The Palms of the Amazon

The Palms

of the Amazon

ANDREW HENDERSON *New York Botanical Garden*

Illustrations by Anthony Salazar

New York   Oxford
OXFORD UNIVERSITY PRESS
1995

Oxford University Press

Oxford   New York   Toronto
Delhi   Bombay   Calcutta   Madras   Karachi
Kuala Lumpur   Singapore   Hong Kong   Tokyo
Nairobi   Dar es Salaam   Cape Town
Melbourne   Auckland   Madrid

and associated companies in
Berlin   Ibadan

Copyright © 1995 by Oxford University Press, Inc.

Published by Oxford University Press, Inc.
200 Madison Avenue, New York, New York 10016

Oxford is a registered trademark of Oxford University Press

All rights reserved. No part of this publication may be reproduced,
stored in a retrieval system, or transmitted, in any form or by any means,
electronic, mechanical, photocopying, recording, or otherwise,
without the prior permission of Oxford University Press.

Library of Congress Cataloging-in-Publication Data
Henderson, Andrew, 1950–
The palms of the Amazon / Andrew Henderson ;
illustrations by Anthony Salazar.
p. cm.   Includes bibliographical references and index.
ISBN 0-19-508311-3
1. Palms—Amazon River Region.
2. Palms—Amazon River Region—Classification.
3. Palms—Amazon River Region—Identification.
I. Title.   QK495.P17H45   1995
584'.5'09811—dc20   93-40717

9 8 7 6 5 4 3 2 1

Printed in the United States of America
on acid-free paper

For Flor and Lidia

Foreword

Although there are only 151 species and 189 taxa of palms in Amazonia, palms are one of the most important groups of plants, both for their abundance and for the large number of uses to which they have been put. It is hard to go into any Amazon habitat and not find a species of palm present. In some areas, palms such as *Mauritia flexuosa* and *Euterpe oleracea* dominate huge areas in almost pure stands.

In spite of their importance, palms have been one of the taxonomically most poorly studied groups of Amazon plants. For information on the palms of the Amazon forest, we have had to rely on the nineteenth century works of Martius (1823-1837), Alfred Russel Wallace (1853), and Richard Spruce (1871). Wallace's remarkable book was written from memory, and notes after his collections were lost in a fire on the ship that was transporting them to England. I am glad to see that many of the species recognized by some of the earlier authors have been united in synonymy.

Palms have been poorly studied because they are difficult to collect and specimens need careful preparation, which requires much patience from the collector. This volume is successful because it is based on extensive field work by an author who has traveled widely in Amazonia and made collections and field observations of most species. I welcome the conservative approach to the taxonomy of the understory genera *Bactris* and *Geonoma,* which have previously been overdivided into microspecies. It should be easy to identify the species that are recognized here. I am sorry that some of my favorite palms, such as the babassu, have needed a change of name from *Orbignya,* as this genus, along with *Maximiliana* and *Scheelea,* is united with *Attalea.* I certainly agree, however, that this is an improvement to the taxonomy. The field work and collections of Andrew Henderson reveal the true relationships of many species and genera.

I am delighted to see that someone whom I introduced to the Amazon forest has produced such a truly useful publication. I hope that this book will both lead to a better conservation plan for the rarer palms and encourage greater employment of palms in sustainable use systems.

Ghillean T. Prance
Director
Royal Botanic Gardens
Kew

Preface

I first became interested in Amazon palms in the mid-1980s, during various collecting trips to Brazil and the western Amazon region. It was obvious then how poorly known the family was. Clearly it was one of the most important, and by far the most abundant, flowering plant families in the region. Virtually every researcher working there, whether botanist, zoologist, entomologist, agronomist, anthropologist or archaeologist, came into contact with palms and wanted to know their names and how to identify them. The only help was the completely out-of-date works of the last century. There was much confusion surrounding the correct names of even the commonest species.

This book, then, attempts to answer some seemingly simple questions. How many genera and species of palms are there in the Amazon region? How can they be told apart? What are their correct names? Where do they grow? What are they called locally, and what are they used for?

The work is based on field, herbarium, and library study between 1988 and 1993. Field work was carried out in the Amazon regions of Bolivia (3 weeks), Brazil (7 months), Colombia (2 weeks), Ecuador (2 months), French Guiana (1 week), Peru (6 weeks), and Venezuela (1 month). Various herbaria in the United States, Europe, and Latin America were also visited. Most herbarium and library work was carried out at the New York Botanical Garden.

The result of my research is a systematic treatment of all the palms that occur naturally in the Amazon region, genus by genus. Keys to genera, species, and varieties are given. Each species is named and described, and its synonyms are given; its distribution, habitat, and ecology (where known) are reviewed; and its common names and uses are given, along with any other relevant facts. The systematic account of the palms is preceded by chapters on the physical setting of the Amazon region (geology, climate, etc.), and notes on the palm family and its biogeography and ecology there. A historical review of botanists who collected and studied palms is also given, followed by a discussion of their species concepts.

My conclusions are that, for its size, the region contains relatively few species of palms. The majority of these are widespread, are for the most part extremely abundant, and present few taxonomic problems. A minority, however, are extremely variable morphologically and very difficult taxonomically, and are considered species complexes. The picture is further complicated by the apparent existence of hybrids between many species. Many taxonomic and biogeographic problems remain. The greatest barrier to solving these is lack of collections. Huge areas of the Amazon region remain virtually uncollected. Far more field work is needed before a complete picture of the palms of the region can be given.

New York A. H.
August 1994

Acknowledgments

This book was born in New York City in 1987 when Dr. Michael Balick suggested the idea of a palm flora of the Amazon. I am grateful to Mike for his continuing support of this project over the years.

I acknowledge World Wildlife Fund's generous support of the field work, which made this publication possible.

Field work in the Brazilian Amazon between 1988 and 1992 was carried out in collaboration with INPA (Instituto Naçional do Pesquisas Amazônicas; with Dr. Marlene Freitas da Silva) in Manaus and with CENARGEN (Centro Nacional de Recursos Genéticos; with Aldicir Scariot) in Brasília. Other field work in Venezuela, French Guiana, Colombia, Peru, Bolivia, and Brazil has been supported by the National Science Foundation, the United States Agency for International Development, the Biological Dynamics of Forest Fragments Project, and Projeto Flora Amazônica. Field work and other aspects of this work have been supported by grants from the Joyce Mertz-Gilmore Foundation.

The manuscript was written at the New York Botanical Garden. It could not have been completed without the resources and magnificent herbarium and library of that institution, and the help of the staff and students.

The curators of the following herbaria made specimens available for study: AMAZ, B, BH, CEN, COL, CUZ, F, INPA, K, LPB, M, MG, MO, NY, P, R, RB, TFAZ, U, UB, US, USM, USZ, VEN. A Short Term Visitor Award from the Smithsonian Institution made possible the study of US specimens.

Many people in many countries have helped me in many ways: in Bolivia, Mónica Moraes, Luis and Papy Moreno, and Michael Nee; in Brazil, Lidio Coradin, Cid Ferreira, Jose Guedes, Evandro Linhares, Eduardo Lleras, Bruce Nelson, Renata Pardini, and Aldicir Scariot; in Colombia, Rodrigo Bernal and Gloria Galeano; in Ecuador, Henrik Balslev; in Peru, Flor Chávez; in the United States, John Butler,

Larry Dorr, Dennis Johnson, Thomas Lovejoy, and Jane McKnight; and in Venezuela, Francisco Guánchez.

The following have reviewed all or parts of the manuscript: Michel Alexiades, William Balée, Michael Balick, Henrik Balslev, Rupert Barneby, Rodrigo Bernal, Ulla Blicher-Mathiesen, Finn Borchsenius, Janet Chernela, Douglas Daly, Jean-Jacques de Granville, John Dransfield, Enrique Forero, Gloria Galeano, Paul Hiepko, Carina Hoorn, Dennis Johnson, Paul Maas, Scott Mori, Michael Nee, Larry Noblick, Heimo Rainer, Anna Roosevelt, Magali Romero Sá, Roger Sanders, David Taylor, Wayt Thomas, David Williams, and Daniel Zarin. Flor Chávez and Amy Berkov prepared many of the distribution maps. Latin descriptions were written by Rupert Barneby, who also advised on questions of style.

Ghillean Prance introduced me to the Amazon in 1977 when I accompanied him on a plant-collecting trip on the newly opened Transamazon Highway, and to the beautiful forests (then) of Carajás and Cachimbo. He subsequently helped me in many ways, and I could hardly have had a better mentor.

Contents

1. *Introduction to the Amazon, 3*
    Size and Limits, *3*
    Geologic History, *4*
    Soils, *7*
    Climate, *9*
    River Systems and Topography, *11*
    Vegetation Types and Phytogeographic Regions, *14*
    Human Occupation, *17*

2. *Introduction to the Palms, 21*
    The Palm Family, *21*
    Diversity, Collecting Density, and Deforestation, *22*
    Local Names and Uses, *24*

3. *Biogeography of Amazon Palms, 29*
    Ecological Biogeography, *29*
    The Refuge Theory, *30*
    Historical Biogeography, *32*

4. *Ecology of Amazon Palms, 37*
    Physical Environment, *37*
    Phenology, *37*
    Reproductive Biology, *38*
    Predation and Dispersal, *39*
    Demography, *42*

5. *Palm Collectors, Palm Botanists, and Species Concepts, 45*
    Historical Review, *45*
    Changing Species Concepts, *52*

6. *The Palms of the Amazon, 57*
    Format, *57*
    Palmae, *58*
    Coryphoideae·Corypheae·Thrinacineae, *62*
    Calamoideae·Calameae·Raphiinae, *68*
    Calamoideae·Lepidocaryeae, *70*
    Ceroxyloideae·Hyophorbeae, *79*
    Arecoideae·Iriarteeae, *88*
    Arecoideae·Areceae·Manicariinae, *100*
    Arecoideae·Areceae·Leopoldiniinae, *103*
    Arecoideae·Areceae·Euterpeinae, *105*
    Arecoideae·Cocoeae·Butiinae, *128*
    Arecoideae·Cocoeae·Attaleinae, *136*
    Arecoideae·Cocoeae·Elaeidinae, *157*
    Arecoideae·Cocoeae·Bactridinae, *160*
    Arecoideae·Geonomeae, *251*
    Phytelephantoideae, *290*

Appendix: Numerical List of Taxa and Specimens Examined, 297
References, 323
Subject Index, 335
Index of Scientific Names, 339
Index of Common Names, 355

1

Introduction to the Amazon

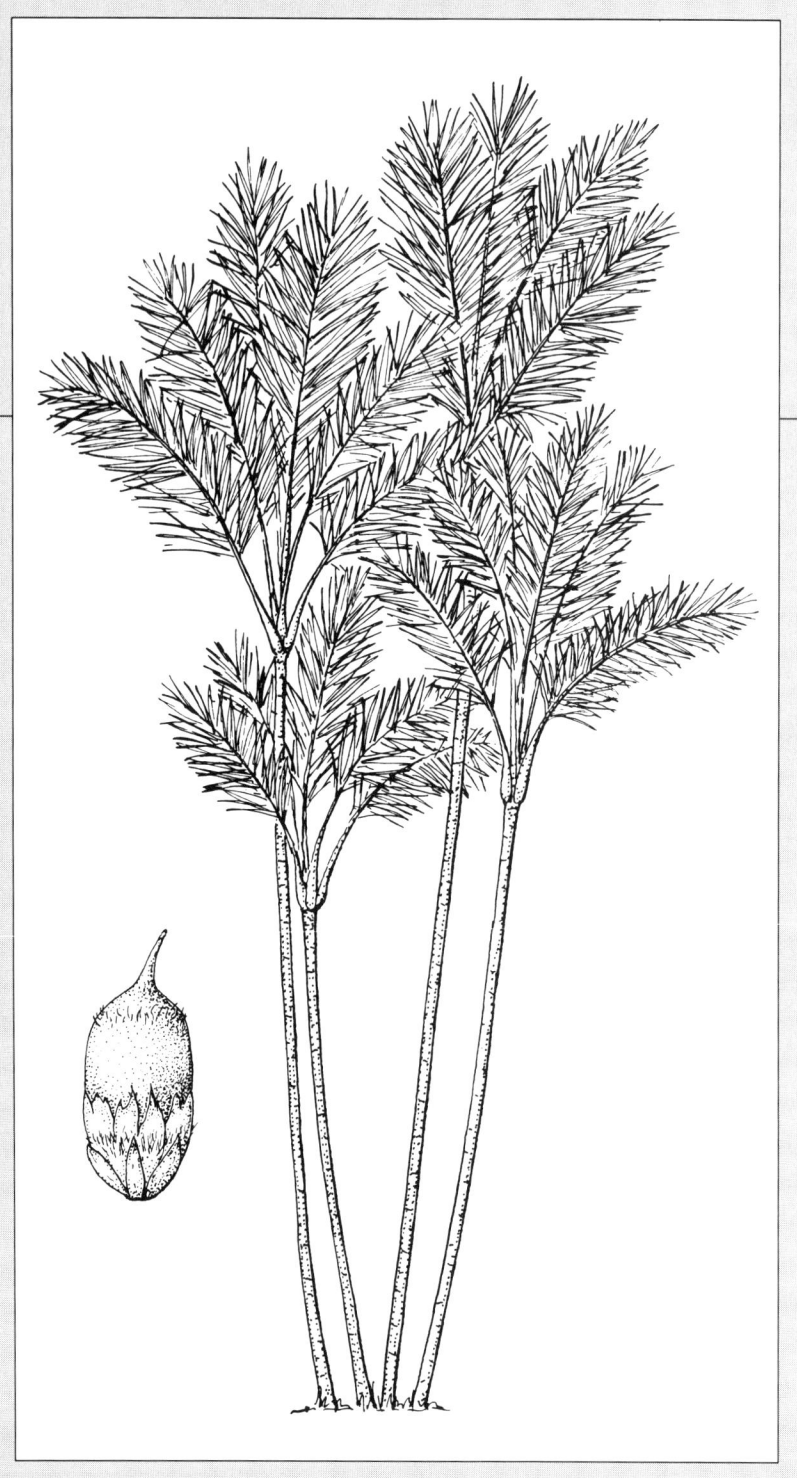

Size and Limits

There are various ways to define and delimit the Amazon in order to convey it as a natural, biogeographic region. Within Brazil it is sometimes defined as Amazônia Legal (Legal Amazon; see Mahar, 1989), but this is a political and economic entity that not only excludes the Amazon regions of adjacent countries, but also includes large areas of non-Amazon vegetation in the south. The watershed of the Amazon river and its tributaries is a better definition, but this excludes large areas of Amazon rain forest in the Guianas and Venezuela that are drained by other rivers, and again includes non-Amazon vegetation in the south. I have therefore followed the floristic criteria of Daly and Prance (1989) and defined the area, henceforth referred to as the Amazon region, as the extent of the lowland rain forest, roughly corresponding to Humboldt's *hylaea* (a term that has come to mean the lowland rain forest, adapted by Humboldt from the Greek word *hyle*, meaning "forest"). This is a phytogeographic definition, and in the distribution maps used here the area is delimited by the dotted line. The western margin is the 500 m contour of the Andes, and the eastern margin is the Atlantic Ocean. As pointed out by Daly and Prance, however, the northern and southern limits of the lowland rain forest are difficult to define, for several reasons. All along the periphery of the region there are transitions to other vegetation types, and the boundaries between these are not always clear. In Colombia, I use the Río Meta as northern boundary rather than the Río Guaviare, in order to include the gallery forests of Vichada. Ducke and Black (1953) noted: "The only natural limits of Amazonia are the Atlantic and the Andes; on the north and south extremes the rain forest is gradually replaced by the flora of the neighboring countries." The Amazon region thus includes, from the northeast, all of French Guiana, Surinam, and Guyana; southern Venezuela; southeastern Colombia; eastern Ecuador; eastern Peru; northeastern Bolivia; and all of northern Brazil (Figure 1.1). This is an enormous area of over 6.5 million km^2.

Not all the vegetation within this area is lowland Amazon rain forest. The region includes large areas of savannas, *campinas,* and

Figure 1.1. Countries of the Amazon region, and Amazonian states of Brazil.

other vegetation types that make up the complex of the Amazon vegetation. These are discussed later in this chapter. A handful of palms occur throughout the area, and the Amazon region is more or less defined by the ranges of *Attalea maripa* (Figure 6.27A), *Mauritia flexuosa* (Figure 6.5A), and *Mauritiella armata* (Figure 6.5C).

Geologic History

The Amazon region comprises parts of four geologic provinces: the Amazon Basin itself, the Guayana Shield to the north, the Brazilian Shield to the south, and the Central Andes to the west. These areas have had a complex and poorly known geologic history. The following account is based on Bigarella (1973), Klammer (1984), Putzer (1984), and Bigarella and Ferreira (1985).

The breaking up of Gondwanaland and the separation of South America from Antarctica, Africa and North America began in the Cretaceous (Pitman et al., 1993). South America and Africa began to drift apart approximately 130 million years ago in the Early Cretaceous. Contact or near-contact between the continents continued for the next 40 to 50 million years as the South Atlantic opened. South America and Antarctica began to separate in the Late Cretaceous, with complete separation accomplished by the end of the Eocene. South and North America began to split much earlier, about 180 million years ago, in the Middle Jurassic. The geologic history of the intervening Caribbean, however, is very complex and poorly understood (Donnelly, 1992), and contact between North and South America may have existed at some time in the Tertiary. Central America finally bridged the gap when the Isthmus of Panama closed approximately 3 million years ago.

South America was thus an island continent throughout much of the Tertiary, when angiosperms were actively diversifying. The core of the continent consists of a craton (a stable, Precambrian continental shield) of metamorphic and igneous rocks that was never submerged below the sea. At some time in the Mesozoic, an intercratonic depression formed between the northern part, known as the Guayana Shield, and the southern part, the Brazilian Shield, and this became the future course of the Amazon river. According to Grabert

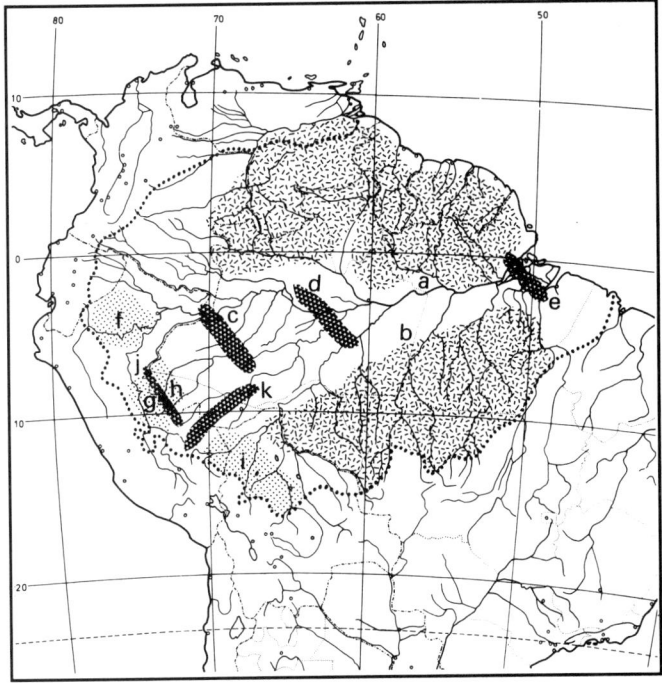

Figure 1.2. Geology of the Amazon region. Shields and arches; **(a)** Guayana Shield; **(b)** Brazilian Shield (extra-Amazonian part of Brazilian Shield not shown); **(c)** Iquitos Arch; **(d)** Purus Arch; **(e)** Gurupá Arch; **(f)** Pastaza-Marañon Basin; **(g)** Ucayali Basin; **(h)** Acre Basin; **(i)** Madre de Dios Basin; **(j)** Serra do Divisor; **(k)** Fitzcarrald Arch. (Adapted from Räsänen et al., 1987; Daly & Prance, 1989; Eden, 1990)

(1983), the continuation of this depression can be traced in Africa. Three ancient areas of uplift, or arches, span this intercratonic depression, running northwest to southeast (Figure 1.2). The Iquitos Arch crosses the basin in Peru; the Purus Arch, near Manaus; and the Gurupá Arch, near Belém. These arches separate basins that themselves have had slightly different stratigraphic histories. According to Caputo (1991), the uplift of the Iquitos Arch was associated with crustal thickening by accretion along the western margin of South America.

The Guayana Shield (Figure 1.2) consists of a tectonically stable base of igneous and metamorphic rocks such as granite, migmatites, gneiss, quartzite, and crystalline schists overlain by the sedimentary Roraima Formation. These sediments reach 3000 m thickness in places, and consist of sandstone probably laid down in Proterozoic times. They are nonfossiliferous, and their age is uncertain. The unequal erosion of the Roraima Formation in the Cretaceous and Early Tertiary led to the characteristic table-top sandstone outcrops, called *tepuís* in Venezuela, such as Roraima, Chimantá, Neblina, and Duida. Basement outcrops occur in the western part, near the Colombia–Venezuela frontier, where the eroded and leveled shield is not covered by sediments. In places the shield is intruded by Precambrian diabasic dikes and sills. Other mountains in the region are the result of granitic rock intrusions, for example the Sierra Parima along the Brazil–Venezuela frontier (Huber et al., 1984). The shield is more or less bounded by the Río Orinoco to the north, the Río Negro to the west, the Rio Amazonas to the south, and the Atlantic Ocean to the west.

The Brazilian Shield (Figure 1.2) is also a Precambrian crystalline platform, sloping gently northward to the intercratonic depression. In the Amazon region it is more or less bounded by the Rio Amazonas in the north and the Rio Tocantins in the east; it reaches its western extreme near the Brazil–Bolivia frontier and is here bounded by the Rio Madeira-Mamoré-Guaporé. Here it is separated from the nearest point of the Andes (at Santa Cruz in Bolivia) by only 200 km. It continues southward outside the Amazon region, into the planalto of central Brazil. The surface is low-lying and supports sediments of varying ages, especially Mesozoic. The basement rocks of the shield outcrop in the region of the upper Rio Tocantins and elsewhere. Unlike the Guayana Shield, the Brazilian Shield has few high mountains. The Serra do Cachimbo and Serra dos Carajás reach 500 m and 400 m elevation, respectively, and the Serra do Roncador reaches a maximum of 630 m. These plateaus, and others such as the Chapada dos Parecís, are capped by deeply eroded sandstones. Active epeirogenic (uplift on a continental scale) elevation of the whole Brazilian Shield in the Late Miocene–Pliocene resulted in intense fluvial deposition in the Amazon Basin.

The Amazon Basin itself is a vast, intercratonic depression filled with sediments dating back as far as the Proterozoic, but principally Cenozoic in origin near the surface. The basin is narrower in the eastern part, and elongate, marginal outcrops of Precambrian, Paleozoic, and Mesozoic rocks occur there. In the western part, the basin is greatly expanded. Both the Guayana and Brazilian Shields were uplifted during the Tertiary, and eroded sediments subsequently were deposited in the basin. In fact, the Tertiary sediments are among the most extensive in the world, although unfortunately the least well understood. The three arches mentioned (Figure 1.2) divide the area into four major sedimentary basins, each having a different stratigraphy. The Marajó Basin lies to the east of the Gurupá Arch. It contains a sequence more than 4000 m thick of Cretaceous, marine Tertiary, and fluvial Quaternary sediments. The middle, Amazonas Basin extends from the Gurupá Arch west to the Purus Arch. It contains a complete sequence of Paleozoic sediments, as well as Cretaceous or Tertiary fluvial sediments of the widespread Barreiras/Alter do Chão Formation. These, consisting of poorly consolidated, sandy, silty, and clayey sediments, are probably of Miocene age and were derived from the shields and deposited under continental conditions. Dissection of these sediments during the Pleistocene resulted in a series of terraces along the middle basin. The Barreiras/Alter do Chão sediments are overlain by the Pliocene to Pleistocene Belterra clays, forming dissected plateaus that are particularly well developed between the Rios Xingu and Tapajós. Theses clays are a uniform, yellowish, kaolinitic material. Sombroek (1966), who first described them, believed they were deposited in a vast, shallow lake during the Nebraskan glacial, when sea levels were higher. Irion (1989), however, considered the Belterra clays much older. The Solimões Basin lies west of the Purus Arch and east of the Iquitos Arch. It contains Silurian to Carboniferous sequences overlain by Tertiary and Quaternary sediments, in particular the Solimões Formation. Finally, to the west of the Iquitos Arch is a complex of smaller basins (Räsänen et al., 1987). From the south, the Madre de Dios Basin is separated by the Fitzcarrald Arch; the Acre Basin is separated from the Ucayali Basin by the Serra do Divisor; and these are separated from the northern Pastaza–Marañon Basin by an area of uplift. The higher areas of the arches are covered with Quaternary, fluvial deposits. This whole western area contains folded and faulted sequences of Carboniferous and Cretaceous sediments, associated with Andean orogeny, overlain by Tertiary deposits. The most widespread Tertiary sequence here is the Solimões Formation, although this is a heterogeneous array of sediments. Frailey and co-workers (1988) discussed the stratigraphic problems of the Late Pleistocene and Holocene sediments in the upper Amazon region, which they interpreted as large-scale lacustrine deposits. Because of supposed equal age and continuity of these deposits, they proposed a large "Lago Amazonas," whose eastern boundary could have been either the Purus or the Gurupá Arch. This hypothesis was not accepted by Tuomisto and colleagues (1992), who pointed out that the sediments of the western Amazon are complex, of different ages, and mostly of fluvial origin. There is, however, evidence of extensive marine inundation in the western Amazon region, but earlier than indicated by supporters of "Lago Amazonas." Hoorn (1990, 1991, in press) made a stratigraphic study of Tertiary sediments in the western Amazon region. She considered that during the Miocene ($\pm$ 16–10 million years ago), marine transgressions from the Pacific or possibly Caribbean, caused by higher sea levels, flooded the area and extensive sedimentation took place. These sediments were laid down in a fluvial or lacustrine, brackish environment. Hoorn called these the Pebas Formation, and considered them equivalent, at least in part, to the Solimões Formation. During the Early to Middle Miocene, the Guayana Shield was replaced by the Andes as the main source of sediments in the region, indicating that Miocene Andean uplift prevented further marine transgressions.

As South America and Africa separated, the continent of South America moved westward. This caused mountain building along the western margin due to colliding plates. The subsequent history of this margin is complex, not only because of various orogenies but also because of Pacific accretions. For example, McCourt and co-workers (1984) interpreted the continental margin of central and southern Colombia as a composite made up of successively accreted oceanic island arcs, dating from the Paleozoic to the Late Cretaceous. Taylor (1991) reviewed the geologic data on Andean orogeny and concluded that mountains have existed in western South America since the Late Cretaceous, formed by westward movement of the continent and accretion from the Pacific. The three major Andean regions—southern, central and northern—have somewhat different histories. In the northern Andes, uplift began in the Late Cretaceous Peruvian orogeny. Central Andean uplift, as a result of accretion, took place at the end of the Cretaceous. In the southern Andes, the first major orogeny was in the Middle Cretaceous, and mountains probably existed from the Late Cretaceous. In general, substantial mountains may have existed throughout the Tertiary but reached their current elevations only recently (van der Hammen, 1989). Most botanical interpretations, however, overlook these earlier mountains and consider that Andean uplift is a relatively recent event (i.e., ±5 million years ago).

Soils

Soils of the Amazon are derived from parent rocks from both inside and outside the region. The Guayana and Brazilian Shields are the source of parent materials inside the region, and the Andes are the main outside source. To give an overview of these soils, it is necessary to choose a classification scheme. Currently several such schemes, both national and international, are in use, and the differences in terminology among them is complex, to say the least. In this book, the World Soil Classification (FAO/UNESCO, 1974) is used because it is internationally recognized and because it links soils to geology. Equivalent names in the U.S. and Brazilian soil classifications, two other commonly used schemes, are given by Sombroek (1984) and Whitmore and Prance (1987).

The FAO classification has two categories: world classes (26) and soil units (106). World classes are separated primarily on formation from parent rock, whereas soil units are separated on the basis of diagnostic horizons. About nine world classes are represented to some extent in the Amazon region. In the following discussion, much general information is taken from Buringh (1979) and local details from Sombroek (1984). Unfortunately, the older FAO/UNESCO (1974) soil map of South America is now out of date and differs substantially from Sombroek's.

Ferralsols are uniform, red or yellow tropical soils consisting of kaolinite and iron or aluminum oxides, with an oxic B horizon. They have a low clay fraction, low cation exchange capacity because of kaolinitic clay minerals, and low base saturation (the percentage of bases on the soil-adsorbing complex). Primary mineral content is low or near zero. These soils can be very deep, and they are usually well drained. They have formed over a very long time, and as a consequence leaching of bases and silica has taken place. In the Amazon region, two kinds are present. Orthic ferralsols occur extensively on the shields north and south of the basin, where they support lowland rain forest. Xanthic ferralsols have a yellowish B horizon because of low iron content. They occur extensively on the lower Amazon terraces, occupying much of the basin between the shields, where they are derived from the Belterra Clay, the upper horizon of the Barreiras/Alter do Chão Formation. They also occur in the northwestern part of the region. Throughout the Amazon, these soils support lowland rain forest in areas with sufficient rainfall.

One interesting feature of ferralsols is the presence of stone lines (i.e., thin layers of quartz or gravel). These were considered by Ab'Sáber (1982) and others to have formed under arid, savanna conditions, such as those found today, for example, in northeastern Brazil. Their presence in the Amazon has been used as evidence of Pleistocene aridity. However, Irion (1989) believed that the formation of these lines was part of the Tertiary formation of the Belterra clays, and thus not evidence in favor of Pleistocene aridity in the Amazon (see also Salo, 1987).

Acrisols are tropical yellow-brown soils with a distinct argillic B horizon, which can limit root growth and drainage. They have low base saturation and low mineral content. Silt content is relatively high. Like ferralsols, they have been formed over long periods of time and have been subject to extensive leaching. They are usually found in areas with a seasonal climate. Acrisols are widespread in the western Amazon, where they support lowland rain forest, but they also occur on both the Brazilian and Guayana Shields. There are three kinds of acrisol in the Amazon region. Ferric acrisols, so called because of the presence of coarse, red mottles or nodules, are widespread in the southwestern part of the region. Orthic acrisols occur in the western Amazon region. Plinthic acrisols contain plinthite, a clayey soil material rich in iron compounds and poor in organic matter. They are principally derived from either marine or Andean sediments, especially in the western and southwestern parts of the region. They support lowland forest as well as savanna and are found, for example, in the Rupununi savannas and Bolivian and Colombian llanos, but they also occur extensively along low-lying terraces of the lower Amazon as well as in the Rio Madeira–Rio Juruá region.

Cambisols have an ochric A horizon (with organic matter) and cambic B horizon (with very fine sand). They are generally young soils and can be considered transitional in a process of development toward acrisols or ferralsols. Eutric cambisols occur on the western margins of the region, and ferralic cambisols are found in Colombia and Ecuador.

Gleysols are hydromorphic soils of permanently poorly drained, low-lying areas. There is no free or dissolved oxygen, and the soil is grayish-blue. Fertility depends on the parent material. In the Amazon region, these soils are found in the floodplains of both white-, black-, and clear-water rivers. Dystric gleysols have a base saturation of less than 50 percent. They are derived from Holocene sediments carried down from the Andes and deposited during periods of falling river levels. A distinctive vegetation develops on these soils. It is known as inundated forest, or *várzea* forest, on white-water rivers and as *igapó* forest on black-water rivers. (*Várzea* is defined as the floodplain of white-water rivers and is usually interpreted as a Holocene depositional plain; *igapó* is the floodplain of black-water rivers [these are further discussed later].) Kahn and Mejia (1990) studied the palm community on these soils in Amazonian Peru. *Mauritia flexuosa* and *Euterpe precatoria* were common, and a total of 11 species of palms were found in a 1 ha plot. The better-drained floodplains of the Orinoco and its tributaries in Colombia have dystric fluvisols, which do not occur in the rest of the Amazon region.

Better drained than the gleysols are the fluvisols. These are young soils developed on recent alluvial deposits, and with little profile development. On floodplains of white-water rivers, they are usually derived from Andean sediments and are nonacidic, whereas on floodplains of black-water rivers they are derived from shield sediments, and are acidic. Fluvisols occupy slightly higher areas than gleysols and are found in broad, elongate stretches along the main rivers, especially in the western parts of the region. Kahn and Mejia (1990) studied palm communities on these soils in Amazonian Peru. Diversity was about the same as on dystric gleysols. Typical palms were *Phytelephas tenuicaulis, Astrocaryum murumuru,* and *Attalea butyracea.*

Podzols are heavily bleached sandy soils with a spodic B horizon (i.e., with a hard layer of organic matter with iron or aluminum or both). The B horizon, a humus iron-pan, displays concentrated leached humus or sesquioxides or both. Gleyic podzols consist of white-gray quartz grains, developed from sandy parent material, itself eroded from the Guayana Shield. They are drained by black-water rivers and are very common north of the Rio Negro in Brazil, extending into adjacent Colombia and Venezuela. Their extent was underestimated by the FAO survey. These soils are poorly drained, are very infertile, and usually support a *campina* vegetation or even open savanna. Due to this infertility, the vegetation forms a mat of roots mixed with raw humus on the surface of the sand, and cycling of nutrients by ectomycorrhizal fungi occurs (Singer & Aguiar, 1986). *Barcella odora* is confined to podzols north of the Rio Negro. As shown by Jordan (1989), however, the soils in these regions can change over a very short distance. Kahn and de Granville (1992) have described the palms of gleyic podzols. In Amazonian Peru, they found 23 species present in a 0.27 ha site, particularly *Oenocarpus bataua* and *Lepidocaryum tenue*.

A complex of different soil types occurs in the Guianas. Arenosols are light gray, coarse-textured, sandy soils with a large proportion of quartz; both albic and ferralic arenosols occur. Nitosols are clayey, red, tropical soils with an argillic B horizon. Both dystric nitosols, with a low base saturation, and eutric nitosols, with a high base saturation, occur in the Guianas and adjacent Bolívar State, Venezuela. Eutric nitosols are also found on the Brazilian Shield, for example in Pará and Rondônia in Brazil. Along the northern Atlantic coast of Brazil (Amapá, Pará, and Maranhão), and particularly on Marajó Island, gleyic solonchaks occur. These are highly saline soils with groundwater levels to within 50 cm of the surface. The palms of these soils have been discussed by Kahn and de Granville (1992); particularly abundant was *Euterpe oleracea*. Wilbert (1976) has described the succession of mangroves, *Manicaria saccifera*, *Mauritia flexuosa*, and *Euterpe oleracea*, on gleyic solonchaks in the Orinoco Delta in Venezuela. In other areas, however, the former two species occur together on dystric gleysols (e.g., Bates, 1863).

Of special interest are *terra preta do índio*, or Indian black-earth soils (Smith, 1980). They are probably formed during long-term human occupation of a site, and are derived from organic waste accumulated in garbage pits. These very fertile, anthropogenic soils are scattered in small to large patches throughout the Amazon region, near prehistoric Indian settlements. Their exact extent is unknown. They have a high content of phosphate, organic carbon, and calcium, and have apparently resisted leaching. Smith considered that the abundance and depth of these soils was evidence that Indian populations were dense and sedentary in pre-Columbian times. Interestingly, these soils are often associated with palms, especially useful species (Balée, 1989). Human beings may have altered soils in other respects. Sandford and colleagues (1985) reported the common presence of charcoal in *terra firme* (any upland area of forest not subject to annual inundation) soils near San Carlos de Río Negro in Venezuela, and considered these to be remains of either natural or human-made Holocene fires.

Climate

In the rainy season the jungle is alive with all these animals. . . . As soon as the dry season sets in things start to change. . . . Animal life in the jungle moves back and forth like the tide of an ocean.

Yungjohann, 1989

The climate of the Amazon region can best be described as hot, wet, and humid, but surprisingly variable depending on locality. The following account of large-scale climatic features is taken from Nobre (1984), Salati and Marques (1984), and Salati (1985).

Average monthly temperature varies little throughout the region. In Belém and Iquitos, the highest average monthly temperature, in November, is 26.9°C and 32°C, respectively; and the coolest, in March and July, is 24.5°C and 30°C. However, cold-air masses occasionally move into the region, particularly in the southwest, and these can cause local cool spells known as *surazos* in Bolivia, *friagens* in Brazil, and *friajes* in Peru. Hurricanes and tornadoes are unknown in the Amazon region, but limited, local windstorms can occur.

There are three important variables in rainfall in the region: total annual amount, seasonal distribution, and between-year variation. Total annual rainfall varies from 3000 mm or more in the western part to 1500 mm in the central-eastern part (Figure 1.3). Locally, the total can reach 5000 to 6000 mm in limited areas near the Andes in Colombia, Ecuador, and Peru, and there are areas in the southwest where the total can be less than 1500 mm. North of the equator, east of 60°W, rainfall is much higher due to the proximity to the Atlantic Ocean.

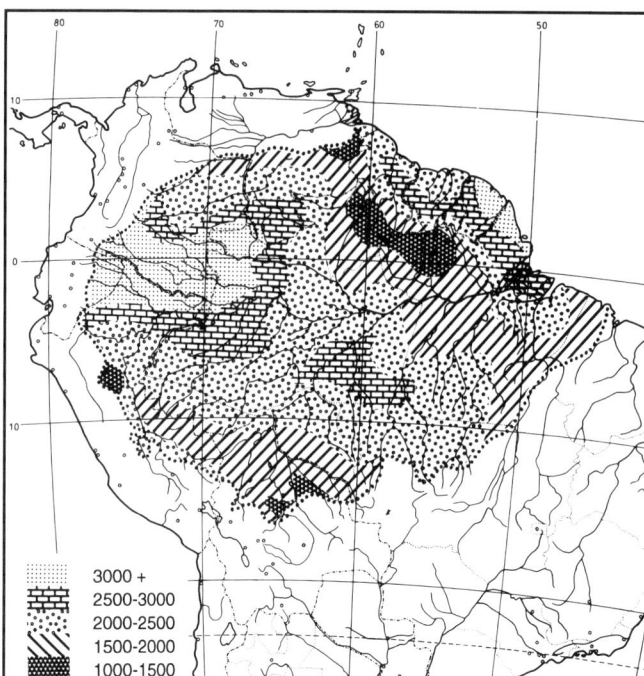

Figure 1.3. Rainfall of the Amazon region. Numbers are in millimeters. (Adapted from Terborgh, 1992)

Also rainfall is higher near the Andes. In the foothills of the Andes in southern Colombia and Peru, warm air is forced to rise and cool, loosing its moisture and giving local areas very high rainfall.

Perhaps more important to vegetation than total amount of rainfall is its seasonal distribution. The annual meridional shift of the position of the sun is responsible for seasonality. The prevailing winds, the trade winds, come from the northeast north of the equator and the southeast south of the equator. These winds pick up moisture from the Atlantic and carry it over the Amazon region. Both trade-wind belts move north or south depending on the zenith position of the sun during its yearly travels. During summer in the southern hemisphere (i.e., December to February), the area of maximum rainfall extends from 0° to 10°S; during summer in the northern hemisphere (June to August), the area of maximum rainfall extends to 10°N. The position of the Intertropical Convergence Zone (ITCZ, formed by the convergence of the two trade-wind zones) is thus moving annually from north to south and back again. So, for example, the rainy season in Rondônia, south of the equator, peaks in January, while north of the equator, in Roraima, the peak is in June. Many local conditions alter this general scheme, however. In many places near to and north of the equator, there are actually two annual peaks of rainfall as a result of the meridional shift in the trade winds. South of the equator, rainfall is generally lower, and dry seasons more pronounced, but again the Andes make the southwest region wetter than the southeast. In the western region, there are no months with less than 60 mm of rain, whereas in eastern parts, there can be three or more months with little or no rainfall.

The third rainfall variable, between-year, is also very marked. For example, large reductions in annual rainfall are thought to be associated with El Niño, the warming of the eastern Pacific Ocean that occurs every few years (so called because it usually comes in December, and thus refers to "The Child", that is, Jesus). Some weather stations recorded a 30 percent reduction in rainfall following the 1983 El Niño (Molion, 1987).

Temperature and rainfall can be combined into a scheme of climate zones, the best known of which is the Köppen system. In the Amazon region, three Köppen categories of humid tropical climate occur, *Af, Am,* and *Aw* (Haffer, 1987a). The *Af* climate is permanently wet with average monthly precipitation never below 60 mm. It is found in the western part of the region and also on the northeast coast. The *Am* climate is transitional and has at least one month with less than 60 mm of rain. It occurs adjacent to the *Af* climate zones. Finally, the *Aw* climate is more strongly seasonal, with several months with less than 60 mm of rain. A broad band, or corridor, of *Aw* climate runs northwest–southeast across the eastern part of the Amazon region, passing through the lower Amazon in the Santarém region. According to van der Hammen (1992), this area was continuous open vegetation in the recent past, thus splitting the Amazon forest into two parts, a northeast and a west-southwest. As we will see later in the discussion on biogeography, this would have had important consequences for plant distributions.

River Systems and Topography

So much rain over such a large area has to go somewhere. Approximately 50 percent of it is immediately returned to the atmosphere by evapotranspiration, but the rest reaches the ground and runs off into rivers. The Amazon river system is by far the largest in the world (Sioli, 1984; Goulding et al., 1988). The catchment area of the river is approximately 6 million km^2; the length of the main river is over 6,500 km; the discharge at the mouth is 175,000 m^3/s, which is between one-fifth and one-sixth of the total discharge of all the world's rivers. An island in the mouth of the river, Marajó, is 48,000 km^2 in size. In the Brazilian Amazon region, there are over 92,000 km^2 of open water (Skole & Tucker, 1993). The gradient of the river is negligible, an average slope of 2 cm/km. A tributary of the Amazon, the Rio Negro, is the second largest river in the world in terms of discharge (and is 100 m deep and nearly 14 km wide near its mouth at Manaus), and both the Juruá and the Madeira are over 3300 km long. In short, the region is dominated by rivers (Figure 1.4).

The typical Amazon river, at least in the western part of the region, meanders as it crosses low-lying, sedimentary areas, and has a large floodplain. For example, although the actual distance between Cruzeiro do Sul and Fonte Boa, two towns at either end of the Rio Juruá, is approximately 1000 km as the crow flies, the downstream trip takes 15 days nonstop by motorboat because of the meanders. The next major tributary to the west, the Rio Purus,

Figure 1.4. Main rivers of the Amazon region.

also has extensive meanders and floodplains. To the east, however, the Rio Madeira, which flows adjacent to the western margin of the Brazilian Shield, has a relatively chanelized course, with few meanders and a narrower floodplain. In Peru, the dynamics of river meandering have been extensively studied. Räsänen and colleagues (1987) proposed that Quaternary and Tertiary tectonics in the western Amazon Basin caused fluvial dynamics leading to a complex mosaic of modern and fossil floodplains. In fact, in this area the differences between *terra firme* and *várzea* are often not clear because of the dynamic history of the river courses.

Rivers of the lower Amazon do not meander nearly so much, and have little floodplain and few chanelized courses (as do the Ríos Orinoco and Guiana). The ones that drain the shields— for example, the Rios Xingu and Tapajós in the south and the Rios Branco and Negro in the north—flow over the crystalline rocks of the shield rather than the sediments of the upper Amazon, and these rivers are characterized by relatively straight courses with many rapids and waterfalls in their upper valleys. A second feature of several lower Amazon tributaries is their lakelike mouths. Good examples are seen on the Rios Tapajós and Xingu. These rivers were formed during Quaternary glacial periods, when sea level was lower, and they dug wide and deep valleys in the soft Tertiary sediments. As the climate warmed, the ice retreated, sea level rose, the valleys were "drowned," and the rivers took on their present-day appearance (Sioli, 1984).

At the western margins of both shields are areas of low elevation where the river system of the Amazon has links with the Orinoco and Paraná systems. In the north, in Venezuela (Amazonas), the Río Casiquiare links the Río Negro to the Río Orinoco; in the south, in Mato Grosso, Brazil, west of the Chapada dos Parecís, the Rio Guaporé almost joins the Rio Paraguai.

An important characteristic of Amazon rivers is their annual rise and fall (Sioli, 1984). Maximum rainfall south of the equator occurs between December and February, so that the southern tributaries of the Amazon are highest at this time. The opposite occurs in the northern hemisphere. Theoretically it would be expected that the main Amazon would not show an annual peak in flooding, but such a peak does occur in May. Water is lowest in October and November. Near the mouth of the Rio Japurá, the difference between high and low water can be as much as 20 m; in Manaus, at the mouth of the Rio Negro, it is 10 to 12 m; and at Santarém, mouth of the Rio Tapajós, it is still 6 to 7 m (Goulding et al., 1988). Below Santarém, tidal influences of the Atlantic are felt. This annual rise of the rivers, and consequent inundation of the *várzea*, has profound consequences for the biota, as we will see later.

Amazon rivers are commonly classified into three color types: white water, black water, and clear water (Furch, 1984; Sioli, 1984; Junk & Furch, 1985) (a division first recognized by Wallace [1889]). These colors are due to a complex combination of chemical and physical factors. The division is not absolute, however. For example, sometimes rivers can change color depending on local or seasonal influences, or an apparent black-water river can in fact be white water chemically. In fact, perhaps it is time for a reevaluation of the classification of Amazon river types (Puhakka et al., 1992).

White-water rivers are a dirty-brown, muddy color, and are turbid as a result of their large sediment load. These rivers—including the Huallaga, Caquetá-Japurá, Madeira, Marañon, Napo, Putumayo-Içá, Ucayali, and Solimões–Amazon itself—flow from the west, draining the Andean region (the Amazon river west of Manaus is referred to in Brazil as the Rio Solimões). The water of these rivers has a pH value of about 7. Annual rise and fall of these rivers, and consequent inundation of the *várzea*, is considerable. Much of their sediment load is deposited on the *várzea*, accounting for the fertility of these soils. Although most of the sediment load carried by white-water rivers originates in the Andes, some white-water rivers drain only lowland areas—for example, the Juruá and Purus—carry sediments eroded from their own floodplains. Typical palms of white-water river margins are *Attalea butyracea, Euterpe precatoria, Mauritia flexuosa, Oenocarpus bataua,* and *Socratea exorrhiza.*

Black-water rivers are the color of tea, and are clearer than white-water rivers. They carry a relatively small sediment load. The largest black-water river is the Rio Negro, but there are numerous small ones, such as most of the right-bank affluents of the Rio Negro and the Rio Cururú, a tributary of the Tapajós; the Rios Urubú and Tefé in the central Amazon region; and the Río Yarí in the Colombian Amazon. Mostly they drain the region between the Andes and the Guayana Shield, but they also occur locally in otherwise white-water regions. The highly weathered rocks of the Guayana Shield release few mineral elements, and thus the rivers are poor in mineral content. Consequently, they are poor in phytoplankton, since photosynthesis is limited by the lack of ions, such as phosphorus and calcium. The water of these rivers is very acidic (pH between 3.8 and 4.9) and is also poor in nutrients. The cause of the color of black-water rivers is still poorly understood, but the color is thought to result from various processes (Goulding et al., 1988). In general, the dark color is caused by dissolved organic substances such as humic and fulvic acid, derived from the breakdown of leaf litter under acidic conditions. This usually occurs on sandy soils, or podzols, especially in the Rio Negro region.

Janzen (1974) suggested that the vegetation of regions drained by black-water rivers had evolved high levels of secondary compounds as a defense against herbivores, and that these compounds, especially tannins and other phenolics, leached out of the vegetation into the water. This, coupled with the nutrient-poor sandy soils, would lead to an ecosystem of low primary production. However, plant communities on podzols do not have exceptionally high levels of secondary compounds (St. John & Anderson, 1982).

Despite its acidic water, humic acid content, and nutrient poverty, the Rio Negro contains the most diverse fish fauna of any freshwater river or lake system in the world (Goulding et al., 1988). Up to 700 species of fish occur in the Rio Negro, compared with, for example, 600 species in all of North America and approximately 350 in all of Western Europe. Some of these fish have important interactions with palms. For example, the fruits of the common Rio Negro palm *Astrocaryum jauari* form an important part of the diet of characins and catfish (Piedade, 1985). Other typical palms of black-water river margins are *Bactris riparia, B. bidentula,* all species of *Leopoldinia,* and *Mauritiella aculeata.* To a greater or lesser extent, however, these palms can also occur on white-water rivers.

Clear-water rivers are generally confined to the lower Amazon region. They have transparent water and a relatively low sediment load, although they can be turbid in the rainy season (and the Xingu and Tocantins can be turbid near their mouths because of tidal influence). Acidity varies, ph being between 4.5 and 7.8. Examples are the Rios Tocantins, Xingu, and Tapajós, and most of the right-bank tributaries of the Madeira. These rivers drain the Brazilian Shield, which has been eroded over very long periods and now consists of relatively resistant formations that release little material into the rivers. The palms of clear-water river margins are very similar to those of white-water rivers, with some minor differences. *Syagrus* species become common along the upper reaches of the southern white-water rivers.

The small-scale topography of the Amazon region is largely defined by its rivers, and the most fundamental division of the landscape is into floodplain and *terra firme*. This topographic difference is important in plant distributions; for palms this has been very well demonstrated by Kahn and de Granville (1992). On a larger scale, however, the region is not one vast, flat, featureless plain covered with rain forest. There is considerable topographic relief in some areas. The most dramatic is the Guayana Highland. The geologic basis of this region, the Guayana Shield and Roraima sediments, has already been discussed in some detail. In southern Venezuela, south of the Río Orinoco, mountains commonly reach 2000 m elevation, while in adjacent western Guyana, southeastern Colombia, and northern Brazil they are smaller. The highest peak in the region, on the Neblina massif, is just over 3000 m elevation. These sandstone mountains, or *tepuis,* have a typical "mesa" shape—that is, a flat top and vertical sides. The most famous are Roraima, Neblina, Duida, and Chimantá. The *tepuis* are notable for their unusual and endemic flora. Steyermark (1986) stated that 8.5

percent of summit genera are endemic. Of much more interest from our point of view are some relict *tepuis* in the Colombian Amazon. For example, Yupatí on the Río Caquetá is a highly eroded hill of scarcely 200 m elevation, but geologically it is still part of the Guayana Highland. Yet no fewer than three species of *Oenocarpus* are found there and only there (*O. circumtextus, O. makeru,* and *O. simplex*), and nearby is the only known locality of *Attalea septuagenata.*

South of the Amazon are several low ranges, none reaching any great elevation. The Serra do Cachimbo and the Serra dos Carajás in southern Pará reach a maximum of 500 m and 400 m elevation, respectively. They are of interest in that certain palms of the *cerrado* of southern and central Brazil just reach the Amazon region on these serras—for example, *Geonoma brevispatha* var. *brevispatha* (Figure 6.61D) and *Syagrus petraea* (Figure 6.22C). Smaller still are the serras of the lower Amazon—for example the Serra Azul and Serra Jauarú on the north bank of the lower Amazon.

Vegetation Types and Phytogeographic Regions

The vegetation cover of the Amazon region is influenced by all of the variables just discussed—geology, soils, climate, river systems, and topography—and their change over time. The Amazon vegetation is thus a constantly changing mosaic of many different types. The following account of the major vegetation types, and the terminology employed, is taken from Prance and colleagues (Pires & Prance, 1985; Prance, 1987). Prance has outlined the most important vegetation types, but has pointed out that these are large-scale divisions and there are many local subdivisions and transition zones. Recent vegetation maps of the Amazon are IBGE (1988) for Brazil, Instituto Geográfico (1985) for Colombia, and Huber and Alarcon (1988) for Venezuela.

Cerrado is the term used for the upland savannas and open forests of central Brazil, and comes, ironically, from the Portuguese word for "closed." It is well described by Eiten (1972). *Cerrado* usually occurs above 500 m elevation and dominates the Planalto region of Brazil. The vegetation is xeromorphic, typically with open woodlands or grasslands, and the trees and shrubs often have thick bark and leathery leaves. The region has a markedly seasonal climate, with 1500 mm to 2000 mm rainfall per year confined to a short wet season. *Cerrado* areas are found all along the southern margin of the Amazon region, and also on the Serra do Cachimbo and Serra dos Carajás in Pará, Brazil. Typical palms of the *cerrado* include *Syagrus, Allagoptera, Butia,* and *Attalea.* In fact, several species of *Syagrus* owe their presence in the Amazon region to the occurrence of *cerrado* or *cerrado*-like vegetation. The only cerrado species of *Attalea* to enter the region is *A. eichleri. Allagoptera leucocalyx* may just reach the southern Amazon region in Brazil and Bolivia, but is not included in this treatment.

Amazon savannas are scattered throughout the region. These open grasslands occur extensively in the Roraima-Rupununi region of the Brazil–Guyana frontier; in the coastal and subcoastal regions of Amapá (Brazil) and the Guianas; in parts of Pará, Brazil, north of the Amazon river; near Humaitá in Amazonas, Brazil; in the frontier region between Madre de Dios, Peru, and La Paz, Bolivia (the Pampas del Heath); in the Gran Sabana region of Bolívar, Venezuela; and to a smaller extent in various other regions. Large, seasonally inundated savanna areas, the *llanos,* border the Amazon region to the northwest, in Colombia and Venezuela, and to the southwest in Bolivia (the Llanos de Moxos). In the estuary of the Amazon, on Marajó Island, there are extensive areas of inundated savanna. Savannas generally occur in areas of less than 2000 mm annual rainfall, but there is still some question as to whether they occur naturally or are maintained artificially by fire. Some savannas are in areas of high rainfall, and their origin is thought to be soil dependent. Huber (1989) has described some of these in Amazonian Venezuela. Typical palms of the Amazon savannas are *Bactris campestris* in the northeastern part of the region, *Acrocomia aculeata* in the coastal savannas of the Guianas and elsewhere, and *Mauritiella armata* in almost all savannas.

Along river margins of savanna regions are found gallery forests. They can extend far from the rain forest into open areas. Typical of these narrow bands of forest is the palm *Mauritia flexuosa,* the main distribution of which is in the Amazon region but which extends into the *cerrado* in gallery forests (Figure 6.4A). Various other palm species are found here, such as species of *Desmoncus* and *Bactris.*

Transition forests mark the boundary of savanna and *cerrado* and tall Amazon forest. Prance (1987) recognized six types of these forests. Semideciduous forest is found in the south of the region between the open vegetation of the south and the closed evergreen forest of the Amazon. Babassu forest, dominated by *Attalea speciosa,* is also found in the southern transition zone. It has greatly expanded its range in various regions, especially in Maranhão and Rondônia, Brazil; where forest has been cleared, babassu palms have come to dominate the landscape (Anderson et al., 1991). Such forests, dominated by one or a few species, are called oligarchic forests by Peters and colleagues (1989), who emphasized their economic potential as managed resources. Liana forest is another type of transition forest. It is best developed in the southeast of the region, between the Rios Tapajós, Xingu, and Tocantins. Bamboo forest is found in Acre, Brazil, and Madre de Dios, Peru. It covers extensive areas, which are dominated by a few species of bamboos (Nelson, in press). These, like other species, undergo extensive dieback after flowering. Few palms occur there, apart from *Bactris maraja.*

Lowland rain forest on *terra firme* is the most extensive forest type in the Amazon region. Here the canopy of the forest is typically 25 to 35 m high, although emergents may be much taller. Species diversity is very high, with much local and regional variation. Large areas of lowland rain forest on *terra firme* occur in the western part of the region. There have been numerous studies of tree diversity in *terra firme* forests; in one site near Iquitos, Peru, Gentry (1988) found more than 300 species of trees over 10 cm diameter at breast height (dbh) in a 1 ha plot. Palm diversity is also high in this western region, particularly in understory species of *Bactris* and *Geonoma.* Many primarily Andean species also occur in the western *terra firme* forests and add to their diversity.

Large areas of pure, leached white sand occur in the Amazon region, mostly derived from erosion of the Guayana Shield (Kubitzki, 1990). Gleyic podzols are derived from these sands, and a distinctive flora is found on them. In Brazil it is called *caatinga* or *campina* (although in other countries it may have other names: e.g., *bana* in Venezuela, *wallaba* in Guyana, white-sand savanna in Surinam). The name *campina* is preferred in order not to confuse it with the desertlike *caatinga* of northeastern Brazil. The largest areas of *campina* vegetation are found in the upper Rio Negro region of Brazil and adjacent Colombia and Venezuela. Patches of *campina* occur, however, throughout the region. The vegetation ranges from open to closed forest, but in general the canopy is somewhat lower than in *terra firme* rain forest, and the plants are adapted to nutrient-poor soils (Anderson, 1981). Inundation is also common due to the presence of a hardpan near the surface. Drainage is usually by black-water streams and rivers. *Campinas* have a distinctive and interesting palm flora; *Barcella odora, Mauritia carana, Mauritiella aculeata, Lepidocaryum tenue* var. *casiquiarense, Euterpe catinga, Bactris* spp., and all species of *Leopoldinia* occur there. Although these palms are common in the upper Rio Negro region, they can occur hundreds of kilometers away in small, isolated patches of *campina.* This raises the question of their origin: by either long-distance dispersal of seeds or previous connections of the *campinas.* Prance and Schubart (1978) considered that some of the smaller campinas in the lower Rio Negro region were anthropogenic, and had previously been cleared by Indians.

Inundated forests, on both *várzea* and *igapó,* are common throughout the region, occurring on both gleysols and fluvisols. Composition of these forests is influenced by the degree and duration of inundation, which in turn depends on rainfall and the consequent rise and fall of rivers. Thus various types of *várzea* are recognized, from annual short inundation to annual long inundation, and the actual area is a mosaic of habitats, from lakes, levees (*restingas* in Brazil), inundated depressions, and mud flats to sand flats (Junk, 1984). The vegetation of the *várzea* is also sensitive to the duration of inundation and varies accordingly. In the western part of the region, particularly in the Peruvian Amazon, the distinction between *várzea* and *terra firme* is somewhat blurred because of the low-lying topography of the whole region and its history of river dynamics. Recently, Worbes and co-workers (1992) have discussed dynamics, floristic subdivisions, and geographic distributions of *várzea* forests in central Amazonia. They conclude that many *várzea* tree species occur on other types of seasonal sites, and their adaptations may not be to inundation but to seasonality in general.

There are various adaptations of plants to seasonally inundated habitats. Many species develop pneumatophores, or breathing roots. These are well developed in such palms as *Mauritia flexuosa* and *Euterpe oleracea*. The fruits of other species develop aerenchymatous tissue (cells with prominent intercellular spaces) in their fruits—for example, all species of *Leopoldinia*. Many riparian palms have the lower parts of their stems submerged for several months. For example, plants of *Astrocaryum jauari* can have parts of their stems submerged from 30 to 340 days per year, depending on site (Piedade, 1985). Worbes (1985) considered that adaptation of trees to inundated habitats was largely physiological rather than anatomical.

The biological diversity of the *várzea* is derived from the nature of the rivers. The high concentration of mineral nutrients in the water and its sediments results in high productivity in both the aquatic and terrestial *várzea* ecosystems. This productivity is often contrasted with nutrient-poor *terra firme* ecosystems. The plant diversity of the *terra firme* is actually much higher per unit area, and a few studies have documented this. Balslev and colleagues (1987) found in Amazonian Ecuador that 1 ha of *terra firme* forest had 228 species of trees over 10 cm dbh, whereas floodplain forest had 149 species. Similar results were obtained by Campbell and co-workers (1986) on the Rio Xingu in Brazil, with 265 species in 3 ha of *terra firme* forest and 40 species in 0.5 ha in adjacent floodplain forest. By contrast, some animal groups have much higher diversity in *várzea* forests. Emmons (1984) studied geographic variation in density and diversity of nonflying mammals at seven sites in Amazonia. She considered that *terra firme* forest north of Manaus had the lowest diversity (but see Malcolm, 1990) and floodplain forest in Amazonian Peru had the highest diversity. This discrepancy is accounted for by the higher productivity of the *várzea*.

Typical palms of the *várzea* are *Attalea butyracea, Euterpe precatoria, Mauritia flexuosa,* and *Socratea exorrhiza*. The vegetation of the *igapó* is closely related to that of the adjacent *campina* forests. Typical of *igapó* are *Astrocaryum jauari, Leopoldinia pulchra,* and *Bactris bidentula*.

The recognition of vegetation types in lowland tropical forests is based on the study of topographic, climatic, and edaphic variables, and not on species composition alone. Phytogeographic regions, however, are distinguished by their constituent species. Both in general terms and specifically for palms, there has been a long history of study of phytogeographic regions in the Amazon. Daly and Prance (1989) reviewed these general schemes and presented maps of the most important ones. Past attempts to define phytogeographic regions for palms are reviewed and are compared with the general schemes to see if there is any agreement.

Spruce (1871) divided the Amazon into five "palm regions". The first was the coastal, or submaritime, region, which included the area of the Amazon estuary under tidal influence as well as the coast of the Guianas. Palms of this region are *Manicaria saccifera, Raphia taedigera,* and *Acrocomia aculeata*. Second was the "granite region of the Casiquiare," which today we would call the upper Rio Negro region. Palms of this region included *Mauritia carana, Mauritiella aculeata, Leopoldinia major,* and *Leopoldinia piassaba*. Spruce considered that *Lepidocaryum* originated in this granite region. The third region was the "diamond region," corresponding to the planalto of central Brazil. Spruce had not visited this region and wrote nothing of its palms. His next, fourth, region was the Amazon region itself, which he divided into middle and upper. Interestingly, Spruce thought that all palms of this region had emigrated from the surrounding areas. The last region was the subandine, or foothills of the Andes in the western Amazon Basin. Here, Spruce wrote, was the headquarters of the stilt-root palms *Iriartea* and *Wettinia;* various species of *Chamaedorea;* and the ivory nut palms, *Phytelephas*.

Drude (1882), himself a prominent biogeographer, gave an extensive account of the biogeography of Brazilian palms. He commented on Spruce's idea that Amazon palms had emigrated from other areas, and cited the example of *Barcella,* a genus endemic to the region (discovered after Spruce's work). He tabulated all genera endemic to each region.

A much more detailed essay on the geography of Amazon palms was written by Barbosa Rodrigues (1903) in his introduction to *Sertum Palmarum Brasiliensium.* He used formal categories, zones, and regions and for each listed the palms found there. Barbosa Rodrigues's six zones were *Amazonina, Montano-Campezina, Marina, Granadina, Platina,* and *Andina.* Here we need only be concerned with the *Amazonina* zone, which Barbosa Rodrigues divided into three regions: *Littoraliae, Planae,* and *Cataractae.* The *Littoraliae* region corresponds directly to Spruce's submaritime region. The *Planae* is the Amazon plain itself, and Barbosa Rodrigues divided it into an eastern and a western part, roughly corresponding to Spruce's middle and upper Amazon. Finally, Barbosa Rodrigues divided his *Cataractae* region into a *Borealiae* and *Australiae,* or northern and southern parts. The northern part roughly corresponds with Spruce's granite region, and the southern one is the area of the Brazilian Shield. Barbosa Rodrigues discussed many of the palms that he had observed on his travels through these regions.

Daly and Prance (1989) reviewed the general phytogeographic subdivisions of the Amazon, on the basis of many different plant families. Four of the most important schemes were illustrated, and here the most recent, that of Prance (1977), is used for comparison with palm distributions. The palm regions of Spruce and Barbosa Rodrigues are too diffuse and too large to correspond with Prance's regions. There is, however, some agreement. The submaritime region of Spruce/the *Littoraliae* of Barbosa Rodrigues is equivalent to the Atlantic coastal sector of Prance; and the granite region of Spruce/the *Cataractae Borealiae* of Barbosa Rodrigues is equivalent to the northwestern sector of Prance. Overall, when one reviews the current distributions of palms (Chapter 3), there is little correspondence between the general phytogeographic regions and palm centers of diversity. There are two main differences. The northeastern region of palms, roughly from Manaus eastward to the Guianas, is split by Prance into three separate regions; the western region of palms, roughly the Peruvian Amazon, is split into two separate regions. However, the northeastern region, roughly corresponding to the Guayana Shield, has a distinctive flora. The flora of the higher elevations have been emphasized (Steyermark, 1986), but more recently Kubitzki (1990) pointed out that the region also has a distinctive lowland flora. One interesting feature that all general schemes recognize is the barrier that the Amazon forms between northern and southern sectors. The lower Amazon acts as a barrier between, for example, the closely related *Oenocarpus bacaba* Mart. and *O. distichus.* The phenomenon of rivers as boundaries of primate species' ranges was first recognized by Wallace (1889), and is discussed recently by Ayres and Clutton-Brock (1992).

Human Occupation

There are no virgin forests today, nor were there in 1492.   Denevan, 1992

Considerable controversy surrounds the timing of the first human occupation of the Americas. While anthropologists generally agree that humans traveled across the Bering Strait from Asia, they do not agree on when. Until recently, the earliest Americans were thought to have arrived some time before 12,000 years ago, a date based on the well-known Clovis site in New Mexico, where stone artifacts have been dated at 12,000 years (Marshall, 1990). However, several other, much older human artifacts have now been claimed. The Pedra Furada site in northeastern Brazil contains rock fragments that were fashioned by humans and are believed to be 50,000 years old (Bahn, 1993).

A second controversy concerns the people themselves. Some researchers (e.g., Greenberg et al., 1986) believe that the region was settled by three separate population movements. These three ancestral groups gave rise to Amerind, Na-Dene, and Aleut-Eskimo peoples. Others disagree with this scenario (e.g., Campbell, 1986).

Whatever the age and origin of the first Americans, the Amazon region has generally not been considered a center of human cultural development. Conventional wisdom is that the resources of the lowland rain forest are too limited or impoverished to have allowed permanent settlement, population expansion, and subsequent cultural development. This view, known as environmental determinism, derived from the work of Julian Steward, who edited a seven-volume work, *Handbook of South American Indians* (1946–1959). His conclusions were that soil infertility and lack of protein in the Amazon region were limiting factors in the development of cultures, and that these limitations led to small, poorly organized populations. Betty Meggers and Clifford Evans carried on this tradition in Amazon anthropology, and effectively excluded others who disagreed with them (Roosevelt, 1991). Meggers's influential book, *Amazonia, Man and Culture in a Counterfeit Paradise* (1971), emphasized dispersal and transience as the most obvious attribute of the Indians. Since current groups practiced shifting agriculture, it was assumed that this had always been the case (see also Balée, 1992). Curiously, both Meggers and Evans carried out field work on Marajó Island, at the mouth of the Amazon. Here they excavated large earthworks and found evidence of a permanently settled culture. They considered this an exception (Meggers & Evans, 1957). In the early 1980s, Anna Roosevelt and colleagues excavated in the Lower Amazon. Using modern techniques, they found evidence of a complex cultural trajectory that began almost 3500 years ago and reached its peak between 500 and 1400 years ago (Roosevelt, 1991). More recently, Roosevelt and colleagues (1991) discovered 7500- to 6000-year-old pottery at sites near Santarém and Monte Alegre. In general, it is now becoming recognized that some prehistoric Amazon cultural groups were considerably more complex than previously supposed, and that the pottery cultures were considerably older than those from the Andes.

Concomitant with these changing views of the age of human culture in the Amazon region has been a reassessment of pre-Columbian population size. An important corollary of this question is the effect of the population on the forest. Again, conventional wisdom considered that the Amazon was a sparsely populated wilderness before European arrival, a terrestial paradise where a few noble Indians lived in harmony with nature and did not alter their forest home. The total population of the Americas was considered modest; Steward and others calculated between 8 and 15 million inhabitants before 1492. Recent estimates, however, are far higher. Denevan (1992) estimated that almost 54 million people lived in pre-Columbian America. Several million of these lived in the lowland rain forests of the Amazon. Furthermore, Denevan believed that such was the size of the population that the current forests of Amazonia are largely anthropogenic in origin. In an article entitled "The Pristine Myth," Denevan argued that human population had greatly altered almost all the New World vegetation before the Europeans arrived.

While the actual size of the population may be debatable, the rapid decline after contact is well established. Approximately 90 percent of the people may have died, most from introduced diseases. The consequence of this decline was the disappearance of the indigenous landscape; the "wilderness" that later explorers found may have been a secondary recovery of the forest, just as the itinerant Indian survivors were remnants of former cultures.

What impact on the vegetation in general, and palms in particular, did this human presence have? Balée (1988, 1989) has pointed out that what appear to be natural palm forests may actually be anthropogenic. Palms are one of the most common families of plants on archaeological sites in Amazonia. Examples are *Astrocaryum vulgare* and *Acrocomia aculeata* on prehistoric Indian mounds on Marajó Island in the Amazon estuary; *Elaeis oleifera* on Indian black-earth sites on the Rio Madeira in Amazonas, Brazil; *Attalea maripa* on archaeological sites on the Rio Xingu in Amazonas, Brazil; and *Attalea speciosa* at numerous sites, especially in the southeast of the region. Balée (1989) goes on to point out that contemporary, foraging Indian societies that exploit these palm resources are not adapted to the "natural" environment, but to a secondary, anthropogenic one, or, "an adaptation to the residue of other cultures, some of which have been long extinct." Balée has estimated that up to 12 percent of *terra firme* forests in Amazonian Brazil may be anthropogenic. It seems very likely that the ranges of the above species, and others, may have been greatly influenced or extended by human activity. But populations of useful species may have been greatly reduced by destructive harvesting.

2

Introduction to the Palms

The Palm Family

There are two main species of palm: in the one, the small leaves are distributed symmetrically along a central stem; in the other, they spread out fanwise.

Lévi-Strauss, 1955

The palm family, Palmae or Arecaceae, has recently been the subject of a magnificent generic monograph (Uhl & Dransfield, 1987). These authors recognize 200 genera worldwide. This number fluctuates somewhat, not only because of new discoveries, but also because of the nebulous concept of the genus. In the present work, several genera recognized by Uhl and Dransfield are placed in synonymy (*Catoblastus* is included in *Wettinia,* and *Maximiliana, Orbignya,* and *Scheelea* in *Attalea*). About 67 genera occur in the New World (for comparison, there are 15 in Africa, 20 in Madagascar, and 97 in the Asian and Pacific tropics). All the New World genera are endemic to the region except for *Elaeis,* with one species in West Africa and one in tropical America, and *Raphia,* with one species in tropical America and numerous species in West Africa.

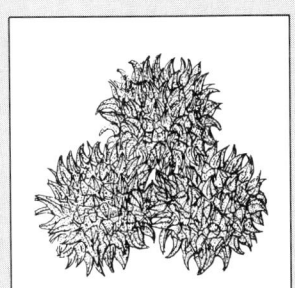

The number of species in the New World is not known, but the available estimates are generally too high. Glassman (1965) estimated 1439 species, and Moore (1973) estimated 1147; in contrast, Henderson and co-workers (in preparation) estimate fewer than 550. The reason for this reduction, as discussed in the following, is not only changing and broadening species concepts, but greatly increased numbers of collections available for study.

In this treatment, 34 genera are recognized in the Amazon region, 8 (24%) of which are endemic. One hundred and fifty-one species occur in the region. Nineteen of these (11 in *Bactris* and *Geonoma*) are widespread and variable and are considered species complexes. They are divided further into 57 varieties. This brings the total number of taxa in the region to 189. Of these, approximately 140 (75%) are endemic. The Amazon region thus contains approximately 50 percent of the New World genera and 30 percent of the species.

Diversity, Collecting Density, and Deforestation

It is a very common observation that the number of species of plants and animals increases as latitude decreases. In other words, the tropics are more diverse than temperate zones. Diversity also tends to increase as elevation decreases. Thus the Amazon region, being at the lowest latitudes and elevations, is regarded, in the popular imagination at least, as being one of the most diverse ecosystems on earth. However, the diversity of most plant and animal groups in the Amazon region is unknown. For flowering plants, many botanists accept an estimated total of 90,000 species for the Neotropics, and Gentry (1982) has estimated that the Amazon region may contain one-third of them. Recently, Henderson and co-workers (1991) made a comparative estimate of the biodiversity of various phytogeographic regions of the Neotropics. On the basis of a sample of 5331 species from 43 volumes of *Flora Neotropica* (i.e., about 6 percent of the estimated total of 90,000), they found that the northern Andes was the most diverse region in the Neotropics. With a surface area one-twentieth that of the Amazon, it contained at least as many species. Furthermore, bryophytes, hepatics, and ferns, surveyed separately from *Flora Neotropica,* were much more diverse in the northern Andes than in the Amazon.

The distribution of palm species supports these findings. One hundred and fifty-one species are recognized from the Amazon region. In Panama, a country of 77,082 km^2, or 1 percent of the surface area of the Amazon, 100 species are present (D'Arcy, 1987); in Ecuador, a country of 283,561 km^2, or 4 percent of the Amazon region, 120 species of palms are known. The reasons for this discrepancy are complex. They certainly have to do with greater topographic and ecological diversity in the Andes and Central American mountains, but there are also historical reasons. This is further discussed in Chapter 3.

If there are relatively few species of palms in the Amazon region, why are they considered so important? The answer lies more in the number of individuals than in the number of species. Palms are extremely abundant. For example, in a 10 ha plot near Manaus, Henderson and collaborators (unpublished data) found 230 adult individual plants of *Astrocaryum gynacanthum*. This is a very common palm and occurs in 259 degree squares (1 degree square is 1 degree longitude by 1 degree latitude, approximately 110 × 110 km), or approximately 3 million km^2 (Figure 6.55A). Thus the potential number of adults is enormous. The amount of fruits produced per year, which feed animals, birds, and insects, must be astronomically large. Thus palms are important in the ecosystem by their sheer abundance.

Within the Amazon region, some areas are much more diverse than others. The degree square with the greatest number of collections (259), near Iquitos, also contains the highest number of taxa (58). The average number of species in squares with more than 30 collections (33 squares) is 30 (range 17–58). There are not enough data, however, to make any assessment of areas of highest diversity at this scale. For all degree squares with one or more collections, the average number of collections for each number of species was calculated and plotted. The slope of the graph was a straight line, indicating that the asymptote was not reached.

One of the biggest obstacles to understanding the diversity of Amazon palms is lack of collections. The Amazon region as a whole is extremely poorly collected, at least for palms (and probably for many other groups). In this study, approximately 6000 specimens have been examined (500 of which have been collected by the author). Not all of these collections are from the Amazon region, but those that are not are probably equaled by specimens that are from the region but have not been examined in this study. Many parts of the region, however, remain uncollected. Figure 2.1 shows degree squares from which two or more palms have been collected in the Amazon region (solid circles), and those with one collection (open circles). This is not to say that no collections exist from the blank areas (although probably they do not), only that such collections have not been seen. There are 170 degree squares without collections, or approximately 30 percent of the total number of 556 degree squares wholly or partly within the Amazon region. Furthermore, over 80 squares have only one collection, and many others have only two

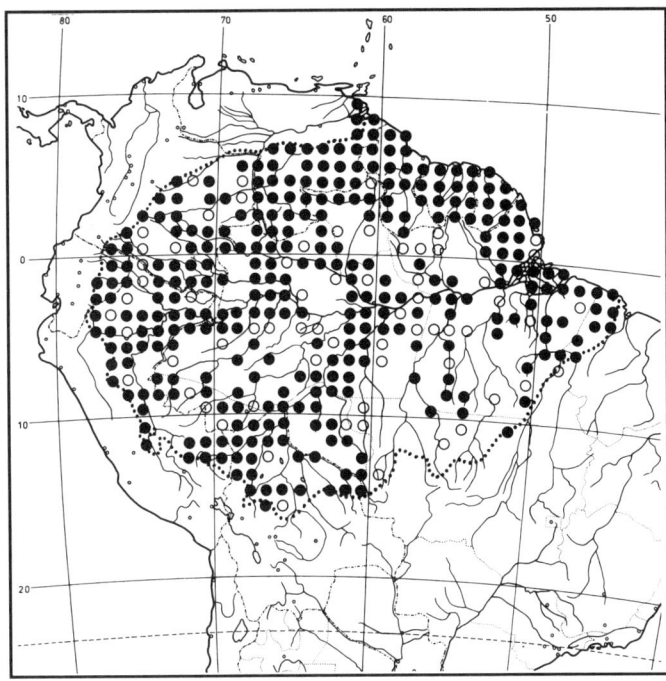

Figure 2.1. Degree squares in which palms have been collected in the Amazon region. Solid circles, more than two collections; open circles, one collection.

or three. This means that about half of the Amazon region is effectively uncollected for palms. These uncollected squares are concentrated in several places, most in the Brazilian Amazon. Even when collections from these regions do exist, they are usually from near margins of major rivers and seldom from the inaccessible *terra firme*.

The Brazilian Shield—including the Brazilian states of southern Pará, northern Mato Grosso, and Tocantins—is virtually uncollected. Nelson and colleagues (1990) also pointed out the lack of general collections of flowering plants from this region. The region is also the area of greatest deforestation in the Amazon region. The only substantial collections of palms from this region are those of Michael Balick and colleagues on a Projeto Flora Amazônica trip in 1977 along the Santarém–Cuiabá highway (BR 163). These collections are extremely interesting, and some are still unidentified. New collections from the area are urgently needed.

The northern part of the Brazilian state of Pará, between the border with Amapá and the Rio Trombetas, is also virtually uncollected. Although the southern part of this region contains savannas and thus few palms, the northern part, bordering on the Guianas, is lowland forest. Judging by the richness of the palms from adjacent French Guiana, it must be an interesting area. To the west of this, in the state of Roraima, the area of the Rio Branco is also completely uncollected. There are perhaps not so many palms in this region, but probably many range extensions from the *campina* forests of the Rio Negro to the west.

The great area of forest between the Rio Negro and the Rio Solimões, from Manaus to the Colombian border, is very poorly collected. This is an extremely interesting area, including as it does large areas of black-water drainage. South of the Rio Solimões is the second largest area with no collections, between the Rio Javari in the west and the Rio Purus in the east, and south to the border with Acre. Here is a huge area of lowland rain forest, more than 500,000 km² in size, which is mostly intact. Finally, in Peru, the department of Ucayali is very poorly collected, but surely contains interesting palms.

The most recent, and probably most accurate, estimate of deforestation in the Brazilian Amazon (i.e, Amazônia Legal) is that of Skole and Tucker (1993). Figures for countries from the rest of the region are scarce and unreliable, but the percentage of deforestation is probably similar to that of Brazil. Skole and Tucker estimated that the original extent of the forest in Amazônia Legal, excluding savannas and water, was just over 4 million km². They calculated that as of 1988, 230,324 km² had been deforested (almost 6 percent of the total). They also pointed out, however, that a further 357,280 km² is adversely affected by either isolation or edge effect. More alarming still are the figures for the amount of deforestation in a recent 10-year period. Skole and Tucker estimated that the amount of deforestation (excluding isolation and edge effect) increased from 78,268 km² in 1978 to 230,324 km² in 1988, while adversely affected areas increased from 129,961 to 357,280 km². Most of this decade's deforestation has been concentrated along the southern and eastern margin of Brazil's Amazon region, precisely the area where there are the fewest collections of flowering plants (Nelson et al., 1990).

The situation in the Amazon region is thus indeed serious, but unfortunately it is often exaggerated. In general terms, no Amazon palm species can be considered in danger of extinction, or even threatened or endangered. Local populations, however, are probably being destroyed. A recent list of the conservation status of New World palms (Dransfield et al., 1988) does not distinguish between species and synonyms and cannot help in assessing the situation. Brown and Brown (1992) reported that in the Atlantic coastal forest of Brazil, which has only about 12 percent of its forests intact, little or no extinction could be detected: "Widely publicized models and predictions have led to a considerable body of mythology, exaggeration and misinformation, spread around the world by a headline-hungry media, often with little factual basis in reliable observations and controlled experimentation." The lowland Amazon rain forests are still largely intact, but do face major threats from colonizing, ranching, mining, damming, road building, and other development schemes.

Local Names and Uses

Man dwells naturally within the tropics and lives on the fruit of the palm tree. He exists in other parts of the world and there makes shift and feeds on corn and flesh.

Linnaeus, quoted in McCurrach, 1960

The local names of palms in the Amazon are very varied and regional, representing a mixture of Spanish, Portuguese, and numerous Indian languages in various combinations. For example, in Brazil the stilt-root palms *Socratea, Iriartea,* and *Iriartella* are called *paxiuba, paxiuba barriguda,* and *paxiubinha,* respectively. *Paxiuba* is derived from *lingua geral* (Jesuit missionaries adopted *lingua geral,* a general language of the Brazilian Amazon, from the coastal Tupinambá language, a member of the Tupi-Guarani family of languages) and comes from the words *paxi,* a thin palm, and *yúa,* a stem. Thus *paxiubinha* is the Portuguese diminutive form of the word, and the modifier *barriguda* is Portuguese for "swollen" (or "potbellied"), in reference to the swollen stem. In Peru, these palms are commonly called *cashapona, huacrapona,* and *ponilla,* respectively. *Pona* is a word of obscure origin, but *casha* is from Quechua and means "spiny" (in reference to the spiny stilt roots); *huacra* is also Quechua and means "horn" (in reference to the inflorescence bud); and *ponilla* is the Spanish diminutive form. So commonly are the stems of *cashapona* used for making floors in the Peruvian Amazon that *el emponado* means any surface covered by *pona,* usually the floor.

The same species of palm can have different names in different places. For example, in the Colombian Amazon, *Attalea maripa* is called *imayá* near Mitú, *guajo* on the Río Caquetá, *guichire* near Villavicencio, *uichira* on the Río Guayabero, *palma real* in Guianía, and *cucurita* in Vichada. Conversely, many different species can have the same name, particularly small, nonuseful species of *Geonoma* and *Bactris*. For example, many different species from these genera are known as *ubim* and *marajá,* respectively, throughout the Brazilian Amazon. Spruce (1908) wrote: "[B]ut I have learnt that it is very unsafe to trust to the native names for the species, these names being, in fact, in some cases generic; I may instance *Assaí, Bacába, Marajá.* . . . The number of *Marajás* is endless."

Botanists have sometimes adopted Indian

names. In many cases, Martius used the local name for a palm as a specific epithet. Thus we have *Astrocaryum paramaca* and *A. chambira; A. tucuma* and *A. munbaca* are both unfortunately now synonyms. Barbosa Rodrigues went even further and used Indian names for sections or subsections of genera. For example, he divided *Astrocaryum* into subsections and called them *Yauary, Chambira, Mumbaca, Mumbacucu, Ayry,* and *Murumuru* (Barbosa Rodrigues, 1903). Later authors have tended to overlook Barbosa Rodrigues's subgeneric classification.

Almost all early travelers and naturalists who visited the Neotropics commented on the importance of palms to indigenous peoples. Columbus himself admired their diversity, and Humboldt, Martius, Spruce, Wallace, and Barbosa Rodrigues were all impressed by their utility. Almost all the material needs of indigenous societies were met by palms, particularly their needs for food, fiber, and construction materials. Balick and Beck (1990) have summarized the economic uses of the family. There are two main reasons for the usefulness of palms, one to do with internal anatomy and chemistry (Tomlinson, 1961), and the other with ecology. The ecological reason is the simple fact that palms are everywhere abundant. The anatomical and chemical reason has three aspects: the anatomy of palm stems, the anatomy of palm leaves, and the chemical composition of palm fruits.

Palm stems are very widely used in the Amazon region in house construction. The highly sclerified vascular bundles in the stems are concentrated in the outer region; the central region consists of soft, parenchymatous material. Short sections from the lower part of the stem, particularly of preferred species such as *Euterpe precatoria* and *Socratea exorrhiza*, are cut and then the outer part is split into short planks. The central pith is removed and discarded.

Perhaps the most common use of palms is of the leaves for thatching. Not only is the large surface area of the leaves or groups of leaves important, but also the internal strength and durability of the leaf segments or pinnae. This strength, in turn, is based on the presence of vascular bundles, silica bodies, and fibers. Palm leaves are astonishingly strong. Niklas (1992) reported that the tensile strength of the mid rib of a coconut leaf rachis was between 0.17 and 0.3 $GN.m^{-2}$ (Giga Newton per square meter), which is close to the tensile strength of annealed aluminium. In the Amazon region, palm thatch is preferred over corrugated tin for several reasons. It is lighter and thus requires less strength in the frame to support it; it makes the interior cooler than tin; rain on thatch is less noisy than rain on tin; and it lasts for up to 15 years. Palm thatch is also cheaper than tin. In some places, such as Rurrenabaque in Bolivia, demand for palm thatch currently outstrips supply (D. Williams, personal communication). It is also interesting to note in passing that humans have learned from animals to use palms; bats of the genera *Artibeus* and *Vampyressa* commonly use palm leaves to make diurnal shelters (Timm, 1987; Foster, 1992).

The edibility of palm fruits for humans is derived from their animal-mediated dispersal strategies; and their importance to humans is based on their abundance and fecundity. Two edible parts of the fruit, the mesocarp and endosperm, have different properties, although caloric values are similar in certain genera. In general, palm fruit mesocarps are high in oleic oils, while endosperm is high in lauric oils. Some indigenous societies depended on palm fruits for their livelihood and have evolved complex ceremonies around the harvest.

Another source of food for indigenous peoples is palm starch, or sago, derived from the soft, central parenchyma of the stem. Heinen and Ruddle (1974) have described the elaborate ritual surrounding the recovery of sago from *Mauritia flexuosa* by the Warao Indians of the Delta Amacuro in Venezuela. The starch content of the palms was greatest between February and April (the beginning of the rainy season) and just before flowering time in May. It decreased rapidly just afterward, increasing again as the fruits formed. When the fruits ripened, the trees provided both starch and edible mesocarp. Heinen and Ruddle considered that the rituals of the harvest and distribution of starch contributed to the preservation of tribal identity and were an important survival mechanism. Wilbert (1976) described the process used by the Warao to extract sago from the stems of *Manicaria saccifera,* although the palm is apparently secondary in importance to *Mauritia.* In the case of both palms, it is extraordinary how closely connected are the lives of the Warao and the annual rhythms of nature. As Wilbert wrote: "As their ethnobotanical lore reveals, the palms have nurtured among the Warao an exquisite partnership between man and nature—a symbiosis that, in addition to a viable socio-economic blue-print, generated an ideological matrix that gave meaning to the world and purpose of life."

Recently, useful palm products have featured prominently in the debate over extractive reserves in the Amazon region. In a list of 28 plants currently collected by Amazon extractivists, nine were palms (Fearnside, 1989). Phillips (1993) has cautioned that most preferred palm fruits are difficult to harvest without destroying the trees. In the Iquitos area of Peru, Vásquez and Gentry (1989) reported that destructive harvesting of two popular palm fruits, *Mauritia flexuosa* and *Oenocarpus bataua,* had greatly reduced natural populations in the region. Pinard (1993) has reported on the impact of stem harvesting of *Iriartea deltoidea* in Acre, Brazil.

Although the uses of palms by indigenous people in the Amazon region are remarkably diverse, they go beyond the utilitarian. Castelnau (1853) described and illustrated the large, beautiful ceremonial masks made from palm leaves and feathers worn by the Karajá Indians for dances. Palms also have symbolic and ritual importance. Reichel-Dolmatoff (1989) discussed the biological and social aspects of the Yuruparí myth of the Indians of the Colombian Amazon, particularly on the Río Vaupés. The basis of this body of teaching is a social code that exhorts exogamy. Of interest to botanists is that various palms are extensively used by the Indians, and the Indians have accumulated a large body of information on them, particularly in relation to reproduction. In fact, since palms are observed by the Indians to be cross-pollinated, they have become a metaphor for exogamy and feature largely in the Yuruparí social code:

> The entire process of pollination is accompanied by an interrelated complex of sensorial phenomena which form part of the sexual physiology of these plants. . . . Palm pollination, as observed and interpreted by the Indians, is equated with human reproductive behavior and from this comparison a large and intricate body of social norms, ritual practices, and mental processes is derived. (Reichel-Dolmatoff, 1989)

The large trumpets and long tubular flutes used in Yuruparí initiation rituals, and the mouthpieces of the trumpets and the flutes, are made from the wood of *Socratea exorrhiza.*

3

Biogeography of Amazon Palms

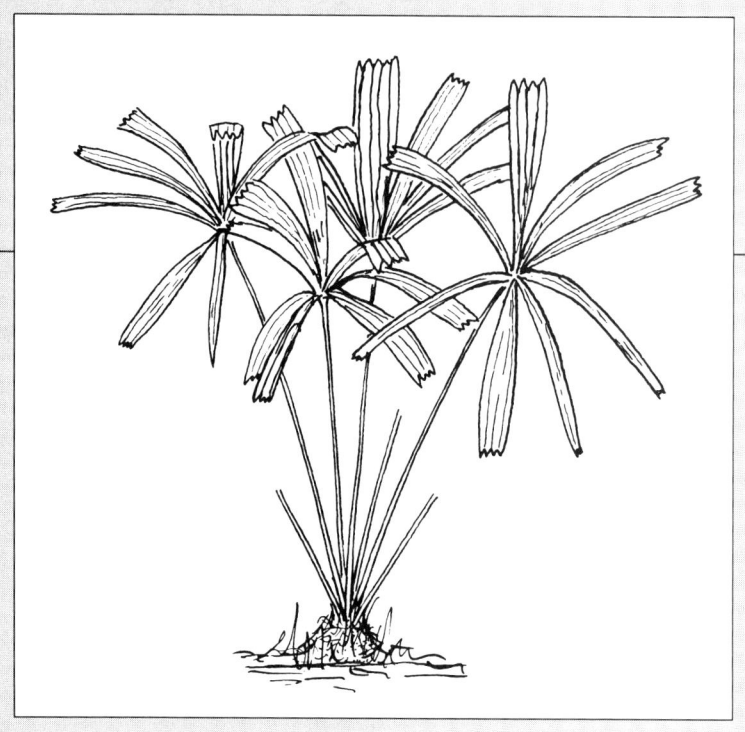

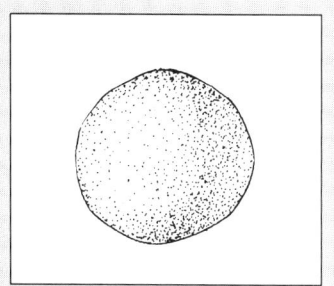

In this chapter, I first bring together the physical factors of the Amazon environment (soils, climate, river systems, topography) and interpret them in relation to palm distributions; that is, I take an ecological biogeography approach. Then I consider geologic history and its relation with palms; that is, I take a historical biogeographic approach, with particular reference to the refuge theory. The main difference between these two approaches is one of scale. The ecological approach is useful at the small scale, whereas the historical biogeographic approach considers distributions on a larger scale. The biotic factors that form part of the ecological biogeography, such as dispersal and dispersers, are discussed as well.

Ecological Biogeography

Few palms are restricted to any one soil type. Widespread species such as *Mauritia flexuosa* almost always occur on inundated soils, but these can be gleysols, solonchaks, or fluvisols. Certain palms are perhaps more abundant on specific soil types, but can also occur on others. For example, Lescure and colleagues (1992) found that *Leopoldinia piassaba* was common on podzols, but also occurred, less abundantly, on other soils. At the small scale, soils are a mosaic of types. For example, Jordan (1989) showed how small variations in topography, which affected the water table, could change soil type over a short distance, with a consequent effect on vegetation. Within a 500 m transect, with an 8 m elevational range, six different communities and associated soil types occurred. Locally, palm distribution is closely tied to both soil type and drainage, as clearly shown by Kahn and de Granville (1992).

Given that the Amazon climate is essentially one of constant warmth, the most important climatic factor is rainfall, including total annual amount, seasonal distribution, and between-year variation. In general, species richness is higher in higher rainfall areas (Gentry, 1982), and this also seems true for palms. There are, however, too few data to make any definite correlations. Although more species of palms are found in the higher rainfall area of the western Amazon Basin, most of these are also found in adjacent drier regions. Numerous species are also found in the high-rainfall coastal region of the Guianas, but, again, many of these cross the dry corridor and occur as far west as Manaus. There is probably a better correlation between palm diversity and climate zones, with the *Af* humid tropical climate having the greatest diversity. Strongly seasonal climates have a less rich but distinctive palm flora. Also of interest is the fact that even very short-term climate changes can cause changes in local abundance and presumably distribution. For example, Condit and colleagues (1992) found that over a 10-year period on Barro Colorado Island in Panama, mortality rates increased dramatically for some species in response to a single, unusually dry season in 1983. The mortality rate of *Oenocarpus mapora* was almost four times greater between 1982 and 1985 than between 1985 and 1990.

The river systems of the Amazon, and their division into white, black, and clear water, also have an effect on palm distributions. As mentioned in Chapter 1, some palms are typical of one type of water or another, but few species are confined to only one water type. Taking the example of *Leopoldinia* again, *L. pulchra* is mostly confined to black-water drainage areas, but also occurs sporadically in white-water areas. The situation is similar with some species of *Bactris*.

Topographic relief on the large scale is influential in the Guayana Shield region, where *tepuis* commonly reach 2000 m elevation. Curiously, few palms occur on these mountains, in contrast to the Andes, and of those that do only *Prestoea tenuiramosa* can be considered endemic. *Dictyocaryum ptarianum* is common on talus slopes between 800 m and 1700 m elevation, but also occurs rarely in the western Amazon Basin at low elevations (Figure 6.7F). *Euterpe catinga* and *Geonoma appuniana* are common components of forested summits and slopes, but are found in the Andes of Ecuador. In general, the impression is that the few palms found on the *tepuis* have migrated from the Amazon lowlands or from the Andes, and this is in agreement with Steyermark's (1986) hypothesis of centripetal dispersal from the surrounding region. Perhaps the distribution of *Euterpe catinga* var. *catinga,* found both in the lowlands and on the *tepuis* (Figure 6.15A), supports this idea.

One interesting aspect of topography, at least in terms of elevation, is the wide elevational range of many species of palms. Many lowland Amazon palms can, even if rarely, be found on eastern Andean slopes to 1200 m elevation, and often in a completely different habitat. For example, the widespread and common *Socratea exorrhiza* is abundant in inundated areas at low elevations throughout the Amazon region. It can also be found on steep, well-drained Andean slopes up to 1000 m elevation. I believe the explanation for this is that rainfall is much higher on eastern Andean slopes, allowing these palms to occur there. Another example is *Oenocarpus bataua,* which occurs in inundated areas at low elevations in areas of lower rainfall, and on *terra firme* in areas with more than 2500 mm annual rainfall.

Finally, the influence indigenous people may have had on palm distributions throughout the history of human occupation of the Amazon region should not be overlooked. Given the considerable economic importance of palms to indigenous societies, it is reasonable to assume that they have widened distributions of some species. An interesting pattern, in this sense, is apparent for three useful species: *Aiphanes aculeata, Bactris macana,* and *Syagrus sancona.* These species occur at moderate elevations along the Andes from Venezuela to Bolivia, whence they spread into the Amazon lowlands. Given the importance of these palms, it seems possible that this is an anthropogenic distribution pattern, and unlikely to represent a natural pattern.

The Refuge Theory

The refuge theory was put forward by Haffer (1969) to explain patterns of bird distribution and diversity in tropical South America. Haffer identified areas of endemism for the neotropical avifauna, and used Quaternary climate change to explain these. Other authors gave supporting evidence from other organisms—for example, plants (Prance, 1982).

The Quaternary as a whole is considered a relatively cold period compared with the Tertiary. It is thought that glacial periods in the polar regions led to drier and cooler tropical climates, and interglacial periods were warmer and wetter. At the same time, sea-level changes were taking place, and these had considerable effects on the landscape. In tropical areas, the changes associated with these glaciations are deduced primarily from radiocarbon-dated pollen cores, deductions being based on the varying levels of different pollen. Other dating methods (e.g., dendrochronology) cannot be used because most tropical trees do not produce annual growth rings (but see Worbes & Junk, 1989). In South America, the most complete work on pollen cores has been carried out in extra-Amazon regions, such as the northern Andes of Colombia; thus climate changes here are best understood (e.g., van der Hammen, 1989). Long sequences of glacial and interglacial periods meant correspondingly lowering and rising temperatures, causing vegetation belts to move either up or down the Andes. For example, Liu and Colinvaux (1985) estimated that vegetation zones on the eastern Andean slopes of Ecuador may have been 700 m lower than today, and temperatures 4.5°C cooler. In the Amazon region, there is far less empirical evidence, and climate changes are generally inferred from current distributions of plants and animals. It is thought that in this low-lying region, open savanna replaced forest in certain regions in arid, glacial periods. Also, temperatures are thought to have been between 2°C and 6°C lower than today.

Various other geomorphological data point to climate change in the Amazon region, but the subject is controversial and not all authors accept the evidence. Salo (1987) reviewed the data for arid Pleistocene cycles and refuges and found them unconvincing. Similarly, Colinvaux (1987) doubted the hypothesis, and both authors emphasized other types of disturbance that could explain diversity patterns. The basic problem is the lack of radiocarbon-dated pollen cores from the region, but recently some have become available. Absy and colleagues (1991) have analyzed a pollen core 6.5 m long, representing 60,000 years of vegetation cover on the Serra dos Carajás in Brazil (Pará). This shows considerable change, with three dry periods.

The authors hypothesized that a reduction of 500 mm in annual rainfall (i.e., from 1500–2000 to 1000–1500) would lead to the replacement of lowland rain forest by other vegetation types. Much more work on fossil-pollen stratigraphy is needed in the Amazon region, although the availability of suitable sites may be a limiting factor (peat bogs, commonly used in the temperate zone, are virtually absent in the Amazon). In North America, for example, radiocarbon-dated pollen profiles have been studied from 250 sites over the last 25 years (Jacobson et al., 1987), and Holocene changes at least are well understood.

Recently, Salo and co-workers (1986) have proposed that small-scale natural forest disturbance and regeneration over a large area of lowland rain forests in Amazonian Peru was caused by lateral erosion and channel shifting of the larger rivers. They considered that this succession on newly deposited river soils was a major mode of regeneration, and that up to 12 percent of forests in this region were in successional stages. They also proposed that the disturbance caused in forests by river dynamics would create and maintain high between-habitat species diversity of this type of forest (see also Puhakka et al., 1992). This itself could have led to species diversity in the region, as opposed to theories of climatic change in the Quaternary. After these new ideas were published, other workers pointed out that the situation was actually more complex. Dumont and colleagues (1990) considered that areas affected by tectonic events are limited to well-defined depressions, and that most upland, *terra firme* areas are unaffected. Also, they considered that diversity of *terra firme* forests was clearly greater than that of the floodplain forest. They attributed the diversity of the former to their great age and to the constant ecological conditions.

Although Haffer's original refuges were explained on the basis of patterns of avifaunal distribution, other explanations have been put forward. The botanical evidence has been criticized by Nelson and co-workers (1990) as an artifact of uneven collecting. Endler (1982; see also Turner, 1982) has proposed that the observed patterns are a result of present-day ecological barriers to gene flow. In this scenario, speciation is parapatric (i.e., speciation with no complete disjunction between daughter populations) rather than allopatric. This model has become known as the hypothesis of parapatric differentiation. Terborgh (1992) considered that the refuge hypothesis was perhaps a special case, applicable only to Pleistocene events.

The Holocene (the last 10,000 years of the Quaternary) has been marked by periods of warming and cooling. For example, studies in North America have shown that climate change was continual and could affect vegetation dramatically in a very short time. One important conclusion of this work is that species behave as individuals and not as communities in response to climate change. Hunter and colleagues (1988) have written:

> Paleoecological information on the distribution of plant taxa in North America, however, indicates that most modern plant communities are less than 8,000 years old and therefore are not highly organized units reflecting long-term co-evolution among species. Rather, they are only transitory assemblages or co-occurrences among plant taxa that have changed in abundance, distribution, and association in response to large climate changes of the past 20,000 years.

And Webb (1987) stated that "such changes are likely to have been a feature of vegetation dynamics for millions of years." If this applies to the Amazon region, and if such changes have been taking place through the Pleistocene and even the Tertiary, then the situation must be considerably more complex than is supposed by the refuge theory. If climate change affects individual species, and not communities, then we would need to look for individual explanations rather than general ones.

Do palms support the refuge theory? There are a few cases where an Amazon palm is confined to one of Prance's (1987, Fig. 3.6) proposed refuges. *Leopoldinia piassaba* occurs partly within the Imerí refuge; *Asterogyne guianensis* and *Syagrus stratincola* occur near the East Guiana and West Guiana refuges, respectively; *Prestoea schultzeana*, *Phytelephas tenuicaulis*, and *Wettinia drudei* occur partly in the Napo refuge; and *Bactris tefensis* occurs within the Tefé refuge. The Napo refuge, in particular, is an area of high palm diversity, but many of the species present are Andean and may owe their presence to the high rainfall in the region. I believe the refuge theory may explain some patterns, but certainly not all. The theory has been so dominant over the last 25 years that the importance of pre-Pleistocene events is sometimes overlooked. Various events during the Cenozoic, however, must have been at least as important in shaping biotic distributions as those in the Quaternary (Cracraft & Prum, 1988). To put Cenozoic events in perspective, angiosperms are known certainly to have existed in the Late Cretaceous (127 million years ago), and some modern orders, families, and genera can be recognized in the remaining 37 million years of the Cretaceous (Raven, 1979). Palm fossils first appear in the Late Cretaceous (about 65 million years ago), become more common in the Paleocene, and by the Eocene are abundant in many parts of the world (Uhl & Dransfield, 1987). To understand Amazon palm biogeography, we need to look beyond the Pleistocene.

Historical Biogeography

The geographical distribution of Palms is not particularly interesting, as these trees are almost equally diffused throughout the Tropics.
　　　　　　　　　　　　　　　　de Candolle, 1857

On a continental scale, it is obvious that there are some recurring patterns of palm distribution in the Amazon. The most striking of these is the endemism of palms in the four geologic provinces discussed in Chpater 1: the Guayana and Brazilian shields, the Amazon Basin, and the Central Andes. Indeed, these geologic provinces appear to be the underlying organizers of Amazon palms.

The Guayana Shield has been recognized as an area of endemism for other groups of plants and animals as well. Mori (1991) cited examples and discussed in detail the Lecythidaceae family. He called the area the Guayana lowland floristic province. Approximately 40 taxa (species and varieties) of palms occur there, but they are not evenly distributed. About six taxa are widespread on the Shield—for example, *Bactris balanophora* and *B. bidentula*. A further seven are more or less confined to the western margin, in upper Rio Negro region. Examples are all species of *Leopoldinia, Mauritiella aculeata,* and *Lepidocaryum tenue* var. *casiquiarense* (but note that some of these Rio Negro palms also occur sporadically in other areas). The largest concentration of taxa, however, is in the eastern part of the shield, in a triangular region roughly bounded on the west by the Rio Branco–Essequibo depression, on the south by the Amazon, and on the east by the Atlantic Ocean. The fact that many of these taxa have their western boundary near Manaus may be a collecting artifact. About 27 taxa occur here—for example, *Attalea attaleoides, Astrocaryum sciophilum,* and eight species of *Bactris*. Within this area is a high concentration of palms in the region of French Guiana, which is also an area of humid tropical *Af* climate. This group can hardly be considered restricted to the *Af* region, however, because many taxa range from French Guiana to Manaus, and thus cross the dry corridor running northwest–southeast across this part of the Amazon region. It is perhaps possible to detect a gap in the range of some species that corresponds to this corridor. *Astrocaryum sciophilum* (Figure 6.57C) is an example, with populations in the wet, coastal part having well-developed aerial stems and those from the drier, Manaus region having short subterranean stems. Because the dry corridor also corresponds to an area with few palm collections (Figure 2.1), the phenomenon may be a collecting artifact. Some members of this group also cross the Amazon and occur in Pará and Maranhão, east of the Rio Tocantins, which marks the eastern boundary of the Brazilian Shield.

The Brazilian Shield species are fewer in number, about nine in total, and consist mainly of species of *Syagrus* and *Attalea. Oenocarpus distichus* also occurs here. Several Brazilian Shield palms reach their western margins along the Rio Madeira—for example, *Attalea eichleri, A. speciosa, Oenocarpus distichus, Syagrus cocoides,* and *S. comosa*. In general, the Brazilian Shield seems poor in species. Scariot and coworkers (1989) found eight species in eight genera in 1 ha of seasonal swamp forest near Serra dos Carajás. A little farther east, at two different sites on the Rio Tocantins, Kahn (1986) found 21 species in nine genera in 10.56 ha of *terra firme* forest. In the present study, only 35 species are known from the whole of the Brazilian Shield (at least the Amazonian part), and these are mostly widespread species that occur throughout the Amazon region. The Brazilian Shield is, however, one of the most poorly collected, and heavily deforested, areas.

The Amazon Basin, principally the western expanded part, is roughly triangular in shape and is bounded on the west by the Andes and on the other two sides by the Rios Japurá-Caquetá and Madeira (i.e., the western margins of the Guayana and Brazilian shields). This area is a region of Cenozoic sediments, intense fluvial dynamics, and *Af* humid tropical climate. Seventy-four taxa occur here. They are not evenly distributed, some being more plentiful in the northern part, others in the south, but there is an extraordinary concentration of species in the central part, in the Iquitos region between the Rios Putumayo and Ucayali (i.e., the Napo refuge). The ranges of some varieties of some species—for example, *Wendlandiella gracilis, Desmoncus mitis,* and *Astrocaryum murumuru*—appear to show some correlation with the small basins of the western Amazon region (Räsänen et al., 1987). Also of interest is that this western region is the area of greatest morphologic diversity in Amazon palms, as measured by the number of varieties present. The western margin of the Brazilian Shield is a boundary for several species, such as *Bactris concinna, B. bifida, Chamaedorea pauciflora, Geonoma laxiflora,* and *Oenocarpus mapora*.

The Andes are a region of high palm diversity. Sixteen taxa that are primarily Andean also reach the Amazon lowlands. Examples are most species of *Chamaedorea, Aiphanes, Wettinia,* and *Pholidostachys*, as well as *Syagrus sancona, Bactris setulosa,* and *B. macana*. There was a possible connection between Amazon and Chocó rain forests through the Huancabamba deflection until the Late Pliocene (Dixon, 1979). This could explain the cross-over of several trans-Amazon palms in the Ecuador region.

Apart from the palms of these four geologic

provinces, there are a few other patterns. About 26 species are widespread in the Amazon region and do not show any western or eastern preference (e.g., *Iriartella setigera*). These often have outliers in Trinidad (e.g., *Bactris simplicifrons*), the Magdalena valley of Colombia (e.g., *Bactris simplicifrons, Geonoma maxima*), or even the Pacific coast of Colombia (e.g., *Bactris brongniartii, Geonoma stricta*). Three species are bimodal. *Raphia taedigera* and *Euterpe oleracea* occur in Central America and northwestern Colombia, and also the estuary of the Amazon. They should perhaps be considered mangroves. *Dictyocaryum ptarianum,* from the Guayana Highland, occurs sporadically in the western lowlands.

How can we explain these patterns? Certainly some palms are found in proposed refuges, as previously discussed. If, however, these refuges were expanded and combined, they would correspond to geologic provinces. For example, the Imataca, W. Guiana, E. Guiana, Imerí, Manaus, and Trombetas refuges are all on the Guayana Shield; the Belém, Tapajós, and Aripuanã refuges are all on the Brazilian Shield; and the Napo, São Paulo de Olivença, Tefé, and E. Peru-Acre refuges are all in the western Amazon Basin. A possible explanation for these patterns has been given by Cracraft and Prum (1988) in their cladistic analysis of neotropical birds. They found a repeating pattern of systematic relationships and geographic areas that they considered a result of a series of vicariance events. They emphasized that these events could have occurred at any time during the Cenozoic, not just in the Pleistocene. The first event was the partitioning of a widespread neotropical biota into two: Chocó–Central America and the rest of the Amazon region. This could have been caused by mountain building in the Andean region as early as the Oligocene–Miocene. The next series of events were the isolation of the Guyanan region (i.e., Guayana Shield) from the rest of the Amazon region, possibly by river valleys or marine transgressions at some time in the Cenozoic; the isolation of the Belém-Pará-Rondônia region (i.e., the Brazilian Shield), also by marine or fluvial action; and finally the splitting of the western Amazon region into the Napo and Inambari refuges.

This scenario makes very good sense for palms, but unfortunately we do not yet have cladistic hypotheses with which to test these ideas. It seems likely, however, that Tertiary geological history as well as the Pleistocene climate changes have been equally important in influencing palm distribution patterns.

The Amazon region is coming to be regarded as an area of continual disturbance, on the large and small scale and over the short and long term. On a small scale and in the short term, this is generally seen as turnover of the forest by tree falls and subsequent light gaps (Denslow, 1987). Other mechanisms, such as inundation on *terra firme,* may also contribute to disturbance (Mori & Becker, 1991; see also Nelson, in press). At the other extreme, on the large scale, are climatic and geologic changes such as Pleistocene changes and Tertiary Andean uplift. Although sometimes described as a "fragile forest," the Amazon ecosystem is extremely robust and has undergone constant change. According to Colinvaux (1987): "The high species richness of Amazonia is a result of numerous opportunities for vicariance because of a very large total area, wide variety of habitats and intermediate levels of disturbance, particularly hydrological processes, that has varied on time scales from years to millennia."

4

Ecology of Amazon Palms

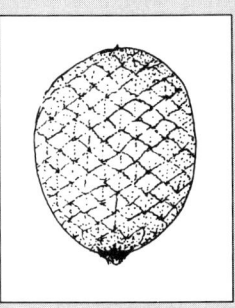

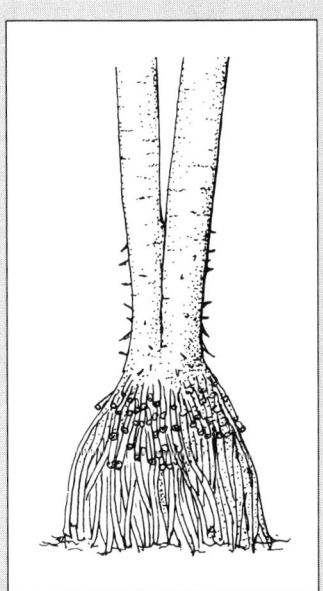

In this chapter, I review some of the ecological studies of palms that have been carried out in the Amazon region. As will become obvious, considering the size of the region and the abundance of the family, we know very little about ecology of Amazon palms. Nevertheless, these studies are of great interest to systematists because they provide explanations for morphologic and geographic patterns.

Physical Environment

Francis Kahn and co-workers have carried out extensive studies on the relationship of palms to their physical environment in the Amazon region (summarized in Kahn and de Granville, 1992). They noted that the palm flora of the western part of the Amazon was richer in species than that of the central or eastern regions, and that diversity of palms was greater in *terra firme* forests than in inundated areas. Furthermore, most of the increase in diversity in the west was due to *terra firme* species; in general, inundated forests have a rather homogeneous palm flora. Overall, they found that most species of palms were small, understory plants; the number of tall, arborescent palms was very low. The reasons for this community structure were to be found in forest architecture and dynamics. Development of arborescent palms depended on high light intensity, and these conditions were found in inundated forests. Here such common arborescent palms as *Mauritia flexuosa* and *Oenocarpus bataua* occurred. These palms were highly specialized for growth in waterlogged soils.

Phenology

As yet, there are no community-level phenological studies of palms in the Amazon, such as exist, for example, in Panama (de Steven and colleagues 1987). Kahn and de Granville (1992) have reviewed the few studies that have been carried out. They concluded that flowering and fruiting periods vary across the Amazon region, depending on wet and dry seasons. Most *terra firme* species in the eastern Amazon flowered in the dry season and fruited during the rainy season.

Various other data can be found scattered in

the literature, and it is interesting to compare the same species from different climate regimes. García (1988) found that in the Ecuadorian Amazon, in an area of approximately 3000 mm annual rainfall, *Oenocarpus bataua* var. *bataua* flowered all year round, with a peak (February–March) at the end of the dry season (December–March). Approximately 300 km to the east, in the Colombian Amazon, also in an area of 3000 mm rainfall, Vélez (1992) found that *O. bataua* var. *bataua* also flowered continually during the year, with peaks at different times in different years, but again with a peak in the dry season (December–February). To the south, in Madre de Dios, Peru, in an area of 2000 mm rainfall, Phillips (personal communication) found that peak flowering took place from December to July 1991. In the central Amazon region, near Manaus in Brazil, in an area of approximately 2200 mm rainfall, Henderson and colleagues (in preparation) found that plants flowered all year round, except for the dry season (July–September). Farther to the east, in French Guiana, in an area of 3000 mm rainfall, Sist (1989a) found that plants of *O. bataua* var. *oligocarpa* flowered all year round, with a peak in the dry season (August–October), but with peaks at other times of the year as well. Although Sist suggested that flowering of var. *oligocarpa* was biennial, evidence from other areas does not support this. The species does appear, however, to flower sporadically, with perhaps a dry-season peak in areas of higher rainfall and a wet-season peak in other areas.

There are some phenological data for another common species, *Mauritia flexuosa*. Heinen and Ruddle (1974) found that in Delta Amacuro, Venezuela, staminate trees flowered together in May, at the start of the rainy season. Pistillate plants also started flowering at this time, but more sporadically. Fruits were ripe from August to October. In the Colombian Amazon (Urrego, 1987; Vélez, 1992), flowering took place in pistillate plants from September to October 1987, and from May to December 1988.

These scattered reports indicate that at least in some palms, flowering and fruiting are not annual events regulated by change of season. It seems that reproductive phenology is supraannual and varies considerably from year to year. This is supported by anecdotal reports that indicate that some palms will flower and fruit profusely in some years and then go for several years with little or no reproduction. The possibility of mast-fruiting in palms should be investigated.

Reproductive Biology

Thus, pollination in many palms is a rather unspecialized process.
Silberbauer-Gottsberger, 1990

Pollination, dispersal, and predation are only recently being investigated in palms. Some previous pollination studies have been reviewed by Henderson (1986a). Three common pollinators were recognized: beetles, bees, and flies. The suite of features associated with each of these was described. Wind pollination was considered rare in palms. Since 1986, a few more pollination studies have been carried out on Amazon palms, and they have shown a much more complicated situation than previously suspected.

García (1988) studied pollination of *Oenocarpus bataua* var. *bataua* in Amazonian Ecuador. He found this species to be protandrous, with diurnal staminate anthesis lasting about 3 weeks. Inflorescences produced a weak scent at this time, and individual staminate flowers lasted a few days. After a pause of about 1 week, pistillate anthesis began, and lasted about 6 days. Inflorescences at this time gave off a stronger, sweet scent between 16:00 and 20:00 hours, accompanied by a rise in inflorescence temperature of 3°C to 5°C. A total of 36 different species of insects visited inflorescences at anthesis, but García considered the most important pollinators to be curculionid beetles of the tribe Derelomini (including *Phyllotrox*), Nitidulidae *(Mystrops),* and Staphylinidae. An unidentified member of the Cydnidae was also considered a possible pollinator. The results of this study are very interesting because the development of the inflorescence appears typical of a bee-pollinated palm (i.e., with protracted diurnal anthesis), but pollination is typical of beetle-pollinated palms. The contracted rachis and closely spaced rachillae seem significant here, considering beetle behavior, compared with the open inflorescence of the related *Euterpe*.

Olesen and Balslev (1990) studied pollination of *Geonoma macrostachys* var. *macrostachys* in Ecuador. This palm is protandrous and diurnal, with staminate and pistillate anthesis lasting 3 to 4 days. Flushes of about 200 staminate flowers opened per day, between 09:00 and 10:00 hours, and produced a strong, sweet smell. Twenty-two species of insects visited staminate flowers, but only 10 visited pistillate flowers, and these were apparently tricked by the resemblance of the pistillate to the staminate flowers. The most probable pollinators were trigonid bees and drosophilid flies. Olesen and Balslev considered that pollination in this case was a result of the pistillate flowers mimicking staminate flowers.

Listabarth (1992) surveyed pollination strategies of seven species of the Bactridinae in the Peruvian Amazon. *Aiphanes aculeata* was found to be protandrous and pollinated by either wind or bugs, beetles or bees. Conversely, *Bactris* spp., *Desmoncus* spp., and *Astrocaryum murumuru* (as *A. gratum*) were all protogynous with brief anthesis and were pollinated by curculionid and nitidulid beetles. Listabarth considered *Aiphanes* less specialized than the other genera, and *Bactris* and *Desmoncus* more specialized than *Astrocaryum*.

Listabarth (1993a) also studied pollination of the dioecious species *Chamaedorea pinnatifrons* and *Wendlandiella simplicifrons*. He found an insect-induced wind pollination in *Chamaedorea*, with thrips and small beetles releasing pollen from the anthers of staminate flowers, which was then carried by the wind to pistillate flowers. *Wendlandiella* was presumed to be wind pollinated, but vegetative reproduction was also considered important.

Finally, Listabarth (1993b) studied four taxa of *Geonoma* in Amazonian Peru. He found that both *Geonoma macrostachys* var. *macrostachys* and *G. macrostachys* var. *acaulis* were pollinated in a way similar to that found by Olesen and Balslev, with protandrous, diurnal inflorescences and pistillate flowers mimicking staminate flowers. However, var. *acaulis* was pollinated by small weevils, whereas *G. brongniartii* was considered to be wind pollinated. *Geonoma interrupta* was also diurnal and protandrous, but staminate and pistillate phases overlapped. Pollination was by meliponid bees, wasps, and muscid flies.

The studies of García, Olesen and Balslev, and Listabarth have shown that pollination modes in palms are more variable than previously thought, and we can begin to see some complex patterns even within genera. What are needed now are comparative studies of closely related species; Listabarth, and Borchsenius (1993), have begun to carry these out. However, we are now at least in a position to doubt the assertion of Silberbauer-Gottsberger (1990) that pollination in palms is an unspecialized process. It also seems likely that in the future many palms, at least those in the understory, will be found to have asexual reproduction rather than sexual reproduction.

Predation and Dispersal

In general, palm dispersal does not appear to involve such mutualistic interactions as pollination (Zona & Henderson, 1989). Although dispersal is often animal-mediated, a whole range of animals can be involved, and these can act as either dispersers or predators. One of the most fascinating and best studied interactions in neotropical palms is between certain cocosoid palms, especially *Attalea* and *Astrocaryum*, and rodents, monkeys, peccaries, macaws, and insects. These two cocosoid palms have fruits with very thick endocarps covering an edible endosperm, and both are often very abundant, evenly distributed, and co-occurring. Only the largest and smallest predators are able to penetrate the endocarp and eat the endosperm.

The story begins in Costa Rica, where Janzen (1971) described the interaction between *Attalea butyracea* (as *Scheelea rostrata*) fruits, bruchid beetles, and rodents. When the fruits fell from the trees, the mesocarp was eaten by rodents and the clean endocarps were discarded beneath the tree. Some endocarps, however, were taken by the rodents and scatter-hoarded for later consumption. Those cleaned endocarps left behind were susceptible to oviposition by bruchid beetles (female bruchids did not oviposit unless the epicarp and mesocarp were removed). The females laid eggs on the endocarp surface; the first instar larvae bored through the very thick and tough endocarp and ate the endosperm. Finally, just before pupating, the larvae cut an exit hole almost to the outer surface, and then the adult finished the hole and emerged. Rodents also chewed open large numbers of endocarps left beneath the tree, in search of either bruchid larvae or intact endosperm. Wilson and Janzen (1972) found that the actions of scatter-hoarders, taking endocarps away from the parent tree, improved the likelihood of the seeds germinating. They noted that adult palms occurred with a density of eight to 20 trees per hectare. Wright (1983) also found that the further the endocarps were from the parent tree, the less the predation by bruchid beetles.

Following these initial studies, more work was carried out in both Costa Rica and Panama. Bradford and Smith (1977) compared two palm populations, one in Costa Rica and one in Panama. Most fruits in both populations contained one seed, but a small proportion contained two to three seeds. In Costa Rica, 61 percent of fruits were attacked by predators, and 84 percent of predators were bruchids. These bruchids had a slight tendency to attack one-seeded fruits, so that two- and three-seeded fruits had a higher chance of survival. In Panama, 81 percent of fruits were attacked, but 78 percent of predators were rodents, and these attacks were independent of the number of seeds. As one might have predicted, researchers found a greater proportion of two- and three-seeded fruits in Costa Rica than in Panama, and this was used by Bradford and Smith to support the hypothesis that bruchid predation in Costa Rica was a selective pressure to maintain multiseeded fruits in the palm population.

In Panama, Smythe (1989) studied seed survival in *Astrocaryum standleyanum*. These palms were extremely fecund and produced up to six infructescences per year, each with 300 to 800 fruits. Fruits of this palm were a staple of the diet of agoutis *(Dasyprocta punctata)* from March to July, the period when the fruits fell from the tree. Although agoutis ate the mesocarp of fresh fruits, they would often peel away the epicarp and mesocarp but not necessarily eat it. Several other animals ate the mesocarp, but only the largest and smallest, peccaries and bruchids, were able to obtain the internal endosperm. The agoutis also scatter-hoarded endocarps for consumption from September to December. Smythe noted that this scatter-hoarding behavior was inherent in agoutis:

> [A]n agouti was raised from the age of about three weeks in a concrete-floored cage, on a diet of soft fruits and vegetables. At the age of seven months the animal was offered the freshly fallen *Astrocaryum* fruits, which it could never have seen before. It peeled the seeds, cleaned them, dug imaginary holes in the concrete, placed the seeds "in" the holes, refilled them with imaginary soil and sometimes placed an imaginary leaf over a "buried" seed.

Smythe hypothesized that since agoutis depend heavily on *Astrocaryum,* and since both have a long evolutionary history in the Neotropics, then perhaps the palm benefited from the agoutis. He carried out some tests and found that if the fruits were not peeled and buried, in the manner of agoutis, then the probability of seed survival was greatly reduced.

Still in Central America, we should note that squirrels also are predators of these palms. Glanz and colleagues (1983) showed that on Barro Colorado Island in Panama, the red-tailed squirrel *(Sciurus granatensis)* was the main arboreal seed predator of *Astrocaryum standleyanum* and *Attalea butyracea*. These palms provided the main diet of the squirrels from March to August, when the squirrels also scatter-hoarded the endocarps. Furthermore, reproduction in the squirrels coincided with the fruiting season of the palms. In fact, in a nearby mainland population of squirrels where *Astrocaryum* was uncommon, reproduction was limited and closely timed to coincide with *Attalea* fruit production.

In Manu National Park, Peru, the same two palm genera are present, but are represented by different species: *Astrocaryum murumuru* and *Attalea phalerata*. These are also large-stemmed palms, and both are very common and important in the diet of monkeys. Terborgh (1983) reported that these palms were surprisingly uniformly distributed in the region, and were used by white-fronted capuchins and brown capuchins *(Cebus albifrons* and *C.*

apella) in a number of ways. Fruits of *Astrocaryum murumuru* matured during late March and early April, when the monkeys fed on the fleshy mesocarp. In early April, the fruits fell from the tree and the mesocarp began to rot. In June, the monkeys fed on these fruits, but only with the help of bruchids. Because an intact endocarp is too hard for a monkey to break open, the monkeys would look for ones partially eaten by a bruchid larva. Most endocarps on the ground were used as oviposition sites by female bruchids, and each soon contained larvae. During June, troops of *Cebus albifrons* would spend hours on the ground beneath *Astrocaryum* trees, selecting endocarps to eat. Terborgh wrote:

> The animals were extremely selective, often checking and rejecting twenty or more nuts before finding one that was suitable. The preliminary screening was conducted by smell. (Presumably byproducts of the bruchid larva and decayed endosperm both emit characteristic odors.) Nuts that passed the smell test were then tapped against a branch or another nut, or bitten to assess their resistance. Many were rejected at this stage, as well as earlier when first sniffed. When an animal found one that seemed to possess the desired qualities, it usually went up onto a low branch for the opening operation. There it would bash the nut vigorously against the branch, and then begin to bite it with its premolars. After a bite or two the nut would be rotated to a slightly different position and bitten again. If it failed to yield, it would be rejected, and the whole selection process would begin anew. If the nut cracked, the endosperm would be laboriously excavated from the shell, using canines or fingernails to pick at the firm material.

Other animals in Manu also compete for fruits of these palms. Emmons (1984) stated that the diets of two large squirrels consisted almost entirely of palm fruits (as in Panama). Presumably, squirrels and other rodents scatter-hoarded endocarps. Kiltie (1981) observed that white-lipped peccaries tended to forage near such objects as trees, logs, or exposed roots. He hypothesized that the peccaries were looking for scatter-hoarded endocarps of *Astrocaryum murumuru*. Macaws are also predators of these palms. Munn and co-workers (1989–1990) reported that hyacinth macaws rely on fruits of *Astrocaryum vulgare* and *Attalea speciosa,* and also *A. phalerata*. With their incredibly powerful beaks, they can shear open the endocarp and eat the endosperm.

Such was the importance of these palms to the larger predators that Terborgh (1986) called them "keystone" species in Manu National Park. Even though a limited set of predators—such as peccaries, capuchins, agoutis, squirrels, and macaws—were able to utilize the palm fruits, these animals were so large that they made up 30 percent of the frugivore biomass. Terborgh considered that *Astrocaryum* and *Attalea* were probably keystone species throughout the Neotropics. This is an interesting idea, because species of both genera are very common and abundant, and are often the most important fruiting trees for many animals. As we have just seen, *Astrocaryum* and *Attalea* have been studied in Costa Rica, Panama, and Peru, where they provide food for bruchids, agoutis, capuchins, macaws, squirrels, and peccaries. In central Amazonia, two different species occur. In a site north of Manaus, Henderson and colleagues (in preparation) found *Astrocaryum sciophilum* and *Attalea attaleoides* dominating the forest understory. Both species are different in having short, subterranean stems, but are evenly spaced in the forest and occur in high densities: 194 and 85 adults per hectare, respectively. Curiously, over 3 years of observations, these two species hardly ever flowered or set fruit, despite their abundance. Emmons (1984) also commented on this, and noted that the two large squirrels present at Manu were absent from this region.

From this brief survey of the dispersers and predators of *Astrocaryum* and *Attalea,* we can see that the selective pressures acting on them, and the competitive interactions between them, are extraordinarily complex. Are the palms in an evolutionary race with their predators to produce endocarps so thick that no animal can break them open or bore through them? Janzen (1971) suggested that the bruchids could play an important role in the evolution of sexuality of *Attalea*. It also has been suggested, though, that host-plant phenology affects sex-pheromone dynamics of bruchids (Pouzat et al.,

1989). The local- and regional-distribution patterns of the palms, and their predators, must also be intricate. Smythe (1989) noted that the demise of any part of the chain—for example, the hunting of agoutis in Panama—could adversely affect the entire forest.

Demography

Demographic studies of palms have recently been reviewed by Pinard and Putz (1992). More than 20 species have been studied in the last 20 years, reflecting the suitability of the family for this type of study. Palms are abundant, and it is possible to estimate their age.

In the Amazon region, only three species have been studied. Sist (1989b) studied the demography of *Astrocaryum sciophilum* in French Guiana. He found a situation common in other plants: a high number of seeds, seedlings, and juveniles, with seedlings making up 60 percent of the population, and a relatively low number of adults. Mortality resulted from both falling branches and predation by squirrels. At the same site, Sist (1987) studied demography of *Oenocarpus bataua* var. *oligocarpa* (as *Jessenia bataua* subsp. *oligocarpa*). Again, the population consisted of many seedlings and juveniles and few adults.

The population structure of a completely different palm in a completely different habitat was studied by Anderson (1983). *Attalea speciosa* (as *Orbignya martiana*) grew in both undisturbed and secondary forests in the Brazilian state of Maranhão. In undisturbed forest, as expected, the population consisted of a large number of seedlings and juveniles and few adults. However, in secondary forest it showed a preponderance of 33-year-old individuals, reflecting the time since disturbance. This demonstrated the remarkable ability of *Attalea speciosa* to flourish on disturbed sites.

5

Palm Collectors, Palm Botanists, and Species Concepts

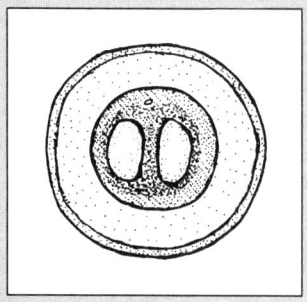

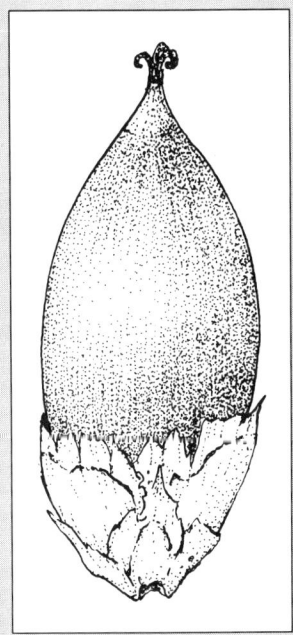

Historical Review

In palmis semperparens juventus; in palmis resurgo (In palms is ever creative youth; in palms I rise again). Martius, 1831–1853

The starting point of modern botanical systematics, the classification of plants, was 1753, the year in which Carl Linnaeus (1707–1778) published his *Species Plantarum*. There he began the practice of using a binomial system to name species and of placing supposedly related species in groups. Linnaeus knew few palms and included only nine species in his work, one of which was a cycad and none of which were from the New World. He at least recognized, however, that palms formed a natural group, the Palmae.

Linnaeus's son Carl (1741–1783) was the first botanist to use the binomial system to describe an Amazon palm. In 1781, he published *Mauritia flexuosa*. He was also, unwittingly, the first botanist to emphasize the importance of field work in palm systematics. He did not see *Mauritia* in the field, and described it as having almost no leaves: "Valde singularis haec arbor fere aphylla" (This tree is alone in being nearly leafless) (Linnaeus, 1781). The first botanists actually to collect Amazon palms, at least those from the northern fringe of the region, were Alexander von Humboldt (1769–1859) and Aimé Bonpland (1773–1858).* These celebrated naturalists traveled in the New World between 1799 and 1804. They reached San Carlos de Río Negro in southern Venezuela and investigated the Río Casiquiare, which links the Orinoco with the Amazon. Humboldt was greatly impressed by palms. Unlike Linnaeus's son, he saw *Mauritia flexuosa* growing in the wild and wrote of the dependence of Indian tribes on it: "The benefits of

---

*It is possible that the first botanist to collect Amazon palms may have been Alexandre Ferreira (1756–1815). He traveled extensively in the Amazon region of Brazil between 1783 and 1792, 10 years before Humboldt and Bonpland (Ferreira, 1972; see also Prance, 1971). To be both patriotic and pedantic, however, I should point out that Sir Walter Ralegh collected a fruit of *Mauritia flexuosa* from South America and brought it back to England in 1595 (Humboldt, 1850).

this life supporting tree are widely celebrated; it alone, from mouth of the Orinoco to north of the Sierra de Imataca, feeds the unsubdued nation of the Guaranis" (Humboldt, 1850). Most of Humboldt and Bonpland's 6000 collections of plants were gathered by Bonpland, but when he returned to Paris he lost enthusiasm when it came to the monumental task of describing them. Finally, in 1816, Bonpland emigrated to Argentina, and Humboldt gave the collections to another botanist, Carl Kunth (1788–1850). Kunth completed the task of describing the plants, and the new genera and species are credited to him, including the Amazon genus *Attalea*. They were published in 1815 in the first volume of *Nova Genera et Species Plantarum,* part of Humboldt and Bonpland's *Voyage aux régions equinoctiales du nouveau continent fait en 1799–1804.*

Two other notable collectors visited South America during the latter part of the eighteenth century, Hipólito Ruíz (1754–1815) and José Pavón (1754–1844). These intrepid botanists endured extreme hardships during their 10 years of collecting in Chile and Peru (Dahlgren, 1940). Although they never reached the Amazon lowlands, they did collect numerous palms. Subsequently, they described two common genera from the western Amazon region, *Iriartea* and *Phytelephas,* as well as nine new species of palms.

Mention should also be made here of two French botanists: Poiteau and Aublet. Jean Aublet (1720–1778) was one of the first botanists to study the flora of the Guianas. His *Histoire des plantes de la Guiane française* was published in four volumes in 1775. The second volume contains descriptions of palms, using pre-Linnaean polynomials, and also an early essay on their uses. Pierre Poiteau (1766–1845) lived for three years in French Guiana, from 1819 to 1822. His article "Histoire des palmiers de la Guiane française," published in 1822 on his return to France, contains descriptions and illustrations of five common species of *Geonoma* (as *Gynestum*). He also was one of the first to write on the difficulty of collecting palms. Although both Aublet and Poiteau published illustrations of Amazon palms, they were preceded in this by Cristovão de Lisboa, who

Figure 5.1. Carl von Martius. (Photograph courtesy of the Hunt Institute for Botanical Documentation, Carnegie Mellon University, Pittsburgh)

illustrated four palms from Maranhão, Brazil, after his journey there between 1625 and 1631 (de Lisboa, 1968).

The first half of the nineteenth century was a golden age for the exploration of the tropics by Westerners. European naturalists, inspired by the works of Humboldt and others, were drawn to South America and especially to the Amazon Basin. One of the greatest of them all was Carl von Martius (1794–1868) (Figure 5.1).

Martius was raised in Germany and developed an interest in botany at an early age (de Candolle, 1857). In 1816, a scientific expedition accompanied the Archduchess Leopoldina of Austria, who was on her way to join her future husband in Brazil. The king of Bavaria, Maximilian Joseph, persuaded the Austrians to take along German scientists, and the botanist Martius and zoologist Spix were chosen. Martius arrived in Brazil in 1817, when he was 23 years old, and spent the next 4 years collecting plants and numerous other natural history specimens, as well as information on the lives of the Indians he encountered. He and Spix traveled overland from Rio de Janeiro and arrived at the mouth of the Amazon, at the city of Belém, in 1819. They spent the next 11 months in the Amazon. They were among the first non-Portuguese scientists allowed into the Brazilian Amazon region. One day, Martius got lost in the forest:

> The joys of that agreeable contemplation of nature were replaced by terror; for I came to a swampy area where I found myself surrounded by impenetrable groves of prickly palms *(Bactris maraja),* where tacky arrowroot bushes became entangled ever more inextricably about me, where the broad-leaved heliconias on which I was trying to gain a foothold concealed a branch of deep water, and where, when I froze and strained my ears, I thought I heard the call of crocodiles, sure of their prey, on their way to make a meal of a stray traveler. Then, to my great horror, I realized that I had wandered into one of those suspicious ponds that the Indians themselves make a point of shunning because they are considered deadly labyrinths and thought to harbour dangerous animals. It was starting to get dark. . . . I clambered up the trunk of a jubatí palm *(Raphia taedigera),* several footstalks of which formed a kind of stairway. I was surely safe from wild animals in the thick branches of this tree. (Spix & Martius, 1831; translated in Gheerbrant, 1992)

Martius continued up the Amazon as far west as Araracuara on the Río Caquetá in Colombia. He collected and described almost all the *várzea* palms of the Amazon, and left for subsequent botanists a few undescribed species on the *terra firme.* On his return to Europe, Martius began what was to become the greatest monograph on palms, or perhaps any plant family, ever produced. The first part of the first volume, in large folio, of his *Historia Naturalis Palmarum* was published in 1823; the final part of the third volume was published 30 years later, in 1853. This work included all the current knowledge on palms from all over the world, including anatomy and palaeobotany. It also included one of the first attempts to provide a subfamilial classification of the family. In the second volume, Martius described his palm collections from the Amazon in magnificent detail and illustrated them with lavish color plates. Here he honored his early patrons with the genera *Leopoldinia* and *Maximiliana.* He described a total of 85 Amazon palms (54 of which are accepted here). In 1840, Martius began a general work on the plants of Brazil, *Flora Brasiliensis,* which was completed in 1906, long after his death. The palms were treated by Drude (1881, 1882). Martius is also remembered for his work on Brazilian ethnography (Hemming, 1987).

Martius also described the specimens collected by other botanists. The Austrian Eduard Poeppig (1798–1868) collected plants in Peru and Brazil between 1829 and 1832. Martius named *Geonoma poeppigiana* for him. Alcide d'Orbigny (1802–1857) was a French naturalist and traveler in southern South America between 1826 and 1833. His journey took him from southern Brazil to Uruguay, Argentina, and Bolivia (Papavero, 1971). In Bolivia, d'Orbigny collected and illustrated numerous palms. On his return to France, he began an eight-volume work on his travels and collections, which was published at the expense of the French government. The seventh volume was written by Martius (1842–1847) and contained descriptions of d'Orbigny's new species from Bolivia. In this work, sometimes known as *Palmetum Orbignianum,* Martius described many Amazon palms, including the genus *Orbignya*.

Two remarkable English botanists, Wallace and Spruce, followed in Martius's footsteps. Alfred Russel Wallace (1823–1913) (Figure 5.2a) was primarily an entomologist, and with his friend and fellow entomologist Henry Bates he left England in 1848 for the Amazon (Wallace, 1889). He spent the next 4 years collecting there, and included palms in his collections, principally from the Rio Negro. He suffered severe malaria and dysentery on this trip, but managed to make valuable observations on the Indians of the region. Although a zoologist, he was impressed by the palms he saw: "In the districts which I visited they were everywhere abundant, and I soon became interested in them, from their great variety and beauty of form and the many uses to which they are applied" (Wallace, 1853). Wallace met Spruce in 1851 in Manaus. Despite an offer from Spruce that they collaborate on palms, they continued their separate ways (Balick, 1980). On Wallace's return to England in 1852, the ship on which he was traveling caught fire and sank. Wallace lost all his plant specimens, but retained a notebook with his sketches of palms. He published these in *Palm Trees of the Amazon and Their Uses* (1853). Here he described 13 new species of Amazon palms (five of which are accepted here, mostly from the upper Rio Negro region). This is a classic although somewhat anecdotal account of the ethnobotany of Amazon palms that showed Wallace's keen interest in the people of the region (he was the first Westerner to record the Yurupari legend of the Rio Vaupés) and how they used palms. In contrast to Spruce's interest in botanical detail, Wallace was more interested in ethnobotany. The book includes the important Rio Negro species *Leopoldinia piassaba* and *Mauritia carana*. Wallace stayed in England for just 2 years after his return from Brazil, and then continued on to Southeast Asia, where he made major contributions to biogeography and the theory of evolution. In fact, Wallace's paper on natural selection, written in Indonesia, prompted Darwin to publish his own studies, *The Origin of Species,* in 1859. Wallace died in Dorset, England, at age 90.

Richard Spruce (1817–1893) (Figure 5.2b) left his native Yorkshire in 1849 (he traveled with Herbert Wallace, who was on his way to join his older brother, Alfred, and who later died of yellow fever in Belém) and spent the next 13 years collecting in the Amazon and Andes of South America. He was extraordinarily industrious and endured great hardship while collecting thousands of plants that he sent back to Kew Gardens in London. He looked back with some envy on the earlier expedition of Martius:

> Protected by the Emperor of Brazil, and provided by the government of that country with all possible aids in the prosecution of his enterprise (rarely lacking Indians to row his boats and to cut down or climb the trees of which he desired to secure specimens) he possessed advantages seldom enjoyed by a solitary botanist travelling and working in so modest a way as myself. (Spruce, 1871)

Like many other early Amazon travelers, Spruce soon became interested in palms. In a letter to England, he wrote: "I can find no one who will talk to me about Palms, and I am now coming among some that are exceedingly interesting" (Spruce, 1908). His major contribution to the knowledge of Amazon palms was his 1871 work *Palmae Amazonicae.* This was written in his house in Yorkshire, using his own specimens sent from Kew. He described 42 new species (10 of which are accepted here), and also produced an essay on the geography of palms. Also at this time, in Yorkshire, he corresponded with Martius, who asked him to contribute to the *Flora Brasiliensis,* but Spruce declined because of his poor health (Spruce, 1908). After Spruce's death, his Amazon and Andean diaries were edited for publication by Wallace (Spruce, 1908). In 1993, the Linnaean Society of London commemorated the centenary of Spruce's death with a symposium in Yorkshire, England.

Two other notable botanists who studied Amazon palms in the latter half of the nineteenth century were Barbosa Rodrigues and Trail. João Barbosa Rodrigues (1842–1909) (Figure 5.2c) was born in Minas Gerais and spent his early years in southern Brazil. In 1872 he was commissioned by the Brazilian government to continue the work of Martius

Figure 5.2. **(a)** Alfred Russel Wallace; **(b)** Richard Spruce; **(c)** João Barbosa Rodrigues; **(d)** James Trail. (Photographs courtesy of the Hunt Institute for Botanical Documentation, Carnegie Mellon University, Pittsburgh)

on palms, and he began a series of collecting trips to the Amazon. One of his earliest publications on the family appeared in 1875 when he described numerous new species in *Enumeratio Palmarum Novarum*. However, his major contribution was the two-volume folio *Sertum Palmarum Brasiliensium* (1903). Here he treated 382 species of palms, 166 of which he had described himself. The work also contained 174 magnificent watercolor paintings done by Barbosa Rodrigues. These are extremely useful because his original specimens are presumed to have been lost in Manaus. The plates in the book thus became the types of his species. Barbosa Rodrigues described 99 new species of Amazon palms (eight of which are accepted here). He relied on native knowledge, and, as mentioned previously, in his subgeneric groupings he used vernacular names for sections and subsections. Barbosa Rodrigues was interested in many aspects of Amazon life; he was a pioneer collector of Indian artifacts and legends and is remembered as an ethnographer as well as a botanist (Hemming, 1987). He was active in contacting the Waimiri-Atroari Indians of the Rio Jauaperi, to the north of Manaus (and from where he described many new species). Brown and Lidstone (1878) provide a portrait of him in the field:

> He . . . was possessed of great energy and ardour almost portentous, when he once got to work. He was usually the foremost in jumping ashore, and drawing aside the first native who chanced to make his appearance. He would seat himself upon a log or other convenient perch, and proceed to note down in his pocket book whatever answers he could obtain to his numerous questions, fixing the unfortunate man to the spot all the time by the earnest way in which he eyed him through his gold-rimmed spectacles.

Barbosa Rodrigues spent the final years of his career as director of the Botanical Garden in Rio de Janeiro. Mori and Ferreira (1987) describe him as one of Brazil's greatest naturalists. He was accompanied on many of his travels by his wife, Constança, and named a remarkable species of *Bactris* for her: *B. constanciae*.

On a palm-collecting trip in the Amazon in 1874, in the small town of Obidos on the north bank of the river, Barbosa Rodrigues met James Trail (1851–1919) (Figure 5.2d), a Scot from the Orkney Islands who had come to Brazil as a doctor on a boat belonging to the Amazon Steam Navigation Company. Trail and Barbosa Rodrigues traveled together for 11 days on the Rio Trombetas, and from then on Trail became interested in palms. After they parted company, Trail went on to collect palms on the Rio Negro and Upper Amazon as far as the Rio Javari, which marks the border with Peru. After his return to England in 1875, he spent a year at Kew Gardens working on his collections. The results of his studies were published in the *Journal of Botany* (Trail, 1876, 1877a, 1877b), in which he described 72 new taxa (nine of which are accepted here). Trail dealt scathingly with Barbosa Rodrigues's *Enumeratio Palmarum Novarum* (1875). For example: "Dr. Barbosa Rodriguez [sic] has given a diagnosis of the species [*Bactris elegans*], with which (along with the name) I furnished him; unfortunately the diagnosis has been so altered (in being printed?) as to be in many places unintelligible even to myself, and therefore of little use." Trail repeated this charge several times. He did not, however, place any of Barbosa Rodrigues's species in synonymy, thus breaking the laws of priority (although he omitted Barbosa Rodrigues's name as author of *Bactris elegans*, implying that he, Trail, was the author).

Hooker sent these journals to Barbosa Rodrigues. So offended was Barbosa Rodrigues that in 1879 he wrote a supplement to his *Enumeratio Palmarum Novarum*, called *Protesto-Appendice*. In it, he ridiculed Trail's early efforts at collecting palms and in great detail defended his priority of publication.

Insult was added to injury when Drude (1881, 1882) followed Trail's scheme. For example, he gave the author of *Bactris elegans* as "Trl. & Rodrig.," even though it was clearly first described by Barbosa Rodrigues and not by Trail. Perhaps Barbosa Rodrigues need not have worried so much; after Trail returned to obscurity in Scotland, Barbosa Rodrigues went on to publish major works on palms, and clearly his contribution is much greater than Trail's.

The nineteenth century ended, not only with Barbosa Rodrigues's work, but also with two important classifications of palms at the generic level, made by Bentham and Hooker (1883) and by Drude (1887). These two generic classifications formed the basis for palm taxonomy for the following 100 years.

Maximilian Burret (1883–1964) was the foremost expert on palms during the first half of the twentieth century. He began working in the Berlin herbarium in 1922, and started to work on palms in 1925, apparently at the urging of the director, Ludwig Diels. Within 10 years, he had monographed several large neotropical genera, including *Attalea, Bactris, Euterpe,* and *Geonoma*. During his 30 years of work in Berlin, Burret published more than 100 papers, describing several hundred new species of palms from all over the world (Potztal, 1965). He had an extremely narrow, typological species concept, based almost entirely on herbarium work. It seems that Burret described as a new species almost every specimen he received. From the Amazon region he described 132 new species (10 of them are accepted here).

Although Burret carried out little field work, many other collectors sent their specimens to him. Philip von Luetzelburg (1880–1948) was a German botanist who collected extensively in Brazil. Between 1928 and 1929, he accompanied Rondon, a famous Brazilian explorer and champion of Indian rights, on a frontier expedition in the states of Amapá, Amazonas, and Roraima. His collections were sent to Munich, and Burret described many of them as new. Günther Tessmann (d. 1926), a German collector, explorer, and ethnographer, collected plants in Peru and sent them also to Berlin. Burret named a magnificent palm after him, *Attalea tessmannii* (the other eponyms *Tessmaniophoenix* and *Tessmaniodoxa* survive only as synonyms).

Jacques Huber (1867–1914), a Swiss botanist, lived in Brazil, in the city of Belém, and was one of the developers of the Museu Goeldi. He collected widely in the Amazon region of Brazil in Acre, Amapá, Amazonas, and Pará. He wrote extensively about his collections and, for example, gave a description of the palms of the upper Rio Purus (Huber, 1906). His palm collections were deposited in the Museu Goeldi, and duplicates were sent to Berlin, where they were described by Dammer and Burret. Ynes Mexia (1870–1938) was an American botanical explorer who collected in Peru and Brazil, and made collections of the larger palms and distributed numerous duplicates. Ellsworth Killip (1890–1968) was an American botanist who worked at the Smithsonian Institution in Washington. He collected intensively, usually with Albert Smith (b. 1906) in Colombia and Peru (Darwin, 1993).

Ernst Ule (1854–1915) was a German botanist who collected in Acre, Amazonas, and Roraima in the early twentieth century. His collections were sent to Berlin and described by Burret. Other German collectors whose collections were described by Burret were Werner Hopp, who traveled in Ecuador, Peru, and western Brazil; and G. Huebner, who also collected in Brazil, partly with Karl Lakó. Burret described *Bactris huberiana, B. luetzelburgii, B. killipii, B. ulei, B. hoppii, B. huebneri,* and *B. lakoi*; only *B. killipii* survived synonymy.

Through the efforts of these and other collectors, Berlin contained, by the late 1930s, the largest palm herbarium in the world, and one of the largest general herbaria in the world. On the night of March 1/2, 1943, direct hits by two phosphorus bombs on the herbarium destroyed the building and many of its 4 million specimens (Merrill, 1943; Hiepko, 1987). Burret's life work, an almost completed manuscript on the world genera of palms, was also lost. Many of the palm specimens, comprising about 1000 large sheets, were in a separate building, and, although damaged, they were not burned (Paul Hiepko, personal communication). After the bombing, these sheets were removed for safekeeping to a small town, Ballenstedt, about 250 km north of Berlin. A second disaster struck when approximately two-thirds of these sheets were burned by Russian soldiers. Of the third that finally returned to the Berlin herbarium, about 44 are types of Amazon palms (the other 140 were lost). Although he continued to work on palms after the war, with Eva Potztal, Burret never really recovered mentally from these losses, and was very ill toward the end of his life.

The American botanist Harold Moore (1917–1980) dominated post–World War II palm studies. He was recruited by Liberty Bailey, himself a formidable collector of palms. Bailey's ambition, stated in an interview with

George Lawrence in 1951, was "And then, I expect to make a classification of all genera, and to describe every genus. . . . It would be a Genera Palmarum." Moore was hired to carry out this task. He worked in the Bailey Hortorium of Cornell University for over 30 years and built up the largest palm herbarium in the world. Moore concentrated most of his collecting efforts in the Old World, but he did collect in the Brazilian Amazon in 1967 with Ghillean Prance. Moore was perhaps most important for palm systematics in that he embraced the ideas of the "new systematics" and began using various lines of evidence for studying palm phylogeny, rather than just morphology. He carried out, or encouraged others to carry out, studies on anatomy, cytology, ecology, biogeography, and chemistry of palms, and used the results to formulate schemes of relationships in palm genera. Moore worked for 30 years on his life's work, *Genera Palmarum,* but died before it was completed. His colleagues completed the task (Uhl & Dransfield, 1987). In this work they describe, illustrate, and discuss the 200 genera of palms.

Contemporary with Moore was another important figure in neotropical palm taxonomy, the Dutch botanist Jan Gerard Wessels Boer (b. 1936) (Görts-van Rijn, 1991; P.J.M. Maas, personal communication). He collected palms in Surinam from 1962 to 1963, and his doctoral dissertation from the University of Utrecht was published as the treatment of the Palmae for the *Flora of Suriname* (1965). This was an influential work, not only because for many years it was the only modern palm flora of a neotropical country, but also because Wessels Boer was the first to consign dozens of obscure names to synonymy and greatly simplify the process of identification. It was also one of the first modern works to be based on extensive field work, and one of the first to recognize the great morphologic variation in palms. After Surinam, Wessels Boer spent one year in Mérida, Venezuela. Unfortunately, the result of this work, a palm flora of Venezuela, was not published until more than 20 years later (Wessels Boer, 1988; but see Wessels Boer, 1971a). On his return from Surinam to the University of Utrecht in 1963, Wessels Boer began a revision of *Geonoma,* which was published 5 years later (Wessels Boer, 1968). It was severely criticized by Harold Moore (1969). After this, Wessels Boer gave up his position at Utrecht and in systematic botany, and became active in environmental protection. He also published a work on hybridization in palms and its significance, and honored Harold Moore by naming a hybrid for him, *Bactris* x *moorei* (Wessels Boer, 1971b). He is also notable for his two-volume, beautifully illustrated, popular work on the plants of Surinam *Fa Joe Kan Tak' Mi No Moi* (Why Can't You Say That I Am Beautiful) (Wessels Boer, 1976).

Finally, mention must be made of several other botanists who have collected palms in the last few decades: Bassett Maguire, Ghillean Prance, and Boris Krukoff, all from the New York Botanical Garden, made extensive collections from the Amazon region. Perhaps one of the most important collecting programs for Brazilian Amazon plants, initiated by Prance, was "Projeto Flora Amazônica" (Prance et al., 1984). Many palms from previously uncollected regions became available for study as a result of this program. A few other prominent botanists have also greatly contributed to our knowledge: José Cuatrecasas, Armando Dugand, and Richard Schultes in Colombia, and Julian Steyermark in Venezuela. Tribute must also be paid to Alwyn Gentry, an outstanding botanist, who collected many palms in the Peruvian Amazon and elsewhere. During the last decade, a new generation of collectors, in many countries, have continued the enormous task of collecting the Amazon's palms.

Changing Species Concepts

Few other subjects in biology have generated as much discussion as species concepts (e.g., Mayr, 1982). Changing concepts in botany are reflected in the work of the palm botanists just discussed. For Linnaeus and his followers, species were created and immutable. The typological species concept was derived from this idea, and species were seen as well-defined and completely constant entities. Plant species could be represented in the herbarium by a single specimen. Darwin's work marked the end of the post-Linnaean century. After 1859, species were regarded not as unchanging creations, but as evolving populations represented by variable individuals.

The new thinking that followed Darwin's work had virtually no effect on palm taxonomy, at least for the next 100 years. Burret and Bailey, and others, had extremely narrow, typological species concepts and continued to describe almost every specimen as a new species. Burret, in particular, studied specimens in the herbarium, and he was criticized for failing to take into account natural variation. His generic revisions have been described as a catalog of herbarium specimens rather than systematic accounts (Wessels Boer, 1965). Ideas on geographic ranges were also narrow. Bailey, for example, believed that different areas had different species, a self-fulfilling concept. He wrote that "we can no longer infer that an insular species is the same as a contiguous continental species or that all the islands of the West Indian chain support the same species. Apparently there are many endemics" (Bailey, 1947).

The turning point in palm taxonomy, at least for the Amazon region, was the mid-1960s, when Wessels Boer began working on the palms of Surinam. He faced many taxonomic problems. First and foremost was the proliferation of names from earlier authors and the problem of their application. The type specimens of Barbosa Rodrigues, Wallace, and Burret were mostly destroyed, and many other types were fragmentary. Second, herbarium specimens were few because of the difficulty of collecting palms. They are large and often spiny plants, and do not fit into the general collecting routine. Those specimens that do exist, especially types, are often fragmentary and not useful for making taxonomic decisions because they lack flowers or fruits, or both. Wessels Boer overcame these problems by consigning dozens of names to synonymy and, more importantly, by carrying out extensive field work.

Wessels Boer's approach was a morphologic–geographic one, usually referred to as a taxonomic species concept. Species are based on multiple, correlative morphologic characters, as well as on cohesive geographic distributions. Currently, this is the most widely used concept, at least by herbarium taxonomists, and is employed in this work. It has been defined by Cronquist (1988): "Species are the smallest groups that are consistently and persistently distinct, and are distinguished by ordinary means."

Two other species concepts are widely used, but neither has yet been applied to palms. The biological species concept defines species as a reproductive community of populations (reproductively isolated from others) that occupies a specific niche in nature (Mayr, 1982). It has been widely used by zoologists, but the problem for botanists is that, in most cases, it is not possible to know whether a population is reproductively isolated or not. A further problem of the biological species concept is that in plants asexual reproduction, polyploidy, and hybridization are widespread. The phylogenetic species concept is derived from phylogenetic systematics, or cladistics. It defines species as the smallest aggregation of populations (sexual) or lineages (asexual) diagnosable by a unique combination of character states in comparable individuals (Nixon & Wheeler, 1990). As yet there have been few cladistic studies of palms.

Although Wessels Boer was criticized for being too radical, perhaps as an overreaction to previous work, he set the pattern for future workers. Recently, palm botanists have adopted a more realistic approach, usually based on extensive field work and appreciation of the plasticity of many characters. A good example of this is the recent revision of *Hyospathe* (Skov & Balslev, 1989). One of the two species of this genus exhibits extreme morphologic variation, but by carefully studying this, Skov and Balslev were able to show a continuum that could not be realistically divided. Once palms are studied in the field, it can be seen that many characters have little taxonomic importance for a given genus. Species can have populations with entire leaves and others with pinnate leaves; tall, aerial stems and short, subterranean stems; solitary and clustered stems; and pilose and glabrous leaves. One very variable character that has been extensively used in the past is fruit size. Fruits vary in size, depending on their age (and, in particular, the mesocarp expands at maturity) and whether they are dry in the herbarium or fresh in the field. Even within the same infructescence, a range can be found that would previously have separated species (and hence the numerous epithets "macrocarpa" and "microcarpa" of Burret, Bailey, and Barbosa Rodrigues). Scattered throughout Amazon palms are examples of species that have populations with fruits larger than normal. In some cases, I recognize these taxonomically (e.g., *Bactris concinna* var. *concinna* and *Desmoncus polyacanthos* var. *prunifer*); in other cases, I do not (e.g., the large-fruited specimens of *Bactris maraja* and *B. simplicifrons*). That palms have the genetic potential to develop larger fruits is shown in exemplary manner by *Bactris gasipaes*. It is interesting to note in passing that several palms in the western Amazon region have larger fruits, as they do in the Chocó. Gentry (1986) noted this for the latter area, and explained it by higher rainfall, an idea that might also apply to the western Amazon region.

The outstanding problem encountered in this work is that 19 species (13 of them in *Bactris* and *Geonoma*) are widespread and extremely variable. Apart from the problem of too few collections, these species are very difficult taxonomically. They are considered complexes, and I have divided them into varieties. Most, but not all, of these varieties have discrete ranges, and vary in vegetative characters only. However, there are others in which the ranges overlap and great variability is encountered. A second problem is hybridization. During the course of this work, it has become increasingly obvious that various forms of hybridization exist and are perhaps quite common.

The solution to these problems, apart from more collections, is new techniques. The next stage in palm taxonomy will, it is hoped, involve studies in cladistics, molecular systematics, and population and reproductive biology. In this way, we will move toward biological and phylogenetic species concepts in palms.

6

The Palms of the Amazon

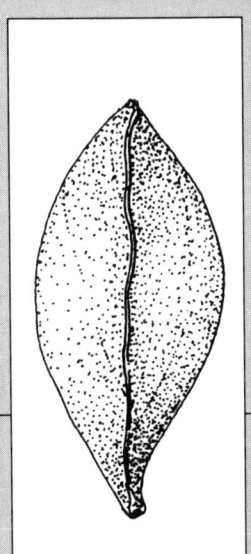

Format

The treatment given here includes all the palms occurring naturally within the Amazon region delimited in Chapter 1. Cultivated palms are excluded; accounts of them are found, for example, in Hoyos and Braun (1984) and in Blombery and Rodd (1982). Some other naturally occurring species, particularly those from drier areas north and south of the Amazon, may just reach into the region. These are excluded, since they can hardly be considered Amazon palms. Examples are, in the north (particularly the savanna areas of northern Bolívar State, Venezuela), *Copernicia tectorum* (Kunth) Mart., *Roystonea oleracea* (Jacq.) O. F. Cook, and *Sabal mauritiiformis* (H. Karst.) Griseb. & H. Wendl.; in the south, *Allagoptera leucocalyx* (Drude) Kuntze, *Astrocaryum campestre* Mart., and *Copernicia alba* Morong. Species found above 500 m elevation in the Guayana Highland are also excluded. These are *Geonoma appuniana* Spruce, *Geonoma simplicifrons* Willd., and *Prestoea tenuiramosa* (Dammer) H. E. Moore. Occasionally, Andean species may just reach below 500 m elevation on eastern Andean slopes in Colombia, Ecuador, Peru, and Bolivia—for example, *Geonoma jussieuana* Mart.—but these are also excluded.

The description of the family applies only to Amazon-region palms, and definitions of terms are given here. The key to the genera also applies only to Amazon palms. Order of genera is taken from Uhl and Dransfield (1987). Generic descriptions given here are for the genus as a whole, and much information, including technical terms, is taken from Uhl and Dransfield. Generic synonyms are given only when they differ from those used by Uhl and Dransfield. The reference(s) following each genus description is to either the most recent or the most useful treatment of the genus, but the number of species given in the reference may not correspond to that in this work. Keys are given only to those species occurring within the Amazon region, and they are designed not only to identify specimens, but also, in most cases, to show relationships among the species. Political units given in the keys are those from the Amazon region only. Species are in alphabetical order, and usually the only synonyms given are those that were described from the region. If a synonym is given that is not from the region, it has been used for Amazon palms. Type specimens are followed by an "n.v." when they have not been examined. If a photograph of a type specimen appears in Dahlgren (1959), this photo is referred to by its "F neg." number (since the original negatives are held at the Field Museum). The species descriptions are based on all specimens available, not just those from the region. Measurements are usually taken from dried herbarium specimens (which shrink on drying). The total range of the species is given by political units (in Ecuador, Napo is still used, rather than the recent division into Sucumbios and Napo). Ecological data are from a variety of sources. Much information is taken from specimen labels, as well as from my own observations. Distribution maps are based on all specimens from the total range of each species. Local names and uses are usually only those given on the labels of the specimens examined from the region.

Author abbreviations are from Brummitt and Powell (1982); serial abbreviations are from Lawrence and co-workers (1968); book abbreviations, and much other information, are from Stafleu and Cowan (1976, 1979, 1981, 1983, 1985, 1986, 1988); and acronyms of herbaria are from Holmgren and colleagues (1990).

PALMAE

Monoecious (with staminate and pistillate flowers) or dioecious (either staminate or pistillate flowers), pleonanthic (stems flowering continuously) or rarely hapaxanthic (stems flowering over a short period and then dying), spiny or nonspiny plants. *Stems* solitary or cespitose, rarely dichotomously branched, short and subterranean to large and aerial, usually erect, sometimes ventricose (swollen), occasionally creeping or climbing, sometimes supported by stilt roots. *Leaves* few to numerous, spirally or rarely distichously (in two ranks) arranged, palmate, costapalmate (palmate but with a short rachis), or pinnate (or pinnately veined if entire); sheaths usually open, sometimes closed and forming a crownshaft (tubular pseudo-stem formed by rolled leaf sheaths); petiole short or absent to elongate; rachis absent in palmate leaves, short in costapalmate leaves, and short to long in pinnate leaves; palmate and costapalmate leaves with few to numerous, induplicate (V-shaped in cross section) or reduplicate (Λ-shaped in cross section) segments (divisions of palmate and costapalmate leaves); pinnate leaves with reduplicate pinnae (leaflets, or divisions of pinnate leaves), the apex usually acute (abruptly tapering to a point), acuminate (tapering to a point), or occasionally praemorse (jagged) apically, often with ramenta (thin, irregular scales) on veins abaxially (on lower surface). *Inflorescences* borne singly, rarely multiply, in leaf axil, interfoliar (borne among the leaves), or infrafoliar (below the leaves) at anthesis, branched to 1–4 orders, or commonly inflorescence spicate; peduncle short to elongate; prophyll (first bract of an inflorescence) short to elongate; peduncular bracts 1–several, papery, fibrous, or woody, often sulcate (grooved on outer surface); rachis short or absent to elongate, occasionally covered with sheathing bracts; rachillae few to numerous; *flowers* bisexual or unisexual, borne singly, in dyads (pairs), acervuli (in lines) or most commonly triads (2 lateral staminate and 1 central pistillate); sepals usually 3, connate, valvate (meeting without overlapping), or imbricate (overlapping); petals usually 3, connate, valvate, or imbricate; stamens (3–)6(–numerous); filaments sometimes connate basally, inflexed or straight at apex; connective (part of stamen that connects anthers) sometimes split and bifid and then the thecae (the locules of an anther) free; staminodes of pistillate flowers digitate or forming a ring, or absent; gynoecium apocarpous (with free carpels) or syncarpous (with fused carpels), 1–several loculate (cavity bearing ovule), 1–several ovulate; pistillode of staminate flowers well developed to absent; *fruits* variously shaped often rostrate (with a short projection); epicarp (outermost layer of fruit) smooth, striate (lined) scaly, or variously spinulose or hairy; mesocarp (middle layer of fruit) fibrous or fleshy; endocarp (inner layer of fruit) thin or thick and bony, with 1–few pores; seed with homogeneous (uniform) or ruminate (with seed coat intruding irregularly) endosperm; eophyll (first seedling leaf) entire, palmate, or pinnate; germination remote-tubular (young plant separated from seed by a short stalk that lacks a ligule), remote-ligular (with the stalk having a ligule), or adjacent-ligular (young plant not separated from seed).

About 200 genera and 2000 species worldwide, mostly in tropical and subtropical areas; 34 genera and 151 species in the Amazon region.

KEY TO THE GENERA OF AMAZON PALMS
1. Leaf segments induplicate; leaves palmate, never spiny; flowers bisexual (Coryphoideae).
    2. Sheath not split at the base; flowers with 4–9 stamens and 1–6 carpels; Colombia (Amazonas), Ecuador (Morona-Santiago), Peru (Loreto, Madre de Dios, Pasco, San Martín, Ucayali), Brazil (Acre, Rondônia), and Bolivia (Beni, Pando) . . . . 1. *Chelyocarpus*.
    2. Sheath with a triangular split at the base; flowers with 5–19 stamens and 1 carpel; Colombia (Amazonas), Peru (Loreto), and Brazil (western Amazonas) . . . . 2. *Itaya*.
1. Leaf segments or pinnae reduplicate; leaves pinnate or pinnately veined, rarely palmate and then with very small spines; flowers unisexual.
    3. Fruits covered with overlapping scales; leaves pinnate or palmate; inflorescences covered with numerous, overlapping bracts (Calamoideae).
        4. Leaves pinnate; plants hapaxanthic and monoecious; Amazon estuary in Brazil (Pará) . . . . 3. *Raphia*.
        4. Leaves palmate; plants pleonanthic and dioecious; widespread.
            5. Large canopy palms with stems 2–20 m tall, 7.5–50 cm diam; fruits 2–5.2(–7) cm diam.
                6. Stems large and solitary (very rarely cespitose), without root spines; leaf green abaxially . . . . 4. *Mauritia*.
                6. Stems moderate and cespitose, less often solitary, with root spines; leaf white-waxy abaxially . . . . 5. *Mauritiella*.
            5. Small understory palms with stems 0.8–4(–6) m tall, 1.5–3 cm diam; fruits 1–2 cm diam. . . . . 6. *Lepidocaryum*.
    3. Fruits not covered with overlapping scales; leaves pinnate, or pinnately veined if entire; inflorescences not covered with numerous, overlapping bracts.
        7. Plants dioecious and with smooth fruits; stigmatic residue basal; peduncle slender, sheathed by (1–)2–6 papery peduncular bracts; leaves completely glabrous; stems green, with prominent internodes (Ceroxyloideae).

8. Staminate flowers not in vertical rows; peduncular bracts 2–6; abaxial surface of sheath and petiole with yellow stripe . . . . 7. *Chamaedorea.*
8. Staminate flowers in vertical rows; peduncular bract 1; abaxial surface of sheath and petiole without yellow stripe. . . . . 8. *Wendlandiella.*
7. Plants monoecious, or rarely dioecious and then with warty fruits; stigmatic residue usually apical; peduncle stout, not sheathed by several papery bracts; leaves usually tomentose or variously pilose or scaly; stems not green with prominent internodes.
    9. Stems with thick stilt roots at base; peduncular bracts 3–many; pinnae praemorse (Iriarteinae).
        10. Inflorescences usually multiple and unisexual; fruits with epicarp rough, hairy, warty, or verrucose . . . . 13. *Wettinia.*
        10. Inflorescences solitary and bisexual; fruits with smooth epicarp.
            11. Stems cespitose and slender, 1–12 m tall, 1–4 cm diam; pinnae entire . . . . 10. *Iriartella.*
            11. Stems solitary (seldom cespitose) and stout, 10–25 m tall, 6.5–30 cm diam; pinnae usually divided into segments.
                12. Stamens 6; seed with basal embryo; pinnae gray-white waxy abaxially . . . . 9. *Dictyocaryum.*
                12. Stamens 10–145; seed with lateral or apical embryo; pinnae green abaxially.
                    13. Stems ± swollen; stilt roots to 100, crowded; stamens 10–25; embryo lateral; peduncular bracts to 15 . . . . 11. *Iriartea.*
                    13. Stems columnar; stilt roots to 30, widely spaced; stamens (17–)30–145; embryo apical to subapical; peduncular bracts to 5 . . . . 12. *Socratea.*
    9. Stems without thick stilt roots at base; peduncular bract 1; pinnae not praemorse (except *Aiphanes,* and then spiny).
        14. Fruits light brown and covered with short, pyramidal projections; leaf partially entire or entire, with serrate margins; peduncular bract fibrous (Manicariinae) . . . . 14. *Manicaria.*
        14. Fruits not as above; leaf without toothed margins; peduncular bract usually not fibrous.
            15. Fruits orange-brown, smooth, and more or less flattened; mesocarp with coarse fibers and air spaces; upper part of stem covered either with long, loose fibers or short, reticulate fibers (Leopoldiniinae) . . . . 15. *Leopoldinia.*
            15. Fruits not as above; mesocarp not as above; upper part of stem not or rarely covered with fibers.
                16. Fruits with thin endocarp; flowers not in pits on rachillae; spines absent; monoecious palms (Euterpeinae).
                    17. Staminate, and occasionally pistillate, flowers pedicellate; inflorescences small; fruits with basal stigmatic residue; stems slender, 1–3 cm diam . . . . 19. *Hyospathe.*
                    17. Staminate and pistillate flowers not pedicellate; inflorescences large; fruits with apical to lateral stigmatic residue; stems stout, (1.5–) 3–45 cm diam.

18. Inflorescences hippuriform, rarely spicate; pinnae usually silvery-gray abaxially. . . . . . 18. *Oenocarpus.*
18. Inflorescences with rachillae spreading in several directions, not hippuriform, never spicate; pinnae green abaxially.
   19. Leaf sheaths forming a distinct crownshaft; rachillae densely covered with trichomes . . . . 16. *Euterpe.*
   19. Leaf sheaths not forming a distinct crownshaft; rachillae not densely covered with trichomes . . . . 17. *Prestoea.*
16. Fruits with thick or thin endocarp; flowers superficial or in pits in rachillae; spines present or absent; monoecious or dioecious palms.
    20. Fruits with a thick, bony endocarp; spines absent or present; monoecious palms (Cocoeae).
        21. Spines absent, or rarely with recurved thorns on the petiole.
            22. Staminate flowers not sunken in rachillae; peduncular bract woody, sulcate.
                23. Inflorescences usually of one kind on the same plant, with both staminate and pistillate flowers. . . . . 20. *Syagrus.*
                23. Inflorescences usually of two kinds on the same plant, either all staminate or staminate and pistillate . . . . . . 21. *Attalea.*
            22. Staminate flowers sunken in the rachillae; peduncular bract fibrous or woody.
                24. Inflorescences with both staminate and pistillate flowers; peduncular bract woody . . . . 22. *Barcella.*
                24. Inflorescences usually unisexual; peduncular bract fibrous . . . . 23. *Elaeis.*
        21. Spines present (occasionally confined to apex of pinnae).
            25. Fruits with epicarp easily cracking when ripe; stems often swollen. . . . . 24. *Acrocomia.*
            25. Fruits with epicarp not easily cracking when ripe.
                26. Apex of pinnae praemorse . . . . 25. *Aiphanes.*
                26. Apex of pinnae not praemorse.
                    27. Rachillae with many triads; pinnae usually green abaxially.
                        28. Nonclimbing palms; pinnae not modified into reflexed hooks . . . . 26. *Bactris.*
                        28. Climbing palms; apical pinnae modified into reflexed hooks . . . . 27. *Desmoncus.*
                    27. Rachillae with few triads; pinnae grayish-white abaxially . . . . . 28. *Astrocaryum.*

20. Fruits with a thin endocarp; spines absent; monoecious or dioecious palms.
29. Fruits 0.5–2.5 cm long, with a smooth epicarp, not densely crowded into a ± globose infructescence; flowers borne in small pits in the rachillae; monoecious palms (Geonomeae).
30. Fruits 1.4–2.5 cm long, 0.8–1.5 cm diam; ovary trilocular at anthesis.
31. Lower lips of floral pits overlapping laterally at anthesis, upper lip absent; western Amazon region in Colombia, Ecuador, Peru, and Brazil . . . . . 29. *Pholidostachys.*
31. Lower lips of floral pits not overlapping at anthesis, upper lip present; French Guiana . . . . 30. *Asterogyne.*
30. Fruits 0.5–1.3 cm long, 0.5–0.9 cm diam; ovary unilocular at anthesis . . . . 31. *Geonoma.*
29. Fruits 6–12 cm long, with warty projections, densely crowded into a ± globose infructescence; flowers not borne in small pits; dioecious palms (Phytelephantoideae).
32. Petioles not elongate, generally less than 2 m long, somewhat flattened adaxially; staminate flowers borne on a short or sessile receptacle, with elongate stamens . . . 32. *Phytelephas.*
32. Petioles elongate, generally more than 2 m long, terete; staminate flowers borne on a well-developed receptacle, with very short stamens.
33. Stems short and subterranean, or aerial and to 1 m tall; sheath not very fibrous; sheath, petiole, and rachis not scaly; Colombia (Caquetá, Putumayo) and Ecuador (Napo) . . . . . 33. *Ammandra.*
33. Stems 3–11 m tall; sheath very fibrous; sheath, petiole, and rachis scaly; Ecuador (Morona-Santiago, Napo, Pastaza), Peru (Loreto, Ucayali), and Brazil (Acre) . . . . 34. *Aphandra.*

CORYPHOIDEAE • CORYPHEAE • THRINACINEAE

1. *Chelyocarpus* **Chelyocarpus** Dammer, Notizbl. Bot. Gart. Berlin-Dahlem 7: 395. 1920.

Small to moderate, hermaphrodite palms. *Stems* solitary or cespitose, erect or procumbent, and then rooting. *Leaves* palmate, induplicate; sheath open and not forming a crownshaft, not split at the base, densely woolly tomentose; petiole long; hastula present adaxially, small or absent abaxially; rachis absent; blade orbicular, divided into two halves, and each half again divided into segments, these segments again divided apically, symmetrically acuminate or acute at apex, green or gray abaxially. *Inflorescences* interfoliar, branched to 1 or 2 orders; peduncular bracts 1–4; rachillae numerous; *flowers* perfect, borne singly; sepals 2–4; petals 2–4, or perianth uniseriate; stamens 4–9; gynoecium apocarpous with 1–6 carpels; *fruits* 1(–2)-seeded, globose, with apical or subapical stigmatic residue; epicarp smooth or corky-tessellate; seed with homogeneous endosperm and lateral embryo; germination remote-tubular; eophyll bifid.

A genus of four species, three of which are distributed in the western Amazon region in Colombia, Ecuador, Peru, Brazil, and Bolivia; the fourth occurs in extra-Amazonian Colombia (Chocó) (Moore, 1972; Kahn & Mejia, 1988).

The species of this genus have an almost linear and mostly nonoverlapping distribution pattern from southeast to northwest. *Chelyocarpus chuco* occurs in western Brazil (southern Acre) and adjacent Bolivia, and was considered by Moore (1972) to be the least specialized species in the genus. *Chelyocarpus ulei* occurs in central Peru, Ecuador, and Colombia and just reaches western Brazil (northern Acre); *C. repens* occurs in northern Peru; and *C. dianeurus* occurs in the Pacific lowlands of Colombia. Two closely related genera continue this pattern: *Cryosophila* occurs from northwestern Colombia through Central America to Mexico (where it meets *Schippia*), and in the south *Trithrinax* occurs in southern Brazil, Bolivia, Paraguay, and Argentina. Thus both the least specialized genus of palms, *Trithrinax* (Uhl & Dransfield, 1987), and the least specialized species of *Chelyocarpus* occur in southern South America, perhaps indicating the origin of the subtribe Thrinacinae.

Moore (1972) considered that the distribution of the species of *Chelyocarpus* known to him correlated with Haffer's (1969) refuges. However, the correlation, especially based on current knowledge, is a poor one; nor do the current ranges of any species coincide with Prance's (1982) refuges.

KEY TO THE SPECIES OF *CHELYOCARPUS*

1. Leaves green abaxially, with prominent cross-veins when dry; flowers with 3 sepals, 3 petals, and 6 stamens; inflorescences with 2 orders of branching; Brazil (Acre, Rondônia) and Bolivia (Beni, Pando) . . . . 1. *C. chuco.*
1. Leaves silvery-gray abaxially, without prominent cross-veins when dry; flowers with 2 sepals, 2 petals, or perianth uniseriate, and 4–8 stamens; inflorescences with 1, rarely 2, orders of branching.
   2. Epicarp of fruit smooth; stems procumbent, to 1 m long; peduncular bract 1; Peru (Loreto). . . . . 2. *C. repens.*
   2. Epicarp of fruit corky-warted; stems erect, 1–8 m tall; peduncular bracts 2; Colombia (Amazonas), Ecuador (Morona-Santiago), Peru (Loreto, Pasco, Madre de Dios, San Martín, Ucayali), and Brazil (Acre) . . . . 3. *C. ulei.*

1. Chelyocarpus chuco (Mart.) H. E. Moore, Principes 16: 73. 1972. *Thrinax? chuco* Mart. in A. D. Orb., Voy. Amérique mér. 7(3). Palmiers 45. 1844. *Trithrinax chuco* (Mart.) Walp., Ann. Bot. Syst. 1: 1005. 1849. *Acanthorrhiza chuco* (Mart.) Drude in Mart., Fl. bras.: Palmae II, fasc. 86, vol. 3(2): 554. 1882. *Tessmanniophoenix chuco* (Mart.) Burret, Notizbl. Bot. Gart. Berlin-Dahlem 10: 400. 1928. *Tessmanniodoxa chuco* (Mart.) Burret, Notizbl. Bot. Gart. Berlin-Dahlem 15: 337. 1941. Type. Bolivia. Beni: Río Guaporé, n.d., *A. d'Orbigny 32* (holotype, P, n.v.) (Figure 6.1a–b).

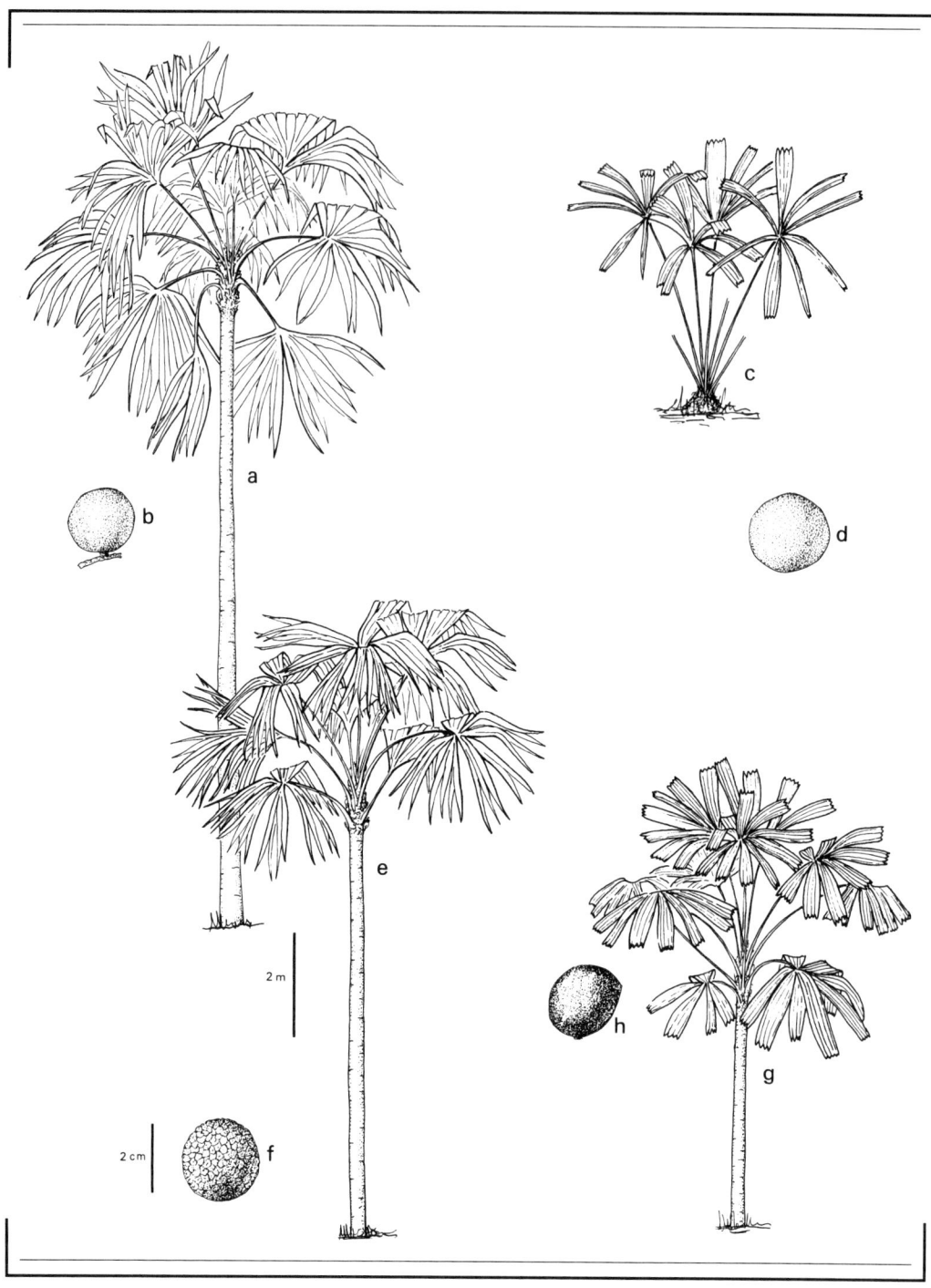

Figure 6.1. **(a)** *Chelyocarpus chuco,* **(b)** fruit (from *G. Prance 5708*); **(c)** *Chelyocarpus repens,* **(d)** fruit (from *F. Kahn 1974*); **(e)** *Chelyocarpus ulei,* **(f)** fruit (from *F. Kahn 1838*); **(g)** *Itaya amicorum,* **(h)** fruit (from *G. Galeano 1654*).

Stems cespitose with 2 stems developed, or solitary, 5–12 m tall, 8–12 cm diam. *Leaves* 14–17; sheath 13–50 cm long, densely whitish-brown woolly tomentose, becoming fibrous apically; petiole 0.8–2.1 m long; hastula small, present adaxially, small or absent abaxially; blade divided almost to the base into 2 equal halves, each half again deeply divided into 11–21 wedge-shaped segments, these again briefly divided apically, the middle ones 77–80 cm long, 12–15 cm wide, green abaxially, with prominent cross-veins when dry. *Inflorescences* interfoliar, branched to 2 orders; peduncle ca. 30 cm long; prophyll not seen; peduncular bracts 2, persistent, densely whitish-brown-tomentose; rachis 30–45 cm long, with persistent bracts; primary branches 6–8; rachillae numerous, 1–8 cm long; *flowers* borne singly, cream colored, 3–4 mm long at anthesis; sepals 3, briefly connate basally, deltate, 2–2.5 mm long; petals 3, briefly imbricate basally, deltate, 2–2.5 mm long; stamens 6, the filaments flattened basally; gynoecium of 3, free carpels; *fruits* 1-seeded, globose, 1.5–2 cm diam; epicarp smooth, yellowish-green or orange-green, becoming dark brown.

Brazil (Acre, Rondônia) and Bolivia (Beni, Pando) around the Rio Madeira and its tributaries (Abunã, Beni, Guaporé, Mamoré) near the Brazil–Bolivia frontier (Figure 6.2A); lowland rain forest in *várzea* or on adjacent *terra firme*, below 200 m elevation.

Bolivia: *hoja grande, hoja redonda*. Brazil: *caranaí, carnaubinha, palha redonda*. In both Bolivia and Brazil, the leaves are used to thatch houses and to weave hats.

Considered by Moore (1972) to be the least specialized species in the genus; see preceding text.

2. **Chelyocarpus repens** Kahn & Mejia, Principes 32: 69. 1988. Type. Peru. Loreto: lower Río Ucayali basin, near Jenaro Herrera, 4°55′S, 73°40′W, 13 Nov 1986, *F. Kahn & K. Mejia 1974* (isotype, NY) (Figure 6.1c–d).

Stems solitary, procumbent, and rooting, to 1 m long, to 9 cm diam. *Leaves* ca. 20; sheath 25–30 cm long, densely whitish-brown woolly tomentose; petiole 0.9–1.8 m long; hastula present adaxially; blade divided into 4–7 segments, these again briefly divided apically, the middle ones 60–70 cm long and 8–14 cm wide, silvery-gray abaxially. *Inflorescences* interfoliar, branched to 1 order, densely tomentose; peduncle 25–28 cm long; prophyll 11–12 cm long, inserted near apex of peduncle; peduncular bract 1, 6–7 cm long, inserted near apex of peduncle; rachis 4–8 cm long; rachillae ca. 25, to 2 cm long; *flowers* borne singly, densely crowded, 3 mm long at anthesis; perianth uniseriate, irregularly lobed, 2.5 mm long; stamens 4–8; gynoecium of (1–)3(–6) free carpels; *fruits* 1(–2)-seeded, globose, to 2.5 cm diam, greenish; epicarp smooth.

Peru (Loreto) in two small localities north and south of Iquitos (Figure 6.2B); lowland rain forest on both *terra firme* and *várzea,* but more abundantly on *terra firme* (Kahn & Mejia, 1988), below 200 m elevation.

The leaves of this species are remarkably like those of *Itaya amicorum*. However the nonsplit sheath and single petiolar phloem strand place it in *Chelyocarpus*, even though its flowers are somewhat similar to those of *Itaya*.

3. **Chelyocarpus ulei** Dammer, Notizbl. Bot. Gart. Berlin-Dahlem 7: 395. 1920. Type. Brazil. Acre: Rio Juruá-Mirim, near Belém, Sep 1901, *E. Ule 5885* (holotype, B) (Figure 6.1e–f).

Tessmanniophoenix longibracteata Burret, Notizbl. Bot. Gart. Berlin-Dahlem 10: 398. 1928. Type. Peru. Pasco: Pozuzo, Palcazú, 0°55′S, 200–300 m, 14 Jul 1913, *A. Weberbauer 6765* (isotype, F, n.v.).

Stems solitary, erect, 1.2–8 m tall, 4–7 cm diam. *Leaves* 10–15; sheath 10–32 cm long, whitish-brown wooly tomentose, fibrous apically; petiole 0.9–2 m long; hastula present adaxially, small or absent abaxially; blade divided almost to the base into 5–12 segments, each segment again divided apically, the middle ones 53–70 cm long and 8–15 cm wide, silvery-gray abaxially. *Inflorescences* interfoliar, branched to 1, rarely 2, orders, tomentose; peduncle 40–53 cm long; prophyll elongate, 30–37 cm long; peduncular bracts 2, ca. 30 cm long; rachis 17–46 cm long; rachillae numerous, 7–20 cm long; *flowers* borne singly, 2.5–3 mm long, yellow or cream-colored, each subtended by an elongate, pilose, linear bract; sepals 2, free, widely ovate, 2–2.5 mm long; petals 2, free, very widely

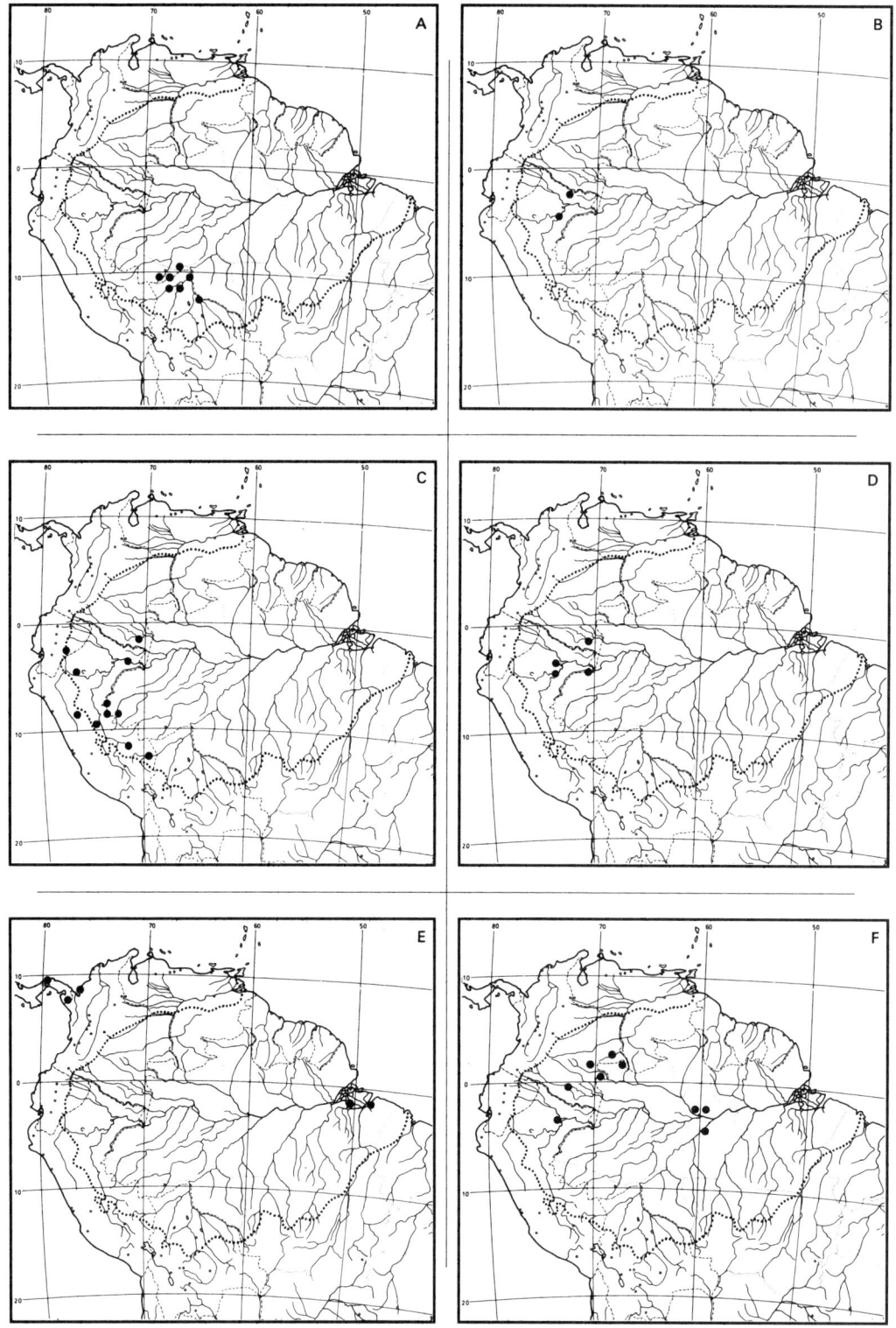

Figure 6.2. **(A)** *Chelyocarpus chuco;* **(B)** *Chelyocarpus repens;* **(C)** *Chelyocarpus ulei;* **(D)** *Itaya amicorum;* **(E)** *Raphia taedigera;* **(F)** *Mauritia carana.*

ovate, 2–2.5 mm long; stamens 5–8; gynoecium of 2 free carpels; *fruits* 1-seeded, globose, 2–2.3 cm diam, brown; epicarp corky-warted (but glabrous early in development); stigmatic residue not apparent.

Colombia (Amazonas), Ecuador (Morona-Santiago), Peru (Loreto, Madre de Dios, Pasco, San Martín, Ucayali), and Brazil (Acre) (Figure 6.2C); lowland rain forest on *terra firme* or *várzea*, at 250–500(–900) m elevation.

Brazil: *xila*. Colombia: *a-ibcom-ba* (Miraña). Peru: *sacha aguajillo*. Salt is extracted from the burned stems by Miraña Indians in Colombia.

2. *Itaya*

Itaya H. E. Moore, Principes 16: 85. 1972.

Moderate, hermaphrodite palms. *Stems* solitary, erect. *Leaves* palmate, induplicate; sheath open and not forming a crownshaft, split proximally; petiole long; hastula present; rachis absent; blade orbicular, split into wedge-shaped segments, these again briefly divided apically, asymmetrically acute at apex. *Inflorescences* interfoliar, branched to 2 orders; peduncle bearing a prophyll and to 9 peduncular bracts; rachis bearing numerous rachillae; *flowers* perfect, borne singly, open within inflorescence bud before anthesis; sepals 3, connate into a 3-lobed cupule; petals 3, connate into a 3-lobed cupule; stamens 15–19; filaments united basally into a cupule; gynoecium unicarpellate; *fruits* 1-seeded, irregularly subglobose to ellipsoid, with apical or subapical stigmatic residue; seed with homogeneous endosperm and basal embryo; eophyll entire.

A monotypic genus (Moore, 1972) occurring in a relatively small area of Colombia, Peru, and Brazil. It differs vegetatively from the closely related *Chelyocarpus* in its split leaf sheaths, thus mirroring the differences between *Thrinax* and *Coccothrinax*. It occurs on the eastern margin of the range of *Chelyocarpus*.

1. Itaya amicorum H. E. Moore, Principes 16: 86. 1972. Type. Peru. Loreto: Prov. Maynas, trail to Omaguas beyond landing on Río Itaya at Varadero de Omaguas, 13 May 1960, *H. Moore et al. 8447* (holotype, BH) (Figure 6.1g–h).

Stems solitary, erect or occasionally slightly leaning, 1–5 m tall, 9–10 cm diam. *Leaves* 11–25; sheath 71–85 cm long, split from the base, densely whitish-brown woolly tomentose; petiole 1.4–2.3 m long, split at base; hastula present adaxially; blade orbicular, divided to the base into 8–16 segments, these again divided briefly and asymmetrically apically, the middle ones 0.6–1.1 m long, 15–16 cm wide, silvery-gray abaxially. *Inflorescences* whitish-brown-tomentose; peduncle ca. 83 cm long; prophyll ca. 15 cm long, split adaxially almost to the base; peduncular bracts 5–9, ca. 33 cm long; rachis 30–47 cm long; rachillae numerous, to 10 cm long; *flowers* borne singly, shortly pedicellate, 4.5–5.5 mm long at anthesis; sepals 1.5–2 mm long; petals 1.5–2 mm long; *fruits* irregularly subglobose to ellipsoid, 2–2.5 cm long, 1.5–2 cm diam, yellowish-brown.

A small area of Colombia (Amazonas), Peru (Loreto), and Brazil (Amazonas) (Figure 6.2D); lowland rain forest in wet areas along rivers and streams as well as on *terra firme*, below 300 m elevation.

Brazil: *xila*. Colombia: *marím ipa* (Miraña). Peru: *falso bombonaje*. Salt is extracted from the burned trunks by Miraña Indians in Colombia, and the leaves are used to thatch temporary shelters.

This species was considered by Moore (1977) to have a narrow range and be in danger of extinction. It is now known to have a wider distribution, although it occurs patchily throughout its range (Henderson & Balick, 1987).

CALAMOIDEAE • CALAMEAE • RAPHIINAE

3. *Raphia*

Raphia P. Beauv., Flore d'Oware et de Bénin, en Afrique 1: 75. 1806.

Large, hapaxanthic, monoecious palms. *Stems* solitary or cespitose, aerial or subterranean, erect. *Leaves* pinnate, reduplicate; sheath open and not forming a crownshaft; petiole either short or long; rachis long; pinnae linear, either regularly arranged and spreading in 1 plane or clustered and spreading in different planes, usually with small spines. *Inflorescences* interfoliar, branched to 2 orders; peduncular bracts several; rachis with numerous rachillae, these distichously arranged, rachis, primary branches, and rachillae covered with sheathing bracts; *flowers* solitary, pistillate proximally and staminate distally on rachillae; staminate flowers with sepals 3, connate into a lobed cupule; petals 3, briefly connate basally, free above, almost woody and greatly exserted; stamens 6–30; pistillode minute or absent; pistillate flowers with sepals 3, connate into a 3-lobed calyx; petals 3, connate into a 3-lobed corolla; staminodes 6–16; gynoecium tricarpellate, triovulate; *fruits* usually 1-seeded, ellipsoid or ellipsoid-oblong, covered with overlapping scales; seed with ruminate endosperm and lateral embryo; germination adjacent-ligular; eophyll pinnate or bifid.

A genus of 28 species (Otedoh, 1982) confined to Africa and Madagascar, except for the neotropical *Raphia taedigera* (Bailey, 1933). Otedoh (1977) considered that even this species had been introduced to the New World from Africa, but this was considered unlikely by Uhl and Dransfield (1987).

Raphia is reported to have the longest leaves of any palm, or flowering plant. In West Africa, the leaves can reach 25.1 m (Hallé, 1977), and in Costa Rica they can exceed 20 m (Anderson & Mori, 1967).

1. Raphia taedigera (Mart.) Mart., Hist. nat. palm. 3: 216. 1839. *Sagus taedigera* Mart., Hist. nat. palm. 2: 54. 1824. *Metroxylon taedigerum* (Mart.) Spreng., Syst. veg. 2: 139. 1825. *Raphia vinifera* var. *taedigera* Drude, Bot. Zeit. 34: 804. 1876. Type. Brazil. Pará: Rio Amazonas, n.d., *C. Martius s.n.* (holotype, M) (Figure 6.3a–b; Plate IIIc).

Stems cespitose, 1–4 m tall, 28–40 cm diam, covered with persistent leaf bases and other debris, with a mound of roots at the base. *Leaves* 5–15, stiffly ascending; sheath 1.5–5.1 m long, with stiff brown fibers on the margins; petiole 1.5–4.2 m long, ± terete; rachis 4.7–8.5 m long, spiny on adaxial ridge; pinnae 136–205 per side, ± regularly inserted or in clusters of 2–3, spreading in different planes, linear, the middle ones 0.6–1.7 m long, 3–5 cm wide, with spines on veins. *Inflorescences* interfoliar; peduncle, prophyll, and peduncular bracts not seen; rachis ca. 1.5 m long, covered with sheathing bracts; rachillae numerous; *flowers* unisexual, the pistillate solitary at the base of the rachillae, the staminate solitary distally; staminate flowers 1.2–1.3 cm long; sepals 3 mm long; petals lanceolate, 11–12 mm long; stamens 9–10; pistillate flowers 10 mm long (postanthesis); sepals 4 mm long; petals 4 mm long; staminodes united into a low ring; *fruits* ellipsoid-oblong, 5–7 cm long, 3–4 cm diam, covered with broad, overlapping, reddish-brown scales; endosperm ruminate.

Central America (Nicaragua, Costa Rica, Panama), the estuary of the Río Atrato in Colombia (Antioquia, Chocó), and the estuary of the Amazon in Brazil (Pará), where it is known at least from Belém to Breves (Figure 6.2E); low-lying, inundated places near rivers in coastal areas, in the Amazon estuary in tidal *várzea* areas. This is one of the least collected of Amazon palms, and it may have a wider distribution. Nigel Smith (personal communication) believes that it occurs near Tucuruí and Carajás, in Pará.

In Central America, it forms huge but isolated stands (Allen, 1965). Anderson and Mori

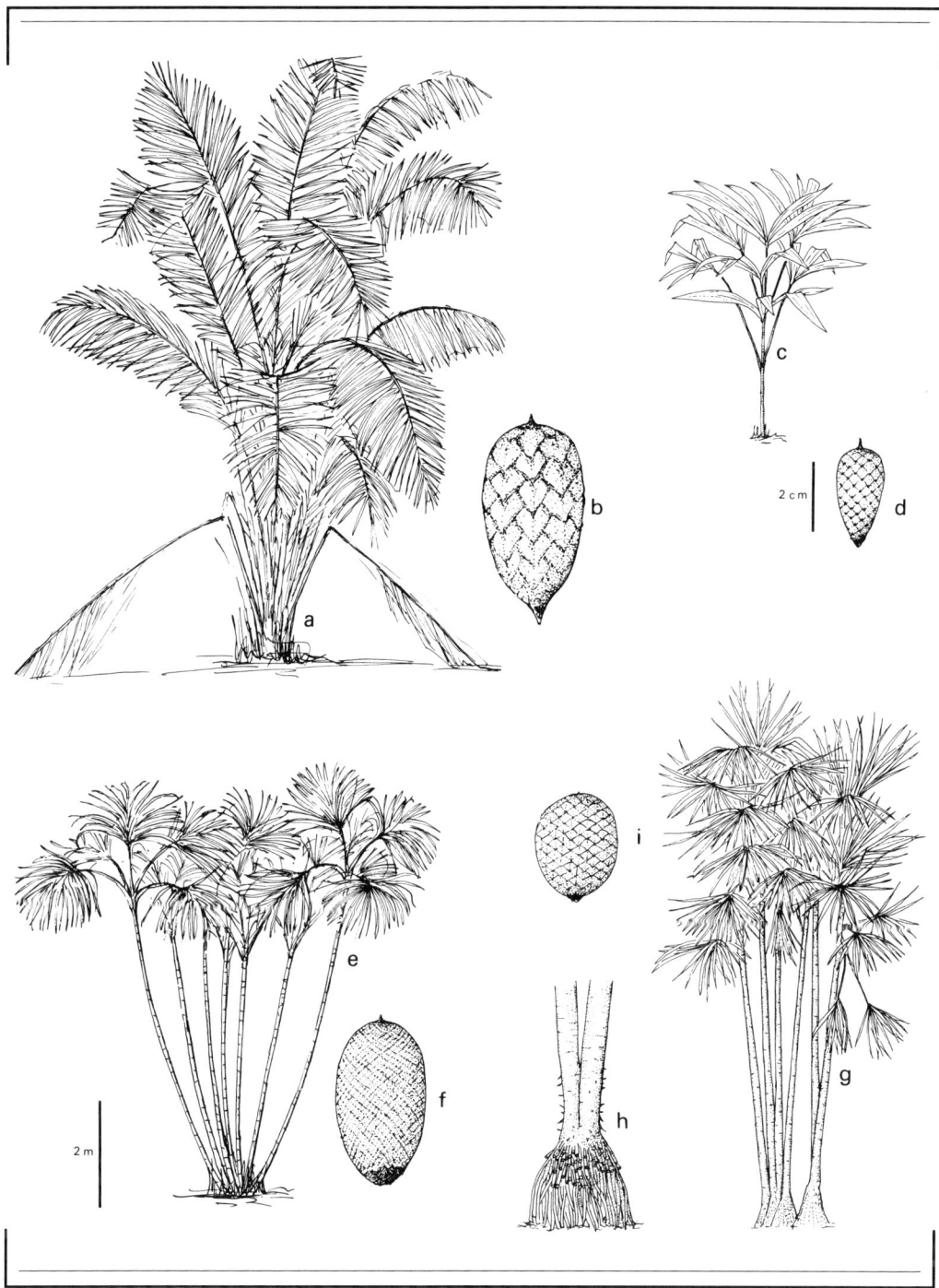

Figure 6.3. **(a)** *Raphia taedigera,* **(b)** fruit (from *A. Henderson 1614*); **(c)** *Lepidocaryum tenue* var. *tenue,* **(d)** fruit (from *E. Lleras P16972*); **(e)** *Mauritiella aculeata,* **(f)** fruit (from *J. Wessels Boer 1916*); **(g)** *Mauritiella armata,* **(h)** detail of stems and root spines, **(i)** fruit (from *J. Wessels Boer 2275*).

(1967) believed these to be pioneer communities, possibly as a result of recent introduction. But Devall and Kiester (1987) proposed that *Raphia*-dominated swamps represented climax communities. A second species, *R. farinifera* from West Africa, is naturalized in the Lesser Antilles of Martinique and Guadeloupe (Read, 1979).

Brazil: *jupatí*. The surfaces of the petioles are used to make shrimp traps, bird cages, and other items. The pith from the petioles is used to cork bottles. Wallace (1853) commented on the usefulness of this species.

This species has an unusual distribution pattern; *Euterpe oleracea* and *Manicaria saccifera* are the only other palms occurring in both the Amazon estuary and northwestern Colombia.

CALAMOIDEAE • LEPIDOCARYEAE

4. *Mauritia*

Mauritia L. f., Suppl. pl. 70. 1782 ("1781").

Large, dioecious palms. *Stems* solitary, very rarely cespitose, erect, smooth. *Leaves* costapalmate, reduplicate; sheath open and not forming a crownshaft, with few to many fibers; petiole long, terete; rachis short; blade divided into halves, each half again divided into numerous, linear segments, these spiny on the margins. *Inflorescences* interfoliar at anthesis and in fruit, horizontally held, branched to 2 orders; peduncular bracts numerous, sheathing the peduncle; rachis, primary branches, and rachillae covered with sheathing bracts; *flowers* crowded on short rachillae; staminate flowers with sepals 3, connate into a 3-lobed cupule; petals 3, connate basally for ca. one-third their length, free and valvate above; stamens 6, 3 longer, 3 shorter; pistillode minute or absent; pistillate flowers with sepals 3, connate into a tubular calyx; petals 3, connate basally, free and valvate above; staminodes 6, stamenlike; ovary trilocular, triovulate; *fruits* usually 1-seeded, globose to ellipsoid, covered with small overlapping scales, with apical stigmatic residue; seed with homogeneous endosperm and basal embryo; germination adjacent-ligular; eophyll palmate.

A genus of two species, widespread in northern South America, east of the Andes; both occur in the Amazon region. Previous authors have recognized several species (Drude, 1881; Beccari, 1918).

Mauritia, Mauritiella, and *Lepidocaryum* have a remarkable, symmetrical distribution pattern. Each has one widespread species (or variety in the case of *Lepidocaryum*) and a second species (or variety) restricted to the upper Rio Negro region of Brazil and adjacent Colombia and Venezuela. These upper Rio Negro taxa, however, can have outliers in other areas.

KEY TO THE SPECIES OF *MAURITIA*

1. Sheaths and basal part of petiole covered with a mass of long, loose fibers; fruit usually globose and covered with small scales 3–4 mm across; Venezuela (Amazonas), Colombia (Amazonas, Guainía), Peru (Loreto), and Brazil (Amazonas) . . . . 1. *M. carana*.
1. Sheaths and basal part of petiole not covered with a mass of fibers; fruit usually ellipsoid and covered with larger scales ca. 6 mm across; widespread . 2. *M. flexuosa*.

1. Mauritia carana Wallace, Palm Trees of the Amazon 53. 1853. *Orophoma carana* (Wallace) Spruce, J. Linn. Soc., Bot. 11: 171. 1871. Lectotype (Wess. Boer, 1988). Wallace, Palm Trees of the Amazon t. 18. 1853 (Figure 6.4a–c).

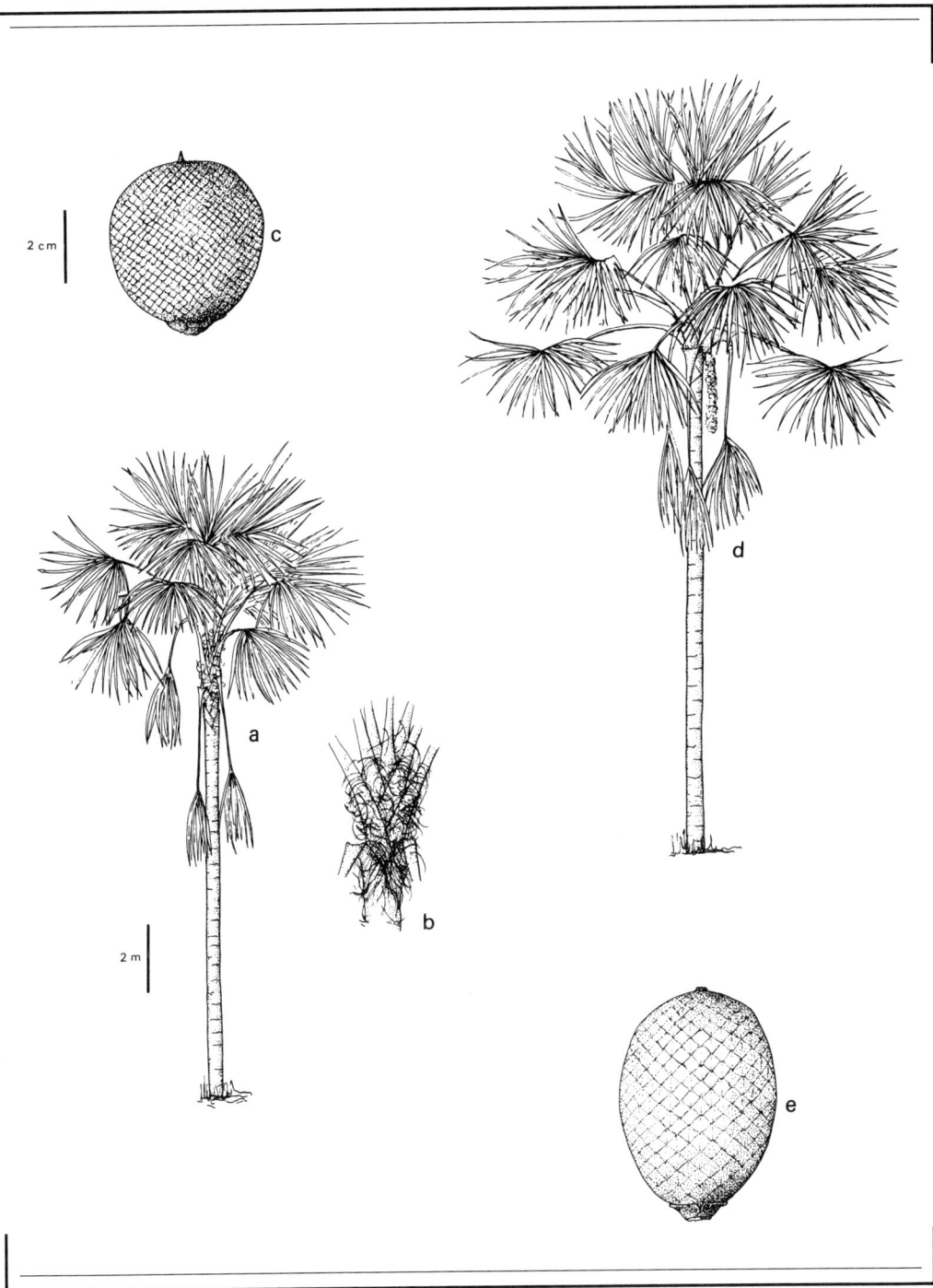

Figure 6.4. (a) *Mauritia carana,* (b) detail of sheath fibers, (c) fruit (from *A. Henderson 483*); (d) *Mauritia flexuosa,* (e) fruit (from *J. Steyermark 87708*).

Stems solitary or very rarely cespitose and then with 2–5 stems, erect, 2–15 m tall, 30–50 cm diam, gray, with large mound of roots at base. *Leaves* 6–12, spreading, dead leaves somewhat persistent; sheath open, 0.6–1.7 m long, with a mass of long, loose fibers on the margins; petiole 0.6–3 m long, fibrous at base; rachis to 60 cm long, recurved; blade divided to one-third or one-half its length into 120–141 or more segments, the middle ones to 3.7 m long, 5–5.5 cm diam, pendulous and shortly bifid apically, occasionally spiny on the margins. *Inflorescences* interfoliar; peduncle 60–80 cm long; prophyll not seen; peduncular bracts to 7; rachis 1.1–1.8 m long; primary branches 24–36, 0.3–1 m long; rachillae numerous, distichously arranged along primary branches; *flowers* borne singly; staminate flowers 5.5 mm long, densely crowded on rachillae; sepals 3 mm long; petals lanceolate, 5 mm long; pistillate flowers not seen; *fruits* 1(–2)-seeded, globose, 4.5–5(–7) cm diam, covered with dark-red or reddish-brown scales 3–4 mm across.

Rio Negro region and adjacent areas from Venezuela (Amazonas) and Colombia (Amazonas, Guainía) to Manaus in Brazil (Amazonas, rarely south of the Amazon), also in Peru (Loreto) (Figure 6.2F); usually on poorly drained, white-sand soils (podzols), either in lowland rain forest or in *campina* forest or open areas, below 200 m elevation.

Brazil: *caraná, caraná do matto, miriti-rnã*. Colombia: *canangucho de sabana, canangucho paso, too'-ee* (Puinave). Peru: *aguaje*. Venezuela: *caraná*. The leaves are commonly used for thatching. Wallace (1853) described how the large and durable leaves were preferred over those of other palms where this species occurred.

Staminate flowers from authentic material of this species should be reexamined; Spruce (1871) described them as having the petals united. On the basis of this, he established the subgenus *Orophoma*.

2. Mauritia flexuosa L. f., Suppl. pl. 454. 1782 ("1781"). Type. Surinam. Without locality, n.d., *C. Dahlberg s.n.* (holotype, S, n.v.) (Figure 6.4d–e; Plates Ia, IIa).
Mauritia vinifera Mart., Hist. nat. palm. 2: 42. 1824. Type. Brazil. Minas Gerais: without locality, n.d., *C. Martius s.n.* (holotype, M).
Mauritia sphaerocarpa Burret, Notizbl. Bot. Gart. Berlin-Dahlem 10: 569. 1929. Type. Brazil. Amazonas: Rio Branco, Dec 1927, *G. Huebner 104* (holotype).
Mauritia minor Burret, Notizbl. Bot. Gart. Berlin-Dahlem 11: 1. 1930. Type. Colombia. Caquetá: Florencia, 6 Aug 1926, *S. Juzepczuk 6458* (holotype, B, n.v.).
Mauritia flexuosa var. *venezuelana* Steyerm., Fieldiana Bot. 28: 90. 1951. Type. Venezuela. Bolívar: between Río Caroni and Ciudad Bolívar, 200 m, 2 Aug 1944, *J. Steyermark 57649* (holotype, F, n.v.).

Stems solitary, erect, 2.8–25 m tall, 23–50(–80) cm diam, gray, with large root mass at base, and with pneumatophores. *Leaves* 8–20, spirally arranged, rarely distichous, pendulous and persistent when dead, especially on younger plants; sheath open, 1–2.1 m long, with few coarse fibers circling younger leaves; petiole 1.6–4 m long; rachis 30–100 cm long, recurved; blade divided almost to the base into (45–)120–236 segments, the middle ones 1.2–2.2 m long, 4.5–5 cm wide, occasionally with small spines on margins, with brown fibers on veins abaxially. *Inflorescences* interfoliar; peduncle 0.7–2.5 m long; prophyll and peduncular bracts not seen; rachis 1.4–2.4 m long, covered with sheathing bracts; primary inflorescence branches 18–46, pendulous, 70–119 cm long; rachillae numerous, distichously arranged, to 5 cm long; *flowers* borne singly; staminate flowers densely crowded on rachillae, to 1.1 cm long, bright orange at anthesis; sepals 3.5–4.5 mm long; petals lanceolate, 1–1.1 cm long; pistillate flowers 8 mm long; sepals 8 mm long, with a small opening at apex; petals lanceolate, 6 mm long; *fruits* 1(–2)-seeded, ellipsoid-oblong or occasionally globose-oblong, 3.7–5.3 cm long, 3–5.2 cm diam, covered with brown or reddish-brown scales ca. 6 mm across.

Widespread over all of northern South America, east of the Andes, in Colombia, Venezuela (Sucre southward), Trinidad, the Guianas, Ecuador, Peru, central and eastern Brazil (to Bahia, Goiás, Mato Grosso, Minas Gerais, Piauí, São Paulo), and Bolivia (Santa Cruz) (Figure 6.5A); gallery forest in savannas, as

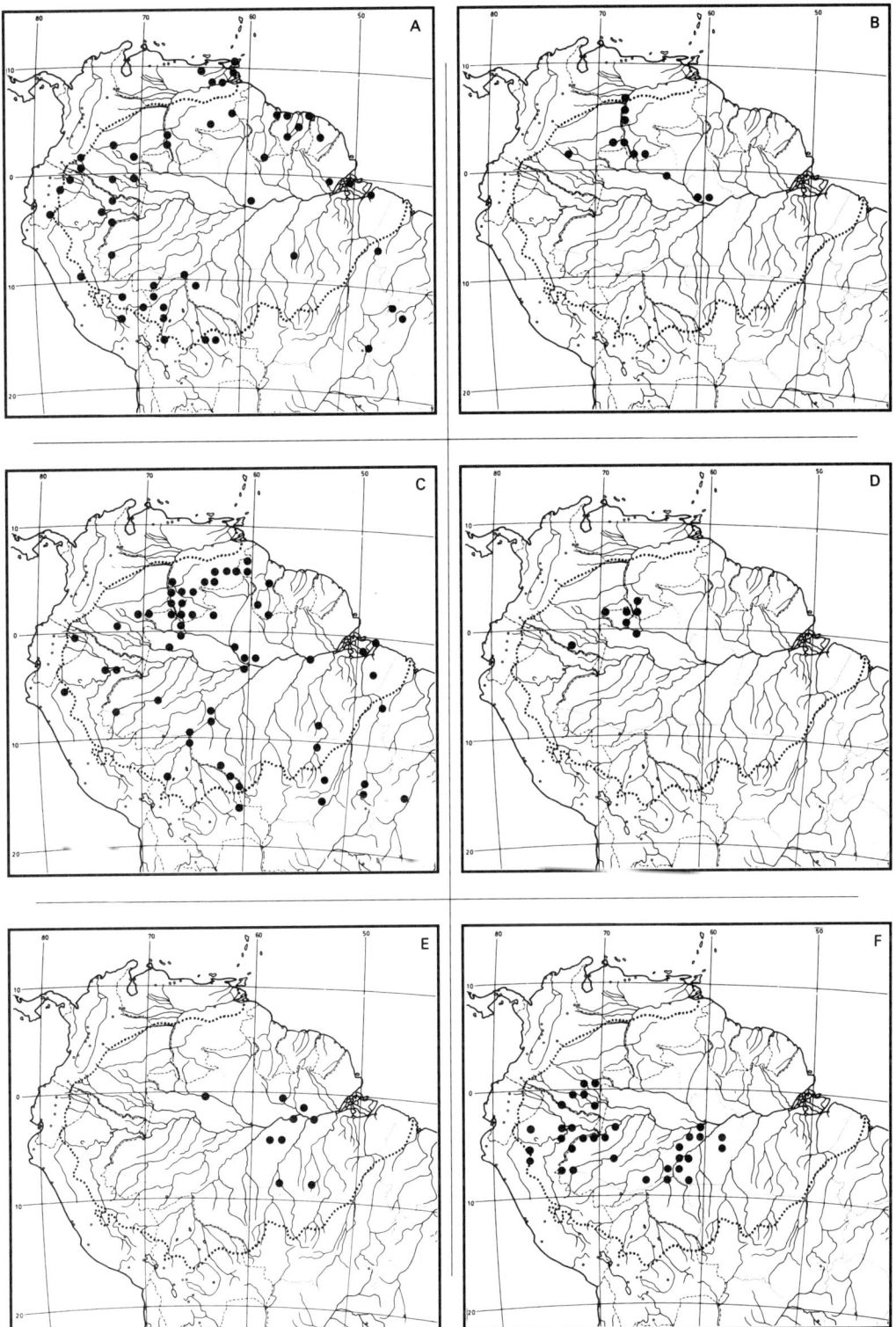

Figure 6.5. (A) *Mauritia flexuosa;* **(B)** *Mauritiella aculeata;* **(C)** *Mauritiella armata;* **(D)** *Lepidocaryum tenue* var. *casiquiarense;* **(E)** *Lepidocaryum tenue* var. *gracile;* **(F)** *Lepidocaryum tenue* var. *tenue.*

well as along stream margins and in wet areas throughout forested regions, often in very large numbers. It generally occurs at low elevations, but occasionally is found up to 1000 m on eastern Andean slopes. In some places, it is extremely abundant and fecund. Near Iquitos, in 2 million ha of forest surveyed, 80 percent of all trees were *M. flexuosa* (ONERN, 1976). In Venezuela, a single pistillate tree may produce between 1500 and 5000 fruits annually (González, 1987). In the Colombian Amazon, trees flowered from August to September, and fruits were mature from February to August (Urrego, 1987; see also Veléz, 1992).

González (1987) has described the ecology of *Mauritia* swamps in the eastern llanos of Venezuela, where huge stands of the palm occur. Goulding (1989) has described the succession of animals that feed on the fruits of this palm. A series of macaws eat the fruits still on the tree (and nest in dead trees), while at least 20 species of fish feed on fruits falling into water, and turtles, tortoises, agoutis, peccaries, deer, pacas, and iguanas feed on fruits falling on the ground.

Bolivia: *kikyura* (Chimane), *palma real.* Brazil: *buriti, miriti, muriti.* Colombia: *canangucha, moriche, nain* (Cubeo), *ne* (Tatuyo), *non* (Puinave). French Guiana: *palmier bache* (Creole). Guyana: *ite palm.* Peru: *aguaje, iñéjhe* (Bora). Venezuela: *moriche.* The large stands of this palm are called *buritales* in Brazil, *canangu-chales* in Colombia, *aguajales* in Peru, and *mori-chales* in Venezuela and adjacent Colombia.

The pulpy mesocarp of the fruit is used to flavor drinks, ices, and other food, and is also used to prepare an alcoholic drink. Oil is also extracted from the mesocarp. The fibers from the leaves are woven into a variety of artifacts.

This species is one of the most widely used palms in the Amazon region. Humboldt (1850) was one of the first European travelers to describe the importance of *Mauritia flexuosa* to indigenous people. He observed the reliance of the Guarani Indians in the Orinoco Delta on this species for almost all their material needs. Heinen and Ruddle (1974) described the rituals surrounding the collection and consumption of stem starch from *M. flexuosa* by the Warao Indians in the Orinoco Delta. Although the use of the palm is declining, it can still be important in local economies. For example, in the Iquitos region of Peru, the fruits are an item of commerce (Padoch, 1988).

The sexual system of this supposedly dioecious palm is still not understood. Isolated pistillate plants have been observed to produce viable seeds. There seems at least a possibility that the species could be parthenocarpic. However, occasionally staminate inflorescences contain pistillate flowers, and possibly the staminodes of pistillate inflorescences can produce viable pollen, and thus the species may not be strictly dioecious. Then the cases of isolated pistillate trees producing seeds could be explained by self-pollination, either in a perfect flower or from a staminate flower. Staminate flowers have a pseudo-ovule.

5. *Mauritiella*

Mauritiella Burret, Notizbl. Bot. Gart. Berlin-Dahlem 12: 609. 1935.

Moderate to large dioecious palms. *Stems* cespitose or less often solitary, often leaning, the nodes and internodes with numerous, stout, root spines, at the base forming a dense cone of roots. *Leaves* costapalmate, reduplicate; sheaths partially open and not forming a crownshaft, sheath and petiole white-waxy; petiole very long, terete; rachis very short; blade divided almost to the base into numerous, linear segments, these white-waxy abaxially and spiny or nonspiny on the margins. *Inflorescences* interfoliar at anthesis and in fruit, branched to 2 orders; peduncular bracts numerous, sheathing the peduncle; rachis bearing numerous primary branches, these alternately and distichously arranged, rachis, primary branches, and rachillae covered with sheathing bracts; rachillae very short and numerous, alternately and distichously arranged; *flowers* crowded; staminate flowers with sepals united into a cupular calyx; petals 3, free, valvate; stamens 6; pistillode minute or absent; pistillate flowers with sepals connate into a tubular calyx; petals 3, connate basally for ca. one-third length, free and valvate above; staminodes 6, stamenlike; ovary trilocular, triovulate; *fruits* usually 1-seeded, ellipsoid or ellipsoid-oblong, or less often globose, covered with overlapping scales, with apical stigmatic residue; seed with homogeneous endosperm and basal embryo; germination adjacent-ligular; eophyll bifid.

A genus of three species distributed throughout northern South America in Colombia (including the Pacific coast), Venezuela, Ecuador, the Guianas, Peru, Brazil (as far south and east as Bahia, Goiás, and Pernambuco), and Bolivia. Earlier authors (Drude, 1881; Beccari, 1918; Burret, 1935) have recognized several species that are not maintained here.

KEY TO THE SPECIES OF *MAURITIELLA*

1. Fruits 4–5 cm long, 3–4.5 cm diam; leaf segments pendulous, 1–2 cm wide, with numerous spines on the margins; stems strongly cespitose, 3–8 m tall, 7–10 cm diam, forming dense colonies; Colombia (Guainía, Vaupés), Venezuela (Amazonas, Apure), and Brazil (Amazonas) . . . . 1. *M. aculeata.*
1. Fruits 2.5–3.5 cm long, 2–3.5 cm diam; leaf segments stiff, 1.5–3.5 cm wide, without spines or with a few spines on the margins; stems solitary or cespitose and then forming small colonies, 2–20 m tall, 7.5–14(-30) cm iam; widespread . . . . 2. *M. armata.*

1. Mauritiella aculeata (Kunth) Burret, Notizbl. Bot. Gart. Berlin-Dahlem 12: 609. 1935. *Mauritia aculeata* Kunth in Humb., Bonpl. & Kunth, Nov. gen sp. 1: 311. 1816. *Lepidococcus aculeatus* (Kunth) H. Wendl. & Drude in Kerch., Les Palmiers 249. 1878. Type. Venezuela. Amazonas: Río Atabapo, n.d., *A. Bonpland s.n.* (holotype, P, n.v.) (Figure 6.3e–f).
Mauritia gracilis Wallace, Palm Trees of the Amazon 57. 1853. Lectotype (Wess. Boer, 1988). Wallace, Palm Trees of the Amazon t. 20. 1853.
Mauritia linophila Barb. Rodr., Enum. palm. nov. 18. 1875. *Mauritia amazonica* Barb. Rodr., Enum. palm. nov. 18. 1875. Lectotype (Glassman, 1972). Barb. Rodr., Sert. palm. brasil. 1: t. 2. 1903.
Mauritiella cataractarum Dugand, Revista Acad. Colomb. Ci. Exact. 8: 385. 1951. Type. Colombia. Vaupés: upper Río Apaporis, basin of Río Macaya, base of Cerro Chiribiquete, 15–16 May 1943, *R. Schultes 5428* (holotype, COL).

Stems strongly cespitose, 3–8 m tall, 7–10 cm diam, forming dense colonies of up to 50 stems, often leaning, covered at least basally with root spines to 3 cm long. *Leaves* 5–9; sheath partially open, to 70 cm long, sheath and petiole white-waxy; petiole 30–55(–100) cm long; rachis 12–14 cm long; blade split almost to the base into 68–80 pendulous segments, the middle ones 65–80 cm long and 1–2 cm wide, with spines on the margins, white-waxy on lower surface, with brown scales on the veins on lower surface. *Inflorescences* interfoliar; peduncle 20–40 cm long; prophyll 2–4.5 cm long; peduncular bracts 6–7, 4–5 cm long; rachis 55–60 cm long; primary branches 9–18, 20–30 cm long, pendulous; rachillae numerous, to 1 cm long; *flowers* subtended by bracts, densely crowded; staminate flowers 5.5 mm long; sepals 3.5 mm long; petals lanceolate, 5 mm long; pistillode not apparent; pistillate flowers not seen; *fruits* oblong-ellipsoid, 4–5 cm long, 3–4.5 cm diam, covered with overlapping, reddish-brown scales.

Upper Río Negro region of Colombia (Guainía, Vaupés), Venezuela (Amazonas, Apure), and Brazil (Amazonas) (Figure 6.5B); inundated margins of black-water streams and rivers, where it can form large colonies.

Brazil: *buritirana, caranaí.* Colombia: *cadanarite* (Curripaco), *moriche.* Venezuela: *morichito.*

2. Mauritiella armata (Mart.) Burret, Notizbl. Bot. Gart. Berlin-Dahlem 12: 611. 1935. *Mauritia armata* Mart., Hist nat. palm. 2: 45. 1824. *Lepidococcus armatus* (Mart.) H. Wendl. & Drude in Kerch., Les Palmiers 249. 1878. Type. Brazil. State?: Rio São Francisco, n.d., C. Martius s.n. (holotype, M, n.v.) (Figure 6.5g–i; Plate IVa).

Mauritia aculeata of Mart., Hist nat. palm. 2: 47. 1824. *Mauritia martiana* Spruce, J. Linn. Soc., Bot. 11: 171. 1871. *Mauritiella martiana* (Spruce) Burret, Notizbl. Bot. Gart. Berlin-Dahlem 12: 611. 1935. *Lepidococcus martianus* (Spruce) H. Wendle. & Drude ex A. D. Hawkes, Arq. Bot. Est. São Paulo, n. s. 2: 174. 1952. Type. Brazil. Pará: without locality, n.d., C. Martius 1720 (holotype, M; F negs. 18583, 18584).

Mauritia pumila Wallace, Palm Trees of the Amazon 59. 1853. *Mauritiella pumila* (Wallace) Burret, Notizbl. Bot. Gart. Berlin-Dahlem 12: 611. 1935. *Lepidococcus pumilus* (Wallace) H. Wendl. & Drude in Kerch., Les Palmiers 249. 1878. Lectotype (Wess. Boer, 1988). Wallace, Palm Trees of the Amazon t. 21. 1853.

Mauritia subinermis Spruce, J. Linn. Soc., Bot. 11: 171. 1871. *Orophoma subinermis* (Spruce) Drude in Mart., Fl. bras.: Cyclanthaceae et Palmae I, fasc. 85, vol. 3(2): 296. 1881. *Lepidococcus subinermis* (Spruce) A. D. Hawkes, Arq. Bot. Est. São Paulo, n.s. 2: 174. 1952. Type. Venezuela. Amazonas: confluence of Casiquiare and Río Guainía, n.d., R. Spruce 39 (holotype, K, n.v.).

Mauritia peruviana Becc., Ann. Roy. Bot. Gard. (Calcutta) 12: 255. 1918. *Mauritiella peruviana* (Becc.) Burret, Notizbl. Bot. Gart. Berlin-Dahlem 12: 611. 1935. *Lepidococcus peruvianus* (Becc.) H. Wendl. & Drude ex A. D. Hawkes, Arq. Bot. Est. São Paulo, n.s. 2: 174. 1952. Type. Peru. Loreto: without locality, n.d., A. Weberbauer 4717 (holotype, B, n.v.).

Mauritia huebneri Burret, Notizbl. Bot. Gart. Berlin-Dahlem 10: 570. 1929. *Mauritiella huebneri* (Burret) Burret, Notizbl. Bot. Gart. Berlin-Dahlem 12: 611. 1935. Lepidococcus huebneri (Burret) H. Wendl. & Drude ex A. Hawkes, Arq. Bot. Est. São Paulo, n.s. 2: 174. 1952. Type. Brazil. Amazonas: Manaus, n.d., G. Huebner 90a (holotype, B).

Mauritia intermedia Burret, Notizbl. Bot. Gart. Berlin-Dahlem 10: 572. 1929. *Mauritiella intermedia* (Burret) Burret, Notizbl. Bot. Gart. Berlin-Dahlem 12: 611. 1935. *Lepidococcus intermedius* (Burret) H. Wendl. & Drude ex A. D. Hawkes, Arq. Bot. Est. São Paulo, n.s. 2: 174. 1952. Type. Brazil. Amazonas: Manaus, Jun 1927, G. Huebner 97 (holotype, B).

Mauritiella duckei Burret, Notizbl. Bot. Gart. Berlin-Dahlem 12: 609. 1935. *Lepidococcus duckei* (Burret) H. Wendl. & Drude ex A. D. Hawkes, Arq. Bot. Est. São Paulo, n.s. 2: 173. 1952. Type. Brazil. Roraima: Caracaraí, Rio Branco, 23 Jul 1933, A. Ducke s.n. (holotype, B, n.v.).

Mauritiella nannostachys Burret, Notizbl. Bot. Gart. Berlin-Dahlem 15: 754. 1942. Type. Brazil. Amazonas: Manaus, Tarumã, n.d., G. Huebner 120b (holotype, B).

Mauritiella campylostachys Burret, Notizbl. Bot. Gart. Berlin-Dahlem 15: 755. 1942. Type. Brazil. Amazonas: Manaus, Tarumã, n.d., G. Huebner 10a (holotype, B, n.v.).

Stems cespitose or less often solitary, 2–20 m tall, 7.5–14(–30) cm diam, basally or entirely covered with root spines to 8 cm long. *Leaves* 4–10; sheath partially open, 35–200 cm long, with a prominent ligule at the apex, sheath and petiole white-waxy; petiole 30–350 cm long; rachis ca. 8 cm long; blade split almost to the base into 60–100 stiff segments, the middle ones 80–120 cm long and 1.5–3.5 cm wide, white-waxy abaxially, nonspiny or spiny on the margins, with brown scales on the veins on lower surface. *Inflorescences* interfoliar at anthesis and in fruit; peduncle 22–75 cm long; prophyll 2–5 cm long; peduncular bracts 6–10, closely sheathing the peduncle, persistent; rachis 20–130 cm long; primary branches 12–34, 28–50 cm long, pendulous, alternately and distichously arranged, each emerging from a rachis bract; rachillae numerous, alternately and distichously arranged, ca. 1.5 cm long; *flowers* subtended by bracts, densely crowded; staminate flowers 3–4 mm long; sepals 2 mm long; petals ovate-lanceolate, 3–4 mm long; pistillode minute or absent; pistillate flowers to 9 mm long; sepals 5 mm long; petals obovate, to 8 mm long; ovary to 2 mm high, covered with overlapping scales; *fruits* globose, ovoid, or ellipsoid-oblong, 2.5–3.5 cm long, 2–3 cm diam, covered with small, overlapping, reddish or reddish-brown scales.

Colombia (Amazonas, Caquetá, Guainía, Guaviare, Meta, Putumayo, Vaupés, Vichada), Venezuela (Amazonas, Bolívar), the Guianas, Ecuador (Napo), Peru (Loreto, San Martín), Brazil (Acre, Amazonas, Bahia, Goiás, Mato Grosso, Minas Gerais, Pará, Pernambuco, Piauí, Rondônia, Roraima, Tocantins), and Bolivia (La Paz, Pando, Santa Cruz) (Figure 6.5C); in a wide variety of habitats such as river margins, savannas, savanna margins, rain forest, and gallery forest. It can occur at up to 1400 m, but is usually found at lower elevations.

Bolivia: *buriticillo, palmilla*. Brazil: *buritirana, caraná, caranaí, caranatinga, jussara*. Colombia: *cananguchillo, caranaí*. Ecuador: *kantine-é* (Siona), *moretillo*. Guyana: *baby-ité*. Peru: *aguajillo*. Venezuela: *caraña* (Maquiritare), *kauwaya, morichito, morichito de tierra firme*. The wood is used to make bows; the petioles are used as corks for bottles; the leaves are woven into baskets; the fruit is edible; and the mesocarp is used to make a drink.

6. *Lepidocaryum* **Lepidocaryum** Mart., Hist. nat. palm. 2: 49. 1824.

Small, dioecious palms. *Stems* cespitose, slender, forming colonies by rhizomes. *Leaves* palmate, reduplicate; sheaths open and not forming a crownshaft; petiole long, terete; rachis very short; blade divided almost to the base into halves, these again divided into few to many linear segments, usually spiny on the margins. *Inflorescences* interfoliar, erect in bud and becoming pendulous in fruit, branched to 2 orders; peduncle elongate with sheathing prophyll and several peduncular bracts; rachis short; rachis, primary branches, and rachillae covered with sheathing bracts; primary branches and rachillae alternately and distichously arranged; rachillae short; *flowers* densely crowded on rachillae; staminate flowers solitary or in pairs, sepals 3, connate into a 3-lobed cupular calyx; petals 3, free, valvate; stamens 6; pistillode minute or absent; pistillate flowers with sepals 3, connate into a 3-lobed cupular calyx; petals 3, connate basally for ca. one-third their length, free and valvate above; staminodes 6, stamenlike, the filaments connate basally; ovary incompletely trilocular, triovulate, covered with overlapping scales; stigmas elongate; *fruits* usually 1-seeded, globose to ellipsoid, covered with overlapping reddish-brown or yellowish scales; stigmatic residue apical; endosperm homogeneous; embryo lateral; germination adjacent-ligular; eophyll bifid.

A genus of one species with three varieties distributed throughout the central and western Amazon region in Colombia, Venezuela, Guyana, Peru, and Brazil. Earlier authors (Drude, 1881; Beccari, 1918) have recognized several species.

1. Lepidocaryum tenue Mart., Hist. nat. palm. 2: 51. 1824. *Mauritia tenuis* (Mart.) Spruce, J. Linn. Soc., Bot. 11: 169. 1871. Type. Brazil. Amazonas: Rio Japurá, n.d., *C. Martius s.n.* (holotype, M; F negs. 18587, 18588) (Figure 6.3c–d).

Stems cespitose, rarely solitary, erect, forming colonies by rhizomes, 0.8–4(–6) m tall, 1.5–3 cm diam. *Leaves* 8–20; sheath 20–80 cm long, sheath and petiole often densely brown-tomentose; petiole 32–130 cm long; rachis very short, to 4 cm long; blade divided into 4(–22) segments, the middle ones 48–75 cm long, (1–)5–8(–13) cm wide, with prominent cross-veins and usually with a few small spines on the margins, especially near the apex. *Inflorescences* interfoliar, erect at anthesis, the pistillate becoming pendulous in fruit; peduncle 30–47 cm long; prophyll 4–8 cm long; peduncular bracts ca. 8, tubular and sheathing the peduncle; rachis 10–20 cm long, covered with tubular, sheathing bracts; primary branches 2–16, 3–10 cm long; rachillae numerous, very short; staminate flowers 6–9 mm long, crowded; sepals 2–4 mm long; petals lanceolate, 6–8 mm long; pistillate flowers 6–7 mm long, crowded; sepals 2–3 mm long; petals lanceolate, 6 mm long; *fruits* ellipsoid or seldom globose, 1.5–3 cm long, 1–2 cm diam, orange-brown, red, or yellowish at maturity, the scales with black, ciliate margins.

Brazil: *burityzinho, caraná, caranai, carana-y, palmeira.* Colombia: *agee* (Miraña), *camanará, caraná, ho-ta-mo-hee'* (Barasana), *kaarú* (Yukuna), *ká-roo* (Yukana), *karugiri* (Yukana), *moo-ee-a* (Tanímuka), *moo-heé* (Makuna), *muin* (Tanimuka), *muiriká* (Tanemuka), *po-ta'-me* (Tukano), *pui, pui ocho hojas, puy, tuee-tee'* (Puinave), *tado* (Andoke), *tegpayaje* (Miraña). Peru: *caraña, irapai, irapay.* Venezuela: *cabaya, cau* (Baré), *morichito, moriquito, tevy* (Baniba). The leaves are very commonly used for thatching; in Peru, three kinds are recognized: *pata de grillo, pata de gallo,* and *cadena.* In Colombia, the stems are used in basketry.

This is a rather complex species. Wessels Boer (1988) recognized one, widespread species. He followed Trail (1877b), who pointed out that *Lepidocaryum tenue:*

> is a most variable species, alike in number of laciniae in the leaves, and in the form and size of fruit, so much so that on the same plant leaves may be found with all numbers from 2 to 10 or even to 14 or more laciniae, while on the same spadix (still more in the same clump of plants) the fruits often vary considerably in form and in size, as well as in form, convexity, ciliation, and colour of scales.

Specimens from the Amazon region, however, can be divided into three forms. The most widespread and consistent form typically has leaves with 4 broad segments, and is here called var. *tenue.* In the upper Río Negro region of Colombia, Venezuela, and Brazil, there is a second form with leaves divided into numerous, narrow segments. This is called var. *casiquiarense.* And in the northern and eastern part of the range of the species, in Guyana and Brazil (eastern Amazonas, northern Mato Grosso, western Pará), a third, very variable form is found with leaves somewhat intermediate between the first two. It is here called var. *gracile.* It was this last form that Trail collected and studied, and I believe this was the reason he considered the species so variable. There appear to be no differences between the flowers and fruits of these three forms, but there are minor differences in the number of primary branches of the inflorescence. Since each form occupies a more or less discrete range, they are here recognized as varieties. Galeano (1991) reported that all three forms were found in the same area of the Colombian Amazon. There could be two explanations for this: either the region is not typical, or all three types are always found together but not sampled. More collecting is needed to better understand this complex species.

KEY TO THE VARIETIES OF *LEPIDOCARYUM TENUE*

1. Leaves typically with 4 segments, occasionally to 8, the middle ones (1–)5–8(–13) cm wide; widespread . . 1c. *L. tenue* var. *tenue.*
1. Leaves typically with 4–22 segments, the middle ones 1–4 cm wide.
   2. Leaves with 16–22 narrow segments; rachis less than 1 cm long; inflorescence primary branches 2–6; Colombia (Amazonas, Guainía), Venezuela (Amazonas, Bolívar), and Brazil (Amazonas) . . . . 1a. *L. tenue* var. *casiquiarense.*
   2. Leaves with 4–19 unequally wide segments; rachis 1–2 cm long; inflorescence primary branches 5–14; Guyana, Brazil (eastern Amazonas, northern Mato Grosso, western Pará). . . . . 1b. *L. tenue* var. *gracile.*

1a. Lepidocaryum tenue var. casiquiarense

(Spruce) Henderson, stat. nov. *Mauritia casiquiarensis* Spruce, J. Linn. Soc., Bot. 11: 173. 1871. *Lepidocaryum casiquiarense* (Spruce) Drude in Mart., Fl. bras.: Cyclanthaceae et Palmae I, fasc. 85, vol. 3(2): 300. 1881. Type. Venezuela. Amazonas: Casiquiare, Dec 1853, *R. Spruce 40* (holotype, K; isotype, NY).

Mauritia guainiensis Spruce, J. Linn. Soc., Bot. 11: 174. 1871. *Lepidocaryum guainiense* (Spruce) Drude in Mart., Fl. bras.: Cyclanthaceae et Palmae I, fasc. 85, vol. 3(2): 298. 1881. Type. Venezuela. Amazonas: Río Guainía, Jun 1854, *R. Spruce 40* (holotype, K).

Leaves 13–18; sheath 30–55 cm long; petiole 50–70 cm long; rachis very short, less than 1 cm long; blade divided into 16–22 segments, the middle ones 41–54 cm long, 1–3 cm wide, spiny on the margins and veins abaxially. *Inflorescences* with 2–6 primary branches.

Upper Río Negro region of Colombia (Amazonas, Guainía), Venezuela (Amazonas, Bolívar), and Brazil (Amazonas) (Figure 6.5D); lowland rain forest on white-sand soils (podzols) near black-water streams and rivers, at low elevations.

1b. Lepidocaryum tenue var. gracile (Mart.) Henderson, stat. nov. *Lepidocaryum gracile* Mart., Hist. nat. palm. 2: 50. 1824. *Mauritia gracilis* (Mart.) Spruce, J. Linn. Soc., Bot. 11: 169. 1871. Type. Brazil. Amazonas: Canumá, n.d., *C. Martius s.n.* (holotype, M; F negs. 18585, 18586).
Lepidocaryum enneaphyllum Barb. Rodr., Enum. palm. nov. 19. 1875. Lectotype (Wess. Boer, 1988). Barb. Rodr., Sert.palm. brasil. 1: t. 4a. 1903.
Lepidocaryum sexpartitum Trail & Barb. Rodr., Enum. palm. nov. 19. 1875. *Lepidocaryum tenue* var. *sexpartitum* (Trail & Barb. Rodr.) Trail, J. Bot. 15: 129. 1877. Lectotype (Wess. Boer, 1988). Barb. Rodr., Sert. palm. brasil. 1: t. 4b. 1903.
Lepidocaryum sexpartitum var. *microcarpum* Drude in Mart., Fl. bras.: Cyclanthaceae et Palmae I, fasc. 85, vol. 3(2): 299. 1881. Type. Brazil. Pará: San Antonio, Rio Tapajós, 17 Mar 1874, *J. Trail & J. Barbosa Rodrigues 1092/XIX* (holotype, K).
Lepidocaryum sexpartitum var. *macrocarpum* Drude in Mart., Fl. bras.: Cyclanthaceae et Palmae I, fasc. 85, vol. 3(2): 299. 1881. *Lepidocaryum macrocarpum* (Drude) Becc., Ann. Bot. Gard. Calcutta 12: 221. 1918. Type. Brazil. Amazonas: Rio Padauiri, 27 Jun 1874, *J. Trail 1095/LXXIX* (holotype, K).
Lepidocaryum gujanense Becc., Ann. Bot. Gard. Calcutta 12: 221. 1918. Type. Guyana. Without locality, n.d, *R. Schomburgk 983* (isotype, NY).

Leaves 6–12; sheath to 38 cm long; petiole 18–50 cm long; rachis 1–2 cm long; blade divided into 4–19 unequal segments, the middle ones 47–54 cm long, 1–4 cm wide, with spines on the margins apically. *Inflorescences* with 5–14 primary branches.

Guyana and Brazil (eastern Amazonas, northern Mato Grosso, western Pará), particularly on the Rios Tapajós, Maués, and Trombetas (Figure 6.5E); lowland rain forest in inundated areas at low elevations. The Guyana specimen is not mapped because its exact locality is unknown.

1c. Lepidocaryum tenue var. tenue
Mauritia quadripartita Spruce, J. Linn. Soc., Bot. 11: 172. 1871. *Lepidocaryum quadripartitum* (Spruce) Drude in Mart., Fl. bras.: Cyclanthaceae et Palmae I, fasc. 85, vol. 3(2): 298. 1881. Type. Brazil. Amazonas: Igarapé da Cachoeira, n.d., *R. Spruce 23* (holotype, K).
Lepidocaryum tessmannii Burret, Notizbl. Bot. Gart. Berlin-Dahlem 10: 771. 1929. Type. Peru. Loreto: Morona, Río Marañon, Tierra Blanca, 160 m, 10 Jan 1925, *G. Tessmann 4906* (holotype, B; isotype NY).
Lepidocaryum allenii Dugand, Caldasia 9: 389. 1944. Type. Colombia. Vaupés: Río Papurí and Río Paca near Bacaricuara, 6 Sep 1943, *P. Allen 3113* (holotype, COL).

Leaves 9–20; sheath 15–80 cm long; petiole 60–130 cm long; rachis 2–5 cm long; blade divided into 4 segments, occasionally to 8, the middle ones (1–)5–8(–13) cm wide. *Inflorescences* with 4–14 primary branches.

Colombia (Amazonas, Vaupés), Peru (Loreto, San Martín), and Brazil (Acre, Amazonas, Pará) (Figure 6.5F); lowland forest, either on *terra firme* or less commonly and abundantly in inundated areas, below 500 m elevation. It can form large colonies in the understory (Kahn & Mejia, 1987). The gap in the middle of the range is a collecting artifact.

CEROXYLOIDEAE • HYOPHORBEAE

7. *Chamaedorea* Chamaedorea Willd., Sp. pl. 4(2): 638. 1806.

Small or rarely moderate, dioecious palms. *Stems* solitary or less often cespitose, slender, erect, or procumbent, seldom climbing, green, with prominent nodes. *Leaves* reduplicate, few, glabrous; sheath partially closed, occasionally forming a crownshaft, abaxial surface of sheath and petiole with yellow stripe; petiole short to moderate; rachis short to moderate; pinnae very variable, from linear to sigmoid, regularly to irregularly arranged, or leaf entire and then usually with praemorse margins. *Inflorescences* interfoliar or infrafoliar, spicate or branched to 1 order, solitary or several per node; peduncle bearing a prophyll and 2–several papery peduncular bracts; rachis short and bearing few to many rachillae, or inflorescence spicate; staminate flowers borne singly or arranged in short rows, occasionally contiguous; sepals 3; petals 3, free, valvate; stamens 6; pistillode small or conspicuous; pistillate flowers borne singly or occasionally almost contiguous; sepals 3; petals 3, valvate; staminodes small or absent; gynoecium trilocular, triovulate; *fruits* 1-seeded, globose or ellipsoid, with basal stigmatic residue; seed with homogeneous endosperm and basal or lateral embryo; germination adjacent-ligular; eophyll bifid or pinnate.

A genus of approximately 100 species or fewer (Hodel, 1992) distributed from Mexico to Bolivia, usually in mountainous areas but also in lowland forest. The greatest concentration of species is in Mesoamerica. In South America, they are usually confined to Andean slopes and the coastal Cordillera of Venezuela. There seems to be a minor center of diversity in the eastern Andean slopes and the adjacent western Amazon region of Peru.

KEY TO THE SPECIES OF *CHAMAEDOREA*
1. Leaves usually pinnate, rarely entire, and then not or scarcely toothed on the margins; inflorescences branched with (2–)6–16 rachillae, 11–30 cm long.
    2. Leaves with 21–39 linear-lanceolate pinnae per side; staminate inflorescences multiple and branched; pistillate inflorescences solitary and branched. . . . . 1. *C. angustisecta.*
    2. Leaves with 4–8 sigmoid pinnae per side, rarely leaf entire; staminate and pistillate inflorescences solitary and branched . . . . 4. *C. pinnatifrons.*
1. Leaves usually entire, toothed on the margins, rarely pinnate; inflorescences spicate or with 2–8 rachillae, these 10–17 cm long.
    3. Inflorescences interfoliar, erect, elongate; peduncle 25–42 cm long; rachillae 25–50 cm long; staminate inflorescences multiple and spicate; pistillate inflorescences solitary and spicate, rarely with 2–3 rachillae . . . . . 3. *C. pauciflora.*
    3. Inflorescences infrafoliar, ± horizontal, short; peduncle 5–10 cm long; rachillae 10–17 cm long; staminate and pistillate inflorescences solitary and branched, rarely spicate . . . . 2. *C. fragrans.*

1. **Chamaedorea angustisecta** Burret, Notizbl. Bot. Gart. Berlin-Dahlem 11: 318. 1932. Type. Peru. Junín: Colonia Perene, ca. 680 m, 14–22 Jun 1929, *E. Killip & A. Smith 25091* (isotypes, F, NY) (Figure 6.6a–b).
Chamaedorea leonis H. E. Moore, Gentes Herb. 12: 30. 1980. Type. USA. Florida: cultivated (seed received from Bolivia) at Parrot Jungle, Red Road, Coral Gables, 1 Apr 1969, *H. Moore 9586* (holotype, BH).

Stems solitary, erect, 1.5–4 m tall, 2–3.5 cm diam. *Leaves* 4–8; sheath partly closed, 30–73 cm long; petiole 35–100 cm long, petiole and rachis mottled green and brown; rachis 1.1–1.5 m long; pinnae 21–39 per side, regularly arranged and spreading in 1 plane, linear-lanceolate, the middle ones 33–52 cm long, 1.5–2.5(–4) cm wide. *Inflorescences* interfoliar, with 1 order of branching, the staminate multiple, 2–7 at each node, the pistillate solitary; peduncle 19–80 cm long; prophyll 1.5–4.5 cm long; peduncular bracts 4–6, 6–27 cm long,

Figure 6.6. (a) *Chamaedorea angustisecta,* **(b)** part of infructescence (from *M. Moraes 610*); **(c)** *Chamaedorea fragrans,* **(d)** infructescence (from *S. Knapp 7422*); **(e)** *Chamaedorea pauciflora,* **(f)** infructescence (from *F. Skov 64725*); **(g)** *Chamaedorea pinnatifrons,* **(h)** infructescence (from *J. Schunke 8537*).

closely sheathing the peduncle; rachis 3.5–7.5 cm long; rachillae 9–16, 11–24 cm long; *flowers* borne singly; staminate flowers 3.5–4 mm long at anthesis; sepals connate into a deeply 3-lobed cupule, 1–1.5 mm long; petals ovate, 3–3.5 mm long; stamens exceeded by the prominent pistillode; pistillate flowers 1 mm long (in bud); sepals connate into a cupular calyx, 1 mm long; petals ovate, 1 mm long; staminodes not apparent; *fruits* ellipsoid, almost falcate, 1.3–1.6 cm long, 0.5–0.8 cm diam, black.

Peru (Ayacucho, Cuzco, Junín, Madre de Dios, Ucayali), Brazil (Acre), and Bolivia (Beni, La Paz, Pando) (Figure 6.7A); lowland rain forest on *terra firme,* in areas below 800 m elevation.

Bolivia: *sialla, siyaiye, siyeyi.* Brazil: *palmeira, palmeirinha.* Peru: *sangapilla.* The fragrant inflorescences are gathered and taken indoors to perfume houses.

Chamaedorea angustisecta is characterized by its pinnate leaves and multiple staminate inflorescences. It is placed, together with *C. pauciflora,* in subgenus *Moreniopsis* (Hodel, 1992). On eastern Andean slopes above 700 m elevation, *C. angustisecta* is replaced by *C. linearis* (Ruiz & Pav.) Mart. (of which *C. poeppigiana* [Mart.] Gentry is a synonym), which also has multiple staminate inflorescences but is distinguished by its grouped staminate flowers and red fruits.

2. **Chamaedorea fragrans** (Ruiz & Pav.) Mart., Hist. nat. palm. 2: 4. 1823. *Nunnezharia fragrans* Ruiz & Pav., Syst. veg. fl. peruv. chil. 294. 1798. Type. Peru. Pasco: Pozuzo, n.d., *J. Pavón s.n.* (holotype, M; F neg. 18539) (Figure 6.6c–d).

Chamaedorea pavoniana H. Wendl. ex Dammer, Gard. Chron., ser. 3, 36: 246. 1904. Type. Cultivated plant, n.d., *H. Wendland s.n.* (holotype, K, n.v.).

Chamaedorea ruizii H. Wendl. ex Dammer, Gard. Chron., ser. 3, 36: 246. 1904. Type. Cultivated plant, n.d., *H. Wendland s.n.* (holotype, K, n.v.).

Chamaedorea cataractarum Hort. (non Mart.) in Guillaum., J. Soc. Natl. Hort. France, ser. 4, 24: 235. 1923.

Chamaedorea verschaffeltii Hort. in Kerch., Les Palmiers 240. 1878.

Chamaedorea gratissima Hort. in Linden, Catalogue 13. 1896.

Stems cespitose, 2–4 m tall, 0.5–1 cm diam. *Leaves* ca. 4; sheath closed, 6–9 cm long; petiole 2–3 cm long; rachis 10–12 cm long; blade entire, narrow, deeply bifid, toothed on margins, the lobes 30–43 cm long, 6–9 cm wide at apex of rachis. *Inflorescences* infrafoliar, spicate or branched to 1 order; peduncle 5–10 cm long; prophyll 0.5 cm long; peduncular bracts 2–3, 1–5 cm long, sheathing peduncle, persistent; rachis 0–2 cm; rachillae 1–8, 10–17 cm long; staminate flowers 3 mm long; sepals briefly connate basally, free and spreading above, depressed ovate, 1 mm long; petals obovate, 3 mm long; pistillode longer than the stamens; pistillate flowers not seen; *fruits* ellipsoid, ca. 1.2 cm long, ca. 0.6 cm diam, black.

Peru (Cuzco, Huánuco, San Martín) (Figure 6.7B); lowland or premontane rain forest on *terra firme* between 400 and 900 m elevation.

Peru: *sangapilla.* The sweet-smelling inflorescences are taken indoors and used to scent houses and clothes, and as a perfume.

This species is apparently rare and seldom has been collected. Spruce (1871), however, reported it to be common on the eastern Andean slopes.

3. **Chamaedorea pauciflora** Mart., Hist. nat. palm. 2: 5. 1823. *Morenia pauciflora* (Mart.) Drude in Mart., Fl. bras. Palmae II fasc. 86 vol 3(2): 526. 1882. *Nunnezharia pauciflora* (Mart.) Kuntze, Revis. gen. pl. 2: 730. 1891. Type. Brazil. Amazonas: Rio Japurá, n.d., *C. Martius s.n.* (holotype, M; F neg. 18545) (Figure 6.6e–f).

Morenia integrifolia Trail, J. Bot. 14: 331. 1876. *Nunnezharia integrifolia* (Trail) Kuntze, Revis. gen. pl. 2: 730. 1891. *Chamaedorea integrifolia* (Trail) Dammer, Verh. Bot. Vereins Prov. Brandenburg 48: 125. 1907. Type. Brazil. Amazonas: Gavião, Rio Jurúa, 11 Nov 1874, *J. Trail 1042/CLVII* (isotype, P).

Morenia integrifolia var. *nigricans* Trail, J. Bot. 14: 331. 1876. Type. Brazil. Amazonas: Tabatinga, 30 Nov 1874, *J. Trail 1044/CLVII* (isotype, P).

Chamaedorea lechleriana H. Wendl. ex Dammer, Gard. Chron. 36(3): 246. 1904. Type.

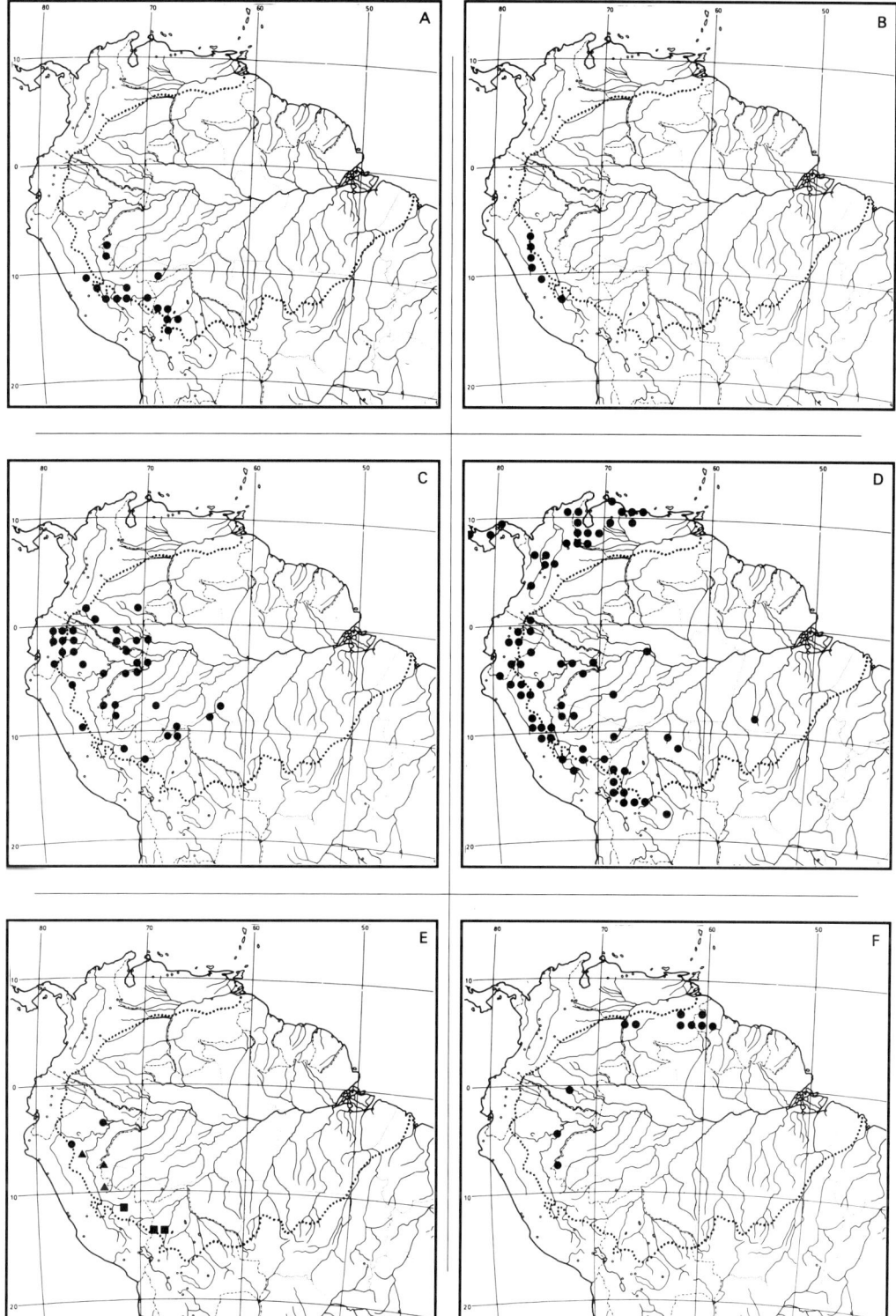

Figure 6.7. **(A)** *Chamaedorea angustisecta;* **(B)** *Chamaedorea fragrans;* **(C)** *Chamaedorea pauciflora;* **(D)** *Chamaedorea pinnatifrons;* **(E)** *Wendlandiella gracilis* var. *gracilis* (triangles), var. *polyclada* (circles), var. *simplicifrons* (squares); **(F)** *Dictyocaryum ptarianum.*

Peru. Department?: St. Garan, n.d., *W. Lechler s.n.* (holotype, GOET).

Chamaedorea amazonica Dammer, Notizbl. Bot. Gart. Berlin-Dahlem 6: 263. 1915. Type. Brazil. Acre: Rio Juruá, Juruá-Mirim, Jun 1901, *E. Ule 5595* (isotypes, K, MG).

Stems solitary or less often cespitose, 0.3–2 (–3) m tall but often less and appearing acaulescent, 1–2 cm diam, with visible roots at the base. *Leaves* 4–9, spreading; sheath 11–25 cm long, open for most of its length; petiole 6–30 cm long; rachis 19–77 cm long; blade entire, becoming split with age, bifid for one-third of its length, strongly plicate, distal margins praemorse, 75–105 cm long, 25–40 cm wide, or occasionally pinnate with 4–7 linear to sigmoid pinnae per side. *Inflorescences* interfoliar, erect at anthesis; staminate inflorescences spicate, multiple, to 7 at each node, pistillate inflorescences solitary, occasionally 2 per node, spicate; peduncle 25–42 cm long; prophyll ca. 3 cm long; peduncular bracts 3–4, 5–15 cm long, tubular and closely sheathing the peduncle; rachilla 25–50 cm long; staminate flowers 2 mm long, borne singly; sepals united into a 3-lobed cupule, 0.5 mm high; petals 1–1.5 mm high; pistillode umbrella-shaped; pistillate flowers borne singly; sepals united into a 3-lobed cupule, 1 mm high; petals slightly imbricate basally, 1 mm high; staminodes 3 or 6?, digitate and very small; ovary globose, to 1 mm high; *fruits* ellipsoid or occasionally 2-lobed, 1.2–1.5 cm long, 5–8 mm diam, black.

Colombia (Amazonas, Caquetá, Putumayo, Vaupés), Ecuador (Napo, Morona-Santiago, Pastaza, Zamora-Chinchipe), Peru (Amazonas, Huánuco, Loreto, Madre de Dios, San Martín, Ucayali), Brazil (Acre, Amazonas, Mato Grosso, Rondônia), and Bolivia (Pando) (Figure 6.7C); lowland or premontane rain forest on *terra firme,* or in *várzea* forest, at elevations below 1000 m.

Colombia: *aacaiba* (Miraña), *citaganomecu* (Mui), *halago, jodaj mena* (Huitoto), *palma gorere, palmita tijereta, ti aacaiba* (Miraña). Ecuador: *suma-yuca* (Quichua), *yaun* (Shuar). Peru: *sangapilla, yaún* (Mayna Jívaro). The leaves are used to wrap and cook fish (Mayna Jívaro, Peru). The inflorescences are used to scent houses and clothes, and a perfume is extracted from them.

The specimens with pinnate leaves are mostly on the periphery of the range of the species.

4. Chamaedorea pinnatifrons (Jacq.) Oerst., Vidensk. Meddel. Dansk. Naturhist. Foren. Kjøbenhavn 1858: 14. 1859. *Borassus pinnatifrons* Jacq., Pl. hort. schoenbr. 2: 65. 1797. *Nunnezharia pinnatifrons* (Jacq.) Kuntze, Revis. gen. pl. 2: 730. 1891. Type. Venezuela. Caracas, n.d., *F. Bredemeyer s.n.* (holotype, W, n.v.; F negs. 29897, 29898) (Figure 6.6g–h).

Martinezia lanceolata Ruiz & Pav., Syst. veg. fl. peruv. chil. 1: 297. 1798. *Chamaedorea lanceolata* (Ruiz & Pav.) Kunth, Enum. pl. 3: 172. 1841. *Nunnezharia lanceolata* (Ruiz & Pav.) Kuntze, Revis. gen. pl. 2: 730. 1891. Type. Peru. Huánuco: Chinchao and Cuchero, Aug 1870, *J. Pavón s.n.* (isotype, M; F neg. 18542).

Chamaedorea boliviensis Dammer, Notizbl. Bot. Gart. Berlin-Dahlem 6: 262. 1915. Type. Bolivia. Pando: Cobija, Jan 1912, *E. Ule 115* (holotype, B, n.v.).

Chamaedorea depauperata Dammer, Notizbl. Bot. Gart. Berlin-Dahlem 6: 263. 1915. Type. Brazil. Acre: Monte Alegre, S. Francisco, Sep 1911, *E. Ule 9155b* (holotype, B, n.v.).

Stems solitary, 1–4 m tall, 1–3 cm diam. *Leaves* 4–10 spreading; sheath 14–30 cm long, partly closed and forming an elongate crownshaft; petiole 10–40 cm long; rachis 16–70 cm long; pinnae 4–8 per side, ± regularly arranged, sigmoid, the middle ones 18–38 cm long, 2.5–8 cm wide, the apical one much broader than the others, or rarely leaf entire. *Inflorescences* infrafoliar at anthesis, solitary at each node; peduncle 20–70 cm long; prophyll 2–4 cm long; peduncular bracts 4–5, 5–15 cm long, closely sheathing the peduncle; rachis 3–8 cm long; rachillae (2–)6–14, 15–30 cm long; staminate flowers 1–2 mm long, borne singly; sepals connate into a shallow, 3-lobed cupule, 0.5–1 mm long, ribbed; petals free, valvate, joined at the apex and opening by lateral slits, 1.5–2 mm long, ribbed; pistillode columnar; pistillate flowers borne singly; sepals 3, imbricate, 1 mm long, ribbed; petals 1 mm long, ribbed; staminodes obscure; ovary pyramidal-shaped; *fruits* globose to ellipsoid, to 1 cm long, 6–8 mm diam, black at maturity.

Widespread from southern Mexico through Central America and then south through the Andes of Colombia, Venezuela, Ecuador, Peru, and Bolivia, and in the adjacent western Amazon region in Colombia (Amazonas), Ecuador (Morona-Santiago, Napo, Pastaza), Peru (Huánuco, Loreto, Madre de Dios, Ucayali), Brazil (Acre, Amazonas, Pará, Rondônia, and one extraordinary occurrence in Pará), and Bolivia (Beni, Cochabamba, La Paz, Santa Cruz) (Figure 6.7D); in a variety of forest types, especially on eastern Andean slopes, where it can occur at over 2500 m elevation. It also occurs in lowland rain forest in the western Amazon region of Ecuador, Peru, Brazil, and Bolivia, where it is found on *terra firme* and occasionally in inundated areas near rivers.

Ataroff and Schwarzkopf (1992) studied leaf production, reproductive patterns, field germination, and seedling survival in *Chamaedorea pinnatifrons* (as *C. bartlingiana*) in Venezuela. Growth rates were lower in juveniles than in adults, and in females than in males. Male inflorescences were produced at the same rate and survival until anthesis was constant. Females produced buds at the same rate, but these buds had a 35 percent probability of becoming ripe infructescences if the plant already had an infructescence, and a 70 percent chance if not. Males took 1 year from bud to inflorescence at anthesis; females, 2 years. Germination showed an annual peak.

Listabarth (1993a) studied reproductive biology in Huánuco, Peru. Both sexes flowered synchronously during the dry season (June to August). Staminate inflorescences were inhabited by thrips, and at anthesis these small insects entered the flowers by the basal slit between the petals. This caused pollen to be shed from the flowers and removed by air currents. Anthesis took place in the evening and was preceded by a fragrance. Pistillate anthesis took place over 3–7 days, with individual flowers lasting about 2 days. Listabarth considered that this type of pollination system was insect-induced wind pollination.

Bolivia: *jatatilla*. Brazil: *ubim*. Colombia: *molinillo*. Ecuador: *chontilla, chontilla blanca, ñucua-ëné* (Siona). Peru: *cashipana, sangapilla sacha*.

This is one of the most wide-ranging species of *Chamaedorea*, both latitudinally and elevationally. Surprisingly, however, it varies relatively little morphologically and is easily recognizable by its few, sigmoid pinnae. Entire-leafed specimens are known from both the northeastern and the southern part of the range.

8. *Wendlandiella*

Wendlandiella Dammer, Bot. Jahrb. Syst. 36: 31. 1905.

Small, dioecious palms. *Stems* cespitose, slender, green, often leaning. *Leaves* pinnate or entire, reduplicate, few, glabrous; sheath closed and forming an elongate crownshaft; petiole short; rachis very short or moderate; pinnae lanceolate, linear-lanceolate, or ovate-lanceolate, few, or leaf entire. *Inflorescences* interfoliar, branched to 1 or 2 orders, rarely the pistillate spicate, recurved or straight at anthesis; peduncle bearing a prophyll and 1 papery peduncular bract; rachis bearing few to many rachillae; *flowers* borne singly, in pairs or rows; staminate flowers borne in short, vertical, biseriate rows of 2–5 flowers; pistillate flowers solitary or rarely paired; staminate flowers with sepals 3, free above, spreading; petals 3, free, valvate; stamens 6; pistillode columnar, 3-fid; pistillate flowers with sepals 3, united into a 3-lobed cupule, this splitting as ovary develops; petals 3, imbricate basally; staminodes 3; gynoecium trilocular, triovulate; *fruits* 1-seeded, globose to ellipsoid, with basal stigmatic residue; seed with endosperm homogeneous and lateral embryo; eophyll ?bifid.

A genus of one species with three varieties distributed in Amazonian Peru,

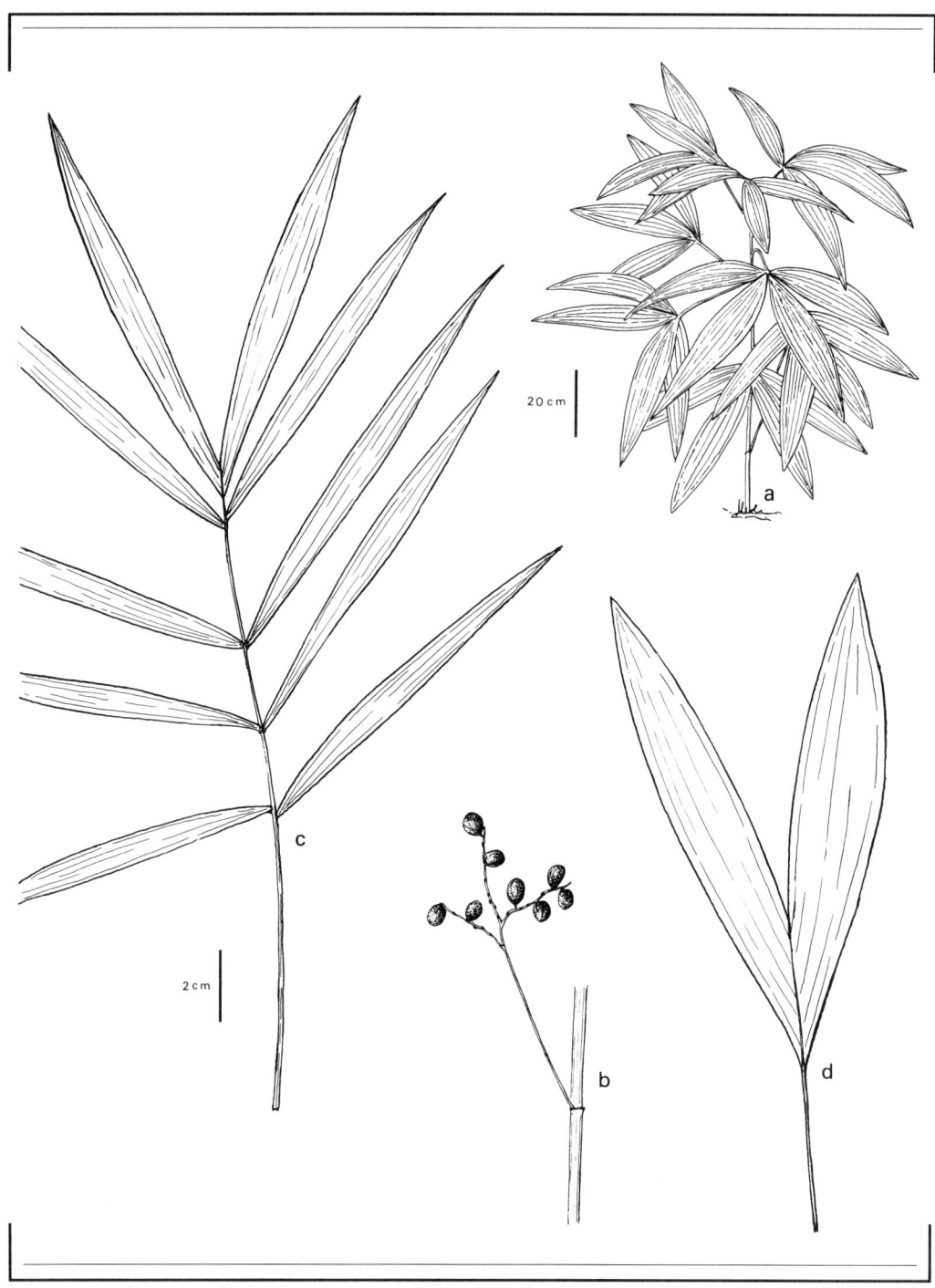

Figure 6.8. (a) *Wendlandiella gracilis* var. *polyclada*; **(b)** infructescence of *Wendlandiella gracilis* var. *simplicifrons* (from *A. Gentry 43421*); **(c)** leaf of *Wendlandiella gracilis* var. *gracilis*; **(d)** leaf of *Wendlandiella gracilis* var. *simplicifrons*.

extreme western Brazil, and northern Bolivia. Macbride (1960) recognized three separate species.

The arrangement of the staminate flowers in this genus, in short, biseriate, vertical rows, is very similar to that of *Synechanthus,* and it seems possible that *Wendlandiella* is actually more closely related to *Synechanthus* than to *Chamaedorea* (Uhl & Dransfield, 1987). If this were the case, then its subtribe, the Hyophorbeae, would have a trans-Pacific linear distribution similar to that of the Ceroxyleae and Butiinae, and all three subtribes would have the "terminal" taxa as the most speciose. In the Hyophorbeae, *Hyophorbe* (5 species) occurs in the Mascarenes, *Wendlandiella* (1) in the western Amazon region, *Synechanthus* (2) in western Ecuador and Colombia and Central America, *Gaussia* (4) in Central America and the Caribbean, and *Chamaedorea* (98) in Central America. In this scenario, *Chamaedorea* has secondarily entered South America.

1. Wendlandiella gracilis Dammer, Bot. Jahrb. Syst. 36: 32. 1905. Type. Peru. San Martín: Huimbayoc, between the Ucayali and Huallaga, Dec 1898, *J. Huber 1541* (holotype, B, n.v.) (Figure 6.8a–d).

Stems cespitose, occasionally solitary, 0.3–1.5 m tall, 0.4–1 cm diam, green, forming clusters of slightly procumbent stems. *Leaves* 4–11, spreading, glabrous; sheath closed and forming an elongate crownshaft 3–13 cm long; petiole 5–19 cm long; rachis 3–17 cm long; blade entire or with to 6 pinnae per side, regularly or somewhat irregularly arranged, lanceolate, linear-lanceolate, or ovate-lanceolate, the middle ones 13–30 cm long, 1–8 cm wide, with a mid-vein and 2 marginal veins. *Inflorescences* interfoliar in bud and decurved or straight and interfoliar at anthesis and in fruit, with 1 or 2 orders of branching, or occasionally spicate; peduncle 9–25 cm long; prophyll 3–7 cm long, inserted well up the peduncle near the point adjacent to top of leaf sheath; peduncular bract 1, 2–12 cm long, inserted near the prophyll; rachis 1–8 cm long; rachillae 1–35, 2–12 cm long; staminate flowers 1 mm long; sepals 0.5 mm long; petals ovate, 1 mm long, recurved at anthesis; pistillode 0.5 mm high; pistillate flowers 1.5–2 mm long; sepals 1 mm long; petals ovate, 1 mm long; *fruits* globose to ellipsoid, 0.8–1 cm long, ca. 0.5 cm wide, red, orange, or orange-red.

Peru: *chontilla, ponilla.*

Surprisingly, this small, obscure genus turns out to have almost exactly the variation described by Macbride (1960), and his key works perfectly. Moreover, the species have cohesive geographic distributions. Since there are so few specimens available, and because some overlap in leaf characters has been observed, I have relegated the three species to varieties pending more detailed study. There is some slight sexual dimorphism in this species; staminate plants are usually larger than pistillate, especially the inflorescences.

There seems to be some correlation between the ranges of the three varieties and three of the basins of the western Amazon region: var. *simplicifrons* in the Madre de Dios Basin, var. *gracilis* in the Ucayali Basin, and var. *polyclada* in the Pastaza-Marañon Basin.

Listabarth (1993a) described reproductive biology in Huánuco, Peru. He found that plants flowered in the dry season (July to October). Staminate flowers opened in the morning and pollen was shed immediately, and blown away from the flowers by air currents. Individual flowers lasted 1 day, but the inflorescence remained at anthesis for up to 70 days. Pistillate flowers were at anthesis for up to 5 days, and an inflorescence remained at anthesis for up to 15 days. No detectable fragrance was produced by either inflorescence type, nor were any insects observed on inflorescences. Listabarth concluded that the species was wind pollinated, but because fruit set was rare he considered that vegetative reproduction was important.

KEY TO THE VARIETIES OF *WENDLANDIELLA GRACILIS*
1. Leaves entire; Peru (Madre de Dios, Pasco) and Bolivia (La Paz, Pando). . . . .
    . 1c. *Wendlandiella gracilis* var. *simplicifrons.*
2. Leaves pinnate.
    3. Leaves with 2 pinnae per side; Peru (Loreto) . . . .
        1b. *Wendlandiella gracilis* var. *polyclada.*
    3. Leaves with 4–6 pinnae per side; Peru (Huánuco) and Brazil (Acre). . .
        . 1a. *Wendlandiella gracilis* var. *gracilis.*

1a. Wendlandiella gracilis var. **gracilis**
Leaves pinnate, with 4–6 linear-lanceolate pinnae per side, the proximal ones somewhat distant from the distal ones, the middle ones 13–18 cm long, 1–2.7 cm wide. *Inflorescences* branched to one order; rachis to 3 cm long; rachillae 5–13.
Brazil (Acre) and Peru (Huánuco) (Figure 6.7E); lowland rain forest on *terra firme* to 700 m elevation.

1b. Wendlandiella gracilis var. **polyclada** (Burret) Henderson, stat. nov. *Wendlandiella polyclada* Burret, Notizbl. Bot. Gart. Berlin-Dahlem 11: 203. 1931. Type. Peru. Loreto: Río Itaya, Soledad, 110 m, 27 Jun 1925, *G. Tessmann 5240* (holotype, B, n.v.).
Leaves pinnate with 2 pinnae per side, lanceolate or linear-lanceolate, 15–30 cm long, 1.5–3.5 cm wide. *Inflorescences* branched to 1 or 2 orders; rachis to 8 cm long; staminate rachillae 13–16 or more; pistillate inflorescence with to 15 rachillae, or with spicate inflorescence and then the flowers in pairs.
Peru (Loreto) (Figure 6.7E); lowland rain forest on *terra firme.*

1c. Wendlandiella gracilis var. **simplicifrons** (Burret) Henderson, stat. nov. *Wendlandiella simplicifrons* Burret, Notizbl. Bot. Gart. Berlin-Dahlem 11: 316. 1932. Type. Peru. Pasco: Puerto Bermudez, ca. 375 m, 14–17 Jul 1929, *E. Killip & A. Smith 26515* (holotype, US; isotype, NY).
Leaves entire, the lobes ovate-lanceolate, 13–15(–30) cm long, 3–3.5(-8) cm wide. *Inflorescences* branched to 1 order; rachis 1–5 cm long; rachillae 4–14.
Peru (Madre de Dios, Pasco) and Bolivia (La Paz, Pando) (Figure 6.7E); lowland rain forest on *terra firme,* occasionally forming large colonies up to 4 m across in the forest understory.

ARECOIDEAE • IRIARTEEAE

9. *Dictyocaryum* **Dictyocaryum** H. Wendl., Bonplandia 8: 106. 1860.

Large, monoecious palms. *Stems* solitary or less often cespitose, stout, erect, columnar or ventricose; stilt roots prominent, numerous. *Leaves* pinnate, reduplicate, few; sheath closed and forming a compact crownshaft; petiole short; rachis long; pinnae praemorse, white-gray waxy abaxially, numerous, divided into segments, these spreading in different planes. *Inflorescences* infrafoliar, erect or pendulous in bud and at anthesis, branched to 1 or 2 orders; peduncle bearing a prophyll and 7–10 peduncular bracts; rachis bearing numerous rachillae; *flowers* in triads; staminate flowers with sepals 3, imbricate; petals 3, valvate; stamens 6; pistillode small; pistillate flowers with sepals 3, imbricate; petals 3, imbricate basally, valvate above; staminodes 6, dentate; gynoecium tricarpellate, triovulate; *fruits* usually 1-seeded, globose to ellipsoid, with subbasal to lateral stigmatic residue; seed with homogeneous endosperm and basal embryo; germination adjacent-ligular; eophyll bifid.

A genus of three species (Henderson, 1990) distributed from eastern Panama to Bolivia, the coastal range of Venezuela, and the Guayana Highland of Venezuela and Guyana; rarely occurring in the western Amazon region at low elevations.

1. **Dictyocaryum ptarianum** (Steyerm.) H. E. Moore & Steyerm., Acta Bot. Venez. 2: 139. 1967. *Dahlgrenia ptariana* Steyerm., Fieldiana, Bot. 28: 82. 1951. Type. Venezuela. Bolívar: Ptari-tepuí, 1585 m, 10–11 Nov 1944, *J. Steyermark 60044* (holotype, F; isotype, BH) (Figure 6.9a–c).

Stems solitary or cespitose, columnar, erect or leaning, 10–15(–20) m tall, 14–20 cm diam, smooth; stilt roots 0.5–1.2 m long, brown with blunt spines. *Leaves* 4–5, spreading; sheath forming a compact crownshaft, 60–70(–125) cm long; petiole 25–50 cm long; rachis 1.7–3 m long; pinnae 23–37 per side, cuneate with entire margins and blunt praemorse apex, split to the base into 2–8 stiff segments, these spreading in different planes, the middle ones 0.7–1 m long, grayish-white-waxy abaxially. *Inflorescences* pendulous and horn-shaped in bud; peduncle curved, 25–35 cm long, at anthesis with prophyll and 8–10 peduncular bract scars; prophyll 9–12 cm long; peduncular bracts 8–10, inserted 2–4 cm apart, to 90 cm long, woody; rachis 20–30 cm long; rachillae 50–80, distal ones branched; *flowers* arranged in triads proximally, paired or solitary staminate distally; staminate flowers 3.5–4 mm long; sepals ovate-deltoid, 1.5 mm long; petals ovate, 3 mm long; pistillate flowers 2 mm long; sepals ovate, 2 mm long; petals ovate, 2 mm long; *fruits* ± globose, 3–3.5 cm long, 2–3 cm diam; epicarp smooth, yellowish-brown at maturity and then splitting irregularly.

Guayana Highland *(tepuis)* of Venezuela (Amazonas and Bolívar) and Guyana (Figure 6.7F); montane rain forest on slopes between (450–)800 and 1700 m elevation. Three lowland-rain-forest populations are known in the western Amazon region of Colombia (Amazonas), Peru (Loreto), and Brazil (Acre), below 360 m elevation.

Brazil: *pifaia, pachiubarana*. Colombia: *bombona paso, jû-kû-fê-nâ* (Huitoto). Peru: *pona colorada*. The leaves are used for thatching, and the stems are split and used to make walls, doors, and especially floors of houses.

The occurrence of lowland-rain-forest populations of this species in the western Amazon region is very unusual; oilbirds are known from the Colombian site and may be the dispersers of this species.

10. *Iriartella*

Iriartella H. Wendl., Bonplandia 8: 103. 1860.

Small to moderate, monoecious palms. *Stems* cespitose, rarely solitary, cylindrical, slender, erect or leaning, forming colonies by rhizomes; stilt roots obscure and few. *Leaves* pinnate, reduplicate; sheaths forming an elongate crownshaft, covered with irritant trichomes; petiole short; rachis short; pinnae few, praemorse, entire. *Inflorescences* interfoliar and erect in bud and at anthesis, becoming infrafoliar and pendulous in fruit, branched to 1 order; peduncle bearing a prophyll and 2–4 peduncular bracts; rachis bearing few rachillae; *flowers* in triads; staminate flowers with sepals 3; petals 3, valvate; stamens 6; pistillate flowers with sepals 3; petals 3; staminodes 6, very small, or staminodes lacking; gynoecium syncarpous, triovulate, tricarpellate; *fruits* 1(–2)-seeded, ellipsoid, with basal stigmatic residue; seed with homogeneous endosperm and apical embryo; germination adjacent-ligular; eophyll entire.

A genus of two species (Henderson, 1990) distributed in the central-western Amazon region of Colombia, Venezuela, Guyana, Peru, and Brazil.

KEY TO THE SPECIES OF *IRIARTELLA*

1. Larger palms with stems 3–12 m tall, (1–)2–4 cm diam; middle pinnae 20–40 cm long; sepals of staminate flowers connate into a shallow cupule; sepals of pistillate flowers connate basally, imbricate above; Colombia (Amazonas, Caquetá, Guainía, Vaupés, Vichada), Venezuela (Amazonas, Bolívar), Guyana, and Brazil (Amazonas, western Pará, Roraima) . . . . 1. *I. setigera*.

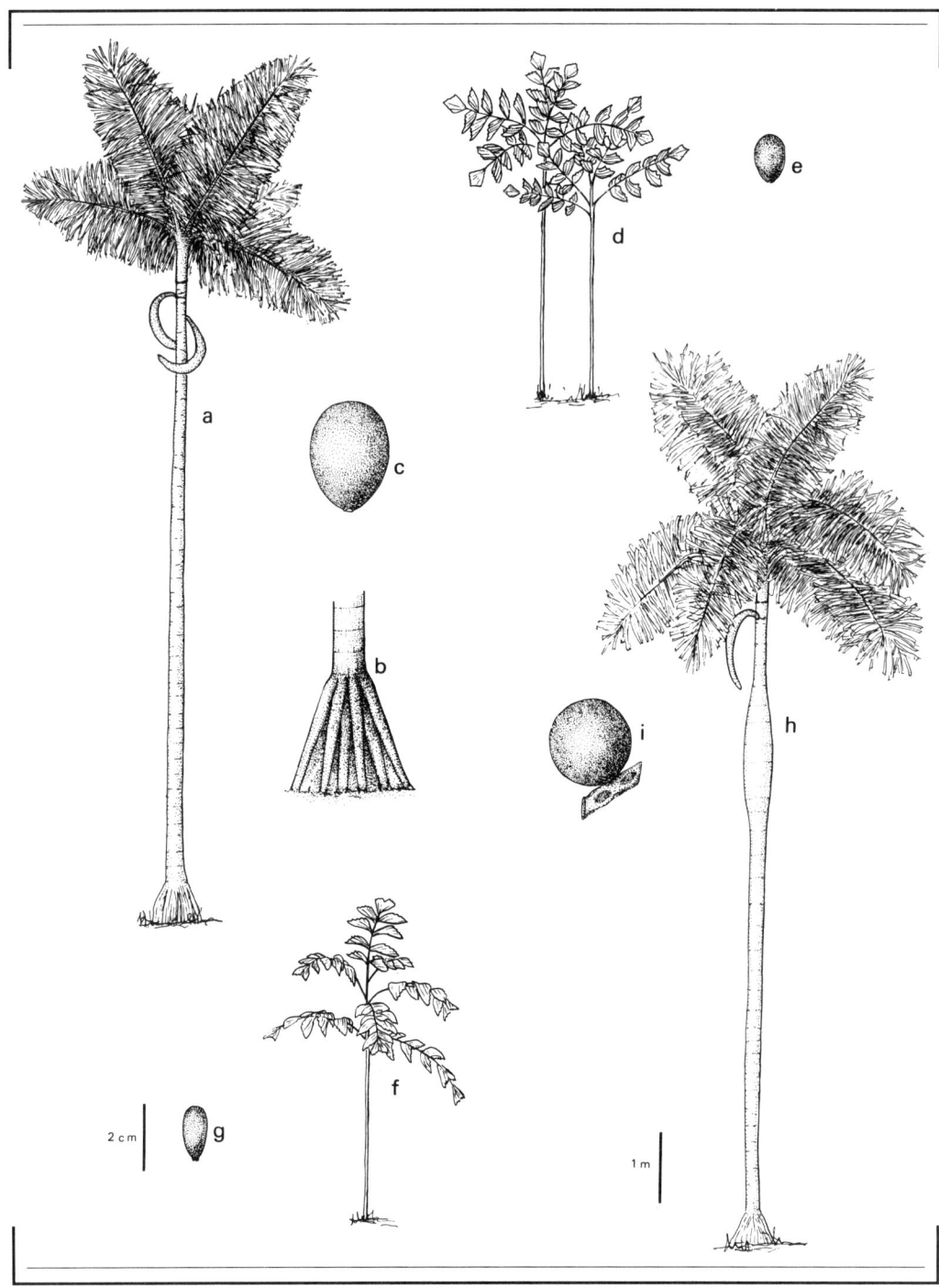

Figure 6.9. **(a)** *Dictyocaryum ptarianum,* **(b)** stilt roots, **(c)** fruit (from *K. Mejia 725*); **(d)** *Iriartella setigera,* **(e)** fruit (from *O. Huber 9411*); **(f)** *Iriartella stenocarpa,* **(g)** fruit (from *J. Ruíz 199*); **(h)** *Iriartea deltoidea,* **(i)** fruit (from *A. Henderson 537*).

1. Smaller palms with stems 1–4(–6) m tall, 1.5–2 cm diam; middle pinnae 14–25(–36) cm long; sepals of staminate flowers briefly connate basally, free and imbricate above; sepals of pistillate flowers distinct and imbricate; Colombia (Amazonas), Peru (Junín, Loreto, Madre de Dios, Pasco, Ucayali), and western Brazil (Acre). . . . . 2. *I. stenocarpa*.

1. Iriartella setigera (Mart.) H. Wendl., Bonplandia 8: 104. 1860. *Iriartea setigera* Mart., Hist. nat. palm. 2: 39. 1824. Type. Brazil. Amazonas: Rio Japurá, [Feb 1820], *C. Martius s.n.* (holotype, M; F negs. 18534, 18534a) (Figure 6.9d–e).

Iriartea spruceana Barb. Rodr., Enum. palm. nov. 13. 1875 ("spruciana"). *Iriartella spruceana* (Barb. Rodr.) Barb. Rodr., Sert. palm. brasil. 1: 18. 1903. *Cuatrecasana spruceana* (Barb. Rodr.) Dugand, Caldasia 2: 72. 1943. Lectotype (Glassman, 1972). Barb. Rodr., Sert. palm. brasil 1: t. 7. 1903.

Iriartella setigera var. *pruriens* Barb. Rodr., Sert. palm. brasil. 1: 18. 1903. *Iriartella pruriens* (Barb. Rodr.) Barb. Rodr., Sert. palm. brasil. 2: 102. 1903. Type not designated.

Cuatrecasea vaupesana Dugand, Rev. Acad. Colomb. Ci. Exact. 3: 392. 1940. Type. Colombia. Vaupés: Mitú, 200 m, 21 Sep 1939, *J. Cuatrecasas 6937* (holotype, COL).

Stems cespitose, occasionally solitary, erect, forming loose clusters of 2–10 main stems, 3–12 m tall, (1–)2–4 cm diam; stilt roots black, spiny, often poorly developed and then not apparent. *Leaves* 6–8; sheath 15–40 cm long, densely reddish-brown-tomentose with dense to moderate clusters of long dark brown, black, or yellow acicular deciduous hairs; petiole 14–30 cm long; rachis 34–96 cm long; pinnae 5–9 per side, entire, cuneate-rhombic in outline, praemorse, the middle ones 20–40 cm long, 4–11 cm wide. *Inflorescences* interfoliar in bud, becoming infrafoliar in fruit; peduncle 25–53 cm long; prophyll to 16 cm long; peduncular bracts 3–4, 8–40 cm long; rachillae 3–25, 8–22 cm long; *flowers* proximally in triads on rachis, near apex paired or solitary staminate; staminate flowers 3 mm long; sepals connate into a 3-lobed calyx, 0.8–1 mm long; petals ovate, 2.5–3.8 mm long; pistillode minute or absent; pistillate flowers 2 mm long; sepals forming a 3-lobed calyx, 0.5–0.6 mm long; petals widely ovate, 1.2–1.4 mm long; staminodes minute or absent; *fruits* ellipsoid, rarely almost globose, 1.4–1.7 cm long, 0.7–1 cm diam; epicarp glabrous, scarlet, orange, or brown at maturity and splitting irregularly from the apex.

Widespread in the central and western Amazon region in Colombia (Amazonas, Caquetá, Guainía, Vaupés, Vichada), Venezuela (Amazonas, Bolívar), Guyana, and Brazil (Amazonas, western Pará, Roraima) (Figure 6.10A); lowland rain forest, on both *terra firme* and in inundated areas, usually below 700 m, rarely reaching 1200 m elevation.

Brazil: *pachiubina, paxiubarana, paxiubinha, ubim do igapó.* Colombia: *boo-han-ñee-kaw-ne* (Makuna), *maá-kan* (Maku). Guyana: *kubina* (Wapisiana). Venezuela: *cola de pava, mabe, mábi, macanilla, palma cola de pescado, yadua* (Maguiritare). The stems were formerly used by indigenous people to make blowguns.

2. Iriartella stenocarpa Burret, Notizbl. Bot. Gart. Berlin-Dahlem 11: 233. 1931. *Iriartea stenocarpa* (Burret) MacBride, Field Mus. Nat. Hist., Bot. Ser. 13: 357. 1960. Type. Peru. Loreto: mouth of Río Napo near Río Amazonas, Mar 1931, *W. Hopp 1110* (holotype, B) (Figure 6.9f–g).

Iriartella ferreyrae H. E. Moore, Gentes Herb. 9: 278. 1963. Type. Peru. Ucayali: Prov. Coronel Portillo, Pampas de Sacramento, a few km SW of Yurac on road to Boquerón del Padre Abad, between Divisoria and Aguaytía, ca. 400 m, 28 Apr 1960, *H. Moore et al. 8367* (holotype, BH; isotype, USM).

Stems cespitose, rarely solitary, erect or leaning, forming loose to dense clusters of up to 10 main stems, 1–4(–6) m tall, 1.5–2 cm diam; stilt roots few, often poorly developed and not apparent, to 20 cm long. *Leaves* pinnate or rarely entire; sheath 11.5–18 cm long, brownish, densely reddish-brown-, or whitish-tomentose and sometimes with scattered to dense dark brown hairs to 2 mm long; petiole 10–19 cm long; rachis 18–50 cm long; pinnae 5–8 per side, entire, rhombic-cuneate to almost

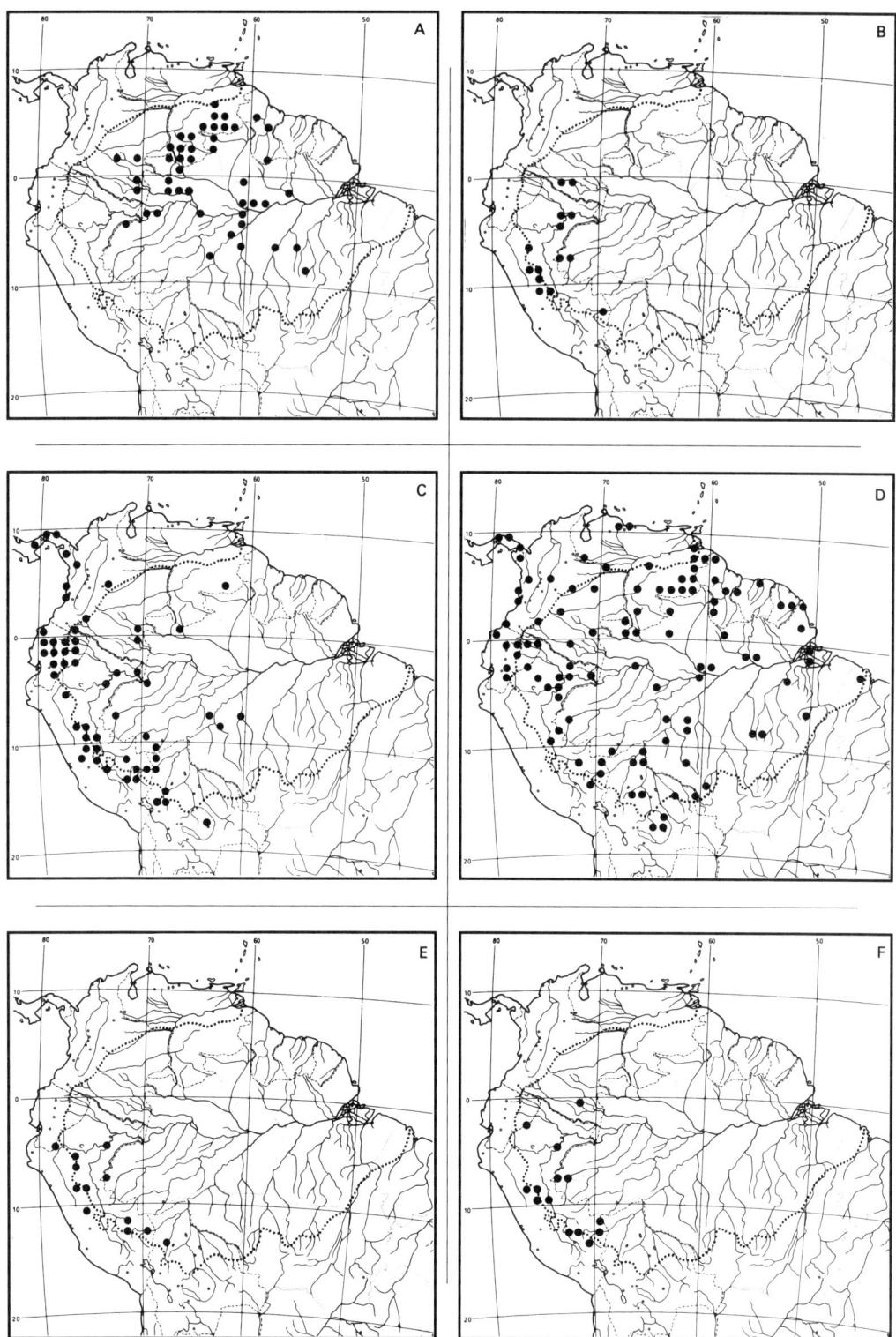

Figure 6.10. **(A)** *Iriartella setigera;* **(B)** *Iriartella stenocarpa;* **(C)** *Iriartea deltoidea;* **(D)** *Socratea exorrhiza;* **(E)** *Socratea salazarii;* **(F)** *Wettinia augusta.*

linear in outline, praemorse, the middle ones 14–25(–36) cm long, 3–10 cm wide. *Inflorescences* interfoliar, becoming infrafoliar in fruit; peduncle to 45 cm long; prophyll inserted near base of peduncle, 4–11 cm long; peduncular bracts 2–3; rachis 1–8 cm long; rachillae 3–20, simple, 4.5–20 cm long; *flowers* in triads, toward end of rachillae paired or solitary staminate; staminate flowers 3–3.5 mm long; sepals very briefly connate basally, imbricate and free above, widely ovate, 1 mm long; petals ovate, 3.2–3.5 mm long; pistillode lacking; pistillate flowers 1.5–1.7 mm long; sepals free or very briefly connate basally, imbricate, widely ovate, 1.5 mm long; petals widely ovate, imbricate, 1.5 mm long; staminodes 6, small, dentate; *fruits* ellipsoid, 0.9–1.5 cm long, 0.5–0.7 cm diam; epicarp glabrous, orange at maturity and splitting irregularly.

Colombia (Amazonas), Peru (Junín, Loreto, Madre de Dios, Pasco, Ucayali), and western Brazil (Acre) (Figure 6.10B); lowland rain forest, usually on *terra firme,* in areas below 500 m elevation. Brazil: *paxiubinha do macaco.* Colombia: *chonta de nutria, fuidorda* (Huitoto), *momogaik-t-t-co* (Muinane), *momoigiak-u* (Muiname). Peru: *camonilla, casha ponita, marrashemapar* (Ameshua), *pona, ponilla.* In Colombia, macerated leaves are used as a shampoo for lice. In Peru (Ameshua), an infusion from the leaves is used for bathing the feet.

11. *Iriartea*

Iriartea Ruiz & Pav., Fl. peruv. prodr. 149. 1794.

Large, monoecious palms. *Stems* solitary, stout, erect, columnar to ventricose; stilt roots prominent, numerous, crowded together. *Leaves* pinnate, reduplicate; sheath forming a compact crownshaft; petiole short; rachis long; pinnae numerous, praemorse, divided into numerous segments, these spreading in different planes. *Inflorescences* infrafoliar and pendulous in bud and at anthesis, branched to 1 or 2 orders; peduncle bearing a prophyll and to 15 peduncular bracts; rachis bearing numerous rachillae; *flowers* in triads, rarely tetrads; staminate flowers with sepals 3, imbricate, very briefly connate basally; petals 3, valvate; stamens (10–)12–15(–17); pistillode minute or absent; pistillate flowers with sepals 3, imbricate; petals 3, imbricate basally, valvate above; staminodes 10–13; gynoecium syncarpous, tricarpellate, triovulate; *fruits* 1(–2)-seeded, globose, with apical to subapical stigmatic residue; seed with homogeneous endosperm and subapical to lateral embryo; germination adjacent-ligular; eophyll entire.

A genus of one species (Henderson, 1990) distributed from Nicaragua south to Bolivia, and east to the western Amazon region of Venezuela and Brazil.

1. Iriartea deltoidea Ruiz & Pav., Syst. veg. fl. peru. chil. 1: 298. 1798. Type. Peru. Pasco: Pozuzo [1784], *J. Pavón s. n.* (holotype, M; isotypes F, P; F neg. 18532a) (Figure 6.9h–i).

Iriartea ventricosa Mart., Hist. nat. palm. 2: 37. 1824. *Deckeria ventricosa* (Mart.) H. Karst., Linnaea 28: 259. 1857 ('1856'). Type. Brazil. Amazonas: between Mutum Coara and Tabatinga, Apr [1821], *C. Martius s.n.* (holotype, M; F neg. 18535).

Stems solitary, columnar to ventricose, to 25 m tall, 10–30 cm in diam at base, 12–70 cm in diam at swelling; stilt roots closely spaced and forming a dense cone, to 2 m long, black. *Leaves* 4–7; sheaths forming a crownshaft, 60–150 cm long; petiole 2–13(–40) cm long; rachis 2.4–3 m long; pinnae 15–27 per side, deltate with praemorse margins, split to the base into numerous (to 18) segments, the proximal segment of a pinna largest and pendulous and all the distal ones smaller and pointing up and away from the axis and giving the leaf a two-ranked appearance. *Inflorescences* infrafoliar, pendulous at anthesis, buds developing below crownshaft and erect at first, soon becoming decurved and eventually horn-shaped; peduncle 20–44 cm long; prophyll 8 cm long; peduncular bracts to 15, caducous as bud elongates, horn-shaped, to 120 cm long; rachis 14–46 cm long; rachillae 23–37, all simple or more often the proximal few bifurcate, 80–140 cm long; *flowers* proximally in triads, distally staminate in pairs or solitary, or often all in an inflorescence staminate; staminate flowers to 7 mm long; sepals depressed ovate, 2–4 mm long, covered with long, stiff, caducous hairs; petals ovate-oblong, 7 mm long; pistillate flowers 4 mm long; sepals widely ovate, 5 mm long; petals widely ovate, 5 mm long; *fruits* globose, 2–2.8 cm diam, greenish-yellow at maturity, the epicarp splitting irregularly from apex.

Central America (Nicaragua, Costa Rica, Panama), Colombia (Amazonas, Antioquia, Caquetá, Chocó, Meta, Putumayo, Valle del Cauca), Venezuela (Amazonas, Bolívar), Ecuador (Cotopaxi, Esmeraldas, Morona-Santiago, Napo, Pastaza), Peru (Cuzco, Junín, Huánuco, Loreto, Madre de Dios, San Martín, Ucayali), Brazil (Acre, Amazonas, Mato Grosso), and Bolivia (Beni, La Paz, Pando) (Figure 6.10C); lowland, premontane, and montane rain forest between sea level and 1200 m elevation. It is very abundant on eastern Andean slopes, but less common in the eastern lowland part of its range, where it is usually confined to areas near streams and rivers. Pollination is by bees (Henderson, 1985; see also Roubik, 1989).

Bolivia: *pachuba, tuamo.* Brazil: *paxiúba barriguda, paxiúba barrigouda, paxiubão.* Colombia: *cachuda barriguda, pona lisa.* Ecuador: *ambakai* (Shuar), *ampakay, ora, pambil, patigua* (Quechua), *pona, tepa* (Waorani). Peru: *camona blanca, huacrapona, pona.* Venezuela: *cadotodek* (Arekuna), *crepísh* (Shiriana).

The outer part of the stems is used throughout its range for building purposes (e.g., floors, posts, poles), as well as for blowguns, bows, harpoons, arrow points, and firewood. A floor can last for up to 25 years. The leaves are used for thatching and basketry. Larger trees with swollen stems near rivers are commonly felled and used as temporary canoes for downstream trips. Pinard (1993) has described the effect of stem harvesting on populations in Acre, Brazil.

12. *Socratea*   **Socratea** H. Karst., Linnaea 28: 263. 1857 ("1856").

Moderate to large, monoecious palms. *Stems* solitary, rarely cespitose, stout, erect, cylindrical; stilt roots prominent, few, widely spaced. *Leaves* pinnate, reduplicate; sheath forming a compact crownshaft; petiole short; rachis long; pinnae regularly arranged, praemorse, numerous, entire or divided into segments and these spreading in different planes. *Inflorescences* interfoliar in bud, infrafoliar and erect at pistillate anthesis, becoming pendulous at staminate anthesis, branched to 1 order; peduncle bearing a prophyll and few peduncular bracts; *flowers* arranged in triads; staminate flowers asymmetrical; sepals 3; petals 3; stamens (17–)30–145; pistillate flowers symmetric, sepals 3, imbricate; petals 3, imbricate; staminodes absent; gynoecium syncarpous, tricarpellate, triovulate; *fruits* 1-seeded, ovoid-ellipsoid, rostrate or not, with apical stigmatic residue; seed with homogeneous endosperm and apical or slightly sub-apical embryo; germination adjacent-ligular; eophyll bifid.

A genus of five species (Henderson, 1990) distributed from Nicaragua south to Bolivia, and east through Venezuela and the Guianas to Pará, Brazil. Two species occur in the Amazon region.

KEY TO THE SPECIES OF *SOCRATEA*

1. Pinnae divided into segments; peduncle to 50 cm long; rachillae to 17; widespread . . . . 1. *S. exorrhiza.*
1. Pinnae entire; peduncle 14–18 cm long; rachillae 3–8; Peru (Amazonas, Loreto, Madre de Dios, Pasco, San Martín, and Ucayali), Brazil (Acre), and Bolivia (La Paz) . . . . 2. *S. salazarii.*

1. **Socratea exorrhiza** (Mart.) H. Wendl., Bonplandia 8: 103. 1860. *Iriartea exorrhiza* Mart., Hist. nat. palm. 2: 36. 1824 ("exorhiza"). Lectotype (Henderson, 1990). Martius, Hist. nat. palm. 2: t. 33 & 34. 1824 (Figure 6.11a–c; Plate IVb).

Iriartea orbigniana Mart., Hist. nat. palm. 3(7): 189. 1838. *Socratea orbigniana* (Mart.) H. Karst., Linnaea 28: 264. 1856 ("orbignyana"). *Iriartea exorrhiza* var. *orbigniana* Drude in Mart., Fl. bras.: Palmae II fasc. 86 vol. 3(2): 540. 1882. Type. Bolivia. Beni: Moxos, n.d., *A. d'Orbigny 33* (holotype, P).

Iriartea philonotia Barb. Rodr., Enum. palm. nov. 13. 1875. *Socratea philonotia* (Barb. Rodr.) Hook. f. in Benth. & Hook. f., Gen. pl. 3: 900. 1883. Lectotype (Wess. Boer, 1965). Barb. Rodr., Sert. palm. brasil. 1: t. 8. 1903.

Socratea microchlamys Burret, Notizbl. Bot. Gart. Berlin-Dahlem 11: 3. 1930. Type. Venezuela. Amazonas: Sacupana, Apr 1896, *H. Rusby & R. Squires 415* (holotype, NY, isotype, M).

Socratea gracilis Burret, Notizbl. Bot. Gart. Berlin-Dahlem 15: 1. 1940. Type. Guyana. Essequibo: basin of Kuyuwini River, ca. 150 mi from mouth, 21–26 Nov 1937, *A. Smith 2619* (isotype, NY).

Socratea albolineata Steyerm., Fieldiana, Bot. 28: 91. 1951. Type. Venezuela. Bolívar: lower portion of Quebrada O-paru-má, tributary of Río Pacairas, below Santa Teresita de Kavanayén, 915–1065 m, 25 Nov 1944, *J. Steyermark 60541* (holotype, F).

Stems solitary, to 20 m tall, usually less, 10–18 cm diam, gray with lichens, smooth; stilt roots to 25, widely spaced and forming an open cone, brown, with spines to 2 cm long. *Leaves* ca. 7; sheath 0.9–1.5 m long; petiole 15–40 cm long; rachis 1.4–2.8 m long; pinnae 15–25 per side, asymmetrically cuneate, the middle ones 40–90 cm long, the margins entire except for praemorse apex, the pinnae split to the base into segments. *Inflorescences* interfoliar and erect in bud, infrafoliar and becoming pendulous at anthesis; peduncle to 50 cm long; prophyll to 11 cm long; peduncular bracts (3–)4(–5), to 61 cm long; rachillae to 17, usually fewer, 30–40 cm long; *flowers* in triads proximally, paired or solitary staminate distally; staminate flowers 9–12 mm long; sepals shortly connate basally, triangular, 2 mm long; petals valvate, very widely ovate, open before anthesis, 9–10 mm long; stamens (17–)30–45(–65); pistillode 1–2 mm long; pistillate flowers 4–8 mm long; sepals very widely ovate, ciliate, 4 mm long; petals similar to sepals; *fruits* ovoid-cylindric, 2.5–3.5 cm long, 1.5–2 cm diam, stigmatic residue subapical and obscure; epicarp yellowish at maturity and splitting irregularly.

Central America (Nicaragua, Costa Rica, Panama) southward across northern South America. In the Amazon region, it occurs in Colombia (Amazonas, Caquetá, Guainía, Guaviare, Meta, Putumayo, Vaupés, Vichada), Venezuela (Amazonas, Bolívar, Delta Amacuro, others), the Guianas, Ecuador (Morona-Santiago, Napo), Peru (Cuzco, Huánuco, Loreto, Madre de Dios, Ucayali), Brazil (Acre, Amapá, Amazonas, Maranhão, Pará), and Bolivia (Beni, Pando, Santa Cruz) (Figure 6.10D); lowland forest, usually below 1000 m elevation. It is common in inundated areas near streams and rivers, but also occurs on *terra firme*. It is pollinated by beetles (Henderson, 1985).

Bolivia: *onipa* (Chacabo), *pachuvilla*. Brazil: *manácam* (Sanama), *manáca-man-ash-quili* (Mayongong), *paxiubinha, paxiuba, paxi'y* (Ka'apor). Colombia: *ñobea, pachuda zancona, zancona*. Ecuador: *anaccu tssatssa'vo'* (Cofán), *bonbon, kúpat* (Shuar), *ñicó* (Siona), *pambil, shiquita* (Quichua), *rayador*. Guyana: *boba, buba*. Peru: *casha pona, huachapona, huacrapona, kupat* (Mayna Jívaro), *palmera*. Venezuela: *buba, cola de pescado, macanilla, upa* (Baniba). The stems are split into planks and very widely used for making floors and walls. The leaves are occasionally used for thatching.

Reichel-Dolmatoff (1989) has described the importance of *Socratea exorrhiza* in the myths and rituals of the Indians of the Vaupés region in Colombia. In particular, the stems were used to make sacred musical instruments such as trumpets.

2. **Socratea salazarii** H. E. Moore, Principes 7: 112. 1963. Type. Peru. Loreto: Prov. Alto Amazonas, km 13–14 on Yurimaguas–Tarapoto road, 24 May 1960, *H. Moore et al. 8517* (holotype, BH; isotype, USM) (Figure 6.11d–e).

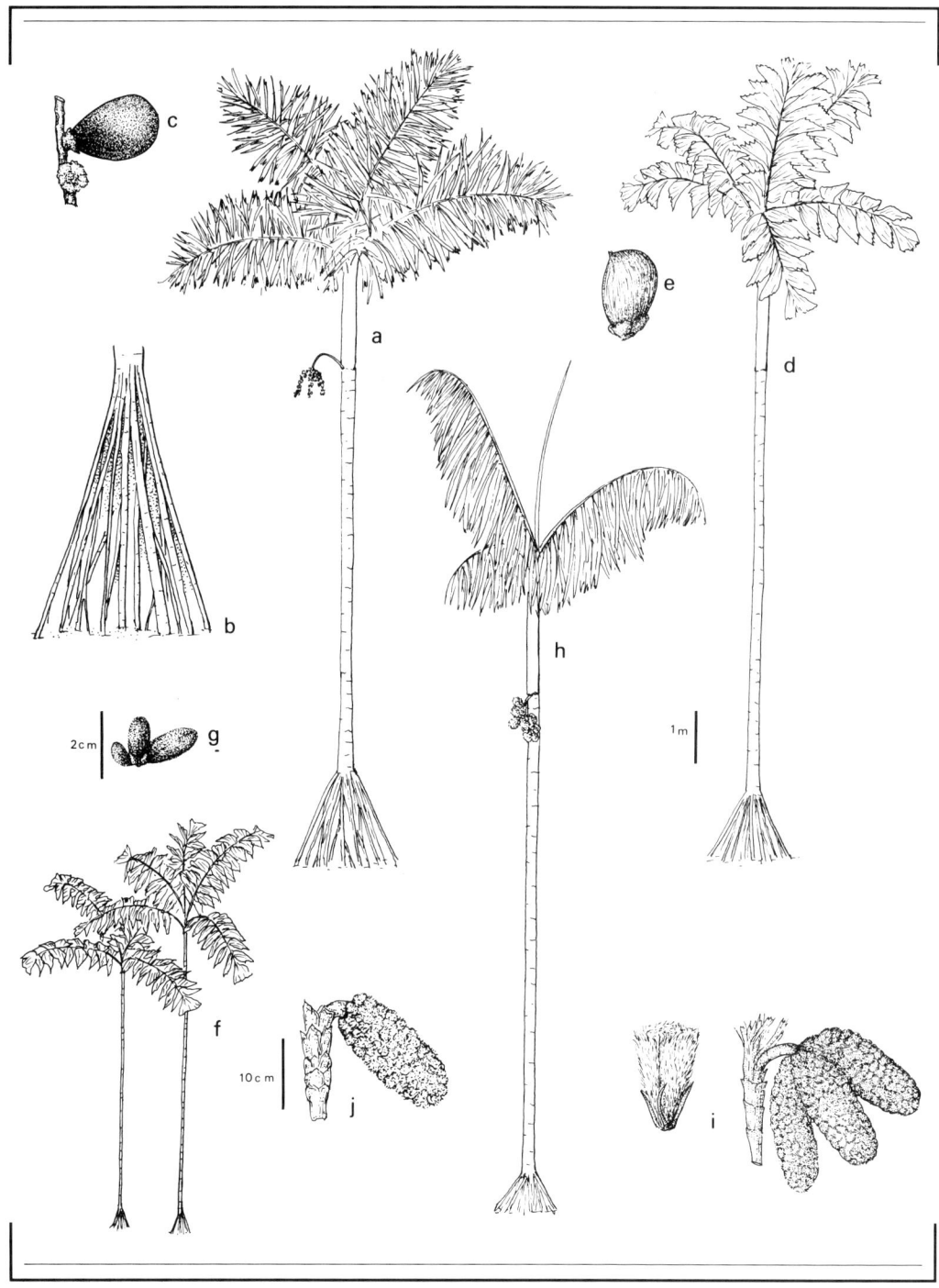

Figure 6.11. (a) *Socratea exorrhiza,* (b) stilt roots, (c) fruit (from *A. Henderson 174*); (d) *Socratea salazarii,* (e) fruit (from *A. Henderson 539*); (f) *Wettinia drudei,* (g) fruit (from *A. Henderson 830*); (h) *Wettinia maynensis,* (i) infructescence and fruit (from *H. Balslev 4265*); (j) infructescence of *Wettinia augusta* (from *F. Kahn 2108*).

Stems solitary or occasionally cespitose, to 16 m tall, 6.5–12 cm diam, smooth; stilt roots forming a cone to 1 m tall, with spines to 3 mm long. *Leaves* 6–7; sheath 0.9–1.2 m long; petiole 30–40 cm long; rachis 1.6–2.5 m long; pinnae 11–16 per side, cuneate, entire, the proximal margins entire, distal margins praemorse. *Inflorescences* erect and interfoliar in bud, becoming pendulous and infrafoliar at anthesis; peduncle 14–18 cm long; prophyll 5–10 cm long; peduncular bracts 2–3, to 30 cm long; rachis 6–7 cm long; rachillae 3–8, 20–35 cm long; *flowers* in triads proximally, for distal ca. 5 cm of rachillae staminate in pairs or solitary, cream-colored; staminate flowers ca. 5 mm long; sepals free, valvate, triangular, 3 mm long; petals free, irregularly shaped, fleshy, 5 mm long; stamens 38–39; pistillode obscure or absent; pistillate flowers ca. 4 mm long; sepals ciliate, very shortly connate basally, broadly triangular, 3–4 mm long; petals similar to sepals; *fruits* ellipsoid-ovoid, 2.5–3.5 cm long, 2–2.5 cm diam, the stigmatic residue prominent and excentrically apical; epicarp yellowish and splitting irregularly.

Amazonian Peru (Amazonas, Loreto, Madre de Dios, Pasco, San Martín, and Ucayali), extreme western Brazil (Acre), and northwestern Bolivia (La Paz) (Figure 6.10E); lowland rain forest on *terra firme* between 300 and 700 m elevation.

Peru: *camonilla, cashapona de altura, imáp, pona de altura.*

13. *Wettinia*

Wettinia Poepp. in Endl., Gen. pl. 243. 1837.
Catoblastus H. Wendl., Bonplandia, 8: 104. 1860.

Moderate to large, monoecious palms. *Stems* solitary or less often cespitose, erect, supported by stilt roots. *Leaves* pinnate, reduplicate, occasionally distichous; sheath closed and forming a crownshaft; petiole short; rachis long; pinnae praemorse, either entire and regularly arranged in 1 plane or split into segments that spread in different planes. *Inflorescences* spicate or branched to 1 order, infrafoliar, usually unisexual, (1–)3–15 per node; peduncle bearing a prophyll and 3–5 peduncular bracts; rachis bearing 1–few rachillae; *flowers* solitary or paired; staminate flowers with sepals 3–4, free; petals 3–4, free; stamens 6–20; pistillode very small or absent; pistillate flowers with sepals 3–4, free; petals 3–4, free; staminodes small or absent; gynoecium syncarpous, uniloculate, or triloculate, uniovulate by abortion of 2 ovules (but occasionally 2 ovules developing); *fruits* 1(-2)-seeded, densely crowded or loosely arranged, irregularly prismatic, globose, ovoid, obovoid, or ellipsoid, with basal stigmatic residue; epicarp rough, hairy, warty, or verrucose; seed with homogeneous or ruminate endosperm and basal embryo; germination adjacent-ligular; eophyll entire.

A genus of 20 species distributed from central Panama and Venezuela to Bolivia, usually in mountainous Andean areas. There is no recent revision, but see Burret (1930) and Moore and Dransfield (1978). *Catoblastus* was, until recently, treated as a separate genus from *Wettinia,* but is now included there (R. Bernal, personal communication). Three species are found in the western Amazon region. A fourth species, *W. praemorsa,* may just enter the northwestern part of the region in Colombia (Meta), but usually occurs above 500 m elevation.

KEY TO THE SPECIES OF *WETTINIA*

1. Fruits densely crowded on the rachillae, prismatic and densely covered with trichomes; stems 3–19 m tall and 3–14 cm diameter.
   2. Stems solitary; inflorescences with 3–7 rachillae; pinnae pilose abaxially . . 3. *W. maynensis.*
   2. Stems cespitose or less often solitary; inflorescences spicate or sometimes bifurcate; pinnae glabrous or scarcely pilose abaxially. . . . . 1. *W. augusta.*
1. Fruits loosely arranged on the rachillae, ellipsoid, brown tomentose; stems slender, 3–6 m tall, 2–3.5 cm diameter. . . . . 2. *W. drudei.*

1. Wettinia augusta Poepp. & Endl., Nov. gen. sp. 2: 39. 1838. *Wettinia poeppigii* Kunth, Enum. pl. 3: 109. 1841. Type. Peru. San Martín: Tocache, *E. Poeppig 2058* (isotype, P; F neg. 29887) (Figure 6.11j).

Stems cespitose or less often solitary, 3–12 m tall, 3–12 cm diam, supported by a cone of spiny stilt roots, 20–50 cm long. *Leaves* 4–8; sheath 0.5–1.5 m long; petiole 15–50 cm long; rachis 1.5–3 m long; pinnae 16–29 per side, narrowly rhombic, entire, regularly arranged and spreading in 1 plane, almost pendulous, the middle ones 45–70 cm long, 6–10 cm wide, praemorse, glabrous or scarcely pilose abaxially. *Inflorescences* infrafoliar, unisexual, solitary or to 8 per node; peduncle 10–15 cm long; prophyll 3–4 cm long; peduncular bracts 3–5; rachis absent or to 2 cm long; rachilla 1 or rarely to 4, 8–19 cm long; staminate flowers 10–11 mm long, densely crowded on rachillae, open in bud before anthesis; sepals 3, narrowly triangular, 1.5–2 mm long; petals 3, lanceolate, 10–11 mm long; stamens ca. 10; pistillode not apparent; pistillate flowers not seen; *fruits* prismatic, densely crowded on the rachilla, 2.5–3 cm long, 1–1.5 cm diam, brownish, densely covered with short, stiff hairs; endosperm homogeneous.

Western Amazon region of Colombia (Amazonas), Peru (Cuzco, Huánuco, Loreto, Madre de Dios, San Martín), and Brazil (Acre), also probably in northern Bolivia (La Paz) (Figure 6.10F); lowland rain forest on *terra firme* between 200 and 800 m elevation. In some places, it occurs in inundated areas, and there has cespitose stems.

Brazil: *paxiúbinha de macaco.* Peru: *ponilla.* The stems are split into planks and used in house construction; the leaves are used for thatching.

The differences between this species and *Wettinia maynensis* are not clearly established; Galeano (1991) described *W. augusta* from the Colombian Amazon with 4 rachillae.

2. Wettinia drudei (O. F. Cook & Doyle) Henderson, comb. nov. *Catoblastus drudei* O. F. Cook & Doyle, Contr. U. S. Nat. Herb. 6: 233. 1913. Type. Not designated (Figure 6.11f–g). *Iriartea pubescens* var. *krinocarpa* Trail, J. Bot. 14: 332. 1876. *Catoblastus pubescens* var. *krinocarpa* (Trail) Drude in Mart., Fl. bras.: Palmae II, fasc. 86, vol. 3(2): 543. 1882. Type. Brazil. Amazonas: Rio Javari, 6 Dec 1874, *J. Trail 1058/CXC* (isotype, P).

Stems cespitose, rarely solitary, erect or occasionally leaning, 3–6 m tall, 2–3.5 cm diam, supported by stilt roots and spreading by rhizomes. *Leaves* ca. 5; sheath 25–43 cm long; petiole 14–47 cm long; rachis to 1.3 m long; pinnae to 14 per side, regularly arranged and in 1 plane, narrowly rhombic, 33–42 cm long. *Inflorescences* often solitary at each node but occasionally multiple; peduncle 9–12 cm long; prophyll 2–3 cm long; peduncular bracts 3, 3–15 cm long; rachis 3 cm long; rachillae 4–8, to 20 cm long; staminate flowers 2.8 cm long; sepals 3, deltate, 1 mm long; petals 3, valvate, linear, 2.8 cm long; stamens 6; pistillode not apparent; pistillate flowers 9 mm long (postanthesis); sepals 3, free, deltate, 1 mm long; petals 3, free, deltate, 4.5 mm long; staminodes ca. 7, stamenlike; stigmas 3, prominent at base of ovary; ovaries 3, only 1 developing, densely brown-tomentose and with long white hairs; *fruits* ellipsoid, 2–3 cm long, 0.5–1 cm diam, loosely spaced on rachillae; epicarp densely brown-tomentose; endosperm homogeneous.

Colombia (Amazonas, Caquetá, Putumayo), Ecuador (Morona-Santiago, Napo), Peru (Amazonas, Loreto), and Brazil (Amazonas) (Figure 6.12A); lowland rain forest in periodically inun-

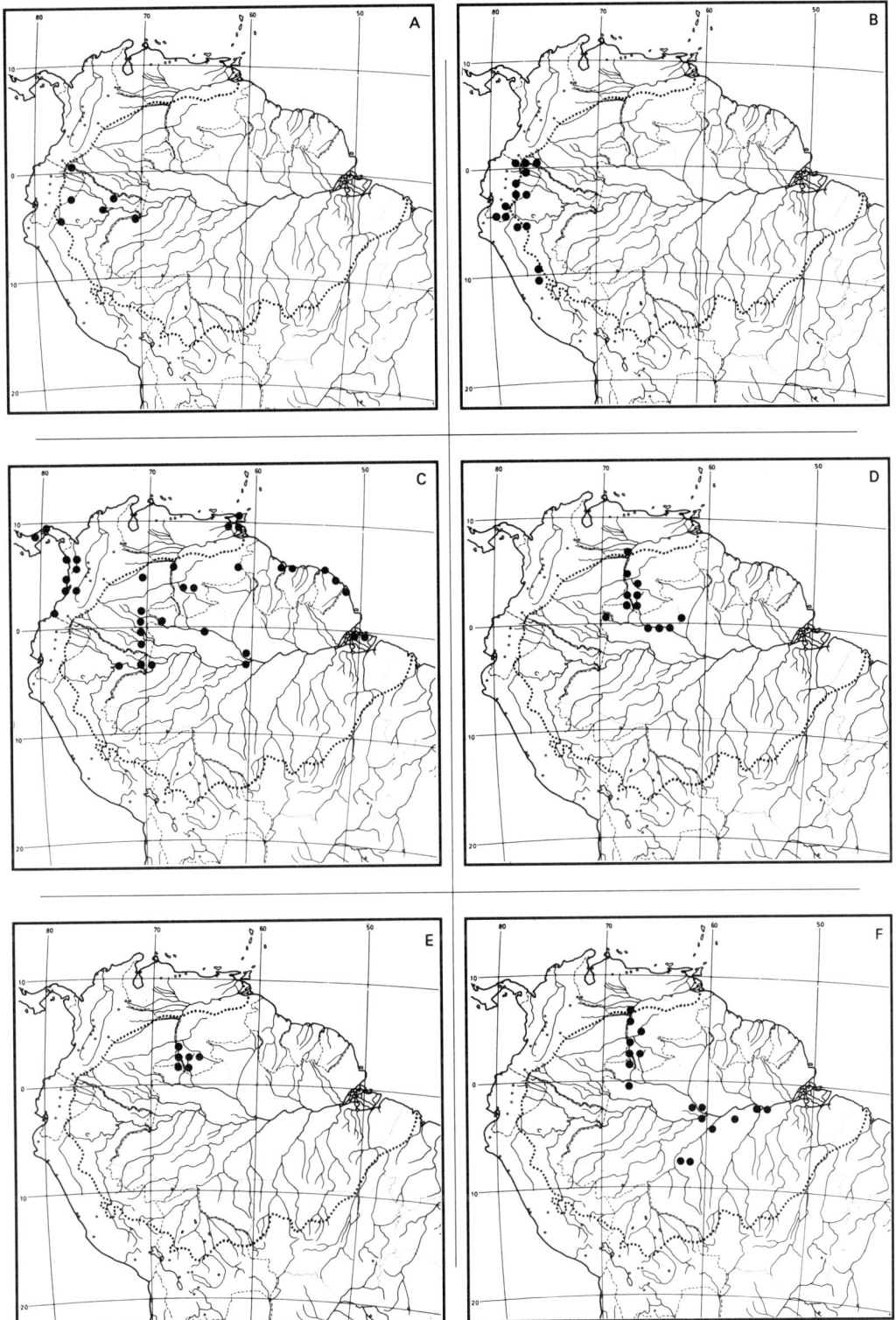

Figure 6.12. (A) *Wettinia drudei;* (B) *Wettinia maynensis;* (C) *Manicaria saccifera;* (D) *Leopoldinia major;* (E) *Leopoldinia piassaba;* (F) *Leopoldinia pulchra.*

dated areas and also on *terra firme,* usually below 600 m elevation.

Brazil: *paxiubinha.* Peru: *pona.*

3. Wettinia maynensis Spruce, J. Proc. Linn. Soc., Bot. 3: 194. 1859. *Catoblastus maynensis* (Spruce) Drude in Mart., Fl. bras.: Palmae II fasc. 86 vol. 3(2): 544. 1882. *Wettinella maynensis* (Spruce) O. F. Cook & Doyle, Contr. U. S. Nat. Herb. 16: 237. 1913. Type. Peru. Without locality, n.d., *R. Spruce 59* (holotype, K, n.v.) (Figure 6.11h–i).

Wettinia illaqueans Spruce, J. Proc. Linn. Soc., Bot. 3: 191. 1859. Nomen nudum.

Stems solitary, 7–19 m tall, 6–15 cm diam, supported by a cone of spiny stilt roots 30–70 cm long. *Leaves* 5–10; sheath 0.9–1.7 m long; petiole 8–30 cm long; rachis 2.1–3.3 m long; pinnae to 43 per side, entire, regularly arranged and spreading in 1 plane, broadly lanceolate, somewhat pendulous, the middle ones 0.5–1 m long, 10–20 cm wide, praemorse marginally, densely pilose abaxially. *Inflorescences* infrafoliar, 5–9 per node, unisexual, the central one pistillate and the lateral staminate (?always); peduncle 16–20 cm long; prophyll 5–7 cm long; peduncular bracts 4–5, from 5 to 25 cm long, outer ones densely hairy; rachis very short or absent; rachillae 3–7, 9–18 cm long; staminate flowers 6–7 mm long, densely crowded on rachillae; sepals 3, 0.5–1 mm long, deltate; petals 3, linear, 6–7 mm long; stamens 7–10; pistillode small or absent; pistillate flowers not seen; *fruits* prismatic, 1.8–2.5 cm long, 1.5–2 cm diam at the middle, densely crowded on rachillae, densely covered with short, stiff, brown trichomes; endosperm homogeneous.

Eastern Andean slopes in Colombia (Putumayo), Ecuador (Morona-Santiago, Napo, Pastaza, Zamora-Chinchipe), and Peru (Amazonas, Huánuco, Loreto, Pasco, San Martín) (Figure 6.12B); lowland, premontane, and montane rain forest on steep slopes between 250 and 2000 m elevation, but most common below 1000 m. In southern part of its range, it occurs at higher elevations.

Ecuador: *ccu'ye* (Cofán), *quirigua* (Quechua), *teren* (Shuar), *walte, wi-ni-co* (Siona). Peru: *camonilla.* The stems are commonly split and the planks used in house construction, especially for floors. The leaves are occasionally used for thatching.

ARECOIDEAE • ARECEAE • MANICARIINAE

14. *Manicaria* **Manicaria** Gaertn., Fruct. sem. pl. 2: 468. 1791.

Moderate to large, monoecious palms. *Stems* solitary or cespitose, occasionally branched (?dichotomously). *Leaves* pinnate, reduplicate; sheath open and not forming a crownshaft; petiole moderate; rachis long; pinnae irregularly split giving an unequally pinnate leaf, or leaf entire, serrate on the margins. *Inflorescences* interfoliar, branched to 1 or rarely 2 orders; peduncle bearing a prophyll and 1(–2) fibrous peduncular bracts; rachis bearing numerous, simple or rarely bifurcate rachillae; *flowers* borne in triads proximally, staminate only distally on rachillae; staminate flowers with sepals 3, connate basally, imbricate above; petals 3, free, valvate; stamens 26–34; pistillode absent; pistillate flowers with sepals 3, free, imbricate; petals 3, free, valvate; staminodes 10–15, linear; gynoecium syncarpous, trilocular, 1–3-ovulate; *fruits* 1–3-seeded, globose or 2–3-lobed, covered with short, pyramidal projections, with subbasal stigmatic residue; seed with homogeneous endosperm and basal embryo; germination adjacent-ligular; eophyll bifid.

A genus of one species distributed from Central America (Belize, Guatemala, Honduras, Nicaragua, Costa Rica, Panama) to Colombia, Venezuela, Trinidad, the Guianas, Ecuador, Peru, and Brazil. There is no recent revision, but see Bailey (1933) or Wessels Boer (1988).

1. **Manicaria saccifera** Gaertn., Fruct. sem. pl. 2: 468. 1791. Lectotype (Glassman, 1972). Gaertn., Fruct. sem. pl. 2: t. 176. 1791 (Figure 6.13a–b).

Manicaria saccifera var. *mediterranea* Trail, J. Bot., 5: 332. 1876. Type. Brazil. Amazonas: Manaus, Tarumã, n.d., *J. Trail 110* (holotype, K, n.v.).

Manicaria martiana Burret, Notizbl. Bot. Gart. Berlin-Dahlem 10: 392. 1928. Type. Brazil. Amazonas: Manaus, n.d., *G. Huebner 2* (holotype, B).

Manicaria atricha Burret, Notizbl. Bot. Gart. Berlin-Dahlem 10: 1013. 1930. Type. Brazil. Amazonas: Jutica, Rio Uaupés, 11 Nov 1928, *P. von Luetzelburg 23031* (holotype, B; isotypes M, R).

Stems solitary or cespitose, occasionally branching, 0.5–10 m tall, 15–20 cm diam, usually covered with persistent leaf bases and other debris. *Leaves* 5–25, rigid and erect, strongly plicate; sheath 60–100 cm long; petiole 0.2–1.3(–3.5) m long; rachis 2.3–4(–8) m long; pinnae 26–55 per side, irregularly arranged and of unequal width, spreading in 1 plane, linear, with serrate margins, the middle ones 1–1.8 m long, occasionally leaf entire but becoming split with age. *Inflorescences* interfoliar at anthesis and in fruit; peduncle 0.4–1.3 m long; prophyll 52–70 cm long; peduncular bract 68–90 cm long, brown, very fibrous, persistent over developing fruits; rachis 33–60 cm long; rachillae 21–56, 8–40 cm long, densely reddish-brown-tomentose, simple or occasionally bifurcate; *flowers* in triads on proximal ca. half of rachillae, paired or solitary staminate distally, subtended by elongate bracteoles; staminate flowers 6 mm long; sepals widely ovate, 3–4.5 mm long; petals ovate, 5–5.5 mm long; pistillate flowers 1–1.2 cm long; sepals ovate, 8–9 mm long; petals triangular, 8–9 mm long; *fruits* globose or 2–3-lobed, 4–6 cm diam, larger when lobed, brownish, covered with short pyramidal projections.

In scattered localities from Belize and Guatemala southward through Central America, Colombia (Amazonas, Antioquia, Chocó, Valle del Cauca, Vaupés, Vichada), Venezuela (Amazonas, Delta Amacuro, Monagas, Sucre), Trinidad, the Guianas, northwestern Ecuador (Esmeraldas), Peru (Loreto), and Brazil (Amazonas, Pará) (Figure 6.12C); usually near the sea in low-lying wet areas, but also far inland near streams and rivers. Although it usually grows at low elevations, rarely it can grow on mountain slopes at up to 1200 m elevation (Cerro Marahuaca, Venezuela). In the Amazon region, it usually grows in lowland rain forest in inundated areas at low elevations.

Brazil: *buçu, bussú, gerua, ubussu*. Colombia: *mecuá-bak* (Miraña), *ubí, wa-heé* (Tanimuka), *wá-hee* (Yukuna). French Guiana: *palmier toulouri, toulouri*. Guyana: *troolie*. Surinam: *truli*. Venezuela: *mavaco, palma temiche, temiche, washí, yajují*. The leaves are commonly used for thatching and are much sought after because they are long-lasting; the peduncular bract, known as *turiri* in Brazil, is used to make hats and bags; the fruits are fed to domestic animals.

Wilbert (1976) described the cultural significance of *Manicaria saccifera* among the Warao Indians of the Orinoco Delta in Venezuela. They used the palm stems not only as a source of sago, or palm starch, but also for many other uses. Weevil grubs (*Rhynchophorus palmarum*) were collected from the fallen stems; the leaves were used for thatching and for boat sails; the peduncular bracts were used as hats; the liquid endosperm from immature fruits was drunk; the endosperm of the germinated seed was eaten; and several parts of the palm were used medicinally.

Manicaria saccifera is widely distributed in scattered localities throughout Central America and northern South America, and it occurs in a variety of habitats. It is consequently variable, and this variability has led to a number of proposed species. None of these is recognized here (but see Wessels Boer, 1988). The occasional occurrence of more than 1 peduncular bract and the branching of the rachillae are unusual.

This palm should have an interesting pollination system. The inflorescence is entirely covered by the bracts at anthesis. Wessels Boer (1965) considered that self-pollination must occur. Insect pollination, however, seems likely. David Furth (personal communication) has collected hundreds of staphylinid and nitidulid beetles from flowers of *M. saccifera* in Colombia.

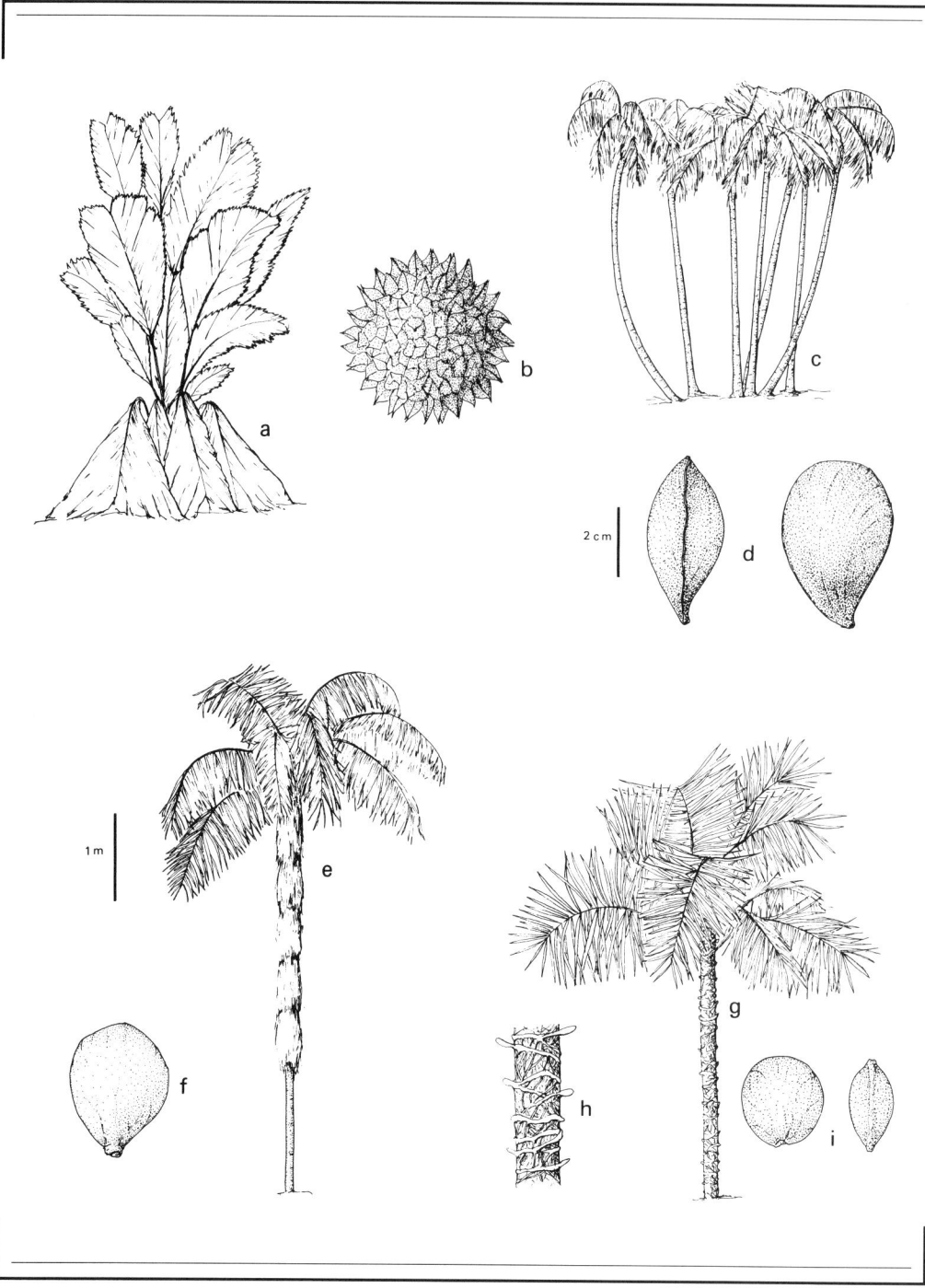

Figure 6.13. **(a)** *Manicaria saccifera,* **(b)** fruit (from *A. Henderson 182*); **(c)** *Leopoldinia major,* **(d)** fruits (from *G. Davidse 27845*); **(e)** *Leopoldinia piassaba,* **(f)** fruit (from *J. Wessels Boer 2309*); **(g)** *Leopoldinia pulchra,* **(h)** detail of fibrous leaf sheath, **(i)** fruits (from *C. Farney 1875*).

ARECOIDEAE • ARECEAE • LEOPOLDINIINAE

15. *Leopoldinia* Leopoldinia Mart., Hist. nat. palm. 2: 58. 1824.

Moderate, monoecious palms. *Stems* solitary or cespitose, erect or slightly leaning. *Leaves* pinnate, reduplicate; sheath open, not forming a crownshaft, fibrous and persistent; petiole long; rachis long; pinnae regularly arranged and spreading in 1 plane. *Inflorescences* interfoliar, branched to 3 or 4 orders; peduncle bearing a prophyll and 1 peduncular bract; rachis with numerous rachillae; *flowers* borne in triads or paired or solitary staminate, very small; staminate flowers with sepals 3, free, imbricate; petals 3, free, valvate; stamens 6; pistillode small; pistillate flowers with sepals 3, free, imbricate; petals 3, free, valvate; staminodes 6; gynoecium syncarpous, trilocular, triovulate before anthesis, pseudomonomerous by anthesis; *fruits* 1-seeded, rounded or reniform, ± flattened, with basal stigmatic residue; mesocarp with coarse fibers and air spaces; seed with homogeneous endosperm and subbasal embryo; germination adjacent-ligular; eophyll bifid.

A genus of three species distributed in the central Amazon region of Brazil, Colombia, and Venezuela. The most recent revision is that of Drude (1882), who recognized four species, but the genus is poorly known and in need of further study. It is of great interest because its phylogenetic and geographic position is both isolated and unexplained. The occurrence of two distinct leaf sheath fiber types in the genus is similar to that in *Oenocarpus:* long and thin in *Leopoldinia piassaba* and *O. bataua*; short and reticulate in *L. major, L. pulchra,* and *O. circumtextus*. The mesocarp of the fruits, with coarse fibers and air spaces, is presumably an adaptation to dispersal by water (Kubitzki, 1991), at least in *L. major* and *L. pulchra*. Spruce (1860) noted the similarity of the leaf sheath fibers to those of the mesocarp.

KEY TO THE SPECIES OF *LEOPOLDINIA*

1. Stems, at least distally, covered with long, loose, brown or black, pendulous fibers, these the persistent remains of the sheaths; pinnae 3.5–5 cm wide; fruits rounded, scarcely flattened . . . . 2. *L. piassaba.*
1. Stems, at least distally, covered with reticulate, stiff, reddish-brown, fibrous, persistent leaf sheaths; pinnae 2–3.2 cm wide; fruits strongly flattened.
    2. Fruits irregularly rounded, widely obovoid or almost reniform, 3–4 cm long, 3–4 cm diam; pinnae pendulous; stems usually cespitose in colonies; main sheath fibers 1–2 mm diam . . . . 1. *L. major.*
    2. Fruits irregularly rounded, 2–2.3 cm diam; pinnae horizontally spreading; stems usually solitary; main sheath fibers 3 mm diam . . . . 3. *L. pulchra.*

1. Leopoldinia major Wallace, Palm Trees of the Amazon 15. 1853. Lectotype (Wess. Boer, 1988). Wallace, Palm Trees of the Amazon t. 5. 1853 (Figure 6.13c–d).

Stems cespitose or rarely solitary, 3–8 m tall, 5–10 cm diam, growing in large, dense colonies. *Leaves* ca. 11; sheath ca. 25 cm long, with persistent, reticulate, brown fibers, the main fibers 1–2 mm diam; petiole ca. 40 cm long; rachis ca. 75 cm long; pinnae 19–20 per side, regularly arranged and spreading in 1 plane, linear, somewhat pendulous, the middle ones 30–40 cm long, 2–2.5 cm diam, with brown scales on veins abaxially. *Inflorescences* interfoliar, branched to 3 or 4 orders; peduncle ca. 30 cm long; prophyll and peduncular bract not seen; rachis ca. 45 cm long; rachillae very numerous, densely reddish-brown-tomentose;

flowers in triads or paired or solitary staminate (some inflorescences all pistillate, others predominantly staminate); staminate flowers 1 mm long (in bud); sepals deltate, 0.6 mm long; petals ovate, 0.8 mm long; pistillode prominent; pistillate flowers 2 mm long; sepals very widely ovate, 1 mm long; petals ovate, 1.5 mm long; staminodes very small; *fruits* irregularly rounded, widely obovoid or almost reniform, 3–4 cm long, 3–4 cm diam, dull orange-red at maturity.

Río Negro region of Colombia (Vaupés), Venezuela (Amazonas), and Brazil (Amazonas, Roraima) (Figure 6.12D); at low elevations along the edges of black-water streams and rivers *(igapó).* Spruce (1871) and Wallace (1853) discussed its distribution. Two specimens, both unfortunately sterile, are known from western Roraima in Brazil. One appears to be *Leopoldinia major,* and the other could be *L. pulchra.*

Brazil: *jará, yará.* Venezuela: *chiquichiquito, morichita, yará.*

The differences between this species and *Leopoldinia pulchra* are still not well established.

2. Leopoldinia piassaba Wallace, Palm Trees of the Amazon 17. 1853. Lectotype (Wess. Boer, 1988). Wallace, Palm Trees of the Amazon t. 6. 1853 (Figure 6.13e–f).

Stems solitary, 4–5 m tall, ca. 15 cm diam (ca. 60 cm diam including fibers). *Leaves* 14–16; sheath ca. 45 cm long, the apex (ligule) consisting of long, loose, pendulous, black or brown fibers, these persistent and covering the stem; petiole 70–150 cm long; rachis 2.7–3.3 m long; pinnae 53–59 per side, linear, regularly arranged and spreading horizontally in 1 plane, the middle ones 70–80 cm long, 3.5–5 cm diam. *Inflorescences* interfoliar; peduncle ca. 80 cm long; bracts not seen; rachis ca. 50 cm long; rachillae numerous, to 15 cm long; staminate flowers 1.5 mm long; sepals widely ovate, 1 mm long; petals ovate, 1.2 mm long; pistillode very small; pistillate flowers not seen; *fruits* irregularly rounded, scarcely flattened, ca. 3.5 cm long, ca. 3 cm diam, orange-brown at maturity.

Upper Río Negro (and Orinoco) region of Colombia (Guainía), Venezuela (Amazonas), and Brazil (Amazonas) (Figure 6.12E); sandy areas near black-water streams and rivers *(igapó),* occasionally by white-water rivers, at low elevations. Lescure and co-workers (1992) reported that the palms grew best on podzols and less well on other soils. Wallace (1853) and Spruce (1860, 1871) discussed the distribution of this species.

Brazil: *piassaba.* Colombia: *chiqui-chiqui, maramapé* (Curripaco), *manamazu* (Puinave). Venezuela: *chiquichiqui.* The fibers from the leaf sheaths are gathered and traded, and at one time were the basis of an important local industry (Wallace, 1853; Spruce, 1860, 1871). They are still traded locally (Putz, 1979). Bernal (1992) reported that in Guainía, Colombia, recent fiber production was between 820 and 850 tons per year, and in May 1991 the price of fiber in Bogotá was $450 per ton. Lescure and colleagues (1992) reported that between 1970 and 1989, production in Brazil was very variable, from 38 to 2359 Mg (megagram) per year. The leaves are occasionally used for thatching.

This species may be dioecious (Wessels Boer, 1988) or monoecious and have unisexual inflorescences, the pistillate with long, thick rachillae and the staminate with short, thin rachillae.

Spruce (1860) wrote: "Nothing that I have seen in Amazonian forests dwells more strongly and pleasantly on my memory than my walk among these strange bearded columns, from whose apex sprang the green interlacing arches which shaded me overhead."

3. Leopoldinia pulchra Mart., Hist. nat. palm. 2: 59. 1824. Type. Brazil. Amazonas: Barra do Rio Negro (=Manaus), *C. Martius s.n.* (holotype, M; F neg. 18493a) (Figure 6.13g–i).

Leopoldinia insignis Mart., Hist. nat. palm. 2: 60. 1824. Type. Brazil. Pará: Canumá, n.d., *C. Martius s.n.* (holotype, M; F neg 18529).

Stems solitary or rarely cespitose, 2.5–7 m tall, 4–10 cm diam, distally covered with persistent, reticulate leaf sheaths. *Leaves* 10–25; sheath 23–30 cm long, persistent and not decaying, fibrous, reddish-brown, reticulate, the main fibers coarse, 3 mm diam; petiole 40–65 cm long, petiole and rachis with brown scales; rachis 1–1.3 m long; pinnae 20–37 per side, linear, long acuminate, regularly arranged and spreading horizontally in 1 plane, the middle ones 53–70 cm long, 2.5–3.2 cm diam, with brown scales on veins abaxially, the cross-veins prominent. *Inflorescences* interfoliar, branched to 3 orders; peduncle 26–35 cm long; prophyll 30–37 cm long; peduncular bract 28–30 cm long; rachis 20–50 cm long; rachillae very numerous, to 8 cm long, densely reddish-brown-tomentose; *flowers* in triads proximally on rachillae, staminate distally; staminate flowers 1 mm long (in bud); sepals depressed ovate, 0.5 mm long; petals widely ovate, 0.7 mm long; pistillode prominent; pistillate flowers 1.3 mm long at anthesis; sepals depressed ovate, 1 mm long; petals depressed ovate, 1 mm long; staminodes obscure; *fruits* rounded, flattened, 2–2.3 cm diam, reddish-brown.

Colombia (Guainía, Vaupés) and Venezuela (Amazonas, Apure) and from there down the Rio Negro to Manaus, rarely south and east of Manaus in Amazonas and Pará (Figure 6.12F); on sandy beaches of black-water streams and rivers *(igapó),* but also on the margins of white-water rivers, at low elevations. The illustration of this species in Martius (1823–1837, t. 52) shows the waterfall at Tarumã near Manaus as background; *Leopoldinia pulchra* still grows there.

Brazil: *jará.* Colombia: *manicoli* (Curripaco), *palmito, yará.* Venezuela: *cucurrito, palmiche.*

ARECOIDEAE • ARECEAE • EUTERPEINAE

16. *Euterpe*

Euterpe Mart., Hist. nat. palm. 2: 28. 1823.

Moderate to large, monoecious palms. *Stems* solitary or cespitose, erect or slightly leaning, often with a dense cone of roots visible at base. *Leaves* pinnate, reduplicate; sheath closed and forming a crownshaft; petiole short or absent; rachis long; pinnae regularly arranged (very rarely somewhat clustered) and spreading in 1 plane. *Inflorescences* infrafoliar, branched to 1 order; peduncle bearing a prophyll and 1 peduncular bract; rachis bearing numerous rachillae, mostly abaxially; *flowers* borne in triads proximally, paired or solitary staminate distally; staminate flowers with sepals 3, free, strongly imbricate; petals 3, free, valvate; stamens 6; pistillode trifid; pistillate flowers with sepals 3, free, imbricate; petals 3, free, imbricate; staminodes absent; gynoecium syncarpous, unilocular, uniovulate; *fruits* 1-seeded, globose or rarely ellipsoid with subapical or lateral stigmatic residue; seed with homogeneous or ruminate endosperm and basal or subbasal embryo; germination adjacent ligular; eophyll bifid or pinnate.

A genus of seven species distributed in Central America, the Lesser Antilles and northern South America. Four species are found in the Amazon region. One variety, *Euterpe catinga* var. *roraimae,* also occurs there, but is found only above 900 m elevation in the Guayana Highland. The most recent revision is that of Burret (1929a), who combined *Euterpe* and *Prestoea.* A new treatment is in progress (Henderson & Galeano, in preparation).

KEY TO THE SPECIES OF *EUTERPE*

1. Crownshaft orange or reddish, rarely green, sometimes with a mass of black, elongate, flimsy scales apically; pinnae usually ± horizontally spreading; fruits 0.8–1 cm diam, with homogeneous endosperm; abaxial surface of petiole and rachis with numerous, large raised scales; staminate flowers 3 mm long; rachillae inserted on all sides of rachis . . . . 1. *E. catinga.*
1. Crownshaft green, dull red-green, yellow-green, dark-brown, or purple, without fibers at apex; pinnae usually ± pendulous; fruits 1–2 cm diam, with ruminate or homogeneous endosperm; abaxial surface of petiole and rachis either glabrous or with scattered, small, appressed scales; staminate

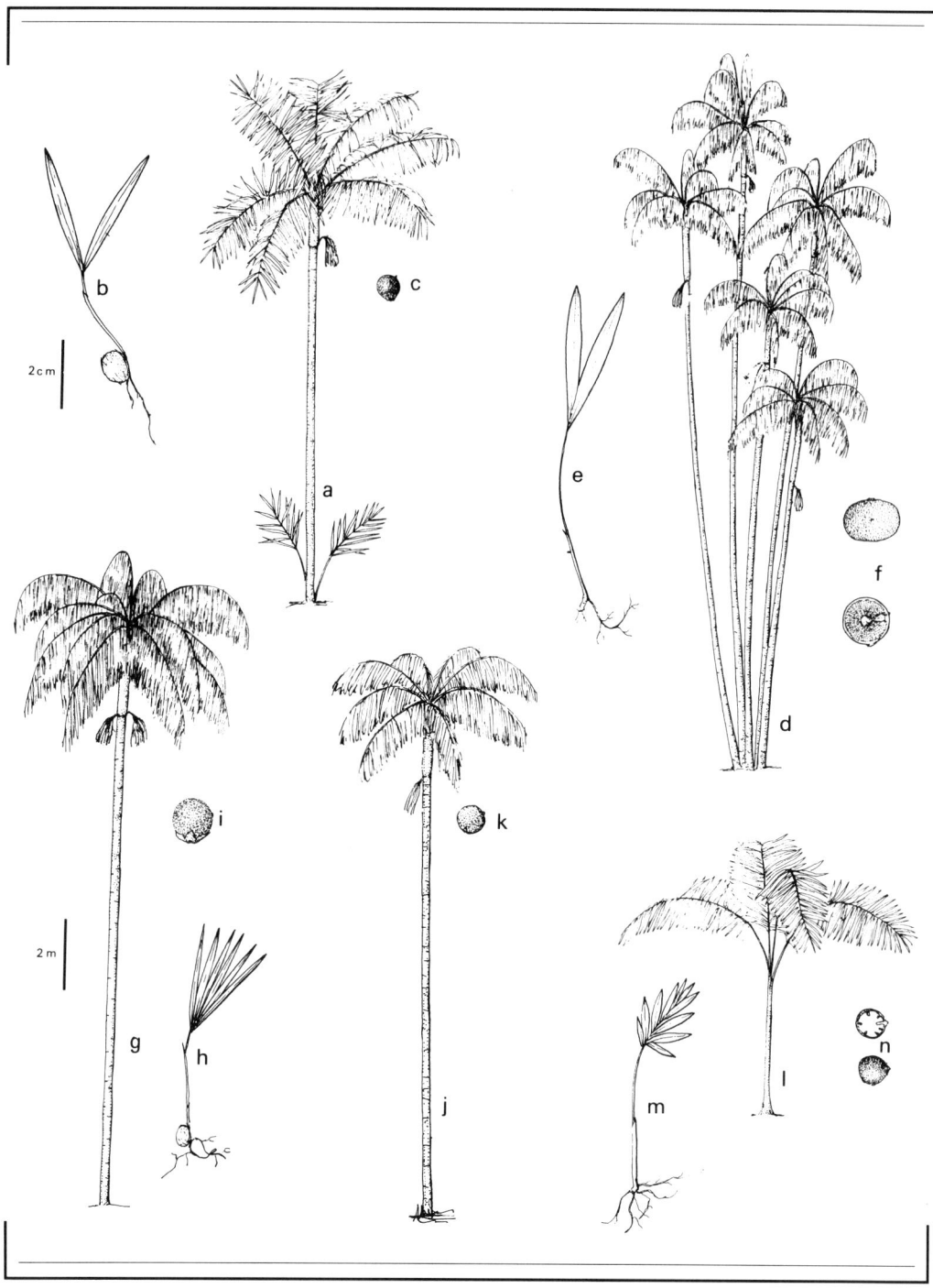

Figure 6.14. (a) *Euterpe catinga* var. *catinga;* **(b)** *eophyll,* **(c)** fruit (from *G. Galeano et al. 1175*); **(d)** *Euterpe oleracea,* **(e)** eophyll, **(f)** fruit and cross section (from *J. Strudwick 4681*); **(g)** *Guterpe precatoria* var. *precatoria,* **(h)** eophyll, **(i)** fruit (from *H. Balslev 4636*); **(j)** *Euterpe longebracteata,* **(k)** fruit (from *M. Jansen-Jacobs 832*); **(l)** *Prestoea schultzeana,* **(m)** eophyll, **(n)** fruit and cross section (from *H. Balslev 62213*).

flowers 4–5.5 mm long; rachillae tending to be absent from adaxial surface of rachis, at least proximally.

2. Fruits with ruminate endosperm; eophyll bifid; stems usually cespitose and forming large clumps . . . . 3. *E. oleracea.*
2. Fruits with homogeneous endosperm; eophyll pinnate; stem usually solitary or if cespitose with only 2–3 stems and not forming large clumps.
   3. Rachillae slender, 1.5–2 mm diam at anthesis, drying brown, covered with very short (less than 0.3 mm long) trichomes; staminate flowers 5.5 mm long with acuminate sepals and petals. . . . . 2. *E. longebracteata.*
   3. Rachillae thicker, 3–5 mm diam at anthesis, drying whitish, covered with longer (0.5 mm long) trichomes; staminate flowers 4–5 mm long with sepals and petals blunt at the apex . . . . 4. *E. precatoria.*

1. Euterpe catinga Wallace, Palm Trees of the Amazon 27, t. 8. 1853. *Euterpe caatinga* Wallace in Spruce, J. Linn. Soc., Bot. 11: 137. 1871. Lectotype (Glassman, 1972). Wallace, Palm Trees of the Amazon t. 8. 1853 (Figure 6.14a–c).

Henderson and Galeano (in preparation) have divided this species into two varieties. Only the following occurs in the lowlands of the Amazon region.

Euterpe catinga var. catinga

Euterpe mollissima Spruce, J. Linn. Soc., Bot. 11: 139. 1871. Type. Brazil. Amazonas: Rio Negro, n.d., *R. Spruce 90* (holotype, K, n.v.).
Euterpe catinga Barb. Rodr., Enum. palm. nov. 15. 1875. *Euterpe catinga* var. *aurantiaca* Drude in Mart., Fl. bras.: Palmae II, fasc. 86, vol. 3(2): 465. 1882. *Euterpe controversa* Barb. Rodr., Les Palmiers 34. 1882. Type. Brazil. Amazonas: Manaus, n.d., *J. Barbosa Rodrigues 225* (destroyed).
Euterpe concinna Burret, Bot. Jahrb. Syst. 63: 69. 1929. Type. Brazil. Amazonas: Manaus, n.d., *J. Huber* (holotype, B).
Euterpe aurantiaca H. E. Moore, Principes 13: 137. 1969. Type. Venezuela. Amazonas: Cerro Sipapo, Caño Grande, 1 km NW of Savanna Camp, 28 Dec 1948, *B. Maguire & L. Politi 28009* (holotype, NY).

Stems cespitose with a few stems or only 1 stem and basal shoots, or solitary, 5–16 m tall, 3.5–9 cm diam. *Leaves* 5–10; sheath closed and forming a crownshaft, 53–87 cm long, orange or reddish, rarely green, with burgundy scales, with a mass of black, elongate, flimsy scales apically; *petiole* 0–10(–17) cm long, petiole and rachis densely covered with black or reddish-brown, raised scales; *rachis* 1.2–2.4 m long; *pinnae* 38–75 per side, regularly arranged and spreading in 1 plane, linear, the middle ones 35–68 cm long, 2–3.5 cm wide. *Inflorescences* infrafoliar at anthesis; peduncle 6–9 cm long; prophyll ca. 54 cm long; peduncular bract ca. 60 cm long; rachis 20–30 cm long; rachillae 48–97, 48–60 cm long, 2.5–4.5 mm diam at anthesis, arranged all around the rachis, densely covered with short, stiff, branched, whitish-brown trichomes; *flowers* in triads almost to ends of rachillae; staminate flowers 3 mm long; sepals very widely ovate, 1.5–2 mm long; petals ovate, 2.5 mm long; pistillode 1–2 mm long; pistillate flowers 2–2.5 mm long; sepals widely ovate, 2 mm long; petals widely ovate, 2–4 mm long; *fruits* globose or depressed-globose, 0.8–1 cm diam, with subapical to lateral stigmatic residue, epicarp purple-black or black, minutely warty; endosperm homogenous; eophyll bifid.

Western Amazon region in Colombia (Amazonas, Caquetá, Guainía, Guaviare, Vaupés), Venezuela (Amazonas), Peru (Loreto), and Brazil (Amazonas) (Figure 6.15A); open or dwarf forest in wet, poorly drained areas on white sand-soil (podzols) with black-water drainage at elevations below 500 m, also in similar habitats in the southwestern Guyana Highland region of Venezuela (Amazonas), in open cloud forest between 1100 and 1500 m elevation. A second variety, *E. catinga* var. *roraimae*, occurs throughout the Guayana Highland at elevations above 900 m.

Brazil: *açaizinho, assaí de caatinga, assaí de catinga, assaí chumbinho, assaí cubinha*. Colombia: *asaí de sabana, asaí paso*. Peru: *huasaí de varillal*. Venezuela: *manaca*. The stems are used

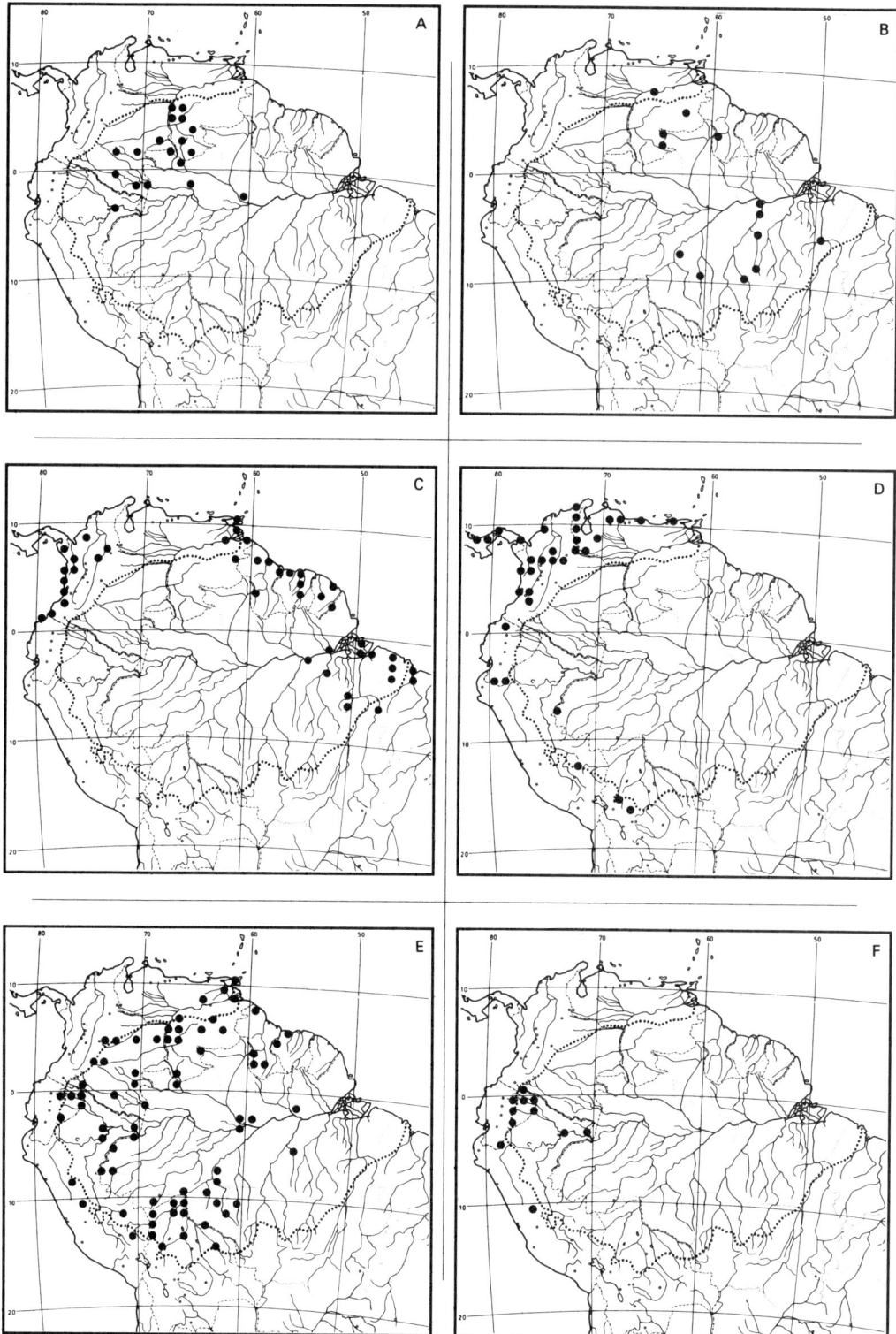

Figure 6.15. **(A)** *Euterpe catinga* var. *catinga;* **(B)** *Euterpe longebracteata;* **(C)** *Euterpe oleracea;* **(D)** *Euterpe precatoria* var. *longevaginata;* **(E)** *Euterpe precatoria* var. *precatoria;* **(F)** *Prestoea schultzeana.*

in house construction; the leaves are used for thatching temporary shelters; and mature fruits are occasionally used to make a drink.

2. Euterpe longebracteata Barb. Rodr., Enum. palm. nov. 17. 1875. *Euterpe longispathacea* Barb. Rodr. ex Huber, Bol. Mus. Paraense Hist. Nat. 6: 133. 1909. Type. Brazil. Pará: Rio Tapajós, n.d., *J. Barbosa Rodrigues 228* (destroyed). Lectotype (Glassman, 1972). Barb. Rodr., Sert. palm. brasil. 1: t. 37. 1903 (Figure 6.14j–k).

Stems solitary or occasionally cespitose, erect, 5–15(–20) m tall, 5–8 cm diam. *Leaves* 8–9; sheath closed and forming a crownshaft, 0.8–1.5 m long, green, with scattered, reddish-brown, woolly scales; petiole 19–41 cm long; rachis 2.5–3 m long; pinnae 70–79 per side, regularly arranged and spreading in 1 plane, pendulous, linear, the middle ones 52–59 cm long, 1.5–2.5 cm wide. *Inflorescences* infrafoliar at anthesis; peduncle 6–10 cm long; prophyll 45–65 cm long; peduncular bract 60–68 cm long; rachis 11–36 cm long; rachillae ca. 60, 21–45 cm long, 1.5–2 mm diam at anthesis, 2.5 mm diam in fruit, absent from all of adaxial surface of rachis, drying distinctively brown, densely covered with stiff, whitish, branched trichomes; staminate flowers 5.5 mm long; sepals very widely ovate, abruptly acuminate, gibbous, 2 mm long, petals ovate, acuminate 5 mm long; pistillode 2 mm long; pistillate flowers 3 mm long; sepals very widely ovate, 2.5 mm long; petals very widely ovate, 2.5 mm long; *fruits* globose, 1–1.2 cm diam, purple-black, with lateral or equatorial stigmatic residue; endosperm homogeneous; eophyll not seen.

Venezuela (Amazonas, Bolívar), Guyana, and Brazil (Amazonas, Mato Grosso, Pará) (Figure 6.15B); lowland forest, usually on *terra firme* but also in inundated areas, at low elevations.

Brazil: *açaí chumbo, açaí da mata, assay da terra firme*.

3. Euterpe oleracea Mart., Hist. nat. palm. 2: 29. 1824. Type. Brazil. Without locality, n.d., *C. Martius 3262* (holotype, M) (Figure 6.14d–f).

Euterpe edulis Mart. in Thurn, Timehri 3: 229. 1884. Misapplied name, = *E. oleracea* Mart.

Euterpe badiocarpa Barb. Rodr., Contr. Jard. Bot. Rio de Janeiro 1: 12. 1901. Lectotype (Wess. Boer, 1965). Barb. Rodr., Sert. palm. brasil. 1: t. 36b. 1903.

Catis martiana O. F. Cook, Bull. Torrey Bot. Club 28: 557. 1901. Nomen nudum.

Euterpe edulis Mart. in Barb. Rodr., Contr. Jard. Bot. Rio de Janeiro 1: 11. 1901. Misapplied name, = *E. oleracea* Mart.

Stems cespitose with to 25 stems per clump, or occasionally appearing solitary and then with shoots at the base, 3–20 m tall, 7–18 cm diam. *Leaves* 8–14; sheath forming a compact crownshaft, 65–150 cm long, dark-brown, purple, green, dull red-green or yellow-green, with few, flat, scattered, brownish scales; petiole 17–50 cm long, petiole and rachis with few, flattened scales or occasionally whitish, scurfy scales adaxially and on upper part of abaxial surface, glabrous abaxially; rachis 1.5–3.7 m long; pinnae 40–80 per side, regularly spaced and spreading in 1 plane, linear, long acuminate, pendulous or less often horizontal (especially on younger plants), the middle ones 64–111 cm long, 2–4.5 cm wide. *Inflorescences* infrafoliar at anthesis; peduncle 5–15 cm long; prophyll 43–66 cm long; peduncular bract 66–95 cm long; rachis 35–68 cm long; rachillae 80–162, 21–75 cm long, 3–4 mm wide at anthesis, thickening in fruit, absent from adaxial, proximal part of rachis, spreading at anthesis, with very short, dense covering of appressed, whitish-brown trichomes; *flowers* in triads proximally, paired or solitary staminate distally; staminate flowers 4–5 mm long; sepals triangular to ovate, 2–3.5 mm long; petals ovate, 3–4 mm long; pistillode 2–3 mm long; pistillate flowers 3 mm long; sepals broadly triangular, 2 mm long, ciliate; petals broadly triangular, 2–3 mm long, briefly valvate at the apex; *fruits* globose or depressed globose, 1–2 cm diam, with subapical to lateral stigmatic residue, purple-black, black, or rarely green, glabrous, minutely verrucose; endosperm deeply ruminate; eophyll bifid.

Pacific coast of Colombia (Cauca, Chocó, Nariño, Valle del Cauca, and some areas of the middle Magdalena valley in Antioquia, Córdoba, Santander) and northern Ecuador (Esmeraldas, Pichincha), Trinidad, Venezuela (Bolívar, Delta Amacuro), the Guianas, and

Brazil (Amapá, Maranhão, Pará, Tocantins) (Figure 6.15C); in large stands of high density in low-lying, tidal areas near the sea and in wet places near rivers, less often occurring inland and then in wet places near streams or rivers.

In the eastern Amazon region, it replaces *E. precatoria* in these habitats. However, in the Pacific coastal region of Colombia and Ecuador, the two species are sympatric. Nevertheless, *E. oleracea* grows in inundated places, while *E. precatoria* grows on noninundated soils. *Euterpe oleracea* can be an aggressive colonizer of disturbed, swampy areas. Oldeman (1969) has discussed the ecology of *E. oleracea* in swamps in French Guiana, and Urdaneta (1981) in Venezuela.

Brazil: *açaí, açaí branco, açaizeiro, assaizeiro* (in Brazil, the tree is *açaizeiro* and the fruit is *açaí*), *ka-be-re* (Apinajé), *juçara, jussara, juzyba* (Assurini), *pinuwa-pihun* (Guajá). French Guiana: *pinot.* Surinam: *baboenpina, kiskis pina, manaka, pina, prasara, wapoe, wapu, wasei.*

This species is important throughout its range because it produces both edible fruit and palmito. In the Brazilian city of Belém, the fruits are an important part of the diet of a large proportion of the inhabitants (Wallace, 1853; Calzavara, 1972; Strudwick & Sobel, 1988). The fleshy mesocarp is mixed with water and made into a drink and, recently, into ice cream. Since the demise of *Euterpe edulis* as a source of palmito, *E. oleracea* is currently the most important species. The canning and sale of palmito was worth $120 million in 1988 (Strudwick & Sobel, 1988). On the Pacific coast of Colombia and Ecuador, there are also canning factories for palmito. Because of its multiple stems, palmito and fruits can be harvested without destroying the tree. This, coupled with the fact that the palms grow in very high-density stands in the Amazon estuary, has recently attracted the attention of researchers interested in sustainable forest products. Anderson (1988) has discussed the use and management of forests dominated by *E. oleracea* near Belém.

Throughout its range, the palm is used for a host of minor items. The stems are used for a variety of construction purposes. The young leaves are mashed, and the sappy residue is applied to stop bleeding (*J. Strudwick et al.* 4681). Fruits and discarded seeds are fed to domestic animals. This palm is also commonly planted as an ornamental throughout the Amazon region in towns and near dwellings.

Near Belém in Brazil, two local forms are known: one with inflorescences branched to 2 orders, and the other with green fruits.

4. Euterpe precatoria Mart. in A. D. Orb., Voy. Amérique mér. 7(3). Palmiers 10. 1842. Type. Bolivia. Moxos, n.d., *A. d'Orbigny 27* (holotype, M) (Figure 6.14g–i).

Stems solitary or less often cespitose and then with 1–2 main stems and basal shoots, erect, (2–)10–20 m tall, (4–)10–23 cm diam, often swollen at the base. *Leaves* (5–)10–20; sheath closed and forming a crownshaft, 0.5–1.6 m long, green or striped lighter green or yellow; petiole (0–)12–57 cm long, petiole and rachis with dense to moderate covering of small, flattened, appressed, irregular, black or reddish-brown scales adaxially and on the margins abaxially, rarely absent (except on juvenile or old leaves); rachis 1.6–3.6 m long; pinnae 43–91 per side, regularly arranged and spreading in 1 plane, linear, pendulous or less commonly horizontal, the middle ones 57–105 cm long, 1–3.5 cm wide, with lateral veins present. *Inflorescences* infrafoliar at anthesis; peduncle 4–20 cm long; prophyll (22–)70–85 cm long; peduncular bract (23–)55–70(–117) cm long; rachis (8–)20–94 cm long; rachillae (24–)70–200, 16–80 cm long, 3–5 mm diam at anthesis, 4–6 mm diam in fruit, absent from adaxial, proximal part of rachis, densely covered with long, whitish or brownish, much branched or stellate, flexuous or stiff trichomes; *flowers* in triads proximally, staminate distally; staminate flowers 4–5 mm long; sepals widely ovate, 2.5–3 mm long; petals narrowly ovate to lanceolate, 3–5 mm long; pistillode 1.5–3 mm long; pistillate flowers 3–4.5 mm long; sepals widely ovate, 3 mm long; petals widely ovate, 4 mm long; *fruits* globose, 0.9–1.3 cm diam, with lateral stigmatic residue, purple-black at maturity, minutely warty; endosperm homogenous; eophyll pinnate with very short rachis.

Widespread and common throughout Central America and northern South America; in forested areas on mountain slopes and ridges at elevations below 2000 m, and in lowland rain forest, very commonly along rivers, below 350 m elevation. This is one of the most widespread and common species in the genus, and perhaps in the family, in the Neotropics. It can be divided into two varieties (Henderson & Galeano, in preparation).

KEY TO THE VARIETIES OF
EUTERPE PRECATORIA

1. Stems moderate, solitary or cespitose; adaxial surface of rachis densely covered with scales, especially near insertion of pinnae; pinnae 2–3 cm wide, with 2 lateral veins on either side of mid-vein, ± pendulous; inflorescences smaller with 3–4 mm diam rachillae, the trichomes ca. 0.1 mm long, erect, stellate; staminate and pistillate flowers not densely pilose, drying brown; low to high elevations in the Andes and Central America . . . .
. . . . 4a. *E. precatoria* var. *longevaginata*.
1. Stems large and solitary; adaxial surface of rachis with few scales; pinnae 1–2 cm wide, with 1 lateral vein on either side of the mid-vein, strongly pendulous; inflorescences larger with 4–6 mm diam rachillae, the trichomes ca. 0.5 mm long, flattened, appressed, much-branched; staminate and pistillate sepals densely pilose abaxially, drying white; low elevations in the Amazon region . . . .
. . . . . . 4b. *E. precatoria* var. *precatoria*.

4a. Euterpe precatoria var. **longevaginata**

(Mart.) Henderson stat. nov. *Euterpe longevaginata* Mart. in A. D. Orb., Voy. Amérique mér. 7(3). Palmiers 11. 1842. Type. Bolivia. Cochabamba, n.d., *A. d'Orbigny 48* (holotype, P, n.v.).

Stems solitary or cespitose, erect, 3–20 m tall, 4–23 cm diam. *Leaves* 5–10(–20); sheath 0.5–1.6 m long, green; petiole (0–)12–49 cm long; rachis 1.6–2.7 cm long, with raised, fimbriate, reddish-brown scales adaxially especially near insertion of pinnae; pinnae 48–73, ± pendulous to horizontal, the middle ones 68–76 cm long, 2–3 cm wide, with prominent mid-vein and 1–2 lateral veins on either side. *Inflorescences* infrafoliar; peduncle 4–13 cm long; prophyll 22–53 cm long; peduncular bract 23–80 cm long; rachis (8–)20–55 cm long; rachillae 24–99, 18–58 cm long at apex, 16–70 cm long at base, 3 mm diam at anthesis, 3–4 mm diam in fruit, brownish, densely covered with ca. 0.1 mm long, stiff, stellate trichomes; *staminate flowers* 5 mm long; sepals very widely ovate, 2.5 mm long, scarcely pilose; petals lanceolate, 4.5 mm long; pistillate flowers 4.5 mm long; sepals very widely ovate, 3 mm long, ± glabrous; petals widely ovate, 4 mm long, scarcely valvate at the apex; *fruits* 0.9–1.1 cm diam.

Central America (Belize, Costa Rica, Guatemala, Nicaragua, Panama), Colombia (Antioquia, Boyacá, Chocó, La Guajira, Norte de Santander, Santander, Valle del Cauca), Venezuela (Barinas, Carobobo, Lara, Miranda, Monagas, Tachira, Yaracuy), Ecuador (El Oro, Esmeraldas, Zamora-Chinchipe), Peru (Amazonas, Loreto, Madre de Dios, San Martín), Brazil (Acre, Serra do Moa), and Bolivia (La Paz) (Figure 6.15D); in forested areas on mountain slopes and ridges, or occasionally in lowland areas, at elevations between sea level and 2000 m.

In the Pacific lowlands of Colombia and Ecuador, this otherwise upland palm occurs at low elevations, as do other Andean palms such as *Chamaedorea linearis*, *Geonoma interrupta*, and *Prestoea decurrens*. This is probably due to the high rainfall in the region.

4b. Euterpe precatoria var. **precatoria**

Cocos venatorum Poepp. ex Mart., Hist. nat. palm. 3: 325. 1853. *Maximiliana venatorum* (Poepp. ex Mart.) H. Wendl. in Kerch., Palmiers 251. 1878. Type. Peru. San Martín: Tocache, n.d., *E. Poeppig 1998* (holotype, W, n.v.; F neg. 31308).

Euterpe oleracea Engel, Linnaea 33: 671. 1865. Nomen nudum.

Euterpe stenophylla Trail ex Thurn, Timehri 3: 229. 1884. Type. Guyana. Corantijn River, *E. Im Thurn* (holotype, K, n.v.).

Euterpe jatapuensis Barb. Rodr., Contr. Jardin Bot. Rio de Janeiro 1: 12. 1901. Lectotype (Glassman, 1972). Barb. Rodr., Sert. palm. bras. 1: t. 36A. 1903 ("yatapuensis").

Euterpe petiolata Burret, Notizbl. Bot. Gart. Berlin-Dahlem 15: 101. 1940. Type. Brazil. Mato Grosso: Amazon floodplain, n.d., *W. Hopp 3044* (holotype, B, n.v.).

Euterpe subruminata Burret, Notizbl. Bot. Gart. Berlin-Dahlem 15: 3. 1940. Type. Guyana.

Basin of Kuyuwini River (Essequibo tributary), about 150 miles from mouth, 21–26 Nov 1937, *A. Smith 2551* (isotypes, K, NY, P, US).

Stems solitary, erect, (7–)10–20 m tall, 10–23 cm diam. *Leaves* 10–20; sheath 0.7–1.6 m long, green or striped lighter green or yellow; petiole (0–)18–57 cm long, with dense to moderate covering of small, flattened, appressed, irregular, black or reddish-brown scales adaxially and on the margins abaxially, rarely absent (except on juvenile or old leaves); rachis 2.1–3.6 m long, with scales similar to those of petiole, especially abaxially, or glabrous; pinnae (43–)48–91 per side, pendulous, the middle ones 62–88 cm long, 1–2 cm wide, with a prominent mid-vein and 1 lateral vein on each side. *Inflorescences* infrafoliar; peduncle 10–15 cm long; prophyll 70–85 cm long; peduncular bract 55–70 cm long; rachis 30–94 cm long; rachillae 70–200, basal rachillae 55–80 cm long, apical ones to 35 cm long, 4–5 mm diam at anthesis, 5–6 mm diam in fruit, densely covered with ca. 0.5 mm long, whitish, much-branched, flexuous trichomes; *staminate flowers* 4–5 mm long; sepals widely ovate, often split apically, 3 mm long, keeled with hairs on keel, unequal, ciliate, white; petals narrowly ovate to lanceolate, 3–5 mm long, white when dry; pistillode 1.5–3 mm long; pistillate flowers 3–4 mm long; sepals widely ovate, 3 mm long, slightly ciliate, with hairs on abaxial surface; petals widely ovate, 4 mm long, scarcely valvate apically; *fruits* 1–1.3 cm diam.

Colombia (Amazonas, Caquetá, Meta, Putumayo, Vaupés, Vichada), Venezuela (Amazonas, Anzoátegui, Bolívar, Monagas, Zulia), Trinidad, the Guianas, Ecuador (Morona-Santiago, Napo), Peru (Amazonas, Cuzco, Loreto, Madre de Dios, Pasco, San Martín), Brazil (Acre, Amazonas, Pará, Rondônia), and Bolivia (Beni, Pando, Santa Cruz) (Figure 6.15E); lowland rain forest, very commonly along rivers, below 350 m elevation, occasionally reaching 600 m in the Andes and Guayana Highland.

Bolivia: *panabí* (Chácobo). Brazil: *açaí, açaí da mata, assaí da mata, juçara*. Colombia: *asaí paso, guasai, ma-na-cáy* (Guahibo), *manaco, ne-e-da* (Huitoto). Ecuador: *ini-bue* (Siona), *palmito, pamiwa, sadke* (Shuar), *sake* (Shuar). Guayana: *manicole, rayhoo, wabo-yaka* (Wapisiana). Peru: *huasai, tunci sake*. Surinam: *nomkie muruku pina, prasara*. Venezuela: *ankú* (Panare), *arimkwe, caruto, guajo* (Yekuana), *manaca, mapora, nenea, palmito manaca*. Trinidad: *manac*.

This is a useful palm throughout its range, but it is not used as much as either *E. edulis* or *E. oleracea*. The fruits are used to make a drink. The stems are commonly used in house construction. The roots are used medicinally for muscle aches and snake bites, and a decoction from the leaves is used to alleviate chest pains (Chacabó, Bolivia). The leaves are used to make brooms and as a source of temporary thatching. In Amazonian Venezuela, the peduncular bract is used to roll cigar tobacco (Zent, label data).

17. *Prestoea*

Prestoea Hook. f. ex Benth. & Hook. f., Gen. pl. 3: 899. 1883.

Small to moderate, monoecious palms. *Stems* solitary or cespitose, erect or procumbent, occasionally short and subterranean. *Leaves* pinnate, sheath open and not forming a crownshaft, or occasionally partly closed and forming a crownshaft; petiole short to medium; rachis long; pinnae regularly arranged and spreading in 1 plane, or occasionally leaf entire. *Inflorescences* interfoliar or infrafoliar, branched to 1 order, or rarely spicate; peduncle bearing a prophyll and 1 peduncular bract; rachis bearing few to many rachillae; *flowers* borne in triads or paired or solitary staminate; staminate flowers with sepals 3, imbricate; petals 3, valvate; stamens 6; pistillode present; pistillate flowers with sepals 3, imbricate; petals 3, imbricate basally, valvate apically; staminodes 6; gynoecium syncarpous, unilocular, uniovulate; *fruits* 1-seeded, globose, with subapical stigmatic residue; seed with ruminate endosperm and basal embryo; germination adjacent-ligular; eophyll bifid or pinnate.

A genus of 11 species (Henderson & Galeano, in preparation) distributed from Central America (Nicaragua) south to Bolivia and also in Venezuela, the Lesser Antilles, and Cuba. The most recent revision is that of Burret (1929a), who did not separate *Prestoea* from *Euterpe*; the genera may indeed be united after further study. Usually confined to mountain areas, but one species just reaches the western Amazon region. It is also possible that another species, *Prestoea ensiformis* (Ruiz & Pav.) H. E. Moore, just enters the region on the eastern Andean slopes of Peru near 500 m elevation.

1. **Prestoea schultzeana** (Burret) H. E. Moore, Gentes Herb. 12: 34. 1980. *Euterpe schultzeana* Burret, Notizbl. Bot. Gart. Berlin-Dahlem 14: 326. 1939. Type. Ecuador. Pastaza: Pacayucu, 200 m, 20 Aug 1937, *H. Schultze-Rhonhof 2433* (holotype, B, n.v.) (Figure 6.14l–n).

Prestoea asplundii H. E. Moore, Gentes Herb. 12: 33. 1980. Type. Ecuador. Napo-Pastaza: Vera Cruz, 900 m, 18 Feb 1956, *E. Asplund 9348* (holotype, S, n.v.).

Stems cespitose, rarely solitary, 0.2–5 m tall, 3–5.5 cm diam, with a cone of roots at base to 70 cm long. *Leaves* 4–10; sheath open and not forming a crownshaft, 37–50 cm long; petiole 80–160 cm long; rachis 1.2–2.2 m long; pinnae (21–)33–38 per side, regularly arranged and spreading in 1 plane, linear-lanceolate, the middle ones (30–)46–58(–81) cm long, 2–4.4(–6) cm wide. *Inflorescences* interfoliar; peduncle 45–80 cm long; prophyll 18–37 cm long; peduncular bract 62–122 cm long; rachis 4–30 cm long; rachillae 5–13, 16–75 cm long; *flowers* arranged in triads proximally on the rachillae, paired or solitary staminate distally; staminate flowers 3–5 mm long; sepals deltate, 1.5–1.7 mm long; petals lanceolate-ovate, 3–5 mm long; pistillode 3.5 mm long; pistillate flowers 3.5–4 mm long; sepals very widely ovate, 3.5 mm long; petals widely ovate, 3.5 mm long; *fruits* globose, black, 0.7–1 cm diam (rarely larger); eophyll pinnate with a long rachis.

Eastern Andean foothills and adjacent Amazon lowlands of Colombia (Amazonas, Putumayo), Ecuador (Morona-Santiago, Napo, Pastaza), and Peru (Amazonas, Loreto, Pasco) (Figure 6.15F); lowland or premontane rain forest in low-lying, flat, inundated areas, rarely on *terra firme*, usually near streams or rivers, usually below 400 m but occasionally to 900 m elevation.

Ecuador: *ca'hue* (Secoya), *chincha, giyikabemo* (Waorani), *na-í* (Secoya), *nai* (Siona), *naicá* (Siona), *palma de pantano*. Peru: *céyacépan* (Amuesha). The leaves are occasionally used to thatch temporary shelters; the seeds are used in blowguns to shoot small birds (Waorani). In Peru, the grated roots are used medicinally (Amuesha).

18. *Oenocarpus* **Oenocarpus** Mart., Hist. nat. palm. 2: 21. 1823.
Jessenia H. Karst., Linnaea 28: 387. 1857.

Moderate to large, monoecious palms. *Stems* solitary or cespitose, erect or leaning. *Leaves* reduplicate, pinnate or pinnately veined if entire; sheath open or partly closed and forming a partial crownshaft, with a prominent fibrous ligule at the apex; petiole short or moderate; rachis long; pinnae regularly or irregularly arranged, spreading in 1 or different planes, rarely leaf entire, silvery-gray or rarely green abaxially. *Inflorescences* infrafoliar, rarely interfoliar, branched to 1 order, rarely spicate or bifurcate, hippuriform (shaped like a horse's tail); peduncle contracted, rarely elongate, bearing a prophyll and 1 peduncular bract; rachis bearing (1–2)–numerous rachillae on lateral and abaxial sides only; *flowers* borne in triads proximally, paired or solitary staminate distally; staminate flowers with sepals 3, very briefly imbricate or connate basally, free and spreading above; petals 3, free, valvate; stamens 6–19; pistillode present; pistillate flowers with sepals 3, widely imbricate; petals 3, widely imbricate; staminodes absent; gynoecium syncarpous, unilocular, uniovulate; *fruits* 1-seeded, globose, oblong, obovoid or ellipsoid, with apical stigmatic residue; seed with ruminate or homogeneous endosperm and large, basal embryo; germination adjacent-ligular; eophyll entire and bifid, or pinnate.

A genus of nine species (Balick, 1986; Bernal et al., 1991) widely distributed over northern South America, mostly north of the Amazon, and reaching Costa Rica in Central America. All species occur in the Amazon region. *Jessenia* is placed in synonymy here and is discussed under *Oenocarpus bataua.* There is an extraordinary concentration of species in the Colombian Amazon; near La Pedrera, Amazonas, seven or eight species occur sympatrically.

The species and clades of this genus have a very interesting distribution pattern in relation to size of the plants. The most widespread, and largest, species is *Oenocarpus bataua.* The next largest, and most widespread, clade comprises *O. bacaba, O. balickii,* and *O. distichus,* which occur in the northern, western, and southern parts of the Amazon region, respectively. The next largest, and widespread, clade comprises *O. mapora* and *O. minor,* occurring in the western and central parts of the Amazon region, respectively. The smallest three species are known only from La Pedrera in the Colombian Amazon.

KEY TO THE SPECIES OF *OENOCARPUS*

1. Leaf sheaths with a few, thick, long, black fibers surrounded by a mass of short, woolly, black fibers; staminate flowers with 8–19 stamens; rachillae glabrous at anthesis; fruits with ruminate endosperm; abaxial surface of pinnae with broad, peltate, multicellular trichomes . . . . 3. *O. bataua.*
1. Leaf sheaths without long, black fibers surrounded by a mass of woolly, black fibers; staminate flowers with 6 stamens; rachillae tomentose at anthesis; fruits with homogeneous, rarely ruminate, endosperm; abaxial surface of pinnae without peltate trichomes.
    2. Leaves entire; inflorescences spicate or bifurcate; Colombia (Amazonas) . . 9. *O. simplex.*
    2. Leaves pinnate; inflorescences with many rachillae.
        3. Fruits with ruminate endosperm; Colombia (Amazonas) . . . . 6. *O. makeru.*
        3. Fruits with homogeneous endosperm.
            4. Peduncle elongate, to 90 cm long; sheath persistent as a reticulate, fibrous covering on the stem; pinnae oblong-lanceolate; Colombia (Amazonas) . . . . 4. *O. circumtextus.*
            4. Peduncle contracted, to 18 cm long; sheath not persistent as a reticulate, fibrous covering on the stem; pinnae linear, lanceolate, or linear-lanceolate.
                6. Leaves distichously arranged; south of the Amazon in Brazil (Maranhão, Mato Grosso, Pará, Rondônia, Tocantins) and Bolivia (Beni, Santa Cruz) . . . . 5. *O. distichus.*
                6. Leaves spirally arranged.
                    7. Stems large and solitary, 7–22 m tall; pinnae irregularly arranged in clusters and spreading in different planes; fruits 1.3–1.8 cm long.
                        8. Peduncle 10–18 cm long; rachis 10–27 cm long; rachillae 103–245, 80–156 cm long; middle pinnae 75–120 cm long; Colombia (Amazonas, Vaupés), Venezuela (Ama-

zonas, Bolívar, Delta Amacuro), the Guianas, and Brazil (Amazonas, Pará) . . . . 1. *O. bacaba.*

8. Peduncle 6–8 cm long; rachis 6–11 cm long; rachillae 46–103, ca. 33 cm long; middle pinnae 60–75 cm long; Colombia (Amazonas), Venezuela (Amazonas), Peru (Loreto, Madre de Dios, San Martín), and Brazil (Acre, western Amazonas, Rondônia) . . . . 2. *O. balickii.*

7. Stems slender and usually cespitose, 2–13.4 m tall; pinnae ± regularly arranged and spreading in 1 plane; fruits 1.5–3 cm long.

9. Peduncle 9–15 cm long; rachis 8–11 cm long; rachillae 79–122, 50–64 cm long; Colombia (Amazonas, Meta, Vaupés), Ecuador (Napo), Peru (Loreto, Madre de Dios, San Martín, Ucayali), Brazil (Acre, western Amazonas), and Bolivia (Pando) . . . . 7. *O. mapora.*

9. Peduncle 3.5–5 cm long; rachis 2.5–6 cm long; rachillae 29–72, 26–56 cm long; Colombia (Amazonas) and Brazil (Amazonas, Pará, Rondônia) . . . . 8. *O. minor.*

1. Oenocarpus bacaba Mart., Hist. nat. palm. 2: 24. 1823. *Oenocarpus bacaba* var. *bacaba* Wess. Boer, Pittieria 17: 130. 1988. Type. Brazil. State?: without locality, n.d., *C. Martius s.n.* (holotype, M, n.v.) (Figure 6.16a–c).

Oenocarpus bacaba var. *xanthocarpa* Trail in Thurn, Timehri 3: 230. 1884. Type. Guyana. Corentyne River, Oct 1879, *G. Jenman 521* (holotype, K).

Oenocarpus baccata in Cuervo Marquez, Trat. elem. Bot. 458. 1913. Nomen nudum.

Oenocarpus hoppii Burret, Notizbl. Bot. Gart. Berlin-Dahlem 11: 1041. 1934. Type. Brazil. Pará: cultivated in Museu Goeldi, Oct 1932, *W. Hopp 8* (holotype, B).

Oenocarpus grandis Burret, Notizbl. Bot. Gart. Berlin-Dahlem 12: 612. 1935. *Oenocarpus bacaba* var. *grandis* Wess. Boer, Pittieria 17: 131. 1988. Type. Brazil. Amazonas: Manicoré, n.d., *W. Hopp 1324* (holotype, B).

Stems solitary, 7–22 m tall, 12–25 cm diam. *Leaves* 8–17; sheath 0.5–1.3 m long, dark purple-green or olive-green, fibrous on margins; petiole 0.3–1.6(–2.3) m long, petiole and rachis whitish-brown-tomentose; rachis 2.2–5.6 m long; pinnae 75–179 per side, irregularly arranged in clusters of 2–9, spreading in different planes, linear, the middle ones 75–120 cm long, 3–5 cm wide, abaxial surface silvery-gray with scattered, brown, punctate scales. *Inflorescences* infrafoliar; peduncle 10–18 cm long; prophyll 50–90 cm long; peduncular bract 0.6–1.9 m long; rachis 10–27 cm long; rachillae 103–245, 80–156 cm long, at anthesis covered with reddish-brown, easily removed, mealy tomentum; *flowers* in triads on basal part of rachillae, paired or solitary staminate distally; staminate flowers 4.5–5 mm long; sepals briefly imbricate basally, triangular, 1–1.5 mm long; petals ovate, 4–4.5 mm long; stamens 6, the filaments somewhat coiled, inflexed at the apex; pistillode 0.5 mm long; pistillate flowers 3–3.5 mm long; sepals very widely ovate, 4 mm long; petals very widely ovate, 2.5 mm long; *fruits* globose-ellipsoid, 1.3–1.5 cm diam, purple-black; endosperm homogeneous; eophyll pinnate.

Central and northeastern Amazon region, usually north of the Amazon river, in Colombia (Amazonas, Vaupés, Vichada), Venezuela (Amazonas, Bolívar, Delta Amacuro), the Guianas, and Brazil (Amazonas, Pará) (Figure 6.17A); lowland rain forest on *terra firme* below 700 m elevation.

Brazil: *bacaba, bacabeira.* Colombia: *milpesillo, patabá.* French Guiana: *comou.* Guyana: *kumu, lu, mapure* (Wapisiana). Surinam: *manni koemboe.* Venezuela: *cudídi* (Banipa), *macába* (Guaica), *seje, seje pequeño, sejito.* The fruits are used to make a drink, although *O. bataua* is usually preferred.

Oenocarpus bacaba forms a monophyletic group with *O. balickii* and *O. distichus.* This

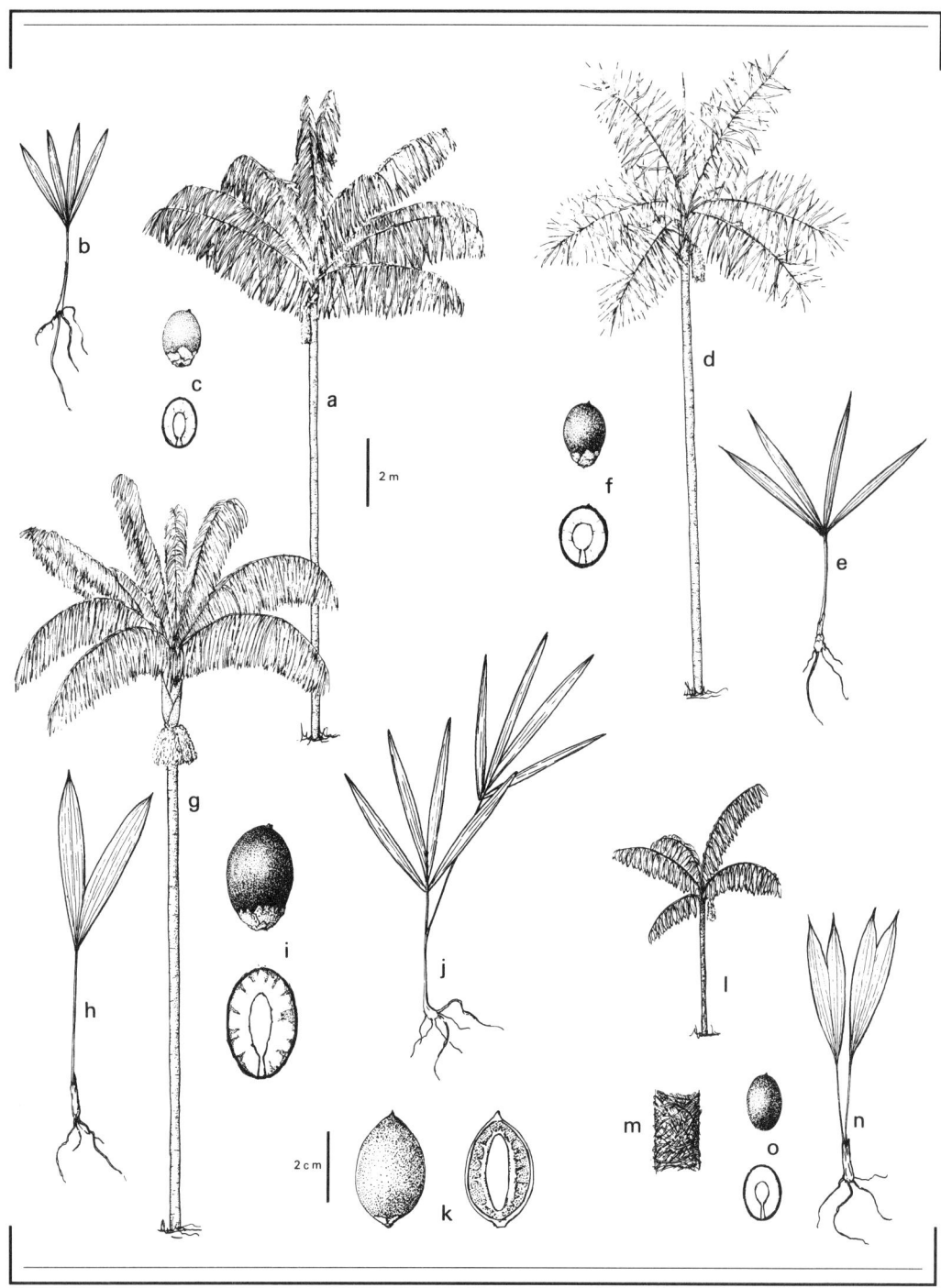

Figure 6.16. (a) *Oenocarpus bacaba,* **(b)** eophyll, **(c)** fruit and cross section (from *M. Jansen-Jacobs 895*); **(d)** *Oenocarpus balickii,* **(e)** eophyll, **(f)** fruit and cross section (from *F. Kahn 1723*); **(g)** *Oenocarpus bataua* var. *bataua,* **(h)** eophyll, **(i)** fruit and cross section (from *D. Smith 12919*); **(j)** *Oenocarpus makeru,* seedling, **(k)** fruit and cross section (j and k from *G. Galeano 2070*); **(l)** *Oenocarpus circumtextus,* **(m)** detail of fibrous sheath, **(n)** eophyll, **(o)** fruit and cross section (from *G. Galeano 1997*).

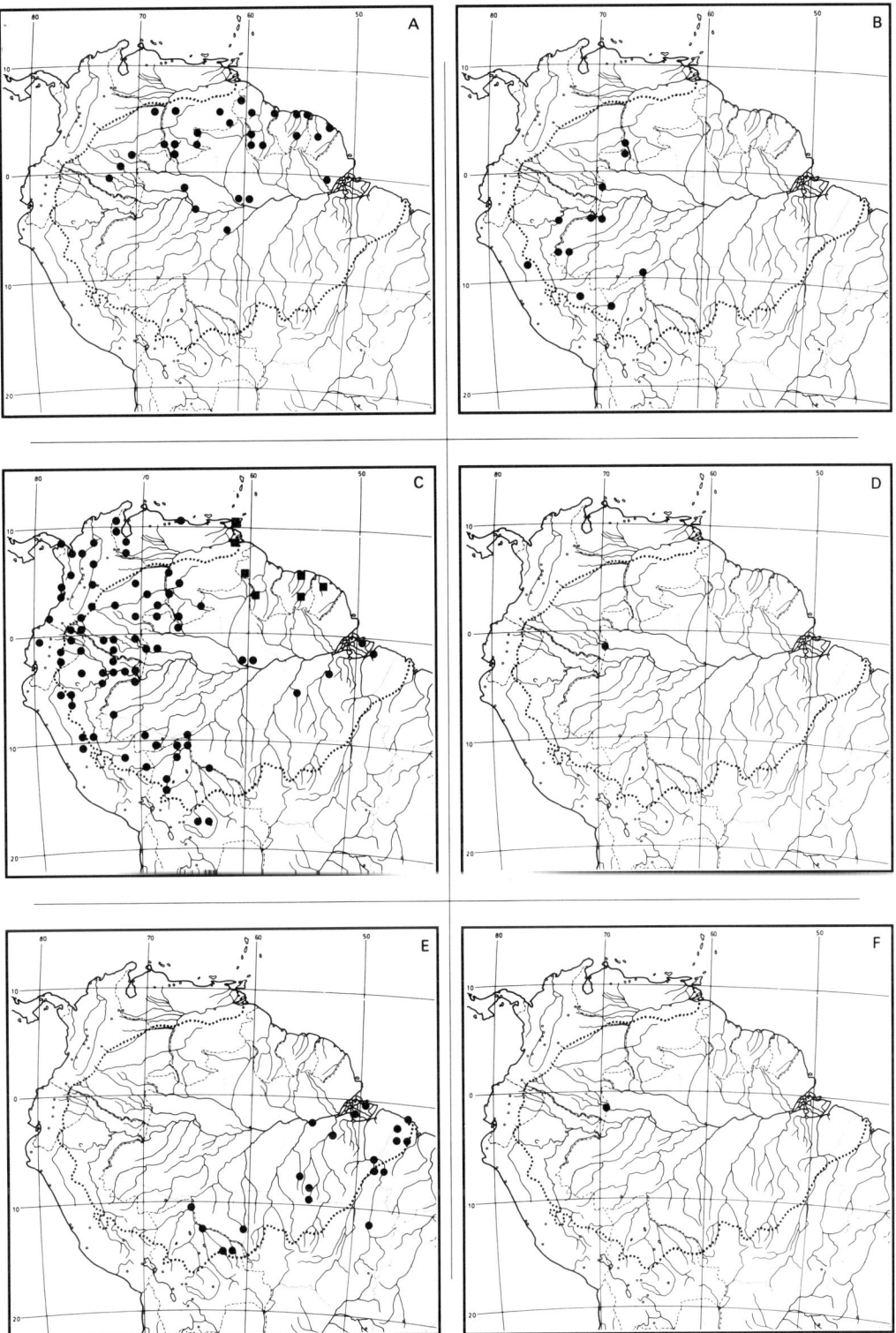

Figure 6.17. (A) *Oenocarpus bacaba;* **(B)** *Oenocarpus balickii;* **(C)** *Oenocarpus bataua* var. *bataua* (circles), var. *oligocarpa* (squares); **(D)** *Oenocarpus circumtextus;* **(E)** *Oenocarpus distichus;* **(F)** *Oenocarpus makeru.*

group is widespread in the Amazon region, but each species has a nonoverlapping distribution: *O. bacaba* north of the Amazon, *O. balickii* in the west, and *O. distichus* south of the Amazon.

Balick (1991) has described a hybrid between this species and *Oenocarpus minor*: *Oenocarpus* x *andersonii* Balick, Bol. Mus. Paraense Hist. Nat. 7: 506. 1991. Type. Brazil. Amazonas: Estrada do Tarumã, 1 km from intersection with Manaus-Itacoatiara Highway, 23 Dec 1979, *A. Anderson 390* (holotype, INPA, n.v.).

2. **Oenocarpus balickii** Kahn, Candollea 45: 351. 1990. Type. Peru. Loreto: Prov. Requena, Bajo Ucayali, Jenaro Herrera, 4°55′S, 73°40′W, 130 m, 25 Jun 1989, *F. Kahn et al. 2380* (isotypes, NY, P) (Figure 6.16d–f).

Oenocarpus bacaba var. *parvus* Wess. Boer, Pittieria 17: 130. 1988. Type. Venezuela. Amazonas: near San Carlos de Río Negro, n.d., *J. Wessels Boer 2276* (holotype, U, n.v.).

Stems solitary (or rarely cespitose?), 7–14(–20) m tall, 6–12 cm diam. *Leaves* 7–11; sheath 45–75 cm long, closed for ca. two-thirds its length and forming a gray-green or green-violet crownshaft, with a fibrous ligule at apex; petiole 10–55 cm long; rachis 2.2–4.2 m long, sheath, petiole, and rachis densely reddish-brown-tomentose or scaly; pinnae 80–179 per side, borne in tight clusters of 2–5, these isolated from one another, spreading in different planes, curved over at the apex, linear, the middle ones 60–75 cm long, 3–4 cm wide, grayish-waxy abaxially, with brown scales on veins abaxially. *Inflorescences* infrafoliar; peduncle 6–8 cm long; prophyll ca. 36 cm long; peduncular bract 30–60 cm long, very densely reddish-brown-tomentose on abaxial surface; rachis 6–11 cm long; rachillae 46–103, 28–33 cm long, densely covered with reddish-brown, granular, easily removed tomentum at anthesis; *flowers* borne in triads on basal part of rachillae, paired or solitary staminate distally; staminate flowers 3.5–4 mm long; sepals very briefly connate basally, narrowly triangular, 1.5 mm long; petals ovate, 3–3.5 mm long; stamens 6, filaments inflexed at the apex; pistillode trifid, 0.5 mm long; pistillate flowers 6 mm long at anthesis; sepals very widely ovate, 3 mm long; petals 4 mm long, very widely ovate; *fruits* globose-ellipsoid, 1.5–1.8 cm long, 1–1.5 cm diam, purple-black; endosperm homogeneous; eophyll pinnate.

Northern and western Amazon region in Colombia (Amazonas), Venezuela (Amazonas), Peru (Loreto, Madre de Dios, San Martín), and Brazil (Acre, western Amazonas, Rondônia) (Figure 6.17B); widespread but rare in lowland rain forest on *terra firme* at low elevations.

Brazil: *bacabão*. Venezuela: *seje*.

This taxon was first recognized by Wessels Boer (1971a) and called *Oenocarpus bacaba* var. *parvus*. At that time, the name was not validly published, but was later validated (Wessels Boer, 1988). Since a name has no priority outside its rank, however, the correct name for the species is *O. balickii* (Bernal et al., 1991). See also discussion under *O. bacaba*.

3. **Oenocarpus bataua** Mart., Hist. nat. palm. 2: 23. 1823. *Jessenia bataua* (Mart.) Burret, Notizbl. Bot. Gart. Berlin-Dahlem 10: 302. 1928. *Jessenia bataua* subsp. *bataua* Balick, Adv. Econ. Bot. 3: 119. 1986. Type. Brazil. Amazonas: without locality, n.d., *C. Martius s.n.* (holotype, M, n.v.) (Figure 6.16g–i).

Stems solitary, 4.3–26 m tall, 15–45 cm diam, columnar or rarely with a swelling, usually with a mound of roots visible at base to 1.2 m tall. *Leaves* (5–)9–20, erect or less often arching; sheath open and not forming a crownshaft, 1–1.9 m long, persistent on young trees, very fibrous on the margins with a few, thick, long, black fibers surrounded by a mass of short, woolly, black fibers, these the remains of the ligule; petiole 0.7–1.6 m long; rachis 3.2–11 m long; pinnae 70–163 per side, linear-lanceolate, regularly arranged and spreading in 1 plane, the middle ones 0.8–1.9 m long, 3.5–14 cm wide, gray-waxy abaxially and with broad, peltate, multicellular trichomes. *Inflorescences* infrafoliar, pendulous at anthesis and in fruit; peduncle 12–23 cm long; prophyll 0.7–1.1 m long; peduncular bract 1.2–2.2 m long, deciduous; rachis 15–54 cm long; rachillae 118–370, 0.7–1.3 m long, yellowish-brown, glabrous; *flowers* borne in triads on proximal part of rachillae, paired or solitary staminate distally; staminate flowers 4–8 mm long; sepals triangular, 1–1.5 mm long; petals ovate-oblong, 3–7 mm long; stamens 8–19; pistillode trifid, 1 mm long; pistillate flowers 6 mm long (in bud); sepals very widely ovate, 4.5–5.5 mm long; petals very widely ovate, 4 mm long; *fruits* ellipsoid or oblong, 2.7–4.5 cm long, 2.2–2.5 cm diam, purple-black; endosperm ruminate; eophyll entire, deeply bifid.

Widely distributed in northern South America and just reaching Central America in eastern Panama, also in Trinidad.

This species has been placed in its own genus, *Jessenia,* by various authors. Most recently, Balick (1986) has emphasized the differences between *Oenocarpus* and *Jessenia:* seeds with homogeneous (*Oenocarpus*) or ruminate (*Jessenia*) endosperm; staminate flowers with 6 (*Oenocarpus*) or 8–19 (*Jessenia*) stamens with inflexed (*Oenocarpus*) or noninflexed (*Jessenia*) filament apex; and pinnae lacking trichomes abaxially (*Oenocarpus*) or with trichomes (*Jessenia*). The discovery by Bernal and colleagues (1991) of a species of *Oenocarpus* with ruminate endosperm, *O. makeru,* has weakened the case for two separate genera, and here they are treated as one.

Like other large palms in the Amazon region, this species varies surprisingly little over its wide range, except for northeastern populations, which, following Balick (1986), are recognized as a separate taxa (at the varietal rank for consistency). The differences, as pointed out by Balick, are not completely consistent, and there may be an intermediate zone.

KEY TO THE VARIETIES OF *OENOCARPUS BATAUA*

1. Staminate part of rachillae flexuous, not cylindrical, with loosely spaced flowers/scars; staminate flowers 7–8 mm long; pistillate flowers confined to proximal half or two-thirds of rachillae; abaxial surface of pinnae not densely gray-waxy, densely covered with trichomes; leaves tending to be erect.
. . . . 3a. *O. bataua* var. *bataua.*
1. Staminate part of rachillae straight, cylindrical, with closely spaced flowers/scars; staminate flowers 4–5 mm long; pistillate flowers confined to proximal quarter or less of rachillae; abaxial surface of pinnae densely gray-waxy with scattered trichomes; leaves tending to be arching . . .
. . . . . . 3b. *O. bataua* var. *oligocarpa.*

3a. Oenocarpus bataua var. bataua

Jessenia polycarpa H. Karst., Linnaea 28: 388. 1857. Type. Colombia. Meta: Llano de San Martín, 1857, *H. Karsten s.n.* (holotype, LE, n.v.; F neg. 29905).

Jessenia weberbaueri Burret, Notizbl. Bot. Gart. Berlin-Dahlem 10: 840. 1929. Type. Peru. San Martín: Moyobamba, 18 Aug 1904, *A. Weberbauer 4561* (isotype, NY).

Amazon region of Colombia (Amazonas, Caquetá, Guaviare, Guainía, Meta, Putumayo, Vaupés, Vichada), Venezuela (Amazonas, Bolívar), the Guianas, Ecuador (Morona-Santiago, Napo), Peru (Junín, Loreto, Huánuco, Madre de Dios, Pasco, San Martín), Brazil (Acre, Amazonas, Pará, Rondônia), and Bolivia (Beni, Pando, Santa Cruz) (Figure 6.17C). In the central Amazon region, it commonly grows in inundated areas along streams in lowland rain forest, but is also occasionally found on *terra firme.* In peripheral areas, it seems to occur more often on *terra firme.* In extra-Amazon regions, especially on eastern Andean slopes, it can grow on steep ridges up to 1400 m elevation.

García (1988) studied pollination of this variety (as *Jessenia bataua*) in Amazonian Ecuador. Palms flowered all year round, with a peak from February to April (see also Veléz, 1992). Inflorescences were protandrous, opening in the afternoon. Staminate flowers were diurnal and not strongly scented, and anthesis lasted for 3 weeks. Pistillate anthesis took place for about 1 week in the fourth and fifth week of anthesis. At this time, a sweet fragrance was given off between 16:00 and 20:00, and inflorescences heated up 3–5°C above ambient. The most important pollinators were considered *Mystrops* (Nitidulidae) and two species of Derelomini (Curculionidae), both of which bred in the flowers.

Bolivia: *itsama* (Chacabo), *majo, mayo.* Brazil: *patoá, patauá, patauá branca, patauá roxa.* Colombia: *batú* (Andoke), *he-bu-ca-nu* (Cubeo), *mil peso, milpesos, palma letchera, patabá, seje, unamo.* Ecuador: *cosá* (Siona), *chapil, milpesos, shiwamuyo* (Quichua), *unguragua, ungurahua.* Peru: *hungurahui, hunguravi, kunkúk* (Mayna Jívaro), *siname, ungurabe morado, ungurahui, ungurawi.* Venezuela: *palma seje, seje, seje grande, seje hembra.* The roots and fruits are used as medicine for various ailments; the leaves are used for thatching; the long, black fibers from the leaf sheath are used as darts for blowguns. The most important use is edible oil. The newly collected fruits are soaked in warm water and then gently massaged to release the oil from the mesocarp. Balick (1986) has discussed the ethnobotany of this palm, as well as the quality of the oil, which is chemically similar to that of olives.

In the southwestern Amazon region of Peru (Madre de Dios) and Brazil (Acre) *Oenocarpus bataua* var. *bataua* apparently forms rare hybrids with *O. mapora*. These are called *ungurahuillo* in Peru and *bacabão* in Brazil, and are always distinguished by local people. Some of these hybrids are unusual in being vegetatively like *O. bataua* and reproductively like *O. mapora*. A brief description is as follows:

Stems solitary or rarely cespitose, 5–15 m tall, 15–30 cm diam. *Leaves* 5–8, erect; sheath to 80 cm long, with fibers like those of *O. mapora*; pinnae 86–88 per side, regularly or slightly irregularly arranged, spreading in 1 plane, abaxially with scales like those of *O. bataua*. *Inflorescences* infrafoliar; peduncle to 10 cm long; peduncular bract to 90 cm long; rachis 15–20 cm long; rachillae 86–93, to 62 cm long; staminate flowers with 6 stamens; *fruits* globose-ellipsoid, 1.5–2.5 cm long, ca. 2 cm diam; endosperm homogeneous with few ruminations or ruminate; eophyll bifid or pinnate with 3–4 pinnae, usually much smaller than that of *O. bataua*.

Michael Balick (personal communication) has discovered a hybrid between *Oenocarpus bataua* var. *bataua* and *Oenocarpus bacaba* from the northwestern Amazon region in Colombia (Vaupés) and Venezuela (Amazonas).

3b. Oenocarpus bataua var. oligocarpa

(Griseb. & H. Wendl.) Henderson, stat. nov. *Jessenia oligocarpa* Griseb. & H. Wendl. in Griseb., Fl. Brit. W. Indies 516. 1854. *Oenocarpus oligocarpa* (Griseb. & H. Wendl.) Wess. Boer, Flora of Suriname 5(1): 58. 1965. *Jessenia bataua* subsp. *oligocarpa* (Griseb. & H. Wendl.) Balick, Adv. Econ. Bot. 3: 126. 1986. Type. Trinidad. Without locality, 1857–1863, *H. Crueger 74* (holotype, GOET, n.v.).

Venezuela (Bolívar), Trinidad, and the Guianas (Figure 6.17C); in lowland rain forest on *terra firme* at low elevations.

French Guiana: *patawa*. Guyana: *turu palm*. Surinam: *patawa*. Venezuela: *palma seje, seje*. The uses are similar to those of var. *bataua*.

Sist (1987) studied regeneration, population dynamics, and dispersal (as *Jessenia bataua* subsp. *oligocarpa*) in French Guiana. Populations consisted of large numbers of seedlings and juveniles and few adults. Light was the most important factor in growth from juvenile to adult. Fruits were eaten by various birds, which ate the mesocarp and discard the seeds. Sist (1989a) considered that in French Guiana this species flowered biennially.

4. Oenocarpus circumtextus Mart., Hist. nat. palm. 2: 26. 1823. Type. Colombia. Amazonas: Cerro de la Pedrera, n.d., *C. Martius s.n.* (holotype, M; F neg. 18554a) (Figure 6.16I–o).

Stems solitary, 3–6 m tall, 6–8 cm diam, densely covered with reticulate, fibrous remains of leaf sheaths. *Leaves* 7–8; sheath 45–47 cm long; petiole 72–75 cm long; rachis 1.7–1.8 m long; pinnae 16–19, regularly arranged and spreading in 1 plane, oblong-lanceolate, abruptly long acuminate, the middle ones 54–66 cm long, 9–11.5 cm wide, glabrous abaxially. *Inflorescences* interfoliar; peduncle ca. 76 cm long; prophyll ca. 42 cm long; peduncular bract ca. 118 cm long; rachis 4–6 cm long; rachillae 21–25, to 36 cm long, sparsely covered with short, brown, crustose hairs; *flowers* in triads proximally on rachillae, paired or solitary staminate distally; staminate flowers 3.5 mm long; sepals briefly connate basally, 1 mm long; petals ovate, 3 mm long; stamens 6; filaments somewhat coiled, inflexed apically; anthers saggitate; pistillode trifid; pistillate flowers 2 mm long (preanthesis); sepals very widely ovate, 2 mm long; petals very widely ovate, 1.5 mm long; staminodes absent; *fruits* ellipsoid-oblong, 1.7 cm long; 1 cm diam; endosperm homogeneous; eophyll entire, bifid at apex.

Known from a small region in eastern Colombia (Amazonas) near La Pedrera on the Río Caquetá (Figure 6.17D); in semiopen savannas on rocky outcrops of granite or quartzite, on sandy soil at low elevations (Bernal et al., 1991).

Colombia: *milpesillo de sabana*.

5. Oenocarpus distichus Mart., Hist. nat. palm. 2: 22. 1823. Type. Brazil. Pará: without locality, *C. Martius 2615* (holotype, M) (Figure 6.18a–c).

Oenocarpus tarampabo Mart. in A. D. Orb., Voy. Amérique mér. 7(3). Palmiers 12. 1842. Lectotype (Balick, 1986). Mart. in A. D. Orb., Voy. Amérique mér. 7(3). Palmiers t. 18b. 1846. Type. Bolivia. Without locality, n.d., *A. d'Orbigny s.n.* (holotype, M).

Oenocarpus discolor Barb. Rodr., Palmae Mattogrossenses 8. 1898. Lectotype (Balick, 1986). Barb. Rodr., Palmae Mattogrossenses t. 3. 1898.

Stems solitary, 5–10(–20) m tall, 8–18(–25) cm diam. *Leaves* 9–12, distichously arranged; sheath 0.5–1 m long, partly closed and forming a crownshaft, olive-green, the margins disintegrating into fibers; petiole 15–40 cm long; rachis 2.5–5.4 m long; pinnae 40–130 per side, irregularly arranged in clusters of 3–10, spreading in different planes, lanceolate, the middle ones 0.7–1.4 m long, 5–6.5 cm wide. *Inflorescences* infrafoliar; peduncle 8–11 cm long; prophyll 53–80 cm long; peduncular bract 1.1–1.4 m long, deciduous; rachis 9–33 cm long; rachillae 51–161, 56–95 cm long, moderately to densely covered with reddish-brown granular tomentum at anthesis; *flowers* arranged in triads proximally on rachillae, paired or solitary staminate distally; staminate flowers 4–5 mm long; sepals briefly connate basally, narrowly triangular, 1–1.6 mm long; petals ovate, 3.5–4 mm long; stamens 6, the filaments inflexed at apex; pistillode trifid, 0.5 mm long; pistillate flowers 5–6 mm long; sepals very widely ovate, 5 mm long; petals very widely ovate, 6 mm long; *fruits* globose to ellipsoid, 1.8–2 cm long, 1.5–1.7 cm diam, purple-black; endosperm homogeneous; eophyll pinnate.

South of the Amazon river in Brazil (Maranhão, Mato Grosso, Pará, Rondônia, Tocantins) and Bolivia (Beni, Santa Cruz) (Figure 6.17E); in a variety of habitats, from lowland rain forest on *terra firme* to savanna margins, savannas, serras, and rocky areas, usually below 350 m elevation, and generally in areas of lower rainfall.

Brazil: *bacaba, pindiwa'y* (Ka'apor), *pinduwa'ywa* (Assurini), *pinuwa-'y* (Guajá), *pinuwa'yw* (Tembé). The fruits are edible, and a beverage is prepared from them. Oil is extracted from the mesocarp. The leaves are used for various purposes, including thatching and weaving.

Balick (1986) recognized three distichous-leaved *Oenocarpus*: *O. distichus, O. discolor,* and *O. tarampabo*. The differences given to separate them—stem size, pinnae size and shape, and staminate petal shape—are not here considered significant, and one widespread species is recognized.

6. Oenocarpus makeru Bernal, Galeano & Henderson, Brittonia 43: 158. 1991. Type. Colombia. Amazonas: Río Caquetá, near the Chorro Córdoba, ca. 250 m, 13 Mar 1990, *G. Galeano et al. 2070* (holotype, COL; isotype, NY) (Figure 6.16j–k).

Stems solitary, 5–8 m tall, 7–8 cm diam, gray with lichens. *Leaves* ca. 12; sheath 65–68 cm long, fibrous; petiole 15–25 cm long; rachis ca. 2.5 m long; pinnae ca. 65 per side, regularly spaced, arranged in 1 plane, linear-lanceolate, the middle ones 73–75.5 cm long, 5–5.3 cm wide, with a wool-like, waxy tomentum abaxially. *Inflorescences* infrafoliar; peduncle 6.2–7.5 cm long; prophyll not seen; peduncular bract with purplish scales; rachis ca. 11 cm long; rachillae 107–125, 56–60 cm long; staminate flowers (young bud) 1 mm long; sepals connate basally, triangular, 0.3 mm long; petals ovate, 1 mm long; pistillate flowers not seen; *fruits* ovoid, 2.2–2.6 cm long, 1.4–1.7 cm diam, purple-black; endosperm ruminate; eophyll pinnate with 4 pinnae.

Known only from the type locality in Colombia (Amazonas) (Figure 6.17F); in lowland forest on *terra firme* at low elevations.

Colombia: *makeru* (Yakuna). Oil is extracted from the mesocarp of the fruits in the same way as for *Oenocarpus bataua*.

This is the second species of *Oenocarpus* to have seeds with ruminate endosperm, and, as previously noted, this character is the main reason for uniting *Jessenia* with *Oenocarpus*. It seems possible, however, that this species is a hybrid between *O. bataua* and *O. minor*.

7. Oenocarpus mapora H. Karst., Linnaea 28: 274. 1857. *Oenocarpus mapora* subsp. *mapora* Balick, Adv. Econ. Bot. 3: 105. 1986.

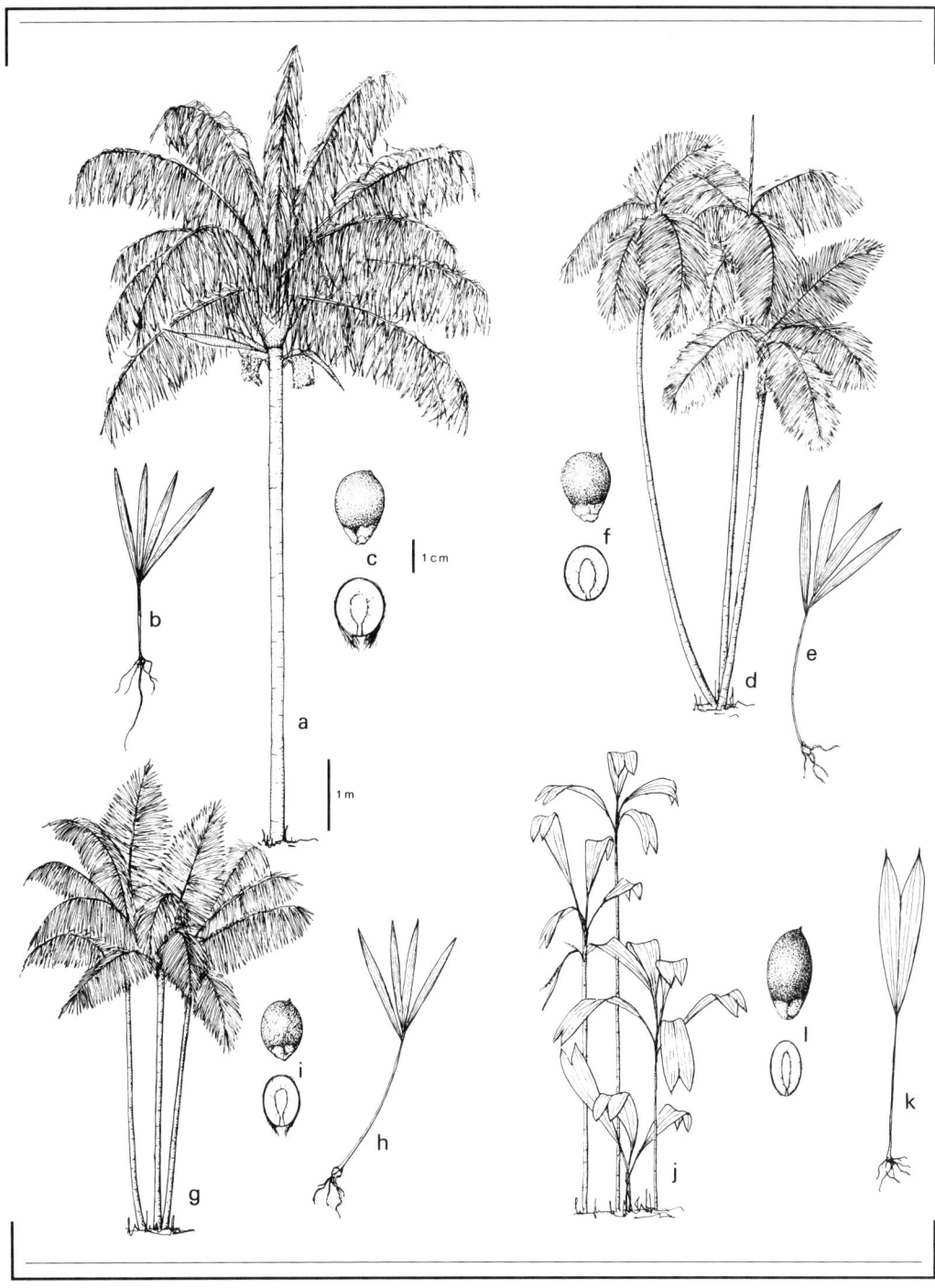

Figure 6.18. (a) *Oenocarpus distichus,* **(b)** eophyll, **(c)** fruit and long section (from *A. Scariot 170*); **(d)** *Oenocarpus mapora,* **(e)** eophyll, **(f)** fruit and long section (from *F. Kahn 1727*); **(g)** *Oenocarpus minor,* **(h)** eophyll, **(i)** fruit and long section (from *A. Henderson 1594*); **(j)** *Oenocarpus simplex,* **(k)** eophyll, **(l)** fruit and long section (from *G. Galeano 2027*).

Type. Venezuela. Zulia: Perija, Maracaibo, n.d., *H. Karsten s.n.* (holotype, LE, n.v.) (Figure 6.18d–f).

Oenocarpus multicaulis Spruce, J. Linn. Soc., Bot. 11: 142. 1871. Type. Peru. San Martín: Tarapoto, n.d., *R. Spruce 63* (holotype, K, n.v.).

Oenocarpus macrocalyx Burret, Notizbl. Bot. Gart. Berlin-Dahlem 11: 1043. 1934. Type. Brazil. Amazonas: Livramento, Rio Madeira and Rio Marmellos, Jan 1932, *W. Hopp 1155* (holotype, B).

Stems 5–15 m tall, 4–17 cm diam, cespitose and forming clumps of to 10 stems, or solitary, with numerous roots visible at the base. *Leaves* 5–12; sheath 0.4–1.4 m long, closed at the base and forming a crownshaft, greenish-brown or olive-green, with a fibrous ligule to 20 cm long at the apex; petiole 10–60(–115) cm long; rachis 2.1–5 m long, sheath, petiole, and rachis densely brown-tomentose, glabrescent; pinnae 40–90 per side, almost regularly arranged or the middle ones irregularly arranged in clusters of 2–5, ± in the same plane but when clustered tending to be arranged in different planes, linear, the middle ones 59–98 cm long, 3.5–7.5 cm diam. *Inflorescences* infrafoliar; peduncle 9–15 cm long; prophyll 31–53 cm long; peduncular bract 50–80 cm long; rachis 8–11 cm long; rachillae 79–122, 50–64 cm long, densely covered with reddish-brown tomentum at anthesis; *flowers* borne in triads proximally on rachillae, paired or solitary staminate distally; staminate flowers 4–5 mm long; sepals briefly connate at the base, narrowly triangular, 1–2 mm long; petals, ovate, 3.5–4 mm long; stamens 6, the filaments inflexed at the apex; pistillode trifid, 0.7 mm long; pistillate flowers 4 mm long (in bud); sepals shallowly triangular, 4 mm long; petals shallowly triangular, 2.5 mm long; staminodes absent; *fruits* globose-ellipsoid, 2–3 cm long, 1.5–2.5 cm diam, purple-black; endosperm homogeneous; eophyll pinnate.

Central America from Costa Rica south to Bolivia and into the western Amazon region of Colombia (Amazonas, Caquetá, Meta, Putumayo, Vaupés), Ecuador (Napo), Peru (Huánuco, Loreto, Madre de Dios, San Martín, Ucayali), Brazil (Acre, western Amazonas), and Bolivia (Pando) (Figure 6.19A); in lowland forest either on *terra firme* or in inundated areas.

There is a possible tendency for plants from inundated areas to have cespitose stems, and those from *terra firme* to have either solitary or cespitose stems.

Bolivia: *bacaba, bacabi, bacabiña, quëboitsama* (Chácobo). Brazil: *bacaba, bacabinha.* Colombia: *milpesillo, posuí, pusuy.* Ecuador: *chimbo* (Quechua), *huicosa* (Siona), *milpesillo, shimbu* (Siona). Peru: *chicyorah* (Bora), *cinamillo, sinamillo, sinami, vacavilla.* The fruits are used to make drinks; oil is extracted from the mesocarp. The palm heart is occasionally eaten, and the stems are used for miscellaneous construction purposes.

In the herbarium, it is sometimes difficult to separate solitary-stemmed specimens of this species from *O. balickii* (and it is possible they hybridize). In the latter, the pinnae tend to be much more closely clustered.

This species and *Oenocarpus minor* form a monophyletic group, with nonoverlapping ranges: the former in the western Amazon region, and the latter in the central region. They are, however, doubtfully distinct (Bernal et al., 1991). See discussion under *O. bataua* for hybrids.

8. Oenocarpus minor Mart., Hist. nat. palm. 2: 25. 1823. *Oenocarpus minor* subsp. *minor* Balick, Adv. Econ. Bot. 3: 112. 1986. Type. Brazil. Amazonas: Manaus, n.d., *C. Martius 3121b* (holotype, M; F neg. 18553a) (Figure 6.18g–i).

Oenocarpus microspadix Burret, Notizbl. Bot. Gart. Berlin-Dahlem 12: 297. 1928. Type. Brazil. Amazonas: Manaus, Oct 1926, *G. Huebner 82* (holotype, B).

Oenocarpus huebneri Burret, Notizbl. Bot. Gart. Berlin-Dahlem 10: 297. 1928. Type. Brazil. Amazonas: Ipiranga, Manaus, Jan 1927, *G. Huebner 87* (holotype, B, n.v.).

Oenocarpus intermedius Burret, Notizbl. Bot. Gart. Berlin-Dahlem 10: 298. 1928. *Oenocarpus minor* subsp. *intermedius* (Burret) Balick, Adv. Econ. Bot. 3: 113. 1986. Type. Brazil. Amazonas: Manaus, Jan 1925, *G. Huebner 1* (holotype, B).

Stems cespitose in clusters of 3–7, or rarely solitary, 2–8 m tall, 4–7 cm diam. *Leaves* 4–13; sheath 40–50 cm long, partially closed and forming a crownshaft, dull-green or maroon,

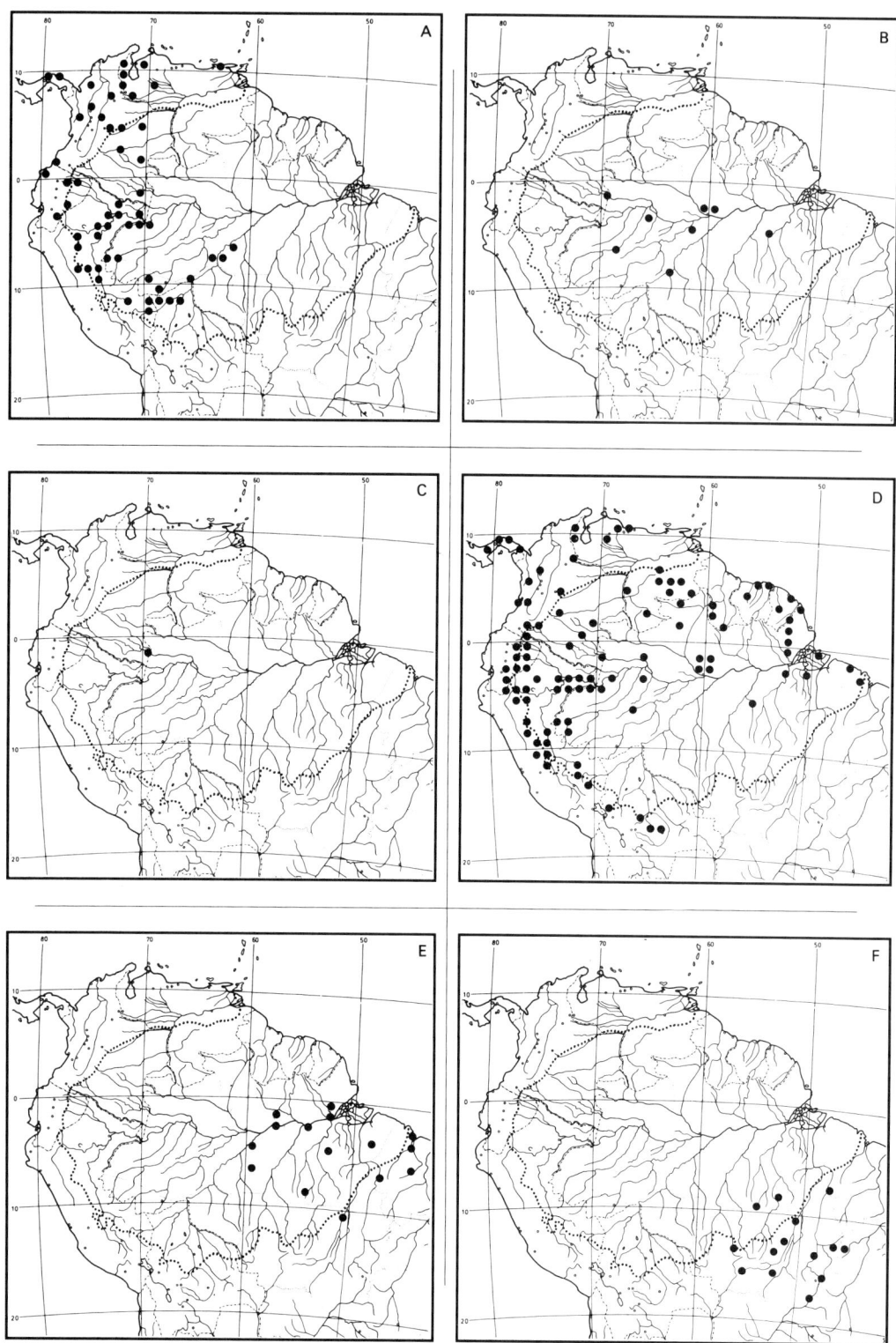

Figure 6.19. (A) *Oenocarpus mapora;* **(B)** *Oenocarpus minor;* **(C)** *Oenocarpus simplex;* **(D)** *Hyospathe elegans;* **(E)** *Syagrus cocoides;* **(F)** *Syagrus comosa.*

fibrous on margins, with a fibrous ligule to 20 cm long; petiole 15–50 cm long; rachis 2–4 m long, rachis, petiole, and sheath covered with reddish-brown tomentum or scales, glabrescent; pinnae 42–79 per side, linear, regularly arranged and spreading in 1 plane, the middle ones 54–69 cm long, 3–6 cm wide, with brown scales on veins abaxially. *Inflorescences* infrafoliar; peduncle 3–5 cm long; prophyll to 30 cm long; peduncular bract not seen; rachis 2.5–6 cm long; rachillae 29–72, 26–56 cm long; densely covered with light-brown or reddish-brown granular tomentum at anthesis; *flowers* borne in triads proximally on rachillae, paired or solitary staminate distally; staminate flowers 3.5–4 mm long; sepals very briefly connate basally, narrowly triangular, 1.5–2 mm long; petals ovate, 3–3.5 mm long; stamens 6, the anthers inflexed at apex; pistillode trifid, 0.6 mm long; pistillate flowers 6 mm long at anthesis; sepals very widely ovate, 5 mm long; petals very widely ovate, 5 mm long, not valvate at the apex; *fruits* globose-ellipsoid, 1.5–2 cm long, 1.3–1.5 cm diam, purple-black at anthesis; endosperm homogeneous; eophyll pinnate.

Central Amazon region in Colombia (Amazonas) and Brazil (Amazonas, Pará, Rondônia) (Figure 6.19B); lowland forest on *terra firme* or occasionally in inundated areas, at low elevations.

Brazil: *bacabinha.*

This taxon is very close to, and probably conspecific with, *Oenocarpus mapora* (in which case, *O. minor* would be the correct name). Bernal and colleagues (1991) have discussed the differences and similarities. A hybrid with *O. bacaba* is known; see discussion under that species.

9. Oenocarpus simplex Bernal, Galeano & Henderson, *Brittonia* 43: 154. 1991. Type. Colombia. Amazonas: ca. 2 km along the trail from La Pedrera to Tarapacá, 230 m, 10 Mar 1990, *G. Galeano et al. 2027* (holotype, COL; isotypes, FTG, NY) (Figure 6.18j–l).

Stems cespitose, 3–4 m tall, 1.5–1.8 cm diam. *Leaves* 5–8; sheath 41–45 cm long; petiole 24–28 cm long; blade entire, bifid at apex, 76–91 cm long, 14–18.6 cm wide, white waxy abaxially. *Inflorescences* interfoliar, spicate or bifurcate; peduncle 16–37 cm long; prophyll 15–20 cm long; peduncular bract 48–66 cm long; rachis absent; rachillae 21–25 cm long; *flowers* in triads on proximal part of rachillae, paired or solitary staminate distally; staminate flowers 3–4.5 mm long; sepals connate basally, triangular, 0.5–0.8 mm long; petals ovate-lanceolate, 3 mm long; stamens 6; filaments inflexed at the apex; pistillode very short; pistillate flowers 2 mm long; sepals suborbicular, 2.5 mm high; petals suborbicular, 1.5 mm high; staminodes absent; *fruits* oblong-ellipsoid, purple-black, 2.2–2.7 cm long, 1.3–1.4 cm diam; endosperm homogeneous; eophyll entire and bifid.

Known only from the type locality in Colombia (Amazonas) (Figure 6.19C); lowland forest on *terra firme* at low elevations.

This is a very distinct species, having entire leaves and spicate or bifurcate inflorescences.

19. *Hyospathe* **Hyospathe** Mart., Hist. nat. palm. 2: 1. 1823.

Small, monoecious palms. *Stems* cespitose or solitary, erect or creeping and then rooting. *Leaves* pinnate, reduplicate; sheath closed and forming an elongate crownshaft; petiole moderate; rachis moderate; pinnae regularly or irregularly arranged, spreading in 1 plane, occasionally leaf entire. *Inflorescences* infrafoliar, branched to 1 order; peduncle bearing a prophyll and 1 peduncular bract; rachis bearing numerous rachillae; *flowers* borne in triads proximally, staminate distally; staminate flowers pedicellate; sepals connate into a 3-lobed tubular calyx, solid basally and forming the pedicel; petals 3, free, valvate; stamens 6, in 2 whorls of 3, the basal whorl with the filaments attached to the base of the pistillode, the upper whorl with the filaments attached to the middle of the pistillode; pistillate flowers pedicellate or sessile; sepals 3, connate into a cupular calyx; petals 3, free, imbricate basally, briefly valvate at the

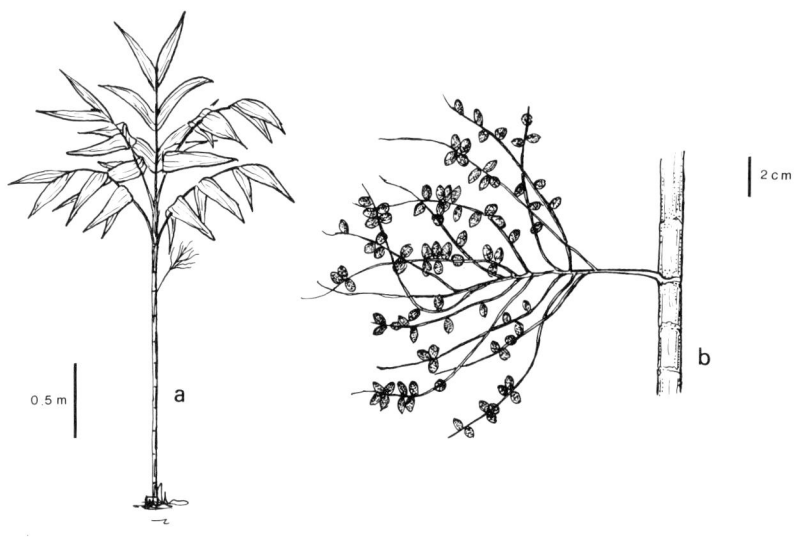

Figure 6.20. (a) *Hyospathe elegans,* **(b)** infructescence (from *I. Amaral 1488*).

apex; staminodes 6; gynoecium syncarpous, unilocular, uniovulate; *fruits* 1-seeded, ellipsoid to ovoid, with basal stigmatic residue; seed with homogeneous endosperm and basal embryo; germination adjacent-ligular; eophyll bifid.

A genus of two species (Skov & Balslev, 1989) distributed from Central America (Costa Rica) south to Bolivia and across northern South America. One species is found in the Amazon region.

1. Hyospathe elegans Mart., Hist. nat. palm. 2: 1. 1823. Type. Brazil. Amazonas: without locality, n.d., *C. Martius 3122* (holotype, M; isotype, P; F neg. 18528a) (Figure 6.20a–b).

Hyospathe filiformis H. Wendl. ex Drude in Mart., Fl. bras.: Palmae II fasc. 86 vol. 3(2): 522. 1882. Type. Brazil. Pará: Yaburu, Rio Yapura, n.d., *C. Martius s.n.* (holotype, M, n.v.; isotype, P).

Hyospathe gracilis H. Wendl. ex Drude in Mart., Fl. bras.: Palmae II fasc. 86 vol. 3(2): 523. 1882. Type. Peru. Without locality, *W. Poeppig 2057* (holotype, W, n.v.).

Hyospathe brevipedunculata Dammer, Verh. Bot. Vereins Prov. Brandenburg 48: 126. 1907. Type. Brazil. Acre: Rio Juruá-Mirim, Sep 1901, *E. Ule 5881* (isotype, MG).

Hyospathe tessmannii Burret, Notizbl. Bot. Gart. Berlin-Dahlem 10: 856. 1929. Type. Peru. Amazonas: mouth of Río Santiago, 160 m, 6 Sep 1924, *G. Tessmann 3980* (holotype, B, n.v.).

Hyospathe micropetala Burret, Notizbl. Bot. Gart. Berlin-Dahlem 10: 857. 1929. Type. Peru. Loreto: San Antonio, Río Marañon, Río Pastaza, 14 Dec 1924, *G. Tessmann 4935* (holotype, B, n.v.).

Hyospathe pallida H. E. Moore, Gentes Herb. 8: 197. 1949. Type. Colombia. Putumayo: Uchupayaco, between Urcusique and Umbria, Río Uchupayaco, 300 m, 22–23 Feb 1942, *R. Schultes 3291* (holotype, BH, n.v.).

Stems 2–8 m tall, 1–3 cm diam, cespitose or appearing solitary, erect or somewhat procumbent and then rooting, sometimes with shoots, with roots often visible. *Leaves* 5–11; sheath 20–50 cm long; petiole 15–28 cm long; rachis 60–102 cm long; pinnae 3–27 per side, narrowly to broadly linear, the middle ones 30–41 cm long, unequally wide, spreading in 1 plane, or occasionally leaf entire. *Inflorescences* infrafoliar; peduncle 3–10 cm long; prophyll 8–22 cm long; peduncular bract 30–41 cm long, prophyll and peduncular bract deciduous; rachis 5–10 cm long; rachillae 3–31, 10–33 cm long; *flowers* borne in triads proximally on the rachillae, staminate at apex; staminate flowers 6 mm long (at anthesis); sepals 2.5 mm long; petals lanceolate, 4 mm long; pistillode 1 mm long; pistillate flowers 2.5 mm long at anthesis; sepals 1 mm long; petals 2 mm long; staminodes 6, digitate; *fruits* ellipsoid to ovoid (immature fruits elongate), 1–1.3 cm long, 0.5–1.2 cm diam, black at maturity.

Central America (Costa Rica, Panama) south through Colombia (Amazonas, Antioquia, Caquetá, Chocó, Cundinamarca, Meta, Putumayo, Valle del Cauca, Vaupés), Venezuela (Amazonas, Aragua, Bolívar, Lara, Tachira, Yaracuy, Zulia), the Guianas, Ecuador (Morona-Santiago, Napo, Pastaza, Santiago-Zamora), Peru (Amazonas, Cuzco, Huánuco, Junín, Loreto, Madre de Dios, Pasco, San Martín), Brazil (Acre, Amapá, Amazonas, Pará), and Bolivia (Cochabamba, Santa Cruz) (Figure 6.19D); in a variety of habitats. In the central Amazon region, it grows at low elevations in forest, in wet areas near small streams; it is seldom found on *terra firme*. In the Andes, it grows in montane forests up to 2000 m elevation on very well-drained slopes.

Colombia: *choó-no-hee* (Barasana), *ñai-cü-r(ü)* (Huitoto), *ña-k-r* (Huitoto), *palmita*. Ecuador: *chontilla de llana-muncu, de-de-hueoco, de-rehue'co* (Siona), *hoja de llana-muncu, kunkupij* (Achuar), *palma de tintas*. Peru: *chellochellpan* (Ameshua), *marrashemapar* (Ameshua), *ñejilla, palmiche, ponilla, saápap, sápap* (Mayna Jívaro), *saupak*. Venezuela: *cutata fedié, san pablo, utata-jididi*. The youngest leaves are chewed by various Indian groups in Colombia, Ecuador, and Peru to protect the teeth against decay. In Peru, the leaves are used as an infusion (Ameshua).

An extremely variable species that has been well documented by Skov and Balslev (1989).

ARECOIDEAE • COCOEAE • BUTIINAE

20. *Syagrus*

Syagrus Mart., Palm. fam. 18. 1824.

Small to large, monoecious palms. *Stems* solitary or cespitose, erect, tall and aerial or less often short and subterranean. *Leaves* pinnate, reduplicate; sheath not forming a crownshaft; petiole moderate; rachis moderate to long; pinnae regularly or irregularly arranged, spreading in 1 or in different planes, or occasionally leaf entire. *Inflorescences* interfoliar, branched to 1 order or rarely spicate; peduncle bearing a prophyll and 1 woody, sulcate peduncular bract; rachis bearing (1–)–few–numerous rachillae; *flowers* borne in triads proximally on rachillae, staminate distally; staminate flowers with sepals 3; petals 3, free, valvate; stamens 6; pistillode small or rarely absent; pistillate flowers with sepals 3, free, imbricate; petals 3, free, imbricate or valvate; staminodial ring present; gynoecium syncarpous, trilocular, triovulate; *fruits* 1-seeded, globose, ovoid, pyriform, or ellipsoid, with apical stigmatic residue; endocarp thick and bony, with 3 basal pores, the cavity rounded or triangular in cross section; seed with homogeneous or ruminate endosperm and basal embryo; germination adjacent-ligular or remote-ligular; eophyll entire.

Glassman (1987) recognized 29 species of *Syagrus* and one each of *Arecastrum, Arikuryroba, Barbosa, Chrysallidosperma,* and *Rhyticocos.* These last five genera were included in the former by Uhl and Dransfield (1987). Even though 34 species are thus recognized, the genus still seems overdescribed. The species are widely distributed from Colombia and Venezuela south to Argentina and east across Brazil, with a concentration in drier regions of central and eastern Brazil. One species reaches the Lesser Antilles. Eight species are found in the Amazon region, and in general they occur all around the margins. It is also possible that *S. flexuosa* just enters the southern part of the region. One specimen from Rondônia in Brazil (*M. Nee 35025*) is considered a possible distinct species (L. Noblick, personal communication).

Syagrus is the terminal genus of a trans-Pacific track formed by genera in the Butiinae. This begins with *Jubaeopsis* in southern Africa, *Voanioala* in Madagascar, *Cocos* in the Pacific, *Jubaea* in Chile, and *Parajubaea* in Bolivia, terminating with the widespread and speciose genus *Syagrus.*

KEY TO THE SPECIES OF *SYAGRUS*

1. Stems short and subterranean; leaves few, 2–7; inflorescences spicate, seldom branched with 2–few rachillae; Brazil (Maranhão, Mato Grosso, Pará, Rondônia) and Bolivia (Santa Cruz) . . . . 5. *S. petraea.*
1. Stems tall and aerial; leaves numerous, 5–22; inflorescences branched with (1–)2–35(–200) rachillae.
    2. Large palms with stems 7–20 m tall and 19–30 cm diam; pinnae 122–170 per side; rachillae (62–)100–200; Colombia (Caquetá, Meta, Putumayo), Venezuela (Bolívar), Peru (Loreto, Madre de Dios, San Martín, Ucayali), Brazil (Acre, southwestern Amazonas), and Bolivia (Beni, La Paz, Santa Cruz) . . . . 6. *S. sancona.*
    2. Smaller palms with stems 1–15 m tall and 6–15 cm diam; pinnae 55–110 per side; rachillae (1–)2–35.
        3. Fruits 6–8 cm long, 3–4 cm diam; endocarp cavity triangular in cross section; endosperm ruminate; Colombia (Amazonas), Peru (Amazonas, Loreto, Ucayali), and Brazil (Acre, western Amazonas). . . . . . 7. *S. smithii.*

3. Fruits 2–5.5 cm long, 1–2.5(–4) cm diam; endocarp cavity rounded, rarely triangular in cross section; endosperm homogeneous.
4. Rachis 0–5 cm long with (1–)2–10(–17) rachillae; pinnae lanceolate or linear-lanceolate, coriaceous, usually with prominent ramenta abaxially; southern Amazon region in Brazil (Mato Grosso, Maranhão, Pará, Tocantins). . . . . 2. *S. comosa.*
4. Rachis 6–50 cm long with 6–35 rachillae; pinnae linear, not coriaceous, usually without prominent ramenta abaxially.
5. Pinnae 1–1.5(–2.5) cm wide; rachillae 4–15; fruits ovoid, elongate, "pear-shaped"; Brazil (Amazonas, Maranhão, Mato Grosso, Pará, Tocantins). . . . . 1. *S. cocoides.*
5. Pinnae 2–3.5 cm wide; rachillae 6–35; fruits ellipsoid, oblong ellipsoid, or subglobose.
6. Fruits subglobose, 3.5–4 cm diam; epicarp striate longitudinally; inflorescence rachis 48–50 cm long; Surinam and French Guiana. . . . . 8. *S. stratincola.*
6. Fruits ellipsoid or oblong ellipsoid, 2–3 cm diam; epicarp not striate; inflorescence rachis 6–30 cm long.
7. Endocarp cavity triangular in cross section; styles slightly or not tomentose; the Guianas and Brazil (Amapá, Amazonas, Maranhão, Pará) . . . . 3. *S. inajai.*
7. Endocarp cavity rounded in cross section; styles densely tomentose; Colombia (Casanare, Meta, Vaupés, Vichada) and Venezuela (Amazonas, Apure, Bolívar). . . . . 4. *S. orinocensis.*

1. Syagrus cocoides Mart., Hist. nat. palm. 2: 130. 1826. *Cocos syagrus* Drude in Mart., Fl. bras.: Cyclanthaceae et Palmae I, fasc. 85, vol. 3(2): 406. 1881. Type. Brazil. Pará: Almeirim, n.d., *C. Martius s.n.* (holotype, M; F neg. 18564) (Figure 6.21a–c).
Syagrus cocoides var. *linearifolia* Barb. Rodr., Enum. palm. nov. 40. 1875. *Cocos syagrus* var. *linearifolia* Barb. Rodr. , Sert. palm. brasil. 1: 103. Lectotype (Glassman, 1987). Barb. Rodr., Sert. palm. brasil. 1: t. 73b. 1903.
Syagrus brachyrhyncha Burret, Notizbl. Bot. Gart. Berlin-Dahlem 13: 686. 1937. Type. Brazil. Pará: Rio Maycurú, Monte Alegre, n.d., *G. Huebner 112* (holotype, B).

Stems solitary or possibly cespitose, 1.3–7(–9) m tall, 6–9 cm diam, covered apically with persistent leaf bases. *Leaves* 14–22; sheath 15–40 cm long, fibrous on the margins; petiole 28–60 cm long; rachis 0.8–1.7(–3) m long; pinnae 55–100 per side, linear, aristate, irregularly arranged in clusters of 2–6, spreading in different planes, the middle ones 35–63 cm long, 1–1.5(–2.5) cm wide, with obscure or prominent cross-veins. *Inflorescences* interfoliar, erect, occasionally all staminate and then with slender rachillae; peduncle 0.8–1.4 m long; prophyll to 35(–100) cm long; peduncular bract 1–1.6 m long, somewhat sulcate; rachis 8–21 cm long; rachillae 4–15, 24–27 cm long; *flowers* in triads on proximal part of rachillae, staminate distally; staminate flowers 1.1–1.6 cm long; sepals briefly connate basally, narrowly triangular, 2 mm long; petals lanceolate, 1.1–1.6 cm long; pistillode small; pistillate flowers 0.8–1.2 cm long, pyramidal; sepals triangular, 6–8 mm long; petals free, valvate, narrowly triangular, 0.7–1.2 cm long; staminodial ring 2 mm high; ovary tomentose; *fruits* ovoid, elongate ("pear-shaped"), 3.5–5 cm long, 2–2.5 cm diam, yellowish-brown; endocarp cavity rounded in cross section; endosperm homogeneous.

Central Brazil (Amazonas, Ceará, Goiás, Maranhão, Mato Grosso, Pará, Tocantins) (Figure 6.19E); usually in open, rocky areas (*cerrado*) at elevations to 500 m, but in the Amazon region it grows in lowland rain forest, gallery forest near rapids, or naturally occurring open areas.

Brazil: *ariry, iriri, piririma, pati, pupunha brava, vo-ti* (Apinajé). The stems are used to make bows and in construction; the seeds are edible.

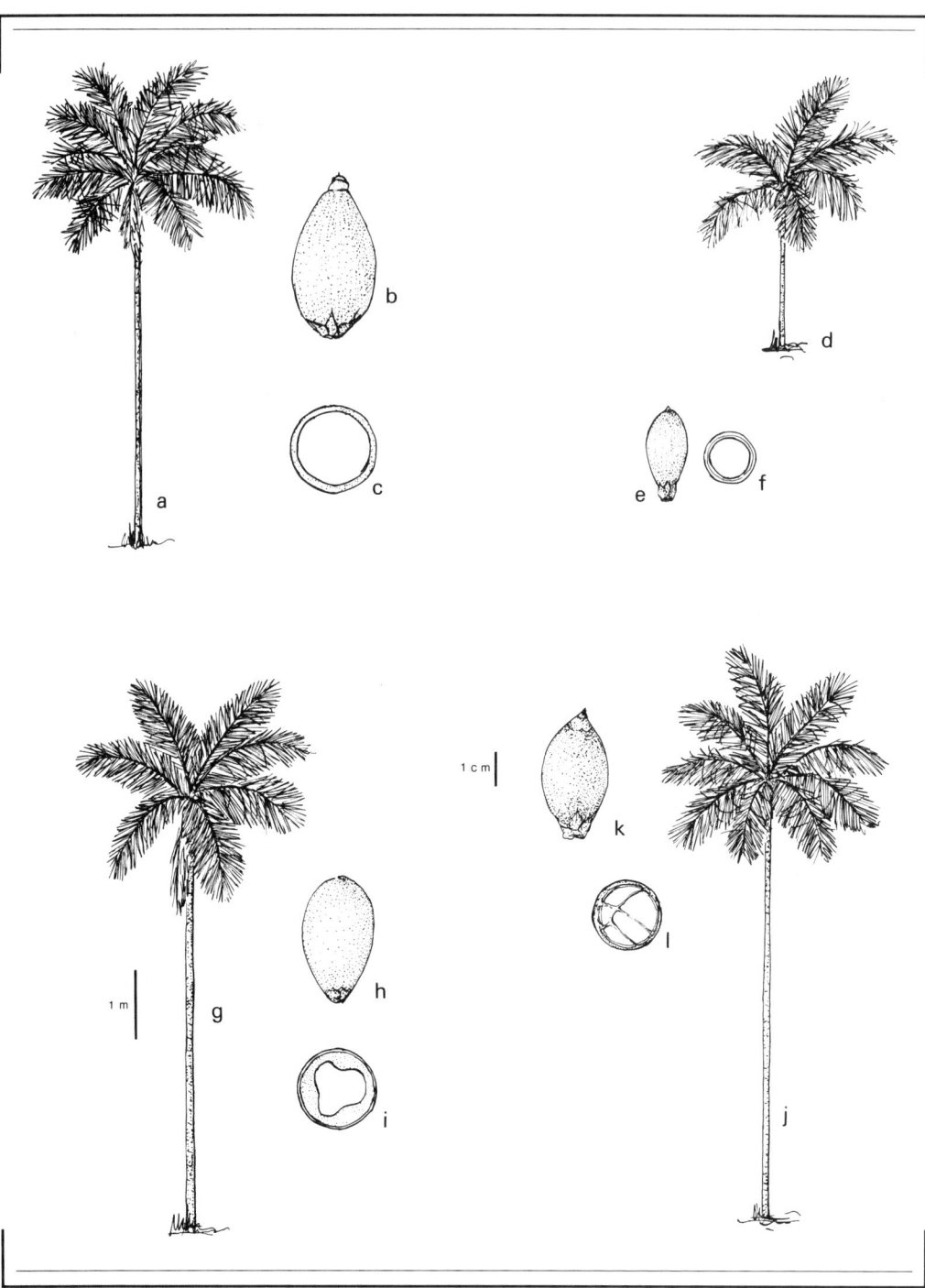

Figure 6.21. (a) *Syagrus cocoides,* (b) fruit, (c) cross section of fruit (b and c from *C. Cid 6413*); (d) *Syagrus comosa,* (e) fruit, (f) cross section of fruit (e and f from *L. Noblick 4658*); (g) *Syagrus inajai,* (h) fruit, (i) cross section of fruit (h and i from *M. Balick 909*); (j) *Syagrus orinocensis,* (k) fruit, (l) cross section of fruit (k and l from *B. Boom 6616*).

2. **Syagrus comosa** (Mart.) Mart. in A. D. Orb., Voy. Amérique mér. 7(3). Palmiers 134. 1847. *Cocos comosa* Mart., Hist. nat. palm. 2: 122. 1826. Type. Brazil. Goiás: without locality, n.d., *C. Martius s.n.* (holotype, M; F neg. 18562) (Figure 6.21d–f).

Stems solitary or occasionally cespitose, 1–3 (–5) m tall, or occasionally stem short and subterranean, 7–11.5 cm diam, rough with old leaf scars. *Leaves* 6–12; sheath 12–33 cm long; petiole 10–27 cm long; rachis 0.7–1.2 m long; pinnae 57–73 per side, lanceolate or linear-lanceolate, asymmetrically bifid at apex, ± coriaceous, irregularly arranged in clusters and spreading in different planes, the middle ones 27–51 cm long, 2–3 cm wide, with prominent cross-veins, often with prominent ramenta abaxially. *Inflorescences* interfoliar; peduncle 30–52 cm long; prophyll 26–35 cm long; peduncular bract 78–90 cm long; rachis absent or very short to 5 cm long; rachillae (1–)2–10(–17), 15–25 cm long; *flowers* in triads on proximal part of rachillae, paired or solitary staminate distally; staminate flowers 0.8–1 cm long; sepals imbricate, deltate, 1 mm long; petals obovate, 8 mm long; pistillode obscure; pistillate flowers 6–7 mm long, rounded, not pyramidal; sepals deltate, 5–6 mm long; petals free, imbricate basally, valvate at tips, deltate, 6 mm long; staminodial ring 3 mm high; ovary glabrous; *fruits* ellipsoid oblong, 2–3 cm long, 1–1.5 cm diam, greenish-brown; endocarp cavity rounded in cross section; endosperm homogeneous.

Central-eastern Brazil (Bahia, Goiás, Mato Grosso, Maranhão, Minas Gerais, Pará, Piauí, Tocantins) (Figure 6.19F); open areas (*cerrado*), especially on rocky slopes, to 1200 m elevation.

Brazil: *arandacê* (Karajá), *gabiroba catolé, pati*.

3. **Syagrus inajai** (Spruce) Becc., Agric. Colon. 10: 467. 1916. *Maximiliana inajai* Spruce, J. Linn. Soc., Bot. 11: 163. 1871. *Cocos inajai* (Spruce) Trail, J. Bot. 6: 79. 1877. *Cocos aequatorialis* Barb. Rodr., Enum. palm. nov. 38. 1875. *Syagrus aequatorialis* (Barb. Rodr.) Barb. Rodr., Protesto-Appendice 33. 1879. Type. Brazil. Amazonas: Rio Negro, n.d., *R. Spruce 83* (holotype, K, n.v.) (Figure 6.21g–i).

Cocos speciosa Barb. Rodr., Enum. palm. nov. 38. 1875. *Syagrus speciosa* (Barb. Rodr.) Barb. Rodr., Protesto-Appendice 49. 1879. Lectotype (Glassman, 1972). Barb. Rodr., Contr. Jard. Bot. Rio de Janeiro 4: t. 24b. 1907.

Cocos chavesiana Barb. Rodr. ex Becc., Malpighia 1: 445. 1887. *Syagrus chavesiana* (Barb. Rodr.) Barb. Rodr., Vellosia 1: 52. 1888. Type. Brazil. Amazonas: Manaus, n.d., *J. Barbosa Rodrigues s.n.* (holotype, FI, n.v.).

Stems solitary or possibly cespitose, 3–15 m tall, 4–15 cm diam. *Leaves* 8–15; sheath 15–75 cm long, fibrous at the margins; petiole 33–100 cm long, abaxial surface of petiole and rachis brown-tomentose; rachis 1.7–3.3 m long; pinnae 51–110 per side, irregularly arranged in somewhat loose clusters of 2–7, spreading in different planes, linear, apex acuminate, the middle ones 50–75 cm long, 2.5–3.5 cm wide, with prominent cross-veins, without ramenta abaxially, young plants with very distinctive, long, narrow, entire leaves. *Inflorescences* interfoliar; peduncle 40–72 cm long; prophyll 20–55 cm long; peduncular bract 75–115 cm long; rachis 6–30 cm long; rachillae 6–35, 10–40 cm long, with a sterile basal part to 10 cm long; *flowers* arranged in triads for most of the rachillae, staminate only distally; staminate flowers 7–8 mm long; sepals connate into a 3-lobed calyx, the lobes 1 mm long; petals oblanceolate-obovate, 6–7 mm long; pistillode trifid, 0.5 mm long; pistillate flowers 6.5–8 mm long (at anthesis), rounded, not pyramidal; sepals strongly imbricate, deltate, 5–7 mm long; petals strongly imbricate, briefly valvate apically, deltate, 5–7 mm long; staminodial ring 2–3 mm high; style slightly or not tomentose; *fruits* ellipsoid, 3.2–4.5(–5.5) cm long, 2–3 cm diam, greenish-brown; endocarp cavity ± triangular in cross section; endosperm homogeneous.

The Guianas and Brazil (Amapá, Amazonas, Maranhão, Pará) (Figure 6.22A); lowland rain forest and gallery forest on *terra firme* at low elevations.

Brazil: *jarevá, marark'y* (Ka'apor), *pirima, piririma, pupunharana, pupunha-brava*. French Guiana: *feuille chasseur*. Surinam: *pëpë* (Trio).

Syagrus inajai is closely related to *S. orinocensis*, and they are sister species, or even the

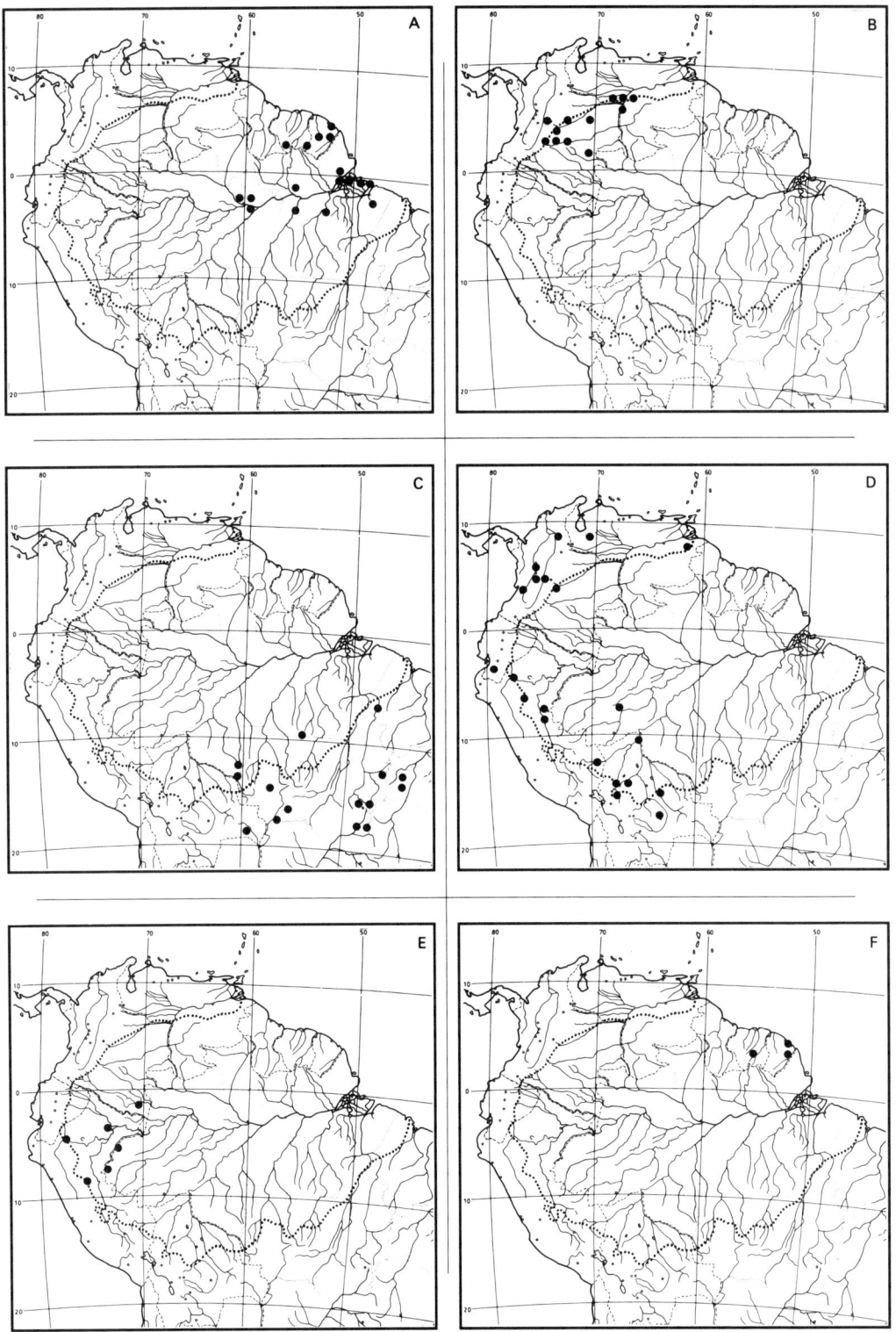

Figure 6.22. **(A)** *Syagrus inajai;* **(B)** *Syagrus orinocensis;* **(C)** *Syagrus petraea;* **(D)** *Syagrus sancona;* **(E)** *Syagrus smithii;* **(F)** *Syagrus stratincola.*

same species. They differ in minor details (see key) and in their ranges. *Syagrus inajai* is another example of a species with a northeastern Amazon region distribution. One specimen from near Manaus (*A. Henderson 634*) is somewhat distinct (L. Noblick, personal communication).

4. **Syagrus orinocensis** (Spruce) Burret, Notizbl. Bot. Gart. Berlin-Dahlem 13: 695. 1937. *Cocos orinocensis* Spruce, J. Linn. Soc., Bot. 11: 161. 1871. Type. Venezuela. Amazonas: Maipurés, Jan 1854, *R. Spruce 49* (holotype, K) (Figure 6.21j–l).
Syagrus allenii Glassman, Fieldiana Bot. 31: 285. 1968. Type. Colombia. Meta: Llanos de San Martín, 21 Oct 1945, *P. Allen 3352* (holotype, COL; isotype, BH).

Stems solitary or cespitose, 1–12 m tall, 8–12 cm diam. *Leaves* 8–12; sheath to 20 cm long; petiole to 50 cm long; rachis 1.6–2.5 m long; pinnae 70–104 per side, irregularly arranged in clusters of 2–4 and spreading in different planes, linear, acute at apex, the middle ones 44–60 cm long, 1.5–3 cm wide, without prominent cross-veins, with ramenta abaxially. *Inflorescences* interfoliar; peduncle 35 cm or more long; prophyll 20–22 cm long; peduncular bract 80–100 cm long; rachis 15–27 cm long; rachillae 23–25, 26–33 cm long; *flowers* in triads on proximal half of rachillae, staminate distally; staminate flowers 0.8–1 cm long; sepals imbricate basally, triangular, 1–1.5 mm long; petals 8–9 mm long; pistillode trifid, very small; pistillate flowers 6.5–8 mm long, rounded, not pyramidal; sepals free, imbricate, deltate, 6–7 mm long; petals free, imbricate basally, valvate apically, triangular, 6–7 mm long; staminodial ring 1 mm high; style densely tomentose; *fruits* oblong-ellipsoid, 3–4 cm long, 2–3 cm diam, brownish, tomentose apically; endocarp rounded in cross section; endosperm homogeneous.

Colombia (Casanare, Meta, Vaupés, Vichada) and Venezuela (Amazonas, Apure, Bolívar) (Figure 6.22B); on granite outcrops in or near rivers, alluvial terraces, savanna margins, generally in places where the soil is very thin or absent, or in forests or gallery forests, at low elevations.

Colombia: *churruái, churubay, churrguay, oró-boto* (Guahibo), *pupunha silvestre.* Venezuela: *cocito, coquito, kopayan* (Panare).

Closely related to *Syagrus inajai*; see discussion under that species.

5. **Syagrus petraea** (Mart.) Becc., Agric. Colon. 10: 467. 1916. *Cocos petraea* Mart. in A. D. Orb., Voy. Amérique mér. 7(3). Palmiers 100. 1844. Type. Bolivia. Santa Cruz: Sierra de Santiago, n.d., *A. d'Orbigny 21* (holotype, M) (Figure 6.23a–c).

Stems solitary or cespitose, short and subterranean. *Leaves* 2–7; sheath 10–18 cm long, fibrous; petiole 6–39 cm long; rachis 50–77 (–130) cm long; pinnae 13–37 per side, linear, asymmetrically bifid apically, irregularly arranged in loose or tight clusters of 2–4, spreading in different planes, the middle ones 16–19(–69) cm long, 0.5–1(–1.9) cm wide, often whitish-tomentose abaxially. *Inflorescences* interfoliar, spicate (rarely with 2–few rachillae); peduncle 15–32 cm long; prophyll 8–13 cm long; peduncular bract (10–)30–45 cm long; rachilla(e) 8–17 cm long; *flowers* borne in triads on proximal part of rachilla, staminate distally; staminate flowers 7–11 mm long; sepals connate into a 3-lobed calyx, 1 mm long; petals free, valvate, obovate, 6 mm long; pistillode 0.2 mm long; pistillate flowers 9–10 mm long (in bud); sepals free, imbricate, narrowly triangular, 8–10 mm long; petals free, imbricate basally for ca. half their length, valvate above, narrowly triangular, 7–8 mm long; staminodial ring 2 mm long; *fruits* ± ellipsoid, 2–2.5 cm long, 1–1.5 cm diam, brown, scurfy-tomentose; endocarp cavity rounded in cross section; endosperm homogeneous.

Brazil (Bahia, Goiás, Maranhão, Mato Grosso, Pará, Piauí, Rondônia), Bolivia (Santa Cruz), and Paraguay (Amambay) (Figure 6.22C); drier, *cerrado* regions in open, grassy areas among rocks, or in low, open forest on sandy soils, to 1000 m elevation.

Brazil: *tucum de indio.*

6. **Syagrus sancona** H. Karst., Linnaea 28: 247. 1857. *Cocos sancona* (H. Karst.) Hook. f., Kew Report 1882: 72. 1884. Lectotype (Glassman, 1987). Colombia. Cundinamarca: Bogotá, n.d., *H. Karsten s.n.* (holotype, LE, n.v.) (Figure 6.23d–f).

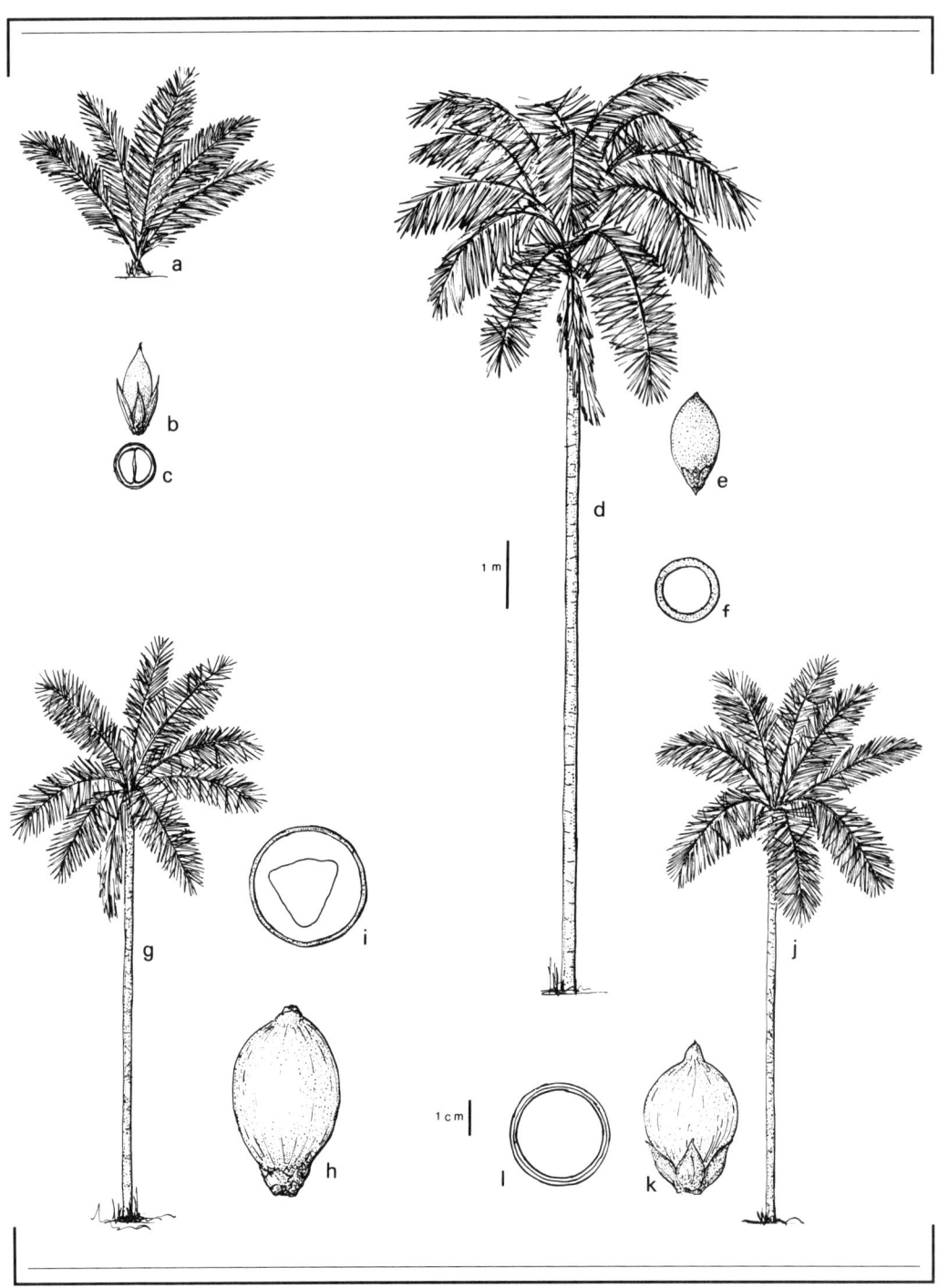

Figure 6.23. (a) *Syagrus petraea,* (b) fruit, (c) cross section of fruit (b and c from *M. Balick 914*); (d) *Syagrus sancona,* (e) fruit, (f) cross section of fruit (e and f from *F. Kahn 2129*); (g) *Syagrus smithii,* (h) fruit, (i) cross section of fruit (h and i from *A. Henderson 1117*); (j) *Syagrus stratincola,* (k) fruit, (l) cross section of fruit (k and l from *J. Wessels Boer 1303*).

Cocos purusana Huber, Bull. Herb. Boissier 6: 271. 1906. *Syagrus purusana* (Huber) Frambach ex Dahlgren, Field Mus. Nat. Hist., Bot. Ser. 14: 268. 1936. Type. Not designated.

Syagrus tessmannii Burret, Repert. Spec. Nov. Regni Veg. Beih. 32: 106. 1933. Type. Peru. Loreto: Río Marañon, near Apaga, 1924, *G. Tessmann 4811* (holotype, B).

Stems solitary, rarely cespitose, 7–20 m tall, 19–30 cm diam. *Leaves* 8–16; sheath and petiole 75–190(–250) cm long, fibrous at margins; rachis 2.1–3 m long; pinnae 122–170 per side, linear, with acute apex, irregularly arranged in clusters of 2–7, spreading in different planes, the middle ones 60–85 cm long, 1.5–4.5 cm wide, cross-veins obscure or visible. *Inflorescences* interfoliar; peduncle 25–45 cm long; prophyll 50–70 cm long; peduncular bract 1–1.5 m long; rachis 46–80 cm long; rachillae (62–)100–200, 15–62 cm long; *flowers* in triads proximally on rachillae, staminate distally, occasionally an inflorescence all staminate; staminate flowers 10 mm long; sepals connate into a 3-lobed calyx, the lobes 1 mm long; petals oblong, 9 mm long; pistillode very small, trifid; pistillate flowers 7–9 mm long, rounded, not pyramidal; sepals deltate, 7 mm long; petals imbricate, deltate, 6 mm long; staminodial ring 1 mm high; styles glabrous; *fruits* ellipsoid, 2.8–3.2 cm long, 1.5–2.3 cm diam; yellowish or orangeish, endocarp rounded in cross section; endosperm homogeneous.

Andean and adjacent regions of Colombia (Caquetá, Meta, Norte de Santander, Putumayo, Risaralda, Valle del Cauca), Venezuela (Barinas, Bolívar), Ecuador (El Oro), Peru (Loreto, Madre de Dios, San Martín, Ucayali), Brazil (Acre, southwestern Amazonas), and Bolivia (Beni, La Paz, Santa Cruz) (Figure 6.22D); lowland or premontane rain forest, or commonly in disturbed areas, on slopes at elevations up to 1200 m, and in gallery forests or river margins at lower elevations. It is occasionally planted as an ornamental. It is a rare palm in the Amazon region, occurring sporadically in the southwestern part. It was considered endangered by Moore (1977). Its distribution is very similar to that of *Aiphanes aculeata* and *Bactris macana*; from the Venezuelan Andes along eastern Andean slopes to Bolivia, except in Ecuador, where it is found on western Andean slopes (and where *Aiphanes eggersii* replaces *A. aculeata,* and *B. macana* is absent). This may be an anthropogenic distribution pattern, since all three palms are useful ones.

Bolivia: *sumuke, sumuqué.* Brazil: *açairana, jaciarana.*

7. Syagrus smithii (H. E. Moore) Glassman, Fieldiana Bot. 32: 231. 1970. *Chrysalidosperma smithii* H. E. Moore, Principes 7: 109. 1963. Type. Peru. Loreto: Alto Amazonas, km 13–14 on Yurimaguas–Tarapoto road, 24 May 1960, *H. Moore et al. 8516* (holotype, BH) (Figure 6.23g–i).

Stems solitary, 4–15 m tall, 5–8 cm diam. *Leaves* 5–18; sheath and petiole 45–90 cm long, fibrous on margins; rachis 1.9–3.6 m long; pinnae 83–94 per side, irregularly arranged in clusters of 2–5 and spreading in different planes, linear, with acute apex, the middle ones 26–84 cm long, 3–3.5 cm wide, or commonly leaf entire (as in juveniles), with prominent cross-veins. *Inflorescences* interfoliar; peduncle 63–100 cm long; prophyll to 30 cm long; peduncular bract 1–1.2 m long; rachis 12–35 cm long; rachillae 9–31, to 30 cm long; *flowers* borne in triads on proximal part of rachillae, paired or solitary staminate distally; staminate flowers 1 cm long; sepals briefly imbricate basally, narrowly triangular, 1 m long; petals valvate, lanceolate, 10 mm long; pistillate flowers 1.2–1.3 cm long (postanthesis), rounded, not pyramidal; sepals broadly imbricate, shallowly deltate, 9–10 mm long; petals broadly imbricate, briefly valvate at the apex, shallowly deltate, 10 mm long; staminodial tube 4 mm long; style tomentose; *fruits* ellipsoid-oblong, 6–8 cm long, 3–4 cm diam, greenish-brown or yellowish; endocarp cavity triangular in cross section; endosperm ruminate.

Colombia (Amazonas), Peru (Amazonas, Loreto, Ucayali), and Brazil (Acre, western Amazonas) (Figure 6.22E); lowland rain forest on *terra firme* at low elevations.

Brazil: *catolé.* Colombia: *coco, toókee* (Miraña). Peru: *kuík.* The leaves are used for thatching temporary shelters, and the seeds are eaten.

8. Syagrus stratincola Wess. Boer, Flora of Suriname 5(1): 170. Type. Surinam. Upper Marowijne Region, Paloemeu River, 19 Apr 1963, *J. Wessels Boer 1303* (holotype, U; isotype, NY) (Figure 6.23j–l).

Stems cespitose and forming clumps of 15–20 stems, rarely solitary, 2–14 m tall, 5–10 cm diam. *Leaves* 6–12; sheath 15–30 cm long, fibrous on margins; petiole 0.4–1.5 m long; rachis 1.5–2.5 m long; pinnae 57–82 per side, linear or linear-lanceolate, with acute apex, irregularly arranged in clusters of 3–4, spreading in different planes, the middle ones 38–70 cm long, 2–3 cm wide, with ramenta on veins abaxially, cross-veins prominent. *Inflorescences* interfoliar; peduncle ca. 70 cm long; prophyll 20–30 cm long; peduncular bract 0.6–1 m long; rachis 48–50 cm long; rachillae 9–15, 20–26 cm long, undulate; *flowers* in triads on proximal part of rachillae, staminate distally; staminate flowers 1.7–1.9 cm long; sepals triangular, 3–5 mm long; petals ovate-lanceolate, valvate, to 1.9 cm long; pistillate flowers to 2.5 cm long, pyramidal; sepals imbricate, ca. 2.5 cm long; petals imbricate at base, valvate apically, triangular, ca. 1.5 cm long; staminodial ring small; *fruits* subglobose, 4–4.5 cm long, 3.5–4 cm diam; epicarp striate longitudinally; endocarp cavity rounded in cross section; endosperm homogeneous.

Surinam and French Guiana, and probably Brazil (Amapá) (Figure 6.22F); lowland rain forest, on granite rocks (inselbergs) or savanna margins, at low elevations.

ARECOIDEAE • COCOEAE • ATTALEINAE

21. *Attalea*   **Attalea** Kunth in Humb., Bonpl. & Kunth, Nov. gen. sp. 1: 309. 1815.
Maximiliana Mart., Palm. fam. 20. 1824.
Orbignya Mart. ex Endl., Gen. pl. 4: 257. 1837.
Scheelea H. Karst., Linnaea 28: 264. 1857.

Small to large, monoecious palms. *Stems* solitary or rarely cespitose, short and subterranean or tall and aerial. *Leaves* pinnate, reduplicate; sheath open and not forming a crownshaft; petiole short or absent to elongate; rachis long; pinnae regularly or irregularly arranged, spreading in 1 or different planes, linear, usually aristate apically but with the main vein ending subterminally, sometimes with an auricle (a short extension of the pinnae margin over the abaxial surface of the rachis), with a thin line of brown tomentum on distal, abaxial margin, this widening at pinnae apex. *Inflorescences* interfoliar, branched to 1 order, either all staminate, staminate and pistillate, or predominantly pistillate, all occurring on same plant; peduncle bearing a prophyll and 1 woody, sulcate peduncular bract; rachis bearing numerous rachillae, these tending to be absent from adaxial surface of rachis, occasionally the pistillate rachillae very short and the inflorescence appearing spicate; *flowers* borne in triads or paired or solitary staminate, tending to be absent from adaxial surface of rachillae; staminate flowers with sepals 3(–4); petals (1–)3(–5), linear or flattened; stamens 6–75, straight or coiled and twisted; pistillode small; pistillate flowers with sepals 3, free, imbricate; petals 3, free, imbricate, mucronate; staminodial ring present; gynoecium syncarpous, 3–several–loculate, 3–several-ovulate; *fruits* 1–several-seeded, globose, ovoid, ellipsoid-oblong or oblong-ovoid with apical stigmatic residue; endocarp thick and bony, with or without internal fibers; endocarp pores subbasal, sunken or sometimes superficial, operculate (with a cover); seeds with homogeneous endosperm and basal embryo; germination remote-tubular; eophyll entire.

A genus of approximately 27 species widely distributed from southern Mexico through Central America and throughout tropical South America, reaching east to Trinidad and south to southern Brazil and Paraguay; one species in Hispaniola. Fourteen species occur in the Amazon region. There is no recent revision, but see Glassman (1977a, 1977b, 1977c, 1978).

Although Uhl and Dransfield (1987) maintained the Attaleinae as consisting of four distinct genera *(Attalea, Maximiliana, Orbignya, Scheelea),* this division is not followed here, and only one genus, *Attalea,* is recognized (see Henderson & Balick, 1991). This is a natural genus, but due to the often large size, species have been poorly collected and misunderstood, nomenclaturally and biologically. There are still many problems in the genus. It is economically important, and is greatly in need of a modern revision. Such revision will be accomplished, however, only on the basis of extensive field work. Wessels Boer (1965) was the first to begin to understand the species, and his work is mostly followed here.

Attalea consists of at least six subgeneric groupings (Henderson & Balick, 1991), five of which occur in the Amazon region. Of these, "Scheelea" contains three Amazon species, all of which are usually found in the western part of the region (I interpret the northeastern *Attalea attaleoides* as being a "Maximiliana" rather than a "Scheelea"). "Orbignya" are mostly either west or east of the Amazon region, with just two species, *A. microcarpa* and *A. spectabilis,* confined to the region. Two other "Orbignya" are found on the Brazilian Shield. Of the three "Attalea" in the region, all are confined to the western part of the region. "Markleya" is in the northeast; "Parascheelea" is in the upper Rio Negro region; and "Maximiliana," with two species, is widespread.

An interesting phenomenon in the genus is the presence of small, peripheral populations with either large and aerial stems or short and subterranean stems conspecific with or close to widespread species with the other type of stem. Examples are the widespread, short-stemmed *Attalea racemosa* and the restricted, tall-stemmed *A. septuagenata*; the widespread, tall-stemmed *A. speciosa* and the restricted, short-stemmed *A. spectablis*; and the restricted, short-stemmed forms of the widespread, tall-stemmed *A. butyracea.*

KEY TO THE SPECIES OF *ATTALEA*
1. Stems usually large and aerial, 2–20 m tall; pinnae 140–318 per side.
    2. Leaves arranged in a few distinctive spirals; petiole elongate, 2.4–3.3 m long; pinnae strongly clustered and spreading in different planes; fruits 4–6 cm long, 2.5–3 cm diam; stamens greatly exceeding the petals. . . . . 7. *A. maripa.*
    2. Leaves not arranged in a few distinctive spirals; petiole not elongate, 0–2.2 m long; pinnae regularly arranged and spreading in 1 plane, or clustered and spreading in different planes; fruits 4.5–13 cm long, 2.5–7 cm diam; stamens shorter than the petals.
        3. Staminate rachillae tending to be arranged on 1 side of rachis; staminate petals flattened, linear or oblanceolate-obovate; stamens 10–75; fruits 10–13 cm long, 5–7 cm diam.
            4. Pinnae regularly arranged and spreading in 1 plane; stamens 36–75; fruits to 10 cm long, 5 cm diam; Colombia (Amazonas) . . 11. *A. septuagenata.*
            4. Pinnae clustered and spreading in different planes; stamens 10–14; fruits 12.5–13 cm long, 6.5–7 cm diam; Peru (Loreto, Madre de Dios, Ucayali) and Brazil (Acre) . . . . 14. *A. tessmannii.*

3. Staminate rachillae arranged all around rachis; staminate petals linear and not flattened, or flattened and then irregularly oblong or lanceolate; stamens 6–28; fruits 4.5–11 cm long; 2.5–7 cm diam.
    5. Stamens 6; petals linear; anthers straight.
        6. Sheath with thick fibers; petiole absent; staminate flowers 1.3–2 cm long; endocarp fibers scattered; Colombia (Amazonas, Caquetá, Putumayo), Venezuela (Amazonas, Bolívar), Ecuador (Napo), Peru (Huánuco, Loreto, Madre de Dios, Ucayali), western Brazil (Acre, Amazonas), and Bolivia (Cochabamba, La Paz, Pando) . . . . 2. *A. butyracea.*
        6. Sheath with thin fibers; petiole to 2 m long; staminate flowers 7 mm long; endocarp fibers in clusters; Peru (Junín, Loreto, Madre de Dios, Ucayali), Brazil (Acre, Maranhão, Mato Grosso, Pará, Rondônia, Tocantins), and Bolivia (Beni, La Paz, Pando, Santa Cruz) . . . . 9. *A. phalerata.*
    5. Stamens 7–28; petals flattened; anthers coiled and twisted.
        7. Pinnae ± regularly arranged and spreading in ± one plane; staminate flowers with 7–10 coiled stamens; Surinam and Brazil (Pará). . . . . 3. *A. dahlgreniana.*
        7. Pinnae regularly arranged and spreading in 1 plane; staminate flowers with 19–28 irregularly coiled and twisted stamens; the Guianas, Brazil (Acre, Amazonas, Maranhão, Pará, Rondônia, Tocantins), and Bolivia (Beni, Pando, Santa Cruz) . . . . 12. *A. speciosa.*
1. Stems usually short and subterranean, rarely to 2 m tall; pinnae 55–148 per side.
    8. Pinnae regularly arranged and spreading in 1 plane.
        9. Staminate rachillae 3–4 mm diam, densely crowded with flowers; staminate flowers with 6–20, coiled and twisted stamens; pistillate rachillae borne all around rachis.
            10. Stamens 6; staminate petals linear; Colombia (Vaupés), southern Venezuela (Amazonas), and western Brazil (Amazonas) . . . . . 6. *A. luetzelburgii.*
            10. Stamens 9–20; staminate petals flattened; widespread north of the Amazon.
                11. Fruits 3.5–4 cm long, 2–3 cm diam; Colombia (Vaupés), Venezuela (Amazonas), the Guianas, Peru (Loreto), and Brazil (Amapá, Amazonas, Pará). . . . . 8. *A. microcarpa.*
                11. Fruits 5–6 cm long, 3–4 cm diam; Brazil (Pará). . . . . . . 13. *A. spectabilis.*
        9. Staminate rachillae 1–3 mm diam, the flowers usually not densely crowded; staminate flowers with 6 or 11–41 straight stamens; pistillate rachillae absent from 1 side of rachis, or borne all around.
            12. Pistillate rachillae borne all round rachis; fruits ovoid, 4.5–5.5 cm long, 2–2.5 cm diam; staminate flowers with linear petals and 6 stamens; French Guiana, Surinam, and Brazil (Amazonas) . . . . 1. *A. attaleoides.*
            12. Pistillate rachillae absent from 1 side of rachis; fruits ovoid, ellipsoid-oblong or globose, 6–9 cm long, 4.5–5 cm diam; staminate flowers with flattened petals and 19–42 stamens; Colombia (Amazonas, Caquetá, Guainía, Guaviare, Vaupés), Venezuela (Amazonas, Bolívar), Peru (Loreto), and Brazil (Amazonas) . . . 10. *A. racemosa.*
    8. Pinnae irregularly arranged in clusters and spreading in different planes.
        13. Apical pinnae partially split and forming "windows"; staminate flowers with linear petals and 6 stamens, arranged all around rachil-

lae; Colombia (Amazonas, Caquetá, Casanare, Meta, Putumayo, Vaupés), Peru (Loreto), and Brazil (Acre, Amazonas) . . . . 5. *A. insignis.*

13. Apical pinnae completely separate; staminate flowers with very widely ovate petals and 17–20 stamens, arranged on 1 side of rachillae; Brazil (Maranhão, Mato Grosso, Pará, Tocantins) and Bolivia (Santa Cruz) . . . . 4. *A. eichleri.*

1. **Attalea attaleoides** (Barb. Rodr.) Wess. Boer, Flora of Suriname 5(1): 157. 1965. *Maximiliana attaleoides* Barb. Rodr., Enum. palm. nov. 41. 1875. *Englerophoenix attaleoides* (Barb. Rodr.) Barb. Rodr., Sert. palm. brasil. 1: 76. 1903. *Attalea transitiva* Barb. Rodr., Protesto-Appendice 49. 1879. Lectotype (Wess. Boer, 1965). Barb. Rodr., Sert. palm. brasil. 1: t. 60A. 1903 (Figure 6.24a–c).

Stems solitary, short and subterranean. *Leaves* 8–11, erect and forming a funnel that fills with litter; sheath partly subterranean, ca. 30 cm long; petiole 0.4–1 m long, petiole and rachis mottled light-green abaxially; rachis 3.3–5(–6) m long; pinnae 75–110 per side, regularly arranged and spreading in 1 plane, linear, lacking an auricle, the middle ones 60–86 cm long, 2–4 cm wide. *Inflorescences* interfoliar; peduncle and prophyll not seen; peduncular bract to 80 cm long, sulcate on outer surface, with a ca. 22 cm long solid apex; rachis 15–20 cm long; staminate rachillae ca. 90, 5–7 cm long, 1–2 mm diam, borne all around rachis, with scattered, silvery-white scales, pistillate rachillae ca. 1 cm long, borne all around rachis; staminate flowers 10–14 mm long, borne all around rachillae; sepals connate for ca. half their length, spreading above, triangular, 0.6 mm long; petals free, linear, 11–12 mm long; stamens 6, 5–6 mm long, much shorter than the petals; pistillode absent; pistillate flowers 2 cm long, 1–2 per rachilla, absent from abaxial surface of rachillae; sepals ovate, 1.2 cm long; petals ovate, 1.2 cm long; staminodial ring present; *fruits* ovoid with an elongate apex, 4.5–5.5 cm long, 2–2.5 cm diam, brown; endocarp pores superficial; endocarp fibers few or absent; seeds 2–3.

Central and eastern Amazon region of French Guiana, Surinam, and Brazil (Amapá, Amazonas) (Figure 6.25A); lowland rain forest on *terra firme,* rarely to 750 m elevation.

Brazil: *coco palha preta, palha branca, palhera,* *palhera branca.* French Guiana: *macoupi blanc* (Creole).

I have followed Wessels Boer (1965) in regarding the original description's statement on clustered pinnae as being a mistake (although it is possible that Barbosa Rodrigues described a hybrid plant [see following]). Staminate flowers of specimens from near Manaus in Brazil (Amazonas) are identical to those from Surinam and French Guiana. There are, however, collections from the latter country with much shorter staminate flowers, and these are reported to grow in inundated areas (e.g., *J.-J. de Granville 2087, 5566).* These may represent an undescribed species, or they may represent the same kind of variation in staminate flowers illustrated by Wessels Boer (1965, figure 8) for *Attalea maripa.*

This species is similar to *Attalea maripa* in some respects, particularly its staminate flowers and fruits, and both have superficial endocarp pores. I believe that both should be placed in "Maximiliana." An apparent hybrid between the two occurs near Manaus in Brazil (Henderson, personal observation). Indeed, Barbosa Rodrigues changed the name of this species from *A. attaleoides* to *A. transitiva* to reflect this similarity, although the laws of nomenclature prevent this name being used.

2. **Attalea butyracea** (Mutis ex L. f.) Wess. Boer, Pittieria 17: 312. *Scheelea butyracea* (Mutis ex L. f.) H. Karst. ex H. Wendl. in Kerch., Palmiers 256. 1878. *Cocos butyracea* Mutis ex L. f., Suppl. plant. 454. 1781. Type. (Glassman, 1977b). Colombia. Tolima: Ibague, n.d., *J. Mutis s.n.* (holotype, ?) (Figure 6.24g–i).

Attalea cephalotes Poepp. ex Mart. in A. D. Orb., Voy. Amérique mér. 7(3). Palmiers 119. 1844. *Scheelea cephalotes* (Poepp. ex Mart.) H. Karst., Linnaea 28: 269. 1857. Type. Peru. Loreto: Maynas, *E. Poeppig s.n.* (holotype, M).

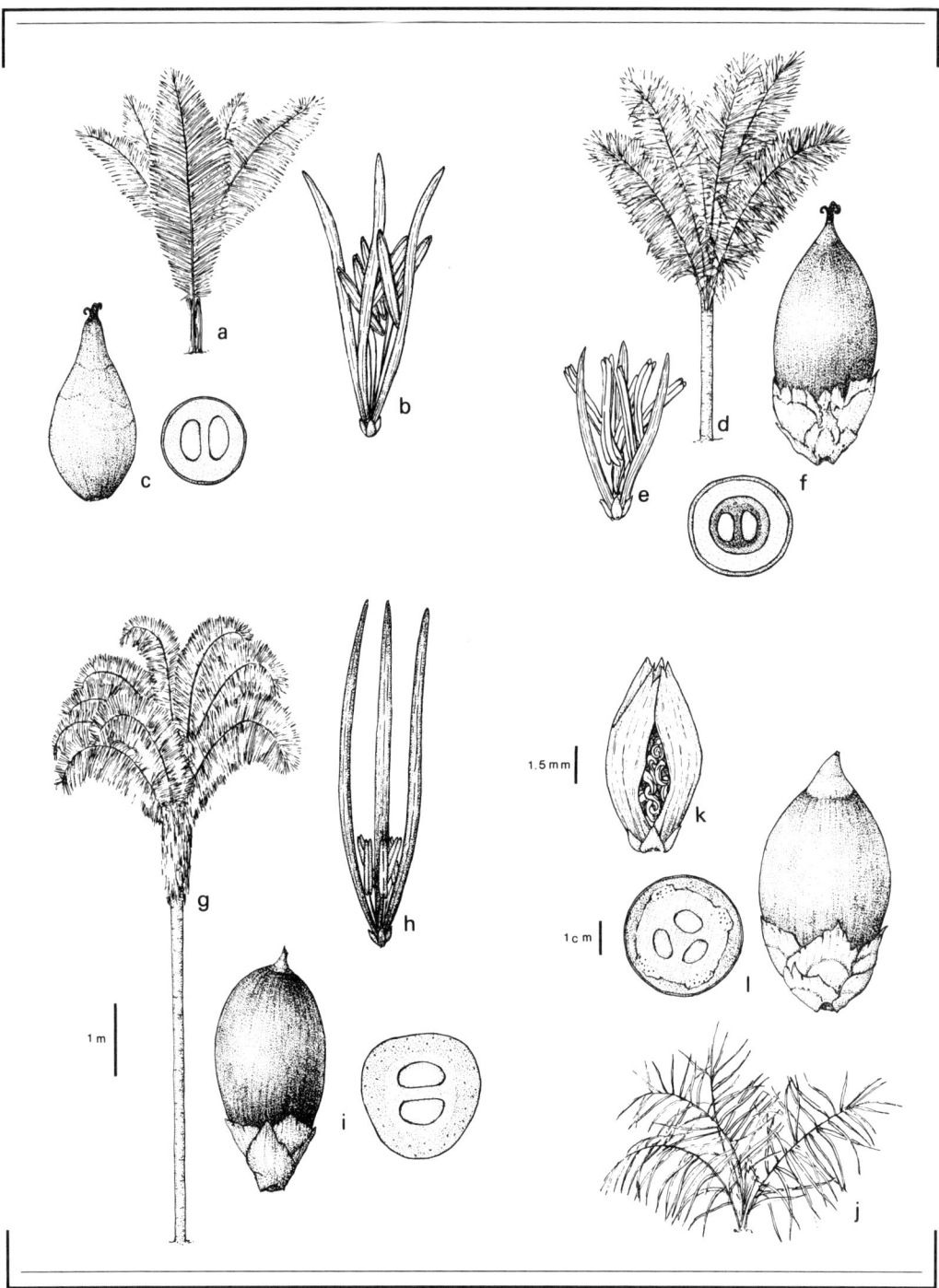

Figure 6.24. (a) *Attalea attaleoides*, (b) staminate flower (from *A. Henderson 656*), (c) fruit and cross section (from *A. Henderson s.n.*); (d) *Attalea dahlgreniana*, (e) staminate flower, (f) fruit and cross section (e and f from *A. Henderson 1734*); (g) *Attalea butyracea*, (h) staminate flower (from *G. Galeano 2068*), (i) fruit and cross section (from *A. Henderson 1644*); (j) *Attalea eichleri*, (k) staminate flower (from *M. Balick 1597*), (l) fruit and cross section (from *A. Henderson 811*).

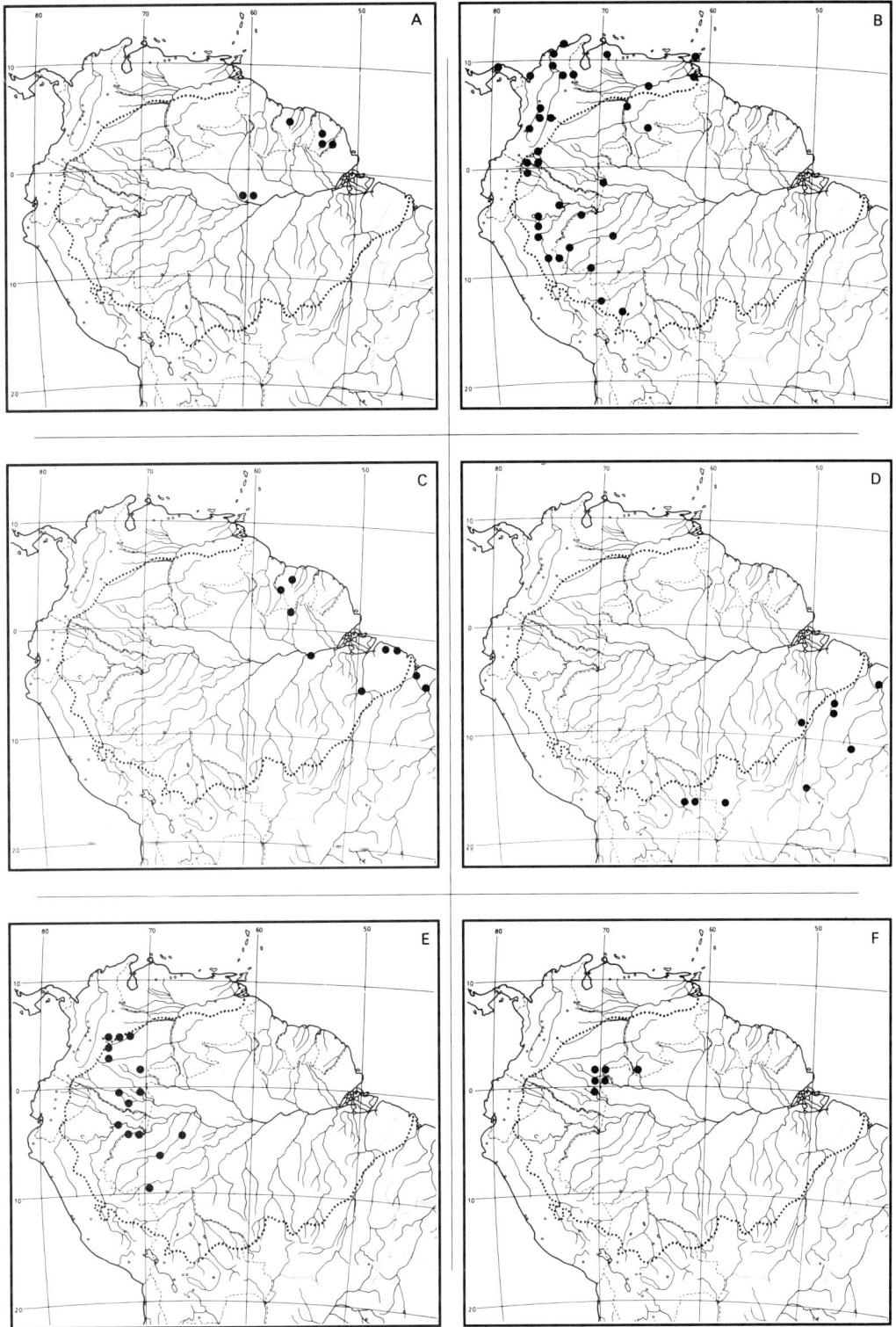

Figure 6.25. (A) *Attalea attaleoides;* **(B)** *Attalea butyracea;* **(C)** *Attalea dahlgreniana;* **(D)** *Attalea eichleri;* **(E)** *Attalea insignis;* **(F)** *Attalea luetzelburgii.*

Attalea humboldtiana Spruce, J. Linn. Soc., Bot. 11: 163. 1871. ?*Scheelea humboldtiana* (Spruce) Burret, Notizbl. Bot. Gart. Berlin-Dahlem 10: 658. 1929. Type. Venezuela. Amazonas: Río Orinoco, Río Casiquiare, Jan 1854, *R. Spruce 43* (holotype, K).

Attalea wallisii Huber, Bull. Herb. Boissier ser. 2, 6(4); 267. 1906. *Scheelea wallisii* (Huber) Burret, Notizbl. Bot. Gart. Berlin-Dahlem 10: 657. 1929. Lectotype (Glassman, 1977). Brazil. Amazonas: Rio Acre and Rio Purus, n.d., *G. Huebner 163* (holotype, B).

Scheelea huebneri Burret, Notizbl. Bot. Gart. Berlin-Dahlem 10: 663. 1929. Type. Brazil. Amazonas: Rio Purus, n.d., *G. Huebner 23a* (holotype, B).

Scheelea bassleriana Burret, Notizbl. Bot. Gart. Berlin-Dahlem 10: 655. 1929. Type. Peru. Loreto: Yarina Cocha, Nov 1925, *G. Tessmann 5490* (isotypes, NY, US).

Scheelea passargei Burret, Notizbl. Bot. Gart. Berlin-Dahlem 10: 671. 1929. Type. Venezuela. "Guayana," n.d., *S. Passarge s.n.* (holotype, B).

Scheelea stenorhyncha Burret, Notizbl. Bot. Gart. Berlin-Dahlem 10: 675. 1929. Type. Peru. Loreto: Río Itaya near Soledad, 110 m, 1 Jul 1925, *G. Tessmann 5256* (holotype, B, n.v.).

Scheelea brachyclada Burret, Notizbl. Bot. Gart. Berlin-Dahlem 10: 680. 1929. Type. Peru. Loreto: Río Itaya, Soledad, Jun 1925, *G. Tessmann 5237* (isotype, NY).

Scheelea tessmannii Burret, Notizbl. Bot. Gart. Berlin-Dahlem 10: 682. 1929. Type. Peru. Loreto: Iquitos, 23 Apr 1925, *G. Tessmann 5085* (holotype, B; isotype, NY).

Scheelea macrolepis Burret, Notizbl. Bot. Gart. Berlin-Dahlem 10: 688. 1929. *Attalea macrolepis* (Burret) Wess. Boer, Pittieria 17: 311. 1988. Type. Venezuela. Bolívar?: Yopal, 14 Feb 1902, *S. Passarge 774* (holotype, B, n.v.).

Attalea pycnocarpa Wess. Boer, Pittieria 17: 299. 1988. Type. Venezuela. Amazonas: near Puerto Ayacucho, n.d., *J. Wessels Boer 1910* (holotype, U, n.v.).

Stems solitary, 2–20 m tall, 25–53 cm diam, sometimes covered apically with persistent, dead leaf sheaths (especially younger plants), occasionally with a mound of roots at base to 40 cm high. *Leaves* 15–30; sheath 1.5–3.2 m long, somewhat persistent on the stem, with distinctive, stiff fibers on margins, these 20–40 cm long and 1–5 mm diam; petiole virtually absent; rachis 6.6–10.3 m long, arching and twisted so that the pinnae appear vertical; pinnae 166–205 per side, regularly arranged and spreading in 1 plane, linear, the middle ones 0.8–1.6 m long, 4–10 cm wide, somewhat glaucous or silvery abaxially, with prominent cross-veins, with an auricle on proximal base. *Inflorescences* interfoliar; peduncle 0.8–1.7 m long; prophyll 1.4–1.5 m long; peduncular bract 2–3.5 m long, woody, sulcate, with a 50–65-cm-long solid apex; rachis 0.7–1.4 m long; staminate rachillae 135–230, 30–53 cm long, borne all around rachis, with silvery-white tomentum; pistillate rachillae 124–300, to 20 cm long, arranged all around rachis, with 1–10 pistillate flowers and an all-staminate section distally; staminate flowers 1.3–2 cm long, arranged all around rachillae, except for a tendency for 1-sided arrangement proximally; sepals very briefly imbricate basally, free and spreading above, triangular, 1 mm long; petals free, linear, 1.3–2 cm long; stamens 6, 2.5–4 mm long; pistillode absent; pistillate flowers 2.5 cm long, to 15 per rachillae, absent from adaxial surface; sepals 2.5 cm long, triangular; petals 1–1.2 cm long, triangular; *fruits* oblong-ovoid or ellipsoid-oblong, 4.5–8.5 cm long, 3–4.5 cm diam; epicarp pinkish-brown, orange-yellow, reddish, or yellowish-brown at maturity; endocarp pores sunken; endocarp fibers scattered, generally not clustered; seeds 1–3.

Southern Mexico south through Central America to Colombia, Venezuela, and Trinidad, and also in the western part of the Amazon region in Colombia (Amazonas, Caquetá, Putumayo), Venezuela (Amazonas, Bolívar), Ecuador (Napo), Peru (Huánuco, Loreto, Madre de Dios, Ucayali), western Brazil (Acre, Amazonas), and Bolivia (Cochabamba, La Paz, Pando) (Figure 6.25B); typical of river margins, but can also occur in forest on *terra firme* or in open areas. It is another example of a widespread species that in the Amazon region grows along rivers, but in other areas grows on sloping, noninundated ground.

Bolivia: *palla*. Brazil: *aricuri, jací*. Colombia: *canambo, mapanaré* (Yucana), *palma real*. Ecuador: *canambo, pa-pa* (Siona). Peru: *shapaha, shapaja, shapajilla, shebon, sheboncita, ümeh* (Bora). Venezuela: *coroba, palma yagua*. The leaves are widely used for thatching, and the seeds are eaten. Beetle larvae from rotting stems are eaten. In Peru, Bora Indians add the burned bark to tobacco paste.

This species has many synonyms in Central America, where in each country it has been described as a separate species.

A few specimens *(U. Blicher 2, U. Blicher 3, G. Galeano 2050, A. Gentry 55997, T. Plowman 6778, R. Vásquez 10834, R. Vásquez 10835)* from the western Amazon region in Colombia (Amazonas) and Peru (Loreto) are similar in some respects (especially staminate flowers and fruits) to *Attalea butyracea,* but are generally smaller and have short and subterranean stems. They may represent an undescribed species.

3. **Attalea dahlgreniana** (Bondar) Wess. Boer, Flora of Suriname 5(1): 158. 1965. *Markleya dahlgreniana* Bondar, Arch. Jard. Bot. Rio de Janeiro 15: 50. 1957. Type. Brazil. Pará: Mun. Bragança, Trocantena, 24 Oct 1956, *G. Bondar 95829* (holotype, RB) (Figure 6.24d–f).

Stems solitary, occasionally partly procumbent but erect apically, 2–15 m tall, 30–55 cm diam, usually covered with persistent, dead leaf bases. *Leaves* 12–20, sheath 0.8–1.6 m long; petiole 0.6–1.1 m long; rachis 4.5–8.2 m long; pinnae 160–321 per side, regularly or slightly irregularly arranged, and then in clusters of 2–4, spreading ± in the same plane, linear, auricle lacking, the middle ones 0.8–1.3 m long, 2.5–6 cm wide, with cross-veins. *Inflorescences* interfoliar; peduncle 0.8–1.7 m long; prophyll 0.8–1 m long; peduncular bract 1.6–2.6 m long; rachis 35–85 cm long; rachillae 162–211, 17–23 cm long, arranged all round rachis, with whitish tomentum; staminate flowers 5–7 mm long, absent from adaxial surface of rachillae; sepals 3, free, narrowly triangular, 1 mm long; petals 3, very briefly connate basally, free and valvate above, lanceolate, 4.5–6 mm long; stamens 7–10, anthers slightly exceeding the petals, somewhat coiled and twisted; pistillode absent; pistillate flowers 2.5 cm long, absent from adaxial surface of rachillae; sepals ovate, 2 cm long; petals widely ovate, mucronate, 2 cm long; staminodial ring 8 mm long; *fruits* ellipsoid-oblong, 6–9 cm long, 2.5–4 cm diam, brownish; endocarp pores slightly sunken; endocarp fibers few, scattered; seeds 1–3.

Surinam and Brazil (Maranhão, Pará) (Figure 6.25C); lowland forest on *terra firme,* on river banks or disturbed places.

Brazil: *dois por dois, perenão, perinão*. Surinam: *majalie, maripa*.

Bondar (1957) speculated that this taxon may be a hybrid between *Attalea speciosa* and *A. maripa,* a view not shared by Wessels Boer (1965).

4. **Attalea eichleri** (Drude) Henderson, comb. nov. *Orbignya eichleri* Drude in Mart., Fl. bras.: Cyclanthaceae et Palmae I, fasc. 85, vol. (3)2: 449. 1881. Type. Brazil. Piauí: Sertão d'Amaroleité, Sep–Oct 1844, *H. Weddell 2705* (holotype, P) (Figure 6.24j–l).

Orbignya humilis Mart. in A. D. Orb., Voy. Amérique mér. 7(3) Palmiers 129. 1847. Type. Bolivia. Santa Cruz: Santa Ana, n.d., *A. d'Orbigny 22* (holotype, P, n.v.).

Stems solitary or rarely cespitose, short and subterranean or occasionally aerial and to 1.5 m tall. *Leaves* 4–8; sheath not seen; petiole 30–50(–73) cm long; rachis 1.4–2 m long; pinnae 66–99 per side, irregularly arranged in clusters of 2–7(–12), spreading in different planes, linear, becoming briefly bifid at apex, the middle ones 20–62 cm long, 1.5–3.6 cm wide, lacking an auricle, cross-veins ± prominent. *Inflorescences* interfoliar; peduncle 5–35 cm long; prophyll not seen; peduncular bract 25–51 cm long; rachis 21–28 cm long; staminate rachillae 20–36, 2.5–15 cm long, absent from 1 side of rachis, with scattered, silvery scales; pistillate rachillae very short (inflorescence almost spicate), absent from 1 side of rachis; staminate flowers 6–7 mm long, absent from adaxial surface of rachillae; sepals 3–4, very briefly connate basally, deltoid, 1 mm long; petals 1–2, very widely ovate, 7 mm long; stamens 17–20, the anthers loosely coiled and twisted; pistillode very small or absent; pistillate flowers 2.5 cm long, 1 per rachilla, but rachillae very short and flowers appearing almost sessile; sepals triangular, 2.5 cm long; petals 3, deltate, 1.5–2 cm long; staminodial ring 3–5 mm long; *fruits* ovoid or ellipsoid-oblong, 7–8 cm long, 4–5.5

cm diam, dark brown; endocarp pores sunken; endocarp fibers scattered, numerous; seeds 3–5.

Southern periphery of the Amazon region in Brazil (western Bahia, Maranhão, Mato Grosso, Pará, Piauí, Tocantins) and Bolivia (Santa Cruz) (Figure 6.25D); drier regions in seasonal forest, rocky slopes, or *cerrado,* often on sandy soils, up to 800 m elevation.

Brazil: *coco-curúa, piaçava, piassava, ruan-rie* (Apinajé).

Although *Orbignya humilis* is the oldest name for this species, the combination *Attalea humilis* has already been made. Balick and co-workers (1987) described *Orbignya teixeriana* Bondar as a hybrid between *A. eichleri* (as *O. eichleri*) and *A. speciosa* (as *O. phalerata*). It seems likely that *O. teixeriana* is a synonym of *O. macrocarpa* Barb. Rodr. (including *O. campestris* Barb. Rodr. and *O. longibracteata* Barb. Rodr.). This species, *O. macrocarpa*, is quite widespread on the southern periphery of the Amazon region in the Brazilian states of Maranhão, Mato Grosso, Piauí, and Tocantins, and it may just enter the region. It appears to form a hybrid zone between *A. eichleri* and *A. speciosa.*

5. **Attalea insignis** (Mart.) Drude in Engl. & Prantl, Nat. Pflanzenfam. 2: 80. 1887. *Scheelea insignis* (Mart.) H. Karst., Linnaea 28: 269. 1857. *Maximiliana insignis* Mart., Hist. nat. palm. 2: 133. 1826. *Englerophoenix insignis* (Mart.) Kuntze, Revis. gen. pl. 3: 322. 1898. Type. Colombia. Caquetá: Río Caquetá, n.d., *C. Martius s.n.* (holotype, M; F neg. 62455) (Figure 6.26a–d).
Scheelea attaleoides H. Karst., Linnaea 28: 265. 1857. Lectotype (Imchanitzkaja, 1987). Colombia. Meta: Llanos de San Martín, 300 m, 1853, *H. Karsten s.n.* (LE, n.v.)
Attalea goeldiana Huber, Bull. Herb. Boissier ser. 2, 6(4): 268. 1906. *Scheelea goeldiana* (Huber) Burret, Notizbl. Bot. Gart. Berlin-Dahlem 10: 658. 1929. Type. Not designated.

Stems solitary, short and subterranean, occasionally aerial and to 2 m tall. *Leaves* 9–11; sheath subterranean, to 48 cm long; petiole 1.6–3.3 m long; rachis 4.3–9 m long; pinnae 114–148 per side, irregularly arranged in distinct clusters of 2–6 and spreading in different planes, linear, acute or blunt at the apex, the middle ones 75–88 cm long, 3.5–4 cm wide, the apical pinnae often partially split and forming "windows," with a prominent auricle at the base of the proximal pinna of a cluster, with prominent cross-veins. *Inflorescences* interfoliar; peduncle to 1.3 m long; prophyll, peduncular bract, and rachis not seen; staminate rachillae 12–30, 15–25 cm long, arranged all around rachis, with silvery-white tomentum; pistillate rachis ca. 16 cm long; pistillate rachillae very short (flowers and fruits appearing almost sessile), arranged all around rachis; staminate flowers 13–15 mm long, arranged all around the rachillae; sepals very briefly connate basally, free and spreading above, triangular, 0.5 mm long; petals free, linear, 13–14 mm long; stamens 6, 4 mm long, included within petals; pistillode absent; pistillate flowers 3.5 cm long, 1–2 per rachilla; sepals triangular, 3 cm long; petals valvate at apex, shallowly deltate, 3.5 cm long; staminodial ring 1 cm high; *fruits* ellipsoid-oblong, 7–7.5 cm long, 3.5–4 cm diam, brown; endocarp pores sunken; endocarp fibers scattered or in small groups; seeds 2–3.

Southeastern Colombia (Amazonas, Caquetá, Casanare, Meta, Putumayo, Vaupés), Ecuador (Napo), Peru (Loreto), and western Brazil (Acre, Amazonas) (Figure 6.25E); lowland rain forest in inundated areas and on *terra firme,* at low elevations. In some areas, it can become a weed of cleared fields (G. Galeano, personal communication).

Brazil: *coco pancha, coquino, palha de flecha, palha rasgada.* Colombia: *coco, yagua.* Peru: *contillo.*

Galeano and Bernal (1989) have given a discussion of this species, and the inclusion of *Scheelea attaleoides* in synonymy.

6. **Attalea luetzelburgii** (Burret) Wess. Boer, Pittieria 17: 303. 1988. *Orbignya luetzelburgii* Burret, Notizbl. Bot. Gart. Berlin-Dahlem 10: 1019. 1930. *Parascheelea luetzelburgii* (Burret) Dugand, Caldasia 3: 24. 1941. Type. Brazil. Amazonas: Jutica, Varadouro, Rio Ayari, Rio Uaupés, 16 Nov 1928, *P. von Luetzelburg 21969* (isotypes, M, R; F neg 18552) (Figure 6.26i–l).

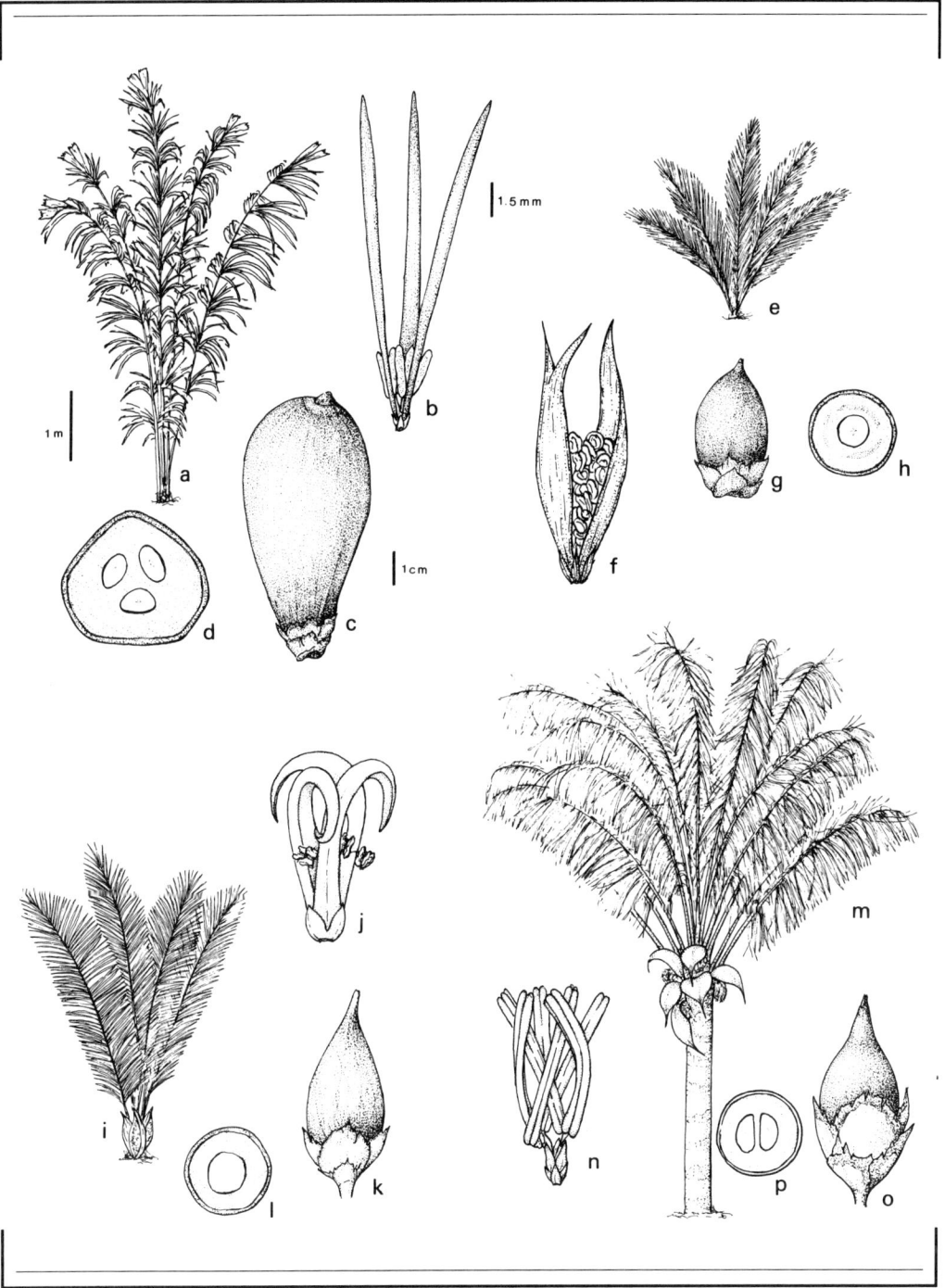

Figure 6.26. (a) *Attalea insignis,* **(b)** staminate flower, **(c)** fruit, **(d)** cross section of fruit (b, c, and d from *A. Henderson 856*); **(e)** *Attalea microcarpa,* **(f)** staminate flower (from *R. Vásquez 7756*), **(g)** fruit, **(h)** cross section of fruit (g and h from *A. Henderson 1742*); **(i)** *Attalea luetzelburgii,* **(j)** staminate flower, **(k)** fruit, **(l)** cross section of fruit (j, k, and l from *J. Wessels Boer 2374*); **(m)** *Attalea maripa,* **(n)** staminate flower (from *H. Balslev 4789*), **(o)** fruit, **(p)** cross section of fruit (o and p from *M. Jansen-Jacobs 2389*).

Parascheelea anchistropetala Dugand, Caldasia 1: 12. 1940. Type. Colombia. Vaupés: Cerro de Circasia, 300–500 m, 10 Oct 1939, *J. Cuatrecasas 7172* (holotype, COL).

Stems solitary, short and subterranean. *Leaves* ca. 6; sheath to 50 cm long; petiole 0.9–1.5 m long, petiole and rachis reddish-brown-tomentose abaxially; rachis 4.6–4.9 m long; pinnae 72–89 per side, regularly arranged and spreading in 1 plane, linear, the middle ones 70–75 cm long, 3.5–4.5 cm wide, auricle lacking, cross-veins obscure. *Inflorescences* interfoliar; peduncle not seen; prophyll to 50 cm long; peduncular bract 70–80 cm long; staminate rachillae to 100, 5–7 cm long, ca. 4 mm diam, arranged all around rachis, lacking silvery scales; pistillate rachillae to 105, 1–5 cm long, arranged all around rachis; staminate flowers 8–10 mm long, mostly absent from adaxial surface of rachillae; sepals 3, briefly connate basally, free and spreading above, triangular, 1.5 mm long; petals 3, connate basally for ca. 3 mm, linear and very narrow, acuminate, curved over at apex, 10 mm long; stamens 6, anthers loosely coiled; pistillode trifid, with 3 spreading arms to 1 mm long; pistillate flowers 2.5 cm long, 1–3 per rachillae, absent from adaxial surface; sepals ovate, 10 mm long; petals very widely ovate, mucronate, 1.5 cm long, with uneven margins apically; staminodial ring 7 mm long; *fruits* ovoid, 5–5.5 cm long, ca. 2.5 cm diam, brownish; endocarp with a few fibers near center; seed 1.

Southeastern Colombia (Vaupés), southern Venezuela (Amazonas), and western Brazil (Amazonas) (Figure 6.25F); open forest or savannas at low elevations on *terra firme,* usually on sandy soils.

Brazil: *curuaraua.* Colombia: *curúa.* Venezuela: *grua.*

7. Attalea maripa (Aubl.) Mart. in A. D. Orb., Voy. Amérique mér. 7(3) Palmiers 123. 1844. *Palma maripa* Aubl., Hist. pl. Guiane 2: 974. 1775. *Palma maripa* Corrêa, Ann. Mus. Natl. Hist. Nat. 8: 75. 1806. *Maximiliana maripa* (Aubl.) Drude in Mart., Fl. bras.: Cyclanthaceae et Palmae I, fasc. 85, vol. 3(2): 452. 1881. *Englerophoenix maripa* (Corrêa) Kuntze, Rev. gen. pl. 2: 728. 1891. Type. French Guiana. Without locality, n.d., *J. Aublet s.n.* (holotype, P, n.v.) (Figure 6.26m–p).

Maximiliana elegans H. Karst., Linnaea 28: 271. 1857. Type. Not designated.

Maximiliana regia Mart., Hist. nat. palm. 2: 132. 1826. *Maximiliana martiana* H. Karst., Linnaea 28: 273. 1857. *Englerophoenix regia* (Mart.) Kuntze, Rev. gen. pl. 2: 728. 1891. *Attalea regia* (Mart.) Wess. Boer, Flora of Suriname 5(1): 150. 1965. Lectotype (Glassman, 1978). Brazil. Maranhão or Pará: without locality, n.d., *C. Martius s.n.* (holotype, M; F neg. 18549a).

Maximiliana tetrasticha Drude in Mart., Fl. bras.: Cyclanthaceae et Palmae I, fasc. 85, vol. 3(2): 455. 1881. *Englerophoenix tetrasticha* (Drude) Barb. Rodr., Sert. palm. brasil. 1: 76. 1903. *Scheelea tetrasticha* (Drude) Burret, Notizbl. Bot. Gart. Berlin-Dahlem 10: 667. 1929. Type. Brazil. Pará: Rio Tocantins, Aug–Sep 1844, *M. Weddell 2331* (holotype, P).

Maximiliana longirostrata Barb. Rodr., Vellosia 1 (ed. 2): 112. 1891. *Englerophoenix longirostrata* (Barb. Rodr.) Barb. Rodr., Sert. palm. brasil. 1: 77. 1903. Lectotype (Glassman, 1978). Barb. Rodr., Vellosia 3: t. 2. 1891.

Maximiliana macrogyne Burret, Notizbl. Bot. Gart. Berlin-Dahlem 10: 692. 1929. Type. Brazil. Maranhão: Tury-assu, 19 Oct 1923, *E. Snethlage 279* (isotype, NY).

Maximiliana stenocarpa Burret, Notizbl. Bot. Gart. Berlin-Dahlem 10: 696. 1929. Type. Peru. Loreto: Iquitos, n.d., *G. Tessmann 5081* (isotype, NY).

Maximiliana macropetala Burret, Notizbl. Bot. Gart. Berlin-Dahlem 10: 699. 1929. *Attalea macropetala* (Burret) Wess. Boer, Flora of Suriname 5(1): 155. 1965. Type. Venezuela. State?: Rosalia, n.d., *S. Passarge 63* (holotype, B, n.v.).

Attalea cryptanthera Wess. Boer, Pittieria 17: 310. 1988. Type. Venezuela. Amazonas: near San Carlos de Río Negro, 22 Jan 1968, *J. Wessels Boer 2323* (isotype, NY).

Stems solitary, 3.5–20 m tall, (9–)20–33(–100) cm diam, often with a basal cone of roots to 70 cm long, erect or rarely somewhat repent. *Leaves* 10–22, erect, arranged in a few (usually five) distinctive spirals; sheath 50–116 cm long, fibrous at margins; petiole to 2.4–3.3 m long; rachis 4.9–9.6 m long; pinnae 152–318 per side, irregularly arranged in clusters of 2–10, spreading in different planes, linear, the middle ones 1–1.5 m long, 4–6.5 cm wide, auricle absent, with prominent cross-veins. *Inflorescences* interfoliar, persistent; peduncle 45–100 cm long; prophyll 0.5–1.6 m long; peduncular bract 1.1–2.5 m long including a 35–50 cm long umbo, deeply sulcate, persistent; rachis 40–100 cm long; staminate and pistillate rachillae very numerous, 254–1000, 15–22 cm long, arranged all around rachis, with silvery-white tomentum; staminate flowers 12 mm long, absent from adaxial surface, at least proximally; sepals 3, briefly imbricate, free and spreading above, triangular to 1 mm long; petals 3, connate basally for ca. 1 mm, free above, lanceolate, 3.5 mm long; stamens 6, the filaments as long as or exceeding the petals, the anthers longer still, to 8 mm; pistillode absent; pistillate flowers 1.5 cm long, 6–10 per rachilla, absent from adaxial surface of rachillae; sepals widely ovate, 12 mm long; petals very widely ovate, 10 mm long; staminodial ring at first adnate to ovary apically, densely tomentose on apical margin, 7 mm long; ovary tomentose; *fruits* ellipsoid-oblong, 4–6 cm long, 2.5–3 cm diam, yellowish-brown, covered almost halfway by fruiting perianth, fruiting staminodial ring fringed on apical margin; endocarp pores superficial; endocarp lacking fibers; seeds 2–3.

Throughout northern South America, east of the Andes, in Colombia (Amazonas, Caquetá, Guaviare, Guainía, Meta, Putumayo, Vaupés, Vichada), Venezuela (Amazonas, Bolívar, Delta Amacuro, Monagas, Sucre), Trinidad, the Guianas, Ecuador (Napo), Peru (Loreto, Madre de Dios, Ucayali), Brazil (Acre, Amazonas, Maranhão, Matto Grosso, Pará, Rondônia), and Bolivia (Beni, Pando, Santa Cruz) (Figure 6.27A); in a variety of habitats. Isolated specimens occur in lowland rain forest on *terra firme* but the species is also found in gallery forest and savanna margins, and can be especially abundant in disturbed areas.

Bolivia: *casicusi, cusi, cusimacho, motacusillo, xëbichoqui* (Chacobo). Brazil: *anaja, anajá, inajá, inajai, inaja-'y* (Guajá), *inaza* (Guajajara), *naja'i* (Arawete). Colombia: *cucurita, guajo, guichire, inayá, imayá, jarive* (Huitoto), *palma real, uichira*. Ecuador: *inayo, oompa* (Waorani), *wa-hó* (Siona). Guyana: *kokerite, kukarit.* Peru: *inayuga, ina yuga.* Surinam: *baba maripa, maripa.* Venezuela: *abadek* (Pemón), *cucurito, mavaco* (Baniba). The leaves are commonly used for thatching; the palm heart is occasionally eaten; the outer part of the petiole and rachis were used to make blowgun darts; the peduncular bract is used for various domestic purposes; the fruits are edible; and salt is extracted from burned infructescences.

This species regularly produces inflorescences with both staminate and pistillate flowers, as well as the more usual all-staminate or all-pistillate inflorescences of other species.

8. Attalea microcarpa Mart. in A. D. Orb., Voy. Amérique mér. 7(3). Palmiers 125. 1844. *Orbignya microcarpa* (Mart.) Burret, Notizbl. Bot. Gart. Berlin-Dahlem 10: 507. 1929. Lectotype (here designated). Martius, Hist. nat. palm. 3: t. 168, f. 2. 1845 (Figure 6.26e–h).
Attalea agrestis Barb. Rodr., Enum. palm. nov. 42. 1875. *Orbignya agrestis* (Barb. Rodr.) Burret, Notizbl. Bot. Gart. Berlin-Dahlem 10: 511. 1929. Lectotype (Glassman, 1972). Barb. Rodr., Sert. palm. brasil 1: t. 55. 1903.
Orbignya sagotii Trail ex Thurn, Timehri 3: 276. 1884. *Attalea sagotii* (Trail ex Thurn) Wess. Boer, Flora of Suriname 5(1): 162. 1965. Type. French Guiana. Karouany, 1856, *P. Sagot 831* (holotype, K; isotype, P).
Orbignya sabulosa Barb. Rodr., Vellosia I. 54. 1888. Lectotype (Glassman, 1972). Barb. Rodr., Sert. palm. brasil. 1: t. 48. 1903.
Orbignya polysticha Burret, Notizbl. Bot. Gart. Berlin-Dahlem 11: 324. 1932. Type. Peru. Loreto: Mishuyacu, near Iquitos, 100 m, Oct–Nov 1929, *G. Klug 205* (isotypes, NY, US).

Stems solitary, short and subterranean. *Leaves* 6–16, stiffly erect and forming a "funnel"; sheath subterranean, 37–50 cm long; petiole very short or absent or to 2 m long, abaxial surface of petiole and rachis lightly to densely reddish-brown-tomentose; rachis 2–5(–7) m long; pinnae 55–115 per side, linear, briefly bifid at the apex, regularly arranged and stiffly spreading in 1 plane, the middle ones 72–104 cm long and 3–5 cm wide, cross-veins prominent abaxially, auricle absent. *Inflorescences* interfoliar; peduncle 50–70 cm long; prophyll to 50 cm long; peduncular bract 0.8–

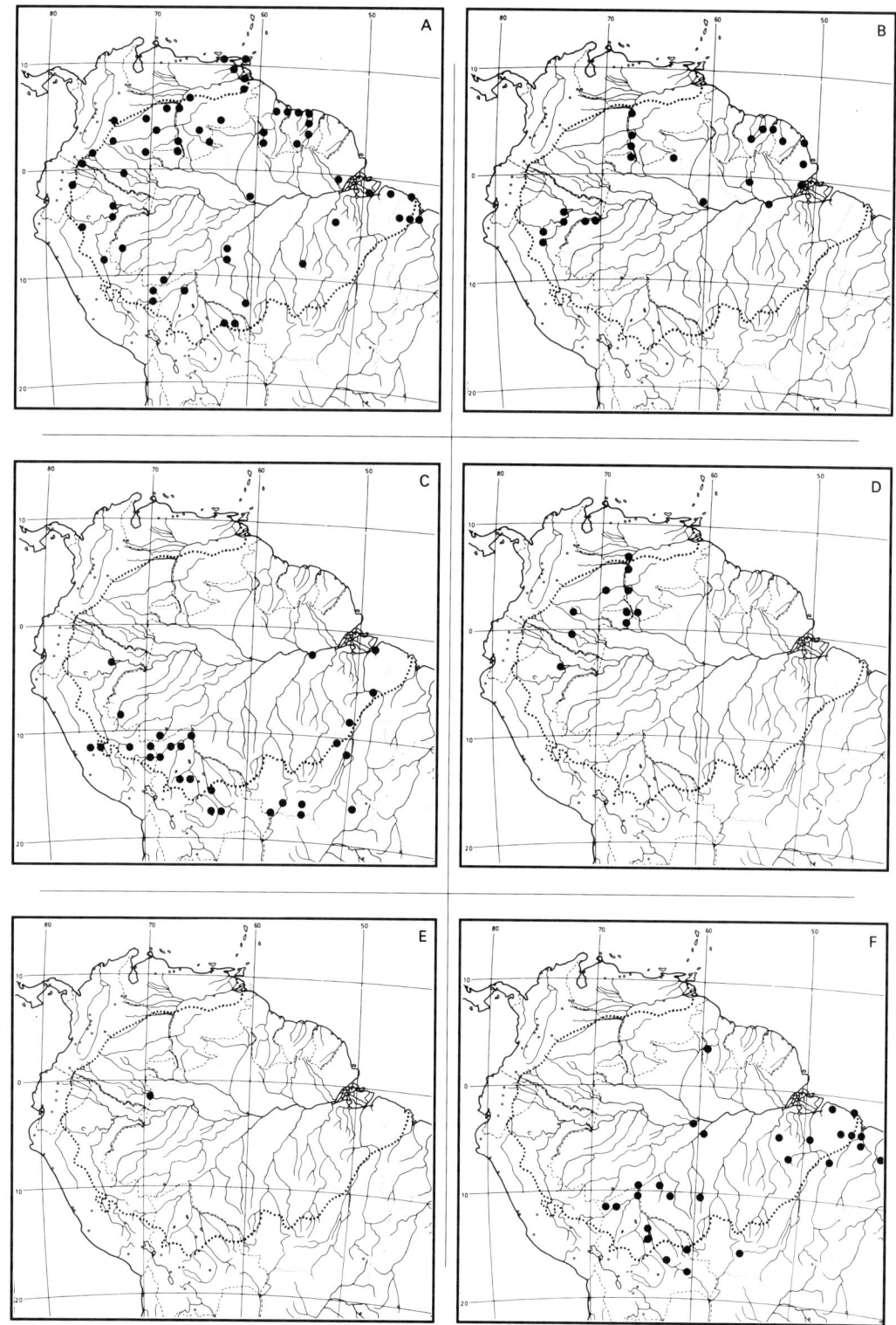

Figure 6.27. (A) *Attalea maripa;* **(B)** *Attalea microcarpa;* **(C)** *Attalea phalerata;* **(D)** *Attalea racemosa;* **(E)** *Attalea septuagenata;* **(F)** *Attalea speciosa.*

1 m long, woody, sulcate abaxially; rachis 20–50 cm long; staminate rachillae 58–73, 3–9 cm long, 3–4 mm diam, arranged all around the rachis, covered with silvery-white scales; pistillate rachillae 1–6 cm long, arranged all around the rachis, glabrous; staminate flowers 7–10 mm long, tending to be arranged all around rachillae but often absent from adaxial surface; sepals very briefly connate basally, free above, deltate, 0.3–1 mm long; petals free, oblanceolate to spathulate, 7–9 mm long; stamens 9–15; anthers coiled and twisted; pistillode minute; pistillate flowers 1.5–2 cm long, 1–7 per rachillae, absent from adaxial surface of rachillae; sepals widely ovate, 1.5 cm long; petals widely ovate, 1.5 cm long; staminodial ring 5 mm high; *fruits* ovoid or obovoid, 3.5–4 cm long, 2–3 cm diam, brown-tomentose; endocarp pores superficial; endocarp thin, with or without fibers; seeds 1–3.

Colombia (Vaupés), Venezuela (Amazonas), the Guianas, Peru (Loreto), and Brazil (Amapá, Amazonas, Pará) (Figure 6.27B); lowland rain forest; open, low forest; or rocky places, usually on sandy soil.

Brazil: *coco curuá, curuaí, kunuana, palha branca, palha vermelha, palhera vermelha, palma vermelha, palmeira, sacurí*. French Guiana: *macoupi*. Peru: *catarina, catherina, shapajilla*. Surinam: *mountain maripa*. Venezuela: *mabaco, mavaco*. The leaves are used for thatching; the mesocarp and seeds are eaten. Carbonized, broken endocarps, probably from this species, are found with Palaeoindian remains in caves near Monte Alegre, Pará, and could be to 7000 years old (A. Roosevelt, personal communication).

I include here all "Orbignya" with short and subterranean stems and small fruits. There are, however, several nomenclatural and systematic problems with this arrangement, and more collections, especially from Brazil (Pará), are needed. This species is very similar to *Attalea spectabilis,* and the differences between the two are not clear; see the discussion under that species.

9. **Attalea phalerata** Mart. ex Spreng., Syst. veg. 2: 624. 1825. *Scheelea phalerata* (Mart. ex Spreng.) Burret, Notizbl. Bot. Gart. Berlin-Dahlem 10: 669. 1929. Type. Not designated (Figure 6.28a–d; Plate IIb).

Attalea excelsa Mart. ex Spreng., Syst. veg. 2: 624. 1825. *Scheelea martiana* Burret, Notizbl. Bot. Gart. Berlin-Dahlem 10: 661. 1929. Lectotype (here designated). Mart., Hist. nat. palm. 2: t. 96, f. 1–3. 1826.

Attalea princeps Mart. in A. D. Orb., Voy. Amérique mér. 7(3). Palmiers 113. 1844. *Scheelea princeps* (Mart.) H. Karst., Linnaea 28: 269. 1857. Type. Bolivia. Santa Cruz: Moxos, n.d., *A. d'Orbigny 16* (holotype, P).

Attalea blepharopus Mart. in A. D. Orb., Voy. Amérique mér. 7(3). Palmiers 116. 1844. *Scheelea blepharopus* (Mart.) Burret, Notizbl. Bot. Gart. Berlin-Dahlem 10: 674. 1929. Type. Bolivia. Department?: Yuracares, n.d., *A. d'Orbigny 34* (holotype, P, n.v.).

Attalea hoehnei Burret, Notizbl. Bot. Gart. Berlin-Dahlem 10: 522. 1929. Type. Brazil. Mato Grosso or Acre: Agua Limpa, n.d., *F. Hoehne 2196* (holotype, B, n.v.).

Scheelea weberbaueri Burret, Notizbl. Bot. Gart. Berlin-Dahlem 10: 659. 1929. Type. Peru. Junín: Prov. Tarma, La Merced, Chanchamayo, 1000 m, Dec 1902, *A. Weberbauer 1848* (holotype, B, n.v.).

Scheelea microspadix Burret, Notizbl. Bot. Gart. Berlin-Dahlem 15: 104. 1940. Type. Brazil. Mato Grosso: without locality, n.d., *W. Hopp 3010* (holotype, B).

Stems solitary, generally short, (0–)2–5(–10) m tall, 25–40(–60) cm diam, often covered with persistent, dead, erect sheaths and petioles. *Leaves* 11–30; sheath 0.7–1.7 m long, with thin fibers on margins; petiole (0–)0.6–2 m long; rachis 3.4–5.9 m long; pinnae 140–205 per side, ± regularly or irregularly arranged in clusters of 2–5, spreading in 1 or different planes (or often the pinnae apex spreading in different planes), linear, the middle ones 0.5–1 m long, 3–5.2 cm wide, with or without auricle basally, with prominent cross-veins. *Inflorescences* interfoliar; peduncle 0.6–1.2 m long; prophyll not seen; peduncular bract (0.5–)1.2–1.7 m long, deeply sulcate abaxially; rachis (15–)35–50 cm long, pistillate rachis often swollen; staminate rachillae very numerous, to 120 or more, 7–10 cm long, arranged all around rachis; pistillate rachillae to 50, 4–8 cm long, arranged all around rachis; staminate flowers 7 mm long, absent from adaxial surface of rachillae; sepals 3, very briefly connate basally, del-

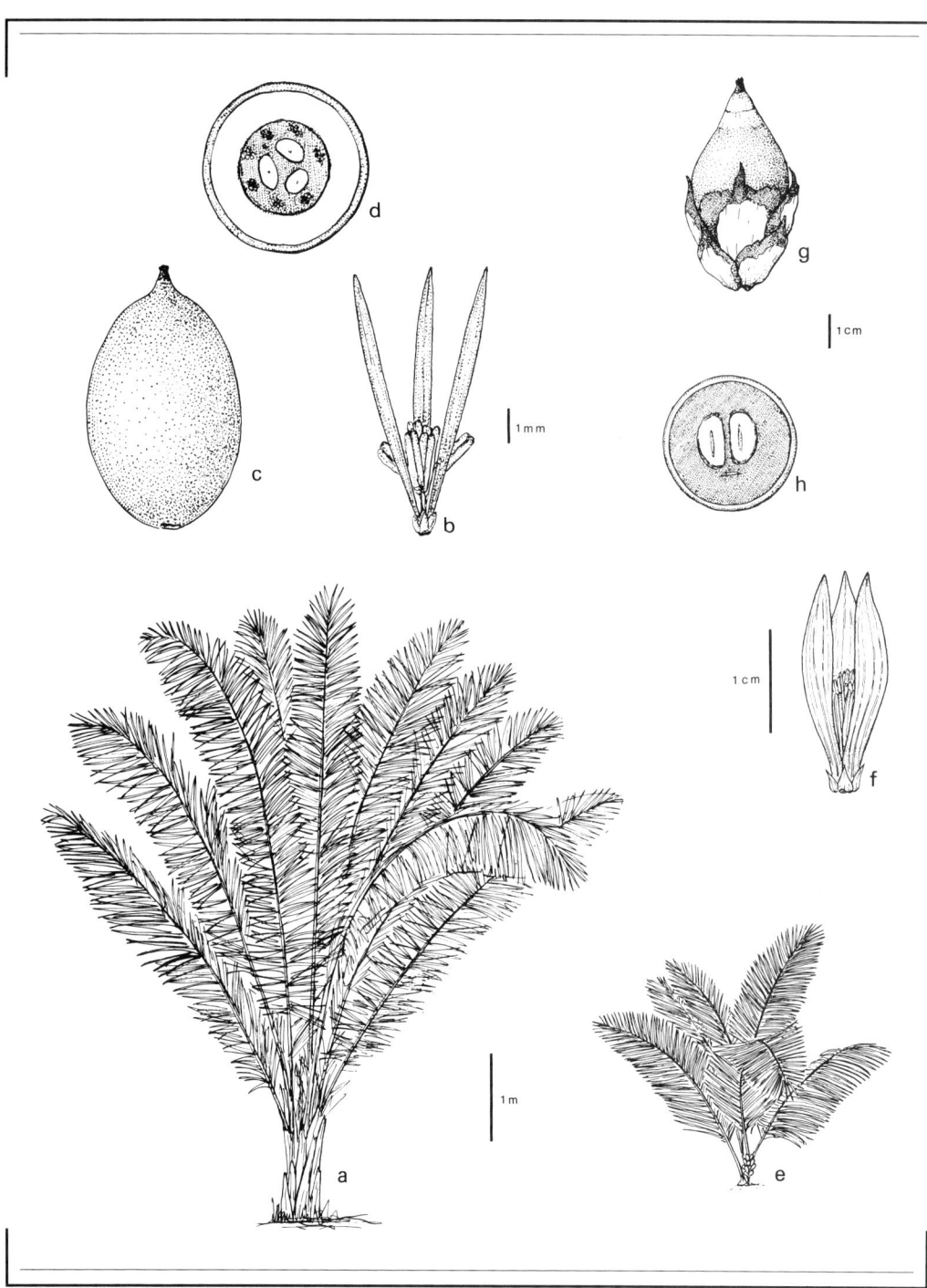

Figure 6.28. (a) *Attalea phalerata,* (b) staminate flower (from *A. Henderson 1643*); (c) fruit, (d) cross section of fruit (c and d from *A. Henderson 1743*), (e) *Attalea racemosa,* (f) staminate flower, (g) fruit, (h) cross section of fruit (f, g, and h from *J. Wessels Boer 1894*).

tate, 0.5 mm long; petals 3, free, linear, ± terete in cross section, 7 mm long; stamens 6; pistillode absent; pistillate flowers (postanthesis) 2.5 cm long, 2–6 per rachillae, absent from adaxial surface of rachillae; sepals deltate, 2 cm long; petals shallowly deltate, 2 cm long; staminodial ring present; *fruits* ellipsoid-oblong, often crowded and then angled or compressed by mutual pressure, 6–11 cm long, 3–5 cm diam, light-brown; endocarp pores sunken; endocarp fibers in distinct clusters; seeds 1–4.

Planalto of Brazil and adjacent countries, and in southern and western parts of the Amazon region in Peru (Junín, Loreto, Madre de Dios, Ucayali), Brazil (Acre, Goiás, Maranhão, Mato Grosso, Pará, Rondônia, Tocantins), and Bolivia (Beni, La Paz, Pando, Santa Cruz) (Figure 6.27C); drier regions at low elevations (occasionally to 1000 m on eastern Andean slopes), in open areas, gallery forest, disturbed forest, forest islands in savanna, and occasionally lowland rain forest. In Bolivia, it is very common on sedimentary soils west of the Brazilian Shield, where it replaces *Attalea speciosa*. William Balée (personal communication) believes this species may be locally allopatric with *A. speciosa*.

Bolivia: *mana'i* (Chimane), *motacú, mutacú, ouricuri, xëbini* (Chácabo). Brazil: *bacurí, urucurí* (fruit), *urucurizeiro* (tree). Peru: *shapaja*. The leaves are used for thatching; a decoction of the leaves is used as a medicine (Chácabo, Bolivia); the mesocarp is eaten; oil from the mesocarp is used as a hair restorative; the roots are used as a medicine (Chimane, Bolivia).

Attalea excelsa is tentatively included in synonymy here, since the only "Scheelea" occurring in Pará and Maranhão (if, indeed, *A. excelsa* is a "Scheelea"; compare Burret, 1929b, with Glassman, 1977b) is *A. phalerata*. The other specimens from the eastern part of the range in Brazil (Mato Grosso, Pará) are smaller than usual. Apart from the synonyms already listed, Hahn (in preparation) has pointed out that there are several other synonyms from Brazil (Mato Grosso) and Paraguay: *Scheelea princeps* var. *corumbaensis* Barb. Rodr., *S. anisitsiana* Barb. Rodr., *S. parviflora* (Barb. Rodr.) Barb. Rodr., *S. quadrisperma* Barb. Rodr., and *S. quadrisulcata* Barb. Rodr.

10. **Attalea racemosa** Spruce, J. Linn. Soc., Bot. 11: 166. 1871. *Orbignya racemosa* (Spruce) Drude in Mart., Fl. bras.: Palmae II 3(2): 448. 1881. Type. Venezuela. Amazonas: between Río Negro and Río Guasié, Oct 1854, *R. Spruce 54* (holotype, K; isotype, P) (Figure 6.28e–h).

Attalea ferruginea Burret, Notizbl. Bot. Gart. Berlin-Dahlem 11: 1044. 1934. Type. Venezuela. Amazonas: San Carlos de Río Negro, Mar 1933, *C. Lakó/G. Huebner 166* (holotype, B).

Stems solitary or occasionally cespitose, short and subterranean, sometimes forming dense colonies. *Leaves* 8–35; sheath largely subterranean, 40–80 cm long, abaxial surface of sheath, petiole, and rachis reddish-brown-tomentose; petiole 0.8–1.4 m long; rachis 2.1–4.6 m long; pinnae 68–126 per side, regularly arranged and spreading in 1 plane, linear, the middle ones 72–90 cm long, 2.5–4.5 cm wide, auricle absent, cross-veins obscure. *Inflorescences* interfoliar; peduncle 1–1.6 m long; prophyll 50–60 cm long; peduncular bract 65–140 cm long, reddish-brown-tomentose abaxially; rachis 17–26 cm long; staminate rachillae 24–35, 7–11 cm long, 1–2 mm diam, arranged all around rachis, lacking scales; pistillate rachillae 6–14, 0.5–5 cm long, borne on 1 side of rachis only; staminate flowers 2 cm long, borne on 1 side of rachillae only; sepals 3, very briefly imbricate basally, free and spreading above, triangular, 2 mm long; petals 3, free, irregularly lanceolate-ovate, 2 cm long, with scattered, whitish scales on abaxial surface, striate abaxially; stamens 19–42, included within petals; pistillode 1 mm long; pistillate flowers 5 cm long, 1–2 per rachilla, appearing almost sessile on rachis, borne on 1 side of rachillae only; sepals ovate, 2 cm long; petals ovate, mucronate, 2 cm long; staminodial ring 8.5 mm long; *fruits* globose, ellipsoid-oblong or ovoid, 6–9 cm long, 4.5–5 cm diam, brownish; endocarp pores deeply sunken; endocarp with numerous, scattered fibers and with a few large clusters; seeds 1–2.

Northwestern Amazon region in Colombia (Amazonas, Caquetá, Guainía, Guaviare, Vaupés), southwestern Venezuela (Amazonas, Bolívar), Peru (Loreto), and Brazil (Amazonas) (Figure 6.27D); open or semiopen areas on

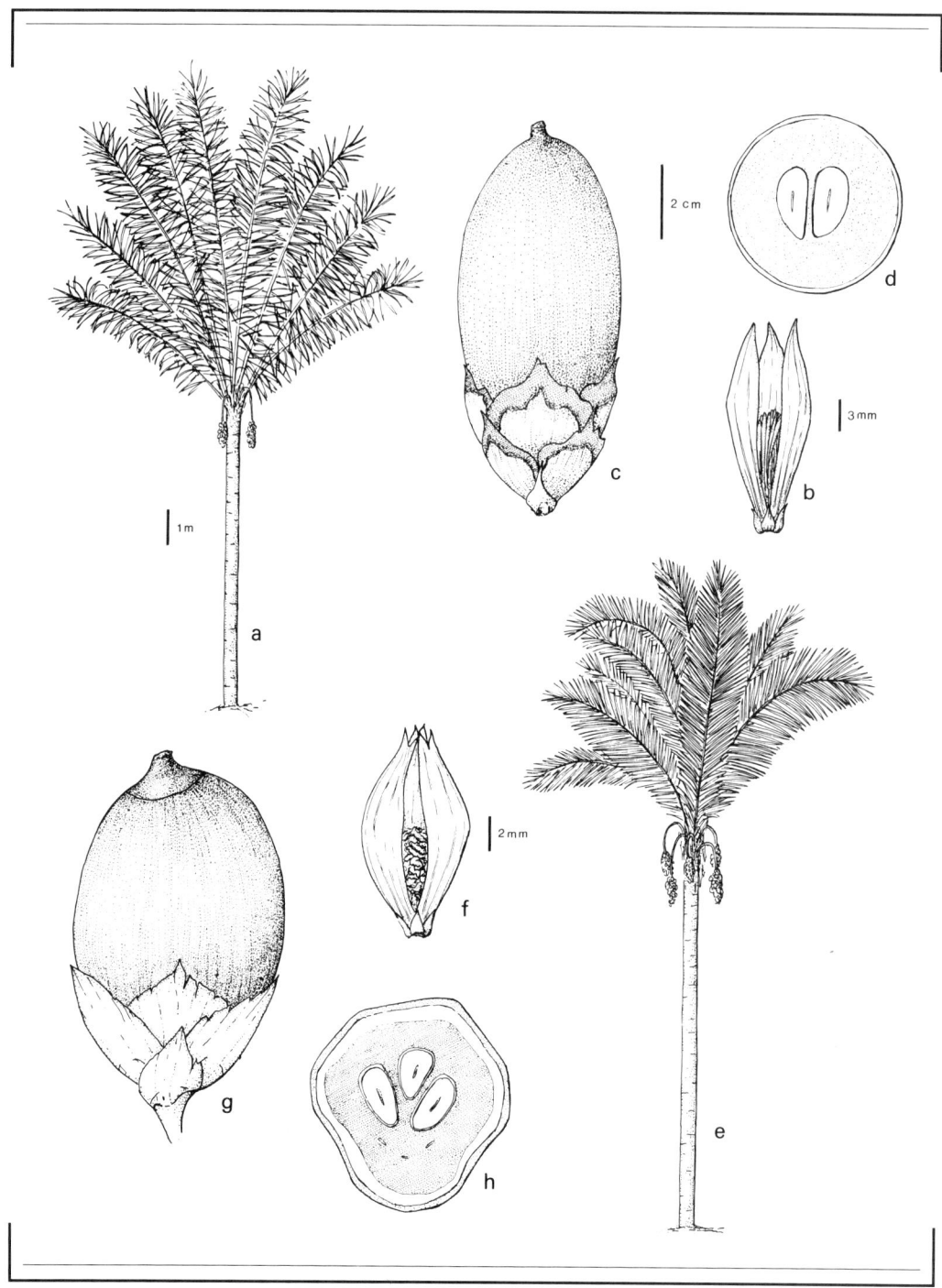

Figure 6.29. (a) *Attalea septuagenata,* (b) staminate flower, (c) fruit, (d) cross section of fruit (b, c, and d from *G. Galeano 2078*); (e) *Attalea speciosa,* (f) staminate flower (from *M. Balick 1432*), (g) fruit, (h) cross section of fruit (g and h from *M. Balick 1946*).

sandy soils, or occasionally in forest or forest margins, at elevations below 500 m.

Brazil: *babassú*. Colombia: *boyon* (Puinave), *coco, codime* (Andoque), *kudina* (Andoque), *novaco, palma de coco*. Peru: *shebon, shebon enano*. Venezuela: *cocito yagua, mabaco, palma mabaca*. The seeds are edible.

There is considerable variation in fruit size and endocarp fiber arrangement. In the eastern part of the range, in Venezuela, fruits are usually smaller and have scattered endocarp fibers; in the western part of the range, in Peru, fruits are larger and endocarp fibers can be in large clusters.

11. **Attalea septuagenata** Dugand, Mutisia 18: 3. 1953. Type. Colombia. Amazonas: Río Miriti-paraná, Mar 1952, *R. Schultes & I. Cabrera 15796* (isotype, BH) (Figure 6.29a–d).

Stems solitary, 7–12 m tall, 25–30 cm diam, with old, persistent leaf sheaths at apex. *Leaves* 18–20; sheath and petiole 1–1.5 m long; rachis to 9 m long, abaxial surface of sheath, petiole, and rachis densely brown-tomentose; pinnae ca. 183 per side, regularly arranged and spreading in 1 plane, linear, the middle ones ca. 1.5 m long, 4.5 cm wide, lacking an auricle, cross-veins obscure. *Inflorescences* interfoliar; peduncle to 1 m long; prophyll not seen; peduncular bract to 1.2 m long, sulcate on abaxial surface; rachis to 84 cm long; staminate rachillae ca. 51, to 17 cm long, tending to be on 1 side of rachis; pistillate rachillae to 5 cm long, arranged on 1 side of rachis, with 1–2 pistillate flower scars per rachilla; staminate flowers 1.7 cm long; sepals 3, free, deltate, 1.5 mm long; petals free, valvate, oblanceolate-obovate, 1.7 cm long; stamens 36–43(–75), included within petals, borne on a solid receptacle at base of petals; anthers straight; pistillode very small; pistillate flowers not seen; *fruits* ellipsoid-oblong, ca. 10 cm long, ca. 5 cm diam, brown or silvery-brown; endocarp pores sunken; endocarp with numerous, scattered fibers; seeds 1–2.

Currently known only from the Río Miriti-paraná region in Colombia (Amazonas) (Figure 6.27E); lowland rain forest in low-lying, wet areas.

Dugand (1953) reported 60–75 stamens, but only about half that number are present in *G. Galeano et al. 2078*.

12. **Attalea speciosa** Mart. ex Spreng., Syst. Veg. 2: 624. 1825. *Orbignya speciosa* (Mart.) Barb. Rodr., Plantas Novas Cultivadas no Jardim Botânico do Rio de Janeiro 1: 32. 1891. *Orbignya barbosiana* Burret, Notizbl. Bot. Gart. Berlin-Dahlem 11: 690. 1932. Type (here designated). Mart., Hist. nat. palm. 2: t. 96, f. 3, 3–6. 1826 (Figure 6.29e–h).

Orbignya phalerata Mart. in A. D. Orb., Voy. Amérique mér. 7(3). Palmiers 126. 1844. Type. Bolivia. Beni: Chiquitos, n.d., *A. d'Orbigny 20* (holotype, P).

Attalea pixuna Barb. Rodr., Enum. palm. nov. 43. 1875. *Orbignya pixuna* (Barb. Rodr.) Barb. Rodr., Protesto-Appendice 49. 1879. *Attalea spectabilis* var. *polyandra* Drude in Mart., Fl. bras.: Palmae II 3(2): 440. 1882. Lectotype (Glassman, 1972). Barb. Rodr., Sert. palm. brasil. 1: t. 49. 1903.

Orbignya lydiae Drude in Mart., Fl. bras.: Palmae II 3(2): 448. 1882. *Attalea lydiae* (Drude) Barb. Rodr., Sert. palm. brasil. 1: 65. 1903. Type. Brazil. Rio de Janeiro, cultivated, 28 Oct 1876, *A. Glaziou 9006* (isotype, P).

Orbignya martiana Barb. Rodr., Palmae Mattogrossensis 68. 1898. Lectotype (Glassman, 1972). Barb. Rodr., Palmae Mattogrossensis t. 22–23. 1898.

Orbignya huebneri Burret, Notizbl. Bot. Gart. Berlin-Dahlem 10: 501. 1929. Type. Brazil. Amazonas: Lago Mondurucú, Rio Manacapurú, n.d., *G. Huebner 64* (holotype, B).

Orbignya macropetala Burret, Notizbl. Bot. Gart. Berlin-Dahlem 10: 507. 1929. Type. Guyana. Rupununi River, n.d., *R. Schomburgk s.n.* (holotype, B).

Stems solitary, erect, 3–15 m tall, 25–41 cm diam. *Leaves* 7–22; sheath to 80 cm long; petiole 0.3–1.6 m long; rachis 5.6–12 m long; pinnae 170–224 per side, regularly arranged and spreading in 1 plane, linear, the middle ones 1–1.1 m long, 4–5.2 cm wide, auricle small or absent, cross-veins prominent. *Inflorescences* interfoliar, erect at anthesis and becoming pendulous in fruit; peduncle not seen; prophyll not seen; peduncular bract 1.4–1.9 cm long, woody, and persistent, sulcate adaxially; predominantly staminate inflorescences with rachis 75–80 cm long; rachillae numerous, to 20

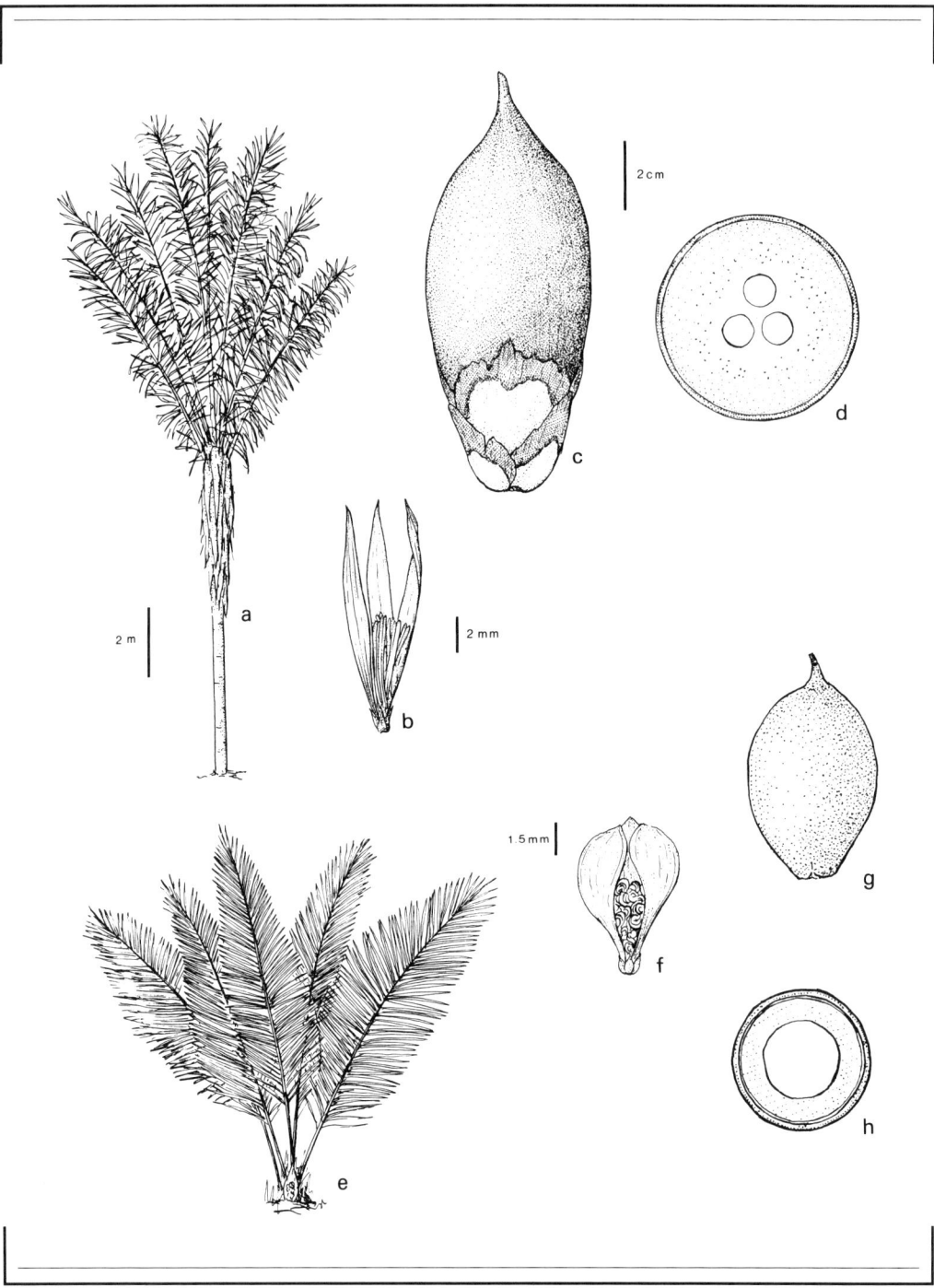

Figure 6.30. **(a)** *Attalea tessmanii,* **(b)** staminate flower (from J. Jangoux 85,nd027), **(c)** fruit, **(d)** cross section of fruit (c and d from A. Henderson 1656); **(e)** *Attalea spectabilis,* **(f)** staminate flower (from A. Henderson 1736), **(g)** fruit, **(h)** cross section of fruit (g and h from M. Lobo 109).

cm long, arranged all around rachis; staminate flowers to 1 cm long, in pairs, these in a single file on abaxial side of rachilla; sepals 3, free, very briefly imbricate basally, free above, triangular, to 3 mm long; petals 2–3 (if 2 by fusion of 2 petals), free, irregularly oblong and curved over and bifid at apex, 12 mm long; stamens 19–28, anthers irregularly coiled and twisted; pistillode small; predominantly pistillate inflorescences with rachis to 75 cm long; rachillae numerous, to 7 cm long; pistillate flowers to 3 cm long, 1–6 per rachilla, absent from abaxial surface; sepals ovate to 3 cm long; petals ovate, to 3 cm long; staminodial ring 6 mm long; stigmas 3; *fruits* ellipsoid-oblong, 7.5–10.7 cm long, 3.5–7 cm diam, brown; endocarp fibers numerous, scattered; seeds 3–6.

Mostly along the southern periphery of the Amazon region in Guyana, Surinam, Brazil (Acre, Amazonas, Bahia, Maranhão, Pará, Rondônia, Tocantins), and Bolivia (Beni, Pando, Santa Cruz) (Figure 6.27F). Its western limit in Bolivia seems to coincide with the margin of the Brazilian Shield. It grows in drier regions in seasonal forest and is especially abundant in areas that have been disturbed by humans, seeming to thrive in such places.

The pollination of this species has been studied by Anderson and co-workers (1988) (as *Orbignya phalerata*). The palm has a complex breeding system, and the most important pollinator is the nitidulid beetle *Mystrops mexicana*.

Bolivia: *cusí*. Brazil: *babaçu, babassu, jetahu-'y* (Ka'apor), *marityna* (Assurini), *wa-'y* (Guajá). This species is the basis of an important local industry in the northeast of Brazil. Large quantities of oil are extracted from the seeds (Anderson et al., 1991). The palm also provides a host of other minor useful products. There is currently an active program for domestication of this species in Brazil.

Anderson and Balick (1988) have discussed the taxonomy of this species (as *Orbignya phalerata*).

13. Attalea spectabilis Mart., Hist. nat. palm. 2: 136. 1826. *Orbignya spectabilis* (Mart.) Burret, Notizbl. Bot. Gart. Berlin-Dahlem 10: 508. 1929. *Attalea spectabilis* var. *typica* Drude in Mart., Fl. bras.: Palmae II 3(2): 440. 1882. Type. Brazil. Pará: Serra de Barú (= Parú), n.d., *C. Martius s.n.* (holotype, M, n.v.) (Figure 6.30e–h).

Attalea monosperma Barb. Rodr., Enum. palm. nov. 42. 1875. *Attalea spectabilis* var. *monosperma* (Barb. Rodr.) Drude in Mart., Fl. bras.: Cyclanthaceae et Palmae I, fasc. 85, vol. 3(2): 440. 1881. Type. Barb. Rodr., Sert. palm. brasil 1: t. 57A. 1903.

Stems solitary but spreading by rhizomes, short and subterranean, rarely aerial and to 1 m long. *Leaves* to 8; sheath and petiole not seen; rachis to 4.4 m long; pinnae to 83 per side, regularly arranged and spreading in 1 plane, linear, aristate at the apex but becoming bifid, the middle ones 45–88 cm long, 3–4 cm wide, auricle small or absent, cross-veins obscure. *Inflorescences* interfoliar; peduncle and prophyll not seen; peduncular bract 60 cm or more long, sulcate abaxially; rachis to 30 cm long; staminate rachillae numerous, 3–5 cm long, 3–4 mm diam, arranged all around rachis, with silvery-white tomentum; pistillate rachillae not seen; staminate flowers to 9 mm long, arranged in two rows on 1 side of rachillae; sepals 3, very briefly imbricate basally, free and spreading above, triangular, 1 mm long; petals 2, free, linear-lanceolate, 1 wider and bifid apically, curved over at apex; stamens 18–20, with anthers coiled and twisted; pistillode small, lobed; pistillate flowers not seen; *fruits* ovoid or ellipsoid, yellowish-brown with brown tomentum (especially on young fruit), 5–6 cm long, 3–4 cm diam; endocarp pores slightly sunken; endocarp thick, fibers few, near outer margin; seeds 1–2.

Brazil (Pará) (Figure 6.31A); usually in savannas or semiopen areas on sandy soils at low elevations. This is one of the few short-stemmed species of *Attalea* that flourishes in disturbed areas (like its close relative, *A. speciosa*). Near Monte Alegre in Pará, this species is a weed of open fields. Pires-O'Brien (1993) has reported that it is one of the most persistent weeds of the Jari timber plantation. She also reports that the species seldom flowers but reproduces by underground rhizomes.

Brazil: *curuá, palha preta, palheira*.

Although Glassman (1977a) considered that the name *Attalea spectabilis* was somewhat doubtful, it is applied here to a few collections of "*Orbignya*" from savanna areas of Brazil

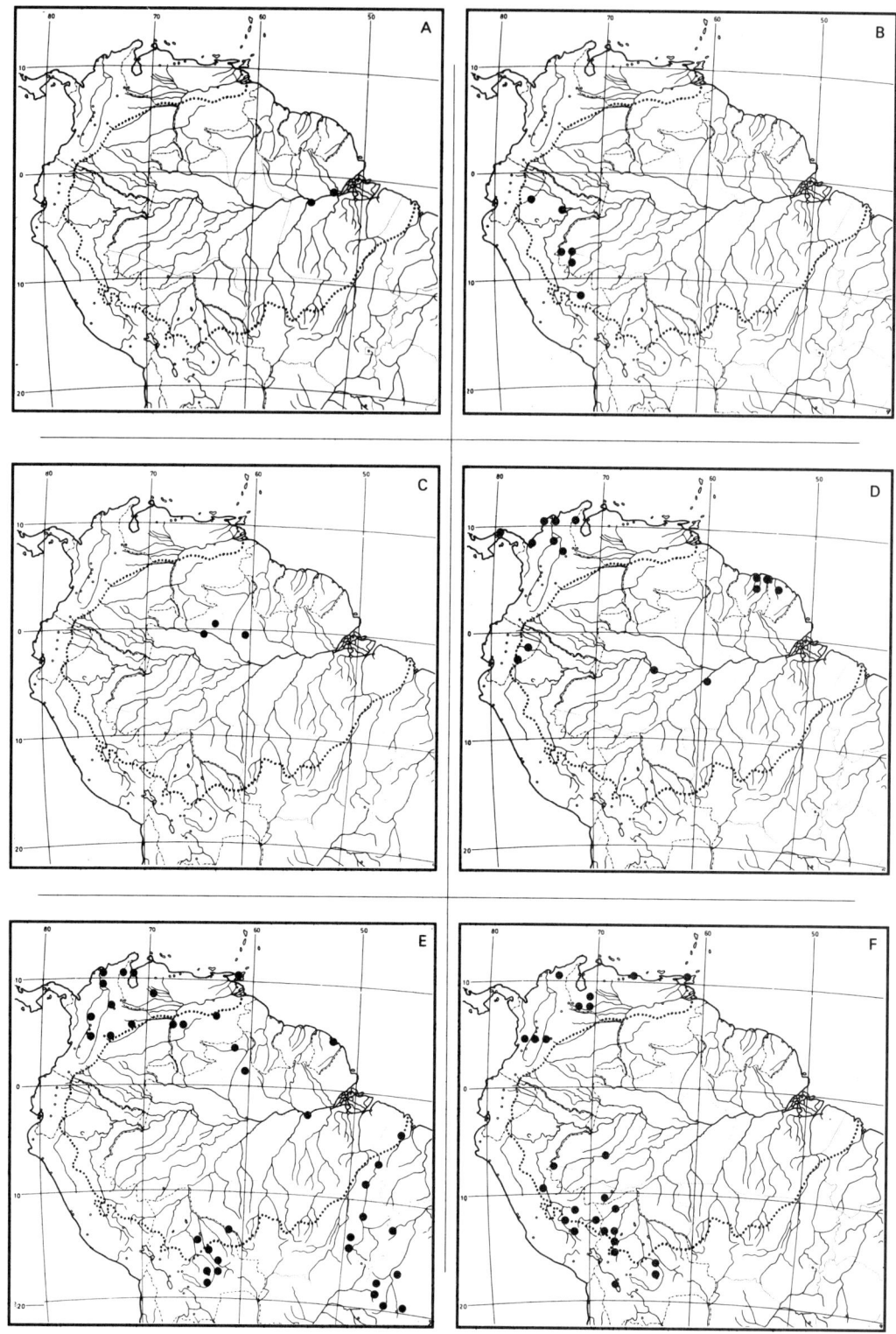

Figure 6.31. (A) *Attalea spectabilis;* **(B)** *Attalea tessmannii;* **(C)** *Barcella odora;* **(D)** *Elaeis oleifera;* **(E)** *Acrocomia aculeata;* **(F)** *Aiphanes aculeata.*

(Pará), with short and subterranean stems and large fruits. The staminate flowers described previously (from *A. Henderson 1736*), with 2 petals and 18–20 stamens, are actually more like those of *A. speciosa*. Other collections of *A. spectabilis* (*M. Pires 1765*) appear to have staminate flowers more like those of *A. microcarpa*, with 3 spathulate petals and 9–15 stamens. It seems possible, then, that *A. spectabilis*, as here conceived, is a mixture of large-fruited *A. microcarpa* and hybrids between *A. microcarpa* and *A. speciosa*. Although the latter has not been collected in the lower Amazon region, it is reported by Barbosa Rodrigues (1903) to occur on the Rio Tapajós. It is also reported to form hybrids with other species (Balick et al., 1987). Alternatively, some specimens of *A. spectabilis* could be conceived of as stemless plants of *A. speciosa*. Wessels Boer (1963) reported *A. spectabilis* from Surinam (with 6–9 stamens), and he also reported *A. speciosa* and *A. microcarpa* (as *A. sagotii*). More collections are needed to solve these problems.

14. **Attalea tessmannii** Burret, Notizbl. Bot. Gart. Berlin-Dahlem 10: 538. 1929. Type. Peru. Loreto: Soledad, Jun 1925, *G. Tessmann 5167* (holotype, B; isotype, NY). (Figuro 6.30a–d).

Stems solitary, erect, 8.5–19 m tall, 31–40 cm diam. *Leaves* ca. 12; sheath 1.4–2 m long, with a few thin fibers on the margins; petiole elongate, 1.3–2.2 m long; rachis 6.5–8.5 m long; pinnae 190–295 per side, ± clustered and spreading in different planes, linear, the middle ones 1.2–1.4 m long, 5–6 cm wide, auricle present. *Inflorescences* interfoliar; peduncle 0.8–1.9 m long; prophyll to 70 cm long; peduncular bract 1.8–2.7 m long, deeply sulcate adaxially; rachis 1–1.6 m long; staminate rachillae numerous, 26–32 cm long, tending to be arranged on 1 side of rachis; pistillate rachillae 103–163, 7–13 cm long, arranged all around rachis, with 1–3 flowers; staminate flowers 1.2–1.5 cm long, arranged on all sides of rachillae; sepals very briefly imbricate at base, free and spreading above, narrowly triangular, 1.2–1.5 mm long; petals 3(–5), free, linear, 1.2–1.5 cm long, with whitish scales abaxially and marginally; stamens 10–14, included within petals; pistillode trifid, ca. 1 mm long; pistillate flowers (immature) 2.5 cm long; sepals triangular, 1.8 cm long; petals triangular, 1.5 mm long; staminodial ring 1 mm long; *fruits* ellipsoid-oblong, 12.5–13 cm long, 6.5–7 cm diam, brownish; endocarp pores sunken; endocarp very thick, 1.7 cm diam, with numerous, small, solitary, ± evenly scattered fibers; seeds 3.

Western Amazon region in Peru (Loreto, Madre de Dios, Ucayali) and Brazil (Acre) (Figure 6.31B); lowland rain forest on *terra firme* at low elevations. It is either uncommon or rarely collected, or both.

Brazil: *cocão*. Peru: *conta, katirina* (Achual Jívaro). In Acre, Brazil, the endocarps are burned to smoke wild-gathered rubber and are used to make charcoal. In Peru, the leaves are occasionally used for thatch; the palmito is eaten; and seeds are eaten by children.

ARECOIDEAE • COCOEAE • ELAEIDINAE

22. *Barcella*

Barcella Drude in Mart., Fl. bras.: Cyclanthaceae et Palmae I, fasc. 85, vol. 3(2): 459. 1881.

Small, monoecious palms. *Stems* solitary, subterranean. *Leaves* pinnate, reduplicate; sheath open and not forming a crownshaft; petiole short; rachis long; pinnae regularly arranged and spreading in 1 plane. *Inflorescences* interfoliar, branched to 1 order; peduncle bearing a prophyll and 1 woody peduncular bract; rachis bearing few rachillae; *flowers* arranged in groups derived from triads but usually borne singly or in pairs, proximal rachillae with pistillate flowers proximally and staminate flowers distally, distal rachillae all staminate,

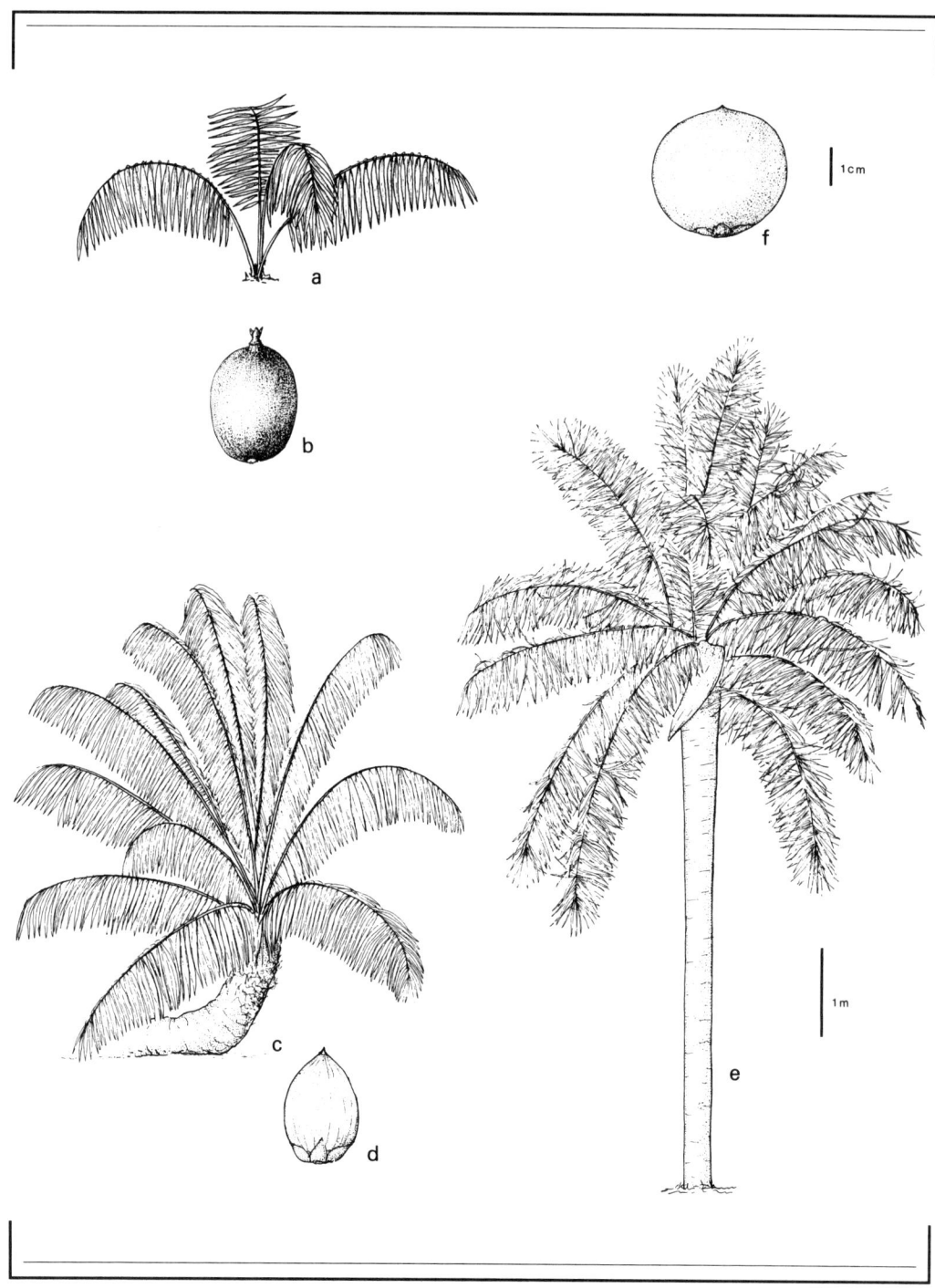

Figure 6.32. (a) *Barcella odora,* (b) fruit (from *A. Henderson 484*); (c) *Elaeis oleifera,* (d) fruit (from *A. Henderson 1508*); (e), *Acrocomia aculeata,* (f) fruit (from *E. Little 13764*).

or occasionally an inflorescence all staminate; staminate flowers with sepals 3, free, imbricate; petals 3, free, valvate; stamens 6, the filaments connate basally into a tube; pistillode trifid; pistillate flowers with sepals 3, free, strongly imbricate; petals 3, free, imbricate; staminodial ring present; gynoecium syncarpous, trilocular, triovulate; *fruits* 1–2 seeded, ellipsoid-oblong with apical stigmatic residue; endocarp thick and bony, with lateral pores; seed with homogeneous endosperm and lateral embryo; germination not known; eophyll entire.

A genus of one species (Drude, 1881) distributed in the central Amazon region of Brazil, north of the Amazon river.

1. **Barcella odora** (Trail) Drude in Mart., Fl. bras.: Cyclanthaceae et Palmae I fasc. 85 vol 3(2): 460. 1881. *Elaeis odora* Trail, J. Bot. 6: 81. 1877. Type. Brazil. Amazonas: Rio Padauari, 26 Jun 1874, *J. Trail 1124/LXXVI* (holotype, K) (Figure 6.32a–b).

Stems solitary, subterranean. *Leaves* 4–7; sheath subterranean, not seen; petiole to 52 cm long, decaying into persistent, black fibers; rachis to 1.8 m long; pinnae to 26 per side, regularly arranged and spreading in 1 plane, linear, the middle ones 40–70 cm long, 3–4 cm wide. *Inflorescences* interfoliar, exserted above the leaf sheaths; peduncle 75–90 cm long; prophyll subterranean, not seen; peduncular bract to 90 cm long, somewhat woody and persistent; rachis 16–25 cm long; rachillae 13–50, 8–13 cm long, 3–4 mm diam, proximal few with both pistillate and staminate flowers (then pistillate flowers basal, staminate above), distal ones with staminate flowers only; *flowers* borne in depressions in rachillae; staminate flowers paired or solitary, 3.5 mm long at anthesis; sepals ovate, 2.5 mm long; petals ovate, 3 mm long; pistillode to 1.5 mm long; pistillate flowers 12 mm long at anthesis; sepals depressed ovate, 6 mm long; petals depressed ovate, 6 mm long; staminodial ring 1 mm long, 6-toothed; *fruits* ellipsoid-oblong, markedly rostrate with persistent stigmas, dark greenish-brown, 3–3.5 cm long, 2.5–2.7 cm diam.

Central Amazon region of Brazil (Amazonas), north of the Amazon River (Henderson, 1986b) (Figure 6.31C); on poorly drained, gleyic podzols, in areas of low, shrubby, *campina* vegetation at low elevations.

Brazil: *piassaba, piassaba preta*. The abundant fibers from the decaying petioles are not strong enough to be of any commercial value.

23. *Elaeis*

Elaeis Jacq., Select. stirp. amer. hist. 280. 1763.

Moderate to large, monoecious palms. *Stems* solitary, erect or procumbent and then rooting. *Leaves* pinnate, reduplicate; sheath open and not forming a crownshaft; petiole moderate, with recurved spines; rachis long; pinnae regularly or irregularly arranged, spreading in 1 or different planes. *Inflorescences* interfoliar, branched to 1 order, usually unisexual but both staminate and pistillate occurring on the same tree; peduncle bearing a prophyll and 1 fibrous peduncular bract; rachis condensed, bearing numerous, short, crowded rachillae; *flowers* borne singly or paired (staminate) in pits formed by the bracts subtending the flowers; staminate flowers with sepals 3, free; petals 3, free; stamens 6; pistillode present; pistillate flowers with sepals 3, free, imbricate; petals 3, free, imbricate; staminodial ring present; gynoecium syncarpous, trilocular, triovulate; *fruits* usually 1-seeded, irregularly ovoid, with apical stigmatic residue; endocarp thick and bony, with subapical pores; seed with homogeneous endosperm and subapical embryo; germination adjacent-ligular; eophyll entire.

A genus of two species (Bailey, 1933; Schultes, 1990). The African oil palm, *Elaeis guineensis* Jacq., originally from West Africa, is now widely cultivated and naturalized throughout tropical areas of the world. It is commonly planted in the Neotropics, both as an ornamental and as a commercial crop. The

American oil palm, *E. oleifera,* is found only in the Neotropics and is generally not cultivated. *Elaeis* is thus the second genus of palms, after *Raphia,* to occur in both Africa and America. Some authors have claimed that the African species of *Elaeis* was introduced there from America (e.g., Cook, 1942; see also Schultes, 1990), although this seems unlikely. Both species are keyed here.

KEY TO THE SPECIES OF *ELAEIS*

1. Stem short or procumbent; pinnae regularly arranged and spreading in 1 plane; bracts subtending pistillate flowers not elongated into a spine . . . . . 1. *E. oleifera.*
1. Stem tall and erect; pinnae ± irregularly arranged in clusters and spreading in different planes; bracts subtending pistillate flowers elongated into a spine . . . . 2. *E. guineensis.*

1. Elaeis oleifera (Kunth) Cortés, Flora de Colombia 1: 203. 1897. *Alfonsia oleifera* Kunth in Humb., Bonpl. & Kunth, Nov. gen. sp. 1: 307. 1815. *Corozo oleifera* (Kunth) L. H. Bailey, Gentes Herb. 3: 59. 1933. Type. Colombia. Atlantico: Cartagena, Mar 1801, *A. Bonpland 5379* (holotype, P) (Figure 6.32c–d).
Elaeis melanococca Mart. non Gaertn., Hist. nat. palm. 2: 64. 1824 (*Elaeis melanococca* Gaertn. = *Elaeis guineensis* Jacq.). Type not designated.

Stems solitary, often curved, procumbent basally and ± erect apically, 1–6 m long, to 40 cm diam, rough, covered in persistent, decaying leaf sheaths and accumulated litter, rooting where procumbent. *Leaves* 20–50; sheath 20–40 cm long, open and not forming a crownshaft, very fibrous; petiole 1.5–3 m long, with stout, recurved spines on margins; rachis 2.9–5.5 m long; pinnae 33–90 per side, regularly arranged and spreading in 1 plane, linear, long acuminate, the middle ones 1–1.2 m long, 4–6 cm wide. *Inflorescences* interfoliar, borne tightly among leaf sheaths; prophyll not seen; peduncular bract fibrous and persistent; peduncle 35–80 cm long; rachis 18–25 cm long; rachillae to 100, 6–20 cm long, the staminate ca. 1 cm diam with the flowers very closely spaced in "pits" formed by a network of bracts and membranes, the pistillate ca. 1.5 cm diam, with the flowers loosely spaced in "pits"; staminate flowers 5 mm long at anthesis; sepals oblanceolate, 3 mm long; petals oblanceolate, 3 mm long; pistillode as long as the anther tube; pistillate flowers 1.2 cm long (postanthesis); sepals ovate, 1.2 cm long; petals ovate, 1 cm long; staminodial ring 1 mm high; *fruit* ellipsoid-oblong, 2.5–3 cm long, 1.8–2 cm diam, orange, orange-yellow, or red, with prominent, apical stigmatic residue.

Central America (Honduras to Panama) and in northern Colombia and Venezuela, also in scattered localities in Brazil, Ecuador, French Guiana, Peru, and Surinam (Figure 6.31D). De Blank (1952) considered it to be introduced into the Amazon region of Brazil, where it is apparently common on the Rio Madeira (Moses, 1962). Andrade (cited in Balée, 1989) found a strict association of this species with *terra preta* soils on the Rio Madeira. Given its sporadic distribution, its occurrence does indeed appear to be anthropogenic in the Amazon region. It grows in low-lying, wet areas, and can form large stands.

Brazil: *caiaué*. The fruits are cooked and used either to make a wine or to extract the oil.

ARECOIDEAE • COCOEAE • BACTRIDINAE

24. *Acrocomia* **Acrocomia** Mart., Hist. nat. palm. 2: 66. 1824.

Small to large, monoecious, spiny palms. *Stems* solitary, either large and occasionally swollen, or small and subterranean. *Leaves* pinnate, reduplicate, usually very spiny; sheath open and not forming a crownshaft; petiole very short;

rachis long; pinnae irregularly arranged, spreading in different planes. *Inflorescences* interfoliar, branched to 1 order, spiny; peduncle bearing a prophyll and 1 peduncular bract; rachis bearing few to numerous rachillae; *flowers* borne in triads proximally on rachillae, paired or solitary staminate distally; staminate flowers with sepals 3, free; petals 3, very briefly connate basally, free and valvate above; stamens 6; pistillode trifid; pistillate flowers with sepals 3, imbricate; petals 3, imbricate; staminodes 6; gynoecium trilocular, triovulate, stigmas large, reflexed; *fruits* 1-seeded, globose, with apical stigmatic residue; endocarp with 3 pores near the equator; seed with homogeneous endosperm and lateral embryo opposite 1 of the pores; germination adjacent-ligular; eophyll simple, linear.

Although Bailey (1941) recognized 25 species of *Acrocomia*, these actually represent one, widespread, variable species: *A. aculeata* (Jacq.) Lodd. ex Mart. Uhl and Dransfield (1987) have added a second species to *Acrocomia* by including in it *Acanthococos*. The one species of this genus, *A. hassleri* Barb. Rodr. (including its synonyms, *A. sericea* Burret, *A. emensis* Toledo, and *A. emensis* var. *pubifolia* Toledo), occurs in northern Paraguay and central eastern Brazil (see also Hahn, 1991). A second monotypic genus, *Gastrococos* from Cuba, is closely related and may not be distinct. Thus conceived, the genus becomes easier to understand; of the three species, one is widespread and the other two are marginal, northern and southern, isolates.

1. **Acrocomia aculeata** (Jacq.) Lodd. ex Mart., Hist. nat. palm. 3: 286. 1849. *Cocos aculeatus* Jacq., Select. stirp. amer. hist. 278. 1763. Lectotype (Glassman, 1972). Jacq., Select. stirp. amer. hist. t. 169. 1763 (Figure 6.32e–f).

Acrocomia sclerocarpa Mart., Hist. nat. palm. 2: 66. 1824. Type. Brazil, Pará: without locality, n.d., *F. Sieber s.n.* (holotype, M; F neg. 18577).

Acrocomia lasiospatha Mart. in A. D. Orb., Voy. Amérique mér. 7(3). Palmiers 81. 1844. Type. French Guiana. Cayenne, n.d., *F. Leprieur s.n.* (holotype, M).

Acrocomia lasiospatha Wallace, Palm Trees of the Amazon 97. 1853. *Acrocomia sclerocarpa* var. *wallaceana* Drude in Mart., Fl. bras.: Cyclanthaceae et Palmae I, fasc. 85, vol. 3(2): 391. 1881. *Acrocomia wallaceana* Becc., Pomona Coll. J. Econ. Bot. 2: 362. 1912. Lectotype (here designated). Wallace. Palm Trees of the Amazon t. 37. 1853.

Acrocomia microcarpa Barb. Rodr., Vellosia 1: 51. 1888. Lectotype (Glassman, 1972). Barb. Rodr., Sert. palm. brasil. 2: t. 82a. 1902.

Acrocomia eriocantha Barb. Rodr., Contr. Jard. Bot. Rio de Janeiro 3: 85. 1902. Lectotype (Glassman, 1972). Barb. Rodr., Sert. palm. brasil. 2: t. 83. 1902.

Stems solitary, erect, 4–12 m tall, 10–35 cm diam, occasionally swollen near middle, occasionally covered with persistent leaf sheath bases, internodes smooth or with spines to 15 cm long (generally younger plants with spines, older plants with smooth stems), often with visible cone of roots to 40 cm long at base. *Leaves* 10–30; sheath 30–90 cm long, fibrous on margins, sheath, petiole, and rachis with moderate to dense covering of black spines to 10 cm long; petiole (3–)28–100 cm long; rachis 1.9–3.7 m long; pinnae 141–190 per side, regularly or irregularly arranged in clusters and spreading in different planes, linear, long acuminate, middle pinnae 60–83 cm long, 2–4 cm wide, densely to scarcely pilose abaxially. *Inflorescences* interfoliar; peduncle 20–90 cm long; prophyll to 70 cm long; peduncular bract 0.8–1.4 m long, densely covered abaxially with brown, velvety tomentum and with scattered, black spines to 2 cm long; rachis 60–90 cm long; rachillae 50–137, 10–36 cm long, distal (staminate) part of rachillae with bracteoles forming a network of shallow pits, and sometimes the bracteoles elongated; *flowers* arranged in a few isolated triads on proximal ca. half of rachillae, paired or solitary staminate distally; staminate flowers to 6 mm long, crowded; sepals spreading, ovate, to 2.5 mm long; petals ovate, to 5.5 mm long; pistillode 2 mm long; pistillate flowers at anthesis 8 mm long; sepals very widely ovate, 3 mm long; petals very widely ovate, 5 mm long; staminodial ring 3 mm long, adnate to petals basally, free above and with 6 minute, stamenlike staminodes; *fruits* globose, yellowish or yellowish-green at maturity, 2.5–3.5 cm diam, spiny, glabrescent, the epicarp easily cracking.

Throughout the Neotropics, from Mexico through Central and South America to Paraguay and Argentina. It also occurs in the Greater and Lesser Antilles (Figure 6.31E). In general, it occurs around the northern, eastern, and southern periphery of the Amazon region from Colombia, Venezuela, the Guianas, Brazil, and Bolivia. A few collections suggest that it also occurs in the dry "corridor" of *Aw* climate that runs northwest–southeast across the lower Amazon (Huber et al., 1984, for example, reported that it occurred on the Sierra Parima on the Venezuela–Brazil frontier). It grows in drier, open or savanna areas, often in disturbed places, at low elevations or occasionally to 2100 m in the northern Andes of Colombia. It has almost certainly had its range extended by humans, but its place of origin is unknown. For example, Lentz (1990) reported that *A. aculeata* (as *A. mexicana*) was probably introduced into sites in Mesoamerica by the Maya.

Phenology, floral biology and pollination of this species were studied by Scariot and colleagues (1991) near Brasilía in Brazil. Flowering took place between August and December, and fruiting between March and June. The principal pollinators were curculionid, nitidulid, and scarab beetles.

Bolivia: *totai*. Brazil: *macaúba, mucuja* (fruits), *mucujazeiro* (tree), *parena* (Guajajara). Colombia: *corozo*. Venezuela: *amankayo* (Panare), *corozo, tucuma*. The fruits are edible, and oil is extracted from the endosperm. The roots are used to make a tea for the cure of hepatitis.

There is some variation in swelling of stems, persistence of leaf sheath bases, and pilose abaxial pinnae surface, none of which has any taxonomic significance.

25. *Aiphanes*

Aiphanes Willd., Mém. Acad. Roy. Sci. Hist. (Berlin) 1804: 32. 1807.

Small to moderate, rarely large, monoecious, spiny palms. *Stems* solitary or cespitose, aerial or short and subterranean, erect or rarely somewhat procumbent. *Leaves* pinnate, reduplicate, spiny; sheath open and not forming a crownshaft; petiole short to moderate; rachis long; pinnae lanceolate to obtriangular, with praemorse apex, regularly or irregularly arranged, or occasionally leaf entire. *Inflorescences* interfoliar, branched to 1 or rarely 2 orders, or rarely spicate, spiny, elongate; peduncle bearing a prophyll and 1 peduncular bract; rachis bearing few to many rachillae; *flowers* borne in triads basally on rachillae with paired or solitary staminate flowers apically; staminate flowers with sepals 3, briefly connate or imbricate basally, free and spreading above; petals 3, free, valvate; stamens 6; pistillode small; pistillate flowers with sepals 3, imbricate basally; petals 3, valvate; staminodes connate into a 6-toothed ring; gynoecium syncarpous, trilocular, triovulate; *fruits* 1-seeded, globose or ellipsoid, usually red, occasionally spinulose, with apical stigmatic residue; seed with homogeneous endosperm and lateral embryo; germination adjacent-ligular; eophyll entire with bifid apex.

A genus of 22 species (Borchsenius & Bernal, in press) distributed in the Antilles and from Panama south through the Andes of Colombia, Venezuela, Ecuador, Peru, and Bolivia; four species just reach the western Amazon region of these countries.

KEY TO THE SPECIES OF *AIPHANES*

1. Rachillae appressed and partially adnate to the rachis . . . . 3. *A. ulei.*
1. Rachillae spreading, not appressed nor adnate to the rachis.
   2. Pinnae abruptly widening near the apex; stems 2–6(–10) m tall, 4–10.5 cm diam . . . . 1. *A. aculeata.*
   2. Pinnae gradually tapering to the apex; stems 0–3 m tall, 2.5–3.5 cm diam.
      3. Petiole 1 m long; rachis 1.3–1.9 m long; rachillae 49–60, slender . . . 2. *A. deltoidea.*
      3. Petiole 20–35 cm long; rachis 0.7–1.1 m long; rachillae 8–34, thickened basally . . . . 4. *A. weberbaueri.*

1. **Aiphanes aculeata** Willd., Samml: Deutsch. Abh. Königl. Akad. Wiss. Berlin 1803: 251. 1806. *Euterpe aculeata* (Willd.) Spreng., Syst. veg. 2: 140. 1825. *Martinezia aculeata* (Willd.) Klotzsch, Linnaea 20: 455. 1847. *Martinezia aiphanes* Mart. in A. D. Orb., Voy. Amérique mér. 7(3). Palmiers 77. 1844. *Marara aculeata* (Willd.) H. Karst., Fl. Columb. 2: 143. 1866. *Caryota horrida* Jacq., Fragm. Bot. 20. 1809. *Aiphanes horrida* (Jacq.) Burret, Notizbl. Bot. Gart. Berlin-Dahlem 11: 575. 1932. Neotype (Borchsenius & Bernal, in press). Venezuela. Miranda: Dist. Brión, Quebrada Agua Bendíta, Río Aricagua, 2.3 km E of Pueblo Seco, 4.6 km E of Aricagua, 75 m, 24–25 Mar 1973, *J. Steyermark & V. Espinoza 106916* (isoneotype, NY) (Figure 6.33a–d).

Martinezia caryotifolia Kunth in Humb., Bonpl. & Kunth, Nov. gen. sp. 1: 305. 1815. *Marara caryotifolia* (Kunth) H. Karst., Fl. Columb. 2: 134. 1866. *Aiphanes caryotifolia* (Kunth) H. Wendl. in Kerch., Palmiers 233. 1878. *Tilmia caryotifolia* (Kunth) O. F. Cook, Bull. Torrey Bot. Cl. 28: 565. 1901. Type. Colombia. Tolima: Ibague, n.d., *A. Humboldt & A. Bonpland s.n.* (holotype, P, n.v.).

Bactris praemorsa Poepp. ex Mart. in A. D. Orb., Voy. Amérique mér. 7(3). Palmiers 66. 1844. *Aiphanes praemorsa* (Poepp. ex Mart.) Burret, Notizbl. Bot. Gart. Berlin-Dahlem 11: 575. 1932. Type. Peru. San Martín: Tocache, Río Huallaga, Jul 1830, *W. Poeppig s.n.* (holotype, M, n.v.).

Martinezia ernesti Burret, Notizbl. Bot. Gart. Berlin-Dahlem 11: 327. 1932. *Aiphanes ernesti* (Burret) Burret, Notizbl. Bot. Gart. Berlin-Dahlem 11: 560. 1932. (based on *Martinezia ulei* Dammer, Notizbl. Königl. Bot. Gart. Berlin 59: 266. 1915. nom. illeg., not *Martinezia ulei* Dammer, Verh. Bot. Vereins Prov. Brandenburg 48: 127. 1907). Neotype (Borchsenius & Bernal, in press). Peru. Madre de Dios: Tambopata, Barsola, Río Piedras, 350 m, 17 Jan 1967, *C. Vargas 18694* (holoneotype, BH).

Aiphanes orinocensis Burret, Notizbl. Bot. Gart. Berlin-Dahlem 11: 560. 1932. Lectotype (Borchsenius & Bernal, in press). Venezuela. State?: Orinoco, n.d., *H. Rusby & R. Squires s.n.* (lectotype, NY).

Stems solitary, 2–6(–10) m tall, 4–10.5 cm diam, spiny at internodes with spines to 6 cm long. *Leaves* 6–15; sheath (15–)26–60 cm long, sheath, petiole (and rachis) moderately to densely covered with flattened, black, brown, or gray spines to 5 cm long; petiole 11–65 cm long; rachis 1–2.3 m long; pinnae 13–40 per side, irregularly arranged in clusters of 2–7, obtriangular, abruptly widening near the apex, bicuspidate, praemorse at the apex, sometimes with spines on abaxial surface, the middle ones 23–43 cm long, 3–5 cm wide at midpoint. *Inflorescences* interfoliar, erect at anthesis, becoming pendulous in fruit; peduncle 0.8–2.3 m long, spiny; prophyll 35–60 cm long; peduncular bract 0.6–1.7 m long; rachis 34–62 cm long; rachillae numerous, 70–81, basal ones 12–64 cm long, apical ones 5–12 cm long, slender; *flowers* in triads, these regularly arranged and isolated basally on rachillae, with paired or solitary staminate apically; staminate flowers 6 mm long; sepals briefly imbricate basally, narrowly triangular, 2 mm long; petals lanceolate, 5 mm long; pistillate flowers 2 mm long (preanthesis); sepals depressed ovate, acuminate, 1.2 mm long; petals deltate, 2 mm long; staminodial ring 1 mm high with dentations;

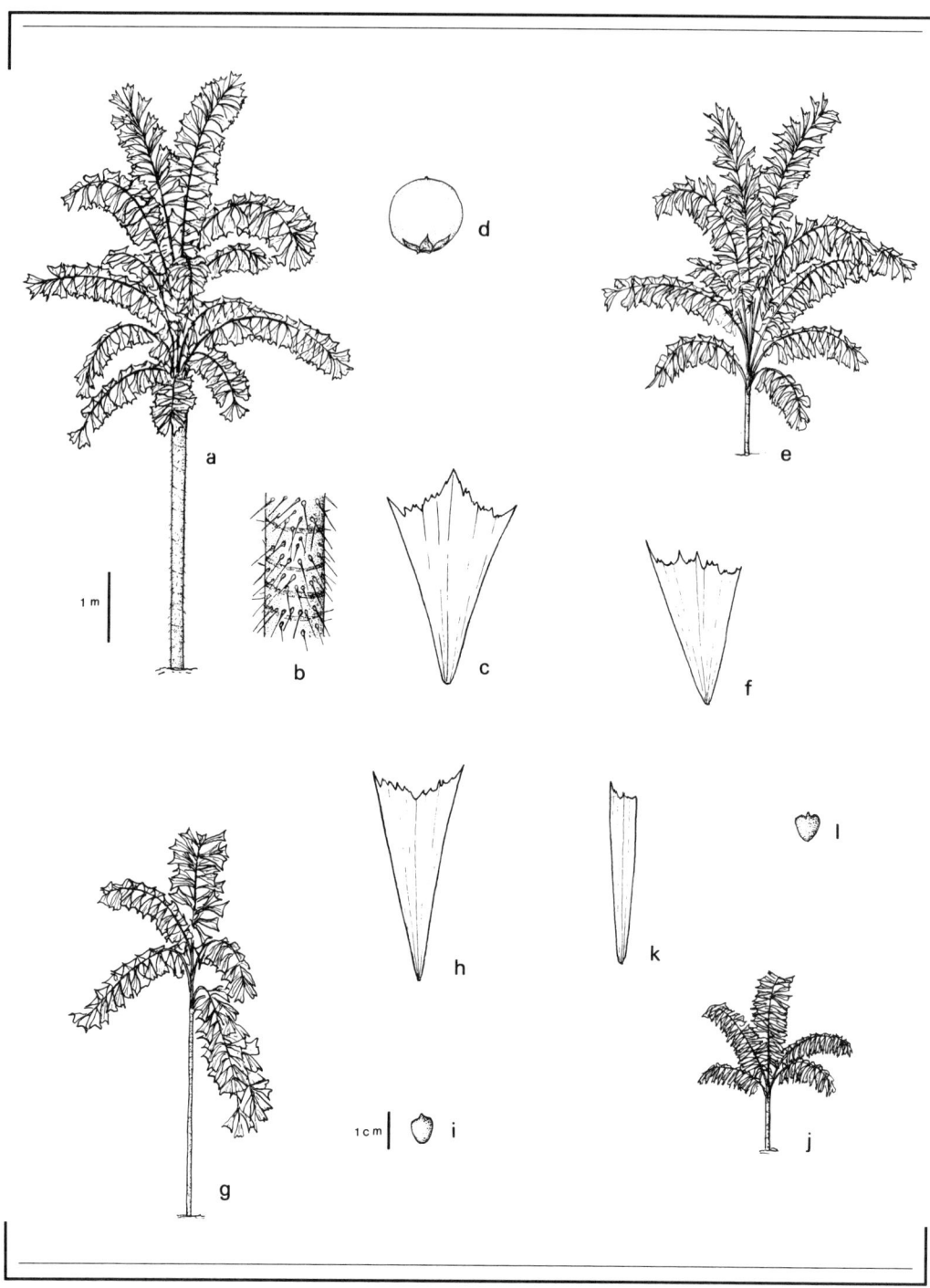

Figure 6.33. (a) *Aiphanes aculeata,* (b) detail of stem, (c) pinna, (d) fruit (c and d from *M. Nee 36870*); (e) *Aiphanes deltoidea,* (f) pinna (from *F. Kahn 2556*); (g) *Aiphanes ulei,* (h) pinna, (i) fruit (h and i from *M. Baker 6805*); (j) *Aiphanes weberbaueri,* (k) pinnae, (l) fruit (k and l from *R. Vásquez 14542*).

fruits globose, 2–2.5 cm diam, orange-red, rarely pink.

Colombia (Antioquia, Casanare, Caldas, Cundinamarca, Meta, Quindio, Tolima, Valle), Venezuela (Barinas, Miranda, Sucre), Trinidad, Peru (Cuzco, Huánuco, Madre de Dios), Brazil (Acre, Amazonas), and Bolivia (La Paz, Pando, Santa Cruz) (Figure 6.31F); lowland or premontane, deciduous or rain forest, or often in disturbed places such as road margins or riverbanks. Commonly planted as an ornamental.

Borchsenius and Bernal (in press) have pointed out that the distribution of this species falls into two disjunct areas: one in Colombia and Venezuela, and the other in Bolivia, Brazil, and Peru. The closely related *Aiphanes eggersii* fills the gap in Ecuador. Listabarth (1992) studied reproductive biology of this species. In Peru, it flowered from February to June and fruited from April to August. Inflorescences were protandrous, and staminate anthesis lasted 15–20 days; pistillate anthesis lasted 2–3 days. Inflorescences were visited by meliponid bees, bugs, and weevils. Listabarth considered that wind pollination was important in this species.

Brazil: *pupunha, pupunha brava*. The fruits and seeds are edible. Balick and Gershoff (1990) reported that the mesocarp of the fruits was rich in vitamin A. A widespread and commonly cultivated species, formerly grown under the name *Aiphanes caryotifolia*.

2. **Aiphanes deltoidea** Burret, Notizbl. Bot. Gart. Berlin-Dahlem 11: 568. 1932. Lectotype (Borchsenius & Bernal, in press). Peru. Loreto: confluence of Río Santiago and Río Marañon, 160 m, 8 Dec 1924, *C. Tessmann 4709* (lectotype, G, n.v.; F neg. 25334) (Figure 6.33e–f).

Stems solitary, short and subterranean or to 2 m tall, 2.5–6 cm diam, spiny on internodes. *Leaves* 10–12; sheath 15–30 cm long, sheath and petiole densely covered with short spines interspersed with longer, black spines to 6 cm long; petiole 0.9–1 m long; rachis 1.3–1.9 m long; pinnae 11–14 per side, narrowly obtriangular, gradually tapering from the apex, irregularly arranged in loose clusters and spreading in different planes, obtriangular, the middle ones 19–32 cm long, 5–8 cm wide at midpoint, paler green abaxially. *Inflorescences* interfoliar; peduncle to 1.7 m long; prophyll 20–25 cm long; peduncular bract not seen; rachis 20–45 cm long; rachillae 49–60, 3–15 cm long, spreading; *flowers* in triads, these regularly arranged almost to apex of rachillae; staminate flowers not seen; pistillate flowers not seen; *fruits* not seen.

Colombia (Amazonas), Peru (Huánuco, Loreto), and Brazil (Amazonas) (Figure 6.34A); lowland or montane rain forest below 700 m elevation.

A poorly known species that is closely related to *Aiphanes weberbaueri*.

3. **Aiphanes ulei** (Dammer) Burret, Notizbl. Bot. Gart. Berlin-Dahlem 11: 568. 1932. *Martinezia ulei* Dammer, Verh. Bot. Vereins. Prov. Brandenburg 48: 127. 1907. Neotype (Borchsenius & Bernal, in press). Peru. Loreto: Prov. Requena, lower Río Ucayali, Jenaro Herrera, n.d., *F. Kahn & K. Mejia 1916* (holoneotype, K, n.v.) (Figure 6.33g–i).

Aiphanes schultzeana Burret, Notizbl. Bot. Gart. Berlin-Dahlem 15: 36. 1940. Type. Ecuador. Pastaza: Mera, 1000 m, 3 Sep 1939, *H. Schultze-Rhonhof 2769* (holotype, B).

Stems solitary, 0.3–4.5 m tall, 2.5–4 cm diam, spiny on internodes. *Leaves* 5–10; sheath 20–30 cm long, sheath, petiole, and rachis with somewhat flattened, black spines to 4 cm long; petiole 21–40 cm long, rachis 0.9–1.2 m long; pinnae 9–14 per side, irregularly arranged in groups of 2, ± in 1 plane, obtriangular, gradually tapering, bicuspidate apically, the middle ones 24–33 cm long, 6–7 cm wide at midpoint. *Inflorescences* interfoliar, erect at anthesis and in fruit; peduncle 75–100 cm long; prophyll 15–35 cm long; peduncular bract 50–87 cm long; rachis 24–30 cm long; rachillae 4–15, adnate to rachis basally, 4–12 cm long; *flowers* borne in congested triads basally, paired or solitary staminate distally, in 2 rows on abaxial side of rachillae; staminate flowers (immature) 1.6 mm long; sepals lanceolate, 1.2 mm long; petals ovate, 1.4 mm long; pistillode trifid, 0.2 mm long; pistillate flowers (immature) 2 mm long; sepals depressed ovate, 1.5 mm long; petals briefly connate basally, ovate, 2 mm long; staminodial ring 0.8 mm long; *fruits* globose, 6 mm diam.

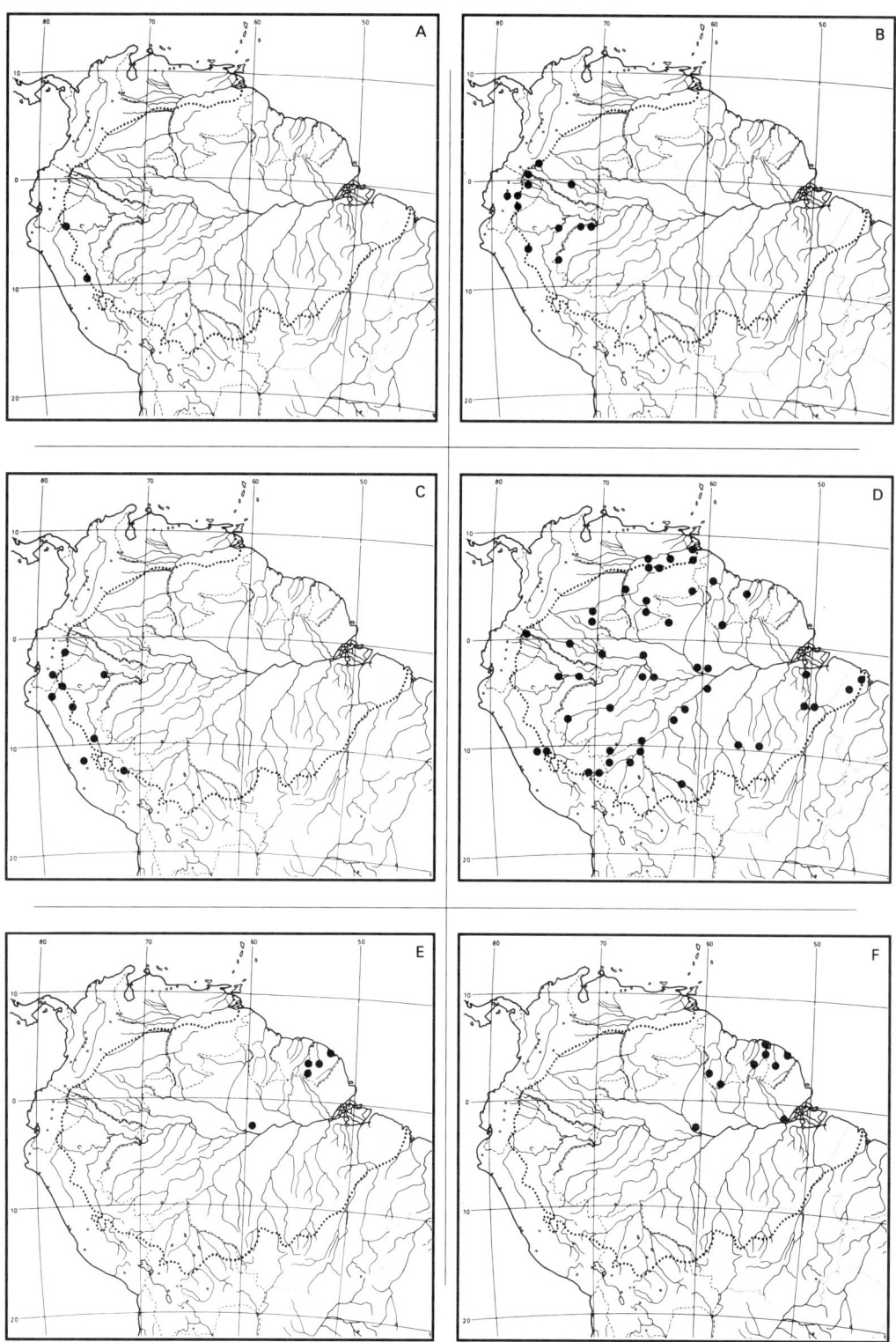

Figure 6.34. **(A)** *Aiphanes deltoidea;* **(B)** *Aiphanes ulei;* **(C)** *Aiphanes weberbaueri;* **(D)** *Bactris acanthocarpa* var. *acanthocarpa;* **(E).** *Bactris acanthocarpa* var. *intermedia;* **(F)** *Bactris acanthocarpoides.*

Colombia (Amazonas, Putumayo), Ecuador (Morona-Santiago, Napo, Pastaza), Peru (Loreto, San Martín), and Brazil (Acre) (Figure 6.34B); lowland or montane rain forest to 1300 m elevation.

Brazil: *pachiubinha de espinha*. Ecuador: *pan si noha* (Kofán).

The specimens from Acre in Brazil have either short or subterranean stems, and the palms appear smaller in all parts. It is strange that this species and *A. weberbaueri* seldom set fruit.

4. **Aiphanes weberbaueri** Burret, Notizbl. Bot. Gart. Berlin-Dahlem 11: 565. 1932. Lectotype (Borchsenius & Bernal, in press). Peru. Huánuco: Río Pozuzo, 1700 m, 20 Jul 1913, *A. Weberbauer 6775* (lectotype, F, n.v.) (Figure 6.33j–l).

Aiphanes tessmannii Burret, Notizbl. Bot. Gart. Berlin-Dahlem 11: 564. 1932. Neotype (Borchsenius & Bernal, in press). Peru. Amazonas: confluence of Río Santiago and Río Marañon, 20 May 1990, *F. Kahn & F. Borchsenius 2546* (isoneotype, NY).

Stems solitary, 0.3–3 m tall, 2.5–3.5 cm diam. *Leaves* 6–13; sheath 15–42 cm long, sheath, petiole, and rachis densely black spinulose, and with rounded, black spines to 5 cm long; petiole 20–35 cm long; rachis 0.6–1.1 m long; pinnae 6–24 per side, irregularly arranged in groups of 3–5, spreading in ± the same plane, obtriangular to almost linear, gradually tapering, cuspidate, the middle ones 16–30 cm long and 2.5–5.5 cm wide at midpoint. *Inflorescences* interfoliar, erect at anthesis; peduncle 0.3–1.1 m long; prophyll 15–32 cm long; peduncular bract 0.9–1 m long; rachis 12–26 cm long; rachillae 8–34, basal ones 16–30 cm long, apical ones to 3 cm long, thicker on pistillate, proximal part; *flowers* in triads for ca. two-thirds the rachillae, paired or solitary staminate distally; staminate flowers 2 mm long (preanthesis); sepals briefly connate basally, widely ovate, 1.5 mm long; petals deltate, 1.5 mm long; pistillode very small; pistillate flowers 1.5 mm long (preanthesis); sepals depressed ovate, 1 mm long; petals depressed ovate, 1 mm long; staminodial ring present; *fruits* ± globose, ca. 6 mm diam.

Ecuador (Morona-Santiago, Pastaza) and Peru (Amazonas, Huánuco, Junín, Loreto, Pasco, Madre de Dios, San Martín) (Figure 6.34C); premontane and montane rain forest on eastern Andean slopes up to 1950 m elevation, and in Amazon lowland rain forest below 350 m elevation.

Peru: *pinglo*.

This is another poorly known species, closely related to *Aiphanes deltoidea*.

26. *Bactris*

Bactris Jacq. ex Scop., Intr. hist. nat. 70. 1777.

Small to large, monoecious, spiny palms. *Stems* cespitose or solitary, usually aerial but also short and subterranean. *Leaves* pinnate or pinnately veined if entire, reduplicate, with various types of spines; sheath open or closed but not forming a crownshaft; petiole moderate to long; rachis short to long; pinnae regularly or usually irregularly arranged, in 1 or different planes, or commonly leaf entire. *Inflorescences* usually interfoliar, branched to 1 order or commonly spicate; peduncle bearing a prophyll and 1 (very rarely 2) peduncular bracts; rachis bearing few to numerous rachillae; *flowers* borne in regularly arranged triads, or triads scattered among paired or solitary staminate flowers; staminate flowers with 3(–4) sepals, these connate basally into a 3-lobed calyx; petals 3(–4), connate basally for ca. half their length, free and valvate above; stamens (3–)6(–12); pistillode minute or absent; pistillate flowers with annular, cupular, or tubular calyx and tubular corolla; staminodes minute or absent, or forming a staminodial ring; gynoecium syncarpous, trilocular, triovulate; *fruits* 1-seeded, smooth, spinulose, or muricate (with fleshy, pyramidal projections), globose, ellipsoid, obovoid, or ovoid, with apical stigmatic residue; mesocarp floury or juicy; endocarp thick and bony, with or without fibers, with 3 pores at or above equator; seed with homogeneous endosperm and embryo opposite 1 of the pores; germination adjacent-ligular; eophyll bifid.

A genus of approximately 60 species distributed throughout the Neotropics; 37 are found in the Amazon region. There is no recent revision, but see Burret (1933–1934) and Sanders (1991).

Some species of *Bactris* are extremely variable morphologically, especially in leaf shape. It is common in various species to find an array of leaf shapes, from entire to pinnate, and pinnae linear to lanceolate to sigmoid. It is this leaf variation that has led to so many names being proposed within the genus. In the following treatment, the six most variable species are divided into varieties.

KEY TO THE SPECIES OF *BACTRIS*

1. Fruits pyriform or ovoid, orange; rachillae densely covered with whitish, flexuous spinules; leaf spines in groups; north of the Amazon in Colombia (Amazonas, Vaupés), Venezuela (Amazonas), Guyana, and Brazil (Amazonas, Pará). . . . . 4. *B. balanophora.*
1. Fruits not pyriform or ovoid and orange; rachillae usually not densely spinulose; leaf spines scattered or rarely in groups.
   2. Stems large and clustered, rarely solitary, 4–18 m tall, 6–25 cm diam; fruits with endocarp fibers flattened and adnate to endocarp, or absent; ocrea small or absent; petiole spines in 3 longitudinal files.
      3. Petiole and rachis spines strongly clustered: pinnae 40–68 per side, the middle ones 3–9 cm wide; a small ocrea present; fruiting corolla with crenulate margins; peduncle ferrugineous; Venezuela (Bolívar) and Surinam. . . . . 31. *B. setulosa.*
      3. Petiole spines not strongly clustered: pinnae 92–135 per side, the middle ones 2–3 cm wide; ocrea absent; fruiting corolla with lobed margins; peduncle yellow-tomentose.
         4. Fruits widely ovoid, to 5 cm long, 3 cm diameter; widespread and always cultivated. . . . . 16. *B. gasipaes.*
         4. Fruits subglobose to obovoid, 1.2–2.3 cm long, 1.1–1.8 cm diam; Peru (Huánuco, Madre de Dios), Brazil (Acre, Rondônia), and Bolivia (Santa Cruz) . . . . 21. *B. macana.*
   2. Stems smaller; fruits with endocarp fibers free, occasionally absent; ocrea present; petiole spines not in 3 longitudinal files.
      5. Inflorescence rachis absent or very short; rachillae few, 1–5(–13), 2.5–7 cm long, with regularly arranged triads throughout; fruits globose, 0.5–0.7 cm diam, or rarely obovoid and to 1 cm diam, red or orange; stems slender, 0.1–2(–3) m tall, 0.3–1(–2) cm diam.
         6. Pinnae 30–35 per side, narrowly linear; rachillae 7–13; Brazil (Pará) . . . . 33. *B. syagroides.*
         6. Pinnae 2–30 per side, linear, linear-lanceolate, or sigmoid, or commonly leaf entire; rachillae 1–5(–8); widespread.
            7. Leaves usually lacking spines except for spinules at pinnae apex, or if spiny then with short, flattened spines to 1.5 cm long; fruits glabrous; pinnae seldom pilose.
               8. Peduncular bract densely spiny; pinnae with prominent cross-veins; French Guiana and Surinam . . . . 3. *B. aubletiana.*
               8. Peduncular bract not spiny; pinnae without cross-veins.

  9. Petiole 20–100 cm long, densely reddish-brown-tomentose; pinnae plicate; peduncle erect at anthesis and in fruit . . . . 20. *B. killipii.*
  9. Petiole 5–26 cm long, not reddish-brown-tomentose; pinnae nonplicate or rarely plicate; peduncle recurved and pointing down at anthesis and in fruit . . . . 32. *B. simplicifrons.*
 7. Leaves usually spiny with terete spines to 5 cm long; fruits spinulose or glabrous; pinnae often pilose.
  10. Fruits spinulose; widespread . . . . 19. *B. hirta.*
  10. Fruits glabrous; French Guiana, Brazil (eastern Amazonas, Pará) . . . . 13. *B. cuspidata.*
5. Inflorescence rachis usually present; rachillae more numerous, 5–89, occasionally 1–3(–7) then usually longer than 2.5–7 cm, the triads scattered or regularly arranged only near the base; fruits variously colored, sized, and shaped, but not as above; stems usually larger.
 11. Rachillae 17–89, filamentous, crowded on a short rachis; triads ± regularly arranged proximally on 1 side of rachillae; endocarp fibers few or absent; fruits usually spinulose, globose to widely obovoid, red, orange-red, or yellowish.
  12. Leaves entire . . . . 36. *B. trailiana.*
  12. Leaves pinnate.
   13. Fruits nonspinulose, glabrous; central Amazon region in Brazil (Amazonas). . . . . 34. *B. tefensis.*
   13. Fruits spinulose.
    14. Pinnae regularly arranged and spreading in 1 plane, linear; French Guiana and Surinam . . . . 29. *B. rhapidacantha.*
    14. Pinnae irregularly arranged and spreading in different planes, linear, linear-lanceolate, or sigmoid; widespread.
     15. Stems usually short and subterranean, rarely to 1 m or more on older plants; pistillate flowers with calyx 1–3 mm long and corolla 2–4 mm long, this usually with a few small spinules; rachillae usually without spinules . . . . 1. *B. acanthocarpa.*
     15. Stems aerial, 1.5–6 m tall; pistillate flowers with calyx 1–1.5 mm long and corolla 3–5 mm long, this densely spinulose; rachillae densely spinulose.
      16. Pinnae linear, the middle ones 61–70 cm long, 2.5–3 cm wide; fruits 1.2–2 cm diam; the Guianas and Brazil (Amazonas, Pará). . . . . 2. *B. acanthocarpoides.*
      16. Pinnae oblanceolate to sigmoid, the middle ones 35–60 cm long, 2.5–7.5 cm wide; fruits ca. 2 cm diam; the Guianas, Peru (Amazonas, Loreto), and Brazil (Acre, Amapá, Amazonas, Pará). . . . . 27. *B. pliniana.*
 11. Rachillae 1–3(–7) or 3–50, not filamentous, not crowded on a short rachis; triads irregularly or occasionally regularly arranged, borne all around and all along rachillae; endocarp fibers present,

rarely absent; fruits seldom spinulose, variously shaped and colored.
17. Fruits orange, red, orange-red, dark red, magenta-purple, yellowish, green, or brown, but not purple-black; rachillae 5–50; endocarp fibers, if present, usually without juice-sacs attached.
    18. Fruits muricate or spinulose.
        19. Fruits muricate . . . . 11. *B. constanciae.*
        19. Fruits spinulose.
            20. Peru (Amazonas) . . . . 9. *B. coloniata.*
            20. Surinam and Brazil (Pará) . . . . 37. *B. turbinocarpa.*
    18. Fruits glabrous.
        21. Leaf spines flattened, grayish; fruits globose, 5–8 mm diam, orange-red; Colombia (Guainía, Vichada), Venezuela (Amazonas), the Guianas, and Brazil (Amapá, Amazonas, Pará, Roraima) . . . . 8. *B. campestris.*
        21. Leaf spines not flattened, black; fruits various but not as above.
            22. Rachillae densely covered with short, brown or white trichomes; leaf spines clustered; Venezuela (Bolívar) and Guyana . . . . 28. *B. ptariana.*
            22. Rachillae not densely covered with trichomes; leaf spines not clustered.
                23. Pinnae lighter green (and drying brown) abaxially, 6–10 per side, lanceolate to almost sigmoid; Venezuela (Bolívar) and the Guianas . . . . 26. *B. oligoclada.*
                23. Pinnae concolorous, 33–58 per side, linear; widespread . . . . 30. *B. riparia.*
17. Fruits purple-black; rachillae 1–5(–7) or 5–50; endocarp fibers usually with juice sacs attached.
    24. Pistillate flowers and fruits without staminodial ring.
        25. Pinnae briefly bifid apically; peduncular bract with few spines; pinnae whitish-brown lepidote abaxially.
            26. Leaf spines to 4 cm long; Colombia (Amazonas), Venezuela (Amazonas), Peru (Loreto), and Brazil (Amazonas). . . . . 5. *B. bidentula.*
            26. Leaf spines to 10 cm long; Brazil (Mato Grosso, Rondônia), and Bolivia (Beni, Santa Cruz) . . . 18. *B. glaucescens.*
        25. Pinnae not briefly bifid apically; peduncular bract with numerous spines (if spines few, then brown tomentose); pinnae green abaxially.
            27. Leaf spines flattened, yellowish in the middle and black at base and apex, or, if black and needlelike, triads regularly arranged.
                28. Inflorescences spicate . . . . 35. *B. tomentosa.*
                28. Inflorescences with 3–17 rachillae . . . . 24. *B. maraja.*
            27. Leaf spines not flattened, black; triads irregularly arranged.
                29. Inflorescences with 1–5(–7) rachillae.
                    30. Fruits obovoid, 2.5–3 cm long; middle pinnae 40–58 cm long . . . . 15. *B. fissifrons.*

30. Fruits globose, 1.2–1.5 cm diam; middle pinnae 9–22 cm long . . . . 14. *B. elegans.*
29. Inflorescences with 4–14 rachillae.
31. Sheath, petiole, and rachis green, not tomentose; endocarp fibers without juice sacs; fruiting corolla usually nonspinulose . . . . 12. *B. corossilla.*
31. Sheath, petiole, and rachis densely brown-tomentose; endocarp fibers with juice sacs attached; fruiting corolla spinulose . . . . 22. *B. macroacantha.*
24. Pistillate flowers and fruits with staminodial ring.
32. Leaf spines strongly flattened, yellowish in the middle and black at the base and apex; rachillae numerous, 15–33 . . . . 7. *B. brongniartii.*
32. Leaf spines not strongly flattened, usually black, occasionally yellowish-brown but then rounded; rachillae fewer, 1–17.
33. Leaves with rachis 0.4–2 m long; pinnae 22–52 per side or leaf entire; leaf drying with a metallic sheen.
34. Leaves entire or rarely pinnate proximally . . . . 6. *B. bifida.*
34. Leaves pinnate with numerous pinnae.
35. Rachillae 1–3(–6) . . . . 10. *B. concinna.*
35. Rachillae (1–)3–17. . . . . 24. *B. major.*
33. Leaves with rachis 8–55 cm long; pinnae 2–10 per side, or leaf entire; leaf drying without a metallic sheen.
36. Leaf entire or with 2–4 pinnae per side; leaf rachis 8–15 cm long; peduncular bract with few spines or nonspiny . . . . 25. *B. oligocarpa.*
36. Leaf pinnate with 8–11 pinnae per side; leaf rachis 45–55 cm long; peduncular bract densely covered with spines. . . . . 17. *B. gastoniana.*

1. Bactris acanthocarpa (Mart.) emend. Henderson. *Bactris acanthocarpa* Mart., Hist. nat. palm. 2: 92. 1826. Type. Brazil. Bahia: Ilheus, n.d., *C. Martius s.n.* (holotype, M; F neg. 18600) (Figure 6.35a–b).

Stems solitary or less often cespitose and then with 2–5 or more stems, 0.1–1 m tall (rarely taller on very old plants), 3–6 cm diam, sometimes procumbent, the internodes very close, usually covered with persistent, decaying leaf bases, seldom with spines. *Leaves* 5–15, erect; sheath 10–50 cm long, sheath, petiole (and rachis) scarcely to densely covered with black spines to 8(–15) cm long (rarely spines lacking); petiole 0.4–1.7 m long, petiole and rachis reddish-brown-tomentose abaxially; rachis 0.6–3.1 m long; pinnae (3–)12–33 per side, irregularly arranged in clusters of 2–4, spreading in slightly different planes, linear-lanceolate or sigmoid, long acuminate, the middle ones 20–60 cm long, 3–10 cm wide, usually with prominent cross-veins, apical pinna usually wider than others. *Inflorescences* interfoliar, often hidden among persistent leaf bases and debris; peduncle 10–20 cm long; prophyll 6–14 cm long; peduncular bract 20–40 cm long, moderately to densely covered with black spines to 1 cm long, sometimes ± glabrous, persistent over infructescence and rotting at base; rachis 2–5 cm long; rachillae 10–46, 4–12 cm long, filamentous, at anthesis with brown, glandular trichomes; *flowers* borne in triads, these ± regularly arranged but often with solitary staminate flowers between the triads, tending to be

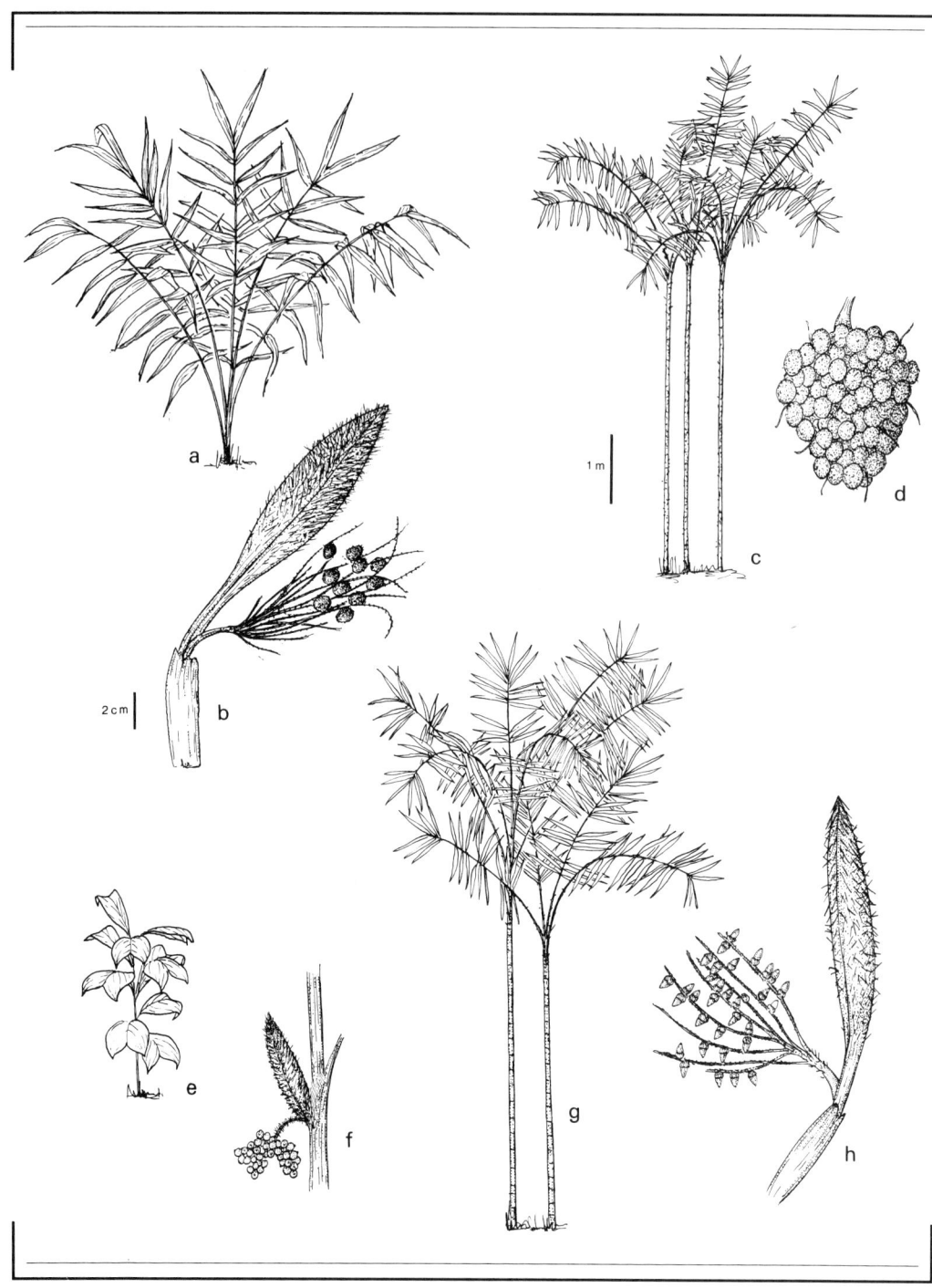

Figure 6.35. (a) *Bactris acanthocarpa* var. *acanthocarpa,* **(b)** infructescence (from *A. Henderson 1559*); **(c)** *Bactris acanthocarpoides,* **(d)** infructescence (from *H. Moore 9529*); **(e)** *Bactris aubletiana,* **(f)** infructescence (from *S. Mori 14731*); **(g)** *Bactris balanophora,* **(h)** infructescence (from *J. Wessels Boer 2310*).

on 1 side of rachillae proximally, paired or solitary staminate flowers distally on rachillae; staminate flowers 3 mm long, deciduous; sepals connate into a 3-lobed cupule, the lobes 0.5–1 mm long; petals oblong, 2.5–3 mm long; pistillode absent; pistillate flowers 2–3 mm long; calyx urceolate, 1–3 mm long, glabrous or with small, brown scales; corolla urceolate, 2–4 mm long, with a few small, brown scales or spinules, glabrescent; ovary with flexuous spinules; staminodes absent; *fruits* very widely obovoid, shortly rostrate, 1–1.8 cm diam, orange or red at maturity, with deciduous, flexuous spinules to 2.5 mm long; mesocarp starchy; endocarp fibers absent; fruiting perianth with short, 3-lobed, glabrous calyx and longer, 3-lobed, glabrous corolla, without staminodial ring.

Martius's original description appears to be based on a mixture of two species, with the stem and leaves incorrectly described. The type consists of an infructescence only. The description is therefore here emended. This species is a member of the Piranga group (Sanders, 1991), characterized by its numerous, filamentous rachillae, regularly arranged triads tending to be on 1 side of rachillae, and spinulose fruits. Wessels Boer (1971b) reported a hybrid, *Bactris* x *moorei* Wess. Boer, between this species (as *Bactris humilis*) and *B. oligoclada* in Venezuela (Type. Venezuela. Bolívar: near El Palmar, 75 m, n.d., *J. Wessels Boer 2092* [isotype, MER]).

Before anthesis, the calyx and corolla of pistillate flowers are subequal, but the corolla lengthens after anthesis, and thus on most herbarium specimens appears longer than the calyx.

Specimens from the northeastern Amazon region, in Surinam and French Guiana, and scattered elsewhere in adjacent areas, have shorter, sigmoid pinnae. These were called *Bactris humilis* by Wessels Boer (1965). They also correspond, in pinnae shape, to *B. interruptepinnata*, but this was described as having a stem 1–2 m tall, and was considered by Wessels Boer to be based on a possible mixture of species. These specimens with sigmoid pinnae are here called var. *intermedia*. Around the margin of the range of this variety, in Venezuela (Amazonas, Bolívar) and Brazil (Maranhão, Pará), occur specimens somewhat intermediate, in pinnae shape and size, with larger plants with longer, linear-lanceolate pinnae from the central and western Amazon region. I have called all these var. *acanthocarpa*.

It is interesting to contrast pinnae shape in the two varieties of *Bactris acanthocarpa* with the situation in two closely related species. *Bactris acanthocarpoides* is also confined to the northeastern Amazon region, but has linear pinnae, while *B. pliniana* is widespread throughout the region, but has sigmoid pinnae.

KEY TO THE VARIETIES OF *BACTRIS ACANTHOCARPA*

1. Leaf rachis 1–3.1 m long; pinnae (3–)22–33 per side, linear-lanceolate or almost sigmoid, 45–60 cm long; widespread. . . . . 1a. *B. acanthocarpa* var. *acanthocarpa*.
1. Leaf rachis 0.7–1.8 m long; pinnae 12–20 per side, sigmoid, 20–37 cm long; northeastern Amazon region in Surinam, French Guiana, and Brazil (Amazonas). . . . . 1b. *B. acanthocarpa* var. *intermedia*.

1a. Bactris acanthocarpa var. **acanthocarpa**

Astrocaryum humile Wallace, Palm Trees of the Amazon 115. 1853. *Bactris humilis* (Wallace) Burret, Repert. Spec. Nov. Regni Veg. 34: 194. 1934. Lectotype (Wess. Boer, 1988). Wallace, Palm Trees of the Amazon t. 45. 1853.

Bactris bicuspidata Spruce, J. Linn. Soc., Bot. 11: 152. 1871. *Pyrenoglyphis bicuspidata* (Spruce) Burret, Repert. Spec. Nov. Regni Veg. 34: 253. 1934. Type. Brazil. Pará: Rio Acará, Sep 1849, *R. Spruce 81* (holotype, K).

Bactris acanthocarpa var. *excapa* Barb. Rodr., Enum. palm. nov. 31. 1875. *Bactris exscapa* (Barb. Rodr.) Barb. Rodr., Sert. palm. brasil. 2: 9. 1903. Type. Destroyed.

Bactris interruptepinnata Barb. Rodr., Enum. palm. nov. 37. 1875. Lectotype. (Glassman, 1972). Barb. Rodr., Sert. palm. brasil. 2: t. 7. 1903.

Bactris aculeifera Drude in Mart., Fl. bras.: Cyclanthaceae et Palmae I, fasc. 85. vol. 3(2): 352. 1881. Type. Brazil. Amazonas: Rio Jutaí, n.d., *J. Trail 892/CCIX* (holotype, K).

Bactris tarumanensis Barb. Rodr., Vellosia 1:

44. 1888. Lectotype (Wess. Boer, 1965). Barb. Rodr., Sert. palm. brasil. 2: t. 5. 1903.

Bactris pinnatisecta Burret, Repert. Spec. Nov. Regni Veg. 34: 191. 1934. Type. Brazil. Amazonas: Rio Madeira, Rio Marmellos, Libramento, 7°S, Jan 1932, *W. Hopp 1157* (holotype, B).

Bactris macrocalyx Burret, Repert. Spec. Nov. Regni Veg. 34: 194. 1934. Type. Guyana. Conawaruk River, Sep 1905, *A. Bartlett 8192* (holotype, B, n.v.).

Bactris microcalyx Burret, Repert. Spec. Nov. Regni Veg. 34: 195. 1934. Type. Brazil. Pará: without locality, Oct 1932, *W. Hopp 5* (holotype, B).

Bactris leptochaete Burret, Notizbl. Bot. Gart. Berlin-Dahlem 12: 620. 1935. Type. Brazil. Amazonas: Rio Madeira, Manicoré, Jun 1934, *W. Hopp 1313* (holotype, B).

Bactris devia H. E. Moore, Gentes Herb. 8: 157. 1949. Type. Colombia. Vaupés: Río Cuduyari, 20 Jan 1944, *P. Allen 3255* (holotype, BH; isotype, MO).

Leaves 5–15, erect; sheath 25–50 cm long, sheath, petiole (and rachis) scarcely to densely covered with black spines to 6(–15) cm long; petiole 0.6–1.7 m long; rachis 1–3.1 m long; pinnae (3–)22–33 per side, irregularly arranged in clusters of 2–4, spreading in slightly different planes, linear-lanceolate or almost sigmoid, long acuminate, the middle ones 45–60 cm long, 3–5.5(–7) cm wide, usually with prominent cross-veins, apical pinna usually wider than others. *Inflorescences* interfoliar; peduncle 11–15 cm long; prophyll 8–14 cm long; peduncular bract 22–40 cm long, densely covered with black spines to 1 cm long, sometimes ± glabrous; rachillae 25–46, 4–12 cm long; *fruits* very widely obovoid, shortly rostrate, 1–1.8 cm diam, orange or red at maturity, with deciduous, flexuous spinules to 2.5 mm long.

Common and widespread throughout the Amazon region of Colombia (Amazonas, Putumayo, Vaupés), Venezuela (Amazonas, Bolívar), Guyana, Ecuador (Napo), Peru (Junín, Loreto, Madre de Dios, Pasco), Brazil (Acre, Amazonas, Maranhão, Pará, Rondônia), and Bolivia (Beni, Pando, Santa Cruz) (Figure 6.34D). It also occurs in the Atlantic coastal forest of Brazil (from Paraíba in the north to Espírito Santo in the south). It grows in lowland rain forest on *terra firme* at low elevations, occasionally to 1000 m.

Bolivia: *canahuaníma* (Chácobo). Brazil: *kwere'i* (Ka'apor), *marajá, munbaca, pupunha de mata, pupunha mança wa'ã* (Guajá). Colombia: *bubúmemëeku* (Muinane), *chontaduro de los peces*. Ecuador: *cui'cui* (Siona). Venezuela: *cubarro*. The fruits are used by the Siona in Ecuador to make necklaces, and by the Chácabo of Bolivia to make a decoction for alleviation of stomach aches.

1b. Bactris acanthocarpa var. **intermedia** Henderson, var. nov. Type. Brazil. Amazonas: BDFF Reserve Km 41, 90 km N of Manaus on BR 174, 41 km E on ZF 3, 11 Sep 1989, *A. Henderson et al. 1070* (holotype, INPA; isotype, NY).

A var. *acanthocarpa rachi breviori pinnisque minus numerosis sigmoideis distans.*

Leaves 5–10(–15); sheath 10–35 cm long, sheath and petiole without spines or with a few black spines to 8 cm long; petiole 0.3–1.1 m long; rachis 0.7–1.8 m long; pinnae 12–20 per side, irregularly arranged in clusters of 2–3, spreading in different planes, sigmoid, glossy green, abruptly pendulous at apex, the middle ones 20–37 cm long, 3.5–10 cm diam. *Inflorescences* interfoliar; peduncle 10–17 cm long; prophyll 8–13 cm long; peduncular bract 20–22 cm long, moderately to densely covered with black spines; rachis 2–5 cm long; rachillae 10–25, 5–10 cm long; *fruits* very widely obovoid, shortly rostrate, 1–1.2 cm diam, bright orangered, covered with short black spinules.

The Guianas and Brazil (Amazonas) (Figure 6.34E); lowland rain forest on *terra firme* at low elevations.

2. Bactris acanthocarpoides Barb. Rodr., Enum. palm. nov. 33. 1875. Lectotype (Wess. Boer, 1965). Barb. Rodr., Sert. palm. brasil. 2: t. 9. 1903 (Figure 6.35c–d).

Bactris acanthocarpa var. *crispata* Drude in Mart., Fl. bras.: Cyclanthaceae et Palmae I, fasc. 85, vol. 3(2): 350. 1881. Type. Brazil. Pará: Almeirim, 19 Feb 1875, *J. Trail 893/CCXXI* (holotype, K).

Stems cespitose, with 2–6 stems per clump, 2–4 m tall, 2–4 cm diam, spiny on internodes and often stems at least partially obscured with persistent, decaying leaf bases. *Leaves* 8–15; sheath 35–50 cm long, sheath, petiole, and rachis moderately to densely covered with somewhat flattened black spines to 7 cm long; petiole 0.9–2 m long; rachis 1.8–2.5 m long; pinnae 20–37 per side, irregularly arranged in clusters of 2–6, spreading in different planes, linear, the middle ones 61–70 cm long, 2.5–3 cm wide, with visible cross-veins, occasionally spiny on the margins. *Inflorescences* interfoliar; peduncle 8–14 cm long, spiny; prophyll 14–18 cm long; peduncular bract to 30 cm long, persistent, covered with blackish or brownish, somewhat soft spines interspersed with straight black spines; rachis to 8 cm long, densely spiny; rachillae 25–89, filamentous, to 14 cm long, densely to moderately covered proximally with spinules; *flowers* arranged in triads, these regularly arranged and on 1 side only of proximal part of rachillae, paired or solitary staminate distally; staminate flowers 2–3 mm long, pedicellate; sepals with deltate lobes, 0.5–1 mm long; petals fleshy, oblong, 2–3 mm long; stamens 0–3–6; pistillode absent; pistillate flowers 4–7 mm long at anthesis; calyx shallow, cupular, 1–1.5 mm long, glabrous; corolla tubular, 3–5 mm long, covered with spinules; staminodes absent; *fruits* very widely obovoid, shortly rostrate, 1.2–2 cm diam, orange, yellowish, or red, covered with soft spinules; mesocarp floury, endocarp fibers absent; fruiting perianth with lobed calyx much shorter than the lobed corolla.

The Guianas and Brazil (Amazonas, Pará) (Figure 6.34F); lowland rain forest on *terra firme*, or occasionally near streams in places liable to inundation, at low elevations.

Guyana: *jawi, ulukpana*. Surinam: *bongie-bongie, haai maba, hanaimaka, maka kow maka, manai maka, nao maka*.

3. Bactris aubletiana Trail, J. Bot. 5: 372. 1876. Type. French Guiana. Without locality, n.d., *J. Aublet s.n.* (holotype, BM, n.v.) (Figure 6.35e–f).

Stems solitary or cespitose, 0.6–2 m tall, 0.3–0.5 cm diam. *Leaves* 10–11; sheath 8–15 cm long, sheath, petiole, and rachis without spines or rarely with a few black spines to 1 cm long on petiole; petiole 5–16 cm long; rachis 3–18 cm long; blade entire, bifid, the lobes 17–30 cm long, 4–9 cm wide, often with soft spines on margins, with prominent cross-veins. *Inflorescences* interfoliar, recurved, becoming infrafoliar in fruit; peduncle 3–3.5 cm long; prophyll 2.5–4 cm long; peduncular bract 5–7 cm long, moderately to densely covered with soft, erect, black spines to 5 mm long; rachillae 1–3, 2.5–3.5 cm long, at anthesis with brown, glandular trichomes and sometimes with short spinules; *flowers* borne in triads, these regularly arranged almost throughout rachillae; staminate flowers 3 mm long, deciduous; sepals with lobes 1 mm long; petals 2.5 mm long; pistillate flowers 2 mm long; calyx cupular, 1 mm long; corolla tubular, 2 mm long, glabrous or spinulose; staminodes absent; *fruits* globose, 0.5–0.7 cm diam, orange, glabrous; mesocarp floury; endocarp fibers absent; fruiting perianth with small, lobed calyx and much longer, 3-lobed corolla, lacking staminodial tube.

Guianas (Figure 6.36A); lowland forest on *terra firme* at low elevations.

French Guiana: *faux wi blanc, wilā si* (Wayápi). Surinam: *yuyba*.

This species belongs to subsection *Amylocarpus* (Sanders, 1991), characterized by its small size, few rachillae, regularly arranged triads, and small fruit. *Bactris aubletiana* is characterized by its entire, bifid leaves with prominent cross-veins and moderately to densely spiny peduncular bract (while the rest of the plant is nonspiny or spinulose).

4. Bactris balanophora Spruce, J. Linn. Soc., Bot. 11: 146. 1871. Type. Venezuela. Amazonas: San Carlos de Río Negro, Oct 1854, *R. Spruce 53* (holotype, K; F negs. 38672, 38673) (Figure 6.35g–h).

Astrocaryum aculeatum of Wallace, Palm Trees of the Amazon 111. 1853.

Stems cespitose, forming clumps of 2–6 stems, 1.5–7 m tall, 2–3 cm diam, spiny on the internodes. *Leaves* 5–8, erect; sheath 30–58 cm long, sheath, petiole, and rachis with somewhat flattened black spines 1.5–3 cm long, these in groups and spreading in different directions; petiole 40–60 cm long; rachis 0.6–1.1 m long; pinnae 15–21 per side, irregularly arranged in loose clusters of 2–5 and often with gaps between the clusters, spreading in slightly different planes, linear-lanceolate, acuminate, the

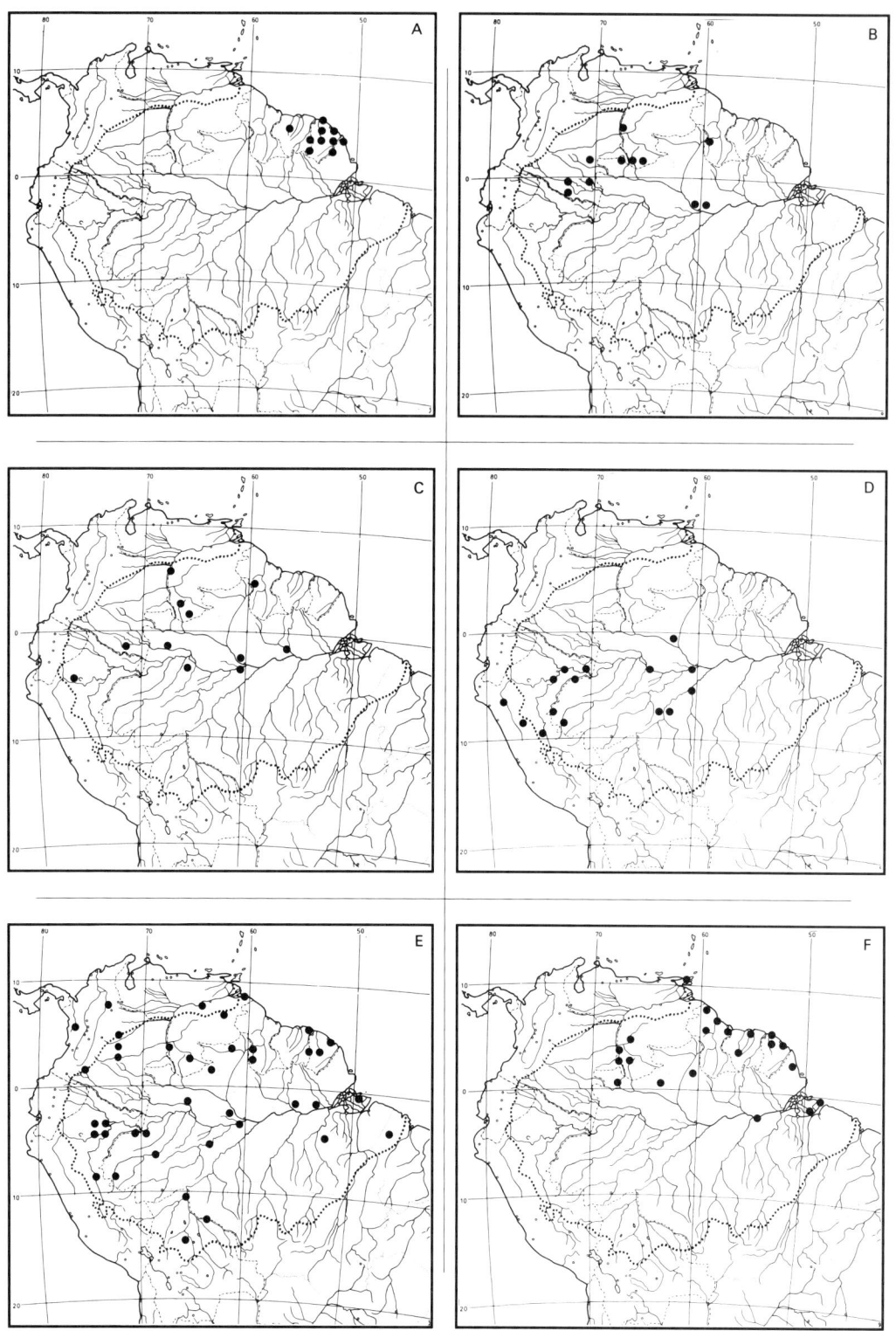

Figure 6.36. (A) *Bactris aubletiana;* **(B)** *Bactris balanophora;* **(C)** *Bactris bidentula;* **(D)** *Bactris bifida;* **(E)** *Bactris brongniartii;* **(F)** *Bactris campestris.*

middle ones 34–58 cm long, 2–4.5 cm wide, glabrous or usually with sparse, fine spinules abaxially. *Inflorescences* infrafoliar among persistent leaf sheaths; peduncle 9–11 cm long; prophyll 12–13 cm long; peduncular bract 21–25 cm long, moderately to densely covered with stiff black spines to 1 cm long; rachis 2–3 cm long; rachillae (5–)7–15, 10–11 cm long, densely covered with soft, whitish, flexuous spinules; *flowers* in triads, these irregularly arranged among paired or solitary staminate flowers; staminate flowers 2.5–3 mm long; sepals with triangular lobes, 1 mm long; petals 3 mm long; pistillate flowers 3 mm long; calyx cupular, 0.8 mm long, glabrous; corolla tubular, 3 mm long, densely whitish-brown-tomentose; staminodes absent; ovary brown-tomentose; *fruits* pyriform or ovoid, erostrate, 1.3–1.5 cm long, 0.7–0.8 cm diam, orange at maturity, glabrous; mesocarp floury; endocarp with flattened, adnate, anastomosing fibers; fruiting perianth with very short, smooth-margined calyx and longer, densely spinulose, smooth-margined corolla, lacking staminodial ring.

North of the Amazon river in Colombia (Amazonas, Vaupés), Venezuela (Amazonas), Guyana, and Brazil (Amazonas, Pará) (Figure 6.36B); lowland rain forest either on *terra firme* or rarely in inundated areas, below 600 m elevation.

Colombia: *chontaduro paso, chontilla, ho-táñe* (Makuna), *oo-chee-an* (Maku), *ya-yo-(e)r(u)* (Huitoto). Venezuela: *ceguera, cubarro, espina de sardina, pijiwau de monte.*

This species is placed in subgenus *Balanophora* and was considered by Sanders (1991) to be the least specialized in the genus because of its pyriform or ovoid fruits and flattened, adnate endocarp fibers. Except for the fruits, it is remarkably similar to *Bactris ptariana,* but the two species are not closely related.

5. **Bactris bidentula** Spruce, J. Linn. Soc., Bot. 11: 146. 1871. Type. Brazil. Amazonas: Manaus, n.d., *R. Spruce 9* (holotype, K) (Figure 6.37a–b).

Bactris sp. Wallace, Palm Trees of the Amazon 83. 1853. Lectotype (here designated). Wallace, Palm Trees of the Amazon t. 31. 1853.
Bactris palustris Barb. Rodr., Enum. palm. nov. 36. 1875. Type. Destroyed.
Bactris nigrispina Barb. Rodr., Palm. hassler. 15. 1900. Lectotype (Wess. Boer, 1988). Barb. Rodr., Sert. palm. brasil. 2: t. 29. 1903.

Stems cespitose, forming dense stands of 4–20 or more stems, 1.7–4 m tall, 3–4.5 cm diam, with rings of spines on the nodes. *Leaves* 3–16; sheath 10–35 cm long, sheath, petiole, and rachis densely whitish-tomentose, glabrescent, moderately covered with somewhat clustered, slightly flattened, 3–4 cm long black spines; petiole 14–53 cm long; rachis 56–120 cm long; pinnae 24–45 per side, irregularly arranged in distant clusters of 2 to 7, spreading in different planes, linear-lanceolate or oblanceolate, the middle ones 16–34 cm long, 2.3–4 cm wide, briefly bifid apically, with few spines on margins, densely whitish-brown-lepidote abaxially. *Inflorescences* interfoliar; peduncle 8–32 cm long; prophyll 9–20 cm long; peduncular bract 22–45 cm long; whitish-brown with few black spines to 2 cm long; rachis 4–12 cm long; rachillae 20–50, 10–12 cm long; *flowers* borne in triads, these irregularly arranged among paired or solitary staminate flowers; staminate flowers 3–4 mm long; sepals with narrowly triangular lobes 1 mm long; petals 3–4 mm long; pistillode minute; pistillate flowers 2.5–3 mm long; calyx cupular, 1 mm long; corolla tubular, 2.5–3 mm long; staminodes absent; *fruits* depressed globose, rostrate, 0.8–1.5 cm diam, purple-black; mesocarp juicy; endocarp densely covered with free fibers; fruiting calyx cupular, 3-lobed, to 0.5 mm long, fruiting corolla tubular, scarcely 3-lobed, to 2 mm long, staminodes absent.

Central Amazon region in Colombia (Amazonas), Venezuela (Amazonas), Peru (Loreto), and Brazil (Amazonas, western Pará) (Figure 6.36C); lowland rain forest, along the margins of small black-water streams or lakes in areas liable to inundation *(igapó).* The lower parts of the stems are often completely submerged during high water.

Bolivia: *chontilla.* Brazil: *marajá do igapó, marajá do jacaré.* Colombia: *boo b(o) me g(e)* (Miraña). The fruits are edible, and are eaten by humans and fish.

Bactris bidentula is distinguished by its clustered leaf spines, densely whitish-brown-lepidote abaxial pinnae surface, bifid pinnae

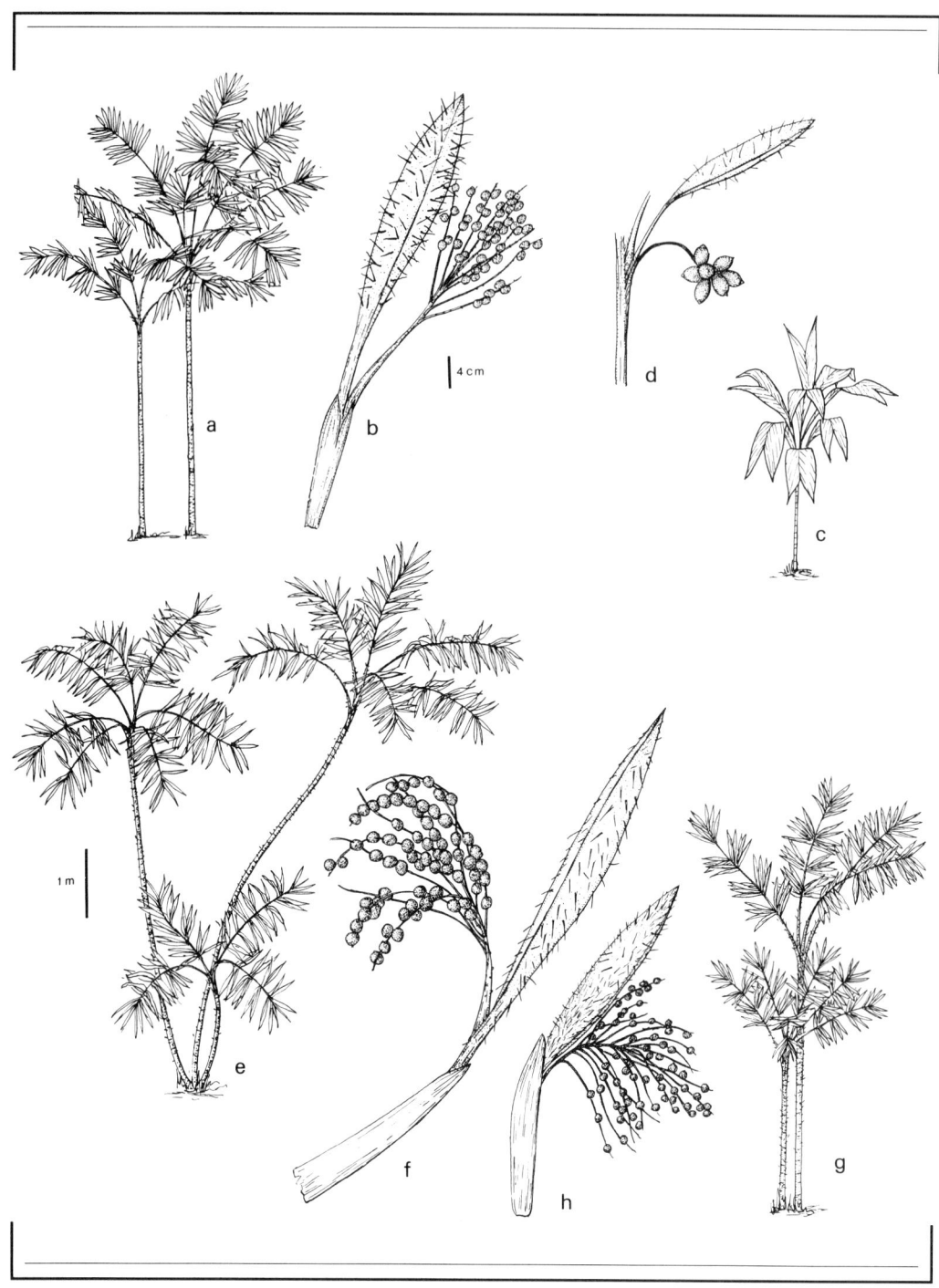

Figure 6.37. (a) *Bactris bidentula,* **(b)** infructescence (from *A. Henderson 1511*); **(c)** *Bactris bifida,* **(d)** infructescence (from *A. Henderson 249*); **(e)** *Bactris brongniartii,* **(f)** infructescence (from *A. Henderson 815*); **(g)** *Bactris campestris,* **(h)** infructescence (from *G. Davidse 17708*).

Plate I *Habits*

- **(a)** *Mauritia flexuosa*
- **(b)** *Bactris major*
- **(c)** *Astrocaryum jauari*
- **(d)** *Geonoma deversa*

Plate II *Flowers*

(a) *Mauritia flexuosa*
(b) *Attalea phalerata*
(c) *Bactris constanciae*
(d) *Geonoma macrostachys* var. *acaulis*

Plate III *Fruits*

(a) *Astrocaryum gynacanthum*
(b) *Geonoma stricta* var. *stricta*
(c) *Raphia taedigera*
(d) *Phytelephas tenuicaulis*

Plate IV *Spines*

(a) root spines on stem of *Mauritiella armata*
(b) spines on stilt roots of *Socratea exorrhiza*
(c) sheath and peduncular bract spines of *Desmoncus giganteus*
(d) stem spines of *Astrocaryum aculeatum*

apex, and purple-black fruits with free endocarp fibers. It is very similar to *B. guineensis,* distributed to the north in Central America and northern South America, and to *Bactris glaucescens,* distributed in the southern part of the Amazon region.

6. **Bactris bifida** Mart., Hist. nat. palm. 2: 105. 1826. *Pyrenoglyphis bifida* (Mart.) Burret, Repert. Spec Nov. Regni Veg. 34: 242. 1934. Type. Brazil. Amazonas: Rio Negro, n.d., *C. Martius s.n.* (holotype, M; F neg. 18604) (Figure 6.37c–d).

Bactris bifida var. *humaitensis* Trail, J. Bot. 6: 47. 1877. *Pyrenoglyphis bifida* var. *humaitensis* (Trail) Burret, Repert. Spec. Nov. Regni Veg. 34: 242. 1934. Type. Brazil. Amazonas: Humaitá, Rio Madeira, 30 May 1874, *J. Trail 909/LII* (holotype, K).

Bactris bifida var. *puruensis* Trail, J. Bot. 6: 47. 1877. *Pyrenoglyphis bifida* var. *puruensis* (Trail) Burret, Repert. Spec. Nov. Regni Veg. 34: 242. 1934. Type. Brazil. Amazonas: Guajaratuba, Rio Purus, 11 Sep 1874, *J. Trail 911/CXXIa* (holotype, K).

Stems cespitose, often leaning, 1–4 m tall, 1–2 cm diam, forming small or large colonies. *Leaves* 4–10; sheath 12–28 cm long, very fibrous on margins, sheath, petiole, and rachis with few to many, blackish-brown, rounded to slightly flattened spines to 7 cm long; petiole (0–)12–22(–100) cm long; rachis 40–70 cm long; blade entire or rarely pinnate proximally, long cuneate basally, deeply bifid apically, 40–100 cm long, 12–14 cm wide at middle, strongly plicate, with a metallic sheen on drying. *Inflorescences* interfoliar; peduncle 15–25 cm long; prophyll 8–13 cm long; peduncular bract 16–25 cm long, sparsely covered with black spines to 5 mm long; rachillae 1–2, 4–6.5 cm long; *flowers* borne in regularly arranged triads, these interspersed with paired or solitary staminate flowers; staminate flowers (immature) 4.5 mm long, deciduous; sepals with narrowly triangular lobes 1 mm long; petals 4 mm long; pistillate flowers 3–4 mm long; calyx tubular, 3–4 mm long; corolla tubular, slightly shorter than the calyx; staminodial ring obscure at anthesis, 0.5 mm long; *fruits* narrowly ellipsoid or ellipsoid-oblong, 2–2.5 cm long, 1–1.5 cm diam, purple-black at maturity; mesocarp juicy; endocarp with numerous, free fibers with juice sacs attached; fruiting perianth with very short calyx and much longer, almost entire-margined corolla, staminodial ring small.

Mostly south of the Amazon river in Colombia (Amazonas), Peru (Huánuco, Loreto, San Martín, Ucayali), and western Brazil (Acre, Amazonas, Mato Grosso) (Figure 6.36D); lowland forest either on *terra firme* or more commonly in areas liable to seasonal inundation, or other wet places, at elevations below 600 m.

Brazil: *ubim de espinho.* Peru: *ñeja negra.*

Bactris bifida is placed in subsection *Pyrenoglyphis* by Sanders (1991), distinguished by its purple-black fruits with staminodial ring. It is distinguished by its long, narrow, entire (rarely partially pinnate), strongly plicate leaves, which are deeply bifid apically and cuneate basally and dry with a metallic sheen, and its usually spicate inflorescence. There is considerable variation in size of leaves. It is closely related to both *B. major* and *B. concinna.*

J. Trail 855/LXVII (incorrectly determined by him as *B. aristata*) from Barcellos is similar to *A. Henderson 176* from near Manaus. Both have regularly pinnate leaves with spicate inflorescence and pistillate flowers with staminodial ring. Both are considered aberrant forms of *B. bifida* (or hybrids?). Neither is included in the description. It is interesting that these two larger specimens, if indeed they are *B. bifida,* are similar to the situation in *B. concinna,* where there are two distinct sizes.

7. **Bactris brongniartii** Mart. in A. D. Orb., Voy. Amérique mér. 7(3). Palmiers 59. 1846. *Pyrenoglyphis brongniartii* (Mart.) Burret, Repert. Spec. Nov. Regni Veg. 34: 251. 1934. Type. Bolivia. Beni: Moxos, n.d., *A. d'Orbigny 25* (holotype, P, n.v.) (Figure 6.37e–f).

Bactris pallidispina Mart. in A. D. Orb., Voy. Amérique mér. 7(3). Palmiers 62. 1846. *Pyrenoglyphis pallidispina* (Mart.) Burret, Repert. Spec. Nov. Regni Veg. 34: 249. 1934. Type. Surinam. Without locality, n.d., *F. Splitgerber 466* (holotype, BR, n.v.).

Bactris rivularis Barb. Rodr., Enum. palm. nov. 36. 1875. *Pyrenoglyphis rivularis* (Barb. Rodr.) Burret, Repert. Spec. Nov. Regni Veg. 34: 251. 1934. Lectotype (Wess. Boer, 1965). Barb. Rodr., Sert. palm. brasil. 2: t. 31. 1903.

Bactris marajaacu Barb. Rodr., Enum. palm. nov. 36. 1875. Lectotype (Glassman, 1972). Barb. Rodr., Sert. palm. brasil. 2: t. 35. 1903.

Pyrenoglyphis microcarpa Burret, Repert. Spec. Nov. Regni Veg. 34: 250. 1934. *Bactris burretii* Glassman, Rhodora 65: 259. 1963. Type. Guyana. Mahaica Creek, Jul 1901, *G. Jenman 7725* (holotype, B, n.v.).

Stems cespitose, often forming large colonies by rhizomes, 3–6(–9) m tall, 3.5–5(–8) cm diam, spiny at internodes. *Leaves* 4–7, stiffly ascending; sheath 30–60 cm long, sheath, petiole, and rachis with moderate to dense covering of flattened spines, these yellowish or brownish in center and black at apex and base (often the ones on the sheath darker); petiole 10–70 cm long; rachis 0.9–1.5 m long; pinnae (10–)23–34 per side, irregularly arranged in clusters of 2–5 and spreading in different planes, linear-lanceolate, strongly plicate, briefly and asymmetrically bifid at apex, the middle ones 45–78 cm long, 3.5–7 cm wide, with small spines on margins. *Inflorescences* interfoliar; peduncle 25–50 cm long; prophyll 15–30 cm long; peduncular bract 25–62 cm long, with sparse to moderate covering of flattened, yellowish spines to 2 cm long; rachis 8–15 cm long; rachillae 15–33, 15–27 cm long, at anthesis densely covered with brown, glandular trichomes; *flowers* borne in triads, these irregularly arranged among paired or solitary staminate flowers; staminate flowers 3.5–4.5 mm long, deciduous; sepals with narrowly triangular lobes 1 mm long; petals 3.5–4 mm long; pistillate flowers at anthesis with subequal, tubular calyx and corolla, 2.5–3 mm long; staminodial ring adnate basally to corolla; *fruits* depressed-globose, 1.3–1.5 cm diam, purple-black, glabrous; mesocarp juicy; endocarp fibers free, numerous, with juice sacs attached; fruiting perianth with irregularly lobed calyx half as long as the irregularly lobed corolla, staminodial ring present.

Throughout the Amazon region of Colombia (Amazonas, Guaviare, Meta), Venezuela (Amazonas, Bolívar, Delta Amacuro), the Guianas, Peru (Loreto, Madre de Dios, Ucayali), Brazil (Acre, Amapá, Amazonas, Maranhão, Pará, Rondônia, Roraima), and Bolivia (Beni). In the north, it extends into Colombia (Antioquia, Chocó, Santander, as *B. tenera* (H. Karst.) H. Wendl.) and Venezuela (Anzoátegui); in the south to southwestern Brazil (Mato Grosso do Sul, as *B. piscatorum* Wedd. ex Drude) (Figure 6.36E). It almost always grows on river margins or in seasonally inundated areas, at low elevations.

Brazil: *marajá, marajá de cacho, marajá pupunha, maraja'i* (Arawaté), *maraja'y* (Ka'apor), *maria-ci* (Guajá), *maria-wa* (Guajá), *tucum bravo*. Colombia: *cachepai montañero, cubarra, maradai*. Guyana: *bango palm*. Peru: *ñejilla*. Venezuela: *caña negra, cubarro, komora* (Yanomami). The stems are used to make walls of houses (Brazil). The sweet mesocarp of the fruits, which are sold in local markets, is edible and is also used to flavor drinks.

Bactris brongniartii is distinguished by its flattened, yellowish or brown leaf spines; its stiffly ascending leaves; its elongate inflorescences with somewhat loosely arranged, numerous, long rachillae; and its relatively small, purple-black fruits with a staminodial ring. Unfortunately, the name that has traditionally been applied to this species, *Bactris maraja,* can no longer be used because it applies to another species.

8. Bactris campestris Poepp. ex Mart., Hist. nat. palm. 2: 146. 1837. Type. Brazil. Pará: Campina Grande de Colares, Jul 1832, *E. Poeppig 3015* (holotype, M; F neg. 18603) (Figure 6.37g–h).

Bactris leptocarpa Trail ex Thurn, Timehri 3: 253. 1884. Type. Guyana. Pomeroon, 1880, *E. Thurn 15* (holotype, K, n.v.).

Bactris lanceolata Burret, Notizbl. Bot. Gart. Berlin-Dahlem 10: 1023. 1930. Type. Brazil. Amazonas: Rio Negro, São Felipe, 22 Sep 1928, *P. von Luetzelburg 22942* (isotype, M).

Stems cespitose, in small clumps, 1–5 m tall, 3–4 cm diam, often covered with persistent leaf bases. *Leaves* 2–5; sheath 24–50 cm long, often gray-tomentose, sheath, petiole, and rachis with reddish-brown or black scales and moderately covered with flattened, gray or gray-black spines to 2(-4) cm long, these black at base and apex; petiole 15–90 cm long; rachis 65–94 cm long; pinnae 17–32, irregularly arranged in clusters of 2–5, spreading in different planes, linear, linear-lanceolate, or oblanceolate, the middle ones 26–52 cm long, 1.5–5 cm wide, lacking spines on margins, bifid at apex. *Inflorescences* interfoliar; peduncle 13–20 cm long, peduncle, rachis, and rachillae covered with dense, reddish-brown tomentum; prophyll 11–20 cm long; peduncular bract 26–40 cm long, densely gray-or brown-tomentose and covered with gray or brown spines to 1.5 cm long; rachis 1–6 cm long; rachillae 8–39, 5–15 cm long; *flowers* borne in triads, these irregularly arranged among paired or solitary staminate flowers; staminate flowers to 4 mm long, deciduous; sepals with deltate lobes 0.5 mm long; petals widely ovate, 3.5 mm long; pistillate flowers 3–3.5 mm long; calyx cupular, 0.5 mm long; corolla tubular, 2.5–3 mm long; staminodes absent; *fruits* globose, 5–8 mm diam, red or orange-red at maturity; mesocarp starchy; endocarp fibers few, free; fruiting perianth with very small calyx and broadly 3-lobed corolla, staminodial ring absent.

Northeastern South America in eastern Colombia (Guainía, Vichada), southern Venezuela (Amazonas), the Guianas, Trinidad (as *B. savannarum* L. H. Bailey), and Brazil (Amapá, Amazonas, Pará, Roraima) (Figure 6.36F); in open areas, white-sand savannas, or in low forest on white sand, generally in poorly drained places, at low elevations (rarely to 800 m).

Brazil: *mumbaca branca*. Venezuela: *cubarro*. The palm heart is mixed with water and placed on the tongue to treat rattlesnake bites (Marajó Island, Pará, Brazil).

Bactris campestris is distinguished by its flattened, grayish leaf spines; reddish-brown tomentum of the inflorescence; and small, globose, red or orange-red fruits. It is further distinguished by its open, savanna habitat. It varies considerably in size of all its parts, and plants from forest are usually much larger than those from open areas.

9. Bactris coloniata L. H. Bailey, Gentes Herb. 3: 106. 1933. Lectotype (Burret, 1934): Panama. Canal Area: Barro Colorado Island, 29 Jun 1931, *L. Bailey 77* (holotype, BH) (Figure 6.38a–b).

Stems cespitose, rarely solitary, forming open colonies, 3.5–7 m tall, 2–5 cm diam, with spiny internodes. *Leaves* 5–7; sheath 27–80 cm long, sheath, petiole, and rachis brown-tomentose abaxially, moderately to densely covered with yellowish-brown to black, somewhat flattened spines; petiole 39–70 cm long; rachis (0.45–)1.2–1.4 m long; pinnae (4–)14–23 per side (or occasionally leaf entire), regularly or irregularly arranged in clusters of 4–12, spreading in 1 or different planes, linear-lanceolate, elliptic or slightly sigmoid, the apex long-caudate and drooping, the middle ones 30–75 cm long, 2.5–8 cm diam, glabrous. *Inflorescences* interfoliar; peduncle 16–25 cm long, curved in fruit; prophyll 15–49 cm; peduncular bract (26–)42–60 cm long, densely covered with appressed, brown, flattened spines; rachis 4–10 cm long; rachillae 9–16, 14–25 cm long, to 3 mm diam in fruit, densely glandular hairy at anthesis; *flowers* borne in triads, these scattered among paired or solitary staminate flowers; staminate flowers 4 mm long; sepals free, linear, 1 mm long; petals 4 mm long; pistillode absent; pistillate flowers 5 mm long; calyx tubular, 4 mm long, glabrous or with spinules; corolla tubular, 2 mm long, with minute spinules; staminodes absent; *fruits* very widely obovoid, markedly rostrate, 1.5–2.5 cm long, 1.5–2 cm diam, yellowish-brown, covered with short brown spinules; mesocarp floury; endocarp fibers free, numerous; fruiting calyx only slightly shorter than the corolla, 4–5 mm long, both with crenulate margins.

Eastern Panama (Canal Area, Colón, San Blas, Darién, Panamá), northwestern Colombia (Antioquia, Chocó) and northeastern Peru (Amazonas) (Figure 6.39A); lowland rain forest on *terra firme* to 700 m elevation.

Peru: *kamancá, úun shigki*. The bark from the trunk is used to pack shotgun cartridges (Amazonas, Peru).

Bactris coloniata is characterized by its large size, open habit of growth, subequal fruiting calyx and corolla, and spinulose fruits that are widest at the middle. This species has a disjunct distribution between Central America and adjacent northwestern Colombia, and the extreme western Amazon region.

10. Bactris concinna Mart., Hist. nat. palm. 2: 99. 1826. *Pyrenoglyphis concinna* (Mart.) Burret, Repert. Spec. Nov. Regni Veg. 34:

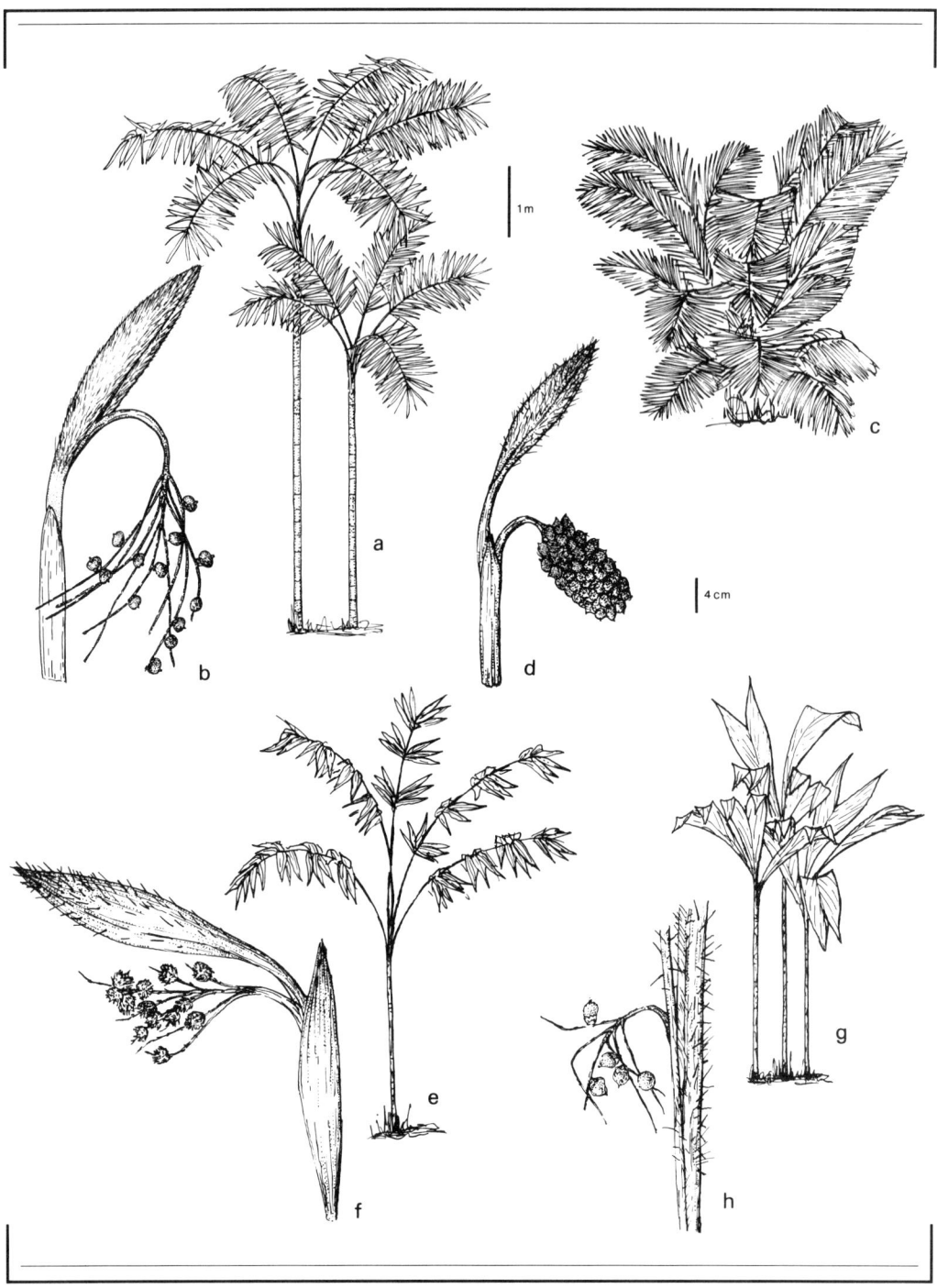

Figure 6.38. (a) *Bactris coloniata,* (b) infructescence (from *T. Croat 15408*); (c) *Bactris concinna* var. *inundata,* (d) infructescence (from *A. Henderson 1510*); (e) *Bactris constanciae,* (f) infructescence (from *S. Mori 19603*); (g) *Bactris corossilla,* (h) infructescence (from *H. Balslev 4790*).

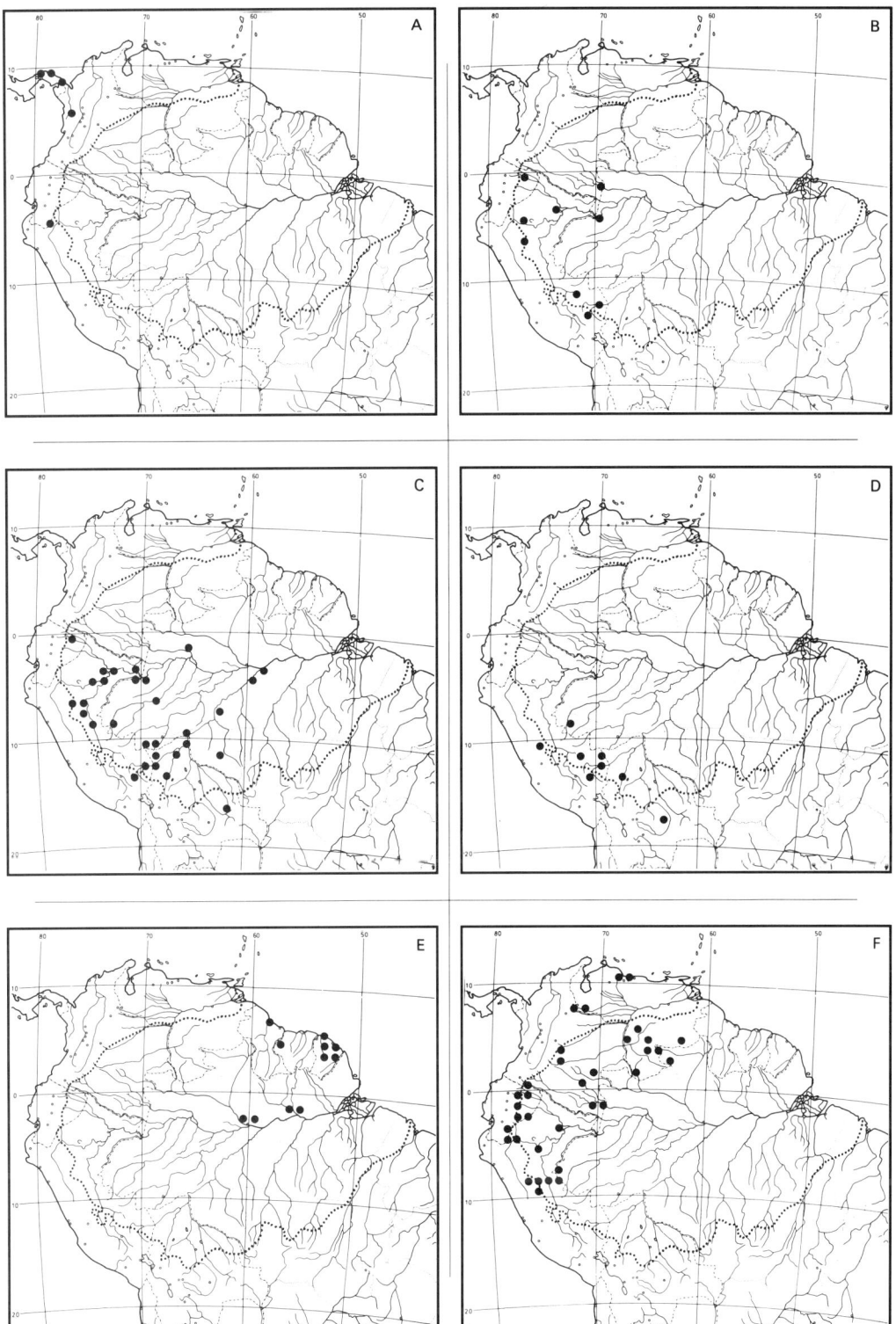

Figure 6.39. (A) *Bactris coloniata;* **(B)** *Bactris concinna* var. *concinna;* **(C)** *Bactris concinna* var. *inundata;* **(D)** *Bactris concinna* var. *sigmoidea;* **(E)** *Bactris constanciae;* **(F)** *Bactris corossilla.*

242. 1934. Type. Brazil. Amazonas: without locality, n.d., *C. Martius 2887* (holotype, M; F negs. 18609, 18610) (Figure 6.38c–d).

Stems cespitose, forming small or large colonies with several hundred stems, (0.5–)2–8 m tall, 1.5–5 cm diam, erect or often leaning, spiny on internodes. *Leaves* 3–10; sheath 18–50 cm long, very fibrous on margins, sheath, petiole, and rachis with a dense covering of black spines to 2 cm long, interspersed with fewer, longer, black or yellowish spines to 10 cm long; petiole 12–74 cm long, with fewer spines than sheath; rachis 0.9–2 m long, not spiny or with few to many spines to 6 cm long; pinnae 16–52 per side, regularly arranged and spreading in 1 plane or irregularly arranged in clusters and spreading in different planes, linear, linear-lanceolate or sigmoid, the middle ones 15–70 cm long, 1–4.5 cm wide, with small spines on margins. *Inflorescences* interfoliar; peduncle 14–30 cm long; prophyll 10–22 cm long; peduncular bract 20–40 cm long, moderately covered with black spines to 1.5 cm long; rachis absent; rachillae 1–3(–6), 5–20 cm long, at anthesis glabrous; *flowers* borne in triads, these regularly arranged on the rachillae; staminate flowers 7–10 mm long, persistent on the rachillae; sepals with narrowly triangular lobes 2.5–4 mm long; petals spathulate, 7–10 mm long; stamens 6–10; pistillate flowers at anthesis to 6 mm long; calyx cupular, to 1 mm long; corolla tubular, to 5 mm long; staminodial ring free from the corolla, to 3.5 mm long; *fruits* very congested on rachillae, irregularly and narrowly obovoid, 2–4.5 cm long, 1–2.5 cm diam, purple-black at maturity; mesocarp fleshy; endocarp fibers free, numerous, with juice sacs attached; fruiting perianth with small, lobed calyx and much longer, scarcely lobed corolla, staminodial ring prominent.

Western Amazon region in Colombia (Amazonas, Putumayo), Ecuador (Napo), Peru (Loreto, Madre de Dios, San Martín, Ucayali), Brazil (Acre, Amazonas, northern Mato Grosso, Pará, Rondônia), and Bolivia (Beni, La Paz, Santa Cruz); lowland rain forest, especially along streams and rivers and in other seasonally inundated areas, where it can form large colonies.

Bolivia: *marajaú*. Brazil: *marajá*. Colombia: *paipigu* (Puinave). Ecuador: *ansepara* (Quichua), *chontilla, nu-que* (Siona). Peru: *ñejilla, niejilla, palmera, síi* (Ese-ejha). The fruits are edible and are sold in local markets and fed to domestic animals; the bark is scraped from the stem and used for loading guns (Brazil, Ecuador); and the stems are used for bows and other small items (Peru).

Bactris concinna is distinguished by its pinnae, which have a metallic sheen when dry; spicate or bifurcate inflorescences (rarely to 6 rachillae); pistillate flowers and fruits with staminodial ring; and purple-black, irregularly and narrowly obovoid, congested fruits.

Bactris concinna and *B. major* are closely related, and appear to be another example of species pairs having nonoverlapping, southwest–northeast distributions in northern South America.

Three varieties of *B. concinna* can be recognized, although there are intermediates between each of them, and they can occur together.

KEY TO THE VARIETIES OF *BACTRIS CONCINNA*

1. Pinnae linear, regularly arranged and spreading in 1 plane; apical pinna the same width as the others.
   2. Fruits 3.5–4.5 cm long, 2–2.5 cm diam; stems 4–8 m tall, 3–5 cm diam; pinnae 46–52 per side, 45–70 cm long, 2–3 cm wide; rachilla 1(–5), 12–20 cm long. . . . . 10a. *B. concinna* var. *concinna*.
   2. Fruits 2–2.5 cm long, 1–1.5 cm diam; stems 1–4 m tall, 1.5–2 cm diam; pinnae 16–52 per side, 15–52 cm long, 1–2 cm wide; rachillae 1–2(–6), 5–12 cm long . . . 10b. *B. concinna* var. *inundata*.
1. Pinnae linear-lanceolate or sigmoid, irregularly arranged in clusters and spreading in different planes; apical pinna usually wider than the others. . . . . 10c. *B. concinna* var. *sigmoidea*.

10a. Bactris concinna var. concinna

Colombia (Amazonas), Ecuador (Napo), Peru (Cuzco, Loreto, Madre de Dios, San Martín), and Brazil (Amazonas) (Figure 6.39B).

This variety usually has 1 rachilla. However, one specimen from Peru *(H. Moore 8540)* has 5 rachillae, and is from the same place as the only specimen of *B. concinna* var. *inundata* that has 6 rachillae *(H. Moore 8535).*

10b. Bactris concinna var. **inundata** Spruce, J. Linn. Soc., Bot. 11: 154. 1871. *Pyrenoglyphis concinna* var. *inundata* (Spruce) Burret, Repert. Spec. Nov. Regni Veg. 34: 243. 1934. Type. Brazil. Amazonas: Rio Negro, n.d., *R. Spruce 8a* (holotype, K).

Bactris concinna subsp. *depauperata* Trail, J. Bot. 6: 48. 1877. *Bactris concinna* var. *depauperata* (Trail) Drude in Mart., Fl. bras.: Cyclanthaceae et Palmae I fasc. 85 vol 3(2): 335. 1881. *Pyrenoglyphis concinna* var. *depauperata* (Trail) Burret, Repert. Spec. Nov. Regni Veg. 34: 243. 1934. Type. Brazil. Amazonas: near Santarém, Rio Jutaí, 4 Feb 1875, *J. Trail 878/CCXVI* (holotype, K; isotype, P).

Colombia (Amazonas, Putumayo), Ecuador (Napo), Peru (Loreto, Madre de Dios, San Martín, Ucayali), Brazil (Acre, Amazonas, Rondônia), and Bolivia (Beni, La Paz, Pando, Santa Cruz) (Figure 6.39C).

10c. Bactris concinna var. **sigmoidea** Henderson, var. nov. Type. Peru. Madre de Dios: Pakitza, right bank of Río Manu, 11°53'S, 71°15'W, 400 m, 7 Nov 1990, *F. Chávez 704* (holotype, CUZ; isotype, NY)

A var. concinna pinnis lineari-lanceolatis vel sigmoideis irregulariter dispositis diversa.

Bolivia (Beni, Santa Cruz), Brazil (Acre), and Peru (Cuzco, Madre de Dios, Pasco) (Figure 6.39D); lowland rain forest on terra firme at low elevations.

11. Bactris constanciae Barb. Rodr., Enum. palm. nov. 37. 1875. Lectotype (Wess. Boer, 1965). Barb. Rodr., Sert. palm. brasil. 2: t. 40. 1903 (Figure 6.38e–f; Plate IIc).

Stems cespitose, 1.5–5 m tall, 2.5–5 cm diam, spiny on internodes. *Leaves* 5–8; sheath 20–70 cm long, sheath, petiole, and rachis lacking spines or with a sparse to moderate covering of black spines to 7 cm long; petiole elongate, 0.8–1.4 m long, brown-tomentose; rachis 1–3.3 m long; pinnae 14–22 per side, irregularly arranged in distant clusters of 2–8, spreading in different planes, oblanceolate to sigmoid, long acuminate, the middle ones 33–38 cm long, 3–5 cm wide, with spiny margins. *Inflorescences* interfoliar; peduncle 22–25 cm long; prophyll 10–22 cm long; peduncular bract 30–45 cm long, with few black spines to 1 cm long; rachis 5–6 cm long; rachillae 12–20, to 12 cm long; *flowers* borne in triads, these irregularly arranged among paired or solitary staminate flowers; staminate flowers not seen, deciduous; pistillate flowers (postanthesis) 5 mm long, purple-red; calyx tubular, 4 mm long; corolla tubular, 5 mm long; staminodes absent; *fruits* globose, 1.5–2.5 cm diam, densely muricate with projections to 3 mm long; rosy, dark-red, or magenta-purple at maturity; mesocarp starchy; endocarp fibers free, numerous; fruiting perianth with deeply 3-lobed calyx almost as long as the deeply 3-lobed corolla, staminodial ring absent.

Eastern Amazon region of the Guianas and Brazil (Amazonas, Pará) (Figure 6–39E); lowland rain forest on *terra firme* at low elevations, rarely to 500 m. At least in the Guianas, it occurs in *wallaba* forest. It seems to be an uncommon species throughout its range, but possibly commoner in the eastern part.

Brazil: *munbaca.*

Bactris constanciae is distinguished by its elongate, brown-tomentose petiole; by its oblanceolate to sigmoid pinnae arranged in distant clusters; and especially by its muricate fruits.

12. Bactris corossilla H. Karst., Linnaea 28: 407. 1857. Type (Glassman, 1972). Venezuela. Carabobo: Puerto Cabello, n.d., *H. Karsten s.n.* (holotype, LE, n.v.; F neg 31312) (Figure 6.38g–h).

Bactris duidae Steyerm., Fieldiana Bot. 28: 73. 1951. Type. Venezuela. Amazonas: Caño Negro, SE base of Cerro Duida, 215 m, 23 Aug 1944, *J. Steyermark 57908* (holotype, F).

Bactris venezuelensis Steyerm., Fieldiana Bot. 28: 80. 1951. Type. Venezuela. Amazonas: Caño Negro, SE base of Cerro Duida, 225 m, 23 Aug 1944, *J. Steyermark 57933* (holotype, F).

Bactris duplex H. E. Moore, Gentes Herb. 8: 160. 1949. Type. Colombia. Vaupés: road from San Martín, vicinity of Río Ocoa, 23 Oct 1945, *P. Allen 3358* (holotype, BH).

Stems cespitose and sometimes forming dense

colonies, or occasionally solitary, 2–4(–6) m tall, 1–3(–4) cm diam, spiny on internodes. *Leaves* 6–8; sheath 20–60 cm long, sheath and petiole with few to many, black, terete spines to 5 cm long, arranged ± in groups; petiole 20–80 cm long; rachis 45–70 cm long; blade entire and bifid, 0.6–2 m long, 23–34 cm wide at apex of rachis, or basal part of leaf pinnate with to 15 pinnae and apical pinna very broad, clustered, and spreading in different planes, the middle ones linear to sigmoid, 45–50 cm long, 3–5 cm wide, aristate, with prominent veins. *Inflorescences* interfoliar; peduncle 14–25 cm long; prophyll 10–20 cm long; peduncular bract 18–40 cm long, sulcate abaxially, densely spiny with ± spreading, flexuous, soft, brown spines; rachis 2–5 cm long; rachillae 4–14, 9–13 cm long, "bumpy" after staminate flowers have fallen; *flowers* arranged in triads, these scattered among paired or solitary staminate flowers; staminate flowers 5–6 mm long; sepals with triangular lobes 1.5–2 mm long; petals 5–6 mm long; pistillode very small; pistillate flowers 5–6 mm long; calyx tubular, 5–6 mm long, glabrous; corolla tubular, 3–3.5 mm long, glabrous or occasionally minutely spinulose; staminodes absent; *fruits* depressed-globose to obovoid, strongly rostrate, 2–2.5 cm long, 1.7–2 cm diam, green becoming yellow and then purple-black; mesocarp floury; endocarp fibers free, numerous; fruiting perianth with small, 3-lobed calyx to 5 mm long and longer, 3-lobed corolla to 1 cm long.

Coastal range, Andes, and adjacent regions of Venezuela (Apure, Carabobo, Mérida, Táchira) and Colombia (Norte de Santander), south to the western Amazon region in Colombia (Amazonas, Meta, Vaupés), Venezuela (Amazonas, Bolívar), Ecuador (Morona-Santiago, Napo, Pastaza), Peru (Amazonas, Loreto, San Martín, Ucayali), and Brazil (Acre) (Figure 6.39F); lowland to montane rain forest on well-drained slopes to 1400 m elevation.

Brazil: *marajá*. Colombia: *coquito, cuparú* (Yucuna). Ecuador: *kamancha* (Shuar). Peru: *ñejilla*. Venezuela: *du* (Piaroa), *juduaro, macanillo*. The fruits are edible.

Bactris corossilla is distinguished by its entire leaves (or, if pinnate, then with a very broad apical pinna), densely spiny, sulcate peduncular bract, few rachillae, and purple-black, obovoid fruits. It is a rather heterogeneous species as conceived here, with an unlikely distribution, and may eventually be split into two taxa. In particular, specimens from Ecuador appear rather different from the others. There may also be hybrids with other species in the specimens examined.

13. **Bactris cuspidata** Mart., Hist. nat. palm. 2: 101. 1826. *Amylocarpus cuspidatus* (Mart.) Barb. Rodr., Contr. Jard. Bot. Rio de Janeiro 3: 72. 1902. Type. Brazil. Amazonas: Rio Japurá, n.d., *C. Martius s.n.* (holotype, M; F neg. 18611) (Figure 6.40a–b).

Bactris mitis Mart., Hist. nat. palm. 2: 102. 1826. *Bactris mitis* subsp. *mitis* Trail, J. Bot. 6: 3. 1877. *Bactris cuspidata* var. *mitis* (Mart.) Drude in Mart. Fl. bras.: Cyclanthaceae et Palmae I, fasc. 85, vol. 3(2): 329. 1881. *Amylocarpus mitis* (Mart.) Barb. Rodr., Contr. Jard. Bot. Rio de Janeiro 3: 72. 1902. Type. Brazil. Amazonas: Rio Japurá, n.d., *C. Martius s.n.* (holotype, M; F neg. 18623).

Bactris floccosa Spruce, J. Linn. Soc., Bot. 11: 146. 1871. *Amylocarpus floccosus* (Spruce) Barb. Rodr., Sert. palm. brasil. 2: 37. 1903. Type. Brazil. Pará: Santarém, n.d., *R. Spruce 37* (holotype, K).

Bactris marajay Barb. Rodr., Enum. palm. nov. 29. 1875. *Bactris cuspidata* var. *marajay* (Barb. Rodr.) Drude in Mart., Fl. bras.: Cyclanthaceae et Palmae I, fasc. 85, vol. 3(2): 329. 1882. *Amylocarpus marajay* (Barb. Rodr.) Barb. Rodr., Contr. Jard. Bot. Rio de Janeiro 3: 72. 1902. Lectotype (Glassman, 1972). Barb. Rodr., Protesto-Appendice t. 2, f. 6. 1879.

Bactris cuspidata var. *coriacea* Trail, J. Bot. 6: 4. 1877. Type. Brazil. Pará: Belém, 1 Jan 1875, *J. Trail 880/ CCXXIV* (holotype, K).

Bactris cuspidata var. *angustipinnata* Trail, J. Bot. 6: 4. 1877. Type. Brazil. Pará: Belém, 2 Mar 1875, *J. Trail 854/CCXXV* (holotype, K).

Stems cespitose, 1–1.5 m tall, 6–8 mm diam. *Leaves* 4–5; sheath 13–17 cm long, sheath, petiole, and rachis whitish-tomentose, moderately to densely covered with black spines to 2 cm long; petiole 35–40 cm long; rachis 30–46 cm long; pinnae 6–13 per side, irregularly arranged in distant clusters, linear-lanceolate to sigmoid or ellipsoid, cuspidate, the middle ones

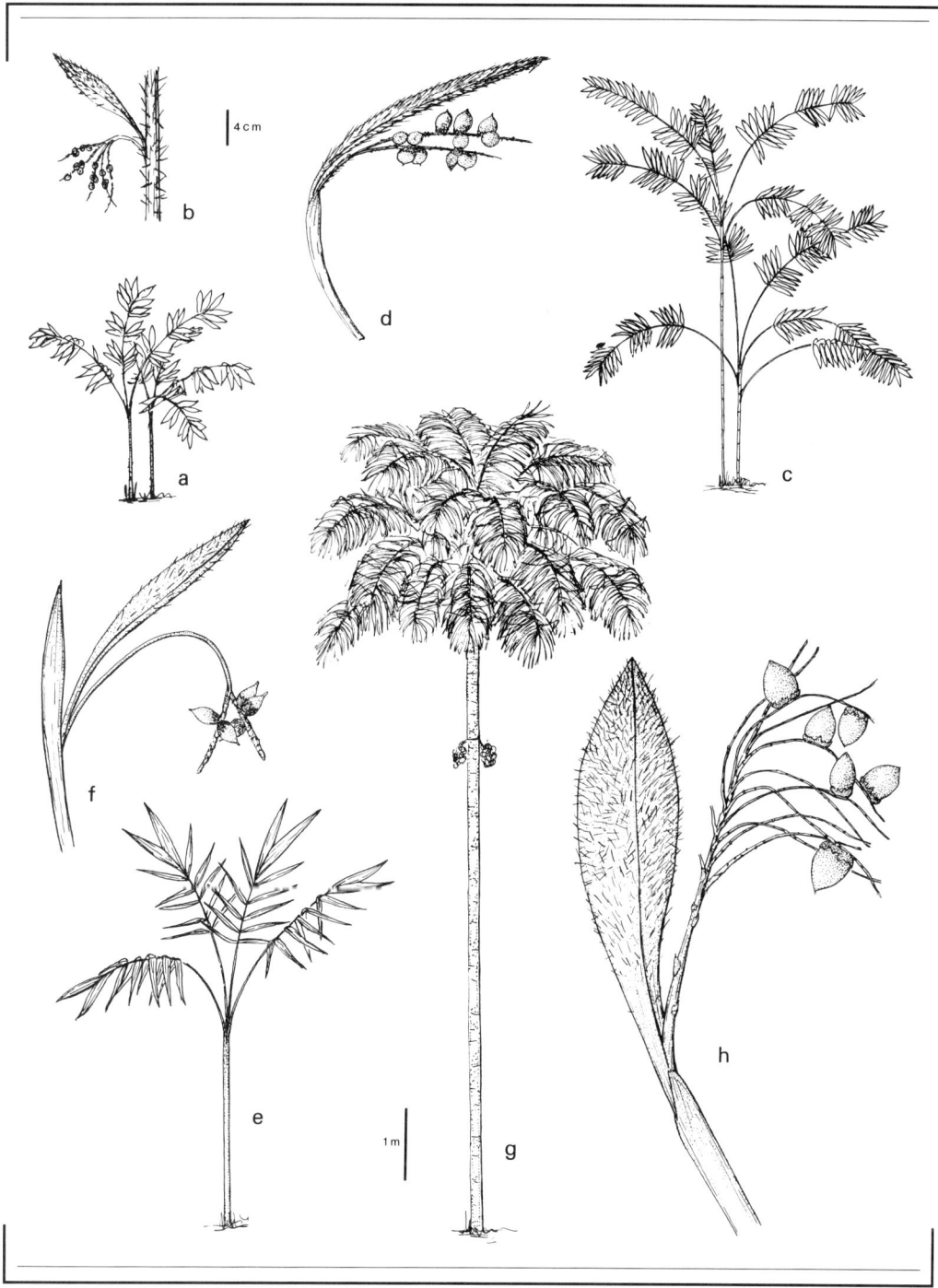

Figure 6.40. (a) *Bactris cuspidata,* **(b)** infructescence (from *G. Prance 30655*); **(c)** *Bactris elegans,* **(d)** infructescence (from *A. Henderson 654*); **(e)** *Bactris fissifrons,* **(f)** infructescence (from *M. Balick 989*); **(g)** *Bactris gasipaes,* **(h)** infructescence (from *M. Saldias 694*).

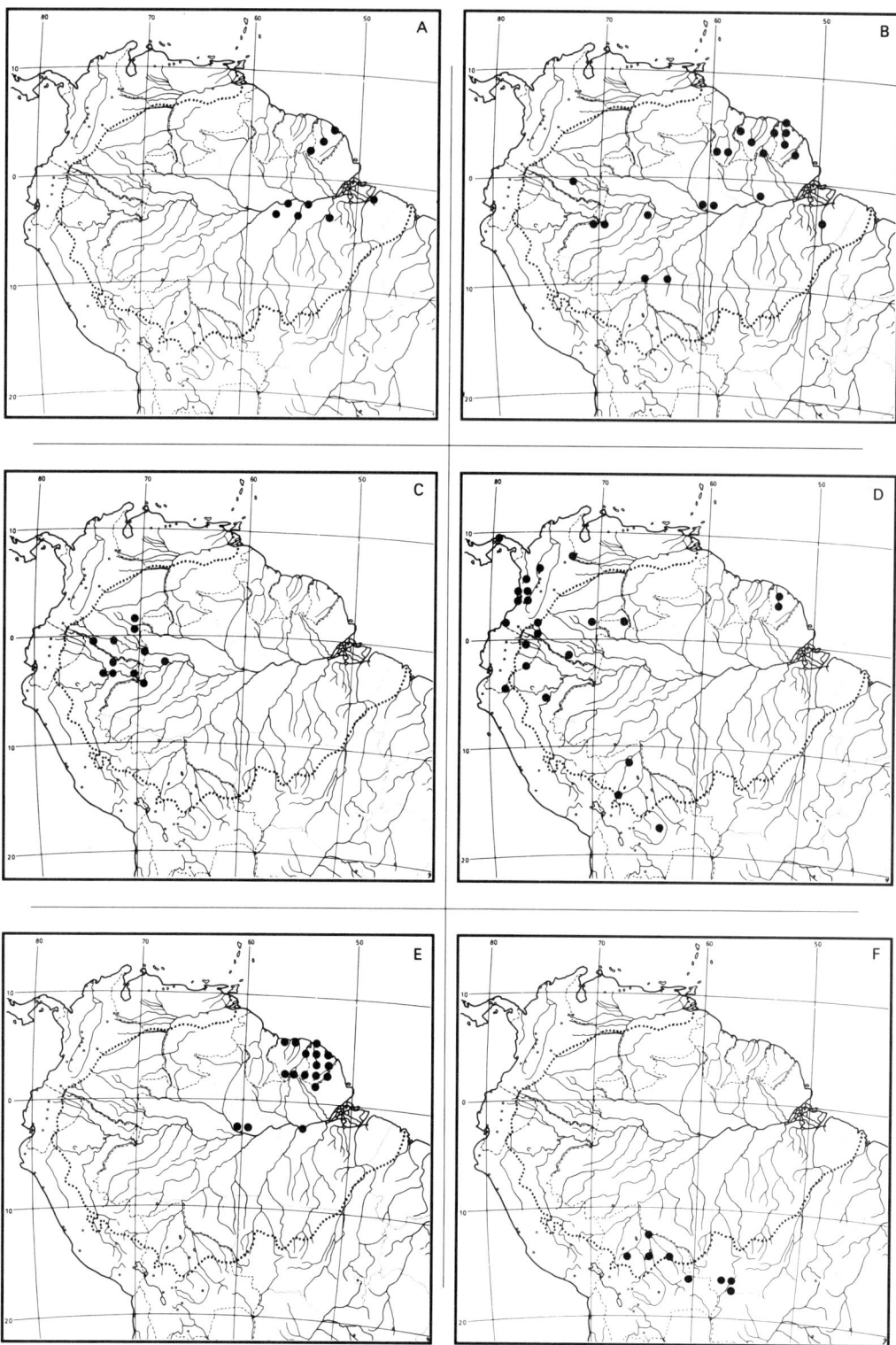

Figure 6.41. **(A)** *Bactris cuspidata;* **(B)** *Bactris elegans;* **(C)** *Bactris fissifrons;* **(D)** *Bactris gasipaes;* **(E)** *Bactris gastoniana;* **(F)** *Bactris glaucescens.*

14–25 cm long, 2–5 cm wide, scarcely pilose abaxially to densely pilose on both surfaces. *Inflorescences* interfoliar, curved down; peduncle 4–6 cm long; prophyll to 8 cm long; peduncular bract 9–12 cm long, densely covered with easily removed, black, soft, straight spines to 1 cm long; rachis absent; rachillae 2–8, 2.5–3.5 cm long, densely covered with short, flexuous, whitish trichomes intermixed with fewer, longer, brown spinules; *flowers* in triads, these close together with a few paired staminate flowers intermixed; staminate flowers not seen; pistillate flowers 2 mm long; calyx annular, 0.2 mm long; corolla tubular, 2 mm long, densely spinulose; *fruits* globose to obovoid, rostrate, 7 mm diam, red.

Central Amazon region in French Guiana and Brazil (Amazonas, Pará) (Figure 6.41A); lowland rain forest on *terra firme* at low elevations, rarely to 780 m.

There are few collections of this species, and these are scattered over a wide area (Martius's collections are not mapped because of lack of exact locality; he may have mixed up his localities, and all his specimens may be from Pará). Thus, as understood here, the species is rather heterogeneous, with populations from the eastern part quite different in some respects from those from the central and western part of the range. It may be the ancestor of the *Bactris hirta-B. simplicifrons* clade, in that it has the smooth fruits of *B. simplicifrons* and annular calyx of *B. hirta* (Sanders, 1991). In fact, some specimens may be hybrids between these two species. It has frequently been confused in the past with pinnate leaved forms of *Bactris simplicifrons* (e.g., the type of *B. mitis* is a mixture of the two). *Bactris floccosa,* included here as a synonym, differs somewhat in its pistillate corolla, and may in fact be a distinct taxon. Many more collections are needed to better understand this species.

14. **Bactris elegans** Barb. Rodr., Enum. palm. nov. 35. 1875. Lectotype (Wess. Boer, 1965). Barb. Rodr., Sert. palm. brasil. 2: t. 24. 1903 (Figure 6.40c–d).
   Bactris elegantissima Burret, Notizbl. Bot. Gart. Berlin-Dahlem 15: 4. 1940. Type. Guyana. Kuyuwini River, Essequibo tributary, ca. 150 mi from mouth, 21–26 Nov 1963, *A. Smith 2533* (holotype, NY).

Stems cespitose, forming loose to dense clusters of to 8 stems, 1.5–3.5 m tall, 0.8–1.5 cm diam, spiny or not spiny at internodes. *Leaves* 6–13, horizontally spreading; sheath 9–30 cm long, closed and not forming a crownshaft, sheath, petiole, and rachis with few black spines to 4 cm long; petiole 11–40 cm long; rachis 47–75 cm long, densely covered with soft, fine, brown bristles to 1 mm long; pinnae 17–32 per side, regularly arranged except for "gaps", spreading in 1 plane, linear-lanceolate to almost sigmoid, asymmetrically toothed and pendulous apically, the middle ones 9–22 cm long, 1–2.5 cm wide, glossy green adaxially, with very small spines on margins. *Inflorescences* interfoliar, at anthesis forming a 45° angle with the stem; peduncle 18–25 cm long; prophyll 10–15 cm long; peduncular bract 30–42 cm long, densely brown-tomentose with few to many black spines to 1.5 cm long; rachis absent; rachillae (1–)2, 8–15 cm long, at anthesis with dense reddish-brown tomentum and brown, glandular trichomes; *flowers* borne in triads, these irregularly arranged among paired or solitary staminate flowers; staminate flowers 4–5 mm long, deciduous; sepals with narrowly triangular lobes 1–2 mm long; petals ovate, 4–5 mm long; pistillate flowers 3–4 mm long; calyx tubular, 2.5 3 mm long; corolla tubular, 3–3.5 mm long; staminodes absent; *fruits* globose, rostrate, 1.2–1.5 cm diam, purple-black at maturity; mesocarp juicy; endocarp fibers free, numerous, with juice sacs attached; fruiting perianth with deeply 3-lobed calyx almost as long as the 3-lobed corolla.

Amazon region of Colombia (Amazonas), the Guianas, Brazil (Amazonas, Pará, Rondônia), and Bolivia (Pando) (Figure 6.41B); lowland rain forest on *terra firme* at low elevations.

Brazil: *marajá*. Colombia: *eérü* (Huitoto). French Guiana: *wili* (Wayâpi). Guyana: *firichí*. Surinam: *akamin-ékoenari, moeroekoe*.

Bactris elegans is distinguished by its small, regularly arranged (but with "gaps"), glossy green pinnae; inflorescences with (1–)2 rachillae; and purple-black, globose fruits. Many plants from the northeastern part of its range have spicate inflorescences.

15. **Bactris fissifrons** Mart., Hist. nat. palm. 2: 103. 1826. Type. Brazil. Amazonas: Rio Japurá, n.d., *C. Martius s.n.* (holotype, M; F neg. 18612) (Figure 6.40e–f).
Bactris aristata Mart., Hist. nat. palm. 2: 97. 1826. *Pyrenoglyphis aristata* (Mart.) Burret, Repert. Spec. Nov. Regni Veg. 34: 242. 1934. Type. Brazil. Amazonas: Rio Japurá, n.d., *C. Martius 3192* (holotype, M; F neg. 18602a).
Bactris fissifrons var. *robusta* Trail, J. Bot. 6: 9. 1877. Type. Brazil. Amazonas: Calderões, Rio Solimões, 10 Dec 1874, *J. Trail 897/CIC* (holotype, K; isotype, P).

Stems cespitose, 2–3 m tall, 1–3 cm diam, forming small clumps of up to 10 stems. *Leaves* 4–8; sheath 22–50 cm long, sheath scarcely to moderately covered with black, terete spines to 4 cm long; petiole 28–40 cm long, only the adaxial surface densely covered with short spines to 2 cm long; rachis 42–110 cm long; pinnae 2–17 per side, irregularly arranged in clusters or almost regularly arranged, spreading in 1 plane, linear-lanceolate, with very long, filiform apex, the apical 1 usually much wider than the others, the middle ones 40–58 cm long, 2.5–10 cm wide, with small spines on margins, with prominent veins adaxially. *Inflorescences* interfoliar; peduncle 10–18 cm long; prophyll 14–22 cm long; peduncular bract 23–34 cm long, sulcate abaxially, moderately covered with black spines to 1 cm long; rachis absent; rachillae 2–5(–7), 13–15 cm long, to 5 mm diam in fruit; *flowers* borne in triads, these irregularly arranged among paired or solitary staminate flowers; staminate flowers 5.5–7 mm long, deciduous; sepals with triangular lobes, 1.5 mm long; petals obovate, 5.5–6.5 mm long; pistillate flowers 5 mm long; calyx tubular, 5 mm long, glabrous; corolla tubular, 3 mm long, minutely spinulose; ovary minutely spinulose; *fruits* obovoid, elongate rostrate, 2.5–3 cm long, 1.5–2 cm diam, yellowish at maturity, glabrous or minutely spinulose at apex; mesocarp juicy; endocarp fibers free, numerous with juice sacs attached; fruiting perianth with irregularly lobed calyx to 5 mm long and irregular lobed, glabrous or minutely spinulose corolla to 1 cm long, staminodial ring absent.

Western Amazon region in Colombia (Amazonas, Putumayo, Vaupés), Peru (Loreto), and Brazil (western Amazonas) (Figure 6.41C); lowland rain forest in well-drained areas, especially on river terraces, at low elevations.

Colombia: *chontilla, cubarro, sitanó* (Andoke), *yachoro.* Peru: *ca-újico* (Bora). The fruits and seeds are edible.

Bactris fissifrons is distinguished by its inflorescences with sulcate peduncular bract and 2–5(–7) thick rachillae; and obovoid, elongate rostrate, yellowish fruits. As noted by Burret (1933–1934), both Trail (1877a) and Drude (1881) considered that *B. aristata* had a staminodial ring. However, Trail based this on his specimen *(J. Trail 855/LXVII),* which is actually an aberrant form of *B. bifida.* Drude followed Trail's mistake. The type of *B. aristata* at M has no fruits, but the locality, leaves, and description match those for *B. fissifrons.*

16. **Bactris gasipaes** Kunth in Humb., Bonpl. & Kunth, Nov. gen. sp. 1: 302. 1815. *Guilielma gasipaes* (Kunth) L. H. Bailey, Gentes Herb. 2: 187. 1930. Type. Colombia. Tolima: Ibagué, n.d., *A. Bonpland s.n.* (holotype, P; F neg. 38701) (Figure 6.40g–h).
Guileilma speciosa Mart., Hist. nat. palm. 2: 82. 1824. Type. Brazil. Maranhão: without locality, n.d., *C. Martius s.n.* (holotype, M, n.v.).
Guilielma insignis Mart. in A. D. Orb., Voy. Amérique mér. 7(3). Palmiers 71. 1844. *Bactris insignis* (Mart.) Baillon, Hist. pl. 13: 305. 1985. Type. Bolivia. Beni: Moxos, n.d., *A. d'Orbigny 18* (holotype, P, n.v.).
Guilielma speciosa var. *flava* Barb. Rodr., Enum. palm. nov. 23. 1875. Lectotype (here designated). Barb. Rodr., Sert. palm. brasil. 2: t. 52b. 1902.
Guilielma speciosa var. *coccinea* Barb. Rodr., Enum. palm. nov. 23. 1875. Lectotype (Glassman, 1972). Barb. Rodr., Sert. palm. brasil. 2: t. 52c. 1902.
Guilielma speciosa var. *ochracea* Barb. Rodr., Vellosia 1: 40. 1888. Lectotype (Glassman, 1972). Barb. Rodr., Sert. palm. brasil. 2: t. 52d. 1902.
Guilielma speciosa var. *mitis* Drude in Mart., Fl. bras.: Cyclanthaceae et Palmae I, fasc. 85, vol. 3(2): 363. 1881. Type. Brazil. Rio de Janeiro: cultivated, 26 Dec 1887, *A. Glaziou 17342* (holotype, P).

Stems cespitose or solitary, 4–18 m tall, 10–25 cm diam, spiny at internodes, rarely without spines. *Leaves* 9–20; sheath lacking ocrea, sheath and petiole 1–1.1 m long, petiole with spines in 3 files, sheath, petiole, and rachis moderately to densely covered with black or brownish spines to 1 cm long; rachis 1.9–2.6 m long; pinnae 92–123 per side, arranged in obscure clusters of 3–5, spreading in different planes, linear, shortly bifid at apex, the middle ones 52–75 cm long, 2–3 cm wide. *Inflorescences* at first interfoliar; peduncle 20–28 cm long; prophyll 20–21 cm long; peduncular bract 47–70 cm long, moderately to densely covered with blackish or brownish spines to 1 cm long; rachis 15–23 cm long; rachillae 46–57, 17–28 cm long, at anthesis densely covered with glandular trichomes, glabrescent; *flowers* in triads, these irregularly arranged among paired or solitary staminate flowers; staminate flowers to 4 mm long, deciduous; sepals with triangular lobes 2 mm long; petals obovate, 3.5 mm long; pistillate flowers 6 mm long; calyx cupular, 2 mm long; corolla tubular, 4.5 mm long; staminodes absent; *fruits* widely ovoid, to 5 cm long, to 3 cm diam, yellow, orange, or red at maturity; mesocarp thick and starchy; endocarp fibers flattened, anastomosing, adnate to endocarp; fruiting perianth with very small calyx with undulate margins and much longer, scarcely lobed corolla, staminodial ring absent.

Widely and commonly cultivated throughout tropical areas of Central and northern South America (Figure 6.41D), and almost always associated with current or past human dwellings. There are many cultivated varieties, including one without spines on the stems and leaves, and another without a seed in the fruit. Its place of origin is not known, but it appears to be a selected form of *Bactris macana*. Indeed, without fruits, the two cannot be told apart (Bernal & Henderson, in press; see also Clement et al., 1989).

Bactris gasipaes is pollinated by a variety of beetles. Scarab beetles were considered the major pollinators by Beach (1984), while Mora Urpí and Solís (1980) emphasized the role played by weevils.

Bolivia: *chima, chonta, chonta de castilla, huanima* (Chacabo), *woy* (Chimane). Brazil: *pupunha*. Colombia: *chontadura*. Ecuador: *chonta pala* (Quichua). Peru: *pijuayo, uwí* (Mayna Jívaro). Venezuela: *pijiguao*. The fruits of *Bactris gasipaes* are an important part of the diet in many parts of the Amazon. Historically, they were much more important to indigenous peoples, not only as a source of food but also in rituals and ceremonies (e.g., Reichel-Dolmatoff, 1989). They are also commonly used to make a drink *(chicha)*. Currently, this species is the object of efforts by agriculturalists to improve yields, and also for use as a source of palm hearts (e.g., Clement & Mora Urpí, 1987).

Bactris gasipaes can be distinguished by its fruits, which are much larger than those of any other species in the Amazon region. It is, however, clearly a selected form of *B. macana* (Bernal & Henderson, in press).

Bernal (1989) has proposed conservation of *Bactris gasipaes* Kunth against the older, overlooked name *B. ciliata* (Ruiz & Pav.) Mart.

17. **Bactris gastoniana** Barb. Rodr., Vellosia 1: 40. 1888. *Pyrenoglyphis gastoniana* (Barb. Rodr.) Burret, Repert. Spec. Nov. Regni Veg. 34: 242. 1934. Lectotype (Wess. Boer, 1965). Barb. Rodr., Sert. palm. brasil. 2: t. 13. 1903 (Figure 6.42a–b).

Stems solitary or cespitose, subterranean or aerial and then 5–60 cm tall, 0.8–2 cm diam, covered with persistent, decaying leaf bases. *Leaves* 4–8, spreading; sheath 11–20 cm long, persistent, fibrous at apex, sheath, petiole, and rachis sparsely covered with black spines to 4.5 cm long; petiole 40–75 cm long; rachis 45–55 cm long; pinnae 8–11 per side, irregularly arranged in clusters of 2–4, spreading in different planes or ± in the same plane, sigmoid, aristate, minutely spiny on the margins, the middle ones 19–28 cm long, 2.5–4 cm wide, apical pinna wider than the others. *Inflorescences* interfoliar; peduncle 13–20 cm long; prophyll 9–12 cm long; peduncular bract 16–25 cm long, densely covered with black or brown spines to 1.5 cm long, glabrescent; rachis absent; rachilla 1, 3–4 cm long; *flowers* borne in triads, these regularly(?) arranged on rachilla; staminate flowers to 1 cm long, persistent and decaying on the rachilla; pistillate flowers not seen; *fruits* ellipsoid or widely ovoid, 2.5–3 cm long, 1.5–2 cm diam, purple-black at maturity, densely crowded on rachilla; mesocarp juicy;

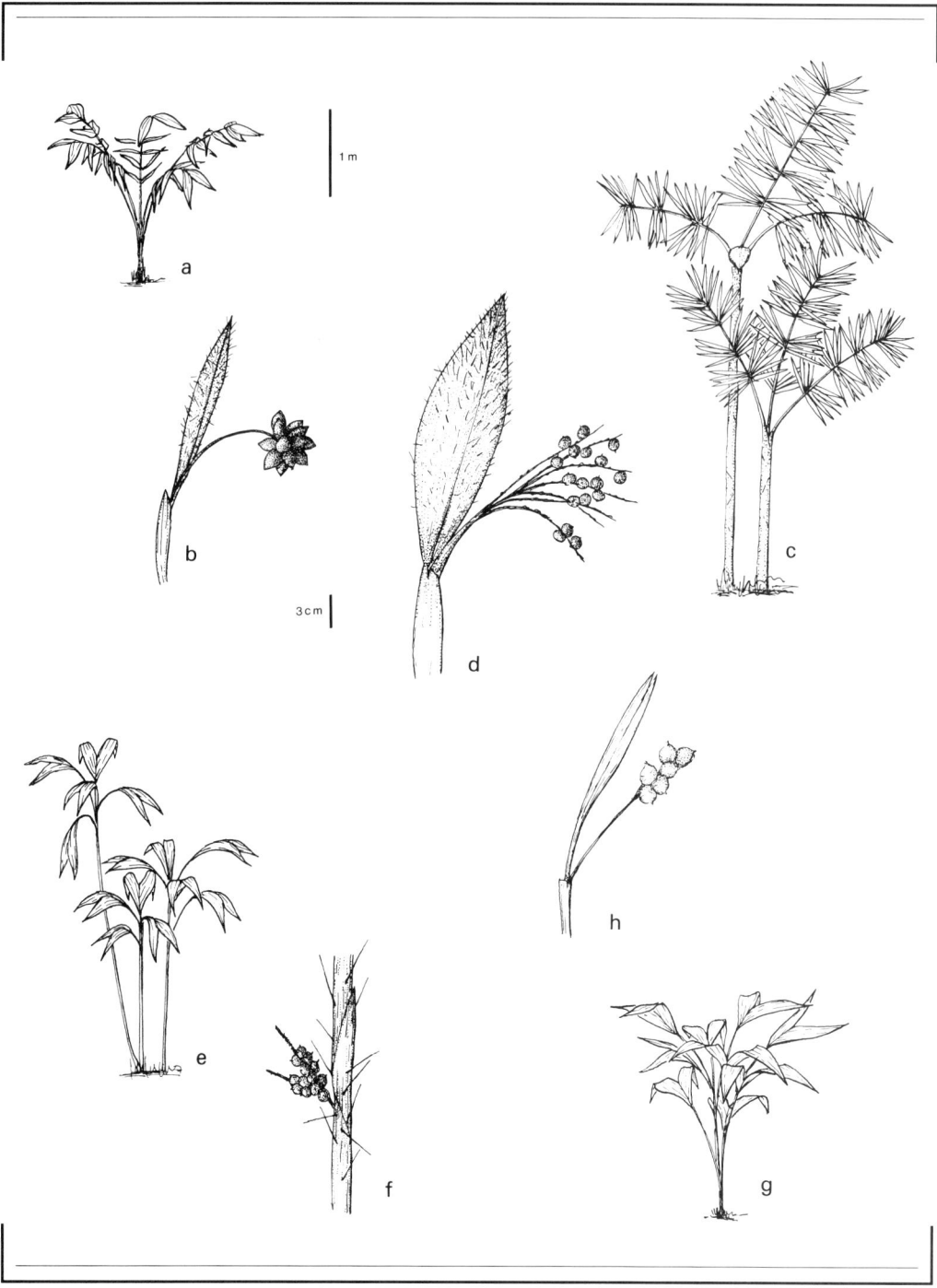

Figure 6.42. (a) *Bactris gastoniana,* **(b)** infructescence (from *A. Henderson 1605*); **(c)** *Bactris glaucescens,* **(d)** infructescence (from *M. Nee 34511*); **(e)** *Bactris hirta* var. *hirta,* **(f)** infructescence (from *W. Balée 1582*); **(g)** *Bactris killipii,* **(h)** infructescence (from *A. Henderson 838*).

endocarp fibers numerous, free, with juice sacs attached; fruiting perianth with very small calyx to 3 mm long and tubular, many-lobed corolla to 6 mm long, staminodial ring to 6 mm long.

Central and eastern Amazon region in the Guianas and Brazil (Amapá, Amazonas, Pará) (Figure 6.41E); lowland rain forest on *terra firme* at elevations below 600 m.

Brazil: *marajá*. French Guiana: *anuya wili* (Wayâpi). Surinam: *hanaimaka, naimacca, tamutubë* (Trio).

Bactris gastoniana is distinguished by its short stem; sigmoid, aristate pinnae; spicate inflorescence; and purple-black fruits with a staminodial ring. It is morphologically similar to *B. oligocarpa* Barb. Rodr.

18. **Bactris glaucescens** Drude in Mart., Fl. bras.: Cyclanthaceae et Palmae I, fasc. 85. vol. 3(2): 345. 1881. Type. Brazil. Mato Grosso: Rio Paraguay, *H. Weddell 3216* (holotype, P) (Figure 6.42c–d).

Stems cespitose, 1.5–4 m tall, 3–4 cm diam, with rings of spines on nodes. *Leaves* 4–16; sheath 26–30 cm long, fibrous at apex, sheath, petiole, and rachis whitish-tomentose, glabrescent, with ± clustered, slightly flattened, black spines to 10 cm long; petiole 8–45 cm long; rachis 56–86 cm long; pinnae 26–45 per side, irregularly arranged in clusters of 2–6 and stiffly spreading in different planes, oblanceolate, briefly bifid at the apex, the middle ones 16–32 cm long, 2–2.5 cm wide, whitish-brown-lepidote abaxially, with few spines on margins. *Inflorescences* interfoliar; peduncle 8–28 cm long; prophyll 9–18 cm long; peduncular bract 22–30 cm long, expanded part hooded over the rachillae, whitish-tomentose, moderately covered with straight, black spines to 1 cm long; rachis 3–8 cm long; rachillae 32–42, 12–14 cm long; *flowers* borne in triads, these irregularly arranged among paired or solitary staminate flowers; staminate flowers 3–4 mm long; sepals with triangular lobes 1 mm long; petals 3–4 mm long; stamens 5–6; pistillode small or absent; pistillate flowers 3 mm long; calyx cupular, 0.5–1 mm long; corolla tubular, 2–2.5 mm long; staminodes 6, digitate; *fruits* depressed globose, rostrate, 1.2–1.3 cm diam, purple-black; mesocarp juicy; endocarp fibers free, numerous, with juice sacs attached; fruiting calyx cupular, with entire margins, fruiting corolla tubular, 3-lobed, staminodial ring absent.

Southwestern part of the Amazon region in Brazil (Goiás, Mato Grosso, Mato Grosso do Sul, Rondônia), Bolivia (Beni, Santa Cruz), and northeastern Paraguay (Amambay, Concepcion, San Pedro) (Figure 6.41F); gallery forest, forest edges, or savannas, along stream margins and other wet places liable to seasonal inundation, at low elevations.

Moraes and Sarmiento (1992) have discussed the reproductive biology of this species in Bolivia. Inflorescences are protogynous, with pistillate anthesis taking place in the first evening of anthesis, followed by a short-lived staminate anthesis approximately 24 hours later. Probable beetle pollinators were species of *Phyllotrox* and *Phytotribus* (Curculionidae).

Bolivia: *chontilla*. Brazil: *coquinha, coquita, tucum*. The fruits are edible.

This species is similar morphologically to two species: *Bactris bidentula* to the north in the Amazon region, and *B. vulgaris* to the south in southern coastal Brazil. It differs from *B. bidentula* in its much longer leaf spines; indeed, it would have been included there except for the geographic separation of the two. This group also contains *B. guineensis* from Central America and northern South America, and is thus another example of a linear, north–south distribution pattern. Sanders (1991) also placed in this group *B. caryotaefolia,* from southeastern Brazil.

19. **Bactris hirta** Mart., Hist. nat. palm. 2: 104. 1826. *Amylocarpus hirtus* (Mart.) Barb. Rodr., Contr. Jard. Bot. Rio de Janeiro 3: 72. 1902. Type. Brazil. Amazonas: Rio Japurá, n.d., *C. Martius s.n.* (holotype, M; F. neg. 18614) (Figure 6.42e–f).

Stems cespitose or solitary, 0.5–3 m tall, 1–2 cm diam, without spines, rarely with spines, or commonly covered with persistent, spiny leaf sheaths. *Leaves* 3–7; sheath 8–26 cm long, partially closed, sheath, petiole, and rachis spinulose or sparsely to very densely covered with black spines 1–5 cm long; petiole 1–89 cm long; rachis 12–81 cm long; blade entire or pinnate, pinnae 7–30 per side, regularly or irregularly arranged, in 1 plane, linear, linear-lanceolate or almost sigmoid, aristate, the middle ones 10–37 cm long, 1–3 cm wide, or often blade is completely or partially entire, 17–80 cm long, 9–20 cm wide at rachis apex, with fine spines on the margins, with soft hairs to 3 mm long abaxially (and occasionally also adaxially), sometimes with prominent cross-veins. *Inflorescences* interfoliar or infrafoliar, usually borne among persistent leaf sheaths, erect in bud and at anthesis; peduncle 6–21 cm long; prophyll 4.5–6(–9) cm long; peduncular bract to 12 cm long, sparsely to densely covered with black spines; rachis 0–0.7 cm long; rachillae 1–4(–5), 5–7 cm long, usually forming a 45° angle with the stem, at anthesis densely covered with brown, glandular trichomes; *flowers* borne in triads, these regularly arranged proximally on rachillae, paired or solitary staminate distally; staminate flowers to 4 mm long, semipersistent or deciduous; sepals to 1.3 mm long; petals to 4 mm long; stamens 6–7; pistillate flowers at anthesis 2–3.5 mm long; calyx annular, 0.5 mm long; corolla tubular, 2–3 mm long, densely covered with long, flexuous, appressed, brown hairs, these exceeding the corolla; staminodes usually absent; *fruits* globose to very widely obovoid, to 1 cm long, usually less, 0.5 cm diam, orange-red or red at maturity, afterward becoming black, covered with fine, brown, deciduous spinules; fruiting perianth with very small, 3-lobed calyx and much longer, hairy, broadly 3-lobed corolla, staminodial ring absent.

Widespread and common in the Amazon region of Colombia (Amazonas, Guainía, Putumayo, Vaupés, Vichada), Venezuela (Amazonas, Bolívar), the Guianas, Peru (Loreto, Madre de Dios), Brazil (Acre, Amapá, Amazonas, Maranhão, Pará, Rondônia), and Bolivia (Pando), also in the Atlantic coastal forest of Brazil (Bahia, Espírito Santo, Pernambuco); usually in lowland rain forest on *terra firme*, occasionally occurring up to 1500 m elevation.

Brazil: *jawpe, marajá, marazaiwa-ran* (Tembé), *ubim rana*. Colombia: *chontaduro de rana*. Surinam: *kiskismaka*. Venezuela: *cubarro, espinita*. The stems are used by Tembé Indians to make arrows.

This species is widespread and variable, particularly in leaf shape and size. The most common and widespread form, which occurs throughout the Amazon region, has a large, entire leaf on an elongate petiole. The abaxial leaf surface is often pilose. This form has been called *Bactris hirta, B. geonomoides,* and *B. integrifolia*. A plicate leaf form found in French Guiana may be a hybrid with *B. aubletiana.* More to the center of the region a second, very common form is found that has pinnate leaves. Pinnae are 7–30 per side, linear, elongate, and are 22–37 cm long and 1–2 cm wide. This form has been called *B. pectinata*. A second pinnate-leaved form is much more local, and is found in the upper Río Negro region of Venezuela (Amazonas, Bolívar) and Colombia (Guainía), often on *tepui* slopes from 400 to 1500 m elevation. It rarely occurs in other places—for example, in Madre de Dios, Peru. It was called *B. turbinata* by Spruce. Pinnae are fewer and shorter, typically 7–10 per side, linear-lanceolate, and 2–3 cm wide. A third pinnate-leaved form known from western Brazil (Amazonas, Rio Juruá) and Peru (Loreto, Río Ucayali) has a much larger inflorescence than usual. All these forms grow together, particularly the first two, and it is not uncommon to find two leaf forms, pinnate and entire, on the same plant. Because of this, and because there are no other differences in vegetative or reproductive characters, I have treated all these forms as var. *hirta*.

In the western Amazon region of Colombia (Amazonas), Peru (Loreto), and Brazil (western Amazonas), a distinctive form occurs that has densely spinulose leaves, with soft spines to 2 cm long. Most plants have pinnate leaves, and the pinnae are of the shorter, linear-lanceolate type. As one moves away from this region, however, forms with entire leaves occur, and also the density of the spinules decreases. This form is here called var. *mollis*.

Scattered throughout the central Amazon region is another form that has a much smaller, entire leaf, which is borne on a very short, usually nonspiny petiole. The leaves commonly have cross-veins. When seen in the field this form, here called var. *pulchra*, appears very distinct from the above forms, but again is essentially similar in reproductive parts.

There are, however, intermediates between these varieties.

KEY TO THE VARIETIES OF BACTRIS HIRTA

1. Leaf entire or pinnate, when entire 34–80 cm long, 16–20 cm wide at apex of rachis, seldom with cross-veins; petiole spiny, (10–) 40–89 cm long.
   2. Sheath, petiole and rachis not densely covered with spinules; widespread . . . . 19a. *B. hirta* var. *hirta*.
   2. Sheath, petiole and rachis densely covered with spinules to 2 cm long; Colombia (Amazonas), Peru (Loreto) and Brazil (western Amazonas) . . . . 19b. *B. hirta* var. *mollis*.
1. Leaf entire, 18–30 cm long, 9–13 cm wide at apex of rachis, commonly with cross-veins; petiole usually non-spiny, 1–10 cm long . . . . . . 19c. *B. hirta* var. *pulchra*.

19a. Bactris hirta var. hirta

Bactris pectinata Mart., Hist. nat. palm. 2: 98. 1826. *Amylocarpus pectinatus* (Mart.) Barb. Rodr., Contr. Jard. Bot. Rio de Janeiro 3: 72. 1902. Type. Brazil. Pará: without locality, n.d., *C. Martius s.n.* (holotype, M; F neg. 18627).

Bactris longipes Poepp. ex Mart., Hist. nat. palm. 2: 145. 1837. Type. Brazil. Amazonas: Tefé, 1831, *E. Poeppig 2750* (holotype, M; F. neg. 18618).

Bactris integrifolia Wallace, Palm Trees of the Amazon 91. 1853. Lectotype (Glassman, 1972). Wallace, Palm Trees of the Amazon t. 35. 1853.

Bactris simplicifrons of Spruce, J. Linn. Soc., Bot. 11: 145. 1871. Type. Brazil. Pará: Tauaú, n.d., *R. Spruce s.n.* (holotype, K, n.v.).

Bactris turbinata Spruce, J. Linn. Soc., Bot. 11: 146. 1871. *Bactris pectinata* subsp. *turbinata* (Spruce) Trail, J. Bot. 6: 7. 1877. Type. Venezuela. Amazonas: San Carlos de Río Negro, Oct 1854, *R. Spruce 51* (holotype, K).

Bactris hylophila Spruce, J. Linn. Soc., Bot. 11: 146. 1871. *Bactris pectinata* subsp. *hylophila* (Spruce) Trail, J. Bot. 6: 6. 1877. *Amylocarpus hylophilus* (Spruce) Barb. Rodr., Contr. Jard. Bot. Rio de Janeiro 3: 72. 1902. Type. Brazil. Amazonas: Rio Negro, n.d., *R. Spruce 15* (holotype, K).

Bactris microcarpa Spruce, J. Linn. Soc., Bot. 11: 146. 1871. *Bactris pectinata* subsp. *microcarpa* (Spruce) Trail, J. Bot. 6: 6. 1877. Type. Brazil. Amazonas: Rio Negro, São Gabriel, Jun 1852, *R. Spruce 31* (holotype, K).

Bactris ericetina Barb. Rodr., Enum. palm. nov. 26. 1875. *Amylocarpus ericetinus* (Barb. Rodr.) Barb. Rodr., Contr. Jard. Bot. Rio de Janeiro 3: 71. 1902. Lectotype (Glassman, 1972). Barb. Rodr., Protesto-Appendice t. 2, f. 4. 1879.

Bactris linearifolia Barb. Rodr., Enum. palm. nov. 31. 1875. *Amylocarpus linearifolius* (Barb. Rodr.) Barb. Rodr., Contr. Jard. Bot. Rio de Janeiro 3: 72. 1902. Lectotype (Glassman, 1972). Barb. Rodr., Sert. palm. brasil. 2: t. 47. 1903.

Bactris setipinnata Barb. Rodr., Enum. palm. nov. 32. 1875. *Bactris pectinata* subsp. *hylophila* var. *setipinnata* (Barb. Rodr.) Trail, J. Bot. 6: 6. 1877. *Amylocarpus settipinnatus* (Barb. Rodr.) Barb. Rodr., Contr. Jard. Bot. Rio de Janeiro 3: 72. 1902. Lectotype (Glassman, 1972). Barb. Rodr., Sert. palm. brasil. 2: t. 48. 1903.

Bactris longipes var. *exilis* Trail, J. Bot. 6: 5. 1877. Type. Brazil. Amazonas: Barreiras de Mutum, Rio Jutahi, n.d., *J. Trail 858/CCX* (holotype, K).

Bactris pectinata subsp. *hylophila* var. *subintegrifolia* Trail, J. Bot. 6: 6. 1877. Type. Brazil. Amazonas: Lages, mouth of Rio Negro, 5 Jul 1874, *J. Trail 869/XCI* (holotype, K).

Bactris pectinata subsp. *microcarpa* var. *nana* Trail, J. Bot. 6: 6. 1877. Type. Brazil. Amazonas: Rio Javari, Camaná, 5 Dec 1874, *J. Trail 861/CXCVI* (holotype, K).

Bactris pectinata subsp. *turbinata* var. *spruceana* Trail, J. Bot. 6: 7. 1877. Type. Brazil. Pará: Belém, 27 Feb 1875, *J. Trail 872/CCXXII* (holotype, K; isotypes, NY, P).

Bactris geonomoides Drude in Mart., Fl. bras.: Cyclanthaceae et Palmae I, fasc. 85, vol. 3(2): 325. 1881. *Amylocarpus geonomoides* (Drude) Barb. Rodr., Contr. Jard. Bot. Rio de Janeiro 3: 72. 1902. Type. Not designated.

Bactris geonomoides var. *setosa* Drude in Mart., Fl. bras. Cyclanthaceae et Palmae I, fasc. 85, vol. 3(2): 325. 1881. Type. Not designated.

Bactris hylophila var. *macrocarpa* Drude in Mart., Fl. bras.: Cyclanthaceae et Palmae I, fasc. 85, vol. 3(2): 332. 1881. *Amylocarpus hylophilus* var. *macrocarpus* (Drude) Barb. Rodr., Sert. palm. brasil. 2: 87. 1903 Type. Not designated.

Bactris hylophila var. *glabrescens* Drude in Mart., Fl. bras.: Cyclanthaceae et Palmae I, fasc. 85, vol. 3(2): 332. 1881. *Amylocarpus hylophilus* var. *glabrescens* (Drude) Barb. Rodr., Sert. palm. brasil. 2: 87. 1903 Type. Not designated.

Bactris hylophila var. *nana* Drude in Mart., Fl. bras.: Cyclanthaceae et Palmae I, fasc. 85, vol. 3(2): 332. 1881. *Amylocarpus hylophilus* var. *nanus* (Drude) Barb. Rodr., Sert. palm. brasil. 2: 87. 1903. Type. Not designated.

Bactris formosa Barb. Rodr., Vellosia 1: 43. 1888. *Amylocarpus formosus* (Barb. Rodr.) Barb. Rodr., Contr. Jard. Bot. Rio de Janeiro 3: 72. 1902. Lectotype (Glassman, 1972). Barb. Rodr., Sert. palm. brasil. 2: t. 46. 1903.

Bactris unaensis Barb. Rodr., Palmae hassler. 14. 1900. Lectotype (Glassman, 1972). Barb. Rodr., Sert. palm. brasil. 2: t. 17. 1903.

Amylocarpus platispinus Barb. Rodr., Contr. Jard. Bot. Rio de Janeiro 3: 72. 1902. *Bactris platyspinus* (Barb. Rodr.) Burret, Repert. Spec. Nov. Regni Veg. 34: 189. 1934. Lectotype (Burret, 1934). Barb. Rodr., Sert. palm. brasil. 2: t. 50b. 1903.

Bactris hoppii Burret, Repert. Spec. Nov. Regni Veg. 34: 181. 1933. Type. Brazil. Pará: Utinga, Oct 1932, *W. Hopp 6* (holotype, B, n.v.).

Bactris huebneri Burret, Repert. Spec. Nov. Regni Veg. 34: 182. 1933. Type. Brazil. Amazonas: Manaus, n.d., *G. Huebner 29* (holotype, B, n.v.).

Bactris atrox Burret, Repert. Spec. Nov. Regni Veg. 34: 182. 1933. Brazil. Pará: without locality, Nov 1932, *W. Hopp 36* (holotype, B).

Sheath moderately to densely covered with stiff spines to 4 cm long; petiole usually spiny proximally; pinnae 7–30 per side, regularly arranged but with "gaps", linear or linear-lanceolate, 22–37 cm long, 1–3 cm wide, usually pilose abaxially, or blade entire or occasionally with 2 broad pinnae, 34–80 cm long, 17–20 cm wide, glabrous or pilose abaxially, rarely adaxially.

Colombia (Amazonas, Guinía, Vaupés), Venezuela (Amazonas, Bolívar), the Guianas, Peru (Loreto, Madre de Dios), Brazil (Amapá, Amazonas, Maranhão, Pará, Rondônia; not yet collected in Acre), and Bolivia (Pando), and also the Atlantic coastal forest of Brazil (Bahia, Espírito Santo, Pernambuco) (Figure 6.43A); occasionally to 1500 m elevation.

19b. Bactris hirta var. **mollis** (Dammer) Henderson, stat. nov. *Bactris mollis* Dammer, Verh. Bot. Vereins Prov. Brandenburg 48: 129. 1907. Type. Colombia. Amazonas: Leticia, Jun 1902, *E. Ule 6221* (isotypes, K, MG).

Bactris lakoi Burret, Repert. Spec. Nov. Regni Veg. 34: 187. 1934. Type. Brazil. Amazonas: Rio Içá, n.d., *C. Lakó 1/G. Huebner 133* (holotype, B, n.v.).

Sheath, petiole, and rachis very densely covered with soft spines to 2 cm long; pinnae 19–30 per side, linear-lanceolate or almost sigmoid, regularly arranged but with "gaps", 11–15 cm long, 1–1.5 cm wide, or less often leaf entire, pilose on both surfaces.

Colombia (Amazonas), Peru (Loreto), and Brazil (western Amazonas) (Figure 6.43B).

This variety has a small range, and on the margins there are intermediates with var. *hirta*. Most plants have pinnate leaves, but two specimens from the edge of the range, both unfortunately sterile but probably belonging here, have either entire leaves or leaves with few, sigmoid pinnae, giving the same variation in leaf shape in the variety that is found in the species.

19c. Bactris hirta var. **pulchra** (Trail) Henderson, stat. nov. *Bactris hirta* subsp. *pulchra* Trail, J. Bot. 6: 4. 1877. *Bactris pulchra* (Trail) Trail ex Drude in Mart., Fl. bras.: Cyclanthaceae et Palmae I, fasc. 85, vol. 3(2): 324. 1882. *Amylocarpus pulcher* (Trail) Barb. Rodr., Contr. Jard. Bot. Rio de Janeiro 3: 72. 1902. Type. Brazil. Amazonas: Manaus, mouth of Rio Negro, Lages, 23 Jul 1874, *J. Trail 876/XCVIII* (holotype, K; isotype, P).

Bactris pulchra var. *inermis* Dammer, Verh. Bot.

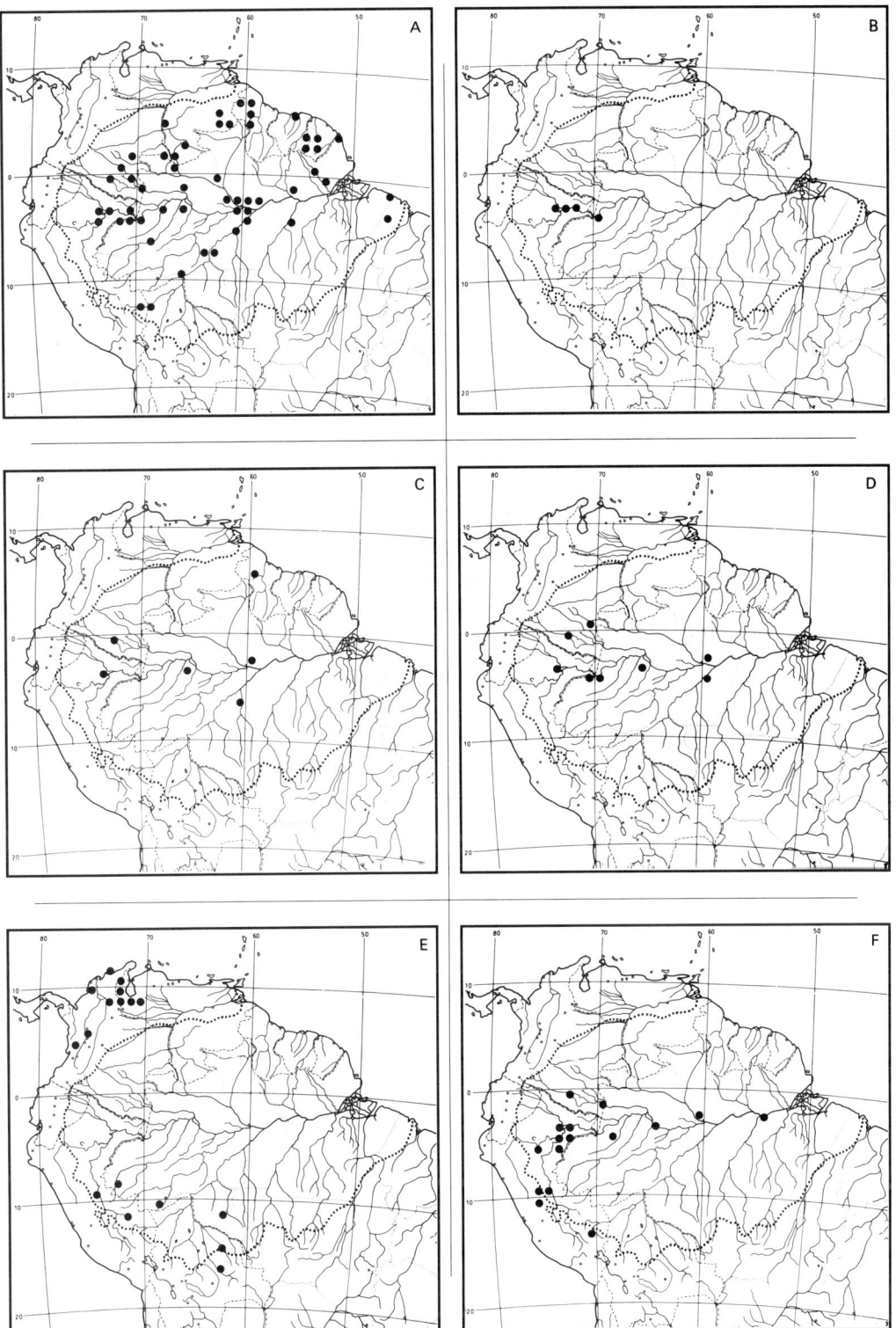

Figure 6.43. (**A**) *Bactris hirta* var. *hirta*; (**B**) *Bactris hirta* var. *mollis*; (**C**) *Bactris hirta* var. *pulchra*; (**D**) *Bactris killipii*; (**E**) *Bactris macana*; (**F**) *Bactris macroacantha*.

Vereins Prov. Brandenburg 48: 128. 1907. *Bactris ulei* Burret, Repert. Spec. Nov. Regni Veg. 34: 177. 1933. Type. Brazil. Amazonas or Acre: Rio Juruá, Bom Fim, Oct 1900, *E. Ule 5283* (holotype, B, n.v.).

Sheath, petiole, and rachis softly spinulose, usually lacking spines; petiole 1–10 cm long; blade entire, deeply to briefly bifid, 18–30 cm long, 9–13 cm wide at apex of rachis, pilose or occasionally glabrous abaxially, with prominent cross-veins.

Colombia (Amazonas), Guyana, Peru (Loreto), and Brazil (Acre, Amazonas) (Figure 6.43C); lowland rain forest on *terra firme* at low elevations, rarely to 1000 m.

In its entire leaf with cross-veins and densely spiny peduncular bract, it resembles *Bactris aubletiana*.

20. **Bactris killipii** Burret, Bot. Jahrb. Syst. 34: 175. 1933. Type. Peru. Loreto: Iquitos, ca. 100 m, 3–11 Aug 1929, *E. Killip & A. Smith 27305* (isotype, US) (Figure 6.42g–h).

Stems solitary or cespitose, 10–60 cm tall, 1–1.5 cm diam, lacking spines. *Leaves* 6–10, stiffly erect, lacking spines; sheath 10–12 cm long; petiole 0.2–1 m long, densely reddish-brown-tomentose; rachis contracted, 6–18 cm long; blade entire (rarely pinnate with 2–4 pinnae per side), deeply bifid, the lobes 30–60 cm long, 6–11 cm wide, lanceolate, stiff and strongly plicate, with small spinules on margins apically. *Inflorescences* interfoliar, lacking spines, erect at anthesis and in fruit; peduncle 15–19 cm long; prophyll 6–8 cm long; peduncular bract 18–24 cm long, densely whitish-brown-tomentose, glabrescent, opening at anthesis and then closing again around developing fruits; rachilla 1, 4–4.5 cm long; *flowers* in triads, these regularly arranged; staminate flowers 5–6 mm long, deciduous; sepals 1 mm long; petals 5–6 mm long; pistillate flowers 2.5–3 mm long; calyx tubular, 2.5–3 mm long; corolla tubular, 2–2.5 mm long; staminodes absent; *fruits* ellipsoid, 1.4–1.5 cm long, 0.8–1 cm diam, orange-red at maturity; mesocarp floury; endocarp fibers lacking; fruiting perianth with deeply 3-lobed calyx equal to the deeply 3-lobed corolla, staminodial ring absent.

Central and western Amazon region of Colombia (Amazonas, Vaupés), Peru (Loreto), and Brazil (Amazonas) (Figure 6.43D); lowland rain forest on *terra firme* at low elevations.

Brazil; *marajá*. Peru: *ñejilla, palmicha*.

Bactris killipii is distinguished by its lack of spines (except on pinnae margins); elongate petiole and contracted rachis; entire (usually), strongly plicate leaves; and spicate, erect inflorescence. It is unusual among *Bactris* in that after anthesis the peduncular bract closes again around the developing fruits.

21. **Bactris macana** (Mart.) Pittier, Manual Plant. Usual. Ven. 276. 1926. *Guilielma macana* Mart. in A. D. Orb., Voy. Amérique mér. 7(3). Palmiers 74. 1844. Type. Venezuela. Zulia: Maraicabo, n.d., *F. Plée s.n.* (holotype, M, n.v.) (Figure 6.44a–b).

Guilielma microcarpa Huber, Bol. Mus. Paraense Hist. Nat. 4: 476. 1906. *Bactris dahlgreniana* Glassman, A Revision of B. E. Dahlgren's Index of American Palms 34. 1972. Type. Not designated.

Stems cespitose or solitary, 9.5–12 m tall, 10–20 cm diam, spiny at internodes, with a basal mound of roots to 50 cm high. *Leaves* 8–17; sheath to 1 m long, lacking an ocrea; petiole to 50 cm long, sheath, petiole, and rachis whitish-tomentose abaxially and moderately to densely covered with black or brownish spines to 1 cm long, those of the petiole in 3 files; rachis 2.5–3.5 m long; pinnae 92–141 per side, arranged in obscure clusters of 3–5, spreading in different planes, linear, shortly bifid at apex, the middle ones 55–100 cm long, 2–3 cm wide. *Inflorescences* at first interfoliar; peduncle to 30 cm long; prophyll 12–25 cm long; peduncular bract to 60 cm long, moderately to densely covered with blackish or brownish spines to 1 cm long; rachis 20–30 cm long; rachillae 40–70, 20–30 cm long, at anthesis densely covered with glandular trichomes, glabrescent; *flowers* in triads, these irregularly arranged among paired or solitary staminate flowers; staminate flowers not seen; pistillate flowers not seen; *fruits* subglobose to obovoid, rostrate, 1.2–2.3 cm long, 1.1–1.8 cm diam, orange; mesocarp starchy; endocarp fibers flattened, anastomosing, adnate to endocarp; fruiting perianth with very small calyx with undulate margins and much longer, lobed corolla, staminodial ring absent.

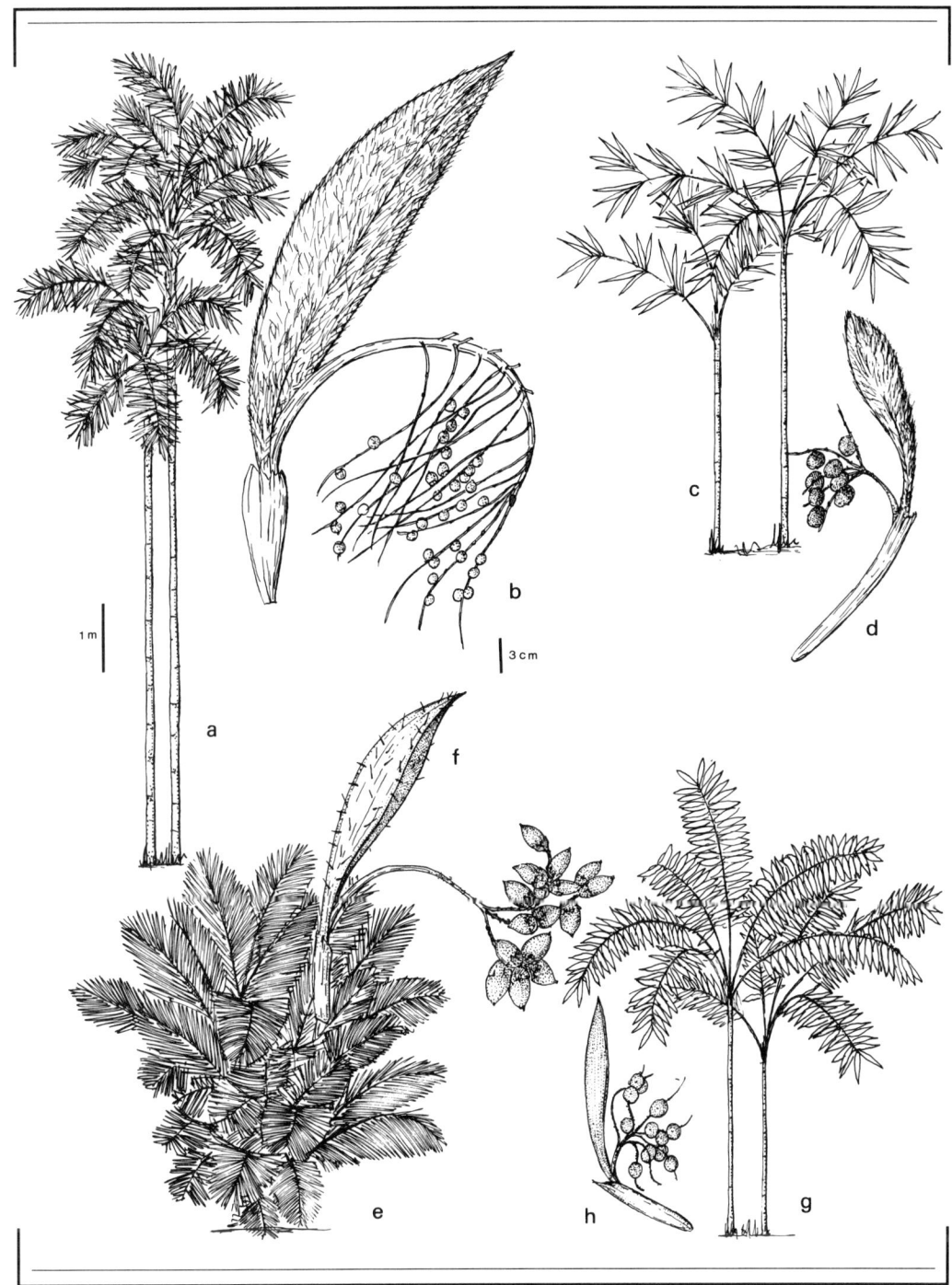

Figure 6.44. **(a)** *Bactris macana,* **(b)** infructescence (from *A. Henderson 1663*); **(c)** *Bactris macroacantha,* **(d)** infructescence (from *F. Kahn 1965*); **(e)** *Bactris major* var. *major,* **(f)** infructescence (var. *socialis*) (from *M. Nee 33730*); **(g)** *Bactris maraja* var. *maraja,* **(h)** infructescence (from *A. Henderson 1500*).

Colombia (Antioquia, Guajira, Norte de Santander, Sucre, Valle), Venezuela (Barinas, Zulia), and Ecuador (Loja, Los Ríos, Manabí, Pichincha), and the western Amazon region of Peru (Huánuco, Madre de Dios), Brazil (Acre, Rondônia), and Bolivia (Santa Cruz) (Figure 6.43E); on Andean slopes, and adjacent lowlands in lowland rain forest on *terra firme*.

Bolivia: *chontilla*. Brazil: *pupunha brava*. Peru: *pijuayo del monte*.

Bernal and Henderson (in press) consider this species to be the wild ancestor of the cultivated *Bactris gasipaes*.

22. Bactris macroacantha Mart., Hist. nat. palm. 2: 95. 1826. Type. Brazil. Pará: Canumá, n.d., *C. Martius 3243* (holotype, M; F neg. 18620, 18621) (Figure 6.44c–d).

Bactris confluens var. *acanthospatha* Trail, J. Bot. 6: 44. 1877. *Bactris acanthospatha* (Trail) Trail ex Drude in Mart., Fl. bras.: Cyclanthaceae et Palmae I, fasc. 85, vol. 3(2): 354. 1881. Type. Brazil. Amazonas: Fonte Boa, 16 Dec 1874, *J. Trail 882/CCII* (holotype, K).

Bactris platyacantha Burret, Notizbl. Bot. Gart. Berlin-Dahlem 14: 265. 1938. Type. Peru. Loreto: Río Huallaga, Lagunas, 1924, *J. Kuhlmann 2329* (holotype, RB).

Bactris setiflora Burret, Notizbl. Bot. Gart. Berlin-Dahlem 14: 328. 1938. Ecuador. Pastaza: Canelos, 350 m, 20 Feb 1937, *H. Schultze-Rhonhof 2233* (holotype, B, n.v.).

Stems cespitose, often in dense clumps, 0.5–4 m tall, 2–4 cm diam, spiny on internodes. *Leaves* 2–6; sheath 25–47 cm long, moderately to densely covered with flattened, but tapering from base, black or occasionally yellowish-brown spines to 8 cm long, sheath, petiole, and rachis densely brown-tomentose (often with a nontomentose abaxial stripe); petiole 25–80 cm long; rachis 1–1.8 m long, lacking spines; pinnae 13–30 per side, either regularly or irregularly arranged and spreading in 1 or different planes, oblanceolate to sigmoid, the middle ones (19–)28–50 cm long, (3–)4.5–9 cm wide, with distinct, spinulose marginal veins. *Inflorescences* interfoliar; peduncle 22–38 cm long; prophyll 12–20 cm long; peduncular bract 35–61 cm long, densely covered with brown or yellowish, strongly appressed, flattened, laciniate spines; rachis 4.5–5 cm long; rachillae 7–12, 8.5–14 cm long, densely covered with globular trichomes at anthesis; *flowers* borne in triads, these irregularly arranged among paired or solitary staminate flowers; staminate flowers 4 mm long; sepals with linear lobes 1.5 mm long; petals 3.5 mm long; pistillate flowers 4–6 mm long; calyx tubular, 4–6 mm long, spinulose or usually glabrous; corolla tubular, 3–4 mm long, densely spinulose; *fruits* very widely obovoid, markedly rostrate, 2.5–3.2 cm long, 2–2.3 cm diam, purple-black; mesocarp juicy; endocarp fibers free, numerous, with juice sacs attached; fruiting perianth with regularly and shortly lobed calyx shorter than the regularly and shortly lobed, spinulose corolla.

Western Amazon region in Colombia (Amazonas), Ecuador (Morona-Santiago, Pastaza), Peru (Amazonas, Cusco, Huánuco, Loreto, Pasco, Ucayali), and Brazil (Amazonas, Pará) (Figure 6.43F); lowland forest on *terra firme* at elevations below 700 m.

Colombia: *ñeeinó* (Andoke). Brazil: *marajá*. Ecuador: *kamancha* (Shuar, Achuar). The fruits are occasionally eaten.

This species is distinguished by its sheath, petiole, and rachis, which are densely brown-tomentose; pinnae with marginal, spinulose veins; peduncular bract densely covered with brown or yellowish, strongly appressed, flattened, laciniate spines; spinulose pistillate corolla shorter than the calyx; and endocarp fibers with juice sacs. It is a rather heterogeneous species, and the specimens are a possible mixture of at least two taxa, or hybrids with other species.

23. Bactris major Jacq. Select. stirp. amer. hist., ed. 2: 134. 1780–81. *Augustinea major* (Jacq.) Oerst., Linnaea 28: 395. 1863. *Pyrenoglyphis major* (Jacq.) H. Karst., Fl. Columb. 2: 141. 1869. Lectotype (Glassman, 1972). Jacq., Select. stirp. amer. hist., ed. 1: t. 171, fig. 2. 1763 (Figure 6.44e–f; Plate Ib).

Stems cespitose, usually growing in large colonies, 1–10 m tall, 2–6 cm diam, spiny on the internodes. *Leaves* 3–10; sheath 22–30 cm long, very fibrous on the margins, sheath, petiole, and rachis with numerous short spines, especially on the sheath, interspersed with long, black or brown spines to 11 cm long; petiole 0.1–1.5 cm long; rachis 0.7–1.8 m long; pinnae 24–48 per side, linear, aristate, regularly or irregularly arranged and spreading in 1 plane, the middle ones 25–62 cm long, 1–3 cm wide, spiny on the margins, with a metallic sheen when dry. *Inflorescences* interfoliar; peduncle 15–40 cm long; prophyll 13–30 cm long; peduncular bract 28–60 cm long, sparsely to moderately covered with black or brown to yellowish spines to 1–2 cm long; rachis 0–4 cm long; rachillae (1–)3–17, 9–23 cm long, ca. 2 mm diam at anthesis, thickening to 3–4 mm in fruit; *flowers* in triads, these scattered among paired or solitary staminate flowers; staminate flowers 3–8 mm long; sepals with narrowly triangular lobes 1.5–3 mm long; petals 3–7 mm long; pistillode obscure or absent; pistillate flowers 4–9 mm long; calyx tubular, 4–8 mm long, minutely spinulose; corolla tubular, 3–5 mm long; staminodial ring 3 mm long; *fruits* ellipsoid to ellipsoid-oblong, obovoid, or ovoid-oblong, 2.5–4.5 cm long, 1.3–3.5 cm diam, purple-black, tomentose or minutely spinulose; mesocarp juicy; fruiting perianth with deeply lobed calyx much shorter than the regularly lobed corolla, staminodial ring present.

It is possible to recognize the following varieties of this species, although there are intermediates between some of them. There are also plants from Trinidad that do not fit into this scheme.

KEY TO THE VARIETIES OF *BACTRIS MAJOR*

1. Rachillae (1–)2–5; expanded part of peduncular bract 16–20 cm long, closing again after anthesis; Venezuela (Amazonas), the Guianas, Brazil (Acre, Amazonas, Maranhão, Mato Grosso, Pará, Rondônia), and Bolivia (Pando) . . . .
. . . . 23a. *B. major* var. *infesta*.
1. Rachillae (3–)5–17; expanded part of peduncular bract 20–33 cm long, remaining open or closing again after anthesis.
    2. Rachillae (3–)5–12(–17); fruits ellipsoid, ellipsoid-oblong, ovoid-oblong, or obovoid, 3.5–4.5 cm long and 2–3 cm diam.
        3. Peduncular bract with yellowish-brown spines, remaining open after anthesis; Bolivia (Beni, Cochabamba, Pando, Santa Cruz) . . . .
. . . . . 23d. *B. major* var. *socialis*.
        3. Peduncular bract with black spines, closing again after anthesis; the Guianas. . . . .
. . . . . 23b. *B. major* var. *major*.
    2. Rachillae 11–17; fruits widely depressed obovoid, 1.5–2 cm long, 2 cm diam; Venezuela (Bolívar), the Guianas, Brazil (Amapá, Pará) . . . .
. . . . . 23c. *B. major* var. *megalocarpa*.

23a. Bactris major var. infesta (Mart.) Drude in Mart., Fl. *bras.*: Cyclanthaceae et Palmae I, fasc. 85, vol. 3(2): 359. 1881. *Bactris infesta* Mart. in A. D. Orb., Voy. Amérique mér. 7(3). Palmiers 54. 1846. *Pyrenoglyphis infesta* (Mart.) Burret, Repert. Spec. Nov. Regni Veg. 34: 248. 1934. Type. Brazil. Rondônia: Principe da Beira, n.d., *A. d'Orbigny 14* (holotype P; F neg. 18631)

Bactris socialis subsp. *gaviona* Trail, J. Bot. 6: 48. 1877. *Bactris gaviona* (Trail) Trail ex Drude in Mart., Fl. bras.: Cyclanthaceae et Palmae I, fasc. 85, vol. 2(3): 360. 1882. *Pyrenoglyphis gaviona* (Trail) Burret, Repert. Spec. Nov. Regni Veg. 34: 246. 1934. Type. Brazil. Amazonas: Gavião, Rio Juruá, 10 Nov 1874, *J. Trail 847/CLVI* (holotype, K).

Bactris exaltata Barb. Rodr., Enum. palm. nov. 32. 1875. *Pyrenoglyphis exaltata* (Barb. Rodr.) Burret, Reppert. Spec. Nov. Regni Veg. 34: 246. 1934. Lectotype (Glassman, 1972). Barb. Rodr., Sert. palm. brasil. 2: t. 20. 1902.

Bactris nemorosa Barb. Rodr., Enum. palm. nov. 32. 1875. *Pyrenoglyphis nemorosa* (Barb. Rodr.) Burret, Repert. Spec. Nov. Regni Veg. 34: 247. 1934. Lectotype (Glassman, 1972). Barb. Rodr., Sert. palm. brasil. 2: t. 18. 1902.

Bactris socialis subsp. *curuena* Trail, J. Bot. 6: 48. 1877. *Bactris curuena* (Trail) Trail ex Drude in Mart., Fl. bras.: Cyclanthaceae et Palmae I, fasc. 85, vol. 3(2): 359. 1882. *Pyrenoglyphis curuena* (Trail) Burret, Repert. Spec. Nov. Regni Veg. 34: 248. 1934. Type. Brazil. Amazonas: Rio Curuem, Rio Jutaí, 5°12′S, 29 Jan 1875, *J. Trail 849/CCVI* (holotype, K).

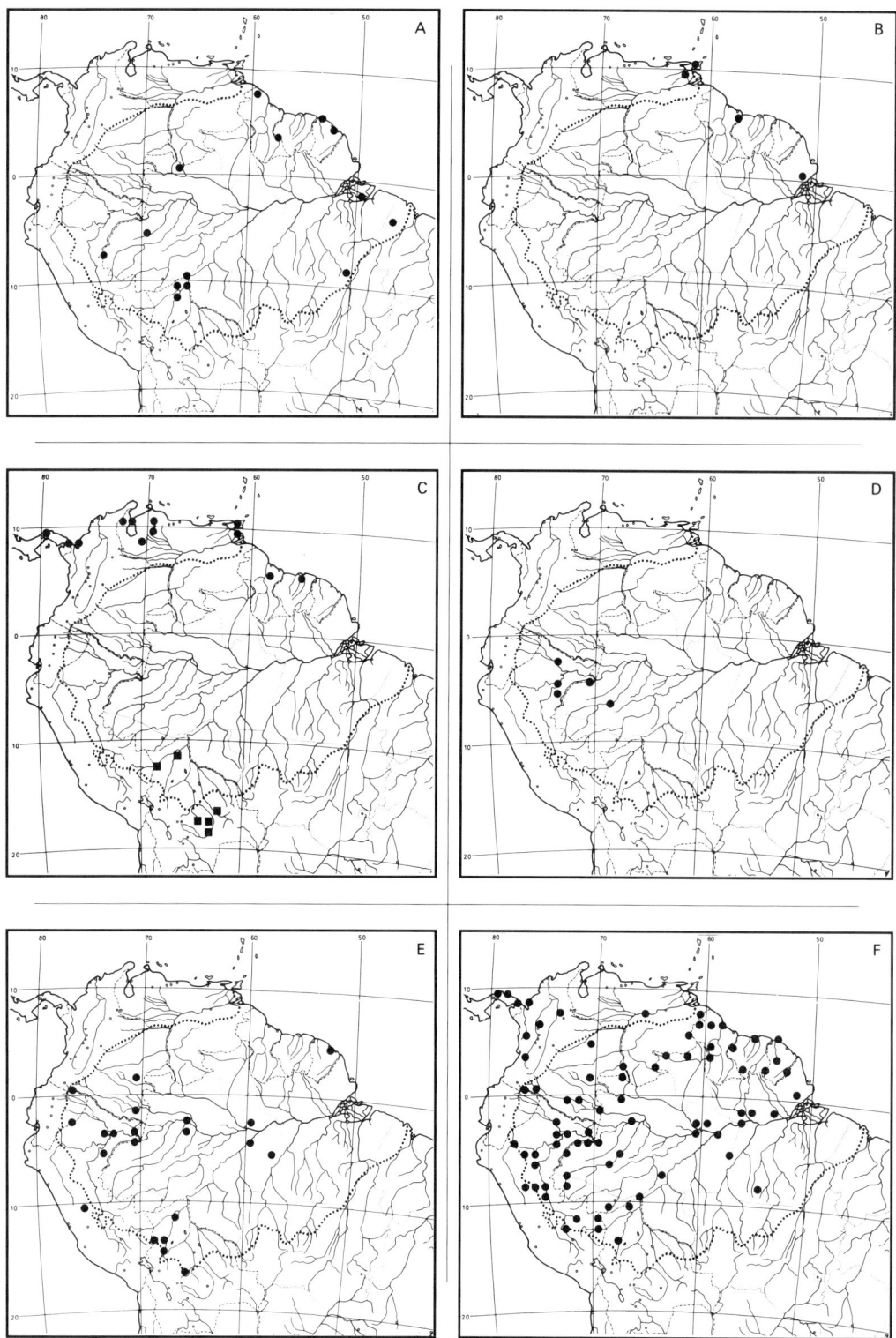

Figure 6.45. (A) *Bactris major* var. *infesta;* **(B)** *Bactris major* var. *megalocarpa;* **(C)** *Bactris major* var. *major.* (circles), var. *socialis* (squares); **(D)** *Bactris maraja* var. *chaetospatha;* **(E)** *Bactris maraja* var. *juruensis;* **(F)** *Bactris maraja* var. *maraja.*

Pyrenoglyphis hoppii Burret, Repert. Spec. Nov. Regni Veg. 34: 246. 1934. *Bactris pyrenoglyphoides* A. D. Hawkes, Arq. Bot. Estado São Paulo 2: 184. 1952. Type. Brazil. Pará: Museu Goeldi, Belém, Oct 1932, *W. Hopp 17* (holotype, B, n.v.).

Bactris mattogrossensis Barb. Rodr., Palmae Mattogrossenses 38. 1898. Lectotype (Glassman, 1972). Barb. Rodr., Palmae Mattogrossenses t. 13b. 1898.

Bactris chapadensis Barb. Rodr., Palmae Mattogrossenses 41. 1898. Lectotype (Glassman, 1972). Barb. Rodr., Palmae Mattogrossenses t. 13a. 1898.

Stems cespitose, forming colonies, 2–3.3 m tall, to 3 cm diam, spiny on internodes. *Leaves* 3–7; sheath 22–30 cm long, very fibrous at apex, sheath, petiole, and rachis moderately to densely covered with short black spines, these intermingled with longer spines to 7 cm long; petiole 0.4–1.2 m long; rachis 0.7–1.6 cm long; pinnae 24–42 per side, ± regularly arranged (occasionally clustered), spreading in 1 plane, linear, aristate, the middle ones 32–57 cm long, 1.5–2.5 cm wide, spiny on margins, somewhat metallic when dry. *Inflorescences* interfoliar; peduncle 15–22 cm long; prophyll to 20 cm long; peduncular bract 30–40 cm long, sparsely covered with black spines to 1 cm long; rachis 0–1 cm long; rachillae (1–)2–5, 9–15 cm long; *fruits* obovoid or ellipsoid, rostrate, 2.5–3 cm long, 1.3–2 cm diam, brown or purple-black, minutely spinulose.

Venezuela (Amazonas), the Guianas, Brazil (Acre, Amazonas, Maranhão, Mato Grosso, Pará, Piauí, Rondônia), and Bolivia (Pando) (Figure 6.45A); lowland rain forest on *terra firme,* or occasionally in inundated areas.

Bolivia: *marayaú.* Brazil: *marajá, piri'a-hu* (Ka'apor). The fleshy mesocarp is edible; in Bolivia it is used to make a drink.

Martius (1842–1847) confused this variety with *B. major* var. *socialis.* A possible hybrid between this variety and *Bactris bifida* occurs near Manaus *(A. Henderson 176);* an unusual specimen from Venezuela, Amazonas *(M. Nee 30829),* could be a hybrid between this variety and perhaps *B. maraja.*

23b. Bactris major var. major

Bactris chaetorachis Mart. in A. D. Orb., Voy. Amérique mér. 7(3). Palmiers 56. 1846. *Pyrenoglyphis chaetorachis* (Mart.) Burret, Repert. Spec. Nov. Regni Veg. 34: 246. 1934. Type. Surinam. Without locality, n.d., *F. Splitgerber 361* (holotype, BR, n.v.).

Bactris minax Miq., Natuurk. Verh. Holl. Maatasch. Wetensch. Haarlem, ser. 2, 7: 207. 1851. Type. Surinam. Paramaribo, n.d., *H. Focke 1212* (holotype, U, n.v.).

Bactris demerarana L. H. Bailey, Gentes Herb. 8: 162. 1949. Type. Guyana. Demerara River, n.d., *L. Bailey 419* (holotype, BH, n.v.).

Stems cespitose, 2–10 m tall, 2–6 cm diam, forming dense or open colonies. *Leaves* 3–10; sheath 22–55 cm long, very fibrous on margins, sheath, petiole, and rachis moderately to densely covered with short black spines, these intermingled with longer, brown or black, ± terete spines to 11 cm long; petiole 0.1–1.5 m long; rachis 0.8–1.8 m long; pinnae 28–46 per side, ± regularly arranged, spreading in 1 plane, linear, aristate, the middle ones 25–60 cm long, 1–3.5 cm wide, minutely spiny on margins, somewhat metallic when dry. *Inflorescences* interfoliar; peduncle 17–40 cm long, densely spiny, recurved; prophyll 13–30 cm long; peduncular bract 28–60 cm long, densely to moderately covered with black spines to 1(–2) cm long; rachis 2–4 cm long; rachillae (3–)5–10(–17), 15–23 cm long, 2 mm diam at anthesis, 3–4 mm diam in fruit; *fruits* irregularly ellipsoid to widely obovoid, 3.3–4.5 cm long, 2.3–3.5 cm diam, brown or purple-black, with minute spinules or small brown scales, glabrescent.

Southern Mexico, Central America (Belize, Costa Rica, El Salvador, Guatemala, Honduras, Panama), Colombia (Atlántico, Antioquia, Bolívar, Córdoba, Meta, Sucre), Venezuela (Apure, Barinas, Monagas, Portuguesa, Yaracuy, Zulia), the Guianas, and Trinidad (Figure 6.45C); in forest or more often in open areas near streams or standing water, often in coastal areas, at low elevations.

French Guiana: *zagrinette.* Venezuela: *ji.*

23c. Bactris major var. megalocarpa (Trail) Henderson, stat. nov.

Bactris megalocarpa Trail ex Thurn, Timehri 3: 256. 1884. Type. Guyana. Corentyne River, 1880, *E. Thurn 14* (holotype, K).

Stems cespitose, forming colonies, 2–6 m tall, to 3 cm diam, spiny on internodes. *Leaves* 6–8; sheath 20–63 cm long, very fibrous at apex, sheath, petiole, and rachis moderately to densely covered with short black spines, these intermingled with longer, brownish spines to 6 cm long; petiole to 10 cm long; rachis to 85 cm long; pinnae 22–33 per side, irregularly arranged in clusters, spreading in the same plane, linear, aristate, the middle ones 38–60 cm long, 2.5–4 cm wide, spiny on margins, somewhat metallic when dry. *Inflorescences* interfoliar; peduncle 20–36 cm long; prophyll 17–23 cm long; peduncular bract to 52 cm long, moderately covered with black spines to 1.5 cm long; rachis 3–4 cm long; rachillae 11–17, 18–22 cm long; *fruits* widely depressed obovoid, shortly rostrate, 1.5–2 cm long, 2 cm diam, purple-black, with minute spinules.

Colombia (Bolívar), Venezuela (Bolívar, Monagas), the Guianas, and Brazil (Amapá, Pará) (Figure 6.45B); in low-lying, inundated areas near streams and rivers. This variety occurs in the area of sympatry of var. *major* and var. *infesta*.

Venezuela: *cucurito*.

The oldest name for this taxon may be *B. leucacantha* Linden & H. Wendl. (Wessels Boer, 1988), but the type has not been located. *Bactris cruegeriana* Griseb. from Trinidad, which Wessels Boer (1988) included here, is distinct in its much larger fruits.

23d. Bactris major var. **socialis** Drude in Mart., Fl. bras.: Cyclanthaceae et Palmae I, fasc. 85, vol. 3(2): 359. 1881. *Bactris socialis* Mart. in A. D. Orb., Voy. Amérique mér. 7(3). Palmiers 56. 1846. *Pyrenoglyphis socialis* (Mart.) Burret, Repert. Spec. Nov. Regni Veg. 34: 246. 1934. Type. Bolivia. Beni: Río Itonamá, n.d., *A. d'Orbigny 30* (holotype, P, n.v.).

Stems cespitose, forming colonies, 3–4 m tall, 2–4 cm diam, spiny on internodes. *Leaves* 4–8; sheath 12–50 cm long, very fibrous on margins, sheath, petiole, and rachis moderately to densely covered with short black spines, these intermingled with longer, brown spines to 6 cm long; petiole 33–60 cm long; rachis 0.9–1.8 cm long; pinnae 38–57 per side, irregularly or regularly arranged, spreading in 1 or slightly different planes, linear, aristate, the middle ones 55–62 cm long, 1.5–2 cm wide, spiny on margins, somewhat metallic when dry. *Inflorescences* interfoliar; peduncle 9.5–40 cm long; prophyll to 33 cm long; peduncular bract to 40 cm long, yellowish-brown on outer surface, sparsely covered with yellowish-brown spines to 1 cm long; rachis 2–3 cm long; rachillae (3–)7–12, to 18 cm long; *fruits* ellipsoid or obovoid, rostrate, 3.5–4 cm long, 2–3 cm diam, purple-black, minutely spinulose.

Bolivia (Beni, Cochabamba, Pando, Santa Cruz) (Figure 6.45C); in open areas or in scrub or secondary forest, always near streams or other wet places, up to 500 m elevation.

Bolivia: *marayaú*. The stems are used for making baskets, and the fruits are eaten.

24. Bactris maraja Mart. Hist. nat. palm. 2: 93. 1826. *Pyrenoglyphis maraja* (Mart.) Burret, Repert. Spec. Nov. Regni Veg. 34: 252. 1934. Type. Brazil. Pará: without locality, n.d., *C. Martius s.n.* (holotype, M; F neg. 18622) (Figure 6.44g–h).

Stems cespitose or solitary, usually in open clusters of 2–15 stems, 1–7(–10) m tall, 1–4 cm diam, with black spines on internodes. *Leaves* 3–10, horizontally spreading; sheath 12–35 cm long, sheath, petiole, and rachis occasionally densely brown-tomentose, with moderate to dense covering of flattened, yellowish (occasionally brown or black) spines to 5(–10) cm long, when yellowish with black base and apex; petiole 13–76 cm long; rachis 0.3–1.3 m long; pinnae 6–22 per side, irregularly arranged in clusters of 2–5 and spreading in different planes, or regularly arranged and spreading in 1 plane, sigmoid or occasionally lanceolate, long acuminate, occasionally pilose abaxially, the middle ones 20–48 cm long, 3–7 cm wide, occasionally leaf entire. *Inflorescences* interfoliar; peduncle 11–18 cm long; prophyll 8–26 cm long; peduncular bract 15–38 cm long, typically velvety brown-tomentose, nonspiny or occasionally with flattened yellowish spines to 8 mm long especially at the apex; rachis 1–5 cm long; rachillae 3–17, 5–15 cm long, at anthesis densely brown-tomentose; *flowers* in triads, these irregularly arranged among paired or solitary staminate flowers; staminate flowers to 3.5 mm long, deciduous; sepals with narrowly triangular lobes 0.5–1 mm long; petals obovate, 3–3.5 mm long; pistillate flowers 3–4 mm long; calyx tubular, 2.5–4 mm long, exceeding the corolla, rarely spinulose; corolla tubular, 2.5–4 mm long, usually spinulose; staminodes absent; *fruits* widely depressed obovoid, rostrate, 1–1.7 cm diam, purple-black at maturity, rarely minutely spinulose; mesocarp juicy; endocarp fibers free, numerous, with juice sacs attached; fruiting perianth with deeply 3-lobed calyx half as long as the deeply 3-lobed, often spinulose corolla, staminodial ring absent.

This species, previously known as *Bactris monticola,* is very common, widespread, and variable, and is perhaps one of the most poorly understood in the Amazon region (especially if the closely related *Bactris tomentosa* is also considered). There are not nearly enough collections, however, to resolve subspecific variation. Spines are typically yellowish at the middle and darker at base and apex, but are often dark throughout, and then often longer and more densely arranged. Pinnae shape varies considerably, and is typically broadly or narrowly sigmoid, but occasionally lanceolate. There are also entire-leaved specimens from scattered localities. Some of these, from the western Amazon region, also have densely brown-tomentose sheath, petiole, and rachis, and spinulose fruits, and are here separated as var. *chaetospatha*. There are also intermediates with *B. tomentosa*. A few specimens appear to be hybrids with other species. The species is here divided into three varieties, but characters from one variety can appear in another. It seems likely that when the species is better understood, it will be divided into numerous taxa.

KEY TO THE VARIETIES OF
BACTRIS MARAJA

1. Leaves entire; fruits spinulose; Peru (Loreto) and Brazil (western Amazonas). . . . .
   . . . . . . 24a. *B. maraja* var. *chaetospatha*.
1. Leaves pinnate, rarely entire; fruits glabrous or rarely spinulose; widespread.
   2. Pinnae (2–)6–11 per side; rachillae 3–6, 5–7 cm long; Colombia (Amazonas, Vaupés), French Guiana, Ecuador (Napo), Peru (Loreto, Madre de Dios, Pasco), Brazil (Acre, Amazonas, Pará), and Bolivia (Beni, Cochabamba, La Paz) . . . . 24b. *B. maraja* var. *juruensis*.
   2. Pinnae 8–22 per side; rachillae (3–)8–17, 6–15 cm long; widespread. . . . .
      . . . . 24c. *B. maraja* var. *maraja*.

24a. Bactris maraja var. **chaetospatha** (Mart.) Henderson, stat. nov. *Bactris chaetospatha* Mart., Hist. nat. palm. 2: 147. 1837. Type. Brazil. Amazonas: Rio Japurá, n.d., *C. Martius s.n.* (holotype, M; F neg. 18605).

Sheath with flattened, black spines, sheath, petiole, and rachis brown-tomentose; blade entire, 55–90 cm long, 14–26 cm wide. *Inflorescences* interfoliar; rachillae 5–9, 5–11 cm long; *fruits* very widely obovoid, spinulose, purple-black, 1.5 cm diam.

Peru (Loreto) and Brazil (western Amazonas) (Figure 6.45D); lowland rain forest at low elevations.

Brazil: *marajá*.

24b. Bactris maraja var. **juruensis** (Trail) Henderson, stat. nov., *Bactris juruensis* Trail, J. Bot. 6: 40. 1877. Type. Brazil. Amazonas: Barreiras de Capiranga, Rio Juruá, 14 Nov 1874, *J. Trail 1115/CLXV* (holotype, K).

Bactris juruensis var. *lissospatha* Trail, J. Bot. 6: 40. 1877. Brazil. Amazonas: Barreiras de Capiranga, Rio Juruá, 14 Nov 1874, *J. Trail 1116/CLXVI* (holotype, K).

Bactris piranga Trail, J. Bot. 6: 41. 1877. Type. Brazil. Pará: Santarém, 4 Feb 1875, *J. Trail 845/CCXVII* (holotype, K; isotypes, NY, US).

Bactris incommoda Trail, J. Bot. 6: 43. 1877. Type. Brazil. Amazonas: Rio Madeira, Paranáquara, 5 Jun 1874, *J. Trail 1119/LX* (holotype, K).

Bactris krichana Barb. Rodr., Vellosia 1: 41. 1888. Type. Brazil. Amazonas: Rio Yauapery, n.d., *J. Barbosa Rodrigues s.n.* (destroyed).

Bactris penicillata Barb. Rodr., Vellosia 1: 42. 1888. Lectotype (Glassman, 1972). Barb. Rodr., Sert. palm. brasil. 2: t. 36. 1903.

Bactris bella Burret, Notizbl. Bot. Gart. Berlin-Dahlem 12: 157. 1934. Type. Brazil. Amazonas: Manaus, May 1934, *W. Hopp 1308* (holotype, B, n.v.).

Bactris pulchella Burret, Notizbl. Bot. Gart. Berlin-Dahlem 12: 621. 1935. Type. Brazil.

Amazonas: Rio Manicoré, Sep 1934, *W. Hopp 1336* (holotype, B).

Bactris chlorocarpa Burret, Notizbl. Bot. Gart. Berlin-Dahlem 12: 622. 1935. Type. Brazil. Amazonas: Manicoré, Rio Madeira, Jun 1934, *W. Hopp 1318* (holotype, B).

Bactris bijugata Burret, Notizbl. Bot. Gart. Berlin-Dahlem 14: 264. 1938. Type. Brazil. Amazonas: Tonantins, Rio Solimões, 24 Jan 1924, *J. Kuhlmann 34793* (holotype, B, n.v.).

Bactris microspadix Burret, Notizbl. Bot. Gart. Berlin-Dahlem 14: 264. 1938. Type. Brazil. Amazonas: Rio Purus, mouth of Juauhiry, n.d., *J. Kuhlmann 889* (holotype, RB).

Sheath and petiole (and rachis) with scattered to moderate covering of slightly flattened, yellowish spines to 3 cm long; pinnae (2–)6–11 per side, irregularly arranged and spreading in slightly different planes, sigmoid, with long, filiform, minutely spiny apex, the middle ones 24–41 cm long, 3.5–4 cm wide, the apical one much wider than the others, occasionally the leaf entire. *Inflorescences* interfoliar; rachillae 3–6, 5–7 cm long; *fruits* very widely obovoid, to 1.5 cm diam, glabrous.

Throughout the central and western Amazon region of Colombia (Amazonas, Vaupés), Ecuador (Napo), Peru (Loreto, Madre de Dios, Pasco), Brazil (Acre, Amazonas, Pará), Bolivia (Beni, Cochabamba, La Paz), and French Guiana (Figure 6.45E); lowland rain forest in *várzea* or *terra firme* at low elevations.

This variety is somewhat intermediate between *Bactris maraja* var. *maraja* and *B. tomentosa* var. *sphaerocarpa*. If the specimens from French Guiana really belong here, they represent a remarkable, isolated eastern occurrence. Exactly the same situation occurs in *Geonoma stricta* var. *trailii*.

24c. Bactris maraja var. maraja

Bactris chloracantha Poepp. ex Mart., Hist. nat. palm. 2: 145. 1837. Type. Peru. Loreto: Maynas, Yurimaguas, n.d., *W. Poeppig 2107* (holotype, M; isotype P; F. neg. 18606).

Bactris elatior Wallace, Palm Trees of the Amazon 81. 1853. Lectotype (Glassman, 1972). Wallace, Palm Trees of the Amazon t. 30. 1853.

Bactris macrocarpa Wallace, Palm Trees of the Amazon 85. 1853. Lectotype (Glassman, 1972). Wallace, Palm Trees of the Amazon t. 32. 1853.

Bactris armata Barb. Rodr., Enum. palm. nov. 27. 1875. *Bactris chaetospatha* var. *macrophylla* Drude in Mart., Fl. bras.: Cyclanthaceae et Palmae I, fasc. 85, vol. 2(3): 338. 1881. Lectotype (Glassman, 1972). Barb. Rodr., Sert. palm. brasil. 2: t. 11. 1903.

Bactris umbrosa Barb. Rodr., Enum. palm. nov. 29. 1875. Lectotype (Glassman, 1972). Barb. Rodr., Sert. palm. brasil. 2: t. 15b. 1903.

Bactris sylvatica Barb. Rodr., Enum. palm. nov. 30. 1875. Lectotype (Glassman, 1972). Barb. Rodr., Sert. palm. brasil. 2: t. 16. 1903.

Bactris monticola Barb. Rodr., Enum. palm. nov. 34. 1875. Lectotype (Wess. Boer, 1965). Barb. Rodr., Sert. palm. brasil. 2: t. 23. 1903.

Bactris umbraticola Barb. Rodr., Enum. palm. nov. 34. 1875. Lectotype (Glassman, 1972). Barb. Rodr., Sert. palm. brasil. 2: t. 15a. 1903.

Bactris paucijuga Barb. Rodr., Enum. palm. nov. 34. 1875. Lectotype (Glassman, 1972). Barb. Rodr., Sert. palm. brasil. 2: t. 34. 1903 ("paucijugata").

Bactris granariuscarpa Barb. Rodr., Enum. palm. nov. 37. 1875. Lectotype (Glassman, 1972). Barb. Rodr., Sert. palm. brasil. 2: t. 39. 1903.

Bactris maraja subsp. *maraja* Trail, J. Bot. 6: 44. 1877. *Bactris maraja* var. *trailii* A. D. Hawkes, Arq. Bot. Estado São Paulo 2: 183. 1952. Type. Brazil. Pará: Obidos, n.d., *J. Trail 4* (holotype, K, n.v.).

Bactris maraja subsp. *sobralensis* Trail, J. Bot. 6: 44. 1877. *Bactris maraja* var. *sobralensis* (Trail) Drude in Mart., Fl. bras.: Cyclanthaceae et Palmae I, fasc. 85, vol. 3(2): 343. 1881. *Bactris sobralensis* (Trail) Barb. Rodr. var. *limnaia* (Trail) Barb. Rodr. Sert. palm. bras. 2: 102. 1903. Type. Brazil. Amazonas: Sobral, Rio Purus, 19 Sep 1874, *J. Trail 838/CXXVII* (holotype, K).

Bactris maraja subsp. *limnaia* Trail, J. Bot. 6: 44. 1877. *Bactris maraja* var. *limnaia* (Trail) Drude in Mart., Fl. bras.: Cyclanthaceae et

Palmae I, fasc. 85, vol. 3(2): 343. 1881. Type. Brazil. Amazonas: Manaus, 13 Aug 1874, *J. Trial 842/CXII* (holotype, K).

Bactris trichospatha Trail, J. Bot. 6: 41. 1877. *Bactris trichospatha* subsp. *trichospatha* Trail, J. Bot. 6: 42. 1877. *Bactris trichospatha* var. *cararaucensis* A. D. Hawkes, Arq. Bot. Estado São Paulo 2: 185. 1952. Type. Brazil. Amazonas: Barcellos, 19 Jun 1874, *J. Trail 835/LXVIII* (holotype, K).

Bactris trichospatha subsp. *trichospatha* var. *elata* Trail, J. Bot. 5: 357. 1876. Type. Not designated.

Bactris trichospatha subsp. *jurutensis* Trail, J. Bot. 6: 42. 1877. *Bactris trichospatha* var. *jurutensis* (Trail) Drude in Mart., Fl. bras.: Cyclanthaceae et Palmae I, fasc. 85, vol. 3(2): 339. 1881. Type. Brazil. Amazonas: Lago Juruty, 4 Apr 1874, *J. Trail 1117/XXX* (holotype, K).

Bactris trichospatha subsp. *trichospatha* var. *robusta* Trail, J. Bot. 6: 42. 1877. *Bactris trichospatha* var. *robusta* (Trail) Drude in Mart., Fl. bras.: Cyclanthaceae et Palmae I, fasc. 85, vol. 3(2): 340. 1881. Type. Brazil. Amazonas: Humaitá, Rio Madeira, 29 May 1874, *J. Trail 833/LI* (holotype, K).

Bactris trichospatha var. *patens* Drude in Mart., Fl. bras.: Cyclanthaceae et Palmae I, fasc. 85, vol. 3(2): 340. 1881. Type. Brazil. Amazonas: Rio Negro, Kahuki, 19 Jun 1874, *J. Trail 834/LXII* (holotype, K).

Bactris actinoneura Drude & Trail in Mart., Fl. bras.: Cyclanthaceae et Palmae I: fasc. 85, vol. 3(2): 344. 1881. Type. Brazil. Amazonas: Santarém, Rio Jutaí, 4 Feb 1875, *J. Trail 1107/CCXVIII* (holotype, K).

Bactris gymnospatha Burret, Notizbl. Bot. Gart. Berlin-Dahlem 10: 1024. 1930. Type. Venezuela. Amazonas: Río Casiquiare, 4 Oct 1928, *P. von Luetzelburg 22346* (isotypes, M, R; F neg. 18613).

Bactris longisecta Burret, Repert. Spec. Nov. Regni Veg. 34: 205. 1934. Type. Brazil. Amazonas: Rio Manacapurú, n.d., *G. Huebner 65* (holotype, B).

Bactris erostrata Burret, Repert. Spec. Nov. Regni Veg. 34: 207. 1934. Type. Brazil. Amazonas: Mocó-Altamira, Rio Japurá, Feb 1926, *G. Huebner 52* (holotype, B, n.v.).

Bactris leptotricha Burret, Repert. Spec. Nov. Regni Veg. 34: 207. 1934. Type. Guyana. Conawaruk River, Sep 1905, *A. Bartlett 8191* (holotype, B, n.v.).

Bactris chaetochlamys Burret, Repert. Spec. Nov. Regni Veg. 34: 208. 1934. Type. Peru. Loreto: Río Napo, Curay, Mar 1931, *W. Hopp 1098* (holotype, B).

Bactris leptospadix Burret, Repert. Spec. Nov. Regni Veg. 34: 210. 1934. Type. Brazil. Amazonas: Igualidade, Rio Japurá, Feb 1926, *G. Huebner 50* (holotype, B, n.v.).

Bactris longicuspis Burret, Notizbl. Bot. Gart. Berlin-Dahlem 15: 5. 1940. Type. Guyana. Kanuku Mountains, Takutu River, 600 m, 4–22 Mar 1938, *A. Smith 3181* (holotype, NY; isotypes, K, P, U, US).

Bactris kamarupa Steyerm., Fieldiana Bot. 28: 75. 1951. Type. Venezuela. Bolívar: Río Pacairao, below Santa Teresita, 915–1065 m, 25 Nov 1944, *J. Steyermark 60542* (holotype, F).

Sheath, petiole, and rachis occasionally densely brown-tomentose and with moderate to dense covering of flattened, yellowish (occasionally brown or black) spines to 5(–10) cm long; pinnae 8–22 per side, irregularly arranged in clusters of 2–5 and spreading in different planes, or regularly arranged and spreading in 1 plane, sigmoid or occasionally lanceolate, long acuminate, occasionally pilose on lower surface, the middle ones 20–48 cm long, 3–7 cm wide, occasionally leaf entire. *Inflorescences* interfoliar; rachillae (3–)8–17 (or rarely inflorescence spicate), 6–15 cm long; *fruits* widely depressed obovoid, rostrate, 1–1.7 cm diam, purple-black at maturity, rarely minutely spinulose.

Central America (Costa Rica, Panama) and throughout northern South America in Colombia (Amazonas, Antioquia, Chocó, Córdoba, Santander, Valle del Cauca, Vaupés, Vichada), Venezuela (Amazonas, Bolívar), the Guianas, Ecuador (Napo), Peru (Amazonas, Cuzco, Huánuco, Loreto, Madre de Dios, San Martín, Ucayali), Brazil (Acre, Amapá, Amazonas, Pará, Rondônia, Roraima), and Bolivia (Beni, La Paz, Pando) (Figure 6.45F); lowland rain forest, or occasionally open forest or secondary forest, usually on *terra firme* but occasionally in inundated areas, at low elevations but occasionally reaching 1500 m.

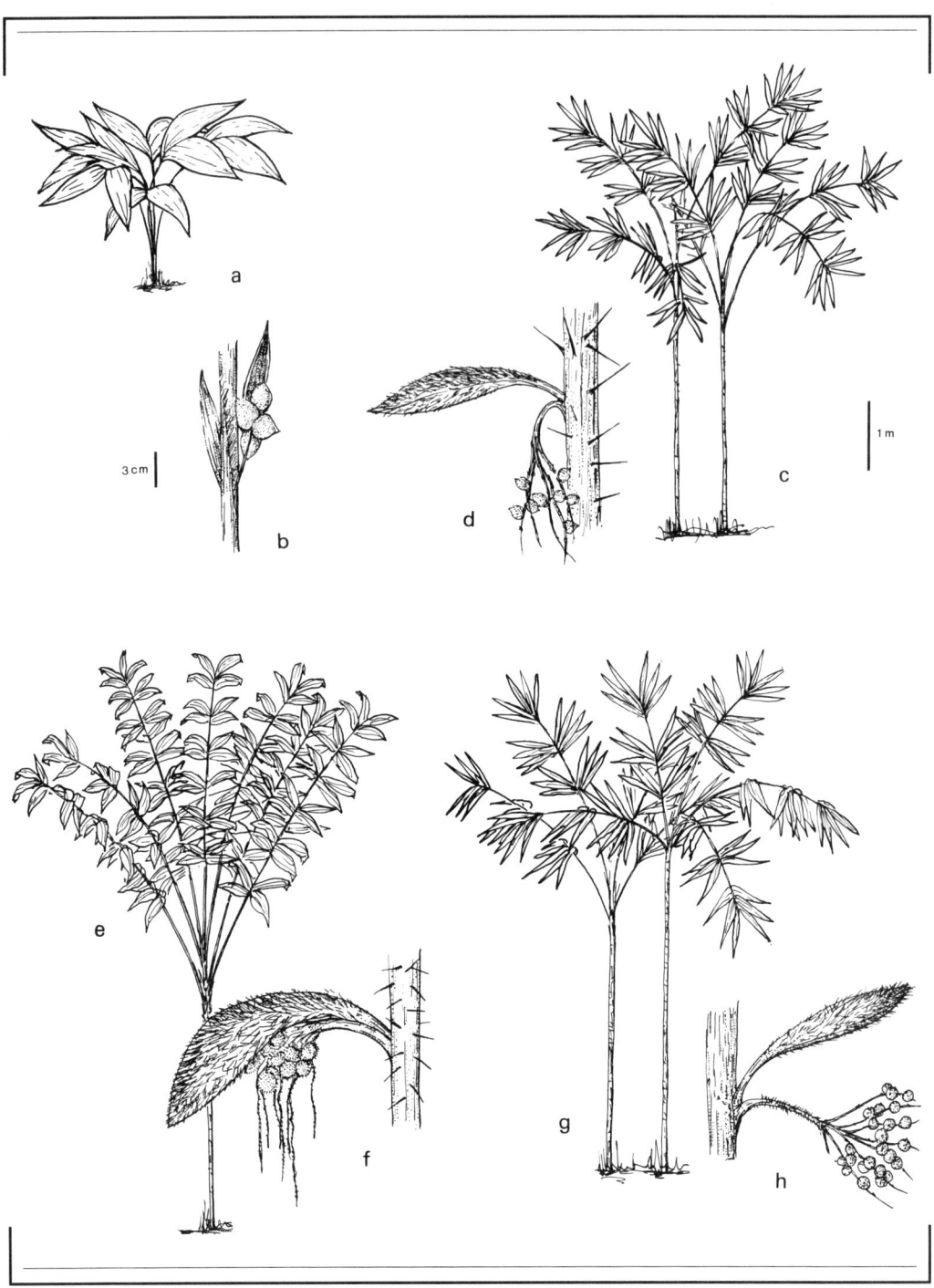

Figure 6.46. (a) *Bactris oligocarpa,* **(b)** infructescence (from *S. Mori 22195*); **(c)** *Bactris oligoclada,* **(d)** infructescence (from *B. Boom 7373*); **(e)** *Bactris pliniana,* **(f)** infructescence; **(g)** *Bactris ptariana,* **(h)** infructescence (from *J. Pipoly 8848*).

Bolivia: *chontilla conguillo, shinishëoxo* (Chácabo). Brazil: *marajá, marajá pupunha, tupina'i* (Araweté), *ubim de espinho*. Colombia: *chontilla, espina*. French Guiana: *anuyawili* (Wayâpi). Guyana: *bunyashiri*. Peru: *ñeja, chambira ñeja, chontilla*. Surinam: *piritu* (Trio), *piritiumë* (Trio). Venezuela: *piritu, uva de montaña*. The fruits are edible.

There is considerable variation in this variety. Forms with entire leaves appear sporadically and have been separated as *Bactris armata*. There are two distinct fruit sizes, one larger and one smaller; the former was separated by Wessels Boer (1988) as *Bactris macrocarpa*. This form is common in Guyana, Colombia (Chocó), and elsewhere. Furthermore, there are forms that occur sympatrically but occupy different habitats—for example, both *terra firme* and *várzea*. The ones in the *várzea*, with strongly sigmoid pinnae (separated by Trail as *Bactris maraja* subsp. *limnaia*), are commonly confused with *B. brongniartii*.

25. **Bactris oligocarpa** Barb. Rodr. & Trail, Enum. palm. nov. 28. 1875. *Pyrenoglyphis oligocarpa* (Barb. Rodr. & Trail) Burret, Repert. Spec. Nov. Regni Veg. 34: 242. 1934. Type. Brazil. Pará: Rio Tapajós, San Antonio, 17 Mar 1874, *J. Trail 1112/XXI* (holotype, K) (Figure 6.46a–b).

Bactris oligocarpa var. *brachycaulis* Trail, J. Bot. 6: 47. 1877. Type. Brazil. Amazonas: Rio Madeira, Exaltacion, 3 Jun 1874, *J. Trail 1114/LVIII* (holotype, K).

Stems solitary or occasionally cespitose, 0.2–1.5 m tall, 0.8–1 cm diam, occasionally short and subterranean. *Leaves* 4–10; sheath 5–15 cm long, sheath, petiole, and rachis with scattered black spines to 3 cm long, or occasionally spines absent; petiole 22–40 cm long; rachis 8–15 cm long; blade usually entire (or commonly 1 apical pinna split into 2) or with 2–4 pinnae per side, the apical pinna much wider than the others, sigmoid, the middle ones 20–33 cm long, 3–8 cm wide. *Inflorescences* interfoliar, occasionally all staminate; peduncle 6–13 cm long; prophyll 5–7 cm long; peduncular bract 10–16 cm long, spines few or absent; rachis absent; rachilla 1, 2–3 cm long; *flowers* in triads, these regularly arranged; staminate flowers 5–6 mm long, persistent; sepals 2 mm long; petals 5–6 mm long; pistillode absent; pistillate flowers (postanthesis) 4 mm long; calyx tubular, 3 mm long; corolla tubular, 3 mm long; staminodial ring adnate to the corolla, 2 mm long; *fruits* widely ovoid, 1.7–2 cm long, 1–1.3 cm diam, purple-black; mesocarp juicy; endocarp fibers free, numerous, with juice sacs attached; fruiting perianth with very short calyx and much longer, irregularly lobed corolla, staminodial ring present.

Central and eastern Amazon region of the Guianas and Brazil (Amapá, Amazonas, Pará) (Figure 6.47A); lowland rain forest on *terra firme* below 400 m elevation.

26. **Bactris oligoclada** Burret, Notizbl. Bot. Gart. Berlin-Dahlem 11: 325. 1931. Type. Guyana. Conawaruk River, Sep 1905, *A. Bartlett 8194* (holotype, B, n.v.) (Figure 6.46c–d).

Stems cespitose, 1–3 m tall, 1–1.5 cm diam, or occasionally stem short and subterranean, spiny or nonspiny. *Leaves* 3–12; sheaths 15–22 cm long, partially closed, sheath, petiole, and rachis with a few, scattered, black, slightly flattened spines to 7 cm long, these often whitish and bulbous basally; petiole 30–55 cm long; rachis 35–45 cm long; pinnae 6–10 per side, in distant clusters, spreading in different planes, lanceolate to almost sigmoid, asymmetrically acuminate, the middle ones 20–32 cm long, 3–6 cm wide, dark green adaxially, lighter green (and drying brown) abaxially, main vein prominent adaxially. *Inflorescences* interfoliar, recurved; peduncle 7–11 cm long; prophyll 3.5–9 cm long; peduncular bract 12–18 cm long, densely covered with very short black spines; rachis 1–2 cm long; rachillae 6–8, 5–8 cm long, at anthesis densely covered with brown trichomes; *flowers* borne in triads, these scattered among paired or solitary staminate flowers; staminate flowers 4.5 mm long; sepals free, 1.5 mm long; petals 4.5 mm long; pistillate flowers at anthesis 3 mm long; calyx annular, 0.5 mm long; corolla tubular, 3 mm long; staminodes absent; *fruits* depressed globose, 1–1.5 cm diam, orange-red at maturity, but green then whitish and yellowish during ripening; mesocarp floury; endocarp pitted, lacking fibers; fruiting perianth with very small calyx and small 3-lobed corolla, staminodial ring absent.

Northeastern Amazon region in Venezuela

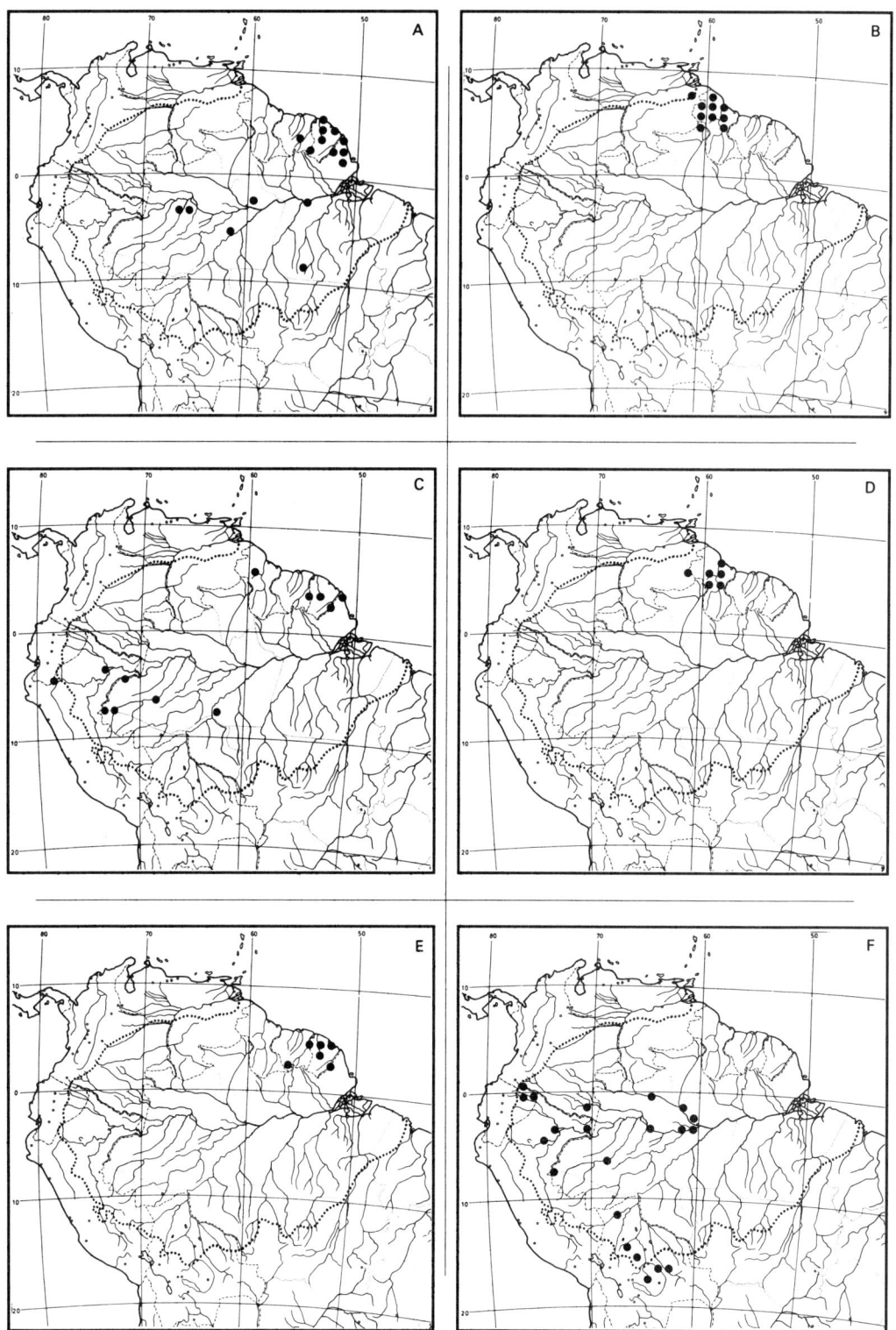

Figure 6.47. (A) *Bactris oligocarpa;* **(B)** *Bactris oligoclada;* **(C)** *Bactris pliniana;* **(D)** *Bactris ptariana;* **(E)** *Bactris rhapidacantha;* **(F)** *Bactris riparia.*

(Bolívar) and the Guianas (Figure 6.47B); lowland rain forest on sandy soils *(wallaba* forest) on *terra firme* at elevations below 800 m.

Guyana: *kidallibanaro, yurua.*

Bactris oligoclada is distinguished by its small stem; elongate petiole with a few, long, black spines; lanceolate pinnae arranged in distant clusters, which are lighter green on the abaxial surface; small inflorescences; and pitted endocarp.

27. **Bactris pliniana** Granv. & Henderson, Brittonia 46:147. 1994. Type. French Guiana. Waki River, Dégrad Somboto, 6 Jul 1973, *J.-J. de Granville 1728* (holotype, CAY; isotype, NY) (Figure 6.46e–f).

Stems cespitose, 1.5–3 m tall, 2.5–5 cm diam, spiny on the internodes. *Leaves* 6–12, erect; sheath 15–80 cm long, sheath and petiole with scattered, black, somewhat flattened spines to 7 cm long; petiole 0.2–1 m long; rachis 0.8–2.5 m long; pinnae 12–30 per side, irregularly arranged in clusters of 2–7, spreading in different planes, oblanceolate to sigmoid, the middle ones 35–60 cm long, 2.5–7.5 cm wide. *Inflorescences* interfoliar; peduncle 9–25 cm long; prophyll 8–15 cm long; peduncular bract 20–38 cm long, densely covered with dark brown or yellowish soft spines to 1 cm long; rachis 2.5–8 cm long; rachillae 20–60, 5–15 cm long, filamentous, spinulose; *flowers* borne in triads proximally, interspersed with solitary staminate flowers, all staminate distally; staminate flowers 2–2.5 mm long (preanthesis); sepals deltate, ca. 0.5 mm long; petals obovate, 2–2.5 mm long; pistillode absent; pistillate flowers (postanthesis) 3–5 mm long; calyx cupular, glabrous, ca. 1 mm long; corolla tubular, spinulose, 3–5 mm long; *fruits* globose to widely obovoid, ca. 2 cm diam, bright orange, densely spinulose; mesocarp starchy; fruiting perianth with glabrous, lobed calyx shorter than the spinulose, lobed corolla.

Guianas, Peru (Amazonas, Loreto), and Brazil (Acre, Amapá, Amazonas, Pará) (Figure 6.47C); lowland rain forest in low, swampy areas at elevations below 500 m.

Brazil: *marajá.* Peru: *uhahík, ujagkit.* Surinam: *hanaimaka, kiskismaka.*

This species is a member of the Piranga group (Sanders, 1991); see the discussion under *Bactris acanthocarpa.* It is similar to, and has been confused with, *B. acanthocarpoides.*

28. **Bactris ptariana** Steyerm., Fieldiana Bot. 28: 77. 1951. Type. Venezuela. Bolívar: SE slopes of Ptari-tepuí, 1585–1600 m, 10–11 Nov 1944, *J. Steyermark 60046* (holotype, F, n.v.) (Figure 6.46g–h).

Stems cespitose or solitary, 2–3 m tall, 3–4 cm diam, spiny on the internodes. *Leaves* 5–9; sheath 20–40 cm long, sheath and petiole with clusters of black spines 1–3 cm long; petiole 34–40(–80) cm long; rachis 1–1.2 m long; pinnae 15–18 per side, linear or linear-lanceolate, acuminate, irregularly arranged in clusters but spreading in ± the same plane, the middle ones 44–60 cm long, 3–5 cm wide, the apical pinna wider than the others, sparsely spinulose abaxially. *Inflorescences* infrafoliar, borne among the persistent leaf bases; peduncle 11–24 cm long; prophyll 7–12 cm long; peduncular bract 12–35 cm long, sparsely covered with appressed black spines to 1 cm long; rachis 2–5 cm long; rachillae 12–18, 6–10 cm long, densely covered with short brown or white trichomes; *flowers* in triads, these irregularly arranged almost to apex of rachillae among paired or solitary staminate flowers; staminate flowers not seen; pistillate flowers 3–3.5 mm long; calyx annular, 0.5 mm long, glabrous or pilose; corolla cupular, 2–2.5 mm long, densely spinulose; staminodes absent; *fruits* very widely obovoid, rostrate, 0.7–1 cm long, 0.7–0.8 cm diam, bright orange or red at maturity; mesocarp floury; endocarp fibers absent; fruiting perianth with obscure calyx and lobed, tomentose corolla, staminodial ring absent.

Venezuela (Bolívar) and Guyana (Figure 6.47D); in open areas near forest margins or in forest, usually on white-sand soils *(wallaba* forest), at elevations between 50 and 800 m.

Guyana: *maswa.*

The preceding description is based on specimens from lowland populations (below 800 m elevation) from Guyana. The specimens from higher elevations (1000–2000 m) in adjacent Venezuela (Bolívar), differ in their plicate linear pinnae, which are densely spinulose abaxially, and their larger inflorescences. There may be two taxa present here.

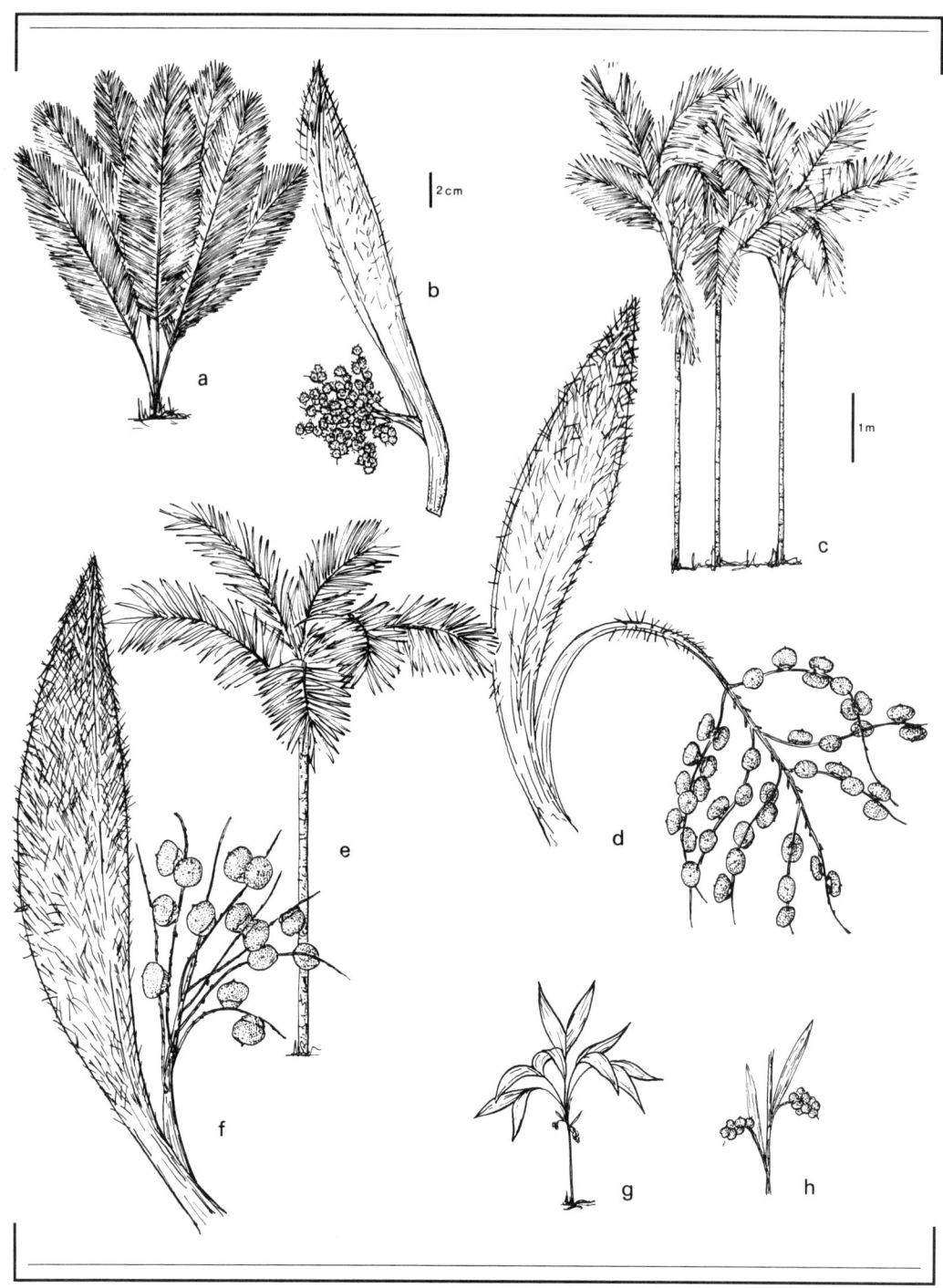

Figure 6.48. (a) *Bactris rhapidacantha,* **(b)** infructescence (from *J.-J. de Granville 11137*); **(c)** *Bactris riparia,* **(d)** infructescence (from A. Henderson 1677); **(e)** *Bactris setulosa,* **(f)** infructescence from *L. Dorr 7182*); **(g)** *Bactris simplicifrons,* **(h)** infructescences (from *T. Plowman 9880*).

29. **Bactris rhapidacantha** Wess. Boer, Flora of Suriname 5(1): 93. 1965. Type. Surinam. Tapanahony River, n.d., *J. Wessels Boer 1183* (holotype, U, n.v.) (Figure 6.48a–b).

Stems solitary or cespitose, 0.3–1.5 m long, 5–8 cm diameter, nonspiny or with a few spines, the internodes very close. *Leaves* 8–15, erect and forming a "funnel"; sheath 40–60 cm long, sheath and petiole densely covered with black spines to 10 cm long; petiole 0.5–1.5 m long; rachis 1.4–3 m long; pinnae 25–40 per side, regularly arranged and spreading in 1 plane, linear, long acuminate, the middle ones 43–85 cm long, 3–4 cm wide. *Inflorescences* interfoliar; peduncle 10–20 cm long; prophyll 8–17 cm long; peduncular bract 15–35 cm long, densely covered with black or brown spines; rachis 5–7 cm long; rachillae 25–40, filamentous, to 8 cm long; *flowers* borne in triads, these ± regularly arranged on proximal part of rachillae and tending to be absent from abaxial surface, paired or solitary staminate flowers distally; staminate flowers (immature) 2 mm long; sepals with triangular lobes 1.5 mm long; petals 2 mm long; pistillate flowers 3 mm long; calyx cupular, 1.5–2 mm long, glabrous; corolla cupular, 2.5–3 mm long, glabrous; *fruits* very widely obovoid, shortly rostrate, 1–1.5 cm diameter, orange-red, with deciduous, short black spinules; mesocarp starchy; endocarp fibers lacking, fruiting perianth with small, glabrous, lobed calyx and larger, glabrous, lobed corolla.

Eastern Amazon region in Surinam and French Guiana (Figure 6.47E); lowland rain forest on *terra firme* at low elevations. Sist (1989a) reported that in French Guiana this species flowered and fruited irregularly.

French Guiana: *koua'hm'* (Oyampi), *petit ouara* (Creole), *piritu* (Trio), *zagrinette, zagrinette forêt*.

This species is a member of the Piranga group (Sanders, 1991); see the discussion under *Bactris acanthocarpa*.

30. **Bactris riparia** Mart., Hist. nat. palm. 2: 97. 1826. Type. Brazil. Amazonas: Rio Japurá, n.d., *C. Martius s.n.* (holotype, M; F neg. 18628) (Figure 6.48c–d).

Bactris longifrons Mart., Hist. nat. palm. 2: 106. 1826. Type. Brazil. Pará: between Rio Tapajós and Rio Tocantins, n.d., *C. Martius s.n.* (holotype, M; F neg. 18617).

Bactris inundata Mart. in A. D. Orb., Voy. Amérique mér. 7(3). Palmiers 58. 1846. Type. Bolivia. Beni: Moxos, n.d., *A. d'Orbigny 24* (holotype, P, n.v.).

Bactris littoralis Barb. Rodr., Enum. palm. nov. 36. 36. 1875. Lectotype (Glassman, 1972). Barb. Rodr., Sert. palm. brasil. 2: t. 32a, 33. 1903.

Guilielma mattogrossensis Barb. Rodr., Palmae Mattogrossenses 33. 1898. *Bactris coccinea* Barb. Rodr., Contr. Jard. Bot. Rio de Janeiro 4: 110. 1907. Type (Glassman, 1972). Barb. Rodr., Contr. Jard. Bot. Rio de Janeiro 4: 24c. 1907.

Stems cespitose, usually forming large colonies, 3–10(–15) m tall, 4.5–10 cm diam, spiny on internodes. *Leaves* 4–18, stiffly spreading horizontally; sheath 30–60 cm long, partially closed, sheath, petiole, and rachis densely covered with slightly clustered black spines to 7 cm long; petiole 10–70 cm long; rachis 0.8–1.6 m long; pinnae 33–58 per side, irregularly arranged in indistinct clusters of 2–7 and spreading in different planes, linear, briefly and asymmetrically bifid apically, the middle ones 40–69 cm long, 1.5–2.5 cm wide, with small marginal spines, usually with soft brown hairs to 2 mm long on veins abaxially. *Inflorescences* interfoliar; peduncle 10–20 cm long; prophyll 8–15 cm long; peduncular bract 30–47 cm long, whitish-tomentose and densely to moderately covered with black spines; rachis 7–8 cm long; rachillae 24–50, to 15 cm long, at anthesis densely covered with brown, glandular trichomes; *flowers* borne in triads, these irregularly arranged among paired or solitary staminate flowers; staminate flowers to 6 mm long, deciduous; sepals 1.5 mm long; petals 5 mm long; pistillate flowers 4.5–6 mm long; calyx tubular, 1–1.5 mm long; corolla tubular, 4–5 mm long; staminodes absent; *fruits* depressed-globose, 1.5–2 cm diam, orange-red or green at maturity, glabrous; mesocarp floury; endocarp fibers numerous, free; fruiting perianth with very small undulate calyx and much longer 3-lobed corolla.

Colombia (Amazonas), Ecuador (Napo), Peru (Loreto, Madre de Dios, Ucayali), Brazil (Acre, Amazonas, Matto Grosso, Pará) and

Bolivia (Beni, Pando, Santa Cruz) (Figure 6.47F); black-water regions at the margins of streams, rivers, and lakes, often where its stems are partially submerged for at least part of the year, usually at low elevations. It is less commonly found on white-water rivers. It can occur in large colonies.

Bolivia: *chontilla, marajaú*. Brazil: *marajá, marajá pupunha*. Colombia: *chontadurillo, ponilla*. Ecuador; *chonta durillo, pa'i ine* (Siona). Peru: *ñeja, ñejilla de canto de cocha*.

31. Bactris setulosa H. Karst., Linnaea 28: 408. 1857. Lectotype (Imchanitzleaya, 1987) Venezuela. Carabobo: Cumbre de Valencia, Puerto Cabello, n.d., *H. Karsten s.n.* (LE, n.v.; F negs. (from W) 31318, 31317) (Figure 6.48e–f).

Stems solitary or cespitose and then forming large clumps, 5–10 m tall, 6–10 cm diameter, densely spiny on internodes. *Leaves* 4–9; sheath 0.5–1.6 m long, sheath, petiole, and rachis densely to moderately (to scarcely) covered with clustered black spines to 4 cm long; petiole 0.3–1.5 m long; rachis 1.3–3 m long; pinnae 40–68 per side, linear-lanceolate, irregularly arranged in clusters but spreading in 1 plane or occasionally different planes, the middle ones 60–91 cm long, 3–9 cm wide, veins prominent and usually spinulose abaxially. *Inflorescences* interfoliar; peduncle 15–21 cm long; prophyll 17–20 cm long; peduncular bract 30–42 cm long, very densely covered in black spines to 1.5 cm long; rachis 11–20 cm long; rachillae 39–60, 15–30 cm long; *flowers* in triads, these scattered among paired or solitary staminate flowers; staminate flowers 4–4.5 mm long; sepals 1 mm long; petals 4 mm long; pistillate flowers 4–5 mm long; calyx cupular, 1 mm long; corolla urceolate, 4 mm long; *fruits* very widely obovoid, 1.8–2 cm long, 1.5–1.8 cm diam, orange-yellow or orange-red; mesocarp floury; endocarp fibers tending to be flattened against the endocarp; fruiting perianth with very small calyx and crenulate margined, undulate corolla.

Andean region of Colombia (Antioquia, Chocó, Nariño, Valle del Cauca), Venezuela (Anzoategui, Barinas, Bolívar, Miranda, Monagas, Táchira, Zulia), Surinam, Trinidad, and Ecuador (El Oro, Esmeraldas, Los Rios, Napo, Pichincha). It just reaches the Amazon region in northwestern Venezuela (Bolívar) and Surinam (Figure 6.49A); in premontane or montane forest on steep slopes between 600 and 1250 m elevation.

Venezuela: *macana*.

Bernal and Henderson (in press) have placed this species in the same group as *Bactris gasipaes* (contra Sanders, 1991). The only specimen from Surinam is tentatively placed here; it is from a different habitat and is much less spiny than usual (as are some specimens from Ecuador, at the other end of the range).

32. Bactris simplicifrons Mart., Hist. nat. palm. 2: 103. 1826. *Amylocarpus simplicifrons* (Mart.) Barb. Rodr., Contr. Jard. Bot. Rio de Janeiro 3: 71. 1902. *Yuyba simplicifrons* (Mart.) L. H. Bailey, Gentes Herb. 7: 146. 1947. Type. Brazil. Amazonas, Rio Negro, n.d., *C. Martius s.n.* (holotype, M; F neg. 18630) (Figure 6.48g–h).

Bactris acanthocnemis Mart. in A. D. Orb., Voy. Amérique mér. 7(3). Palmiers 67. 1844. *Bactris simplicifrons* var. *acanthocnemis* (Mart.) Drude in Mart., Fl. bras.: Cyclanthaceae et Palmae I, fasc. 85, vol. 3(2): 322. 1881. *Yuyba simplicifrons* var. *acanthocnemis* (Mart.) A. D. Hawkes, Arq. Bot. Estado São Paulo 2: 193. 1952. *Amylocarpus acanthocnemis* (Mart.) Barb. Rodr., Contr. Jard. Bot. Rio de Janeiro 3: 71. 1902. Type. French Guiana. Without locality, n.d., *E. Mélinon s.n.* (isotype, M).

Bactris tenuis Wallace, Palm Trees of the Amazon 87. 1853. *Bactris mitis* subsp. *tenuis* (Wallace) Trail, J. Bot. 6: 3. 1877. *Bactris cuspidata* var. *tenuis* (Wallace) Drude in Mart., Fl. bras.: Cyclanthaceae et Palmae I, fasc. 85, vol. 3(2): 329. 1881. Lectotype (Glassman, 1972). Wallace, Palm Trees of the Amazon t. 33. 1853.

Bactris brevifolia Spruce, J. Linn. Soc., Bot. 11: 144. 1871. *Bactris simplicifrons* var. *brevifolia* (Spruce) Trail, J. Bot. 6: 1. 1877. Type. Brazil. Amazonas: São Gabriel, Rio Negro, n.d., *R. Spruce 35* (holotype, K, n.v.).

Bactris carolensis Spruce, J. Linn. Soc., Bot. 11: 145. 1871. *Bactris simplicifrons* var. *carolensis* (Spruce) Trail, J. Bot. 6: 1. 1877. *Bactris negrensis* var. *carolensis* (Spruce)

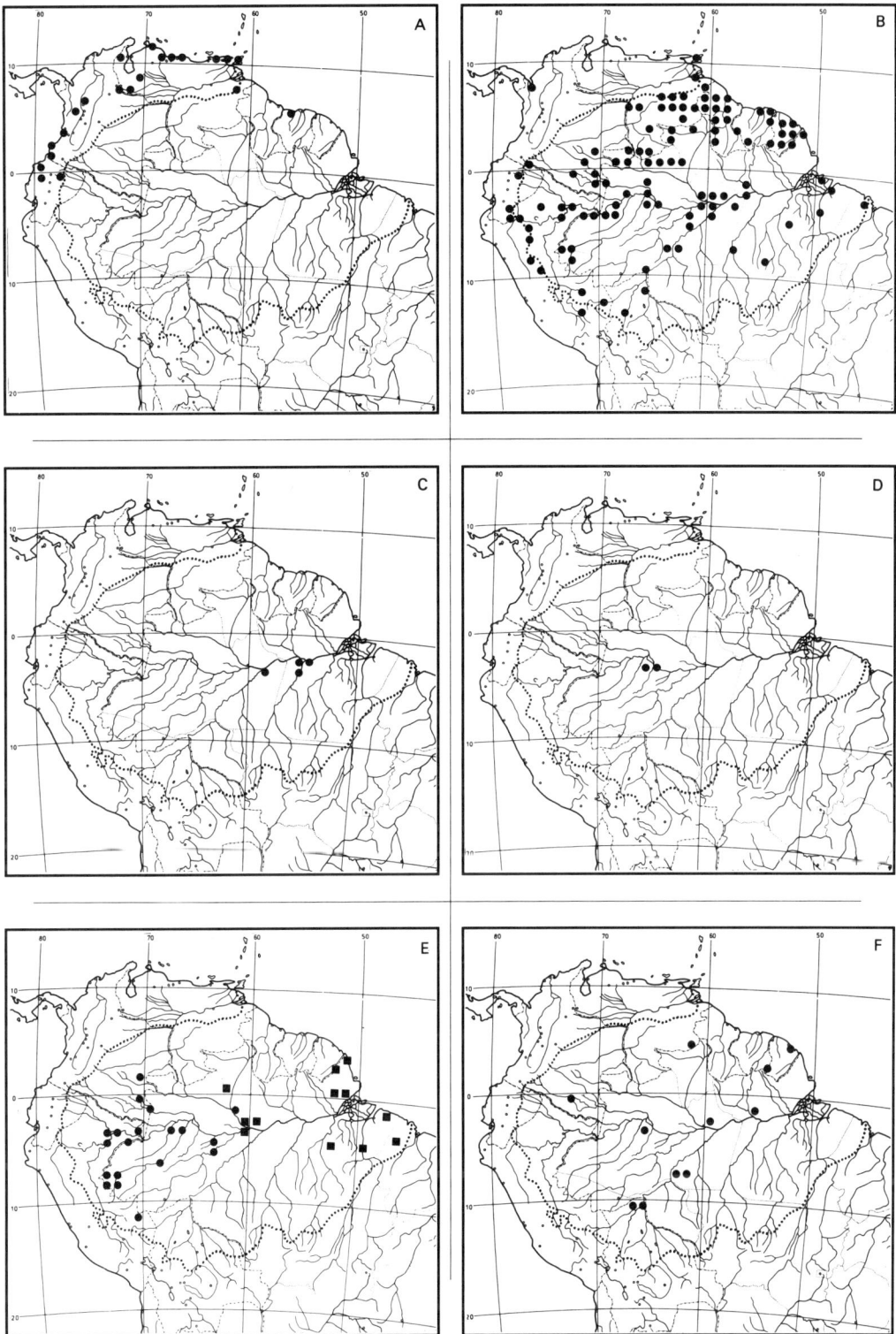

Figure 6.49. (A) *Bactris setulosa;* **(B)** *Bactris simplicifrons;* **(C)** *Bactris syagroides;* **(D)** *Bactris tefensis;* **(E)** *Bactris tomentosa* var. *sphaerocarpa* (circles), var. *tomentosa* (squares); **(F)** *Bactris trailiana.*

Burret, Repert. Spec. Regni Veg. 34: 174. 1933. Type. Venezuela. Amazonas: San Carlos de Río Negro, n.d., *R. Spruce 52* (holotype, K; isotype, P).

Bactris negrensis Spruce, J. Linn. Soc., Bot. 11: 145. 1871. *Bactris simplicifrons* var. *negrensis* (Spruce) Trail, J. Bot. 6: 1. 1877. Type. Brazil. Amazonas: Rio Negro, n.d., *R. Spruce 17* (holotype, K).

Bactris uaupensis Spruce, J. Linn. Soc., Bot. 11: 145. 1871. *Bactris mitis* subsp. *uaupensis* (Spruce) Trail, J. Bot. 6: 3. 1877. Type. Brazil. Amazonas: Rio Uaupés, n.d., *R. Spruce 77* (holotype, K).

Bactris negrensis var. *minor* Spruce, J. Linn. Soc., Bot. 11: 148. 1871. Type. Brazil. Amazonas: Tarumã, Feb 1855, *R. Spruce 70* (holotype, K).

Bactris microspatha Barb. Rodr., Enum. palm. nov. 26. 1875. *Amylocarpus microspathus* (Barb. Rodr.) Barb. Rodr., Contr. Jard. Bot. Rio de Janeiro 3: 72. 1902. Type. Not designated.

Bactris gracilis Barb. Rodr., Enum. palm. nov. 27. 1875. Type. Not designated.

Bactris arenaria Barb. Rodr., Enum. palm. nov. 29. 1875. *Amylocarpus arenarius* (Barb. Rodr.) Barb. Rodr., Contr. Jard. Bot. Rio de Janeiro 3: 72. 1902. Lectotype (Wess. Boer, 1965). Barb. Rodr., Sert. palm. brasil. 2: t. 44. 1903.

Bactris xanthocarpa Barb. Rodr., Enum. palm. nov. 30. 1875. *Amylocarpus xanthocarpus* (Barb. Rodr.) Barb. Rodr., Contr. Jard. Bot. Rio de Janeiro 3: 71. 1902. Lectotype (Wess. Boer, 1965). Barb. Rodr., Sert. palm. brasil. 2: t. 41. 1903.

Bactris inermis Trail ex Barb. Rodr., Enum. palm. nov. 30. 1875. *Amylocarpus inermis* (Trail ex Barb. Rodr.) Barb. Rodr., Sert. palm. brasil. 2: t. 45a. 1903. *Bactris tenuis* var. *inermis* (Trail ex Barb. Rodr.) Burret, Repert. Spec. Nov. Regni Veg. 34: 172. 1933. Lectotype (Glassman, 1972). Barb. Rodr., Sert. palm. brasil. 2: t. 45a. 1903.

Bactris inermis var. *tenuissimis* Barb. Rodr., Enum. palm. nov. 30. 1875. *Amylocarpus tenuissimus* (Barb. Rodr.) Barb. Rodr., Contr. Jard. Bot. Rio de Janeiro 3: 72. 1902. *Bactris tenuissimis* (Barb. Rodr.) Barb. Rodr., Sert. palm. brasil. 2: t. 10d. 1903. Lectotype (Glassman, 1972). Barb. Rodr., Sert. palm. brasil. 2: t. 10d. 1903.

Bactris mitis subsp. *inermis* Trail, J. Bot. 5: 355. 1876. Lectotype (here designated). Brazil. Pará: Rio Trombetas, Lago Tapagem, 28 Feb 1874, *J. Trail 914/X* (holotype, K).

Bactris simplicifrons var. *subpinnata* Trail, J. Bot. 6: 2. 1877. *Amylocarpus simplicifrons* var. *subpinnata* (Trail) Barb. Rodr., Sert. palm. brasil. 2: 87. 1903. Type. Brazil. Amazonas: Barcellos, 19 Jun 1874, *J. Trail 940/LXIII* (holotype, K).

Amylocarpus angustifolius Huber, Bol. Mus. Paraense Hist. Nat. 7: 285. 1913. *Bactris huberiana* Burret, Repert. Spec. Regni Veg. 34: 174. 1933. Type. Colombia. Amazonas: Río Caquetá, Cerro de Yupatí, 24 Nov 1912, *A. Ducke 12302* (isotypes, MG, RB).

Bactris luetzelburgii Burret, Notizbl. Bot. Gart. Berlin-Dahlem 10: 1022. 1930. *Amylocarpus luetzelburgii* (Burret) Burret, Notizbl. Bot. Gart. Berlin-Dahlem 10: 1023. 1930. Type. Venezuela. Amazonas: Río Casiquiare, Solano, 7 Oct 1928, *P. von Luetzelburg 23150* (isotypes, M, R).

Bactris obovata Burret, Notizbl. Bot. Gart. Berlin-Dahlem 11: 16. 1930. Type. Colombia. Caquetá: Getuchá, Río Orteguaza, 21 Jul 1926, *G. Woronow & J. Juzepczuk 6121* (holotype, B, n.v.).

Bactris paucisecta Burret, Repert. Spec. Nov. Regni Veg. 34: 171. 1933. Type. Guyana. Conawaruk River, Sep 1905, *A. Bartlett 8190* (holotype, B, n.v.).

Bactris luetzelburgii var. *anacantha* Burret, Repert. Spec. Nov. Regni Veg. 34: 174. 1933. Type. Brazil. Amazonas: Tarumã Grande, n.d., *G. Huebner 75* (holotype, B).

Bactris naevia Poepp. ex Burret, Repert. Spec. Nov. Regni Veg. 34: 179. 1933. Type. Peru. Loreto: Yurimaguas, Jun 1831, *W. Poeppig 2475* (holotype, M).

Bactris simplex Burret, Repert. Spec. Nov. Regni Veg. 34: 179. 1933. Type. Brazil. Amazonas: Rio Catrimany, Serra do Pacu, 800–900 m, Nov 1929, *C. Lakó/ G. Huebner 129* (holotype, B, n.v.).

Bactris amoena Burret, Repert. Spec. Nov. Regni Veg. 34: 180. 1933. Type. Brazil. Amazonas: Rio Içá, 4 Apr 1930, *C. Lakó 4/G. Huebner 136* (holotype, B, n.v.).

Bactris kuhlmannii Burret, Notizbl. Bot. Gart. Berlin-Dahlem 14: 262. 1938. Type. Brazil. Acre or Amazonas: Rio Purus, Juauhiny, 27 Nov 1923, *J. Kuhlmann 888* (holotype, RB).

Bactris kuhlmannii var. *aculeata* Burret, Notizbl. Bot. Gart. Berlin-Dahlem 14: 262. 1938. Type. Not designated.

Yuyba stahelii L. H. Bailey ex Maguire, Bull. Torrey Bot. Club 75: 106. 1948. *Bactris stahelii* (L. H. Bailey) Glassman, Rhodora 65: 259. 1963. Type. Surinam. Zandery Island, 10 Mar 1946, *G. Stahel s.n.* (holotype, NY; isotype, BH).

Yuyba maguirei L. H. Bailey ex Maguire, Bull. Torrey Bot. Club 75: 106. 1948. *Bactris maguirei* (L. H. Bailey) Steyerm., Fieldiana Bot. 28: 80. 1951. Type. Surinam. Tafelberg, 29 Aug 1944, *B. Maguire 24555* (holotype, NY; isotypes, BH, US).

Yuyba essequiboensis L. H. Bailey ex Maguire, Bull. Torrey Bot. Club 75: 107. 1948. *Bactris essequiboensis* (L. H. Bailey) Glassman, Rhodora 65: 259. 1963. Type. Guyana. Essequibo River, Kamuni Creek, Groete Creek, 14 Apr 1944, *B. Maguire & D. Fanshawe 22835* (holotype, NY).

Yuyba dakamana L. H. Bailey ex Maguire, Bull. Torrey Bot. Club 75: 108. 1948. *Bactris dakamana* (L. H. Bailey ex Maguire) Glassman, Rhodora 65: 259. 1963. Type. Surinam. Tafelberg, 3 Sep 1944, *B. Maguire 24614* (holotype, NY; isotype, BH).

Yuyba gleasonii L. H. Bailey, Gentes Herb. 8: 174. 1949. *Bactris gleasonii* (L. H. Bailey) Glassman, Rhodora 65: 259. 1963. Type. Guyana. Kangaruma-Potaro Landing, 25–27 Jun 1921, *H. Gleason 220* (holotype, NY).

Yuyba schultesii L. H. Bailey, Gentes Herb. 8: 174. 1949. *Bactris schultesii* (L. H. Bailey) Glassman, Rhodora 65: 259. 1963. Type. Colombia. Putumayo: Río San Miguel or Sucumbios, Conejo, opposite Quebrada Conejo, 300 m, 2–5 Apr 1942, *R. Schultes 3519* (holotype, BH).

Bactris soropanae Steyerm., Fieldiana Bot. 28: 78. 1951. Type. Venezuela. Bolívar: Río Karuai, Sororopán-tepuí, W of Laja, 1220 m, 29 Nov 1944, *J. Steyermark 60781* (holotype, F).

Stems cespitose or solitary, 0.5–2 m tall, 0.3–1 cm diam, erect or leaning. *Leaves* 5–9; sheath 5–20 cm long, closed but not forming a crownshaft, sheath and petiole without spines or occasionally with black, flattened spines to 1(–1.5) cm long; petiole 5–26 cm long; rachis 8–25 cm long; blade typically entire and shaped like a whale's fin, the lobes 24–58 cm long, 1.5–10 cm wide, or oblanceolate, plicate, deeply bifid apically and cuneate basally, or leaf pinnate, the pinnae 2–6(–20) per side, linear, linear-lanceolate or sigmoid, the middle ones 10–33 cm long, 1.5–15 cm wide, irregularly or regularly arranged, glabrous except for a few small spines distally on margins. *Inflorescences* interfoliar but appearing infrafoliar because exserted through leaf sheath; peduncle 3–4 cm long; prophyll 3–7 cm long; peduncular bract 6–12 cm long, erect at anthesis and forming an angle of ca. 30° with stem, usually nonspiny; rachis 0–0.5 cm long; rachillae 1(–5), 3.5–5.5 cm long, pendulous at anthesis, densely whitish-brown-tomentose or almost glabrous; *flowers* borne in triads, these regularly arranged almost throughout rachilla(e); staminate flowers 3.5–4.5 mm long, deciduous; sepals 1 mm long; petals 3–4 mm long; pistillate flowers 3.5–5.5 mm long at anthesis; calyx tubular, 3.5–5 mm long; corolla tubular, 3.5–5.5 mm long; staminodes absent; *fruits* globose, rostrate, 5–8 mm diam, rarely obovoid then 0.8–1.2 cm long, 0.7–1 cm diam, red or orange at maturity, glabrous; mesocarp floury; endocarp fibers absent; fruiting perianth with deeply 3-lobed calyx as long as the deeply 3-lobed corolla.

Common and widespread throughout the Amazon region and adjacent areas in Colombia (Amazonas, Córdoba, Guainía, Putumayo, Vaupés), Venezuela (Amazonas, Bolívar), Trinidad, the Guianas, Ecuador (Napo), Peru (Amazonas, Cusco, Huánuco, Loreto, Madre de Dios, Pasco, San Martín), Brazil (Acre, Amazonas, Maranhão, Mato Grosso, Pará, Rondônia, Roraima), and Bolivia (Beni, La Paz) (Figure 6.49B); lowland forest on *terra firme,* but also in open, sandy areas at both low and high (to 1800 m) elevations.

Brazil: *maraja, ubim-mirim.* Colombia: *chontaduro de rana de rastrojo, jo-da-jime-ru* (Huitoto), *oo-ree-ñaw* (Barasana), *u-ma-chú-ku-su* (Kofan). Ecuador: *chonta duro de tintin, huaso-e-ne* (Siona), *wa-tos-uné* (Siona). Guyana: *parapi-balli.* Peru: *ceyacepan* (Ameshua), *chon-*

tilla, kamanchá (Mayna Jívaro), *pijuaito, rotetenpar* (Ameshua), *uwinim, uyainim*. Surinam: *kiskismaka, yuyba*. Venezuela: *cubarillo*. Scrapings from the stem are used to stuff gun cartridges (Siona, Ecuador), and the plant is used medicinally (Ameshua, Peru).

This is one of the most common, widespread, and variable palms in the Amazon region. Although several forms can be recognized, there are too many intermediates between them to make any meaningful separation into varieties. Here I point out some of the morphologic tendencies in the species. Most variation is in the leaves; the inflorescence is generally consistent and typically consists of an erect, nonspiny peduncular bract and a pendulous rachilla(e).

The most widespread form, occurring throughout the Amazon region and beyond, in lowland rain forest on *terra firme*, is a plant without spines (except for spinules on the leaf apex) and an entire leaf with sigmoid venation (Figure 6.59n). The lobes of the leaf are typically 17–26 cm long and 4–10 cm wide. Some plants from the western Amazon region in Peru (Loreto) and Brazil (western Amazonas) have very small, entire leaves with the lobes 13–17 cm long and 1.5–3 cm wide. Less often, pinnate-leaved plants occur with 2–20 sigmoid pinnae per side, these 11–27 cm long and 1.5–7 cm wide (Figure 6.59m). The inflorescence of this form has a rachilla(e) that is 3–4 cm long, and the globose fruits are 5–8 mm diam.

Throughout the central Amazon region, mostly north of the Amazon river, a larger form occurs, often sympatrically with the first, and is found in lowland rain forest on *terra firme*, rarely to 1000 m elevation. The leaves are usually nonspiny (except for apex), but spiny plants are occasionally found. The leaf blade is commonly entire, with sigmoid venation, and the lobes are 24–58 cm long and 9–10 cm wide. Less often, pinnate-leaved plants are found with 2–6 sigmoid pinnae per side, these 15–23 cm long and 3–7 cm wide. Inflorescences are larger than those of the first form, with the rachilla(e) 4–6 cm long. Fruits are usually globose, to 8 mm diameter, but in the western Amazon region forms with obovoid fruits, 0.8–1.2 cm long and 0.7–1 cm diam, occur. These have been called *Bactris amoena* and *B. schultesii*.

A third form occurs north (rarely south) of the Amazon river on podzols in poorly drained places, and in forest, *campina*, or savanna at both low and high (to 1700 m) elevations. The center of its distribution is the Guayana Highland region of Venezuela and adjacent countries. The sheath and proximal part of the petiole is covered with few to many, flattened black spines to 1(–1.5) cm long. This form has been called *B. acanthocnemis* and *B. tenuis*. Leaf shape shows the same variation as the first two forms, but linear venation is also common. Then the leaf is oblanceolate and deeply bifid, and lobes are linear, strongly plicate, 27–44 cm long and 1.5–5 cm wide (Figure 6.59o). Plants with this leaf shape have been called *Bactris huberiana* and *B. luetzelburgii*.

33. **Bactris syagroides** Barb. Rodr. & Trail, Enum. palm. nov. 33. 1875 ("cyagroïdes"). *Amylocarpus syagroides* (Barb. Rodr. & Trail) Barb. Rodr., Contr. Jard. Bot. Rio de Janeiro 3: 72. 1902. Type destroyed. Lectotype (here designated). Brazil. Pará: Rio Tapajós, 17 Mar 1874, *J. Trail 890* (K) (Figure 6.50a–b).

Bactris multiramosa Burret, Notizbl. Bot. Gart. Berlin-Dahlem 14: 263. 1938. Type. Brazil. Pará: Itapacurú, Rio Tapajós, 4 Apr 1924, *J. Kuhlmann 1908* (holotype, RB).

Stems cespitose, 0.6–1.5 m tall, 0.8–2 cm diam, nonspiny, covered with persistent leaf bases. *Leaves* 5–8; sheath 10–20 cm long, sheath and petiole nonspiny or moderately covered with somewhat flattened black spines to 1 cm long; petiole 45–90 cm long; rachis 50–58 cm long; pinnae 30–35 per side, ± regularly arranged and spreading in 1 plane, narrowly linear, the middle ones 25–35 cm long, 0.7–1 cm wide, with lines of spinules on veins abaxially. *Inflorescences* interfoliar; peduncle to 4.5 cm long; prophyll to 5 cm long; peduncular bract 10–12 cm long, densely covered with soft black spines to 1 cm long, with white, swollen base; rachis 0–1.5 cm long; rachillae 7–13, 4–6 cm long; *flowers* borne in triads, these regularly arranged almost throughout rachillae; staminate and pistillate flowers not seen; *fruits* globose, ca. 7 mm diam, orange-yellow; fruiting perianth with 3-lobed calyx and corolla.

Eastern Amazon region of Brazil (Pará),

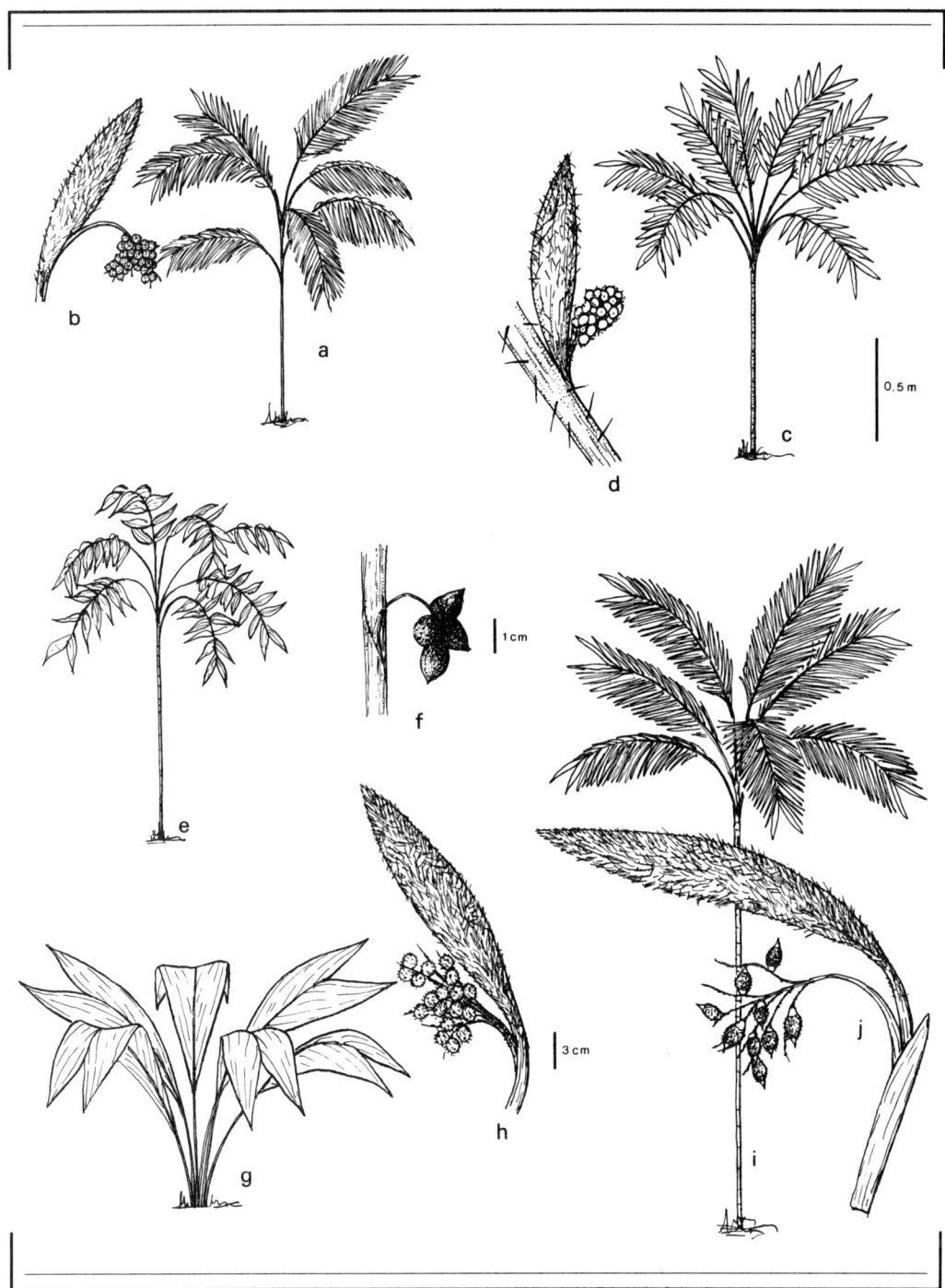

Figure 6.50. (a) *Bactris syagroides*, (b) infructescence (from *J. Kuhlmann 1908*); (c) *Bactris tefensis*, (d) infructescence (from *A. Henderson 1584*); (e) *Bactris tomentosa* var. *tomentosa*, (f) infructescence (from *A. Henderson 1501*); (g) *Bactris trailiana*, (h) infructescence (from *A. Henderson 1613*); (i) *Bactris turbinocarpa*, (j) infructescence (from *J. Wessels Boer 1392*).

known mostly from the Rio Tapajós (Figure 6.49C); lowland rain forest on *terra firme*.

This species is very poorly known, and there are no recent collections. The lectotype matches very well with the type of *Bactris multiramosa*, and the latter is thus placed in synonymy. Burret was somewhat puzzled by Kuhlmann's collection because, although it appeared to be related to the *Amylocarpus* group, it had many more rachillae than usual. In some respects, this species is reminiscent of *B. cuspidata*.

34. Bactris tefensis Henderson, sp. nov. Type. Brazil. Amazonas: Mun. Tefé, near mouth of Rio Tefé, 19 Jan 1991, *A. Henderson & J. Guedes 1584* (holotype, INPA, isotype, NY) (Figure 6.50c–d).

A congeneribus omnibus corolla foeminea spinulis flexuosis brunneis.

Stems solitary or cespitose, 0.8–1.5 m tall, 1.5–2 cm diam, covered with persistent leaf bases. *Leaves* 4–10; sheath 15–20 cm long, sheath, petiole, and rachis with a few, scattered, black spines to 4 cm long; petiole 0.6–1 m long; rachis 40–76 cm long; pinnae 9–13 per side, irregularly arranged in clusters of 2–3 and spreading in different planes, sigmoid, light green, the middle ones 22–31 cm long, 4.5–6 cm wide. *Inflorescences* interfoliar; peduncle to 10 cm long; prophyll not seen; peduncular bract 18–20 cm long, moderately covered with soft, brownish spines to 1 cm long; rachis to 3 cm long, very spiny; rachillae numerous, filamentous, crowded on rachis, to 3 cm long; *flowers* borne in triads, these regularly arranged on proximal part of rachillae, solitary or paired staminate flowers distally; staminate flowers not seen; pistillate flowers 4 mm long (postanthesis), borne on 1 side and at base of rachillae; calyx cupular, 1 mm long, with a few, flexuous, brown spinules to 2 mm long; corolla tubular, 3 mm long, densely covered with flexuous, brown spinules to 3 mm long; staminodes very small, digitate; *fruits* widely obovoid, shortly rostrate, ca. 1 cm diam, orange-red, lacking spinules; meoscarp starchy; endocarp fibers few; fruiting perianth with 2 mm long lobed calyx and 3 mm long densely spinulose corolla.

Rio Tefé region of Brazil (Amazonas) (Figure 6.49D); lowland rain forest on *terra firme* at low elevations.

This species is a member of the Piranga group (Sanders, 1991); see the discussion under *Bactris acanthocarpoides*. It shares with another member of the group, *Bactris glandulosa* from Central America, nonspinulose fruits, and is also similar in habit. The spinules on the pistillate corolla are remarkably like those of *Astrocaryum gynacanthum*.

35. Bactris tomentosa Mart., Hist. nat. palm. 2: 100. 1826. *Bactris tomentosa* subsp. *tomentosa* Trail, J. Bot. 6: 4. 1877. *Bactris tomentosa* var. *negrensis* A. D. Hawkes, Arq. Bot. Estado São Paulo 2: 184. 1952. Type. Brazil. Amazonas: Rio Solimões, n.d., *C. Martius s.n.* (holotype, M; F negs. 18633, 18634) (Figure 6.50e–f).

Stems cespitose, 0.3–3 m tall, 0.8–1.6 cm diam, spiny on internodes. *Leaves* 9–13; sheath 11–18 cm long, sheath and petiole (and rachis) brown-tomentose, sparsely to moderately covered with slightly flattened, yellowish spines or needlelike, black spines to 3 cm long; petiole 10–28 cm long; rachis 18–55 cm long; pinnae 2–9 per side, irregularly arranged and spreading in slightly different planes, sigmoid, with long, filiform, minutely spinulose apex, the middle ones 18–27 cm long, 3–4 cm wide, the apical ones wider than the others, occasionally pilose abaxially, or blade entire or with a few pinnae, then deeply bifid apically and narrowly cuneate basally, to 50 cm long and 15 cm wide at middle. *Inflorescences* interfoliar; peduncle ca. 6 cm long; prophyll 8–9 cm long; peduncular bract 10–17 cm long, nonspiny or spiny with short, black spines to 0.3 cm long or flexuous, yellowish spines to 0.8 cm long; rachilla 1, 3–5 cm long; *flowers* in triads, these regularly arranged almost throughout rachilla; staminate flowers 6 mm long; sepals 1–2 mm long; petals 6 mm long; pistillode absent; pistillate flowers 4 mm long; calyx tubular, 3–4 mm long, occasionally spinulose; corolla tubular, 3–4 mm long, occasionally spinulose; *fruits* very widely obovoid, to 1.5 cm long, to 1.5 cm diam, purple-black, occasionally spinulose; mesocarp juicy; fruiting perianth with lobed calyx half as long as the lobed corolla, occasionally spinulose, staminodial ring absent.

I have included here all specimens of *Bactris* with regularly arranged triads on a spicate inflorescence, pistillate flowers with subequal calyx and corolla, no staminodial ring, and purple-black fruits. The resultant assemblage is rather heterogeneous and may not represent a real taxon. There are, however, too few collections to resolve the matter. Specimens from the western Amazon region in Colombia, Peru, western Brazil, and possibly Bolivia—with needlelike, black spines and entire or partially entire leaves—are here called var. *sphaerocarpa.* Specimens from the central and eastern part of the region in Brazil and French Guiana—with somewhat flattened, yellowish spines and pinnate leaves—are very variable. Near Manaus are specimens that have pilose pinnae and spinulose pistillate perianth and fruits. These were called *B. tomentosa* by Martius. In the same region are similar plants, but the leaves are less pilose and the perianth and fruits are nonspinulose. These were called *B. eumorpha* by Trail. Toward the east, plants only rarely have minutely spinulose fruits. All these central and eastern forms are here called var. *tomentosa.*

KEY TO THE VARIETIES OF BACTRIS TOMENTOSA

1. Sheath and petiole with needlelike, black spines to 3 cm long; leaves entire or pinnate with 2–4 pinnae per side; Colombia (Amazonas, Vaupés), Peru (Loreto, Madre de Dios) and Brazil (Acre, Amazonas). . . . . 35a. *B. tomentosa* var. *sphaerocarpa.*
1. Sheath and petiole with slightly flattened, yellowish spines; leaves pinnate with to 8 pinnae per side; French Guiana, Brazil (Amapá, Amazonas, Maranhão, Pará) . . . . 35b. *B. tomentosa* var. *tomentosa.*

35a. Bactris tomentosa var. **sphaerocarpa**

(Trail) Henderson, stat. nov. *Bactris sphaerocarpa* Trail, J. Bot. 6: 8. 1877. Type. Brazil. Amazonas: Tabocal, Rio Purus, 11 Sep 1874, *J. Trail 898/CXIX* (holotype, K).

Bactris sphaerocarpa var. *minor* Trail, J. Bot. 6: 8. 1877. Type. Brazil. Amazonas: Barreiras de Catatiha, Rio Purus, 28 Sep 1874, *J. Trail 902/CXXX* (holotype, K).

Bactris sphaerocarpa var. *ensifolia* Trail, J. Bot. 6: 8. 1877. Type. Brazil. Amazonas: Barreiras de Carurú, Rio Jutaí, 5 Feb 1875, *J. Trail 904/CXIXa* (holotype, K).

Bactris sphaerocarpa var. *platyphylla* Trail, J. Bot. 6: 8. 1877. Type. Brazil. Amazonas: Gavião, Rio Juruá, 10 Nov 1874, *J. Trail 844/CLIV* (holotype, K).

Bactris sphaerocarpa subsp. *pinnatisecta* Trail, J. Bot. 6: 9. 1877. *Bactris sphaerocarpa* var. *pinnatisecta* (Trail) A. D. Hawkes, Arq. Bot. Estado São Paulo 2: 184. 1952. Type. Type. Brazil. Amazonas: Barreiras de Catatiha, Rio Purus, n.d., *J. Trail 906/CXXXII* (holotype, K; isotype, P).

Bactris sphaerocarpa var. *schizophylla* Drude in Mart., Fl. bras.: Cyclanthaceae et Palmae I, fasc. 85, vol. 3(2): 326. 1881. Type. Brazil. Amazonas: Barreiras da Cupana, Rio Purus, 4 Oct 1874, *J. Trail 901/CXXXIX* (holotype, K).

Bactris angustifolia Dammer, Verh. Bot. Vereins Prov. Brandenburg 48: 128. 1907. Type. Brazil. Acre: Rio Juruá-mirim, Jun 1901, *E. Ule 5596* (holotype, B; isotypes, K, MG).

Brazil: *marajá, marajazinha.* Peru: *kamáncha* (Achual Jívaro), *ñejilla.* The fruits are edible.

Throughout the western Amazon region of Colombia (Amazonas, Vaupés), Peru (Loreto, Madre de Dios), and Brazil (Acre, Amazonas) (Figure 6.49E); lowland rain forest on *terra firme* at low elevations.

35b. Bactris tomentosa var. **tomentosa**

Bactris sp. Wallace, Palm Trees of the Amazon 79. 1853. Lectotype (here designated). Wallace, Palm Trees of the Amazon t. 29. 1853.

Bactris tomentosa subsp. *capillacea* Trail, J. Bot. 6: 5. 1877. *Bactris capillacea* (Trail) Drude in Mart., Fl. bras.; Cyclanthaceae et Palmae I, fasc. 85, vol. 3(2): 336. 1881. Type. Brazil. Amazonas: Barreiras de Pariti, Rio Purus, 5 Oct 1874, *J. Trail 884/CXL* (holotype, K; isotype, P).

Bactris eumorpha Trail, J. Bot. 6: 9. 1877. *Bactris eumorpha* subsp. *eumorpha* Trail, J. Bot. 6: 9. 1877. Type. Brazil. Amazonas: Lages at mouth of Rio Negro, 2 Jul 1874, *J. Trail 1109/C* (holotype, K; isotype, P).

Bactris eumorpha subsp. *arundinacea* Trail, J. Bot. 6: 9. 1877. *Bactris arundinacea* (Trail) Drude in Mart., Fl. bras.: Cyclanthaceae et Palmae I, fasc. 85, vol. 3(2): 333. 1881. Type. Brazil. Amazonas: Barreiras de Cariwacanga, Rio Purus, 10 Oct 1874, *J. Trail 1110/CLXI* (holotype, K).

Bactris capinensis Huber, Bol. Mus. Paraense Hist. Nat. 6: 60. 1909. Type. Brazil. Amazonas: Rio Cupim, Approaga, 17 Jun 1897, *J. Huber 743* (holotype, MG).

French Guiana and Brazil (Amapá, Amazonas, Maranhão, Pará) (Figure 6.49E); lowland rain forest on *terra firme* at low elevations.

36. Bactris trailiana Barb. Rodr., Enum. palm. nov. 27. 1875. *Bactris acanthocarpa* subsp. *trailiana* (Barb. Rodr.) Trail, J. Bot. 6: 46. 1877. Lectotype (here designated). Barb. Rodr., Sert. palm. bras. 2: t. 4. 1903 (Figure 6.50g–h).

Stems solitary, short, and subterranean, rarely 1–2 m tall, 5–7 cm diam. *Leaves* 7–10; sheath ca. 25 cm long, sheath and petiole sparsely covered with black spines to 5 cm long, or occasionally plants lacking spines; petiole 0.7–1(–2) m long, petiole and rachis reddish-brown-tomentose abaxially; rachis 65–70 cm long; blade entire, slightly praemorse on the margins, 0.7.1.2(–2) m long, 22–34 cm wide. *Inflorescences* interfoliar; peduncle 12–20 cm long; prophyll 6–7 cm long; peduncular bract 22–26 cm long, densely covered with black spines to 1 cm long; rachis 4–8 cm long; rachillae ca. 23, filamentous, to 10 cm long; *flowers* in triads proximally, these tending to be on abaxial side of rachillae, paired or solitary staminate distally; staminate flowers 2.5 mm long; sepals connate into a 3-lobed cupule, the lobes 0.5 mm long; petals widely ovate, 2.5 mm long; pistillate flowers 3 mm long; calyx cupular, 2 mm long, glabrous; corolla tubular, 2.5–3 mm long, glabrous; *fruits* very widely obovoid, shortly rostrate, 1.5–2 cm diam, orange-red, covered with black spinules; meoscarp starchy; endocarp fibers few; fruiting perianth with 2 mm long, lobed calyx and 3 mm long lobed corolla.

Colombia (Amazonas), Venezuela (Bolívar), French Guiana, Brazil (Amazonas, Pará), and Bolivia (Beni, Pando) (Figure 6.49F); lowland rain forest on *terra firme* at elevations below 520 m.

Brazil: *ubussuhy.*

This species is a member of the Piranga group (Sanders, 1991); see the discussion under *Bactris acanthocarpoides.* I initially considered it a variety of *B. acanthocarpa,* following Trail, but the insect pollinators are so different (Henderson et al., in preparation) that I now maintain it as a separate species. *Bactris cuesco* Engel is a synonym of *B. corossilla* H. Karst., not of *B. trailiana* (contra Sanders, 1991).

37. Bactris turbinocarpa Barb. Rodr., Enum. palm. nov. 23. 1875. *Pyrenoglyphis turbinocarpa* (Barb. Rodr.) Burret, Repert. Spec. Nov. Regni Veg. 34: 248. 1934. Lectotype (Wess. Boer 1965). Barb. Rodr., Sert. palm. brasil. 2: t. 22. 1903 (Figure 6.50i–j).

Stems cespitose, 1–1.7 m tall, 2–2.5 cm diam, with a few spines on the internodes. *Leaves* 6–13; sheath 19–24 cm long, sheath, petiole, and rachis densely reddish-brown-velutinose, with scattered to numerous, black spines to 9 cm long; petiole 50–57 cm long; rachis 0.7–1 m long; pinnae 15–22 per side, linear-lanceolate, aristate, regularly arranged and spreading in 1 plane, the middle ones 36–45 cm long, 2.5–4 cm wide, sparsely tomentose abaxially, cross-veins evident. *Inflorescences* interfoliar; peduncle 22–30 cm long; prophyll 15–20 cm long; peduncular bract 30–45 cm long, densely covered with appressed, black and brown spines to 1 cm long; rachis 3–5 cm long; rachillae 8–12, 6–12 cm long; *flowers* in triads, these scattered among paired or solitary staminate flowers; staminate and pistillate flowers not seen; *fruits* obovoid, strongly rostrate, 2.5–3.5 cm long, 1.5–2 cm diam, densely brown-tomentose, covered with short spinules; mesocarp starchy; fruiting perianth with deeply lobed, spinulose calyx 5–7 mm long and briefly lobed, spinulose corolla to 1 cm long, staminodial ring absent.

Northeastern Amazon region in Surinam and Brazil (Pará) (Figure 6.51A); lowland rain forest on *terra firme* at low elevations.

This species is very rare, and rarely collected. It is known from only two localities that are at least 500 km apart. Another specimen, possibly this species but differing in its entire leaf, is known from the Pacific coast of Colombia (Chocó). Roger Sanders (personal communication) considers *B. turbinocarpa* a relict species,

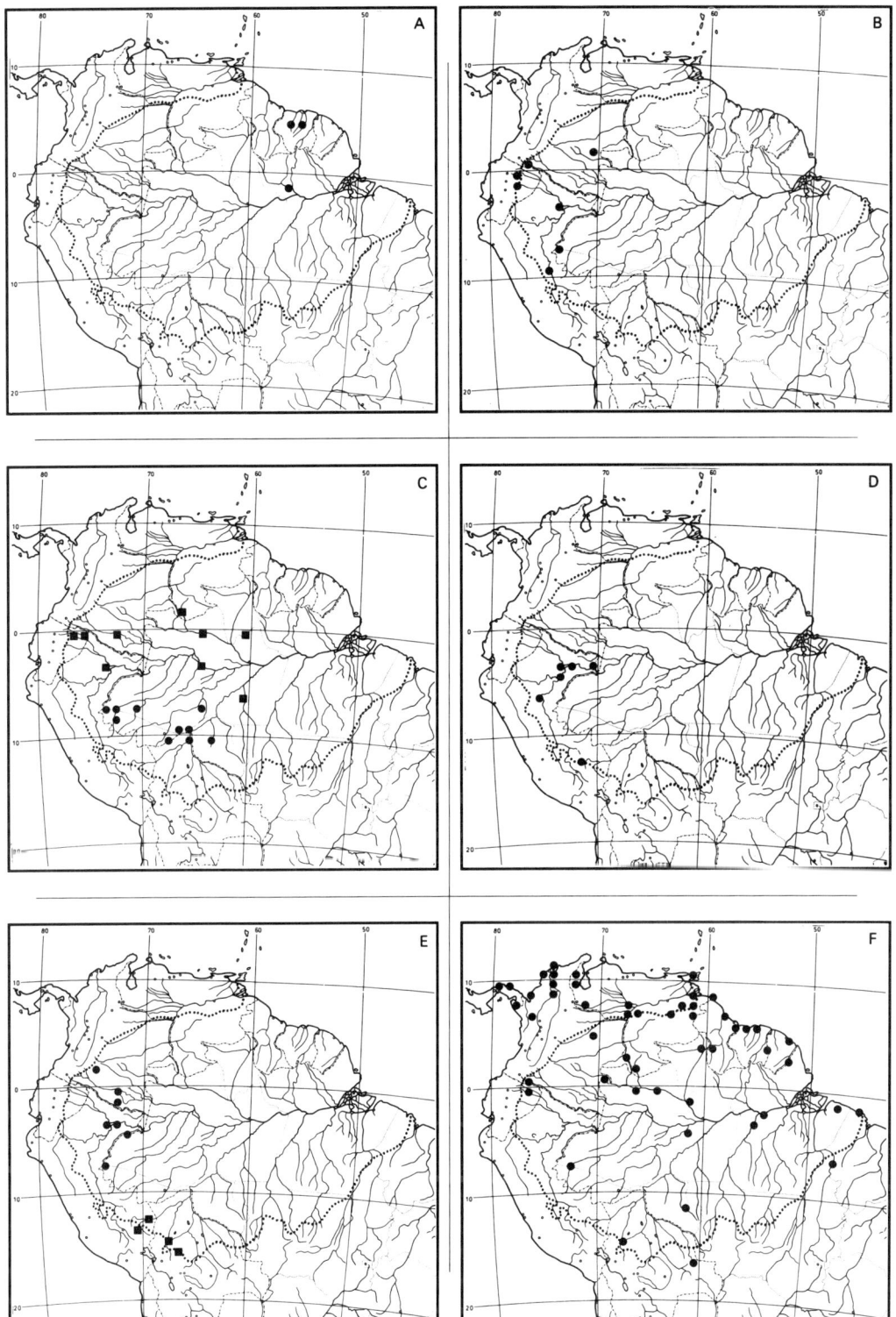

Figure 6.51. **(A)** *Bactris turbinocarpa;* **(B)** *Desmoncus giganteus;* **(C)** *Desmoncus mitis* var. *leptoclonos* (circles), var. *mitis* (squares); **(D)** *Desmoncus mitis* var. *leptospadix;* **(E)** *Desmoncus mitis* var. *rurrenbaquensis* (squares), var. *tenerrimus* (circles); **(F)** *Desmoncus orthacanthos*.

and this might account for this unusual distribution.

Trail (1877a) considered that the fruiting corolla of this species had a staminodial ring; on the basis of this, Burret (1933–1934) transferred the species to *Pyrenoglyphis*. However, no staminodial ring appears to be present, although the apical part of the corolla is spinulose internally, which is unusual in *Bactris*.

Uncertain Names

Bactris vexans Burret, Notizbl. Bot. Gart. Berlin-Dahlem 14: 266. 1938. Type. Brazil. Amazonas: Tocantins, 24 Jan 1924, *J. Kuhlmann 1239* (holotype, RB).

Bactris megistocarpa Burret, Notizbl. Bot. Gart. Berlin-Dahlem 12: 623. 1935. Type. Brazil. Amazonas: Manicoré on Rio Manicoré, 1934, *W. Hopp 1334* (holotype, B).

27. *Desmoncus*

Desmoncus Mart., Palm. fam. 20. 1824.

Small to large, spiny, monoecious palms. *Stems* elongate, climbing, scrambling, or rarely free-standing, solitary or cespitose. *Leaves* pinnate, reduplicate, variously spiny, arranged all along upper part of the stem, often distichously; sheath closed, extending above the petiole into a prominent and persistent ocrea; petiole short or absent; rachis elongate, the apical part extended into a cirrus; pinnae regularly or irregularly arranged and spreading in 1 plane, rarely the proximal ones with tendril-like apex, the distal few usually modified into short, reflexed acanthophylls (spines derived from pinnae). *Inflorescences* interfoliar, branched to 1 order, solitary or rarely multiple at each node; peduncle bearing a prophyll and 1 spiny, or nonspiny and tomentose, peduncular bract; rachis bearing few to many simple rachillae; *flowers* borne in triads or paired or solitary staminate; staminate flowers with sepals connate into a shallow, 3-lobed cupule; petals 3, connate basally for ca. half their length, free and valvate above; stamens 6–11; pistillode small or absent; pistillate flowers with a cupular calyx, tubular corolla, and 6 staminodes; gynoecium syncarpous, trilocular, triovulate; *fruits* 1-seeded, globose, obovoid or ellipsoid, with apical stigmatic residue; endocarp thick, black and bony with 3 lateral pores, covered with flattened, appressed fibers; seed with homogeneous endosperm and lateral embryo; germination adjacent-ligular; eophyll bifid.

Burret (1934a) recognized at least 41 species, and several more were added subsequently. However, there are far fewer. The large number of names and small number of species, together with the large amount of variation and small number of specimens, have led to great confusion. In addition to the five species recognized from the Amazon region, there are two others. One, *Desmoncus stans* Grayum & de Nevers, occurs in Costa Rica; the other, *D. cirrhiferus* Gentry & Zardini, in the Pacific coastal region of Colombia and Ecuador. Variation has been discussed by Wessels Boer (1966). The genus, however, is still poorly understood and in need of a modern revision. Matters are further complicated by the occurrence of apparent hybrids.

Desmoncus contains two of the most widespread species in the Neotropics: *D. orthacanthos* and *D. polyacanthos*. If *D. giganteus* is related to the former, then it can be considered a western peripheral isolate. It is distinguished by its larger fruits. Of remarkable note is that two other varieties from the same general region also have larger fruits: *D. mitis* var. *tenerrimus,* and *D. polyacanthos* var. *prunifer.*

Listabarth (1992) studied reproductive biology of two species in Peru: *Desmoncus mitis* var. *leptospadix* and *D. polyacanthos* var. *polyacanthos*. Both were protogynous with inflorescences at pistillate anthesis in the evening, with fragrant flowers and inflorescences that heated up. Approximately 24 hours later, staminate flowers opened, and were very short-lived. Probable pollinators were curculionid and nitidulid beetles, as occurs in *Bactris*.

One interesting feature of the genus is its phenology. Plants can grow without flowering for years. Then several inflorescences at different nodes will develop together at the same time, and will flower and fruit at the same time. It is common to find numerous sterile juveniles in the forest, but fertile adults are rarely encountered.

KEY TO THE SPECIES OF *DESMONCUS*

1. Petiole and rachis with ± straight spines to 5.5 cm long, cirrus without spines; inflorescences with (8–)24–31 rachillae; pinnae 16–48 cm long, often with spines on the lower surface, especially proximally; spines of peduncular bract ± straight.
   2. Ocrea very fibrous; fruits 3–4 cm long, with a corolla up to one-third the length of the fruit; Colombia (Vaupés), Ecuador (Napo), Peru (Huánuco, Loreto), and Brazil (Acre) . . . . 1. *D. giganteus*.
   2. Ocrea not very fibrous; fruits 1.5–2 cm long, the corolla very small; widespread . . . . 3. *D. orthacanthos*.
1. Petiole, rachis, and cirrus with strongly recurved spines to 0.5 cm long, or spines absent; inflorescences with 3–17 rachillae; pinnae 8–25 cm long, rarely with spines on the lower surface.
   3. Inflorescences stout with 5–17 rachillae; peduncular bract usually densely spiny, wide; sheath and ocrea usually spiny.
      4. Peduncular bract with recurved spines, 3–6 mm long; inflorescences with 5–17 rachillae, these 3–12 cm long; fruits ellipsoid to obovoid, 1–2.2 cm long and 0.8–1.8 cm diam; widespread . . . . 5. *D. polyacanthos*.
      4. Peduncular bract with straight spines, 2–4 mm long; inflorescences with 8–11 rachillae, these 2–6 cm long; fruits globose to obovoid, 0.8–1 cm long, 0.8–1 cm diam; Venezuela (Amazonas), the Guianas, and Brazil (Amapá, Amazonas, Maranhão, Pará) . . . . 4. *D. phoenicocarpus*.
   3. Inflorescences slender with 3–7 rachillae; peduncular bract usually not spiny, brown-tomentose, narrow; sheath and ocrea often nonspiny . . . . . . 2. *D. mitis*.

1. Desmoncus giganteus Henderson, sp. nov. Type. Brazil. Acre: Rio Moa near mouth of Rio Azul, 7°25′S, 73°15′W, 14 Feb 1992, *A. Henderson et al. 1688* (holotype, INPA; isotype, NY) (Figure 6.52a–c; Plate IVc).

Ocrea fibrosissima et corolla tertiam longitudinem fructuum longiorum usque attingenti ab aliis speciei varietatibus distans.

Stems solitary or cespitose, 10–25 m or more long, 1.5–2 cm diam, 4–8 cm diam with sheaths, covered with persistent leaf sheaths. *Leaves* distichously arranged, to 50; sheath to 60 cm long, densely spiny with black spines to 3 cm long; ocrea very fibrous and spiny; petiole 20–30 cm long; rachis 1.1–2 m long; cirrus 0.8–1.5 m long; pinnae 5–10 per side, elliptic, the middle ones 30–48 cm long, 7–10 cm wide, glaucous abaxially. *Inflorescences* interfoliar; peduncular bract at least 40 cm long, densely covered with black or brown spines to 2 cm long; *flowers* not seen; *fruits* ellipsoid, 3–4 cm long, 1.5–2.5 cm diam, red, the persistent thick petals covering almost one-third the length of the fruit.

Western Amazon region of Colombia (Vaupés), Ecuador (Napo), Peru (Huánuco, Loreto), and Brazil (Acre) (Figure 6.51B); lowland rain forest on *terra firme* at low elevations.

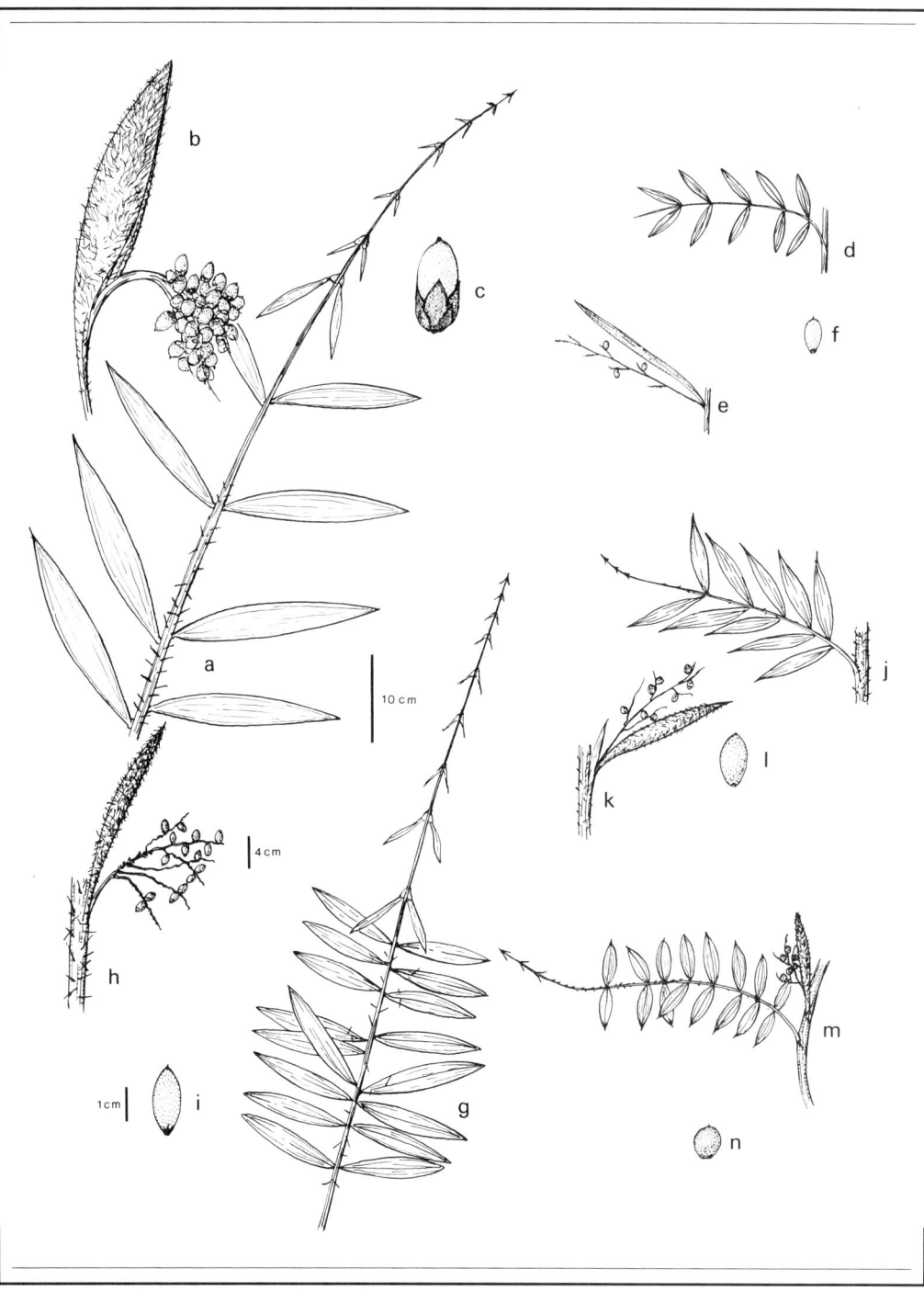

Figure 6.52. **(a)** *Desmoncus giganteus,* part of leaf, **(b)** infructescence (from *A. Henderson 1688*), **(c)** fruit; **(d)** *Desmoncus mitis* var. *leptospadix,* leaf, **(e)** infructescence (from *E. Killip 28807*), **(f)** fruit; **(g)** *Desmoncus orthacanthos,* part of leaf, **(h)** infructescence (from *A. Henderson 1696*), **(i)** fruit; **(j)** *Desmoncus polyacanthos* var. *polyacanthos,* leaf, **(k)** infructescence (from *A. Henderson 1675*), **(l)** fruit; **(m)** *Desmoncus phoenicocarpus,* leaf and infructescence (from *T. McDowell 4104*), **(n)** fruit.

Brazil: *jacitara*. Ecuador: *urpi-chunda* (Quichua). Peru: *vara-casha*. In Iquitos, Peru, the stems (and those of other species) are used to weave furniture (Henderson & Chávez, 1993).

2. **Desmoncus mitis** Mart., Hist. nat. palm. 2: 90. 1824. *Atitara mitis* (Mart.) Kuntze, Revis. gen. pl. 2: 727. 1891. Type. Brazil. Amazonas: Rio Negro, n.d., *C. Martius s.n.* (holotype, M; F neg. 18593) (Figure 6.52d–f).

Stems cespitose or solitary, 1–10 m long, 0.5–1 cm diam, climbing or scandent, completely covered with sheaths. *Leaves* 6–30; sheath 15–30 cm long, with a prominent ocrea, nonspiny or with scattered or numerous, bulbous-based, recurved or straight spines; petiole 1–15 cm long; rachis 17–27 cm long, with scattered, recurved spines; cirrus to 36 cm long, but often not developed; pinnae 2–24 per side, elliptic, linear or lanceolate, 6–25 cm long, 0.5–6 cm wide, with a prominent mid-vein and obscure lateral veins. *Inflorescences* interfoliar, slender and elongate, solitary or rarely multiple (3–4) at each node; peduncle 12–30 cm long; prophyll 8–15 cm long, inserted ± at middle of peduncle; peduncular bract 12–28 cm long, inserted near apex of peduncle, without spines or rarely with a few, short spines, narrow, brown-tomentose; rachis 7–15 cm long; rachillae 3–7, 3–9 cm long, very slender; *flowers* arranged in triads on proximal part of rachillae, staminate distally; staminate flowers 6–7 mm long at anthesis; sepals 0.5–1 mm long; petals lanceolate-ovate, 6–7 mm long; stamens 6; pistillode not apparent; pistillate flowers 2 mm long at anthesis; calyx cupular, 0.5 mm long; corolla tubular, 3-lobed, 1.5 mm long; staminodes not apparent; *fruits* ellipsoid to obovoid, 1–2.2 cm long, 0.5–1.5 cm diam, bright red or yellowish-green at maturity.

Western Amazon region in Colombia (Amazonas, Caquetá), Venezuela (Amazonas), Ecuador (Napo), Peru (Cusco, Huánuco, Loreto, Madre de Dios), Brazil (Acre, Amazonas, Rondônia, Roraima), and Bolivia (Beni, La Paz).

This is the most complex species of *Desmoncus*, but there are still too few collections to understand the considerable variation. Here it is divided into five varieties.

KEY TO THE VARIETIES OF *DESMONCUS MITIS*

1. Fruits 1.8–2.2 cm long, 1–1.5 cm diam; Colombia (Amazonas, Caquetá), Peru (Loreto), and Brazil (Acre, Amazonas) . . . . 2e. *D. mitis* var. *tenerrimus*.
1. Fruits 0.8–1.5 cm long, 0.5–1 cm diam.
   2. Pinnae 10–24 per side, linear or lanceolate, 6–14 cm long, 1–2.5 cm wide.
      3. Leaf-sheath and ocrea with numerous, short, straight spines; pinnae linear, 22–24 per side; Peru (Cusco, Madre de Dios) and Bolivia (Beni, La Paz) . . . . 2d. *D. mitis* var. *rurrenabaquensis*.
      3. Leaf-sheath and ocrea without spines; pinnae lanceolate, 10–11 per side; Colombia (Amazonas), Venezuela (Amazonas), Ecuador (Napo), Peru (Loreto), and Brazil (Amazonas, Roraima) . . . . 2c. *D. mitis* var. *mitis*.
   2. Pinnae 2–8 per side, elliptic to lanceolate, 12–25 cm long, 3–6 cm wide.
      4. Pinnae 2–3 per side, 21–25 cm long, 3–6 cm wide; cirrus often not developed; Brazil (Acre, Amazonas, Rondônia) and Bolivia (Beni, La Paz) . . . . . . 2a. *D. mitis* var. *leptoclonos*.
      4. Pinnae 5–8 per side, 14–15 cm long, 3–4 cm wide; cirrus well-developed; Colombia (Amazonas), Peru (Huánuco, Loreto, Madre de Dios) . . . . 2b. *D. mitis* var. *leptospadix*.

2a. Desmoncus mitis var. **leptoclonos** Henderson, nom. nov. *Desmoncus leptoclonos* Dammer, Verh. Bot. Vereins Prov. Brandenburg 48: 129. 1907 (not *D. leptoclonos* Drude). Type. Brazil. Acre: Juruá-mirim, May 1901, *E. Ule 5515* (isotypes, MG, K).

Leaves with rachis and cirrus 15–41 cm long, often the cirrus not developed; pinnae 2–3 per side, elliptic, 21–25 cm long, 3–6 cm wide. *Inflorescences* with peduncular bract 20–23 cm long, narrow, brown-tomentose, nonspiny; *fruits* ellipsoid, 0.8–1 cm long, 0.5 cm diam, bright red.

Brazil (Acre, Amazonas, Rondônia) and Bolivia (Beni, Pando). (Figure 6.51C); lowland rain forest at low elevations.

2b. Desmoncus mitis var. **leptospadix** (Mart.) Henderson, stat. nov. *Desmoncus leptospadix* Mart. in A. D. Orb., Voy. Amérique mér. 7(3). Palmiers 52. 1844. *Atitara leptospadix* (Mart.) Kuntze, Revis. gen. pl. 2: 727. 1891. Type. Peru. Loreto: Maynas, n.d., *E. Poeppig 2207* (holotype, G, n.v.).

Leaves with pinnae 5–7 per side, elliptic to lanceolate, 14–15 cm long, 3–4 cm wide. *Inflorescences* with peduncular bract to 25 cm long, narrow, nonspiny, brown-tomentose; *fruits* ellipsoid or obovoid, 1.3–1.5 cm long, 0.6–1 cm diam, bright red.

Colombia (Amazonas) and Peru (Huánuco, Loreto, Madre de Dios) (Figure 6.51D); lowland rain forest at low elevations.

Peru: *barahuasca, vara casha.*

2c. Desmoncus mitis var. **mitis**
Desmoncus pumilus Trail, J. Bot. 5 353. 1876. *Atitara pumila* (Trail) Kuntze, Revis. gen. pl. 2: 727. 1891. Type. Brazil. Amazonas: Rio Padauiri, 26 Jun 1874, *J. Trail 1086/LXXV* (holotype, K; isotype, P).
Desmoncus setosus var. *mitescens* Drude in Mart., Fl. bras.: Cyclanthaceae et Palmae I fasc. 85 vol. 3(2): 316. 1881. Type. Brazil. Amazonas: without locality, n.d., *J. Trail s.n.* (holotype, K, n.v.).

Leaves with pinnae 10–11 per side, lanceolate, 6–14 cm long, 0.5–2.5 cm wide. *Inflorescences* with peduncular bract to 12 cm long, brown-tomentose, slender, nonspiny or with a few spines; *fruits* ellipsoid, 0.8 cm long, 0.6 cm diam, bright red.

Colombia (Amazonas), Venezuela (Amazonas), Ecuador (Napo), Peru (Loreto), and Brazil (Amazonas, Roraima) (Figure 6.51C); lowland rain forest, forest margins, or *campina*, or in open vegetation on podzols, at low elevations.

2d. Desmoncus mitis var. **rurrenabaquensis** Henderson, var. nov. Type. Peru. Madre De Dios: Explorer's Inn on Río Tambopata, 12°50′S, 69°17′W, ca. 250 m, 6 Nov 1991, *A. Henderson & F. Chávez 1640* (holotype, USM; isotypes, CUZ, NY).
A var. miti vagina et ocrea spinis rectis numerosis armatis pinnisque numerosis linearibus diversa.

Leaves with sheath and ocrea with numerous, short, straight spines; pinnae linear, 22–24 per side, to 10 cm long, to 1 cm diam. *Inflorescences* with peduncular bract, nonspiny, brown-tomentose; *fruits* ellipsoid, 1 cm long, 0.5 cm diam, yellowish-green.

Peru (Cusco, Madre de Dios) and Bolivia (Beni, La Paz) (Figure 6.51E); lowland rain forest on *terra firme* at low elevations. Its range is centered on the Madre de Dios Basin.

2e. Desmoncus mitis var. **tenerrimus** (Mart.) Henderson, stat. nov. *Desmoncus tenerrimus* (Mart.) Mart. ex Burret, Repert. Spec. Nov. Regni Veg. 34: 236. 1934. *Bactris tenerrima* Mart. in Drude, Fl. bras.: Cyclanthaceae et Palmae I fasc. 85 vol. 3(2): 328. 1881. Type. Colombia. Amazonas: Río Caquetá, Araracuara, n.d., *C. Martius s.n.* (holotype, M; F neg. 18632).
Desmoncus vacivus L. H. Bailey, Gentes Herb. 8: 186. 1949. Type. Colombia. Amazonas: Río Igaraparaná, near La Chorrera, ca. 180 m, n.d., *R. Schultes 3941* (holotype, BH).

Leaves with pinnae 3–5 per side, elliptic, the middle ones 12–15 cm long, 3–5 cm diam. *Inflorescences* with peduncular bract brown-tomentose, nonspiny, slender; *fruits* ellipsoid to obovoid, 1.8–2.2 cm long, 1–1.5 cm diam, bright red.

Western Brazil (Acre, Amazonas), southern Colombia (Amazonas, Caquetá), and Peru (Loreto) (Figure 6–51E); in lowland rain forest either on *terra firme* or in inundated areas.

Brazil: *jacitara.* Colombia: *foodá* (Andoke).

Two varieties of *Desmoncus mitis* occur in Araracuara, in the Colombian Amazon, and were called by Galeano (1991) *D. setosus* and *D. pumilus.* I believe that *D. setosus* of Galeano corresponds to *D. mitis* var. *mitis,* and *D. pumilus* of Galeano corresponds to *D. mitis* var. *tenerrimus.* Even though the type of the latter is sterile, the pinna shape is characteristic.

A large-fruited variety of *Desmoncus polyacanthos,* var. *prunifer,* also occurs in the same general region.

3. Desmoncus orthacanthos Mart., Hist. nat. palm. 2: 87. 1824. *Atitara orthacantha* (Mart.) Barb. Rodr., Contr. Jard. Bot. Rio de Janeiro 3: 76. 1902. Type. Brazil. Bahia: Rio Mucuri, n.d., *M. Neuwied s.n.* (holotype, M; F negs. 18594, 18595) (Figure 6.52g–i).

Desmoncus rudentum Mart. in A. D. Orb., Voy. Amérique mér. 7(3). Palmiers 48. 1844. Type. Bolivia. Santa Cruz: Río Piray, near Palometa, n.d., *A. D'Orbigny 26* (holotype, P).

Desmoncus horridus Splitg. ex Mart. in A. D. Orb., Voyage Amérique mér. 7(3). Palmiers 51. 1844. *Atitara horrida* (Splitg. ex Mart.) Kuntze, Revis. gen. pl. 2: 727. 1891. Type. Surinam. Paramaribo, n.d., *F. Splitgerber 61* (holotype, BR, n.v.).

Desmoncus ataxacanthus Barb. Rodr., Enum. palm. nov. 25. 1875. *Atitara ataxacantha* (Barb. Rodr.) Kuntze, Revis. gen. pl. 2: 727. 1891. Lectotype (Wess. Boer, 1965). Barb. Rodr., Sert. palm. brasil. 2: t. 55. 1903.

Desmoncus palustris Trail, J. Bot. 5: 353. 1876. *Atitara palustris* (Trail) Kuntze, Revis. gen. pl. 2: 727. 1891. Type. Brazil. Amazonas: Rio Padauiri, 28 Jun 1874, *J. Trail 1087/LXXXI* (holotype, K).

Desmoncus melanacanthos Drude in Mart., Fl. bras.: Cyclanthaceae et Palmae I fasc. 85 vol. 3(2): 306. 1881. Type not designated.

Desmoncus orthacanthos var. *trailianus* Drude in Mart., Fl. bras.: Cyclanthaceae et Palmae I fasc. 85 vol. 3(2): 311. 1881. Type. Brazil. Amazonas: Rio Purus, Jauaria, 10 Sep 1874, *J. Trail 1079/CXXVI* (holotype, K).

Desmoncus macrocarpus Barb. Rodr., Vellosia 1: 34. 1888. *Atitara macrocarpa* (Barb. Rodr.) Barb. Rodr., Contr. Jard. Bot. Rio de Janeiro 3: 75. 1902. Lectotype (Wess. Boer, 1965). Barb. Rodr., Sert. palm. brasil. 2: t. 53. 1903.

Desmoncus angustisectus Burret, Notizbl. Bot. Gart. Berlin-Dahlem 10: 1025. 1930. Type. Brazil. Amazonas: Rio Negro, Trinidade, 12 Sep 1928, *P. Luetzelburg 22171* (isotypes, M, R; F neg. 18589).

Desmoncus luetzelburgii Burret, Notizbl. Bot. Gart. Berlin-Dahlem 10: 1025. 1930. Type. Brazil. Amazonas: Rio Papuri, Rio Uaupés, Trinidade, 13 Dec 1928, *P. Luetzelburg 23833* (isotypes, M, R; F neg. 18591).

Desmoncus huebneri Burret, Repert. Spec. Nov. Regni Veg. 36: 200. 1934. Type. Brazil. Roraima: Serra do Frechal, Rio Branco, n.d., *G. Huebner 80* (holotype, B, n.v.).

Desmoncus kuhlmanii Burret, Notizbl. Bot. Gart. Berlin-Dahlem 14: 267. 1938. Type. Bolivia. Beni: Riberalta, 25 Sep 1923, *J. Kuhlmann 522* (isotype, R).

Desmoncus demeraranus L. H. Bailey & H. E. Moore, Gentes Herb. 8: 181. 1949. Type. Guyana. River Demerara, near Craig village, between Georgetown and Atkinson Field, n.d., *L. Bailey 418* (holotype, BH).

Desmoncus multijugus Steyerm., Fieldiana Bot. 28: 85. 1951. Type. Venezuela. Bolívar: Tumeremo, 18 Dec 1944, *J. Steyermark 60968* (holotype, F, n.v.).

Stems 2–12 m long, 1.5–2 (–4 including leaf sheaths) cm diam, cespitose, climbing, covered with persistent leaf sheaths. *Leaves* to 10, often appearing distichously arranged; sheath 40–55 cm long, whitish-brown-tomentose, moderately to densely covered with short black spines to 1.5 cm long; ocrea 13–23 cm long; petiole 3–7 cm long, petiole and rachis with scattered, ± straight black spines to 5.5 cm long; rachis 0.9–1.9 m long; cirrus 35–70 cm long; pinnae (7–)20–28 per side, regularly or irregularly arranged, elliptic to linear, the middle ones 16–35 cm long, 2.5–6.5 cm wide, often with a few long black spines on veins abaxially, occasionally with a cluster of spines at pinnae base adaxially. *Inflorescences* interfoliar; peduncle elongate, 54–60 cm long, mostly included within the subtending sheath; prophyll 21–40 cm long; peduncular bract 22–37 cm long, moderately to densely covered with black or brown straight spines to 1 cm long, often white and bulbous at the base; rachis 11–12 cm long; rachillae (8–)24–31, 5–12 cm long; *flowers* arranged in triads, these regularly arranged along proximal part of the rachillae, staminate distally; staminate flowers to 11 mm long (at anthesis); sepals 1 mm long; petals ellipsoid, to 11 mm long; stamens 6–11, the filaments short, connate to the petals; pistillode absent; pistillate flowers 2.5–3 mm long (at anthesis); calyx cupular, 0.5–0.7 mm long; corolla tubular, 1.5–2 mm long; staminodes 6, digitate; *fruits* ellipsoid or obovoid to almost globose, 1.5–2 cm long, 1–1.5 cm diam, orange, yellow-orange, or red at maturity.

Atlantic slope of southern Mexico (Oaxaca, Veracruz) and Central America (Guatemala, Belize, Honduras, Nicaragua, Costa Rica, Panama), across tropical South America in Colombia (Atlántico, Bolívar, Guainía, Magdalena,

Vichada), Venezuela (Amazonas, Apure, Barinas, Bolívar, Delta Amacuro, Zulia), Trinidad and Tobago, the Guianas, Ecuador (Napo), Brazil (Acre, Amazonas, Bahia, Espírito Santo, Maranhão, Pará, Rio de Janeiro, Rondônia, Roraima, Tocantins), and Bolivia (La Paz, Santa Cruz) (Figure 6.51F). It seems to be particularly common in coastal regions. It grows in a variety of habitats; lowland rain forests, secondary forests, river margins, low-lying areas near the coast, and especially in disturbed areas.

Brazil: *jacitara*. Colombia: *ca-mu-vé* (Guahibo), *enredadera, kamawa*. Guyana: *karwari*. Surinam: *bambamaka*. Venezuela: *camuare, camuari, volador*.

This is a widespread and variable species, characterized by its nonspiny cirrus, ± straight leaf spines and large inflorescences. It is possible to recognize various local forms. Populations in the Atlantic coastal rain forest of Brazil are generally smaller and less spiny, the peduncular bracts often lack spines, and staminate flowers have 6–7 stamens. Those from the central part of the range are usually large and spiny and have ellipsoid fruits. Populations from northern Central America are large, robust, spiny plants; the staminate flowers have 9–11 stamens; and fruits are almost globose.

4. **Desmoncus phoenicocarpus** Barb. Rodr., Enum. palm. nov. 24. 1875. *Atitara phoenicocarpa* (Barb. Rodr.) Kuntze, Revis. gen. pl. 2: 727. 1891. Lectotype (Glassman, 1972). Barb. Rodr., Enum. palm. nov. t. 1, f. 3–4a-d (Figure 6.52m–n).

Desmoncus macrodon Barb. Rodr., Vellosia 1: 39. 1888. *Atitara macrodon* (Barb. Rodr.) Barb. Rodr., Contr. Jard. Bot. Rio de Janeiro 3: 75. 1902. Lectotype (Glassman, 1972). Barb. Rodr., Sert. palm. brasil. 2: t. 59. 1903.

Desmoncus nemorosus Barb. Rodr., Vellosia 1: 36. 1888. *Atitara nemorosa* (Barb. Rodr.) Barb. Rodr., Contr. Jard. Bot. Rio de Janeiro 3: 75. 1902. Lectotype (Glassman, 1972). Barb. Rodr., Sert. palm. brasil. 2: t. 56. 1903.

Desmoncus kaieteurensis L. H. Bailey, Bull. Torrey Bot. Club 75: 115. 1948. Type. Guyana. Trail from Tukeit to Kaiatuk Plateau, 29 Apr 1944, *B. Maguire & D. Fanshawe 23093* (isotypes, K, NY).

Desmoncus parvulus L. H. Bailey, Bull. Torrey Bot. Club 75: 115. 1948. Type. Guyana. Tumatumari, 18 Jun–8 Jul 1921, *H. Gleason 164* (isotype, NY).

Stems 2–3(–9) m long, 0.7–1 cm diam, trailing or climbing. *Leaves* numerous; sheath and ocrea with scattered, short, straight spines, or sometimes spines absent; petiole 3–10 cm long, petiole, rachis, and cirrus with short recurved spines especially on lower surface; rachis and cirrus 56–71 cm; pinnae 6–7 per side, elliptic, the middle ones 8–15 cm long, 2.5–4.5 cm wide, often with small spines on veins abaxially. *Inflorescences* interfoliar; peduncle not seen; prophyll 7–12 cm long, inserted ± halfway along peduncle; peduncular bract 13–21 cm long, moderately to densely covered with ± straight, 2–4 mm long, brown spines with white bulbous bases; rachis 4–7 cm long; rachillae 8–11, 2–6 cm long; *flowers* borne in triads on proximal parts of the rachillae, staminate distally; staminate flowers 7.5 mm long (at anthesis); sepals deltate, 1 mm long; petals ovoid, acuminate, 7 mm long; stamens 6; pistillode 0.5 mm long; pistillate flowers 2 mm long (at anthesis); calyx cupular, 1 mm long; petals corolla tubular, 2 mm long; staminodes 6, digitate; *fruits* globose to obovoid, rostrate, 0.8–1 cm long, 0.8–1 cm diam, orange or bright red at maturity.

Northeastern Amazon region in Venezuela (Amazonas), the Guianas, and Brazil (Amapá, Amazonas, Maranhão, Pará) (Figure 6.53A); forest margins or in clearings in lowland rain forest or secondary forest, often on white-sand soil, at elevations below 700 m.

Brazil: *iraparpukwaha* (Ka'apor), *jacitara, pijacicara*. French Guiana: *yasita* (Wayâpi). Guyana: *kamuari*.

The name that has traditionally been applied to this species, *Desmoncus macroacanthos* (e.g., Wessels Boer, 1965), can no longer be used. Although the type is sterile, it clearly belongs to *D. polyacanthos*.

Two specimens from Brazil (Amazonas), near Manaus (Mun. Careiro, Manaus–Porto Velho highway, km 22, 2 km on road to Purupuru, 3°30′S, 60°W, 1 Apr 1985, *A. Henderson 178* (NY); Rio Solimões, 3°23′S, 60°16′W, 16 Jun 1992, *S. Mori 22378* (NY)), are intermediate

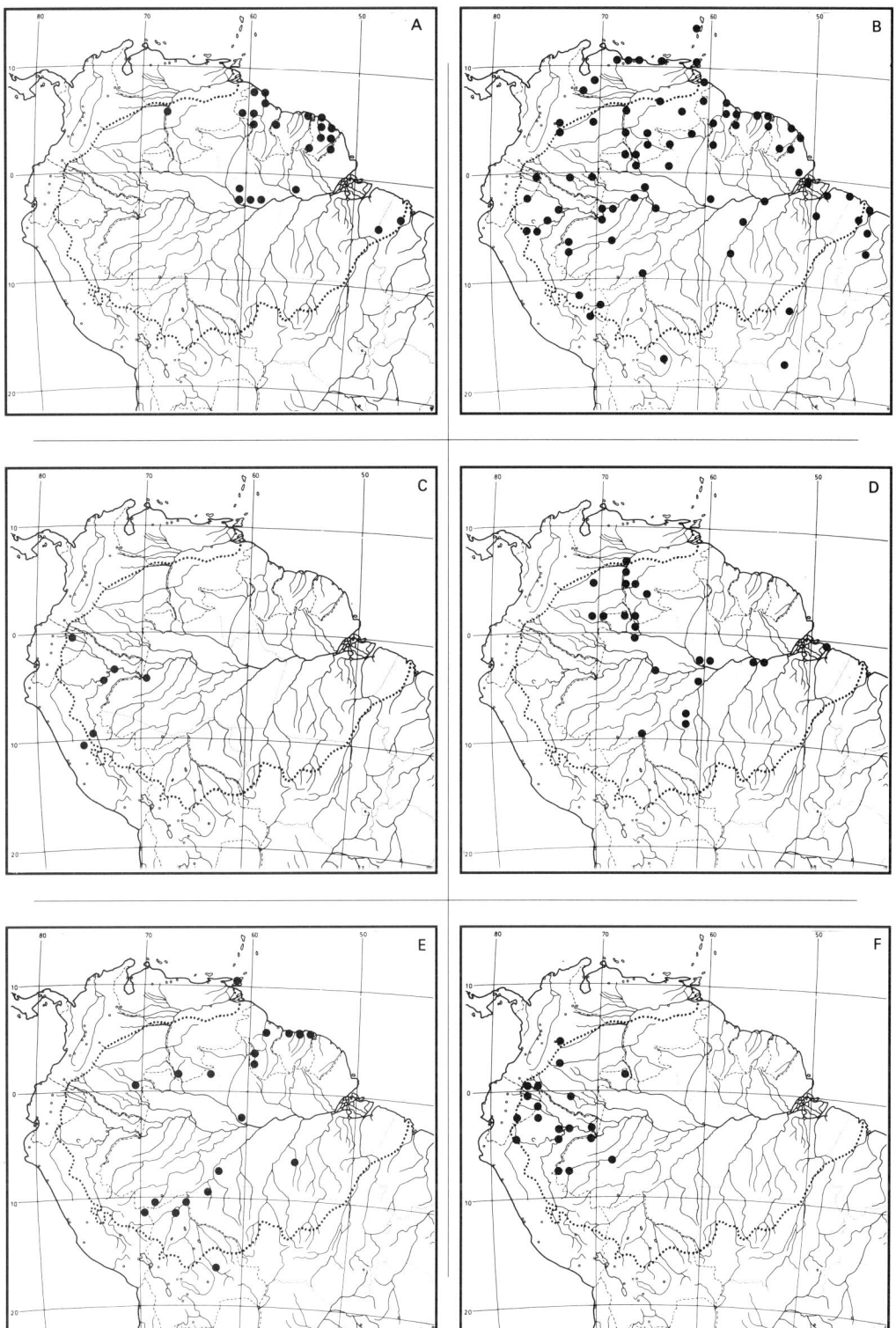

Figure 6.53. (A) *Desmoncus phoenicocarpus;* **(B)** *Desmoncus polyacanthos* var. *polyacanthos;* **(C)** *Desmoncus polyacanthos* var. *prunifer;* **(D)** *Astrocaryum acaule;* **(E)** *Astrocaryum aculeatum;* **(F)** *Astrocaryum chambira.*

between *D. phoenicocarpus,* and *D. mitis.* The specimens have linear-lanceolate pinnae, like *D. mitis;* inflorescences with densely spiny peduncular bracts, like *D. phoenicocarpus,* and 4–9 short rachillae on a slender rachis, like *D. mitis.* The fruits and endocarp fibers are intermediate between the two species. It seems possible that these plants, and the type of *D. macrodon,* are hybrids. These proposed hybrids occur at the margins of the range of both parents.

5. Desmoncus polyacanthos Mart., Hist. nat. palm. 2: 85. 1824. *Atitara polyacantha* (Mart.) Kuntze, Revis. gen. pl. 2: 726. 1891. Type. Brazil. Bahia: without locality, n.d., *C. Martius s.n.* (holotype, M; F neg. 18597) (Figure 6.52j–l).

Stems 2–15 m long, 0.5–2 cm diam, climbing. *Leaves* 17–26, usually distichous; sheath 18–48 cm long, closed, with a prominent ocrea, ± densely covered with to 2 cm long, black or brown, straight or curved spines with white, bulbous base; petiole 2–7 cm long, petiole, rachis, and cirrus with short recurved spines, especially on lower surface; rachis 40–100 cm long; cirrus 30–42 cm long; pinnae 4–14 per side, lanceolate to elliptic, regularly or irregularly arranged, the middle ones 12–35 cm long, 2.5–6 cm wide. *Inflorescences* interfoliar; peduncle 40–60 cm long, mostly included within the subtending leaf sheath; prophyll 16–20 cm long, inserted ca. halfway up the peduncle; peduncular bract 18–27 cm long, inserted near apex of peduncle, outer surface sparsely to moderately covered with brownish, recurved, 3–6 mm long, white bulbous-based spines; rachis 10–15 cm long; rachillae 5–17, 3–12 cm long; *flowers* borne in triads on proximal part of rachillae, staminate distally; staminate flowers 4.5 mm long (at anthesis); sepals deltate, 1 mm long; petals irregularly obovate, apiculate, 5 mm long; stamens 6; pistillate flowers 4 mm long at anthesis; calyx cupular, to 1.5 mm long; corolla cupular, 2.5 mm long; staminodes obscure; *fruits* ellipsoid to obovoid, 1–2.2 cm long, 0.8–1.8 cm diam, orange or red at maturity.

The name *Desmoncus polyacanthos* var. *angustifolia* Drude in Mart., complete with citation (Fl. bras.: Cyclanthaceae et Palmae I fasc. 85 vol. 3(2): 313. 1881), appears in Glassman (1972), and there is a type (Brazil. Pará: Santarém, Mar 1850, *R. Spruce 627* (holotype, M)), but this name is apparently not published. The type is determined as *Desmoncus nemorosus.*

Specimens from the western Amazon region with large fruits are here separated as a variety, but there are some intermediates.

KEY TO THE VARIETIES OF *DESMONCUS POLYACANTHOS*

1. Fruits 1–1.4 cm long, 0.8–1 cm diam; peduncular bract usually densely spiny; widespread. . . . .
   5a. *Desmoncus polyacanthos* var. *polyacanthos.*
1. Fruits 1.5–2.2 cm long, 1.3–1.8 cm diam; peduncular bract with few spines, or spines absent; Colombia (Amazonas), Ecuador (Napo) and Peru (Huánuco, Loreto, Pasco) . . . . 5b. *Desmoncus polyacanthos* var. *prunifer.*

5a. Desmoncus polyacanthos var. **polyacanthos**

Desmoncus macroacanthos Mart., Hist. nat. palm. 2: 86. 1824. *Atitara macroacantha* (Mart.) Kuntze, Revis. gen. pl. 2: 727. 1891. Type. Brazil. Pará: without locality, n.d., *C. Martius s.n.* (holotype, M; F neg. 18592).

Desmoncus setosus Mart., Hist. nat. palm. 2: 89. 1824. *Atitara setosa* (Mart.) Kuntze, Revis. gen. pl. 2: 727. 1891. Type. Brazil. Amazonas: Rio Japurá near Tefé, n.d., *C. Martius s.n.* (holotype, M; F neg. 18599).

Desmoncus riparius Spruce, J. Linn. Soc., Bot. 11: 156. 1871. *Atitara riparia* (Spruce) Kuntze, Revis. gen. pl. 2: 727. 1891. Type. Venezuela. Amazonas: Río Negro, San Carlos de Río Negro, Oct 1854, *R. Spruce 46* (holotype, K).

Desmoncus oligacanthus Barb. Rodr., Enum. palm. nov. 24. 1875. *Atitara oligacantha* (Barb. Rodr.) Kuntze, Revis . gen. pl. 2: 727. 1891. Lectotype (Glassman, 1972). Barb. Rodr., Sert. palm. brasil. 2: t. 58. 1903.

Desmoncus aereus Drude in Mart., Fl. bras.: Cyclanthaceae et Palmae I fasc. 85 vol. 3(2): 307. 1881. *Atitara aerea* (Drude) Barb. Rodr., Contr. Jard. Bot. Rio de Janeiro 3: 75.

1902. Type. Brazil. Amazonas: Rio Negro, Ayrão, 4 Jul 1874, *J. Trail 1088/LXXXVIII* (holotype, K).

Desmoncus phengophyllus Drude in Mart., Fl. bras.: Cyclanthaceae et Palmae I fasc. 85 vol. 3(2): 314. 1881. *Atitara phenogophylla* (Drude) Kuntze, Revis. gen. pl. 2: 726. 1891. Type. Brazil. Amazonas: Rio Jutahi, 27 Jan 1875, *J. Trail 1075/CCV* (holotype, K).

Desmoncus philippianus Barb. Rodr., Vellosia 1: 38. 1888. *Atitara philippiana* (Barb. Rodr.) Barb. Rodr., Contr. Jard. Bot. Rio de Janeiro 3: 76. 1902. Lectotype (Glassman, 1972). Barb. Rodr., Sert. palm. brasil. 2: t. 61. 1903.

Atitara paraensis Barb. Rodr., Contr. Jard. Bot. Rio de Janeiro 3: 76. 1902. *Desmoncus paraensis* (Barb. Rodr.) Barb. Rodr., Sert. palm. brasil. 2: 57. 1903. Lectotype (Wess. Boer, 1965). Barb. Rodr., Sert. palm. brasil. 2: t. 57. 1903.

Desmoncus ulei Dammer, Verh. Bot. Vereins Prov. Brandenburg 48: 129. 1907. Type. Brazil. Amazonas: Manaus, Feb 1901, *E. Ule 5388* (isotype, MG).

Desmoncus brevisectus Burret, Repert. Spec. Nov. Regni Veg. 36: 215. 1934. Type. Brazil. Pará: without locality, 15 Nov 1932, *W. Hopp 35* (holotype, B, n.v.).

Desmoncus maguirei L. H. Bailey, Bull. Torrey Bot. Club 75: 108. 1948. Type. Surinam. Kwakoegron, 19 Oct 1947, *B. Maguire & G. Stahel 25011* (holotype, NY; isotypes, BH, U).

Desmoncus duidensis Steyerm., Fieldiana Bot. 28: 85. 1951. Type. Venezuela. Amazonas: Caño Negro, SE of Cerro Duida, n.d., *J. Steyermark 57944* (holotype, F, n.v.).

Stems 2–15 m long, 0.5–1.25 cm diam. *Leaves* with pinnae 4–14 per side, the middle ones 12–25 cm long, 2.5–6 cm wide. *Inflorescences* with peduncular bract usually densely spiny; *fruits* 1–1.4 cm long, 0.8–1 cm diam.

East of the Andes in Colombia (Amazonas, Guainía, Meta, Vichada), Venezuela (Amazonas, Anzoategui, Apure, Barinas, Bolívar, Carabobo, Delta Amacuro, Miranda), Trinidad, St. Vincent, the Guianas, Peru (Cusco, Loreto, Madre de Dios), Brazil (Acre, Amazonas, Amapá, Bahia, Goiás, Maranhão, Mato Grosso, Pará, Roraima), and Bolivia (Pando, Santa Cruz) (Figure 6.53B). According to Noblick (1991), it occurs throughout the Atlantic coastal forest of Brazil. It grows in a variety of habitats, often along river banks but also in lowland rain forest, forest gaps, secondary forest, forest margins, and disturbed places, usually at low elevations or occasionally to 1000 m on eastern Andean slopes.

Brazil: *espera-i, jacitara, jassitara, titara.* Colombia: *bejuco alcalde, ca-mu-vé* (Guahibo), *enredadera, yasitara.* Guyana: *kamawarri.* Peru: *pishuallo rojo, vara casha.* Surinam: *bambakka, bamba maka.* Venezuela: *voladora.* The stems are occasionally used to weave baskets and other items.

Two specimens from Brazil (Amazonas), near Tefé (Mun. Maraã, Rio Japurá, Ati Paraná, 6 Nov 1982, *I. Amaral 368* (NY); Mun. Tefé, near mouth of Rio Tefé, 19 Jan 1991, *A. Henderson 1588* (NY)), are unusual. They have slender stems; sheaths densely covered with short straight spines to 0.5 cm long; linear pinnae, to 15 per side, to 20 cm long, 1 cm wide; densely spiny peduncular bracts with short, recurved, 3 mm long, white, bulbous-based spines; and 14–17 rachillae, 2–8 cm long. These specimens match very well, in both morphology and locality, *D. setosus* (although the type is sterile). They are somewhat intermediate between *Desmoncus polyacanthos* var. *polyacanthos* and *D. mitis,* and it is possible that they are hybrids.

5b. Desmoncus polyacanthos var. **prunifer** (Mart.) Henderson, stat. nov. *Desmoncus prunifer* Poepp. ex Mart., Hist. nat. palm. 2: 148. 1837. *Atitara prunifera* (Poepp. ex Mart.) Kuntze, Revis. gen. pl. 2: 727. 1891. Type. Peru. Loreto: Maynas, n.d., *E. Poeppig 2148* (holotype, G, n.v.; F neg. 31323).

Stems to 20 m long, to 2 cm diam, including leaf sheaths. *Leaves* with pinnae 8–11 per side, 20–35 cm long, 3–6 cm wide. *Inflorescences* with peduncular bract with few spines, or spines absent; *fruits* 1.5–2.2 cm long, 1.3–1.8 cm diam.

Colombia (Amazonas), Ecuador (Napo), and Peru (Huánuco, Loreto, Pasco) (Figure 6.53C); lowland rain forest on *terra firme* at low elevations.

28. *Astrocaryum*   **Astrocaryum** G. Mey., Prim. fl. esseq. 265. 1818.

Small to large, densely spiny, monoecious palms. *Stems* solitary or cespitose, short and subterranean or aerial and then moderate to large. *Leaves* pinnate, reduplicate; sheath open and not forming a crownshaft; petiole long; rachis long; pinnae either regularly arranged and spreading in 1 plane and occasionally leaf ± entire, or irregularly arranged in clusters and spreading in different planes, grayish-white abaxially. *Inflorescences* interfoliar, branched to 1 order; peduncle bearing a prophyll and 1 peduncular bract; rachis bearing numerous rachillae, the distal part thickened; *flowers* in triads on proximal part of rachillae, staminate flowers only distally; staminate flowers with sepals 3, briefly connate basally, free above; petals 3, free or connate basally, valvate; stamens (3–)6(–12); pistillode small; pistillate flowers with sepals connate into a 3-lobed, cupular or tubular calyx; petals connate into a 3-lobed, tubular or urceolate corolla; staminodes borne in a ring, or rarely digitate or absent; gynoecium syncarpous, trilocular, triovulate, styles and stigmas usually large; *fruits* 1-seeded, globose to obovoid or irregularly shaped, spinulose or nonspinulose, with apical stigmatic residue, dehiscent or indehiscent; endocarp black, thick, and bony with 3 lateral pores; seed with homogeneous endosperm and lateral embryo; germination adjacent-ligular; eophyll bifid.

Burret (1934b) recognized 45 species, but actually only about 15 exist (see also Kahn & Millán, 1992). They are widely distributed from Mexico south to Bolivia and across Brazil; not in the Antilles. Ten species are found in the Amazon region. One other species, *A. campestre* Mart., from savanna areas of Bolivia and Brazil, may also just reach the southern Amazon region.

The genus is usually divided into two subgenera whose characters are given in couplets number 1 in the following key. Subgenus *Pleiogynanthus* includes six Amazon species. The two tall, solitary-stemmed species have almost non-overlapping ranges, with *Astrocaryum chambira* in the west and *A. aculeatum* in the east. There is some question, however, as to the amount of human influence on these distributions, and the occurrence of *A. chambira* on the Rio Negro should be considered in this context (see Wallace, 1853). A second species pair with nonoverlapping ranges is the eastern *A. vulgare* and the southern *A. huaimi*. They seem closely related and perhaps conspecific. Both grow in drier, more open areas. Of the two other species in this subgenus, *A. jauari* occurs north of the Amazon river along river margins, and is extremely abundant (but not collected) on the Rio Negro; and *A. acaule* is also common on the Rio Negro, but occurs in other places. At least three other extra-Amazonian species of subgenus *Pleiogynanthus* are known: *A. malybo* H. Karst. in the Magdalena valley of Colombia, and the trans-Andean *A. confertum* H. Wendl. ex Burret and *A. standleyanum* L. H. Bailey.

Subgenus *Monogynanthus* includes four Amazon species. Section *Munbaca* contains two species: the widespread *Astrocaryum gynacanthum,* and the northeastern *A. paramaca*. Similarly, section *Ayri* also contains two species: the widespread *A. murumuru,* and the northeastern *A. sciophilum*. However, *A. murumuru* is variable morphologically, and has several, western, peripheral isolates (here treated as varieties). Both *A. paramaca* and *A. sciophilum* appear absent from the dry "corridor" of *Aw* climate that runs northwest–southeast across the lower Amazon (and *A. vulgare* is present there). There are several other species of this subgenus outside the Amazon, notably *A. aculeatissimum*

(Schott) Burret in the Atlantic coastal forest of Brazil; *A. triandrum* Galeano, Bernal & Kahn in the Magdalena valley in Colombia; *A. alatum* Loomis in southern Central America and Pacific Colombia and Ecuador; and *A. mexicanum* Liebm. ex Mart. in northern Central America.

Distribution patterns of the species of both subgenera correspond well with those of some birds (Haffer, 1987b; Cracraft & Prum, 1988).

KEY TO THE SPECIES OF *ASTROCARYUM*

1. Pinnae usually irregularly arranged in clusters, spreading in different planes, strongly reduplicate; peduncular bract deciduous at or before anthesis, its scar near apex of peduncle; pistillate flowers and fruits (or scars) 2–5 (rarely 0–1) per rachilla, the rachillae loosely spaced; fruits usually glabrous or scurfy, indehiscent; subgenus *Pleiogynanthus*.
   2. Stem solitary, short, and subterranean, rarely to 1 m tall; sheath, petiole, rachis, and peduncular bract mottled brown-white-tomentose; Colombia (Guainía, Vaupés, Vichada), Venezuela (Amazonas, Bolívar), the Guianas, and Brazil (Amazonas, Pará, Rondônia) . . . . 1. *A. acaule.*
   2. Stem solitary or cespitose, 5–20 m tall; sheath, petiole, rachis, and peduncular bract not mottled.
      3. Stems large and solitary, 3.5–22 m tall; fruits 4.5–6 cm long, 3.5–4.5 cm diam.
         4. Fruits with glabrous epicarp; mesocarp fleshy; fruiting calyx and corolla lobed; staminodial ring in fruit about half as long as corolla; staminate flowers yellowish; ?Colombia, Venezuela (Amazonas), the Guianas, Brazil (Acre, Amazonas, Mato Grosso, Pará, Rondônia), and Bolivia (Beni, Pando, Santa Cruz) . . . . 2. *A. aculeatum.*
         4. Fruits with whitish-brown-scurfy epicarp with minute spinules; mesocarp fibrous; fruiting calyx fimbriate, corolla with entire margin; staminodial ring in fruit as long as corolla; staminate flowers purplish; Colombia (Meta), Venezuela (Amazonas), Ecuador (Morona-Santiago, Napo), Peru (Amazonas, Loreto), and Brazil (Acre, Amazonas) . . . . 3. *A. chambira.*
      3. Stems moderate to large, cespitose or rarely solitary, 4–13 m tall; fruits 2.5–5 cm long, 1.7–3.7 cm diam.
         5. Fruits 2.5–3.5(–5) cm long, 1.7–2.5(–3) cm diam; stamens 7–9; Colombia (Amazonas, Caquetá, Guaviare, Putumayo), Venezuela (Amazonas, Bolívar), the Guianas, Ecuador (Napo), Peru (Loreto), and Brazil (Acre, Amapá, Amazonas, Pará, Roraima) . . . . 6. *A. jauari.*
         5. Fruits 3.5–5 cm long, 2–3.7 cm diam; stamens 6.
            6. Fruits 3.5–4.5 cm long, 2–3 cm diam; Peru (Madre de Dios), Brazil (Goiás, Mato Grosso), and Bolivia (Beni, La Paz, Santa Cruz). . . . . 5. *A. huaimi.*
            6. Fruits 4–5 cm long, 3–3.7 cm diam; Surinam, French Guiana, and Brazil (Maranhão, Pará, Tocantins) . . . . 10. *A. vulgare.*
1. Pinnae regularly arranged, spreading in 1 plane, not strongly reduplicate but ± flat; peduncular bract ± persistent after anthesis, sheathing the peduncle for ± half its length; pistillate flowers and fruits (or scars) 1 per rachilla, the rachillae densely crowded; fruits usually spiny, dehiscent or indehiscent; subgenus *Monogynanthus*.

7. Fruits 2.5–4 cm long, 1.2–2 cm diam, smooth, or spinulose on "shoulders" only, the epicarp splitting regularly at maturity to expose the orange or yellow mesocarp; pistillate calyx and/or proximal part of rachillae with flexuous black spines.
8. Stems cespitose, 2–6(–12) m tall, 3–6(–10) cm diam; fruits glabrous; Colombia (Amazonas, Vichada), Venezuela (Apure, Amazonas, Bolívar), the Guianas, Brazil (Amapá, Amazonas, Maranhão, Pará, Rondônia), and Bolivia (Pando) . . . . 4. *A. gynacanthum.*
8. Stems solitary, short and subterranean, or rarely to 8 m tall and 11 cm diam; fruits with ring of spinules on "shoulder"; the Guianas and Brazil (Amapá, Amazonas, Pará) . . . . 8. *A. paramaca.*
7. Fruits 3–9 cm long, 2.5–4.5 cm diam, spinulose or scurfy-tomentose, brown at maturity, the epicarp not splitting at maturity, or splitting irregularly to expose the brown mesocarp; pistillate calyx and proximal part of rachillae without flexuous, black spines.
9. Petiole (and rachis) spines not in oblique rows; rachis not contracted, 42–70 cm long; pinnae 1-veined, not strongly plicate; widespread . . 7. *A. murumuru.*
9. Petiole (and rachis) spines in oblique rows; rachis contracted, 11–15 cm long; pinnae several-veined, plicate; the Guianas and Brazil (Amapá, Amazonas) . . . . 9. *A. sciophilum.*

1. Astrocaryum acaule Mart., Hist. nat. palm. 2: 78. 1824. Type. Brazil. Amazonas: without locality, n.d., *C. Martius s.n.* (holotype, M) (Figure 6.54a–b).

Astrocaryum giganteum Barb. Rodr., Contr. Jard. Bot. Rio de Janeiro 3: 82. 1902. Lectotype (Glassman, 1972). Barb. Rodr., Contr. Jard. Bot. Rio de Janeiro 3: t. 10C. 1902.

Astrocaryum luetzelburgii Burret, Notizbl. Bot. Gart. Berlin-Dahlem 10: 1021. 1930. Type. Brazil. Amazonas: Rio Uaupés, Jutica, 16 Nov 1928, *P. Luetzelburg 23045* (isotypes, M, R; F neg. 18571a).

Astrocaryum huebneri Burret, Repert. Spec. Nov. Regni Veg. 35: 128. 1934. Type. Brazil. Amazonas: near Manaus, n.d., *G. Huebner 8* (holotype, B, n.v.).

Stems short and subterranean, rarely to 1 m tall. *Leaves* 5–9; sheath 0.8–1 m long, mostly subterranean, sheath, petiole, and rachis mottled brownish-white-tomentose, with dense to moderate covering of flattened black spines to 9 cm long; petiole 0.9–3 m long; rachis 1.7–4 m long; pinnae 55–103 per side, linear, irregularly arranged in clusters of 3–7, spreading in different planes, strongly folded, the middle ones 30–90 cm long, 2–2.5 cm wide. *Inflorescences* interfoliar, erect at anthesis and in fruit; peduncle 0.4–1.3 m long; prophyll ca. 25 cm long; peduncular bract 37–46 cm long, inserted near apex of peduncle, moderately to densely covered with black spines, mottled brownish-white-tomentose; rachis 10–25 cm long; rachillae numerous, ca. 14 cm long, densely whitish-brown-tomentose, glabrescent; staminate flowers solitary, borne close together in pits in the distal part of the rachillae, 3.5 mm long at anthesis; sepals triangular, 0.7 mm long; petals ovate, 2.5 mm long; stamens 6, the filaments inflexed at the apex; pistillode very small; pistillate flowers 2–3 at base of each rachillae, 1 cm long at anthesis; calyx 3-lobed, 6 mm long; corolla 3-lobed, 7 mm long; staminodial tube connate to the corolla; *fruits* ± obovoid, indehiscent, glabrous, 2.5–3 cm long, 1.5–2 cm diam, green at first, becoming yellow-green, yellow-orange, and finally orange at maturity.

Amazon region and peripheral areas in Colombia (Guainía, Meta, Vaupés, Vichada), Venezuela (Amazonas, Bolívar), the Guianas, and Brazil (Amazonas, Pará, Rondônia) (Figure 6.53D). It is especially common in the upper Rio Negro region. It grows in open areas, stream banks, river islands, savanna margins, rocky outcrops, road margins or other disturbed areas, or rarely forest, often on sandy soils at low elevations.

Brazil: *tucuma-í.* Colombia: *corocito, espina, ma-tav-icú-li* (Guahibo). Guyana: *unabai* (Wap-

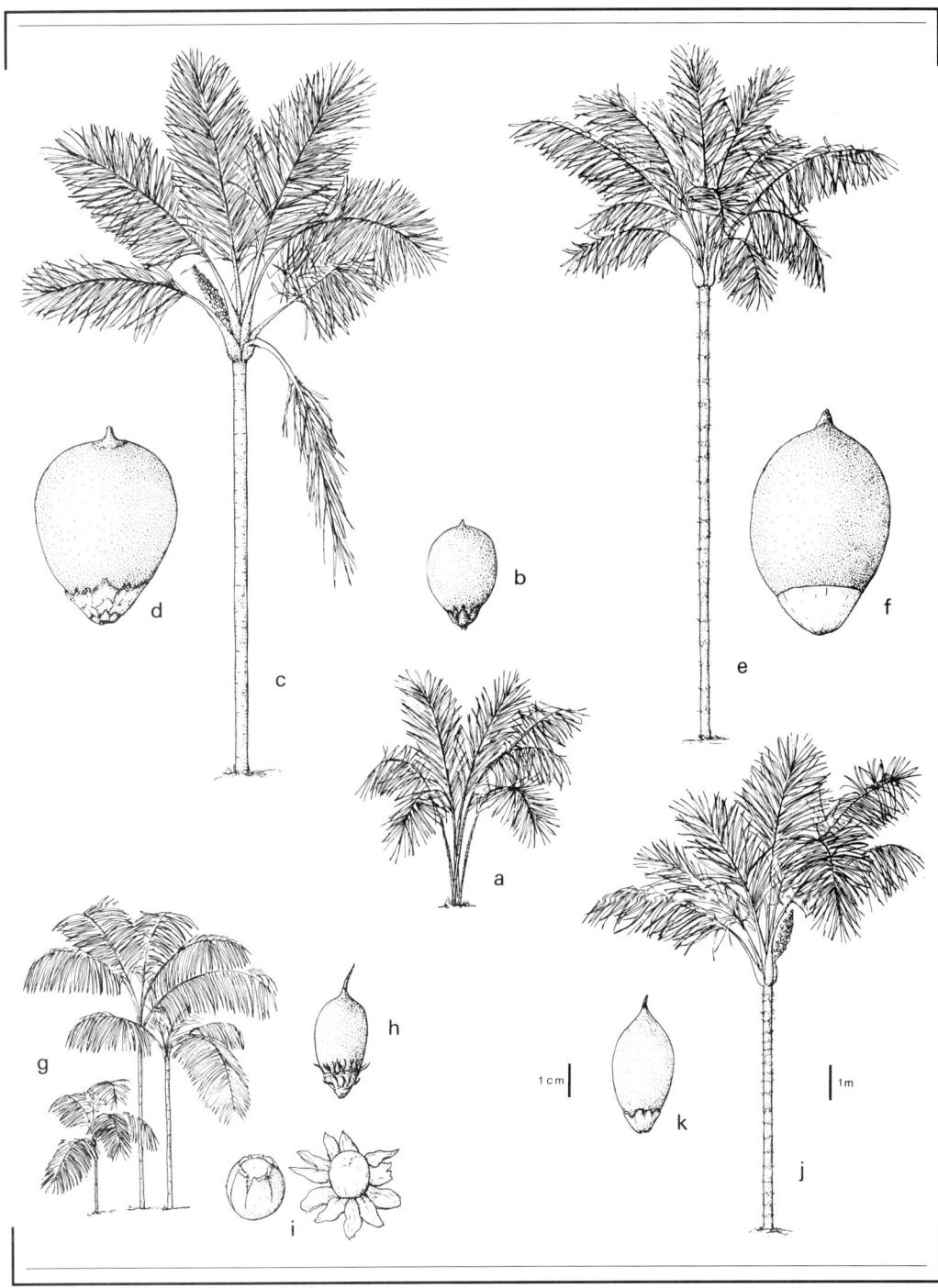

Figure 6.54. **(a)** *Astrocaryum acaule,* **(b)** fruit (from *J. Poole 2114*); **(c)** *Astrocaryum aculeatum,* **(d)** fruit (from *B. Boom 4159*); **(e)** *Astrocaryum chambira,* **(f)** fruit from (*S. Mori 14683*); **(g)** *Astrocaryum gynacanthum,* **(h)** fruit (from *J. Steyermark 125695*), **(i)** epicarp splitting; **(j)** *Astrocaryum huaimi,* **(k)** fruit (from *M. Nee 40133*).

isiana). Venezuela: *corozo, quidíja*. The ripe fruits are edible, and fibers are occasionally extracted from the leaves.

2. Astrocaryum aculeatum G. Mey., Prim. fl. esseq. 266. 1818. Type. Guyana. Essequibo River, n.d., *E. Rodschied s.n.* (holotype, GOET, n.v.) (Figure 6.54c–d; Plate IVd).

Astrocaryum tucuma Mart., Hist. nat. palm. 2: 77. 1824. Type. Brazil. Pará: without locality, n.d., *C. Martius s.n.* (holotype, M; F neg. 18575).

Astrocaryum princeps Barb. Rodr., Enum. palm. nov. 22. 1875. Lectotype (Wess. Boer, 1965). Barb. Rodr., Sert. palm. brasil. 2: t. 72. 1903.

Astrocaryum caudescens Barb. Rodr., Enum. palm. nov. 22. 1875. Lectotype (Glassman, 1972). Barb. Rodr., Sert. palm. bras. 2: t. 66. 1903.

Astrocaryum princeps var. *sulphureum* Barb. Rodr., Vellosia 1: 48. 1888. Lectotype (Glassman, 1972). Barb. Rodr., Vellosia 1: t. 80B. 1888.

Astrocaryum princeps var. *aurantiacum* Barb. Rodr., Vellosia 1: 49. 1888. Lectotype (Glassman, 1972). Barb. Rodr., Vellosia 1: t. 79B. 1888.

Astrocaryum princeps var. *flavum* Barb. Rodr., Vellosia 1: 50. 1888. Lectotype (Glassman, 1972). Barb. Rodr., Vellosia 1: t. 80C. 1888.

Astrocaryum princeps var. *vitellinum* Barb. Rodr., Vellosia 1: 48. 1888. Lectotype (Glassman, 1972). Barb. Rodr., Vellosia 1: t. 79C. 1888.

Astrocaryum manaoense Barb. Rodr., Vellosia 1 (ed. 2): 105. 1891. Lectotype (Glassman, 1972). Barb. Rodr., Vellosia 4 (ed. 2): t. 1. 1891.

Astrocaryum macrocarpum Huber, Bull. Herb. Boissier 6: 271. 1906. Type not designated.

Stems solitary, erect, 8–20 m tall, 12–25(–40) cm diam, the internodes covered with black spines to 15 cm long. *Leaves* 6–15, stiffly ascending; sheath and petiole, 1.8–3.7 m long, sheath, petiole and rachis covered with long, somewhat flattened, black or gray spines to 10 cm long; rachis 1.4–6.4 m long; pinnae 73–130 per side, linear, irregularly arranged in obscure or obvious clusters of 2–5, spreading in different planes, the middle ones 1–1.4 m long, 4–6 cm wide. *Inflorescences* interfoliar, erect at anthesis and in fruit; peduncle 0.3–0.7 m long; prophyll to 45 cm long, persistent; peduncular bract 1.2–2.2 m long, inserted near apex of peduncle, densely black or brown spiny abaxially; rachis 0.8–1 m long; rachillae 194–278, to 26 cm long, proximal ca. half with pistillate flowers, distal ca. half with staminate flowers, glabrous or brownish-white-tomentose, laxly spreading at anthesis; *staminate flowers* to 5 mm long, yellowish, paired or solitary in crowded pits in the rachillae; sepals narrowly triangular, 1 mm long; petals briefly connate basally, 4 mm long; stamens 6, inflexed apically; pistillode deeply trifid, 1 mm long; pistillate flowers 1.5 cm long, 2–4 per rachillae; calyx cupular, 1 cm long; corolla cupular, 1 cm long; staminodial ring 2 mm high; *fruits* globose or obovoid, 4.5–6 cm long, 3.5–4.2 cm diam, yellow-orange or orange-green, glabrous; mesocarp fleshy; calyx 3-lobed, corolla several-lobed, staminodial ring about half as long as the corolla.

Central and eastern Amazon region in ?Colombia, Venezuela (Amazonas), Trinidad, the Guianas, Brazil (Acre, Amazonas, Mato Grosso, Pará, Rondônia), and Bolivia (Beni, Pando, Santa Cruz) (Figure 6.53E). Although Wessels Boer (1965) stated that this species is found only near past or present human settlements, and other authors have followed this, this is not always the case. This species is found sporadically in apparently undisturbed forest (e.g., Peres, in press). It is also common at the boundaries of savannas and forest—for example, on the Ilha do Maracá, Roraima, Brazil.

Bolivia: *chonta, panima* (Chácabo), *tucum*. Brazil: *tucumã*. Guyana: *awara*. Surinam: *cemau*. Venezuela: *tucuma*. The most useful part of this species is the fruits. The mesocarp is edible and is highly prized, and fruits are sold in markets in Manaus. The endosperm is also eaten.

3. Astrocaryum chambira Burret, Repert. Spec. Nov. Regni Veg. 35: 122. 1934. Type. Peru. Loreto: Iquitos, 7 Apr 1925, *G. Tessmann 5079* (holotype, B) (Figure 6.54e–f).

Astrocaryum vulgare of Wallace, not Mart., Palm Trees of the Amazon 105. 1853.

Stems solitary, erect, 3.5–22 m tall, 19–35 cm diam, with a mound of roots at the base, the internodes covered with gray or black spines to 20 cm long. *Leaves* 8–15, stiffly ascending; sheath and petiole 2.9–5 m long, sheath, petiole, and rachis densely covered with somewhat flattened, 3–15 cm long, yellowish-brown or gray spines; rachis 3.9–5.5 m long; pinnae 125–175 per side, linear, irregularly arranged in obscure to obvious clusters, spreading in different planes, the middle ones 1.2–1.4 m long, 3.5–5 cm wide. *Inflorescences* interfoliar, erect at anthesis and in fruit; peduncle 1.2–2.5 m long; prophyll 0.8–1.2 m long, persistent; peduncular bract 0.9–1.9 m long, inserted near apex of peduncle, densely black or brown spiny abaxially; rachis 0.8–1.5 m long; rachillae 150–300, 23–50 cm long, proximal ca. half with pistillate flowers, distal ca. half with staminate flowers, glabrous or whitish-brown-tomentose, laxly spreading at anthesis; *staminate flowers* ca. 6 mm long (in bud), purple, paired or solitary in crowded pits in the rachillae; sepals narrowly triangular, 2 mm long; petals briefly connate basally, obovate, 5 mm long; stamens 6, the filaments flattened, not inflexed apically; pistillode deeply trifid, 1 mm long; pistillate flowers ca. 2 cm long, 2–4 (rarely 0–1) per rachillae; calyx cupular, 1.5 cm long; corolla cupular, inturned at the apex, with a moderate covering of spinules, 1.3 cm long; staminodial tube as long as and adnate to the corolla; *fruits* obovoid, 5–6 cm long, 4–4.5 cm diam; epicarp yellowish-green, whitish-brown-scurfy with minute spinules; mesocarp fibrous; the persistent calyx fimbriate, the corolla with entire apical margin, with the staminodial ring as long as the corolla.

Western Amazon region in Colombia (Amazonas, Meta, Putumayo), Venezuela (Amazonas), Ecuador (Morona-Santiago, Napo), Peru (Amazonas, Loreto), and Brazil (Acre, Amazonas) (Figure 6.53F); lowland rain forest on *terra firme* at low elevations. It is also found in disturbed areas—for example, near Iquitos. Wallace (1853) reported that this species was planted by Indians on the upper Rio Negro in Brazil in areas where it did not occur naturally.

Brazil: *tucúm, tucuma*. Colombia: *corombolo, cumare, palma coco, palma de cumare, takone* (Andoke). Ecuador: *chambira, coco de mono, kumri* (Shuar), *tuinfa* (Cofán). Peru: *batái, chambira* (Quechua), *mataa* (Achual Jívaro). Venezuela: *cumare*. The leaf fibers of this species are commonly used to weave various articles. The youngest, unexpanded leaves are removed from the tree and the fibers pulled away from the pinnae margins. After drying, they are twisted into string, which is used for fishing lines and to weave hammocks, fishing nets, and bags. The mid-veins of the pinnae are used to make brooms. The liquid endosperm of the fruits is drunk. The leaves are occasionally used for thatching.

4. Astrocaryum gynacanthum Mart., Hist. nat. palm. 2: 73. 1824. Type. Brazil. Amazonas: Manaus, n.d., *C. Martius n.d.* (holotype, M; F neg. 18570) (Figure 6.54g–i; Plate IIIa).

Astrocaryum munbaca Mart., Hist. nat. palm. 2: 74. 1824. *Astrocaryum gynacanthum* var. *munbaca* Trail, J. Bot. 6: 78. 1877. Type. Brazil. Pará: without locality, n.d., *C. Martius 3280* (holotype, M; F; negs. 18573, 18573a).

Astrocaryum minus Trail, J. Bot. 6: 78. 1877. *Astrocaryum rodriguesii* var. *minus* (Trail) Barb. Rodr., Protesto-Appendice 28. 1879. Type. Brazil. Amazonas: Barreiras de Mutum, Rio Jutaí, 2 Feb 1875, *J. Trail 1071/CCXIII* (holotype, K).

Astrocaryum minus var. *terrafirme* Drude in Mart., Fl. bras.: Cyclanthaceae et Palmae I fasc. 85 vol. 3(2): 374. 1881. Type. Brazil. Amazonas: Maués, 1861, *G. Wallis s.n.* (holotype, B, n.v.).

Astrocaryum vulgare of Warburg, not Mart., Pflanzenwelt. 3: 409. 1922.

Astrocaryum gymnopus Burret, Notizbl. Bot. Gart. Berlin-Dahlem 10: 1020. 1930. Type. Venezuela. Amazonas: Esmeralda, 5 Oct 1928, *P. Luetzelburg 22841* (isotypes, M, R; F neg. 18569a).

Astrocaryum gynacanthum var. *dasychaetum* Burret, Repert. Spec. Nov. Regni Veg. 35: 141. 1934. Type. Brazil. Amazonas: Livramento, Rio Madeira, Marmellos, n.d., *W. Hopp 1152* (holotype, B, n.v.).

Stems cespitose or occasionally solitary, 2–6(–12) m tall, 3–6(–10) cm diam, with rings of flattened black spines to 12 cm long on the

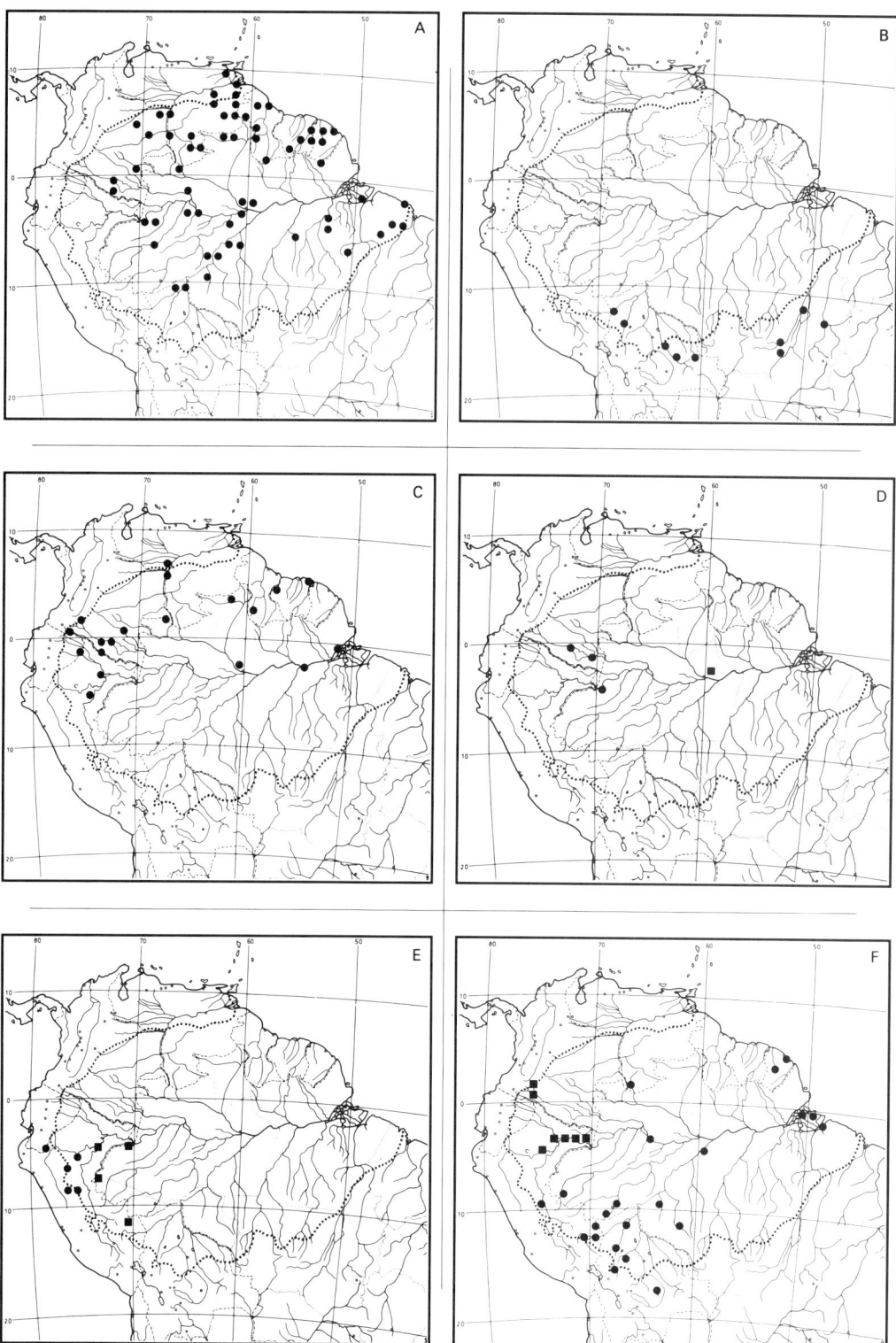

Figure 6.55. **(A)** *Astrocaryum gynacanthum;* **(B)** *Astrocaryum huaimi;* **(C)** *Astrocaryum jauari;* **(D)** *Astrocaryum murumuru* var. *ciliatum* (circles), var. *ferrugineum* (squares); **(E)** *Astrocaryum murumuru* var. *huicungo* (circles), var. *javarense* (squares); **(F)** *Astrocaryum murumuru* var. *murumuru* (circles), var. *macrocalyx* (squares).

internodes. *Leaves* 6–13, horizontally spreading; sheath partly closed, 15–80 cm long, sheath, petiole, and rachis densely to moderately covered with flattened black spines to 14 cm long, these tending to be in clusters; petiole 20–60 cm long; rachis 1.2–2.5 m long; pinnae 21–40 per side, linear, regularly arranged and spreading in 1 plane, the apical one much broader than the others and occasionally with a wider one interspersed among the others, flat, not bifid at the apex, the middle ones 54–60 cm long, 2.5–6 cm wide. *Inflorescences* interfoliar, pendulous at anthesis and in fruit; peduncle 35–80 cm long, very spiny; prophyll 30–40 cm long; peduncular bract 60–80 cm long, densely black spiny on abaxial surface, inserted ca. halfway along peduncle, both bracts persistent; rachis 10–25 cm long; rachillae numerous, ca. 4 cm long, the staminate part deciduous after anthesis; staminate flowers densely crowded along rachillae, borne singly in pits, 3 mm long; sepals narrowly triangular, 1 mm long; petals connate basally for ca. half their length, reflexed at the apex at anthesis, obovate, 3 mm long; stamens 6; filaments briefly connate basally, free above, not inflexed at apex; pistillode prominent, trifid; pistillate flowers solitary and ± sessile on rachis, 1 per rachillae, 1 cm long; calyx cupular, 4 mm long, densely covered on outer surface with persistent, black, flexuous, flattened spinules to 8 mm long; corolla cupular, apical margins smooth, 3 mm long; staminodial ring adnate to the corolla, 1 mm high; *fruits* obovoid, densely clustered on rachis, 2.5–3.5 cm long, 1.2–1.5 cm diam, epicarp bright orange at maturity, splitting ± regularly to expose the orange mesocarp and black endocarp.

Central and eastern Amazon region of Colombia (Amazonas, Guainía, Vaupés, Vichada), Venezuela (Apure, Amazonas, Bolívar, Monagas), the Guianas, Brazil (Acre, Amapá, Amazonas, Maranhão, Pará, Rondônia), and Bolivia (Pando) (Figure 6.55A); very common and abundant in lowland rain forest on *terra firme*, generally at low elevations, seldom reaching 850 m.

Brazil: *ju* (Ka'apor), *jupihu* (Ka'apor), *marajá, maraju, marayua* (Guajajara), *munbaca, munbaca da preta, spinho preto, yu-'y* (Guajá). Colombia: *cubarra, rui-re'-gö* (Huitoto). French Guiana: *counana agouti, ti-wara* (Creole). Guyana: *urishi, wulo*. Surinam: *awëke* (Trio), *highland kiskismaka*. Venezuela: *amaint-nak* (Shiriana), *cubarro, devéke*. The fruits are edible.

A palm notorious for its long, sharp spines, especially those on the stem. Spruce (1871) noted that the local name in Brazil, *munbaca*, came from the Tupi word meaning "awakeners," in reference to the spines. Spruce also wrote that he once had a *munbaca* spine in his finger joint, and even 16 years later his finger would occasionally be paralyzed.

5. Astrocaryum huaimi Mart. in A. D. Orb., Voy. Amérique mér. 7(3) Palmiers 86. 1847. Type. Bolivia. Beni: Ascensio de Guayaros, n.d., *A. d'Orbigny 23* (holotype, P, n.v.) (Figure 6.54j–k).

Stems cespitose or solitary, 3–7 m tall, 5–15 cm diam, densely covered with long, flat, black spines. *Leaves* 6–12, erect; sheath 30–35 cm long, sheath, petiole, and rachis whiteish-tomentose, with dense to moderate covering of flattened black spines to 7 cm long; petiole 1.4–2 m long; rachis 2.2–2.5 m long; pinnae 50–72 per side, irregularly arranged in clusters and spreading in different planes, linear, briefly bifid at the apex, the middle ones 58–77 cm long and 1.5–3 cm wide, spiny on the margins. *Inflorescences* interfoliar, erect; peduncle to 60 cm long; prophyll 45–54 cm long; peduncular bract 70–84 cm long, mottled brown with few black spines; rachis 46–51 cm long; rachillae numerous, to 72, to 20 cm long; *staminate flowers* 6–7 mm long, densely crowded on apical part of rachillae; sepals connate into a 3-lobed cupule, the lobes acute, 2 mm long; petals connate basally, free above, recurved at anthesis, obovate, 6 mm long; stamens 6; pistillode trifid, 1.5 mm long; pistillate flowers 10 mm long, 2–3 per rachillae; calyx tubular, 7 mm long; corolla urceolate, 7 mm long; staminodial ring 3 mm high; *fruits* ellipsoid, 3.5–4.5 cm long, 2–3 cm diam, glabrous, orange or orange-yellow.

Peru (Madre de Dios), Brazil (Goiás, Mato Grosso), and Bolivia (Beni, La Paz, Santa Cruz)(Figure 6.55B); forest islands in savannas (e.g., Pampas del Heath in Madre de Dios, Peru), semideciduous forest, or rocky places, at low elevations.

Bolivia: *chontilla*.

A very poorly known species, perhaps not distinct from *A. vulgare*, and of which *Astrocaryum leiospatha* Barb. Rodr. is probably a synonym.

6. Astrocaryum jauari Mart., Hist. nat. palm. 2: 76. 1824. Type not designated (Figure 6.56a–b; Plate lc).

Astrocaryum guara Burret, Notizbl. Bot. Gart. Berlin-Dahlem 11: 15. 1930. Type. Colombia. Caquetá: Getuchá, Río Orteguaza, 21 Jul 1926, *G. Woronow & S. Juzepczuk 6121* (holotype, B, n.v.).

Stems cespitose or less often solitary, usually leaning, 5–13 m tall, 9–30 cm diam, with flattened black spines 10–14 cm long on internodes, with a mound of roots at base. *Leaves* 6–10, stiffly ascending; sheath 0.5–1.7 m long, closed at base, sheath, petiole, and rachis moderately to densely covered with flattened black or gray spines; petiole 0.6–1.5 cm long; rachis 1.5–2.6 m long; pinnae 56–148 per side, irregularly arranged in clusters of 3–7, spreading in different planes, linear, bifid at apex, the middle ones 0.6–1.1 m long, 2–3 cm wide. *Inflorescences* interfoliar, erect at anthesis and in fruit; peduncle 65–115 cm long; prophyll 40–75 cm long, persistent; peduncular bract ca. 1.1 m long, inserted near the apex of the peduncle, deciduous at anthesis; rachis 50–85 cm long; rachillae 49–85, 15–60 cm long, with triads proximally and staminate flowers distally, the staminate part densely reddish-brown-tomentose; *staminate flowers* 4.5 mm long (in bud), paired or solitary in pits on distal part of rachillae; sepals free, narrowly triangular, 3 mm long; petals very briefly connate basally, valvate above, reflexed at anthesis, 4.5 mm long; stamens 7–9, the filaments inflexed apically; pistillode trifid, 0.5 mm long; pistillate flowers 7 mm long, 3–6 on basal part of rachillae; sepals forming a cupular calyx, 3 mm long, apical margins ciliate, 3-lobed; petals forming a cupular corolla, 3 mm long, apically ciliate; staminodial ring free from the corolla, 2 mm high; *fruits* globose, globose-ellipsoid, or obovoid, 2.5–3.5(–5) cm long, 1.7–2.5(–3) cm diam, glabrous, green at first, becoming yellowish-green or orange at maturity, indehiscent.

Throughout the Amazon region in Colombia (Amazonas, Caquetá, Guaviare, Putumayo), Venezuela (Amazonas, Bolívar, Anzoátegui, Apure), the Guianas, Ecuador (Napo), Peru (Loreto), and Brazil (Acre, Amapá, Amazonas, Pará, Roraima) (Figure 6.55C); very abundant and common along forested margins of larger rivers, occasionally in low-lying, open, seasonally inundated places away from rivers. The stems are usually cespitose but can be solitary, especially in peripheral parts of its range (e.g., Ecuador).

Piedade (1985) studied the ecology of *A. jauari* in the Arquipélago das Anavilhanas on the Rio Negro in Brazil. She found that the reproductive cycle of the palm was closely synchronized with the rise and fall of the Rio Negro. Palms were very abundant along a narrow zone on the edges of river channels and lakes. They flowered in the dry season, in August and September, as the level of the river was falling. Approximately 9 months later, in May and June, toward the end of the rainy season when the river was at its highest level, fruits were mature and began falling from the trees. Many fruits were destroyed by predators. Those that fell on the ground were attacked by beetles and fungi. Those that fell in the water were eaten by at least 16 different species of fish, especially the large *tambaqui* (*Colossoma macropomum*) and *pirapitinga* (*Colossoma bidens*). These fish have very powerful jaws and are able to crack the endocarps and eat the seeds. However, some seeds were dispersed by fish. As the waters began to fall, in August, surviving seeds germinated.

Brazil: *jauarí*. Colombia: *ko-rü'-ne* (Huitoto), *tupí, yavarí*. Ecuador: *huiririma, oco-be-to* (Siona). Guyana: *sauarai* (Wapisiana). Peru: *chambirilla, huiririma* (Quechua). Surinam: *liba awara*. Venezuela: *albarico*. The leaf rachis is used for weaving; and the endocarps are used for necklaces. The fruits are eaten by fish and are used by fishermen as bait (especially the endosperm). There is currently a commercial palm heart–canning business based on *A. jauari* in Barcellos in Brazil. The hearts come from the large populations of palms that occur on the Rio Negro.

7. Astrocaryum murumuru Mart., Hist. nat. palm. 2: 70. 1824. Type. Brazil. Pará: without

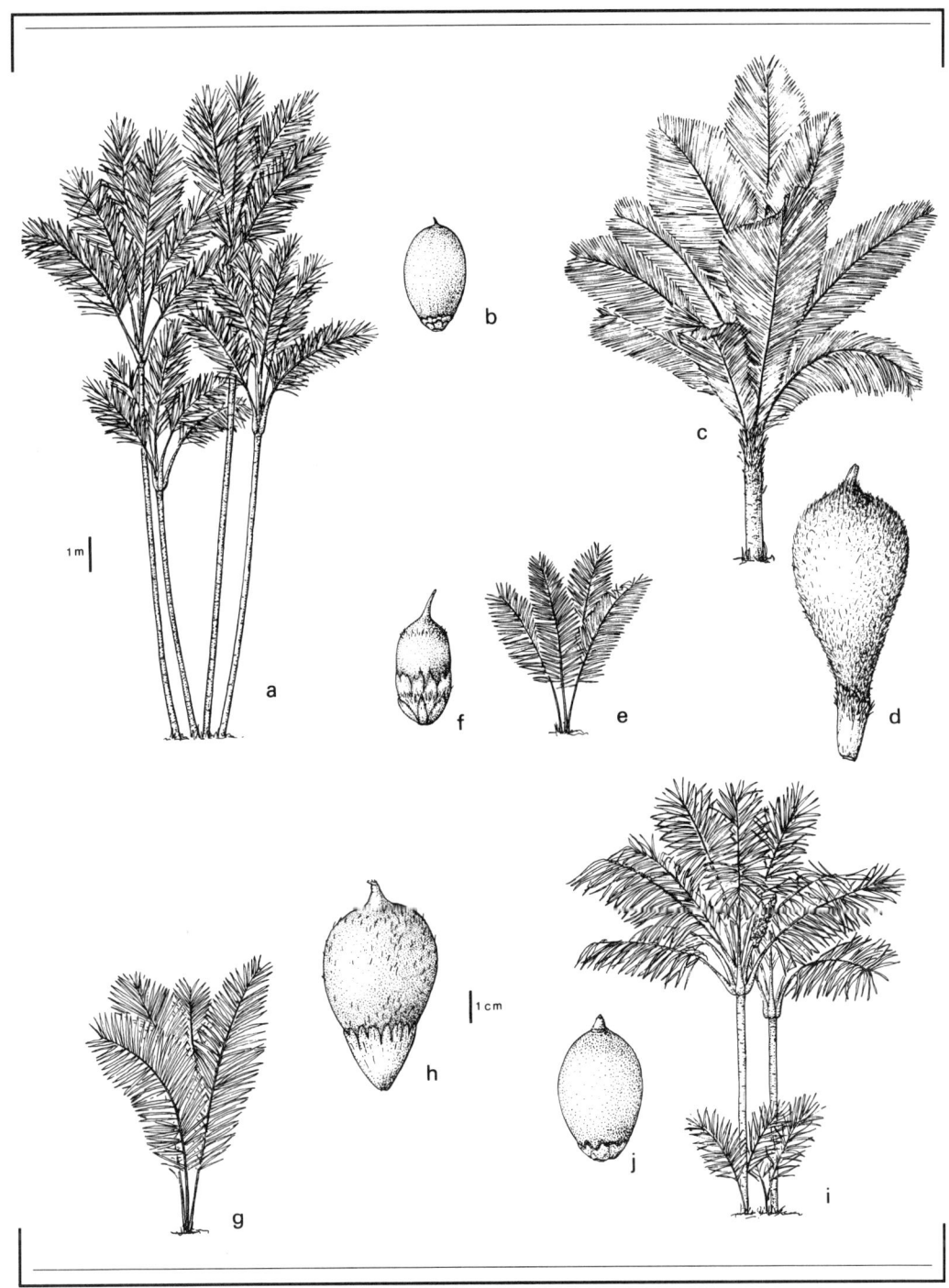

Figure 6.56. **(a)** *Astrocaryum jauari,* **(b)** fruit (from *H. Balslev 4313*); **(c)** *Astrocaryum murumuru* var. *murumuru,* **(d)** *Astrocaryum murumuru* var. *huicungo,* fruit (from *E. Killip 28840*); **(e)** *Astrocaryum paramacca,* **(f)** fruit (from *J. Wessels Boer 1226*); **(g)** *Astrocaryum sciophilum,* **(h)** fruit (from *M. Jansen-Jacobs 2390*); **(i)** *Astrocaryum vulgare,* **(j)** fruit (from *M. Balick 1626*).

locality, n.d., *C. Martius s.n.* (holotype, M; F negs. 18572, 18572a) (Figure 6.56c–d).

Stems solitary or cespitose, short and subterranean or 1.5–15 m tall, 10–30 cm diam, at least partly covered with persistent, decaying, spiny leaf sheaths. *Leaves* 6–15(–25), spreading; sheath 1–1.7 m long, sheath, petiole, and rachis moderately to densely covered with flattened black spines to 30 cm long, these tending to be clustered; petiole 1–2(–5) m long; rachis 4–6 m long; pinnae (38–)92–133 per side, regularly arranged and stiffly spreading in 1 plane, linear, briefly bifid at apex, the middle ones 1–1.2 m long, 4.5–7 cm wide, usually without prominent lateral veins. *Inflorescences* interfoliar, erect at anthesis and in fruit; peduncle (0.6–)1–1.7 m long; prophyll 65–88 cm long; peduncular bract 0.8–1.5 m long, inserted on upper half of peduncle, persistent, densely velutinous with longer spines abaxially; rachis 42–70 cm long; rachillae 100–700, 13–19 cm long; *staminate flowers* 2.5–3 mm long, borne singly in crowded pits along distal part of rachillae; sepals triangular, 0.5–0.7 mm long; petals free, valvate, oblong, 2 mm long, thickened and curved over at apex; stamens 6, anthers slightly inflexed at apex; pistillode very small and obscure; pistillate flowers 1–1.4 cm long, usually sessile on rachis at base of each rachilla; calyx 3-lobed, cupular or tubular, 6–14 mm long, glabrous or spinulose; corolla scarcely 3-lobed, cupular or tubular, 6–14 mm long, usually with few to many spinules; staminodial ring usually adnate to corolla, occasionally reduced to 6 small teeth; *fruits* densely crowded on rachis, obovoid, oblong-obovoid, or elongate-obovoid but misshapen by mutual pressure, 3.5–9 cm long, 2.5–4.5 cm wide, scurfy-tomentose or densely to moderately covered with black or brown black spinules, the epicarp sometimes splitting irregularly and longitudinally at maturity to expose the mesocarp; mesocarp fleshy or fibrous.

Colombia (Amazonas, Caquetá, Putumayo), Venezuela (Amazonas), the Guianas, Ecuador (Morona-Santiago, Napo), Peru (Amazonas, Loreto, Madre de Dios, Pasco, San Martín), Brazil (Acre, Amazonas, Pará, Rondônia), and Bolivia (Beni, Pando, Santa Cruz); typically in lowland rain forest in inundated areas, floodplain forest, near rivers or in swamps, at low elevations, but also on *terra firme* on eastern Andean slopes to 900 m elevation.

Bolivia: *chonta, chonta loro, pani* (Chácabo). Brazil: *mormuru, murumuru*. Colombia: *chuchana*. Ecuador: *chuchana, iika* (Waorani), *sirá* (Siona), *usahua* (Quichua). French Guiana: *mourou-mourou*. Peru: *chonta, huicungo, masanke*. Venezuela: *orocori*. The trunks are sometimes used in house construction as corner poles; the leaves are occasionally used for thatching; the liquid endosperm of young fruits is drunk; the fruits are used as food for domestic animals; the mesocarp of some varieties is eaten; the palmito is occasionally eaten; and the endocarps are used in necklaces.

Astrocaryum murumuru, as here understood, is a widespread, variable species, occurring throughout the Amazon region and containing various local varieties. Kahn and Millán (1992) have studied this variation in some detail and have concluded that 13 different species exist. The treatment given here differs mostly in ranking. Since all plants are practically indistinguishable vegetatively, and since Kahn and Millán's scheme depends on pistillate flowers at anthesis for identification, I prefer to recognize various varieties, on the basis of morphology and geography.

This species is of interest in that it drops its ripe fruits en masse over a very short period. The endosperm is then an important food source for any terrestrial animal that can crack open, or bore through, the thick endocarp, such as peccaries, agoutis, some monkeys, and bruchid beetles. Other animals may eat the mesocarp and also act as dispersers, such as deer and tapirs.

KEY TO THE VARIETIES OF
ASTROCARYUM MURUMURU

1. Pistillate flowers with spinulose calyx as long as or almost as long as the spinulose corolla.
    2. Pinnae with reddish-brown hairs abaxially; Brazil (central Amazonas). . . . .
        . . . 7b. *A. murumuru* var. *ferrugineum*.
    2. Pinnae without reddish-brown hairs abaxially.
        3. Fruits with corolla margins entire and spinulose; Colombia (Amazonas)
            . . . 7a. *A. murumuru* var. *ciliatum*.

3. Fruits with lobed, nonspinulose corolla margin.
   4. Stems solitary; fruits 3.5–6.5 cm long; mesocarp dry; Peru (Loreto, Madre de Dios) and Brazil (Acre, western Amazonas) . . . .
   . . 7d. *A. murumuru* var. *javarense*.
   4. Stems cespitose; fruits 6.5–7.5 cm; mesocarp fleshy; long; Peru (Amazonas, western Loreto, San Martín) . . . .
   . . 7c. *A. murumuru* var. *huicungo*.
1. Pistillate flowers with glabrous calyx as long as or shorter than the spinulose corolla.
   5. Fruits scurfy-tomentose, not densely spinulose; Venezuela (Amazonas), the Guianas, Peru (Huánuco, Loreto, Madre de Dios), Brazil (Acre, Amazonas, Pará, Rondônia), and Bolivia (Beni, La Paz, Pando, Santa Cruz). . . . .
   . . . . 7f. *A. murumuru* var. *murumuru*.
   5. Fruits not scurfy-tomentose, usually densely spinulose; western Amazon region.
   6. Fruits greatly narrowed basally; Peru (Pasco). . . . .
   7g. *A. murumuru* var. *perangustatum*.
   6. Fruits not greatly narrowed basally.
   7. Stems cespitose; Ecuador (Morona-Santiago, Napo) and Peru (Loreto) . . . .
   . 7h. *A. murumuru* var. *urostachys*.
   7. Stems solitary; Colombia (Amazonas, Caquetá, Putumayo) and Peru (Loreto) . . . .
   . 7e. *A. murumuru* var. *macrocalyx*.

7a. Astrocaryum murumuru var. **ciliatum**
(Kahn & Millán) Henderson, stat. nov. *Astrocaryum ciliatum* Kahn & Millán, Bull. Inst. fr. études andines 21: 497. 1992. Type. Colombia. Amazonas: Araracuara, 24 Sep 1987, *G. Galeano & J. Huitoto 1315* (isotype, NY).

Stems solitary, short and subterranean or to 5 m tall, to 10 cm diam. Pistillate flowers with spinulose calyx as long as the corolla, this spinulose at the apex; *fruits* obovoid, 4.5–5 cm long, 2.5–3 cm diam, moderately spinulose, with the corolla margin entire and spinulose.

Colombia (Amazonas) (Figure 6.55D); lowland rain forest on *terra firme*.

Pinnae venation is somewhat different from other varieties.

7b. Astrocaryum murumuru var. **ferrugineum**
(Kahn & Millán) Henderson, stat. nov. *Astrocaryum ferrugineum* Kahn & Millán, Bull. Inst. fr. études andines 21: 497. 1992. Type. Brazil. Amazonas: Reserva Florestal Adolfo Ducke, 26 km NE of Manaus on road to Itacoatiara, 10–21 Nov 1986, *A. Henderson 674* (holotype, INPA; isotype, NY).

Stems solitary, to 4 m tall, to 25 cm diam. Pinnae with long reddish-brown hairs abaxially. Pistillate flowers with spinulose calyx as long as the spinulose corolla; *fruits* elongate-obovoid, 4.5–5 cm long, 3–3.5 cm diam, densely spinulose.

Known only from a small area near Manaus, Brazil (Amazonas) (Figure 6–55D); lowland rain forest on *terra firme*.

A specimen of var. *murumuru* from near Humaitá in Amazonas, Brazil (*B. Krukoff 1607*) approaches var. *ferrugineum* in the pilose abaxial leaf surface.

7c. Astrocaryum murumuru var. **huicungo**
(Dammer) Henderson, stat. nov. *Astrocaryum huicungo* Dammer ex Burret, Repert. Spec. Nov. Regni Veg. 35: 146. 1934. Type. Peru. San Martín: Moyobamba, 800–900 m, *A. Weberbauer s.n.* (holotype, B).
Astrocaryum scopatum Kahn & Millán, Bull. Inst. fr. études andines 21: 503. 1992. Type. Peru. Amazonas: Río Marañon, Bagua, Santa María de Nieve, 22 May 1990, *F. Kahn & F. Borchsenius 2563* (holotype, P).
Astrocaryum carnosum Kahn & Millán, Bull. Inst. fr. études andines 21: 504. 1992. Type. Peru. San Martín: upper Huallaga valley, near Uchiza, Dec 1985, *F. Kahn 1839* (holotype, P; isotype, K).

Stems cespitose, short and sometimes subterranean, or to 2 m tall, 15–18 cm diam. Pistillate flowers with spinulose calyx as long as the spinulose corolla; *fruits* obovoid, 6.5–7.5 cm long, 3.5–4.5 cm diam, densely covered with short, dark spinules.

Andean foothills of Peru (Amazonas, western Loreto, San Martín) (Figure 6.55E); lowland or premontane rain forest in periodically inundated areas near streams and rivers, at 240–900 m elevation. This variety, whose range is centered on the Ucayali Basin, is very similar to the following, which is centered on the adjacent Acre Basin.

7d. Astrocaryum murumuru var. javarense
(Trail) Henderson, stat. nov. *Astrocaryum paramaca* var. *javarense* Trail, J. Bot. 6: 77. 1877. *Astrocaryum javarense* (Trail) Drude in Mart. , Fl. bras.: Cyclanthaceae et Palmae I fasc. 85 vol 3(2): 372. 1881. Type. Brazil. Amazonas: Rio Javari, Camaná, 4 Dec 1874, *J. Trail 1073/CLXXXV* (holotype, K).

Astrocaryum horridum Barb. Rodr., Vellosia 1 (ed.2): 104. 104. 1891. Lectotype (Glassman, 1972). Barb. Rodr., Sert. palm. brasil. 2: t. 81A. 1903.

Stems usually solitary, short and subterranean or 2–5 m tall, 15–18 cm diam. Pistillate flowers with spinulose calyx as long as the spinulose corolla; *fruits* elongate-obovoid, 3.5–6.5 cm long, 2–3 cm diam, densely covered with short dark spinules; mesocarp dry.

Peru (Loreto, Madre de Dios) and Brazil (Acre, Amazonas), principally in the region of the Rio Javari but also south through Acre, Brazil, to Cocha Cashu in Madre de Dios, Peru (Figure 6.55E); lowland rain forest in wet places liable to inundation, at elevations below 250 m.

In Manu National Park in Madre de Dios, Peru, bats *(Vampyressa macconnelli)* make roosts in the leaves of *A. murumuru*. The bats severed the basal part of the fused apical pinnae and then bend these over to make a "tent" (Foster, 1992).

7e. Astrocaryum murumuru var. macrocalyx
(Burret) Henderson, stat. nov. *Astrocaryum macrocalyx* Burret, Repert. Spec. Nov. Regni Veg. 35: 150. 1934. Type. Peru. Amazonas: Río Marañon, May 1931, *W. Hopp 1126* (holotype, B, n.v.).

Astrocaryum cuatrecasanum Dugand, Caldasia 1: 18. 1940. Type. Colombia. Caquetá: between Florencia and Venecia, 400 m, 31 Mar 1940, *J. Cuatrecasas 8957* (holotype, COL).

Stems solitary, 2–10 m or more tall, 18–20 cm diam. Pistillate flowers with glabrous calyx as long as the spinulose corolla; *fruits* elongate-obovoid, 7–9 cm long, 3–3.5 cm diam, densely covered with short dark spinules; mesocarp fleshy.

Colombia (Amazonas, Caquetá, Putumayo) and Peru (Loreto) (Figure 6.55F); lowland rain forest in inundated or noninundated areas below 400 m elevation.

7f. Astrocaryum murumuru var. murumuru
Astrocaryum chonta Mart. in A. D. Orb., Voy. Amérique mér. 7(3) Palmiers 85. 1847. Type. Bolivia. Santa Cruz: Moxos, n.d., *A. d'Orbigny s.n.* (holotype, P, n.v.).

Astrocaryum yauaperyense Barb. Rodr., Vellosia 1: 48. 1888. Lectotype (Wess. Boer, 1988). Barb. Rodr., Sert. palm. brasil. 2: t. 80A. 1903.

Astrocaryum ulei Burret, Repert. Spec. Nov. Regni Veg. 35: 147. 1934. Type. Brazil. Acre: Rio Acre, n.d., *E. Ule 106* (holotype, B).

Astrocaryum gratum Kahn & Millán, Bull. Inst. fr. études andines 21: 503. 1992. Type. Peru. Madre de Dios: Puerto Maldonado, 1 Oct 1987, *F. Kahn & J. Llosa 2147* (holotype, P).

Stems solitary or cespitose, 1.5–15 m tall, 12–30 cm diam. Pistillate flowers with nonspinulose calyx as long as or usually shorter than the spinulose corolla; staminodial ring normal or reduced and then membranous or digitate; *fruits* oblong-obovoid, 3–8.5 cm long, 2–4.5 cm diam, scurfy-tomentose, usually not or scarcely spinulose, mesocarp soft and fleshy.

Venezuela (Amazonas), the Guianas, Peru (Huánuco, Loreto, Madre de Dios) Brazil (Acre, Amazonas, Pará, Rondônia), and Bolivia (Beni, La Paz, Pando, Santa Cruz) (Figure 6.55F); in periodically inundated areas or tidally inundated areas near the sea, along river margins, or occasionally in lowland rain forest on *terra firme*. Since the fruits are edible, it is likely that its range has been extended by humans.

Listabarth (1992) studied reproductive biology of this variety in Peru. He found that the palms flower from October to December, and the protogynous inflorescences are pollinated by small curculionid and nitidulid beetles.

7g. Astrocaryum murumuru var. perangustatum
(Kahn & Millán) Henderson, stat. nov.

Astrocaryum perangustatum Kahn & Millán, Bull. Inst. fr. études andines 21: 503. 1992. Type. Peru. Pasco: Prov. Oxapampa, Villa Rica–Iscozacin road, San Juan de Cacazu, 13 Jan 1992, *F. Kahn 3232* (holotype, P; isotype, NY).

Stems solitary, to 4 m tall, 20–23 cm diam. Pistillate flowers with the glabrous calyx as long as the spinulose corolla; *fruits* elongate-obovoid, greatly narrowed at the base, 7–8.5 cm long, 2–5 cm diam, densely covered with short dark spinules; mesocarp fleshy.

Peru (Pasco) (Figure 6.57A); premontane rain forest on Andean foothills on steep slopes at 380–800 m elevation.

7h. Astrocaryum murumuru var. urostachys
(Burret) Henderson, stat. nov. *Astrocaryum urostachys* Burret, Repert. Spec. Nov. Regni Veg. 35: 151. 1934. Type. Peru. Loreto: Río Napo, 200 m, Feb 1931, *W. Hopp 1078* (holotype, B).

Stems cespitose, 3–10 m tall, 12–18 cm diam. Pistillate flowers with the glabrous calyx as long as the spinulose corolla; *fruits* elongate-obovoid, 5–8 cm long, 3–4 cm diam, densely covered with short dark spinules.

Ecuador (Morona-Santiago, Napo, Pastaza) and just reaching Peru (Loreto) (Figure 6.57A); lowland rain forest at low elevations along river margins.

8. Astrocaryum paramaca Mart. in A. D. Orb., Voy. Amérique mér. 7(3) Palmiers 88. 1844. *Bactris paraensis* Splitg. ex Vriese, Jaarb. Kon. Ned. Maatsch. Aanm. Tuinb. 1848: 10. 1848. Type. Surinam. Without locality, 1839, *F. Splitgerber 507* (holotype, M; F neg. 18574) (Figure 6.56e–f).
Astrocaryum acanthopodium Barb. Rodr., Enum. palm. nov. 20. 1875. Lectotype (Wess. Boer, 1965). Barb. Rodr., Sert. palm. brasil. 2: t. 76B. 1903.
Astrocaryum paramaca var. *platyacantha* Drude in Mart., Fl. bras.: Cyclanthaceae et Palmae I fasc. 85 vol 3(2): 371. 1881. Type. Brazil. Amazonas: Maués, 1861, *G. Wallis s.n.* (holotype, B, n.v.).
Astrocaryum aculeatum of Barb. Rodr., not Meyer, Enum. palm. nov. 20. 1875.

Astrocaryum rodriguesii Trail, J. Bot. 6: 79. 1877. Type. Brazil. Amazonas: José Açu, Villa Bella de Imperatriz, n.d., *J. Barbosa Rodrigues 320* (destroyed). Lectotype (Wess. Boer, 1965). Barb. Rodr., Sert. palm. brasil. 2: t. 76A. 1903.

Stems solitary, short and subterranean, or less often to 8 m or more tall and 11 cm diam, then clean and without persistent leaf bases. *Leaves* 10–15; sheath ca. 40 cm long, sheath, petiole, and rachis with strongly flattened, winged, black spines to 5 cm long, occurring together in obscure clusters; petiole 0.9–1.3 m long; rachis 2–4.4 m long; pinnae 55–100 per side, regularly arranged and spreading in 1 plane, linear or linear-lanceolate, aristate, the middle ones 64–83 cm long, 2–3 cm wide. *Inflorescences* interfoliar, erect at anthesis, erect or pendulous in fruit; peduncle 0.6–1.5 m long; prophyll to 30 cm long; peduncular bract 69–95 cm long, densely brown-velutinous abaxially; rachis 18–46 cm long; rachillae numerous, to 8 cm long, with black flexuous spines proximally; *staminate flowers* 3 mm long, densely crowded, paired or solitary in pits on the distal part of the rachillae; sepals briefly connate basally, free and spreading above, narrowly triangular, 1 mm long; petals free, valvate, obovate, 2 mm long; stamens 6, the filaments not inflexed at the apex, the thecae almost free; pistillode very small; pistillate flowers 2 cm long, solitary, ca. 1 cm from the base of each rachilla; calyx 3-lobed, cupular, 1.3 cm long, with a dense covering of spinules; corolla 3-lobed, cupular, 1.1 cm long, with dense covering of spinules; staminodial ring 2 mm high, adnate to the corolla for 1 mm; *fruits* oblong, prominently rostrate, 3.5–4 cm long, 1.5–2 cm diam, with ring of spinules on "shoulder," glabrescent; epicarp orange-brown at maturity, splitting ± regularly to expose the yellow mesocarp.

Northeastern Amazon region in Surinam, French Guiana, and Brazil (Amapá, Amazonas, Pará) (Figure 6.57B); very abundant in the understory of lowland forests on *terra firme* in Surinam and French Guiana, but less common in the western part of its range. de Granville (1977) has described how the "funnel-shaped" leaves of this species act as a litter trap, and falling leaves and other debris accumulate in the center of the palm. Sist (1989a) reported

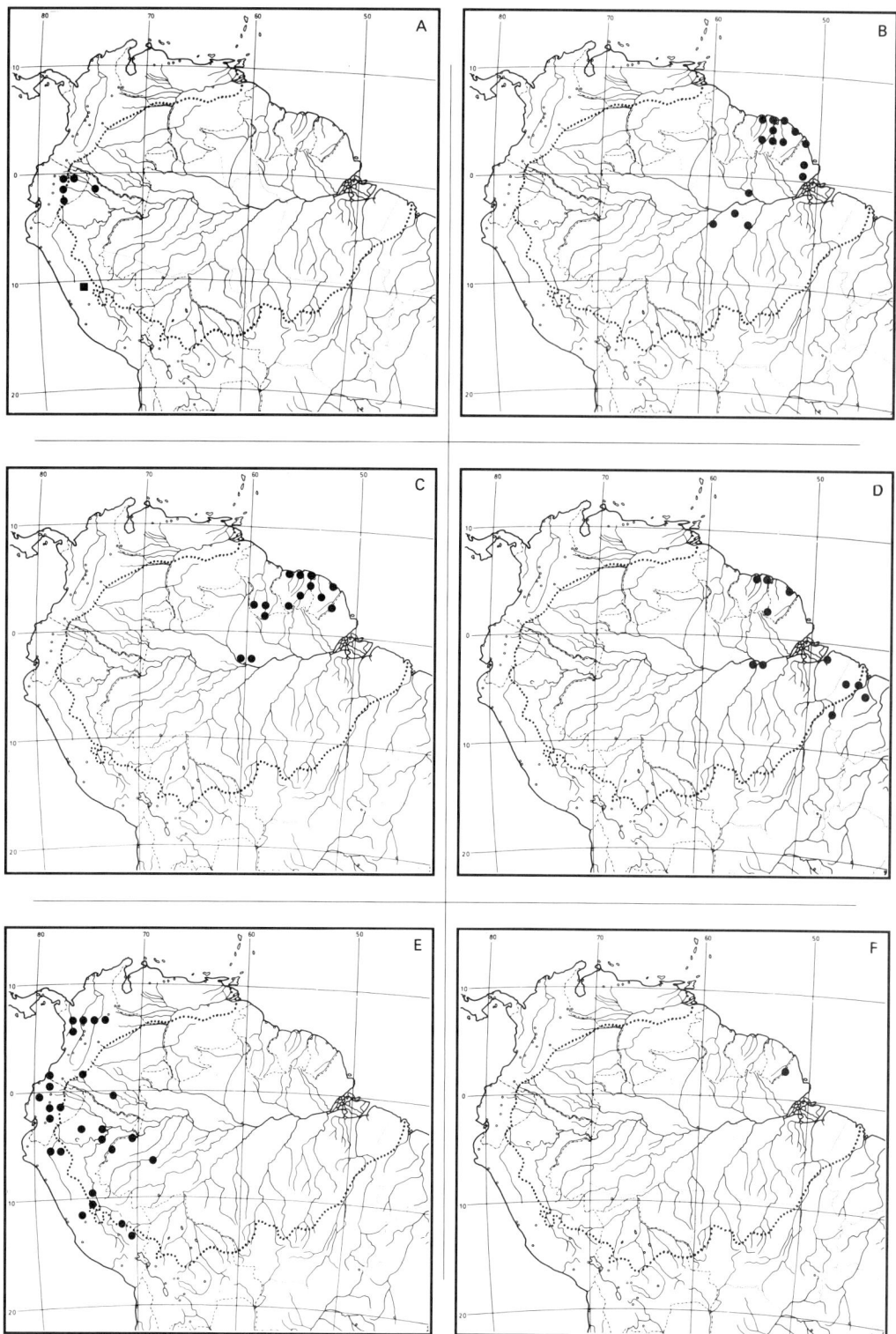

Figure 6.57. (A) *Astrocaryum murumuru* var. *perangustatum* (square), var. *urostachys* (circles); **(B)** *Astrocaryum paramaca;* **(C)** *Astrocaryum sciophilum;* **(D)** *Astrocaryum vulgare;* **(E)** *Pholidostachys synanthera;* **(F)** *Asterogyne guianensis.*

that in French Guiana this species showed a peak of flowering in November.

Brazil: *murumuru da terra firme, tucumã branco*. French Guiana: *counana* (Creole), *counana agouti, taki-taki* (Boni), *tiemaka* (Boni). Surinam: *guie, paramaka, paramaka x boefro maka*.

This species is still somewhat problematic. The original description is of a palm from Surinam with a short and subterranean stem (but Martius also wrote: "teste cl. Splitgerber in 20 ad 30 pedum altitudinem evehitur"). It has been re-collected several times in Surinam, French Guiana, and Brazil (Amapá), and is characterized by its short and subterranean stem and erect inflorescence and infructescence. Trail (1877a) described *Astrocaryum rodriguesii* from Brazil (but his collection, from the Rio Trombetas, is not the type) and stated that this species had a stem to 16 m tall, but also wrote that "the stem varies much in height, and is sometimes wanting." The lectotype of the species (illustrated with a pendulous inflorescence), from eastern Amazonas, is identical to a specimen from near Borba, also in eastern Amazonas, and both are identical (except for the pendulous inflorescence and aerial stem) to the specimens from Surinam and French Guiana. For this reason, I include *A. rodriguesii* as a synonym. Unfortunately, there is another, similar, palm occurring sporadically in Surinam and French Guiana, which has been interpreted by both Wessels Boer (1965) and Kahn and Millán (1992) as *A. rodriguesii*. I believe, however, this may be an undescribed taxa, differing at least in its spiny peduncular bract and larger, and spinier, pistillate calyx and corolla. More collections are needed to solve this problem.

9. Astrocaryum sciophilum (Miq.) Pulle, Enum. vasc. pl. Surinam 73. 1906. *Bactris sciophila* Miq., Natuurk. Verh. Kon. Maatsch. Wetensch. Haarlem 7: 208. 1851. Type. Surinam. Blaauwe Berg, n.d., *H. Focke 922* (holotype, L, n.v.) (Figure 6.56g–h).

Astrocaryum plicatum Drude in Mart., Fl. bras.: Cyclanthaceae et Palmae I fasc. 85 vol 3(2): 375. 1881. Type. French Guiana. Acarouany, 1858, *P. Sagot 594* (holotype, P; isotypes, K, US).

Astrocaryum sociale Barb. Rodr., Vellosia 1: 48. 1888. Lectotype (Wess. Boer, 1965). Barb. Rodr., Sert. palm. brasil. 2: t. 79A. 1903.

Astrocaryum farinosum Barb. Rodr., Enum. palm. nov. 21. 1875. Lectotype (Wess. Boer, 1965). Barb. Rodr., Sert. palm. bras. 2: t. 78. 1903.

Stems solitary, short and subterranean or less often aerial and then 0.3–2 m or more tall and 14(–30) cm diam, covered with persistent leaf bases. *Leaves* 7–12; sheath and petiole 1.1–2.5 m long, petiole and rachis with oblique rows of black spines, these to 8(–15) cm long and not strongly flattened; rachis 2.1–4.7 m long; pinnae 58–90 per side, regularly arranged and spreading in 1 plane, linear or linear-lanceolate, slightly bifid at apex, the apical one much broader than the others, the middle ones 74–140 cm long, 3–5 cm wide, strongly plicate with several prominent veins. *Inflorescences* interfoliar, erect at anthesis; peduncle 0.4–1.5 m long; prophyll 35–40 cm long; peduncular bract 46–75 cm long, inserted ± halfway up peduncle, densely black spiny abaxially; rachis 11–15 cm long; rachillae numerous, to 7 cm long; *staminate flowers* 2.5 mm long, paired or more often solitary in pits in the distal part of the rachillae; sepals forming a shallow 3-lobed ring, the lobes narrowly triangular, 0.5 mm long; petals briefly connate basally, free and valvate above, curved over at the apex, irregularly oblong, 2.5 mm diam; stamens 6, filaments not inflexed at the apex; pistillode not apparent; pistillate flowers 1 cm long, sessile on rachis at base of rachillae; calyx 3-lobed, cupular, 5 mm long; corolla 3-lobed, cupular, 5 mm long, staminodial ring adnate to corolla; *fruits* obovoid but appearing almost globose on the rachis, somewhat misshapen by mutual pressure, 3–6 cm long, 2.5–4 cm diam, rostrate, brown, moderately covered with short black spinules.

Northeastern Amazon region in the Guianas and Brazil (Amapá, Amazonas) (Figure 6.57C); lowland rain forest on *terra firme* at low elevations, in some places extremely abundant. Henderson and co-workers (in preparation) found almost 2000 adult plants in a 10 ha plot near Manaus. De Granville (1977) described how the "funnel-shaped" leaves of this species act as a litter trap, and falling leaves and other debris accumulate in the center of the palm. Sist (1989b) has described the demography and dispersal of this species in French Guiana. Populations are characterized by a high proportion of small, immature individuals. Seedlings and young juveniles suffer the highest level of mortality. Seeds suffer high predation rates from squirrels *(Sciurus aestuans)*. Rodents store endocarps singly near objects, and so create a random distribution of individuals. Sist (1989a) reported that in French Guiana this species flowered and fruited irregularly throughout the year.

Brazil: *murumuru*. French Guiana: *counana* (Creole), *mulumulusi, muru muru*. Guyana: *bulishi* (Wapisiana), *mumu*. Surinam: *boegroe maka, murumuru* (Trio). The fruits are occasionally eaten.

Kahn and Millán (1992) considered *Astrocaryum sciophilum* to have an aerial stem, and thus differ from the short-stemmed *A. sociale* and *A. farinosum*. This stem character is not considered taxonomically significant. The populations in the Guianas probably produce stems as a result of higher rainfall in that region. Apart from this the species is evidently homogeneous, and has a typical northeastern-Amazon-region distribution.

10. **Astrocaryum vulgare** Mart., Hist. nat. palm. 2: 74. 1824. Type. Brazil. Pará: without locality, n.d., *C. Martius 2565* (holotype, M; F neg. 18576) (Figure 6.56i–j).

Astrocaryum awarra de Vriese, Jaarboek van de Koninklijke Nederlandsche Maatschappij. 1848: 12. *Astrocaryum guianense* Splitg. ex Mart., Hist. nat. palm. 3: 323. 1853. Type. Surinam. Paramaribo, n.d., *F. Splitgerber 60* (holotype, L, n.v.).

Astrocaryum tucuma of Wallace, not Martius, Palm Trees of the Amazon 107. 1853.

Astrocaryum segregatum Drude in Mart., Fl. bras.: Cyclanthaceae et Palmae I fasc. 85 vol. 3(2): 382. 1881. Type. French Guiana. Without locality, n.d., *P. Sagot 593* (isotype, K).

Astrocaryum tucumoides Drude in Mart., Fl. bras.: Cyclanthaceae et Palmae I fasc. 85 vol. 3(2): 381. 1881. Type. Brazil. Pará: without locality, n.d., *A. Glaziou 8060* (holotype, P; F neg. 21149).

Stems cespitose or less often solitary, erect, 4–10 m tall, 10–20 cm diam, the internodes covered with black spines 4–22 cm long. *Leaves* 8–16, stiffly ascending; sheath and petiole 1–2 m long, sheath, petiole, and rachis covered with long, somewhat flattened, black spines to 10 cm long; rachis 1.7–4 m long; pinnae 73–120 per side, irregularly arranged in clusters of 2–6, spreading in different planes, linear, the middle ones 0.9–1.2 m long, 3.5–5 cm wide. *Inflorescences* interfoliar, erect at anthesis and in fruit; peduncle 0.9–1 m long; prophyll to 50 cm long, persistent; peduncular bract 1–1.3 m long, inserted near apex of peduncle, densely black spiny abaxially; rachis 45–75 cm long; rachillae 214–244, to 30 cm long, proximal ca. half with 2–4 triads, distal ca. half with staminate flowers; *staminate flowers* to 4 mm long, paired or solitary in crowded pits in the rachillae; sepals connate basally, free above, triangular, 1.5 mm long; petals connate basally, free and valvate above, ovate, 4 mm long; stamens 6, the filaments flattened, inflexed apically; pistillode 3 mm long; pistillate flowers 1.3 cm long, 2–4 per rachillae; sepals connate basally for ca. one-third their length, widely ovate, 7 mm long; corolla cupular, 8 mm long; staminodial tube 2 mm high; *fruits* globose to ellipsoid, 4–5 cm long, 3–3.7 cm diam, orange, glabrous.

Eastern Amazon region in Surinam, French Guiana, and Brazil (Maranhão, Pará, Tocantins) (Figure 6.57D); lowland rain forest, but frequent in disturbed areas and old fields (*capoeiras* in Brazil), especially on sandy soils.

Brazil: *rohn-di* (Apinajé), *takamã* (Guajá), *tucumã, tucuma* (fruit), *tucumazeiro* (tree), *tucum, tukumã'y* (Ka'apor). French Guiana: *aroira palm, awara*. Surinam: *awarra*. The fruits are edible and are sold in local markets. Fibers from the young leaves are occasionally used to make twine. The endocarps are carved into rings for fingers.

Wessels Boer (1965) considered that this species was very similar to *Astrocaryum aculeatum*, and the two could be conspecific. They are kept separate here on the strength of the clustered stems and smaller fruits of the former, but intermediates do occur. This species is also very similar to *A. huaimi*.

ARECOIDEAE · GEONOMEAE

29. *Pholidostachys* **Pholidostachys** H. Wendl. ex Hook. f. in Benth. & Hook. f., Gen. pl. 3: 915. 1883.

Small to moderate, monoecious palms. *Stems* solitary, erect or somewhat procumbent. *Leaves* pinnate, reduplicate; sheath open and not forming a crownshaft; petiole long; rachis moderate; pinnae regularly or irregularly arranged, spreading in 1 plane. *Inflorescences* interfoliar, branched to 1 order or spicate; peduncle bearing a prophyll and 1 peduncular bract; rachis bearing 1–few rachillae; *flowers* borne in triads, these sunken in pits along the rachillae; staminate flowers with sepals 3, free; petals 3, connate basally, free and valvate above; stamens 6, the filaments connate basally into a hollow tube, free above; pistillode present; pistillate flowers with sepals 3, free, imbricate; petals 3, connate basally, free and valvate above; staminodial tube digitately lobed; gynoecium syncarpous, trilocular, triovulate; *fruits* 1-seeded, obovoid, with basal stigmatic residue; endocarp thin; seeds with homogeneous endosperm and basal embryo; germination adjacent-ligular; eophyll bifid.

A genus of four species distributed from Central America (Costa Rica, Panama) south through Colombia and Ecuador to Peru, and just reaching extreme western Brazil. Wessels Boer (1968) recognized three species and one more has since been rediscovered (Greg de Nevers, personal communication); one species occurs in the Amazon region.

1. **Pholidostachys synanthera** (Mart.) H. E. Moore, Taxon 18: 231. 1969. *Geonoma synanthera* Mart., Hist. nat. palm. 2: 13. 1823. *Calyptrogyne synanthera* (Mart.) Burret, Bot. Jahrb. Syst. 63: 137. 1930. *Calyptronoma synanthera* (Mart.) L. H. Bailey, Gentes Herb. 4: 166. 1938. Type. Peru. Department?: Chicoplaya, n.d., *H. Ruíz Lopez & J. Pavón s.n.* (holotype, M, n.v.; F neg. 18531) (Figure 6.58a–b).
Calyptronoma robusta Trail, J. Bot. 14: 330. 1876. *Calyptrogyne robusta* (Trail) Burret, Bot. Jahrb. Syst. 63: 137. 1930. Type. Brazil. Amazonas: Rio Javari, 5 Dec 1876, *J. Trail 961/CLXXXVI* (isotypes, NY, P).

Stems solitary, 1.8–5 m tall, 3–8 cm diam. *Leaves* 10–25; sheath 14–45 cm long, open and not forming a crownshaft, fibrous on margins, sheath, petiole, and rachis ferrugineous-tomentose; petiole 25–98 cm long; rachis 47–124 cm long; pinnae 5–17 per side, regularly arranged and spreading in 1 plane, ± linear, the middle ones 43–60 cm long, 3.5–13 cm wide. *Inflorescences* interfoliar; peduncle 31–70 cm long; prophyll 31–51 cm long; peduncular bract 40–57 cm long, both bracts somewhat woody, flattened, ferrugineous, tomentose; rachis 11–40 cm long; rachillae 7–17, the proximal few sometimes bifid, 40–64 cm long, 6–10 mm diam, with the pits arranged in close vertical files, the basal lip prominent, with entire, rounded, recurved margin, and covering the pit before anthesis, the upper lip absent; staminate flowers 6 mm long at anthesis; sepals spreading at anthesis, oblanceolate, 3 mm long; petals oblanceolate-obovate, 5 mm long; pistillode gynoeciumlike, covered by the filament tube, 3.5 mm long; pistillate flowers 3.5 mm long (in bud); sepals oblanceolate, 3.5 mm long; petals oblanceolate-obovate, 3 mm long; staminodial tube with lobes spreading at anthesis; *fruits* obovoid, 1.4–1.7 cm long, 8–10 mm diam, black at maturity.

Andes and western Amazon region of Colombia (Amazonas, Antioquia, Caquetá, Nariño, Santander), Ecuador (Carchi, Napo, Pastaza, Pichincha), Peru (Amazonas, Cusco, Huánuco, Junín, Loreto, Madre de Dios, Puno, San Martín), and Brazil (Amazonas) (Figure 6.57E). It is another example of a cis-trans-Andean distribution pattern. It usually grows in premontane and montane rain forest on steep

Figure 6.58. (a) *Pholidostachys synanthera,* **(b)** infructescence (from G. Galeano 1968); **(c)** *Asterogyne guianensis,* **(d)** infructescence (from J.-J. de Granville 7124).

slopes, below 1500 m elevation, but in the western Amazon it grows on *terra firme* in lowland rain forest.

Brazil: *ubim uassú, ubim.* Colombia: *pe-co-r* (Huitoto). Peru: *palmicha, palmiche, palmiche grande, wayúr* (Mayna Jívaro). The leaves are used for thatching.

30. *Asterogyne* **Asterogyne** H. Wendl. & Hook. f. in Benth. & Hook. f., Gen. pl. 3: 914. 1883.

Small to moderate, monoecious palms. *Stems* solitary or rarely cespitose, short and subterranean or longer and aerial. *Leaves* entire, pinnately veined, reduplicate; sheaths open and not forming a crownshaft; petiole short or medium; rachis medium; blade entire, bifid at apex. *Inflorescences* interfoliar, branched to 1 order, or spicate; peduncle bearing a prophyll and 1(–2) peduncular bracts; rachis bearing 1–few rachillae; *flowers* borne in triads, these borne in pits along the rachillae; staminate flowers with sepals 3, free, imbricate; petals 3, connate basally for ca. half their length; stamens 6–26, with the filaments connate basally; pistillode small; pistillate flowers with sepals 3, free; petals 3, connate basally for ca. half their length, valvate above; staminodial tube digitately lobed; gynoecium syncarpous, trilocular, triovulate; *fruits* 1-seeded, ellipsoid or obovoid, with basal stigmatic residue; seed with homogeneous endosperm and subbasal embryo; germination adjacent-ligular; eophyll bifid.

A genus of five species (Wessels Boer, 1968; Henderson & Steyermark, 1986; de Granville & Henderson, 1988) distributed from Central America (Belize) to Colombia and Venezuela, with an outlying species in French Guiana.

1. **Asterogyne guianensis** Granv. & Henderson, Brittonia 40: 76. 1988. Type. French Guiana. Camopi River, ca. 1.5 km NE of Mont Belvédère, 150 m, 4 Dec 1984, *J.-J. de Granville 7124* (isotype, NY) (Figure 6.58c–d).

Stems solitary, 1.5–2 m tall, 3.4–5 cm diam, with a mound of roots at the base. *Leaves* 15–18; sheath and petiole ca. 60 cm long; rachis 80–90 cm long; blade entire, cuneate at the base, bifid at the apex, 100–110 cm long, 35–40 cm wide. *Inflorescences* interfoliar, spicate; peduncle 50–55 cm long; prophyll ca. 20 cm long; peduncular bract 40–42 cm long; rachillae 1, 26–30 cm long, ca. 1.5 cm diam, the flower pits spirally arranged, each one ca. 5 mm apart, with a reflexed lower lip; staminate flowers 10 mm long; sepals lanceolate, 7 mm long; petals 8 mm long; stamens 25–26; pistillode 2 mm long, trifid; pistillate flowers 5 mm long (in bud); sepals lanceolate, 4 mm long; petals 4 mm long; staminodial ring with ca. 21 lobes; *fruits* ellipsoid, ca. 2.5 cm long, ca. 1.5 cm diam, red at maturity.

A small area in French Guiana (Figure 6.57F); lowland rain forest in a flat, inundated area with *Euterpe oleracea* and *Hyospathe elegans.*

This species is geographically isolated; its nearest congener occurs in northern Venezuela (de Granville & Henderson, 1988).

31. *Geonoma* **Geonoma** Willd., Sp. pl. 4: 174. 1805.

Small to moderate, monoecious palms. *Stems* cespitose or solitary, short and subterranean or longer and aerial. *Leaves* pinnate, or pinnately veined if entire, reduplicate; sheaths open and not forming a crownshaft; petiole short to medium; rachis medium; pinnae regularly or irregularly arranged, spreading in 1 plane, or often leaf entire. *Inflorescences* interfoliar or infrafoliar, branched to 1–3 orders, or commonly spicate; peduncle bearing a prophyll and (0–)1(–2) peduncular bracts; rachis bearing 1–many rachillae; *flowers* borne in triads, sunken in pits along the rachillae, the pits spirally, verticillately, or decussately arranged (alternating in pairs at right angles), with upper and lower lips or upper lip absent; staminate flowers with sepals 3, free, imbricate; petals 3, connate basally for ca. half their length, free and valvate above; stamens (3–) 6(–more) with filaments partly connate basally, the connective often bifid with free thecae; pistillode small; pistillate flowers with sepals 3, free, imbricate; petals connate basally for ca. two-thirds their length, free and valvate above; staminodial tube either blunt, crenate, or digitately lobed at the apex; gynoecium syncarpous, unilocular, uniovulate; style basifixed; *fruits* 1-seeded, globose to ellipsoid, with basal stigmatic residue; seed with homogeneous endosperm and subbasal embryo; germination adjacent-ligular; eophyll bifid.

A genus of approximately 80 species or fewer distributed throughout the Neotropics, from Mexico south to Bolivia and across to eastern Brazil; also in the Lesser and Greater Antilles (Haiti) (Wessels Boer, 1968). Twenty-one species occur in the Amazon region. Two other species may just enter the region. *Geonoma jussieuana* Mart. occurs on eastern Andean slopes between 700 and 1700 m elevation, but may occur near 500 m; and *G. appuniana* Spruce (and possibly also *G. simplicifrons* Willd.) occurs in the region, but on *tepuis* of the Guayana Highland above 1800 m elevation. *Geonoma pauciflora* Mart. does not enter the region.

Geonoma is one of the most complex genera of neotropical palms. Despite a relatively recent revision (Wessels Boer, 1968), there still remain many problems, both taxonomic and nomenclatural (perhaps not least of which is that Wessels Boer listed 59 names of uncertain application). Some species are extremely variable in leaf shape and division, and are not well understood. Wessels Boer divided these into several species. He had far fewer specimens available for study, however, and eight species in his treatment were represented by 10 or fewer specimens. In many cases, new collections have tended to obscure the boundaries between species because they are morphologically intermediate. Skov (1989), in his treatment of *Geonoma* for Ecuador, found many intermediate states in characters, and he treated variable species as consisting of several varieties and at the same time noted the existence of intermediates. I have followed Skov's scheme here.

Leaf variation is very interesting, and in the three most variable species (*Geonoma macrostachys*, *G. maxima*, and *G. stricta*) the same type of variation occurs (Figure 6.59). Depending on the angle between the vein and rachis, various leaf forms occur. Leaves with a wide vein angle of 40–60° have pinnate leaves with sigmoid pinnae (e.g., *G. macrostachys* var. *acaulis* [Figure 6.59a], *G. maxima* var. *maxima* [Figure 6.59e–f], and *G. stricta* var. *trailii* [Figure 6.59i]). As the angle narrows, leaves become less pinnate (and the pinnae more linear) and more entire, and then leaf shape goes from broadly (e.g., *G. macrostachys* var. *macrostachys* [Figure 6.59b]) to narrowly obovate. Finally, at about 10°, leaves become entire, narrow, and strongly plicate (e.g., *G. macrostachys* var. *macrostachys* [Figure 6.59c] and *G. maxima* var. *spixiana* [Figure 6.59h]). In general, vein angle of the most proximal pinna is wider than that of the middle ones, and this proximal pinna is often contracted at the point of insertion. Interestingly, the same type of leaf shape variation occurs in *Bactris hirta*, *B. simplicifrons* (Figure 6.59m–o), and *B. tomentosa*.

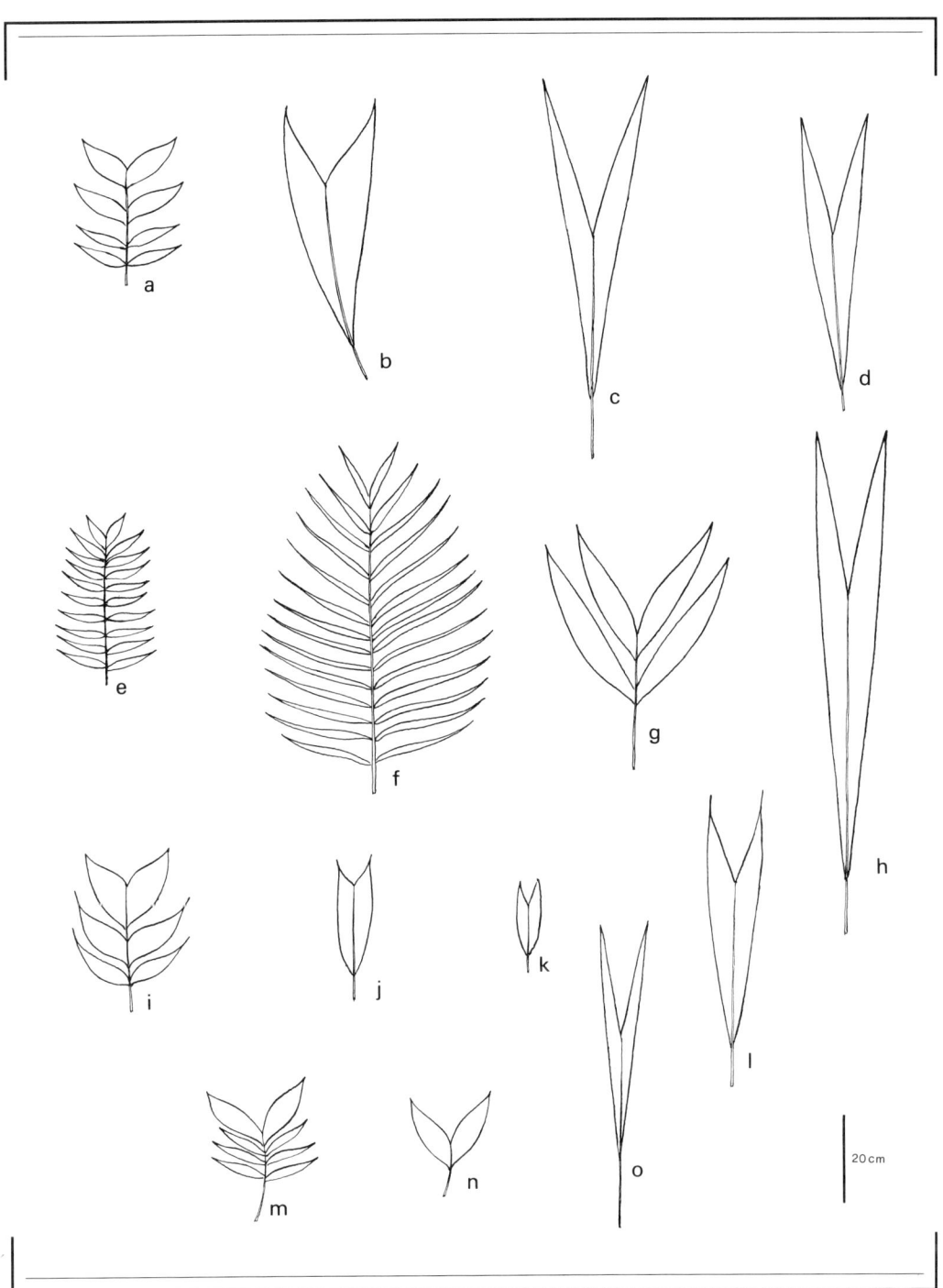

Figure 6.59. Leaf shapes: **(a)** *Geonoma macrostachys* var. *acaulis* (F. Chávez 710); **(b)** *Geonoma macrostachys* var. *macrostachys* (F. Chávez 695); **(c)** *Geonoma macrostachys* var. *macrostachys* (J. Ruíz 1248); **(d)** *Geonoma macrostachys* var. *poiteauana* (H. Irwin 47627); **(e)** *Geonoma maxima* var. *maxima* (J. Torres 3132); **(f)** *Geonoma maxima* var. *maxima* (J. Steyermark 125642); **(g)** *Geonoma maxima* var. *chelidonura* (A. Henderson 633); **(h)** *Geonoma maxima* var. *spixiana* (A. Henderson 1053); **(i)** *Geonoma stricta* var. *trailii* (A. Henderson 303); **(j)** *Geonoma stricta* var. *stricta* (H. Balslev 60584); **(k)** *Geonoma stricta* var. *stricta* (B. Boswezen 6508); **(l)** *Geonoma stricta* var. *piscicauda* (P. Cazalet 7775); **(m)** *Bactris simplicifrons* (J. Steyermark 88054); **(n)** *Bactris simplicifrons* (S. Mori 18164); **(o)** *Bactris simplicifrons* (G. Galeano 1975).

Staminate and pistillate flowers show considerable variation and can be used to divide the genus into various groups. However, they are too small, ephemeral in the case of staminate flowers, and inaccessible (in the pits) to be of much practical value in identifying the species. They are included in the key because, if present, they are diagnostic.

KEY TO THE SPECIES OF *GEONOMA*
1. Inflorescences spicate.
    2. Peduncle much longer than the ± glabrous rachilla; peduncular bract present; dry fruits with tuberculate epicarp.
        3. Flower pits close together on a straight rachilla; prophyll and peduncular bract inserted close together at base of peduncle; staminodial tube digitately lobed apically; rachilla usually green at anthesis; mesocarp 0.3–0.6 mm thick, with numerous short, erect fibers; widespread.
            4. Stems short and subterranean; fruits 8–12 mm long, 7–8 mm diam; each filament inflexed apically with the connective not split; widespread. . . . . 12. *G. macrostachys.*
            4. Stems aerial, (0–)0.3–2 m tall and 2–7 cm diam; fruits 1–1.5 cm long, 0.6–1.3 cm diam; each filament very briefly inflexed apically with the connective split.
                5. Leaves pinnate with 2–25 unequally wide pinnae; veins forming a 40–70° angle with the rachis.
                    6. Pinnae bicolorous, sigmoid; stamens 6; Colombia (Amazonas, Putumayo), Ecuador (Morona-Santiago, Napo), Peru (Loreto, Madre de Dios), and Brazil (Acre, Amazonas). . . . . 6. *G. camana.*
                    6. Pinnae concolorous, linear to somewhat sigmoid; stamens more than 6; Colombia (Caquetá) and Ecuador (Napo) . . 18. *G. polyandra.*
                5. Leaves usually entire; veins forming a 15° angle with the rachis; French Guiana and Brazil (Pará) . . . . 15. *G. oldemanii.*
        3. Flower pits loosely arranged, not close together, on an often curved rachilla; prophyll and peduncular bract inserted far apart (3.5–4.5 cm) at base of peduncle; staminodial ring crenate apically; rachilla reddish-brown at anthesis; mesocarp 0.2–0.3 mm thick, without fibers; Colombia (Amazonas, Caquetá, Meta, Putumayo), Ecuador (Napo), Peru (Ayacucho, Huánuco, Junín, Loreto, Madre de Dios, Pasco, Puno, San Martín, Ucayali), Brazil (Acre, Amazonas), and Bolivia (Beni, Cochabamba, La Paz, Santa Cruz). . . . . 5. *G. brongniartii.*
    2. Peduncle shorter than the often pilose rachilla; peduncular bract absent or present; dry fruits with striate epicarp.
        7. Peduncular bract absent or much reduced; rachilla 0.3–1.2 cm diam, straight within bud; widespread . . . . 19. *G. stricta.*
        7. Peduncular bract present; rachilla 1.5–2.5(–3) mm diam, folded and twisted in bud; Colombia (Amazonas, Putumayo), Ecuador (Napo, Pastaza), Peru (Amazonas, Huánuco, Loreto, Pasco, Puno), and Brazil (Amazonas). . . . . 1. *G. arundinacea.*
1. Inflorescences branched.
    8. Fruits 0.8–1.6 cm long, 0.7–1.2 cm diam; staminodial ring digitately lobed apically.

9. Inflorescences branched to 1 order; peduncle 25–50 cm long; rachillae 2–8, 3–13 cm long, 3–13 mm diam; leaves drying brown, with alternate veins brown-tomentose abaxially.
   10. Rachillae 10–30 cm long, (3.5–)8–13 mm diam; flower pits closely inserted; stems 0.5–4 m tall, 3–5 cm diam; Colombia (Caquetá), Ecuador (Morona-Santiago, Napo), and Peru (Junín, Madre de Dios, Pasco) . . . . 20. *G. triglochin.*
   10. Rachillae 15–23 cm long, 3–3.5 mm diam; flower pits loosely inserted; stems 0.6–2 m tall, 2–2.5 cm diam; the Guianas and Brazil (Amapá, Pará) . . . . 21. *G. umbraculiformis.*
9. Inflorescences usually branched to 2 orders; peduncle 5–19 cm long; rachillae 4–43, 6–24 cm long, 1.5–6 mm diam; leaves drying yellowish-green, alternate veins not brown-tomentose abaxially . . . . 13. *G. maxima.*
8. Fruits 0.5–1(–1.2) cm long, 0.5–0.8 cm diam; staminodial ring crenate or blunt apically, rarely digitately lobed.
  11. Prophyll and peduncular bract short, 1–10(–14) cm long, ± swollen, usually deciduous; stems slender, 0.4–2 cm diam.
    12. Flower pits usually arranged in raised, alternating whorls of 3; inflorescences branched to 1 or 2 orders, with 3–25 rachillae.
      13. Rachillae 3–8, 25–35 cm long; flower pits ca. 2.5 mm apart; blade entire, wedge-shaped; fruits globose-ellipsoid, 8–10 mm long, 5–7 mm diam; Colombia (Amazonas), Ecuador (Napo), Peru (Loreto), Brazil (Acre, Amazonas), and Bolivia (Pando) . . . . 9. *G. laxiflora.*
      13. Rachillae 3–25, 7–32 cm long; flower pits ca. 1 mm apart; blade pinnate, typically with 3 pinnae per side, rarely entire and then not wedge-shaped; fruits globose, 5–7 mm diam; widespread . . . . 7. *G. deversa.*
    12. Flower pits not in alternating whorls; inflorescences branched to 1 order, with 2–10 rachillae.
      14. Flower pits widely spaced, 5–6 mm apart, with prominent basal lower lip almost forming a cupule with upper lip, leaves often entire and with almost parallel margins . . . . 10. *G. leptospadix.*
      14. Flower pits closely spaced, 1–2 mm apart, the two lips not forming a cupule; leaves seldom entire and then not with parallel margins.
        15. Rachillae 3–10, 20–30 cm long; fruits globose, 5 mm diam; several buds often visible on the stem below the leaves; dry fruits with tuberculate epicarp; Colombia (Amazonas), Venezuela (Amazonas), and Brazil (Amazonas) . . . . 16. *G. oligoclona.*
        15. Rachillae 2–5, 5–41 cm long; fruits globose-ellipsoid or ellipsoid, 8–10 mm long, 5–6 mm diam; without several buds visible on the stem; dry fruits with striate epicarp.
          16. Blade entire or with 2–3(–4) pinnae, oblong in outline; rachillae 10–41 cm long, 1.5–2.5(–3) mm diam, folded and twisted within the bud; Colombia (Amazonas, Putumayo), Ecuador (Napo, Pastaza), and Peru (Amazonas, Huánuco, Loreto, Pasco) and Brazil (Amazonas) . . . . 1. *G. arundinacea.*
          16. Pinnae 2–6 per side, the blade not oblong in out-

line; rachillae 5–26 cm long, 3–5 mm diam, straight within the bud.
- 17. Rachillae 5–14 cm long, (2–)3–4 mm diam; staminodial tube digitately lobed apically; Colombia (Amazonas), Guyana, and Brazil (Amazonas) . . . . 2. *G. aspidiifolia.*
- 17. Rachillae 18–26 cm long, 4–5 mm diam; staminodial tube crenate apically; Peru (Amazonas, Loreto) . . . . 19. *G. stricta.*
- 11. Prophyll and peduncular bract elongate, (5–)10–37 cm long, tubular and flattened and usually persistent; stems stout, 1–5(–7) cm diam.
- 18. Flower pits alternating in pairs at right angles (decussate); Brazil (Mato Grosso, Pará, Rondônia, Tocantins), Bolivia (La Paz, Pando, Santa Cruz), and Peru (Madre de Dios). . . . . 4. *G. brevispatha.*
- 18. Flower pits not decussate, spirally arranged.
- 19. Upper lip of flower pit present; Venezuela (Amazonas, Bolívar), the Guianas, and Brazil (Amapá, Amazonas, Maranhão, Pará, Roraima). . . . . 3. *G. baculifera.*
- 19. Upper lip of flower pit obscure or absent.
- 20. Inflorescences infrafoliar, branched to 2 orders; rachis 14–16 cm long; rachillae 30–50.
- 21. Peduncle ca. 6 cm long; rachillae 28–30 cm long, straight within the bud; Brazil (Acre) . . . . . . 14. *G. myriantha.*
- 21. Peduncle 5–40 cm long; rachillae twisted and folded within the bud; Colombia (Caquetá, Putumayo), Ecuador (Morona-Santiago, Napo, Pastaza), Peru (Huánuco, Madre de Dios, San Martín), Brazil (Pará, Rondônia), and Bolivia (Beni, La Paz) . . . 8. *G. interrupta.*
- 20. Inflorescences interfoliar, branched to 1 order; rachis 5–16 cm long; rachillae 2–10.
- 22. Leaf rachis not brown-tomentose; peduncle densely whitish-brown-tomentose; rachillae 3–5.5 mm diam . . . . 17. *G. poeppigiana.*
- 22. Leaf rachis densely brown-tomentose; peduncle not densely tomentose; rachillae 2–3 mm diam . 11. *G. longepedunculata.*

1. Geonoma arundinacea Mart., Hist. nat. palm. 2: 17. 1823. Type. Brazil. Amazonas: Rio Japurá, n.d., *C. Martius 3127* (holotype, M; F neg. 18502) (Figure 6.60d–f).

Geonoma uleana Dammer, Verh. Bot. Vereins Prov. Brandenburg 48: 122. 1907. Type. Brazil. Amazonas or Acre: Rio Juruá, Cachoeira, May 1901, *E. Ule 5521* (isotype, MG; F neg. 5467).

Stems cespitose or solitary, 0.5–1.5(–4) m tall, 0.5–1 cm diam, erect or leaning. *Leaves* 6–10; sheath 4–9 cm long; petiole 5–17 cm long; rachis 8–25 cm long; leaf entire, oblong, or with 2–3(–4) unequal pinnae per side, ± sigmoid, the middle ones 12–33 cm long, (1–)5–11(–20) cm wide, the apical one usually much wider, veins forming a 40–55° angle with the rachis. *Inflorescences* interfoliar, becoming infrafoliar in fruit, spicate or branched to 1 order; peduncle 2–9 cm long; prophyll 4–7 cm long; peduncular bract 4.5–6.5 cm long, both bracts early deciduous; rachis absent; rachillae 1–3(–5), folded and twisted within the bud, with a very uneven surface when dry, densely to scarcely covered with wooly, brown hairs, 10–41 cm long, 1.5–2.5(–3) mm diam, with flower pits loosely arranged in vertical files, the lower lip prominent and bifid, the upper lip prominent; staminate flowers 3–4 mm long (in bud); sepals ovate, 3–3.5 mm long; petals ovate, 2–

Figure 6.60. (a) *Geonoma myriantha,* **(b)** infructescence (from *A. Henderson 1684*), **(c)** fruits; **(d)** *Geonoma arundinacea,* **(e)** infructescence (from *R. Pardini 14*), **(f)** fruits; **(g)** *Geonoma aspidiifolia,* **(h)** infructescence (from *S. Tillett 44931*), **(i)** fruits; **(j)** *Geonoma baculifera,* **(k)** infructescence (from *G. Aymard 7290*), **(l)** fruits.

3 mm long; filaments very briefly inflexed apically, with two long, free thecae; pistillode 1 mm long; pistillate flowers (in bud) 3–3.5 mm long; sepals ovate, 2.5–3.5 mm long; petals ovate, 2–3 mm long; staminodial tube crenate apically; *fruits* ellipsoid, to 8 mm long, to 6 mm diam.

Colombia (Amazonas, Putumayo), Ecuador (Napo, Pastaza), Peru (Amazonas, Huánuco, Loreto, Pasco), and Brazil (Amazonas) (Figure 6.61A); lowland rain forest on *terra firme,* at low elevations but up to 1500 m on eastern Andean slopes.

Brazil: *ubim.* Peru: *kamáncha.*

Geonoma arundinacea is a seldom collected and poorly known species, and is apparently a rare plant (Galeano, 1991). It is distinguished by its elongate inflorescences with 1–3(–5) rachillae, presence of both prophyll and peduncular bract, rachilla(e) folded and twisted within the bud, and loosely arranged flower pits with a bilobed lower lip. The few specimens examined are rather heterogeneous and the species remains equivocal.

2. Geonoma aspidiifolia Spruce, J. Linn. Soc., Bot. 11: 112. 1871. Type. Brazil. Amazonas: Rio Tarumá, Feb 1855, *R. Spruce 75* (holotype, K; isotype, NY) (Figure 6.60g–i).

Geonoma fusca Wess. Boer, Mem. New York Bot. Gard. 23: 93. 1972. Type. Guyana. Upper Mazaruni River Basin, Mt. Ayanganna, 700–800 m, 5 Aug 1960, *S. & C. Tillett 45047* (holotype, NY).

Stems cespitose, 1–2(–3) m tall, 0.5–1 cm diam. *Leaves* 7–12; sheath 7–11 cm long; petiole 9–22(–30) cm long; rachis 16–21 cm long; pinnae 2–4(–6) per side, usually 3, sigmoid, of varying width, the middle ones 13–20 cm long, 3–6 cm wide, veins forming a 40–60° angle with the rachis. *Inflorescences* infrafoliar, branched to 1 order; peduncle 3–7 cm long; prophyll 5–8 cm long; peduncular bract ca. 4 cm long, included within the prophyll; rachis virtually absent; rachillae 2–5, 5–14 cm long, (2–)3–4 mm diam, stiff and erect, straight within the bud, reddish-brown-tomentose, with flower pits arranged in loose spirals ca. 1 mm apart, lower lip slightly raised, upper lip scarcely raised; staminate flowers to 5 mm long at anthesis (including exserted stamens); sepals oblanceolate-obovate, 2.5 mm long; petals obovate, 3 mm long; filaments inflexed apically with 2 free connectives and 2 free thecae; pistillode to 1 mm long; pistillate flowers 4.5–6 mm long at anthesis (including exserted stigmas); sepals oblanceolate-obovate, 3.5–4 mm long; petals obovate, 3.5–4 mm long; staminodial tube digitately lobed apically; *fruits* globose-ellipsoid, 0.8–1 cm long, 5–6 mm diam, dark red or reddish-purple.

Central Amazon region of Colombia (Amazonas), Guyana, and Brazil (Amazonas) (Figure 6.61B); lowland rain forest on *terra firme* and also in inundated areas, at low elevations but occasionally to 1350 m.

Geonoma aspidiifolia is distinguished by its small size; leaves typically with 3 sigmoid pinnae; small inflorescences with few, short, stiff, reddish-brown-tomentose rachillae; staminate flowers with free, inflexed thecae; and pistillate flowers with digitately lobed staminodial tube. The lobes of the staminodial tube are not so deep as those of other species, nor do they spread at anthesis. The only specimen examined from the western part of the range is considerably smaller than the others. Wessels Boer (1968) confused two species in his treatment of *Geonoma aspidiifolia*; he included in synonymy Spruce's *Geonoma chelidonura,* which is quite distinct. Wessels Boer's *G. fusca* clearly belongs here.

3. Geonoma baculifera (Poit.) Kunth, Enum. pl. 3: 233. 1841. *Gynestum baculiferum* Poit., Mém. Mus. Hist. Nat. 9: 389. 1822. Type. French Guiana. Without locality, n.d., *A. Poiteau s.n.* (holotype, P) (Figure 6.60j–l).

Geonoma acutiflora Mart., Hist. nat. palm. 2: 10. 1823. Type. Brazil. Pará: without locality, n.d., *C. Martius s.n.* (holotype, M, n.v.; F neg. 18503).

Geonoma macrospatha Spruce, J. Linn. Soc., Bot. 11: 105. 1869. *Geonoma baculifera* var. *macrospatha* (Spruce) Drude in C. Martius, Fl. bras.: Palmae II 3(2): 490. 1882. Type. Venezuela. Amazonas: Río Casiquiare, Dec 1853, *R. Spruce 42* (holotype, K; isotype, P; F neg. 38652).

Geonoma estevaniana Burret, Notizbl. Bot. Gart.

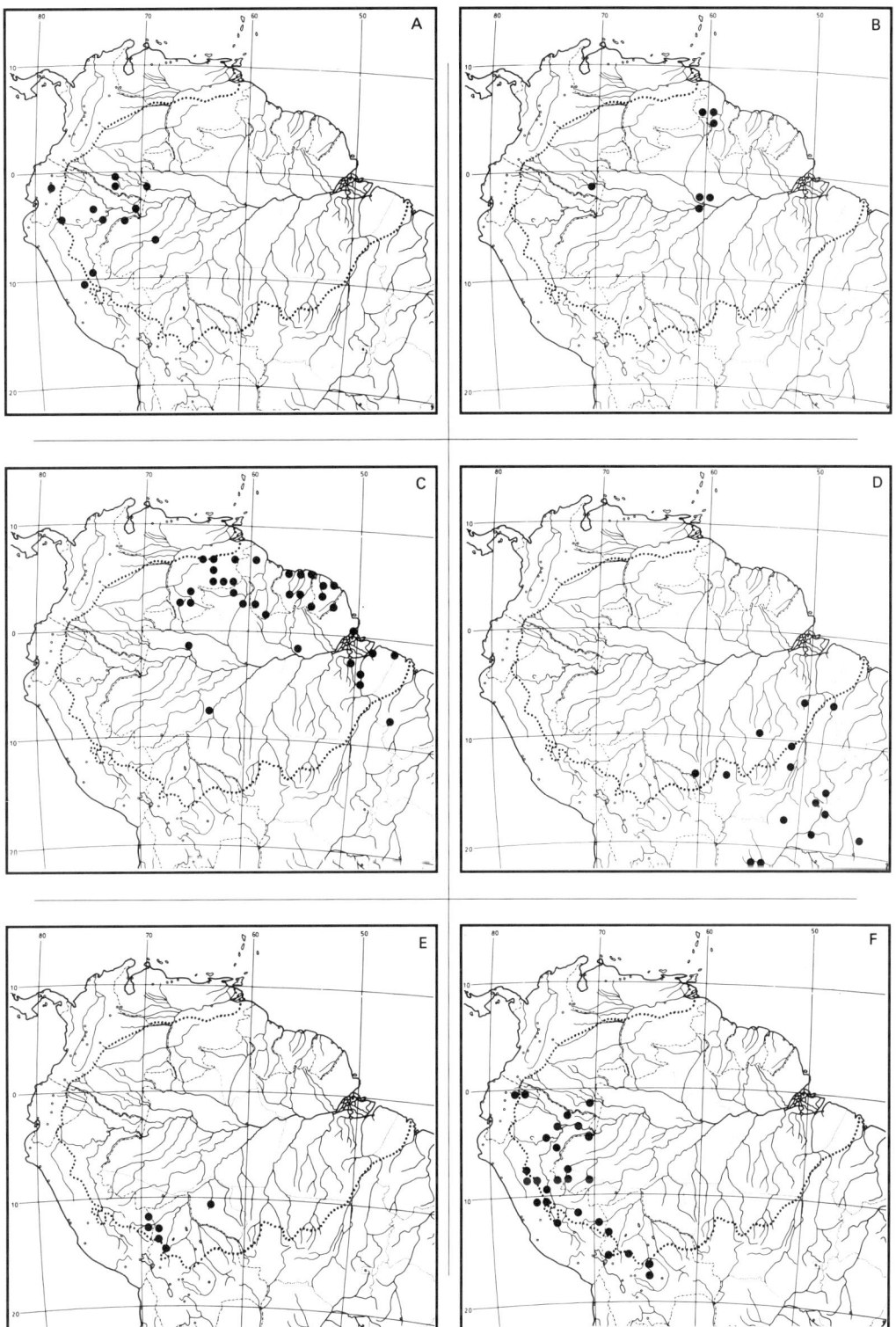

Figure 6.61. (A) *Geonoma arundinacea;* **(B)** *Geonoma aspidiifolia;* **(C)** *Geonoma baculifera;* **(D)** *Geonoma brevispatha* var. *brevispatha;* **(E)** *Geonoma brevispatha* var. *occidentale;* **(F)** *Geonoma brongniartii.*

Berlin-Dahlem 14: 256. 1938. Type. Brazil. Pará: Utinga, n.d., *M. Burret 208* (holotype, B, n.v.).

Stems cespitose, 1.3–4 m tall, 1–3 cm diam, erect or partly procumbent, often forming large colonies. *Leaves* 7–12; sheath 14–21 cm long; petiole 10–30 cm long; rachis 23–60 cm long; blade entire or irregularly pinnate, pinnae 3–several per side, the middle ones 40–46 cm long, 1–5 cm wide, veins forming a 30–35° angle with the rachis. *Inflorescences* interfoliar, erect at anthesis, branched to 1 or rarely 2 orders; peduncle 21–44 cm long; prophyll 16–32 cm long; peduncular bract 18–28 cm long, both bracts strongly flattened and persistent; rachis 1–10(–16) cm long; rachillae 3–10, 12–30 cm long, ca. 4 mm diam, simple or occasionally the basal ones bifurcate, the pits arranged in loose spirals 1–1.5 mm apart, lower lip entire and prominent, upper lip present; staminate flowers to 4 mm long (in bud); sepals narrowly elliptic, 3 mm long; petals elliptic, 3.5 mm long; filaments briefly inflexed apically, with 2 long, free thecae; pistillode at top of tube, very small; pistillate flowers 6 mm long at anthesis (including exserted stigmas), often persistent; sepals elliptic, 4 mm long; petals lanceolate-ovate, 4 mm long; staminodial tube crenate apically; *fruits* ovoid or ellipsoid, 0.9–1.2 cm long, 0.5–0.8 mm diam, black.

Northeastern part of the Amazon region in southern Venezuela (Amazonas, Bolívar), the Guianas, and Brazil (Amapá, Amazonas, Maranhão, Pará, Roraima) (Figure 6.61C); lowland rain forest in low, flat, swampy areas near rivers, usually at low elevations but occasionally up to 650 m.

Brazil: *haijowa-'i* (Guajá), *owi* (Ka'apor), *ubi, ubim, ubím, ubím grande*. French Guiana: *waï* (Creole). Guyana: *meena*. Venezuela: *baraboro, baroboro, palma cola de pescado, palma san pablo, san pablo*. Surinam: *tas, tassie*. The leaves are commonly used for thatching, and the seeds are occasionally used as beads.

Geonoma baculifera is distinguished by its inflorescences with elongate, flattened, subequal prophyll and peduncular bract; elongate peduncle, short rachis, and few, long, simple, rarely bifurcate rachillae; flower pits with upper lip; and ovoid or ellipsoid fruits.

4. **Geonoma brevispatha** Barb. Rodr., Protesto-Appendice 41. 1897. Lectotype (Wess. Boer, 1968). Barb. Rodr., Sert. palm. brasil. 1: t. 22. 1903 (Figure 6.62a–c).

Stems 0.7–4 m tall, 1.5–2.3(–7) cm diam, cespitose and usually forming clumps, rarely appearing solitary. *Leaves* 7–18; sheath 12–19 cm long; petiole 20–46 cm long; rachis 30–43 cm long; pinnae 3–19 per side, linear to sigmoid, regularly arranged or commonly narrow, linear pinnae interspersed with 3–more broader pinnae, often only 3 broad pinnae present, the middle ones 13–42 cm long, 0.5–10 cm wide, often with ramenta on abaxial veins, veins forming a 35–45° angle with the rachis. *Inflorescences* infrafoliar, branched to 1 or 2 orders; peduncle 5–27 cm long; prophyll 6–15 cm long; peduncular bract 5–13 cm long, both bracts flattened, deciduous or ± persistent; rachis 5–14 cm long; rachillae numerous, to 23 or more, (4–)15–33 cm long, 2–3 mm diam; flower pits alternating in pairs at right angles (decussate), the pairs loosely spaced 2–4 mm apart, lower lip prominent, entire or bifid, upper lip smaller; staminate flowers 2–2.5 mm long (in bud); sepals lanceolate-ovate, 2 mm long; petals ovate, 2–2.5 mm long; stamens 6; filaments very briefly inflexed apically, with 2 long, free thecae; pistillode very small; pistillate flowers 3.5 mm long (at anthesis); sepals ovate, 2 mm long; petals ovate, 2 mm long; staminodial tube crenate apically; *fruits* globose-ellipsoid, 7–8 mm long, 6–6.5 mm diam, black.

As here understood, *Geonoma brevispatha* is a heterogeneous species (and is close to *G. schottiana*, which is the older name). Specimens from the open regions of central Brazil (e.g., Goiás, Mato Grosso, Minas Gerais) tend to have regularly pinnate leaves with numerous, linear pinnae (although there is a tendency for plants from the north and west of this part of the range to have wider pinnae), and interfoliar, long-pedunculate inflorescences with ± persistent bracts. These are here called var. *brevispatha*. Specimens from forested regions of the northwest part of the range (e.g., Peru [Madre de Dios], Brazil [Rondônia], and Bolivia [La Paz, Pando]) tend to have a few, broad, almost sigmoid pinnae, and infrafoliar, short-pedunculate inflorescences with deciduous

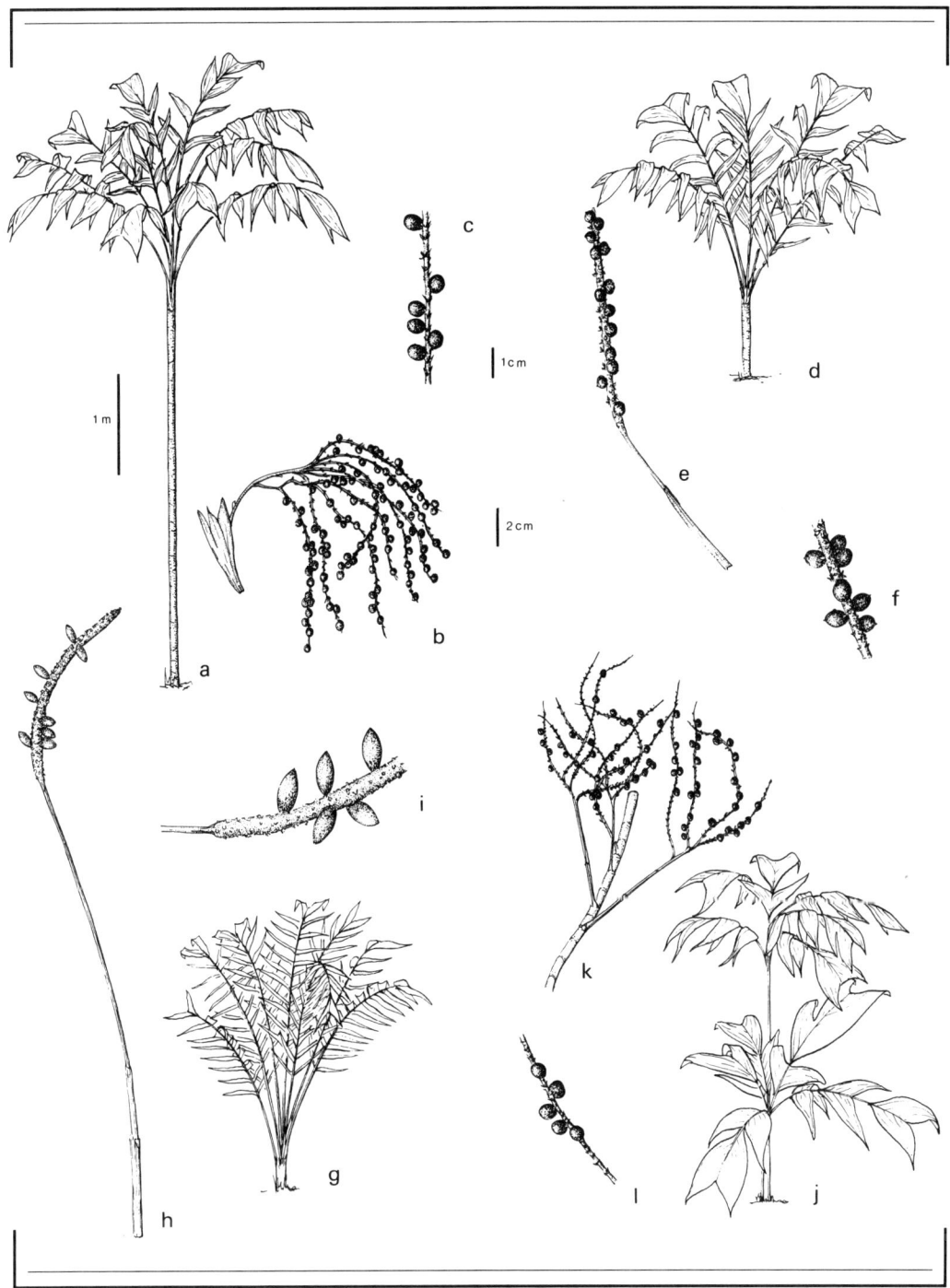

Figure 6.62. **(a)** *Geonoma brevispatha* var. *occidentale,* **(b)** infructescence of var. *brevispatha* (from G. Prance 59110), **(c)** fruits; **(d)** *Geonoma brongniartii,* **(e)** infructescence (from F. Chávez 705), **(f)** fruits; **(g)** *Geonoma camana,* **(h)** infructescence (from A. Henderson 1661), **(i)** fruits; **(j)** *Geonoma deversa,* **(k)** infructescences (from B. Maguire 29347), **(l)** fruits.

bracts. These northwestern plants are here called var. *occidentale*. They may be a separate species; unfortunately, inflorescence bracts are lacking from all specimens and their absence makes a decision difficult.

KEY TO THE VARIETIES OF *GEONOMA BREVISPATHA*

1. Leaves typically with numerous, linear pinnae; inflorescences interfoliar; peduncle (8–)11–27 cm long . . . .
. . . . 4a. *G. brevispatha* var. *brevispatha*.
1. Leaves typically with few, broad, ± sigmoid pinnae; inflorescences infrafoliar; peduncle 5–7 cm long. . . . .
. . . . 4b. *G. brevispatha* var. *occidentale*.

4a. Geonoma brevispatha var. brevispatha

Leaves typically regularly pinnate with 13–19 stiff, linear pinnae per side, occasionally pinnae wider and fewer or very rarely leaf almost entire, the middle ones 13–30 cm long, 0.5–1(–8) cm wide, with prominent ramenta abaxially. *Inflorescences* interfoliar, slender; peduncle (8–)11–27 cm long; prophyll (6–)9–15 cm long; peduncular bract 7–13 cm long, shorter than prophyll and inserted some distance from it, both bracts tubular (rarely swollen) and ± persistent.

Planalto *(cerrado)* region of central Brazil (Bahia, Distrito Federal, Goiás, Mato Grosso do Sul, Minas Gerais, Pará, Tocantins), Bolivia (Santa Cruz), and northern Paraguay (Amambay, Canendiyu, Alto Parana), reaching the southern Amazon region in Brazil (Pará) on the Serra do Cachimbo and Serra dos Carajás (Figure 6.61D). It grows in *cerrado,* commonly in gallery forest or forest patches at elevations up to 1600 m, and is almost always found in wet places near streams.

Brazil: *icaí, te-ere* (Apinajé), *ubim.* A tea is made from the palmito and drunk for stomach pains (Apinajé).

4b. Geonoma brevispatha var. occidentale
Henderson, var. nov. Type. Peru. Madre de Dios: Río Tambopata, Explorer's Inn at junction with Río La Torre, 12°50'S, 69°17'W, 3 Nov 1991, *A. Henderson & F. Chávez 1633* (holotype, USM; isotypes, CUZ, NY).

A var. brevispatha pinnis minus numerosis latioribus pedunculoque breviori differt.

Leaves typically with 3 broad, ± sigmoid pinnae per side, often with a few, linear pinnae at base of rachis, the middle ones 25–33 cm long, (1–)5–10 cm wide, lacking ramenta abaxially. *Inflorescences* infrafoliar, stouter; bracts not seen, early deciduous (probably shorter than var. *brevispatha);* peduncle 5–7 cm long.

Southwestern part of the Amazon region in Peru (Madre de Dios), Brazil (Rondônia), and Bolivia (La Paz, Pando), and also just reaching Andean foothills in Bolivia (Figure 6.61E); lowland rain forest on periodically inundated soils at low elevations.

Bolivia: *jatata.* Peru: *palmiche.* The leaves are seldom used for thatching in the southwestern Amazon region, since they are much less durable than the sympatric, and more common, *Geonoma deversa.*

5. **Geonoma brongniartii** Mart. in A. D. Orb., Voy. Amérique mér. 7(3). Palmiers 24. 1843. Type. Bolivia. Cochabamba: Prov. Carrasco, n.d., *A. d'Orbigny 39* (holotype, P; F neg. 38642) (Figure 6.62d–f).

Geonoma werdermannii Burret, Bot. Jahrb. Syst. 63: 173. 1930. Type. Bolivia. Beni: Mission Todos Santos, 300 m, 2 Aug 1926, *E. Werdermann 2183* (holotype, B, n.v.).

Geonoma cuneifolia Burret, Notizbl. Bot. Gart. Berlin-Dahlem 11: 199. 1931. Type. Peru. Loreto: Río Ucayali, n.d., *G. Tessmann 3317* (isotype, NY).

Stems solitary or cespitose, short and subterranean, or aerial and 0.3–1 m long, 2.5–3.5 cm diam. *Leaves* 5–13; sheath 7–10 cm long; petiole 19–65 cm long; rachis 30–84 cm long; blade entire, or with 5–14 pinnae per side, these irregularly spaced and of varying width, the middle ones 27–37 cm long, 3–9 cm wide, veins forming a 40–60° angle with the rachis. *Inflorescences* interfoliar, spicate or rarely branched to 1 order; peduncle 19–28 cm long; prophyll 17–21 cm long; peduncular bract ca. 16 cm long, inserted 3.5–4.5 cm above base of peduncle; rachilla 1 (very rarely 5), 14–40 cm long, 2.5–3.5(–5) mm diam, often curved, red at anthesis, with a 1–2 cm pointed apex, with pits arranged in loose (rarely close) spirals, 1.5–2.2 mm apart, individual pits about 2 mm diam, with a prominent, bifid lower lip and obscure upper lip; staminate flowers (immature) 2.5 mm long; sepals lanceolate-ovate, 2 mm long; petals ovate, 2 mm long; stamens 6; filaments very briefly inflexed apically, with 2 long, free thecae; pistillode very small; pistillate flowers at anthesis (including stigmas) 4.5 mm long; sepals lanceolate-ovate, 3 mm long; petals lanceolate-ovate, 3 mm long; staminodial tube crenate apically; *fruits* globose-ellipsoid, 5–7 mm diam, black.

Western Amazon region in Colombia (Amazonas, Caquetá, Meta, Putumayo), Ecuador (Napo), Peru (Ayacucho, Huánuco, Junín, Loreto, Madre de Dios, Pasco, Puno, San Martín, Ucayali), Brazil (Acre, Amazonas), and Bolivia (Beni, Cochabamba, La Paz, Santa Cruz) (Figure 6.61F); lowland rain forest in areas liable to inundation, occasionally on *terra firme*, below 500 m elevation. According to Wessels Boer (1988), it also occurs in Venezuela (Apure).

Colombia: *San Pablo*. Ecuador: *ní-ní* (Secoya). Peru: *cullulí*. The leaves are occasionally used for thatch, but are not considered very durable.

Geonoma brongniartii is distinguished by its subterranean or short stem; long, pedunculate, spicate inflorescences (one specimen from Madre de Dios, Peru [*A. Henderson 1636*] has an unusual, branched inflorescence, and may be a hybrid with *G. poeppigiana*); long rachilla(e); spirally arranged flower pits, anthers with free, inflexed thecae; and crenate staminodial ring. There is a tendency for leaf division to change from north to south. Plants from Colombia and Ecuador, in the northern part of the range, tend to have pinnate leaves with numerous, narrower pinnae. Plants from the middle part of the range have pinnate leaves with fewer, broader pinnae. Those from the southern part of the range, in Bolivia and southern Peru, tend to have entire leaves.

This species is very heterogeneous, and is part of a complex of species (Group 5 of Wessels Boer, 1968) from eastern Andean slopes and adjacent areas that includes *Geonoma jussieuana* (which may just reach below 500 m elevation on Andean slopes) and *G. lehmanii*. As pointed out by Skov (1989), there are intermediates between all three species. Furthermore, this complex is poorly distinguished from the trans-Andean *G. cuneata* complex (including *G. gracilis, G. procumbens,* and *G. sodiroi*; Skov, 1989). This latter complex has the peduncular bract inserted close to the prophyll at the base of the peduncle, whereas in the *G. brongniartii* complex it is inserted some distance from the prophyll. Some specimens of *G. brongniartii*, from scattered localities, with large inflorescences to 30 cm long and 5 mm diameter, actually resemble more *G. cuneata,* and it seems possible that this also occurs on western Andean slopes. Alternatively, the specimens of *G. brongniartii* with smaller inflorescences may actually be *G. gracilis* or *G. jussieuana*. Unfortunately, the Amazon region specimens are incomplete (i.e., lack the base of the peduncle), and the problem cannot be solved without complete material.

6. Geonoma camana Trail, J. Bot. 5: 324. 1876. *Taenianthera camana* (Trail) Burret, Bot. Jahrb. Syst. 63: 270. 1930. Type. Brazil. Amazonas: San Antonia da Boa Vista, Rio Javari, 4 Dec 1874, *J. Trail 977/CLXXXII* (holotype, K; isotype, P; F neg. 38643) (Figure 6.62g–i).

Geonoma lagesiana Dammer, Verh. Bot. Vereins Prov. Brandenburg 48: 121. 1906. *Taenianthera lagesiana* (Dammer) Burret, Bot. Jahrb. Syst. 63: 270. 1930. Type. Brazil. Amazonas: Rio Juruá, Juruá-mirim, Aug 1901, *E. Ule 5745* (isotype, MG).

Stem solitary, short and subterranean or more often aerial and 0.3–2 m tall, (2–)4–6 cm diam, with a basal cone of roots to 20 cm long. *Leaves* 5–14; sheath and petiole (48–)70–175 cm long; rachis 0.5–1.3 m long; pinnae 2–31 per side, occasionally leaf entire, regularly arranged and spreading in 1 plane, elongate-sigmoid, the middle ones 18–45 cm long, usually of equal (ca. 2 cm) width but often with wider (ca. 5 cm) pinnae interspersed, apical pinna much wider than the others, lighter green abaxially, the narrower pinnae with prominent central and two lateral, submarginal veins, veins forming a 40–70° angle with the rachis, the veins often with brown tomentum abaxially. *Inflorescences* interfoliar, spicate; peduncle 54–96 cm long; prophyll 8–17 cm long; peduncular bract 28–70 cm long; rachilla 1, 12–30 cm long, 7–9 mm diam, with flower pits arranged in close, vertical files, with prominent, bifid lower lip and obscure upper lip; staminate flowers 3 mm

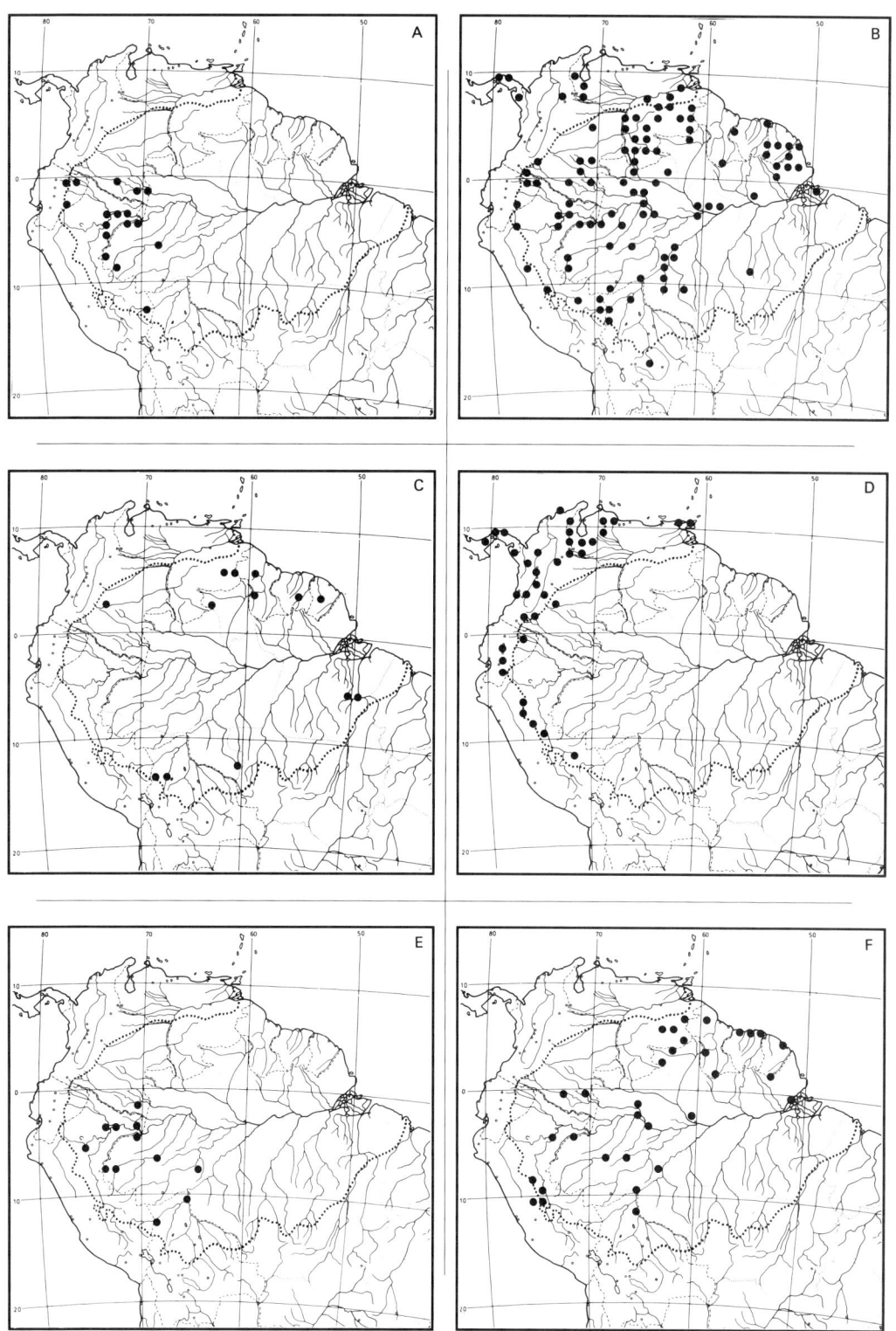

Figure 6.63. **(A)** *Geonoma camana;* **(B)** *Geonoma deversa;* **(C)** *Geonoma interrupta* var *euspatha;* **(D)** *Geonoma interrupta* var. *interrupta;* **(E)** *Geonoma laxiflora;* **(F)** *Geonoma leptospadix.*

long (immature); sepals oblanceolate, 3 mm long; petals obovate, 3 mm long; stamens 6; filaments with 2 long, free thecae; pistillate flowers (immature) 2.5 mm long; sepals oblanceolate, 2.5 mm long; petals obovate, 2 mm long; staminodial tube digitately lobed apically; *fruits* ellipsoid, 1–1.2 cm long, 6–8 mm wide, black or blue-black.

Western Amazon region in Colombia (Amazonas, Putumayo), Ecuador (Morona-Santiago, Napo), Peru (Loreto, Madre de Dios), and Brazil (Acre, Amazonas) (Figure 6.63A); lowland rain forest on *terra firme* or in wet places, at elevations below 300 m.

Brazil: *assai-rana, ubim*.

Geonoma camana is distinguished by its short, stout stem; its elongate, sigmoid pinnae with their characteristic color and venation; spicate inflorescences; staminate flowers with short, united thecae in line with the filament; and pistillate flowers with digitately lobed staminodial tube.

7. Geonoma deversa (Poit.) Kunth, Enum. pl. 3: 321. 1841. *Gynestum deversum* Poit., Mém. Mus. Hist. Nat. 9: 390. 1822. Type. French Guiana. Without locality, n.d., *A. Poiteau s.n.* (holotype, P) (Figure 6.62j–l; Plate Id).

Geonoma paniculigera Mart., Hist. nat. palm. 2: 11. 1823. Type. Brazil. without locality, n.d., *C. Martius s.n.* (holotype, M, n.v.; F neg. 18517).

Geonoma desmarestii Mart. in A. D. Orb., Voy. Amérique mér. 7(3). Palmiers 23. 1847. Type. Bolivia. Cochabamba: without locality, n.d., *A. d'Orbigny 50* (holotype, P; F neg. 38647).

Geonoma rectifolia Wallace, Palm Trees of the Amazon 67. 1853. Lectotype (Glassman, 1972). Wallace, Palm Trees of the Amazon t. 25. 1853.

Geonoma microspatha Spruce, J. Linn. Soc., Bot. 11: 108. 1871. *Geonoma paniculigera* var. *microspatha* (Spruce) Trail, J. Bot. 5: 327. 1876. Type. Brazil. Amazonas: Rio Negro, Serra do Gama, São Gabriel, Feb 1853?, *R. Spruce 28* (holotype, K; isotype, P; F neg. 38653).

Geonoma microspatha var. *pacimoensis* Spruce, J. Linn. Soc., Bot. 11: 108. Type. Brazil. Amazonas: Río Casiquiare. n.d., *R. Spruce 41* (holotype, K; isotype, P).

Geonoma paniculigera var. *papyracea* Trail, J. Bot. 5: 326. 1876. Type. Brazil. Amazonas: Rio Javari, 8 Dec 1874, *J. Trail 943/CXCIII* (holotype, K).

Geonoma paniculigera var. *cosmiophylla* Trail, J. Bot. 5: 326. 1876. Type. Brazil. Amazonas: Tabatinga, 30 Nov 1874, *J. Trail 956/CLXXVII* (holotype, K; isotype, P; F neg. 38656).

Geonoma paniculigera subvar. *gramineifolia* Trail, J. Bot. 5: 327. 1876. Type. Brazil. Amazonas: Tabatinga, 30 Nov 1874, *J. Trail 958/CLXXVII* (holotype, K; isotype, P).

Geonoma trijugata Barb. Rodr., Enum. palm. nov. 12. 1875. Lectotype (Wess. Boer, 1968). Barb. Rodr., Sert. palm. brasil. 1: t. 14. 1903.

Geonoma yauaperyensis Barb. Rodr., Contr. Jard. Bot. Rio de Janeiro 3: 88. 1902. Lectotype (Wess. Boer, 1968). Barb. Rodr., Sert. palm. brasil. 1: t. 30. 1903.

Geonoma tessmannii Burret, Bot. Jahrb. Syst. 63: 181. 1930. Type. Peru. Amazonas: Río Marañon, 7 Oct 1924, *G. Tessmann 4225* (holotype, B, n.v.).

Geonoma bartlettii Dammer ex Burret, Bot. Jahrb. Syst. 63: 183. 1930. Type. Guyana. Conawaruk River, n.d., *A. Bartlett 8195* (holotype, B, n.v.).

Geonoma leptostachys Burret, Notizbl. Bot. Gart. Berlin-Dahlem 10: 1014. 1930. Type. Brazil. Amazonas: Rio Negro, Camanaos, 26 Sep 1928, *P. Luetzelburg 23072* (holotype, B, n.v.; F neg. 18511).

Geonoma macropoda Burret, Notizbl. Bot. Gart. Berlin-Dahlem 10: 1015. 1930. Type. Brazil. Amazonas: Manaus, 26 Aug 1928, *P. Luetzelburg 22089* (isotype, R; F neg. 18513).

Geonoma major Burret, Notizbl. Bot. Gart. Berlin-Dahlem 10: 1016. 1930. Type. Brazil. Amazonas: Rio Negro, Serra do Cucui, 25 Sep 1928, *P. Luetzelburg 22273* (isotype, R; F neg. 18515).

Geonoma killipii Burret, Notizbl. Bot. Gart. Berlin-Dahlem 11: 320. 1932. Type. Peru. Junín: Puerto Bermudez, ca. 375 m, 14–17 Jul 1929, *E. Killip & A. Smith 26594* (isotypes, NY, US).

Stems cespitose, or appearing solitary, 0.8–3 m tall, 0.8–3 cm diam. *Leaves* 7–16; sheath 8–13(–22) cm long; petiole 12–28(–35) cm long, rachis 31–50 cm long; pinnae very variable, typically 3 per side, but sometimes leaf entire, or regularly pinnate with to 18 pinnae per side, these broad or narrow, somewhat sigmoid, the middle ones 14–30 cm long, 0.5–6.5 cm wide, veins forming a 30–40° angle with the rachis, alternate veins brown-tomentose abaxially. *Inflorescences* infrafoliar, branched to 1 or 2, rarely 3, orders; peduncle 5–15 cm long; prophyll 5–10 cm long; peduncular bract included within prophyll, 4–9 cm long, both bracts early deciduous, not flattened, somewhat inflated; rachis 4–13 cm long; rachillae 3–25, 7–32 cm long, 1–2 mm diam, folded and twisted in bud; flower pits raised and arranged in alternating whorls of 3, occasionally the whorls very close and then the pits appearing spirally arranged; staminate flowers 2.5 mm long (in bud); sepals ovate, 2 mm long; petals ovate, 2.2 mm long; stamens 6; filaments inflexed apically with 2 free connectives and 2 free thecae; pistillode at base of tube; pistillate flowers 4 mm long at anthesis (including exserted stigmas); sepals ovate, 2.5 mm long; petals ovate, 3 mm long; staminodial tube crenate apically (it and style becoming detached together as ovary swells); *fruits* globose, 5–7 mm diam, purple-black.

Central America (Belize, Nicaragua, Costa Rica, Panama) and throughout northern South America in Colombia (Amazonas, Antioquia, Caquetá, Chocó, Meta, Norte de Santander, Putumayo, Vaupés, Vichada), Venezuela (Amazonas, Apure, Bolívar, Delta Amacuro, Mérida, Zulia), the Guianas, Ecuador (Morona-Santiago, Napo, Pastaza), Peru (Junín, Loreto, Madre de Dios, San Martín), Brazil (Acre, Amapá, Amazonas, Pará, Rondônia, Roraima), and Bolivia (Beni, Cochabamba, La Paz, Beni) (Figure 6.63B); lowland or premontane rain forest, usually on *terra firme* but occasionally near streams in areas liable to inundation, at elevations up to 1200 m.

Bolivia: *jatata*. Brazil: *ubim, ubim comum*. Colombia: *go-güi-re de centro de monte* (Huitoto), *San Paulo, vá-va-ra* (Guahibo). Ecuador: *chontillo, chontillo de loma, chontillo de monte, huasoëne* (Siona), *ñu-quan-ne* (Siona), *suté-dé-dé* (Siona), *turuji* (Shuar). French Guiana: *ohau-imon* (Oyampi), *vaí* (Creole). Peru: *kampana* (Huambisa Jívaro), *palmiche*. Venezuela: *baro, manása, manasha, palma San Pablo, varo-varo*. The thin, straight stems are used for various purposes. The leaves are widely used for thatching, and occasionally for lining baskets. This species is one of the most important thatch palms in the Amazon region. Rioja (1992) has discussed the marketing of *jatata* in eastern Bolivia.

Most specimens from the northeastern Amazon region in Venezuela (Bolívar), Guyana, Surinam, French Guiana, and Brazil (Amapá) are somewhat distinct. They have smaller leaves and small inflorescences with 3–8 simple rachillae. The type of the species is of this form, as is *Geonoma bartletti* (although Wessels Boer [1968] considered this a separate species and erroneously placed it next to *G. stricta*). This smaller form occasionally occurs in the western part of the Amazon region, and also in the Andes and Central America. Some populations of these smaller plants form apparent hybrids with *G. leptospadix* (e.g., *J. Pires 50759, H. Irwin 48082,* both from Amapá, Brazil; and *J.-J. de Granville 2592,* from French Guiana). The larger, widespread form that occurs throughout the rest of the Amazon region and beyond is very variable in leaf division, but is essentially uniform in its inflorescences with 4–25 slender rachillae with alternate whorls of 3 raised flower pits. A group of specimens from the western Amazon region in Colombia (Amazonas, Caquetá), Peru (Loreto), and Brazil (western Amazonas) have densely tomentose rachillae and the whorls are close together, and the pits appear almost spirally arranged.

8. Geonoma interrupta (Ruiz & Pav.) Mart., Hist. nat. palm. 2: 8. 1823. *Martinezia interrupta* Ruiz & Pav., Syst. veg. fl. peruv. chil. 298. 1798. Type. Peru. Pasco: between Pozuzo and Cuchero, n.d., *H. Ruíz Lopez & J. Pavón s.n.* (holotype, M, n.v.) (Figure 6.64a–c).

Stems usually solitary, 0.1–5 m tall, 2–7 cm diam. *Leaves* 6–23, spreading; sheath 8–40 cm long; petiole 9–115 cm long; rachis 0.3–2 m long; pinnae 3–41 per side (rarely leaf entire), linear to falcate or sigmoid, unequally wide, closely or widely spaced, strongly plicate, the middle ones 36–75 cm long, 3–15 cm wide, veins forming a 30–45° angle with the rachis, alternate veins tomentose abaxially. *Inflorescences* usually infrafoliar, branched 1–3 orders;

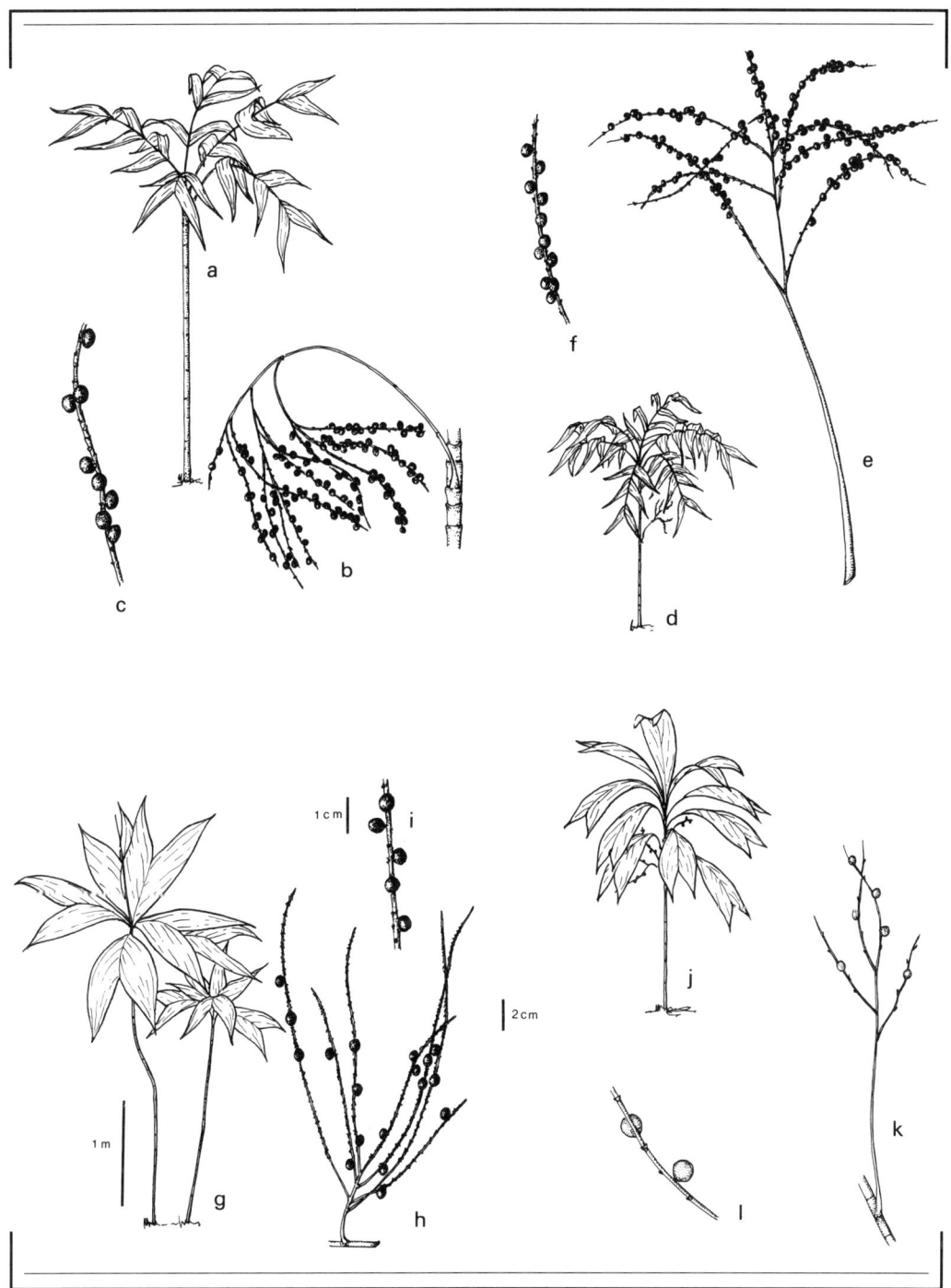

Figure 6.64. (a) *Geonoma interrupta* var. *interrupta,* (b) infructescence (from *L. Dorr 7861*), (c) fruits; (d) *Geonoma longepedunculata,* (e) infructescence (from *H. Balslev 4303*), (f) fruits; (g) *Geonoma laxiflora,* (h) infructescence (from *J. Solomon 10875*), (i) fruits; (j) *Geonoma leptospadix,* (k) infructescence (from *J. Steyermark 74750*), (l) fruits.

peduncle 5–40 cm long; prophyll 12–30 cm long; peduncular bract 11–29 cm long, both bracts tubular, early deciduous; rachis 2–55 cm long; primary branches to 14; rachillae numerous, 3–50, 5–25 cm long, 2–3 mm diam, tomentose, twisted and folded within the bud; pits in loosely arranged spirals, lower lip small, entire, upper lip absent; staminate flowers 2.5 mm long (in bud); sepals lanceolate, 2–2.5 mm long; petals narrowly elliptic, 2 mm long; stamens 6; filaments very briefly inflexed apically, with 2 long, free thecae; pistillate flowers 3–3.5 mm long; sepals free, lanceolate or linear-elliptic, 2–2.5 mm long; petals 2–2.5 mm long; staminodial tube crenate apically; *fruits* globose or almost ellipsoid, 5–7 mm long, to 5 mm diam, black.

This widespread species is very difficult taxonomically. Wessels Boer (1965) recognized a widespread, variable species, and included *Geonoma pinnatifrons* and *G. oxycarpa* in synonymy. He wrote that "a series of slightly different palms which have been described subsequently as a large number of slightly different species. . . . The differences are very slight indeed but tend to show some correlation with the distribution. This correlation, however, is weak and does not hold consistently." Later, Wessels Boer (1968) divided the specimens into several different species:

> The emphasis on slight but clearly detectable differences in bracts, pubescence of the rachillas, lips, pubescence of the flower pits, and last but not least, on shape, size and colour of the fruits permits a further separation. This separation proves to be natural as it coincides with geography and, as far as can be seen from the labels, with ecology.

However, there is now much more material available, and study of this has shown that the characters used by Wessels Boer are not consistent (see Moore, 1969). Here one widespread species is recognized. However, there are still problems with lowland, Amazon populations. Three populations from the Amazon region are known, and plants of these are smaller in all parts. They occur in the northeastern part of the region in Venezuela (Bolívar), the Guianas, and Brazil (Pará); in Colombia (Meta); and in the southwestern part of the region in Brazil (Rondônia) and Bolivia (Beni, La Paz). Most of these appear to be smaller versions of typical *G. interrupta,* and I have called them var. *euspatha.*

KEY TO THE VARIETIES OF *GEONOMA INTERRUPTA*

1. Stems 0.1–3 m tall; leaves 6–12; pinnae 3–10 per side or rarely leaf entire; rachillae 3–15 . . . . 8a. *G. interrupta* var. *euspatha.*
1. Stems 3–5 m tall; leaves 10–23; pinnae 4–41 per side; rachillae numerous, to 50. . . . . 8b. *G. interrupta* var. *interrupta.*

8a. Geonoma interrupta var. **euspatha** (Burret) Henderson, stat. nov. *Geonoma euspatha* Burret, Notizbl. Bot. Gart. Berlin-Dahlem 11: 10. 1930. Type. Colombia. Caquetá: Sucre, 10 Jul 1926, *G. Woronow & S. Juzepczuk 5885* (holotype, LE, n.v.).

Geonoma karuaiana Steyerm., Fieldiana Bot. 28: 88. 1951. Type. Venezuela. Bolívar: Río Karuai, Sororopán-tepuí, W of La Laja, 29 Nov 1944, *J. Steyermark 60789* (holotype, F).

Stems solitary, 0.1–3 m tall, 2.5–4 cm diam, erect or occasionally procumbent. *Leaves* 6–12; sheath 8–15 cm long; petiole 10–70 cm long; rachis 34–65 cm long; blade irregularly pinnate with 3–10 unequal pinnae, linear or somewhat sigmoid, of unequal width, the middle ones 36–46 cm long, 4–5.5 cm wide, or rarely leaf entire, veins forming a 35–40° angle with the rachis. *Inflorescences* interfoliar, erect at anthesis, branched 1 or 2 orders; peduncle 5–40 cm long; prophyll 12–20 cm long; peduncular bract 11–19 cm long, both bracts tubular; rachis 2–10 cm long; rachillae 3–15, 5–25 cm long, lower ones usually branched, with the flower pits loosely arranged in spirals 1.5–2.5 mm apart, lower lip entire, prominent, upper lip absent and inside of pit pilose.

Colombia (Meta), Venezuela (Bolívar), the Guianas, Brazil (Pará, Rondônia, Roraima), and Bolivia (Beni, La Paz) (Figure 6.63C). It has a rather unlikely distribution. In the northeastern part of its range (Venezuela [Bolívar], the Guianas, and Brazil [Roraima]), it grows in forest on mountain slopes between 400 and 1200 m elevation. These populations were called *G. interrupta* by Wessels Boer (1965). The specimens from the southwestern part of the range, in Bolivia, also grow in forest, but can occur at

200 m elevation. Although these populations are very similar to the eastern ones, phytogeographically they are separate. There are also isolated collections from southern Pará and Rondônia, Brazil.

Venezuela: *san pablo*. The leaves are used for thatching.

8b. Geonoma interrupta var. interrupta

Stems usually solitary, 3–5 m tall, 2–7 cm diam. *Leaves* 10–23, spreading; sheath 15–40 cm long; petiole 9–115 cm long; rachis 1–2 m long; pinnae 6–41 per side, falcate, unequally wide, closely or widely spaced, strongly plicate, the middle ones 46–75 cm long, 3–15 cm wide, veins forming a 30–45° angle with the rachis, alternate veins tomentose on lower surface. *Inflorescences* usually infrafoliar, branched to 1–3 orders; peduncle 22–35 cm long; prophyll 13–30 cm long; peduncular bract 20–29 cm long, both bracts early deciduous; rachis 30–55 cm long; primary branches to 14; rachillae numerous, to 50, 9–22 cm long, 2–3 mm diam, tomentose, twisted and folded within the bracts; pits in loosely arranged spirals, lower lip small, entire, upper lip absent.

Atlantic slope in southern Mexico (Chiapas, Oaxaca, Tabasco, Veracruz) and Central America (Belize, Costa Rica, Guatemala, Nicaragua, Panama), and then Colombia (Antioquia, Chocó, La Guajira, Magdalena, Meta, Norte de Santander, Risaralda, Santander, Valle del Cauca), northern Venezuela (Barinas, Lara, Mérida, Monagas, Portuguesa, Sucre, Tachira, Yaracuy, Zulia), Trinidad, Haiti, the Lesser Antilles, and the Amazon region of Colombia (Caquetá, Putumayo), Ecuador (Morona-Santiago, Napo, Pastaza), and Peru (Huánuco, Madre de Dios, San Martín) (Figure 6.63D); lowland, premontane or montane rain forest, usually on slopes on *terra firme* or rarely inundated areas, at elevations of 200–1500(–1850) m.

Ecuador: *palma real de loma, púi* (Siona). Peru: *palmiche*.

9. **Geonoma laxiflora** Mart., Hist. nat. palm. 2: 12. 1823. Type. Brazil. Amazonas: Rio Japurá, n.d., *C. Martius s.n.* (holotype, M; F neg. 18510) (Figure 6.64g–i).

Geonoma laxiflora var. *depauperata* Trail, J. Bot. 5: 326. 1876. Type. Brazil. Amazonas: Ananaá, N bank of Rio Solimões, 6 Sep 1874, *J. Trail 1024/CXVI* (holotype, K).

Geonoma beccariana Barb. Rodr., Vellosia 1: 33. 1888. Lectotype (Wess. Boer, 1968). Barb. Rodr., Sert. palm. brasil. 1: t. 17. 1902.

Stems cespitose, bamboolike, sometimes forming large colonies, erect or leaning, 1.8–4 m tall, 0.4–1.5 cm diam. *Leaves* 6–10, clustered at top of stem; sheath 6–10 cm long; petiole 4–10 cm long; rachis 15–36 cm long; blade entire, wedge-shaped, deeply bifid at apex, cuneate at base, 25–40 cm long, veins forming a 25° angle with the rachis. *Inflorescences* infrafoliar, branched to 1, rarely 2, orders; peduncle 3.5–5 cm long; prophyll 3–5 cm long; peduncular bract similar to and slightly shorter than prophyll, both bracts somewhat inflated and early deciduous; rachis 4–8 cm long; rachillae 3–8, 25–35 cm long, simple or occasionally the basal ones bifurcate, twisted and folded within the bracts, with pits arranged in loose, vertical files or commonly tristichous, ca. 2.5 mm apart, the upper and lower lips prominent, entire; staminate flowers to 3.5 cm long at anthesis (including exserted stamens); sepals obovate-oblanceolate, 2 mm long; petals obovate, 2.5 mm long; stamens 6; filaments inflexed apically with 2 free connectives and 2 free thecae; pistillode ca. 1 mm long; pistillate flowers 3 mm long at anthesis (including exserted stigmas); sepals obovate-lanceolate, 2 mm long; petals obovate, 2 mm long; staminodial tube crenate apically; *fruits* globose-ellipsoid, 8–10 mm long, 5–7 mm diam, blue-black.

Western Amazon region in Colombia (Amazonas), Ecuador (Napo), Peru (Loreto), Brazil (Acre, Amazonas), and Bolivia (Pando) (Figure 6.63E); lowland rain forest on river margins in areas subject to inundation, at low elevations.

Brazil: *ubim da várzea*. Peru: *ponilla*.

Geonoma laxiflora is distinguished by its bamboolike stems forming large colonies; entire, wedge-shaped leaves; strongly infrafoliar inflorescences with short, inflated, subequal prophyll and peduncular bracts; and short peduncle and elongate rachillae with often tristichous flower pits.

10. **Geonoma leptospadix** Trail, J. Bot. 5: 327. 1876. Type. Brazil. Amazonas: Rio Tocantins, 24 Nov 1874, *J. Trail 963/CLXXII* (holotype, K; isotypes, NY, P) (Figure 6.64j–l).

Geonoma saramaccana L. H. Bailey, Bull. Torrey Bot. Club 75: 104. 1948. Type. Surinam. Saramacca River, 9 Jul 1944, *B. Maguire 24095* (holotype, NY).

Stems solitary or cespitose, 0.5–1.2(–2) m tall, 0.4–1 cm diam. *Leaves* 6–17; sheath 4–9 cm long; petiole 4–15(–28) cm long; rachis (13–)25–43 cm long; blade entire or sometimes regularly to irregularly divided into a few broad pinnae, the blade 30–52 cm long, 8–12 cm wide, veins forming a 20–25° angle with the rachis, alternate veins brown-tomentose abaxially. *Inflorescences* interfoliar, becoming infrafoliar, branched to 1 order; peduncle 6–20 cm long; prophyll 4–10 cm long; peduncular bract 3–8 cm long, rachis 0–7 cm long; rachillae 2–5, (5–)8–20 cm long, 1–1.5 mm diam, with flower pits loosely and spirally arranged 5–6 mm apart, lower lip prominent and almost forming a cupule with the upper lip; staminate flowers at anthesis to 4 mm long (including exserted stamens); sepals oblanceolate-obovate, 2 mm long; petals ovate, 1.5–2.5 mm long; filaments very briefly inflexed apically, with 2 long, free thecae; pistillode prominent, 1.5 mm long; pistillate flowers (in bud) 2 mm long; sepals obovate, 1.5 mm long; petals obovate; staminodial tube crenate apically; *fruits* globose, 5–8 mm diam, purple-black.

Amazon region of Colombia (Amazonas), Venezuela (Bolívar), the Guianas, Ecuador (Pastaza), Peru (Huánuco, Junín, Loreto, Pasco, Ucayali), Brazil (Acre, Amapá, Amazonas, Maranhão, Rondônia, Roraima), and Bolivia (Beni) (Figure 6.63F); lowland rain forest on *terra firme* at low elevations, rarely to 1200 m.

Brazil: *owi-ran* (Ka'apor), *ubim, ubim brava*. Colombia: *go-güi-n(e)* (Huitoto). Ecuador: *sápap* (Achuar Jivaro), *uksha* (Quechua). French Guiana: *awilâ* (Wayâpi). Peru: *sangapilla masha*. The leaves are used to thatch houses, especially the sides.

Geonoma leptospadix is distinguished by its narrow, linear, entire leaves with alternate veins tomentose abaxially; and its small inflorescences with slender rachillae and distantly spaced, spirally arranged flower pits. It has an unusually long pistillode. This species is also unusual in that many inflorescences or infructescences at the same stage can be present on individual plants at the same time (Trail, 1876). There are two possible specimens from the Magdalena Valley in Colombia. These are much smaller than usual, and may represent either this species or *G. deversa*. There are also apparent hybrids with *G. deversa*; see discussion under that species.

11. **Geonoma longepedunculata** Burret, Notizbl. Bot. Gart. Berlin-Dahlem 11: 8 1930. Type. Colombia. Caquetá: Getuchá, Río Orteguaza, 21 Jul 1926, *G. Woronow & S. Juzepczuk 6157* (holotype, LE, n.v.) (Figure 6.64d–f).

Stems solitary, 0.1–1 m tall, 3–5 cm diam, often somewhat procumbent and covered with debris. *Leaves* 9–10, erect; sheath 10–15 cm long; petiole 37–70 cm long; rachis 34–65 cm long, densely brown-tomentose, glabrescent; pinnae 3–11 per side (rarely leaf entire), elongate-sigmoid, well spaced, unequally wide, ± irregularly arranged and spreading in 1 plane, the middle ones 30–40 cm long, 2.5–8 cm wide, bicolorous, veins forming a 40–45° angle with the rachis. *Inflorescences* interfoliar, branched to 1 order, slender, erect; peduncle 28–40 cm long; prophyll 14–19 cm long; peduncular bract 13–19 cm long, both bracts flattened; rachis 5–11 cm long; rachillae 4–7, straight and stiff, 12–25 cm long, 2–3 mm diam, sparsely pilose; pits arranged in loose spirals, lower lip entire, upper lip obscure or absent; staminate flowers (in bud) 2 mm long; sepals lanceolate, 1.5–2 mm long; petals narrowly elliptic, 2 mm long; filaments very briefly inflexed at the apex with 2 long, free thecae; pistillate flowers (immature) 1.5 mm long; sepals lanceolate, 1.5 mm long; petals 2 mm long; staminodial tube blunt apically; *fruits* globose, 4–5 mm diam, black.

Colombia (Caquetá, Meta, Putumayo), Ecuador (Morona-Santiago, Napo, Pastaza), and Peru (Junín, Loreto, Pasco) (Figure 6.65A); lowland or premontane rain forest on Andean foothills, on *terra firme* at 200–700 m elevation.

Peru: *palmiche, sápap* (Achual Jívaro), *turúji* (Mayna Jívaro). The leaves are used for thatching.

Skov (1989) called the Ecuadorean specimens of this species *Geonoma euspatha*. I believe that the name should be applied to a

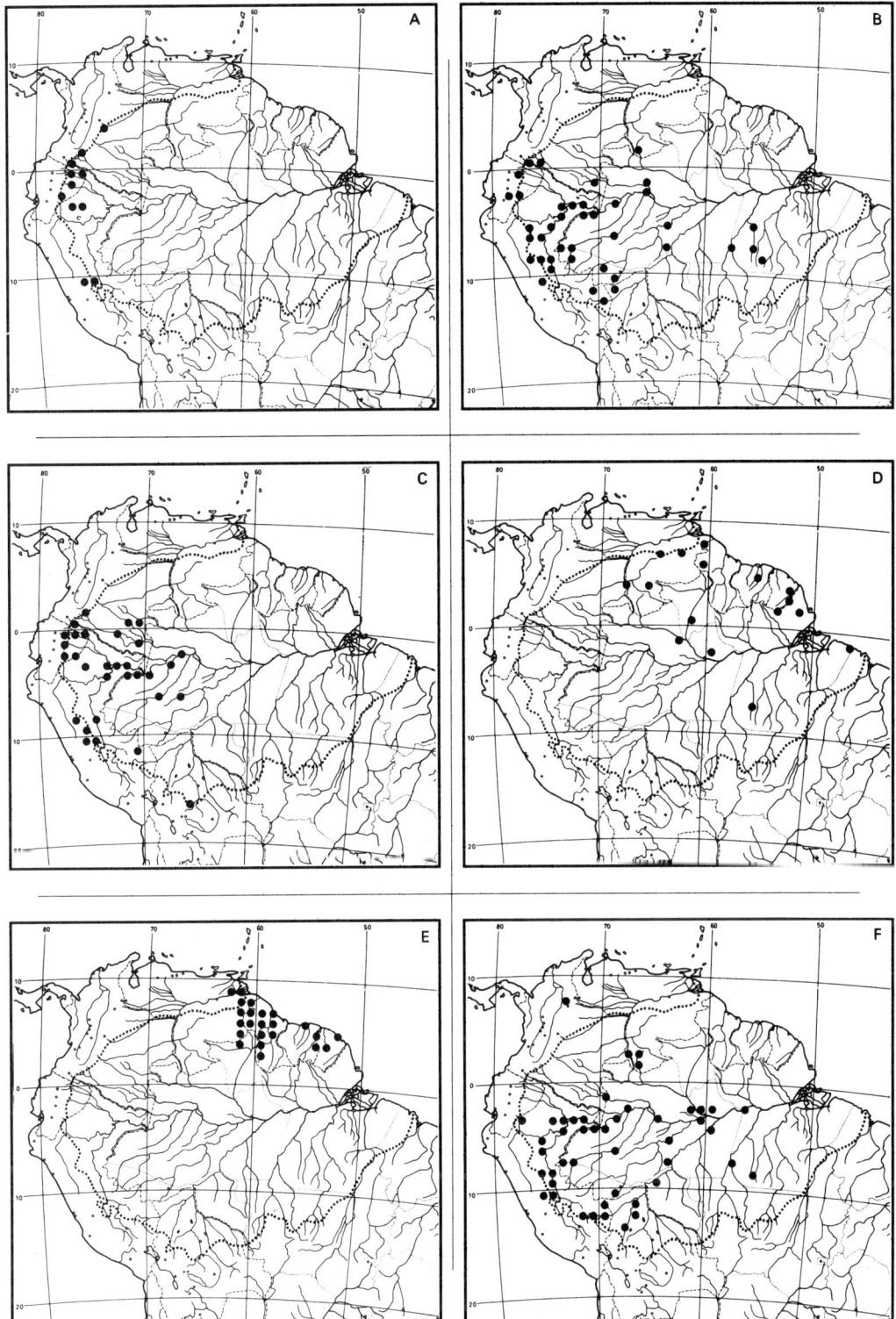

Figure 6.65. (A) *Geonoma longepedunculata;* **(B)** *Geonoma macrostachys* var. *acaulis;* **(C)** *Geonoma macrostachys* var. *macrostachys;* **(D)** *Geonoma macrostachys* var. *poiteauana;* **(E)** *Geonoma maxima* var. *ambigua;* **(F)** *Geonoma maxima* var. *chelidonura.*

variety of *G. interrupta,* and the correct name for the Ecuadorean, and other, specimens is *G. longepedunculata.* This species is superficially similar to *G. dicranospadix,* and the latter just reaches below 500 m elevation in Peru, but is generally from higher elevations and so is excluded from this flora. There is also some similarity to *G. poeppigiana,* and the differences are given in the key.

12. Geonoma macrostachys Mart., Hist. nat. palm. 2: 19. 1823. *Taenianthera macrostachys* (Mart.) Burret, Bot. Jahrb. Syst. 63: 268. 1930. Type. Brazil. Amazonas: Rio Japurá, n.d., *C. Martius s.n.* (holotype, M; F neg. 18514) (Figure 6.66a–c; Plate IId).

Stems solitary or rarely cespitose, short and subterranean, rarely aerial and 10–50 cm long, procumbent or erect. *Leaves* 4–14; sheath 13–25 cm long; petiole 0.6–1 m long; rachis 0.2–1 m long; blade entire, often long cuneate at the base and deeply bifid apically, narrow to broader, or leaf pinnate and then the pinnae 2–12 per side, linear to sigmoid, the middle ones 15–54 cm long, 2.5–7 cm wide, veins forming a 10–60° angle with the rachis. *Inflorescences* interfoliar, spicate; peduncle 0.3–1.1 m long; prophyll 5–16 cm long; peduncular bract 15–35 cm long; rachilla 1, 6–28 cm long (elongating between anthesis and fruiting), 3–10 mm diam, straight, green at anthesis, with flower pits arranged in close spirals almost touching one another or loose spirals ca. 2 mm apart, lower lip bifid, upper lip small; staminate flowers 3.5–6 mm long; sepals ovate, 2–3.5 mm long; petals ovate, 3–6 mm long; filaments inflexed apically with the connective not split and 2 free thecae; pistillode very small, at apex of filament tube; pistillate flowers 3–6 mm long; sepals lanceolate, 3–3.5 mm long; petals 3–3.5 mm long; staminodial ring digitately lobed apically; *fruits* globose to ellipsoid, apiculate, 8–12 mm long, 7–8 mm diam, black.

This is a widespread, variable species, which is here divided into three varieties. Needless to say, there are intermediates between all these, which are further discussed later. Variation in leaf shape in the three varieties is similar to that of *Geonoma maxima* and *G. stricta.*

KEY TO THE VARIETIES OF GEONOMA MACROSTACHYS

1. Leaves usually entire, or if pinnate then with linear pinnae; veins forming a 10–35° angle with the rachis.
    2. Rachilla 13–28 cm long and 7–10 mm diam; flower pits arranged in dense spirals almost touching one another; Colombia (Amazonas, Caquetá, Putumayo, Vaupés), Ecuador (Morona-Santiago, Napo), Peru (Huánuco, Junín, Loreto, Pasco, Madre de Dios, San Martín, Ucayali), Brazil (Acre, Amazonas), and Bolivia (Santa Cruz) . . . .
        . 12b. *G. macrostachys* var. *macrostachys.*
    2. Rachilla 8–17 cm long and 3–5 mm diam; flower pits arranged in loose spirals ca. 2 mm apart; Colombia (Vichada), Venezuela (Amazonas, Bolívar), the Guianas, and Brazil (Amapá, Amazonas, Pará) . . . .
        . . 12c. *G. macrostachys* var. *poiteauana.*
1. Leaves usually pinnate with sigmoid pinnae; veins forming a 40–60° angle with the rachis. . . 12a. *G. macrostachys* var. *acaulis.*

12a. Geonoma macrostachys var. **acaulis** Skov, Aarhus University PhD Diss. 112. 1989. *Geonoma acaulis* Mart., Hist. nat. palm. 2: 18. 1823. *Taenianthera acaulis* (Mart.) Burret, Bot. Jahrb. Syst. 63: 269. 1930. Type. Colombia. Amazonas: Río Caquetá, Cerro Yupatí, n.d., *C. Martius s.n.* (holotype, M; F neg. 18501).

Geonoma acaulis subsp. *tapajotensis* Trail, J. Bot. 5: 324. 1876. *Geonoma tapajotensis* (Trail) Drude in Mart., Fl. bras.: Palmae II fasc. 86 vol 3(2): 508. 1882. *Taenianthera tapajotensis* (Trail) Burret, Bot. Jahrb. Syst. 63: 269. 1930. Type. Brazil. Pará: Aramanahy, Rio Tapajós, 10 Jan 1874, *J. Trail 1017/IX* (holotype, K).

Taenianthera gracilis Burret, Notizbl. Bot. Gart. Berlin-Dahlem 11: 14. 1930. Type. Brazil. Amazonas: Rio Iça, Apr 1930, *C. Lakó 10* (holotype, B, n.v.).

Taenianthera oligosticha Burret, Notizbl. Bot. Gart. Berlin-Dahlem 11: 201. 1931. Type. Peru. Loreto: Río Nanay, May–Jun 1929, *L. Williams 737* (holotype, F, n.v.).

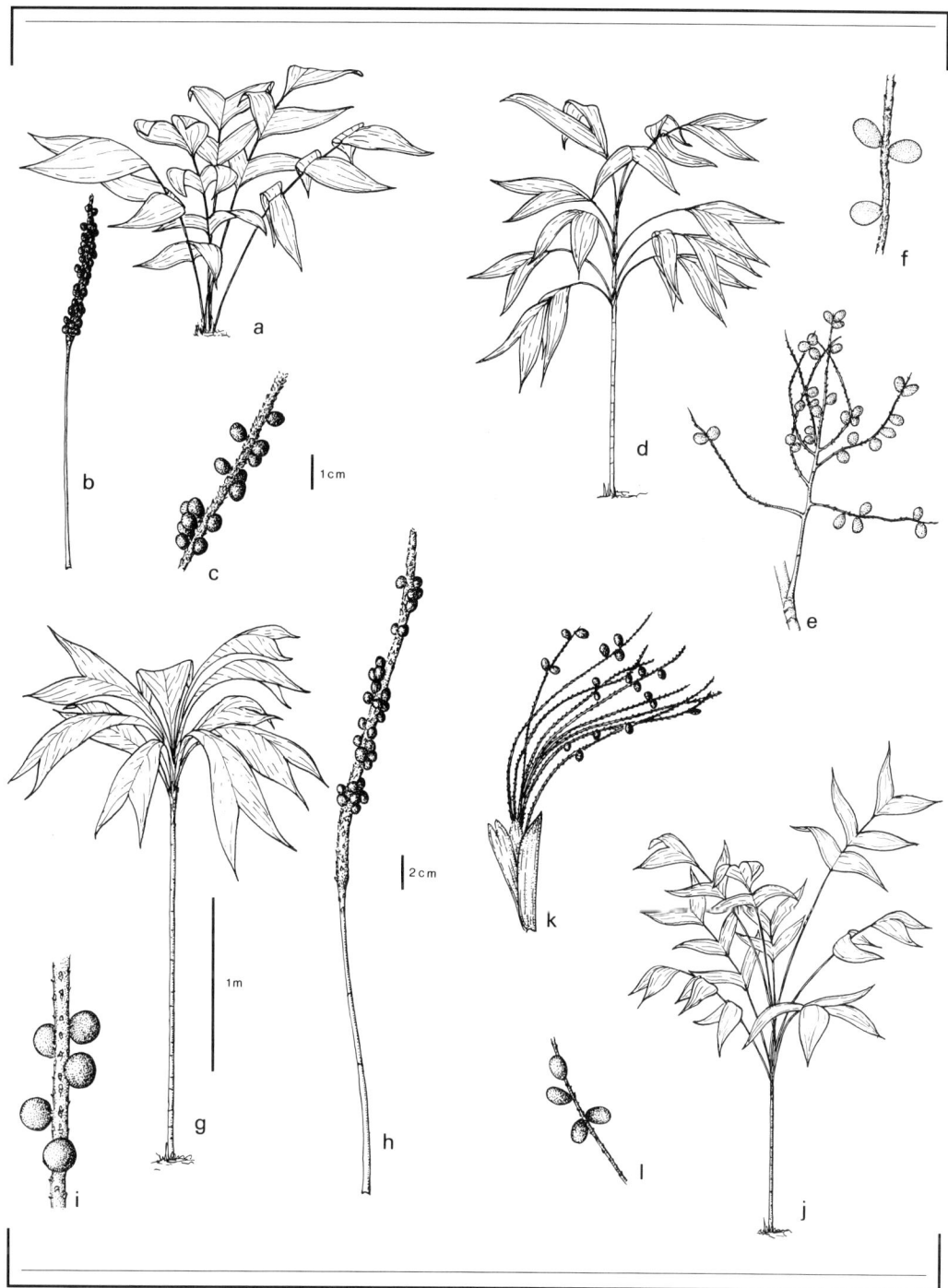

Figure 6.66. (a) *Geonoma macrostachys* var. *acaulis,* **(b)** infructescence (from *F. Casas 8169*), **(c)** fruits; **(d)** *Geonoma maxima* var. *chelidonura,* **(e)** infructescence (from *A. Henderson 1122*), **(f)** fruits; **(g)** *Geonoma oldemanii,* **(h)** infructescence (from *H. Beck 163*), **(i)** fruits; **(j)** *Geonoma oligoclona,* **(k)** infructescence (from *B. Maguire 28800*), **(l)** fruits.

Taenianthera minor Burret, Notizbl. Bot. Gart. Berlin-Dahlem 14: 324. 1939. Type. Ecuador. Pastaza: Canelos, 350 m, 12 Jan 1937, *H. Schultze-Rhonhof 2106* (holotype, B, n.v.).

Leaves 4–13; sheath 13–25 cm long; petiole 62–100 cm long; rachis 25–70 cm long; pinnae 3–12 per side, sigmoid, the middle ones 15–26 cm long, 2.5–5 cm wide, veins forming a 40–60° angle with the rachis. *Inflorescences* interfoliar; peduncle 82–100 cm long; prophyll 10–16 cm long; peduncular bract 15–35 cm long; rachilla 1, 6–22 cm long (elongating between anthesis and fruiting), 3–5 mm diam, with flower pits arranged in close spirals almost touching one another.

Western and central Amazon region of Colombia (Amazonas, Putumayo), Venezuela (Amazonas), Ecuador (Morona-Santiago, Napo), Peru (Huánuco, Loreto, Madre de Dios, Pasco, San Martín, Ucayali), Brazil (Acre, Amazonas, Pará), and Bolivia (Pando) (Figure 6.65B); lowland rain forest in areas subject to inundation and also on *terra firme,* rarely reaching 1250 m elevation on eastern Andean slopes in Ecuador and Peru. Wessels Boer (1968) stated that this variety (as *Geonoma acaulis*) also occurred in Zulia, Venezuela, but I believe the specimens he cited probably belong to another species.

Brazil: *ubim, ubim galope, ubimzinho.* Peru: *irapaillo, ponilla.* Venezuela: *barubaru.*

From the east, this variety first appears in western Venezuela and western Pará, Brazil, and from there is common westward to Colombia, Ecuador, Peru, and Bolivia. It always has pinnate leaves, sigmoid venation, and a wide vein angle. It is not uniform, however, and there are two widespread forms that sometimes occur together. The first form is a delicate plant with a short, procumbent, rhizomelike stem; small leaves with narrow, sigmoid pinnae contracted at the point of insertion; and a short, slender rachilla without an elongate, sterile tip. This form grows in the lowlands and appears to be common on *terra firme.* The second form is a robust plant with a short, erect subterranean stem; larger leaves with the broader, more linear pinnae not contracted basally (Figure 6.59a); and a longer, thicker rachilla often with an elongate, sterile tip (and occasionally with a few staminate flowers at the apex of the tip). This form grows in lowlands and uplands, usually in areas liable to inundation near streams.

There are intermediate specimens that do not fit easily into one form or the other, and for this reason these forms are not recognized taxonomically. According to Heimo Rainer (personal communication), there may also be a difference between the two in the filament position at anthesis. Field work is needed to investigate this problem further.

12b. Geonoma macrostachys var. macrostachys

Geonoma tamandua Trail, J. Bot. 5: 323. 1876. *Taenianthera tamandua* (Trail) Burret, Bot. Jahrb. Syst. 63: 268. 1930. Type. Brazil. Amazonas: Rio Javari, 4 Dec 1874, *J. Trail 976/CLXXXIII* (holotype, K; isotype, P; F neg. 38665).

Geonoma woronowii Burret, Notizbl. Bot. Gart. Berlin-Dahlem 11: 6. 1930. Type. Colombia. Caquetá: Río Orteguaza, 21 Jul 1926, *G. Woronow & S. Juzepczuk 6119* (holotype, B, n.v.).

Stems solitary, short and subterranean, rarely aerial and to 50 cm long. *Leaves* 7–14, erect; sheath and petiole 15–77 cm long; rachis 0.3–1 m long; leaf usually entire, broad and obovate or narrow and oblanceolate, then long cuneate at base and deeply bifid at apex, 0.4–1.4(–2) m long, occasionally pinnate with 2–7 pinnae, these linear, veins forming a 10–35° angle with the rachis. *Inflorescences* interfoliar; peduncle 35–115 cm long; prophyll 5–13 cm long; peduncular bract 20–25 cm long; rachilla 13–28 cm long, 7–10 mm diam, with flower pits arranged in close spirals almost touching one another, lower lip bifid, upper lip obscure.

Western Amazon region in Colombia (Amazonas, Caquetá, Putumayo, Vaupés), Ecuador (Morona-Santiago, Napo), Peru (Huánuco, Junín, Loreto, Pasco, Madre de Dios, San Martín, Ucayali), Brazil (Acre, Amazonas), and Bolivia (Santa Cruz) (Figure 6.65C); lowland rain forest on *terra firme* at low elevations, rarely to 1200 m.

Brazil: *ubim, ubim rasteiro.* Colombia: *ñaigú-ru* (Huitoto). Ecuador: *calson-panga, geke* (Waorani), *guacamaya-panga, guacamayopanga, huan-só-dé-dé* (Secoya), *mo* (Waorani), *naya-*

huë-dadu (Siona), *o-có-pui* (Secoya), *yampuna ujuk* (Shuar), *yija-déré* (Siona). Peru: *calzón panga, palmiche, sápap* (Mayna Jívaro). The leaves are occasionally used for thatching.

This variety first appears west of Tefé and is common from there to Colombia, Ecuador, and Peru and south to Bolivia. The typical form has entire, broad leaves (Figure 6.59b). A second form, very similar to the first but restricted to Colombia, Ecuador, and northern Peru, has much narrower leaves. Both the broad and narrow leaf forms can occasionally have pinnate leaves, but do not have the wide vein angle and sigmoid venation of var. *acaulis* (only rarely do pinnate-leaved forms of var. *macrostachys* approach var. *acaulis*). A third form from Ecuador and northern Peru has very long, narrow, strongly plicate leaves (Figure 6.59c), but is connected by intermediates to other forms. Both Wessels Boer (1968) and Skov (1989), following Trail (1876), considered this to be a distinct species, *Geonoma tamandua*. The first two authors stated that the staminate flowers had long, free thecae reflexed in relation to the filament (as opposed to *G. macrostachys* with short, united thecae in line with the filament). The type specimen of *G. tamandua,* however, clearly has staminate flowers of the latter type, as do all other specimens examined. I believe the confusion arose because Wessels Boer included specimens of *G. oldemanii* in his concept of *G. tamandua* (which do have the former type of staminate flower).

12c. **Geonoma macrostachys** var. **poiteauana** (Kunth) Henderson, stat. nov. *Geonoma poiteauana* Kunth, Enum. pl. 3: 233. 1841. *Gynestum acaule* Poit., Mém. Mus. Hist. Nat. 9 391. 1822. *Geonoma poiteana* Mart. in A. D. Orb., Voy. Amérique mér. 7(3). Palmiers 39. 1843. *Geonoma acaulis* (Poit.) Burret, Jahrb. Bot. Syst. 63: 162. 1930. Type. French Guiana. Without locality, n.d., *A. Poiteau s. n.* (holotype, P; isotype, US).
Geonoma dammeri Huber, Bol. Mus. Paraense Hist. Nat. 3: 409. 1902. Type. Brazil. Pará: Furo Macujubim, 6 Oct 1901, *M. Guedes 2241* (holotype, MG).
Taenianthera lakoi Burret, Notizbl. Bot. Gart. Berlin-Dahlem 11: 11. 1930. Type. Brazil. Roraima: Rio Catrimany, Aamaro, Cachoeira do Mirity, Nov 1929, *C. Lakó/G. Huebner 128* (holotype, B, n.v.).
Geonoma chaunostachys Burret, Bull. Torrey Bot. Club 58: 318. 1931. Type. Venezuela. Amazonas: Mount Duida, ca. 250 m, 18 Nov 1928, *G. Tate 394* (holotype, NY).

Stems solitary, short and subterranean or to 10 cm tall, procumbent. *Leaves* 8–12; sheath and petiole 25–50 cm long; rachis 40–92 cm long; pinnae usually 2–3, or sometimes leaf entire, long cuneate at base, veins forming a 10–20° angle with the rachis. *Inflorescences* interfoliar; peduncle 54–105 cm long; prophyll 10–16 cm long; peduncular bract 15–30 cm long; rachilla 1, 8–17 cm long, 3–5 mm diam, with flower pits arranged in loose spirals ca. 2 mm apart, lower lip prominent and bifid, upper lip obscure.

Eastern Amazon region in Colombia (Vichada), Venezuela (Amazonas, Bolívar), the Guianas, and Brazil (Amapá, Amazonas, Pará, Roraima) (Figure 6.65D); lowland rain forest on well-drained soils on *terra firme,* usually below 400 m elevation.

French Guiana: *waï* (Creole). Guyana: *dalibana*. In French Guiana, the leaves are used for thatching.

This variety occurs in the northeastern Amazon region, principally on the Guayana Shield. It is distinguished by its pinnate leaves with a narrow vein angle (Figure 6.59d), and inflorescences with somewhat loosely spaced flower pits. Occasionally, entire-leaved specimens are found. It appears to be rare compared with the other varieties.

13. **Geonoma maxima** (Poit.) Kunth, Enum. pl. 3: 229. 1841. *Gynestum maximum* Poit., Mém. Mus. Hist. Nat. 9: 388. 1822. Type. French Guiana. Without locality, n.d., *A. Poiteau s.n.* (holotype, P) (Figure 6.66d–f).

Stems solitary, or appearing solitary with basal shoots, or cespitose and sometimes forming large colonies, 1–7 m tall, 0.5–4.5 cm diam. *Leaves* 4–20; sheath 4–30 cm long; petiole 2–89 cm long; rachis 6–120 cm long; pinnae 2–31 per side (or occasionally the leaf entire and then deeply bifid), regularly arranged and spreading in 1 plane, sometimes strongly plicate, linear, falcate or almost sigmoid, often narrow pinnae interspersed between wider ones, the apical one often wider, the middle ones 30–74 cm long, 1–10 cm diam, veins forming a 5–60° angle with the rachis. *Inflorescences* generally infrafoliar or sometimes interfoliar, branched to 1 or 2 orders; peduncle 5–19 cm long; prophyll 6–22 cm long; peduncular bract 6–18 cm long, both bracts deciduous, ± equal at the apex; rachis 5–15 cm long; rachillae 4–43, 6–24 cm long, 1.5–6 mm diam, the pits arranged in loose or close spirals, lower lip prominent and bifid, upper lip scarcely raised; staminate flowers to 4.5 mm long at anthesis (including the exserted stamens); sepals ovate or linear-lanceolate, 1.5–2 mm long; petals 2.5 mm long; stamens 6; filaments very briefly inflexed apically with 2 long, free thecae; pistillode at top of filament tube; pistillate flowers to 5 mm long at anthesis (including the exserted stigmas); sepals ovate-lanceolate or lanceolate, 2–2.5 mm long; petals ovate, 2–2.5 mm long; staminodial tube digitately lobed apically, the fingers spreading at anthesis; *fruits* globose to ellipsoid, 0.8–1.6 cm long, 6–12 mm diam, yellowish-green or orange-green, usually becoming purple-black at maturity.

Common and widespread in the Amazon region (and also occurring in the Magdalena valley in Colombia); usually in lowland rain forest on *terra firme* or inundated areas at low elevations, but reaching 1300 m on the *tepuis* of the Guayana Highland and 900 m on eastern Andean slopes.

Bolivia: *jatata grande, tananë* (Chácabo). Brazil: *ubim, ubim açaí, ubim cavalo, ubim no céu ubimrana, ubi-ran* (Tembé). Ecuador: *oomawe* (Waorani). Guyana: *dhalli, dhallibanna, hikuripipia.* Peru: *palmiche negro, ponilla, taco de altura, yunkúp* (Mayna Jívaro). Surinam: *mantas.* Venezuela: *baru baru, palma de San Pablo.*

The stems were at one time used for spears (Waorani, Ecuador) and arrow shafts (Chácabo, Bolivia). A decoction made from the pulverized palm heart is taken orally as a remedy for measles (Tembé, Brazil). The leaves are occasionally used for thatching, but are not durable.

Geonoma maxima (like *G. macrostachys* and *G. stricta*) is a widespread and extremely variable species. There are at least 10 distinct leaf forms in the species, but all are linked by intermediates. On the basis of leaf morphology, and to a lesser extent inflorescence morphology, it is possible to recognize the following four varieties. There are intermediates between all of these, and in some cases specimens do not fit easily into any of them. Leaf shape variation is similar to that in *Geonoma macrostachys* and *G. stricta* (Figure 6.59).

KEY TO THE VARIETIES OF *GENOMA MAXIMA*

1. Leaves regularly pinnate with 9–31, equally wide, ± sigmoid, nonplicate pinnae per side; veins forming a 40–60° angle with the rachis. . . . .
. . . . 13c. *G. maxima* var. *maxima.*
1. Leaves irregularly pinnate with 2–15, unequally wide, ± linear, plicate pinnae per side, or leaf entire; veins forming a 5–50° angle with the rachis.
  2. Leaves typically with 2 wide (3.5–13 cm) pinnae per side, often irregularly interspersed with narrow (1–2 cm wide) pinnae, or leaf entire, plicate; veins forming a 25–50° angle with the rachis.
    3. Rachillae slender, elongate, 11–15 cm long, 1.5–2 mm diam; the Guianas, Venezuela (eastern Bolívar, Delta Amacuro), and Brazil (Roraima). . . . .
. . . 13a. *G. maxima* var. *ambigua.*
    3. Rachillae stout, short, 6–15 cm long, 1.5–4 mm diam; Colombia (Amazonas, Vaupés), Venezuela (Amazonas), Peru (Amazonas, Huánuco, Loreto, Madre de Dios, Pasco), Brazil (Acre, Amazonas, Pará, Rondônia), and Bolivia (Beni, La Paz, Pando). . . . .
. . 13b. *G. maxima* var. *chelidonura.*
  2. Leaves typically entire, sometimes irregularly pinnate with linear pinnae interspersed, strongly plicate; veins forming a 5–20° angle with the rachis; Brazil (Amazonas) and Colombia (Amazonas).
. . . . . 13d. *G. maxima* var. *spixiana.*

13a. Geonoma maxima var. **ambigua** (Spruce) Henderson, stat. nov. *Geonoma ambigua* Spruce, J. Linn. Soc., Bot. 11: 111. 1869. Type. Guyana. Without locality, n.d., *C. Appun 566* (holotype, K).

Geonoma schomburgkiana Spruce, J. Linn. Soc., Bot. 11: 111. 1869. Type. Guyana. Without locality, 1837, *R. Schomburgk 705* (holotype, K, excluding leaf).

Geonoma robusta Burret, Bot. Jahrb. Syst. 63: 259. 1930. Type. Guayana. Canawaruk River, Sep 1905, *A. Bartlett 6/8189* (holotype, B, n.v.).

Stems tending to be solitary, 2–5 m tall. *Leaves* typically with basal and apical pinnae much wider (9–13 cm) than the middle ones, these narrowly linear (1.5–2 cm), middle ones 38–48 cm long, veins forming a 30–35° angle with the rachis. *Inflorescences* with slender, elongate rachillae, these 11–15 cm long, 1.5–2 mm diam, with the flower pits in loose spirals; *fruits* ellipsoid, 1–1.2 cm long, 0.7–0.8 cm diam, purple-black.

Venezuela (Bolívar, Delta Amacuro), the Guianas, and Brazil (Roraima) (Figure 6.65E); lowland rain forest on *terra firme* at low elevations.

This variety can be considered a northeastern form of var. *maxima,* to which it is linked by intermediates, especially in French Guiana, where both varieties occur. Spruce's description of *Geonoma ambigua* states that the leaf is entire; however, it is more likely that the piece on the sheet is only the apical pinna and that the leaf is actually pinnate. This then becomes a good name for a somewhat doubtfully distinct variety.

13b. Geonoma maxima var. chelidonura

(Spruce) Henderson, stat. nov. *Geonoma chelidonura* Spruce, J. Linn. Soc., Bot. 11: 111. 1871. Type. Brazil. Amazonas: Rio Uaupés, Nov 1852, *R. Spruce 73* (holotype, K).

Geonoma densiflora Spruce, J. Linn. Soc., Bot. 11: 112. 1869. Type. Brazil. Amazonas: São Gabriel, Garnás, Mar 1852, *R. Spruce 30* (holotype, K).

Geonoma personata Spruce, J. Linn. Soc., Bot. 11: 112. 1869. Type. Brazil. Amazonas: Serra of São Gabriel, Jun 1852, *R. Spruce 34* (holotype, K).

Geonoma tuberculata Spruce, J. Linn. Soc., Bot. 11: 112. 1871. Type. Brazil. Amazonas: Rio Negro, n.d., *R. Spruce 18* (holotype, K).

Geonoma densiflora var. *monticola* Spruce, J. Linn. Soc., Bot. 11: 118. 1869. Type. Brazil. Amazonas: São Gabriel, Jun 1852, *R. Spruce 33* (holotype, K).

Geonoma speciosa Barb. Rodr., Enum. palm. nov. 9. 1875. Lectotype (Wess. Boer, 1968). Barb. Rodr., Sert. palm. brasil. 1: t. 18. 1903.

Geonoma brachyfoliata Barb. Rodr., Enum. palm. nov. 10. 1875. Lectotype (Wess. Boer, 1968). Barb. Rodr., Sert. palm. brasil. 1: t. 33. 1903.

Geonoma bijugata Barb. Rodr., Enum. palm. nov. 10. 1875. Lectotype (Wess. Boer, 168). Barb. Rodr., Sert. palm. brasil. 1: t. 16. 1903.

Geonoma falcata Barb. Rodr., Enum. palm. nov. 10. 1875. Lectotype (Wess. Boer, 1968). Barb. Rodr., Sert. palm. brasil. 1: t. 19. 1903.

Geonoma palustris Barb. Rodr., Enum. palm. nov. 11. 1875. Lectotype (Wess. Boer, 1968). Barb. Rodr., Sert. palm. brasil. 1: t. 27. 1903.

Geonoma furcifolia Barb. Rodr., Enum. palm. nov. 11. 1875. Lectotype (Glassman, 1972). Barb. Rodr., Sert. palm. brasil. 1: t. 15. 1903.

Geonoma spruceana Trail, J. Bot. 5: 328. 1876. *Geonoma spruceana* subsp. *spruceana* var. *spruceana* Trail, J. Bot. 5: 329. 1876. Type. Brazil. Pará: Lago Juriti, 3 Apr 1874, *J. Trail 1002/XXIV* (holotype, K).

Geonoma spruceana subsp. *spruceana* var. *heptasticha* Trail, J. Bot. 5: 329. 1876. Type. Brazil. Amazonas: Rio Negro at Assutuba, 6 Jul 1874, *J. Trail 1007/XCIII* (holotype, K).

Geonoma spruceana subsp. *spruceana* var. *micra* Trail, J. Bot. 5: 329. 1876. Type. Brazil. Amazonas: Lago Juruty, n.d., *J. Trail 29* (holotype, K, n.v.).

Geonoma spruceana subsp. *intermedia* var. *tuberculata* Trail, J. Bot. 5: 329. 1876. Type. Brazil. Amazonas: Rio Marmellos, Rio Madeira, 2 Jun 1874, *J. Trail 983/LIV* (holotype, K).

Geonoma spruceana subsp. *intermedia* var. *major* Trail, J. Bot. 5: 330. 1876. Type. Brazil. Amazonas: Rio Solimões, Coary, 16 Oct 1874, *J. Trail 984/CXLIV* (holotype, K; isotype, NY).

Geonoma juruana Dammer, Verh. Bot. Vereins Prov. Brandenburg 48: 119. 1906. Type. Brazil. Acre: Rio Juruá, Juruá-mirim, Aug 1901 *E. Ule 5744* (isotype, MG).

Geonoma lakoi Burret, Bot. Jahrb. Syst. 63: 253. 1930. Type. Brazil. Amazonas: Rio Manacaparu, 10 May 1929, *C. Lakó/ G. Huebner 116* (holotype, B, n.v.).

Geonoma longisecta Burret, Bot. Jahrb. Syst. 63: 257. 1930. Type. Peru. Loreto: Iquitos, n. d., *G. Tessmann 5087* (isotype, NY).

Geonoma parvisecta Burret, Notizbl. Bot. Gart. Berlin-Dahlem 10: 1018. 1930. Type. Brazil. Amazonas: Rio Negro, São Pedro do Uaupés, 23 Sep 1928, *P. Luetzelburg 22278* (holotype, B; isotypes, NY, R; F neg. 18518).

Stems cespitose or occasionally solitary, 1.7–7 m tall, 1–3 cm diam. *Leaves* with 2–4(–10) pinnae, per side, typically with 2 broad pinnae, but often these interspersed with narrower, linear ones, the middle ones 35–65 cm long, (1–)7–10 cm wide, or leaf entire and then deeply bifid, veins forming a 25–50° angle with the rachis. *Inflorescences* interfoliar, becoming infrafoliar, initially erect in leaf axil, branched to 1–3 orders; rachillae 4–36, 6–15 cm long, 1.5–4 mm diam, the pits arranged in close to loose spirals; *fruits* ellipsoid or globose, 0.8–1.6 cm long, 0.7–1 cm diam, yellowish or orange-green, maturing black.

Western Amazon region in Colombia (Amazonas, Putumayo, Vaupés, and also Magdalena Valley in Norte de Santander), Venezuela (Amazonas), Peru (Amazonas, Huánuco, Loreto, Madre de Dios, Pasco), Brazil (Acre, Amazonas, Pará, Rondônia), and Bolivia (Beni, La Paz, Pando) (Figure 6.65F); lowland rain forest on *terra firme* and also *várzea,* at low elevations, or to 900 m elevation on eastern Andean slopes.

This variety is extremely variable and has several forms. Most plants from the western Amazon region in Colombia, Peru, western Brazil, and Bolivia occur on *terra firme* and are robust plants with stems to 7 m tall and 3 cm diameter, and with relatively large leaves (Figure 6.59g). They replace var. *maxima* in this western region, where they have often been called *Geonoma juruana.* There is a second robust form known only from Acre in Brazil, with large, entire leaves (and called locally *ubim do céu*). Along the Amazon river itself, at least from Iquitos to Manaus, mostly in the *várzea,* is a smaller form with more slender stems to 4 m tall and 2 cm diameter, and smaller leaves. Trail collected many of these and described several of them as distinct. A third form occurs in the upper Río Negro region of Venezuela (Amazonas). Here plants occur on gleyic podzols, and tend to be smaller still and have entire leaves. Wessels Boer (1968) confused this form and *G. aspidiifolia.*

Geonoma densiflora, G. personata, and *G. densiflora* var. *monticola* are here interpreted as possible hybrids between var. *maxima* and var. *chelidonura.* The few specimens available *(M. Balick 942, W. Lewis 10122, G. Prance et al. 7596),* with 2–10 broad pinnae and few, short, thick rachillae (8–13 cm long and 5–6 mm diameter) come from widely scattered localities in Brazil (Acre, Amazonas, Pará) and Peru (Loreto), and seem unlikely to represent a separate variety. They are not mapped.

13c. Geonoma maxima var. maxima

Geonoma multiflora Mart., Hist. nat. palm. 2: 7. 1823. Type. Brazil. Pará: without locality, n.d., *C. Martius s.n.* (holotype, M; F negs. 18516, 18516a).

Geonoma discolor Spruce, J. Linn. Soc., Bot. 11: 110. 1869. Type. Brazil. Pará: Rio Tapajós, n.d., *R. Spruce 36* (holotype, K, n.v.).

Geonoma hexasticha Spruce, J. Linn. Soc., Bot. 11: 110. 1869. Type. Brazil. Amazonas: Gamás, São Gabriel, n.d., *R. Spruce 29* (holotype, K; isotype, NY).

Geonoma paraensis Spruce, J. Linn. Soc., Bot. 11: 112. 1869. Type. Brazil. Pará: near Belém, n.d., *R. Spruce 69* (holotype, K).

Geonoma negrensis Spruce, J. Linn. Soc., Bot. 11: 113. 1869. Type. Venezuela. Amazonas: Río Negro, San Carlos, Sep 1853, *R. Spruce 70* (holotype, K).

Geonoma capanemae Barb. Rodr., Enum. palm. nov. 9. 1875. Lectotype (Wess. Boer, 1968). Barb. Rodr., Sert. palm. brasil. 1: t. 29. 1903.

Geonoma uliginosa Barb. Rodr., Enum. palm. nov. 11: 1875. Lectotype (Wess. Boer, 1968). Barb. Rodr., Sert. palm. brasil. 1: t. 28. 1903.

Geonoma spruceana subsp. *intermedia* var. *intermedia* Trail, J. Bot. 5: 329. 1876. Type. Brazil. Amazonas: Lago Cerrado, Rio Juruá, 30 Oct 1874, *J. Trail 989/CXLVII* (holotype, K).

Geonoma spruceana subsp. *intermedia* var. *compta* Trail, J. Bot. 5: 329. 1876. Type. Brazil. Amazonas: Barcellos, 30 Jun 1874, *J. Trail 997/LXXXIV* (holotype, K).

Geonoma huebneri Burret, Bot. Jahrb. Syst. 63: 254. 1930. Type. Colombia. Amazonas: Sierra de Yupatí near La Pedrera, n.d., *G. Huebner 43* (holotype, B).

Geonoma latisecta Burret, Bot. Jahrb. Syst. 63: 255. 1930. Type. Brazil. Amazonas: near Manaus, n.d., *G. Huebner 30* (holotype, B, n.v.).

Geonoma camptoneura Burret, Notizbl. Bot. Gart. Berlin-Dahlem 11: 210. 1931. Type. Peru. Loreto: Río Huallaga, Yurimaguas, Mar 1930, *L. Williams 7836* (holotype, B, n.v.).

Stems tending to be solitary, occasionally cespitose and forming large colonies, 1–5 m tall, 0.5–2.5 cm diam. *Leaves* regularly pinnate with 9–31, linear to sigmoid pinnae per side, the middle ones 10–47 cm long, 0.5–4.5 cm wide, the apical one slightly wider than the others, veins forming a 40–60° angle with the rachis. *Inflorescences* branched to 1 or 2 orders; rachillae slender, elongate, 7–16 cm long, 1–4 mm diameter; *fruits* ellipsoid, 8–12 mm long, 5–10 mm diam, purple-black.

Colombia (Amazonas, Caquetá, Guainía), Venezuela (Amazonas), French Guiana, Ecuador (Morona-Santiago, Napo, Pastaza), Peru (Amazonas, Loreto), Brazil (Amapá, Amazonas, Pará, Rondônia) and Bolivia (Beni, Pando) (Figure 6.67A); lowland rain forest on *terra firme* or *várzea,* or the transition zone between them, at low elevations, occasionally reaching 1300 m on *tepui* slopes in the Guayana Highland.

This variety shows some variation in pinnae shape and number and in rachillae diameter. Plants from the upper Río Negro region of Venezuela (Amazonas) and adjacent areas have fewer pinnae (Figure 6.59f) and shorter, thicker rachillae. These were called *G. hexasticha* and *G. negrensis* by Spruce. Plants from the western Amazon region tend to have narrower, more linear pinnae, whereas those from the eastern part of the range have wider, more sigmoid pinnae. These latter occur with var. *ambigua* in French Guiana. Several specimens from Colombia (Amazonas) have few (10–18), short (10–22 cm long), strongly sigmoid pinnae (Figure 6.59e) and very narrow rachillae; these were called *G. multiflora* by Galeano (1991).

13d. Geonoma maxima var. **spixiana** (Mart.) Henderson, stat. nov. *Geonoma spixiana* Mart., Hist. nat. palm. 2: 15. 1823. Type. Brazil. Amazonas: Rio Japurá, n.d., *C. Martius s.n.* (holotype, M; F negs. 18525, 18525a).

Geonoma grandisecta Burret, Bot. Jahrb. Syst. 63: 258. 1930. Type. Brazil. Amazonas: Manaus, Aug 1928, *G. Huebner 106* (holotype, B, n.v.).

Stems cespitose, sometimes appearing solitary, 2–5 m tall, 1.5–4 cm diam. *Leaves* stiff and erect; blade entire, 1–1.6 m long, 16–24 cm wide, or pinnate with 2–3 broad pinnae per side, the middle ones ca. 74 cm long, (1–)7 cm wide, occasionally with narrow, linear pinnae interspersed, deeply bifid at apex, long cuneate at base, veins forming a 5–20° angle with the rachis. *Inflorescences* interfoliar; rachillae 8–14, (6–)10–24 cm long, (3–)4–5 mm diam, the pits arranged in close spirals; *fruits* ellipsoid or ovoid, 1.2–1.5 cm long, 1–1.2 cm diam, yellowish-green, becoming black at maturity.

Central and western Amazon region in Colombia (Amazonas) and Brazil (Amazonas) (Figure 6.67B); lowland rain forest in areas subject to inundation or on *terra firme,* at low elevations.

This variety is distinguished by its long, narrow, strongly plicate leaves (Figure 6.59h)

14. Geonoma myriantha Dammer, Verh. Bot. Vereins Prov. Brandenburg 48: 120. 1907. Type. Brazil. Acre: Rio Juruá, Juruá-mirim, Sep 1901, *E. Ule 5882* (isotype, MG) (Figure 6.60a–c).

Stems cespitose, 2–5 m tall, 2.5–3 cm diam. *Leaves* 9–14; sheath 20–23 cm long; petiole 31–60 cm long; rachis 70–95 cm long; pinnae 3–19 per side, sigmoid, stiffly spreading in 1 plane, the middle ones 26–33 cm long, 3–4 cm wide, veins forming a 40–70° angle with the

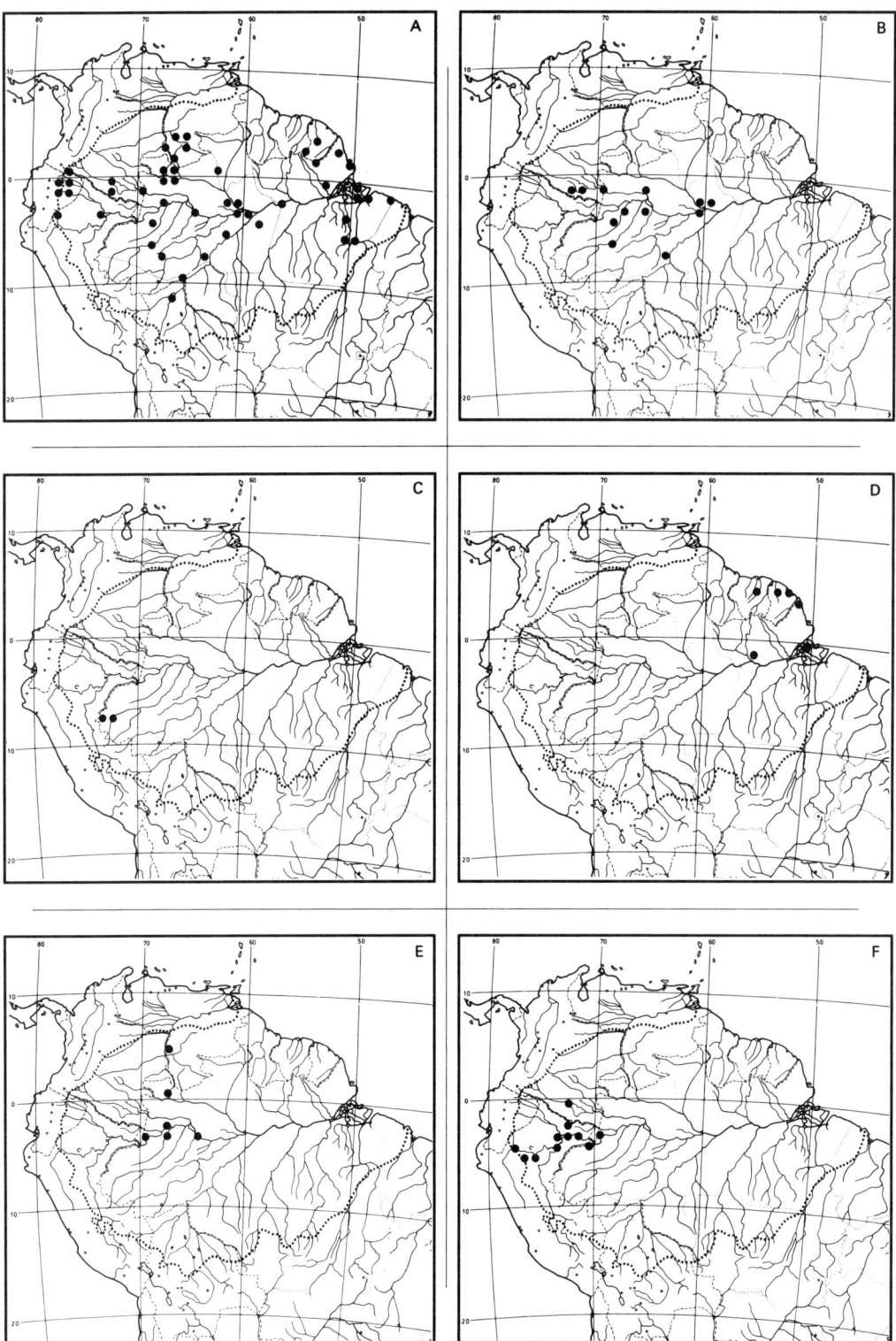

Figure 6.67. (A) *Geonoma maxima* var. *maxima;* **(B)** *Geonoma maxima* var. *spixiana;* **(C)** *Geonoma myriantha;* **(D)** *Geonoma oldemanii;* **(E)** *Geonoma oligoclona;* **(F)** *Geonoma poeppigiana.*

rachis. *Inflorescences* infrafoliar, branched to 2 or 3 orders; peduncle ca. 6 cm long; prophyll and peduncular bract subequal, to 18 cm long; rachis 14–16 cm long; rachillae at least 30, 28–30 cm long, 2.5 mm diam, thickening to 3.5 mm in fruit, straight within the bud; flower pits tristichous or in loose spirals, with a pronounced lower lip and obscure upper lip; staminate flowers (in bud) 2.5 mm long; sepals lanceolate, 2 mm long; petals lanceolate, 2 mm long; stamens 6; filaments very briefly inflexed apically with long, free thecae; pistillate flowers (in bud) 2 mm long; sepals ovate, 2 mm long; petals 2 mm long; staminodial tube crenate apically; *fruits* globose to ellipsoid, 6 mm long, 5.5 mm diam, black.

Brazil (Acre) (Figure 6.67C); lowland rain forest on *terra firme* at low elevations.

Brazil: *ubim*.

15. **Geonoma oldemanii** Granv., Adansonia ser. 2, 14: 553. 1975. Type. French Guiana. Region of Saint-Georges-d'Oyapock, Gabaret creek, 1 Oct 1973, *J.-J. de Granville 1992* (isotypes, P, US) (Figure 6.66g–i).

Stems cespitose or solitary, 1.5–2 m tall, 2–7 cm diam. *Leaves* 12–13; sheath 10–24 cm long; petiole 10–17 cm long; rachis 1–1.2 cm long; blade entire, 1.2–1.5 m long, 15–20 cm wide, long cuneate at the base, bifid at the apex, veins forming a 15° angle with the rachis. *Inflorescences* interfoliar, spicate; peduncle 35–45 cm long; prophyll 24–36 cm long; peduncular bract 19–27 cm long, slightly exceeding the prophyll at the apex, both bracts tubular and semipersistent, a third bract sometimes present on peduncle; rachilla 22–45 cm long, 0.8–1 cm diam, elongating and widening during anthesis; pits arranged in 10–12 vertical files, lower lip large and bifid, upper lip slightly smaller; staminate flowers 3.5 mm long (in bud); sepals lanceolate, 3 mm long; petals lanceolate; stamens 6; filaments very briefly inflexed apically, with 2 long, free thecae; pistillate flowers 5 mm long (at anthesis); sepals lanceolate, 4 mm long; petals 4 mm long; staminodial tube digitately lobed apically; *fruits* globose to ellipsoid, 1.2–1.5 cm long, 8–13 mm diam, black.

Eastern Amazon region in French Guiana and Brazil (Pará) (Figure 6.67D); lowland rain forest in permanently wet areas, occasionally on *terra firme*, at low elevations. In French Guiana, it flowers from June to December, with a peak in September and October, this coinciding with the peak dry season. Fruits are borne 9 months later, between March and July, with a peak in May and June, the peak of the rainy season (Sist, 1989a).

Brazil: *ubimaçu*. The leaves are used for thatching (Brazil).

16. **Geonoma oligoclona** Trail, J. Bot. 5: 325. 1876. Type. Brazil. Amazonas: Barreira Branca, Rio Jutaí, 31 Jan 1875, *J. Trail 1019/CCI* (holotype, K; isotype, P F neg. 38654) (Figure 6.66j–l).

Stems cespitose, 1–2 m tall, 1.2–1.5 cm diam. *Leaves* ca. 10; sheath ca. 10 cm long; petiole 30–45 cm long; rachis 22–37 cm long; pinnae 3 per side, sigmoid with long filiform apex, the middle one 20–26 cm long, 4–8 cm wide, veins forming a 40° angle with the rachis. *Inflorescences* infrafoliar, often with several buds visible on the stem below the leaves, branched to 1 order; peduncle 4–7 cm long; prophyll 6–10 cm long; peduncular bract 6–9 cm long, both bracts ± persistent; rachis 0–1 cm long; rachillae 3–10, 20–30 cm long, 3–4 mm diam, cylindrical and not narrowed between the pits, folded and twisted with the bud; flower pits arranged in loose spirals, lower lip small and bifid, upper lip small; staminate flowers 2.5–3 mm long (in bud); sepals oblanceolate, 2–2.5 mm long; petals oblanceolate, 2 mm long; stamens 6; filaments inflexed apically with 2 free connectives and 2 free thecae; pistillate flowers 2–2.5 mm long (in bud); sepals ovate, 2 mm long; petals ovate, 2 mm long; staminodial tube crenate apically; *fruits* globose, ca. 5 mm diam, black.

Western Amazon region in Colombia (Amazonas), Venezuela (Amazonas), and Brazil (Amazonas) (Figure 6.67E); lowland rain forest on *terra firme* at elevations up to 1000 m. Despite a relatively wide range, it is rare or rarely collected.

17. **Geonoma poeppigiana** Mart. in A. D. Orb., Voy. Amérique mér. 7(3) Palmiers 35. 1847. Type. Peru. Loreto: Yurimaguas, n.d., *E. Poeppig 2295* (holotype, M; F neg. 18522) (Figure 6.68a–b).

Figure 6.68. (a) *Geonoma poeppigiana,* **(b)** infructescence (from *G. Galeano 1323*); **(c)** *Geonoma polyandra,* **(d)** infructescence (from *H. Balslev 60739*); **(e)** *Geonoma stricta* var. *stricta,* **(f)** infructescence of var. *stricta* (from *A. Henderson 1578*); **(g)** *Geonoma triglochin,* **(h)** infructescence (from *H. Balslev 62071*); **(i)** *Geonoma umbraculiformis,* **(j)** infructescence (from *J.-J. de Granville 8584*).

Geonoma oligoclada Burret, Notizbl. Bot. Gart. Berlin-Dahlem 11: 9. 1930. Type. Brazil. Amazonas: Rio Içá, n.d., *C. Lakó 7/G. Huebner 138* (holotype, B).

Stems solitary, 0.1–2 m tall, 2–5 cm diam, erect or occasionally procumbent, aerial or occasionally short and subterranean. *Leaves* 8–16; sheath and petiole 64–70 cm long, densely brown-tomentose, glabrescent; rachis 39–80 cm long; pinnae 4–several per side, of varying width, the middle ones 30–58 cm long, 2–5 cm wide, or often blade entire and then deeply bifid apically for ca. 30 cm, veins forming a 20–50° angle with the rachis, the angle smaller on entire leaves. *Inflorescences* interfoliar, erect, branched to 1 order, densely whitish-brown-tomentose especially on the peduncle, glabrescent; peduncle 30–60 cm long; prophyll 20–33 cm long; peduncular bract 20–37 cm long, both bracts flattened; rachis 0–15 cm long; rachillae 2–5, 12–40 cm long, 3–5.5 mm diam; pits arranged in close spirals, with obscure lower and upper lips, these ca. 1 mm diam at anthesis; staminate flowers 4.5 mm long at anthesis (including the exserted stamens); sepals oblanceolate-obovate, 2 mm long; petals oblanceolate-obovate, 2.5 mm long; stamens 6, greatly exserted at anthesis; filaments very briefly inflexed apically, with 2 long, free thecae; pistillode very small; pistillate flowers 2–3 mm long; sepals oblanceolate-obovate, 2 mm long; petals obovate, 1.5–2 mm long; staminodial tube crenate apically; *fruits* globose, 5–6 mm diam, black.

Western Amazon region in Colombia (Amazonas), Peru (Amazonas, Loreto), and Brazil (Amazonas) (Figure 6.67F); lowland rain forest on *terra firme* at elevations below 500 m.

Colombia: *jiyui* (Muinane), *ü-da-ti-ño-rü* (Huitoto). Peru: *palmiche*. The leaves are used for thatching.

Geonoma poeppigiana is distinguished by its short, stout stems; inflorescences that are densely tomentose initially; flattened bracts; elongate peduncle, short rachis, and few, elongate, thick rachillae; and staminate flowers with inflexed, bifid connectives. It appears similar to both *G. longepedunculata* and *G. baculifera*.

18. Geonoma polyandra Skov, Nord. J. Bot. 14: 39. 1994. Type. Ecuador. Napo: Añangu, 0°32′N, 76°23′W, 300 m, 19 Jul 1985, *H. Balslev et al. 60536* (isotypes, CAY, NY) (Figure 6.68c–d).

Stems solitary, 0.4–2.3 m tall, 2.5–5 cm diam. *Leaves* 8–15; sheath and petiole 45–54(–92) cm long; rachis 0.5–1.1 cm long; pinnae 8–22 per side, linear to somewhat sigmoid, long acuminate, of varying width, the middle pinnae 25–35 cm long, 1–3 cm wide, veins forming a 40–50° angle with the rachis. *Inflorescences* interfoliar, spicate, erect; peduncle 45–49 cm long; prophyll 27–29 cm long; peduncular bract 12–22 cm long; rachilla 21–30 cm long, ca. 1 cm diam in fruit, with spirally arranged flower pits, lower lip deeply bifid, upper lip obscure; flowers not seen; stamens 10–12; *fruits* globose to ellipsoid, ca. 1 cm long, 6 mm diam, black.

Colombia (Caquetá) and Ecuador (Napo) (Figure 6.69A); lowland rain forest on *terra firme* at low elevations.

Ecuador: *dédé-bui* (Siona), *dí-dí* (Siona), *huasipanga* (Quichua), *tsao-he-tsi* (Kofan). The leaves are used for thatching.

Geonoma polyandra is distinguished by its numerous stamens. It appears most similar to *G. chococola* Wess. Boer, which is known from western Colombia and eastern Panama.

19. Geonoma stricta (Poit.) Kunth, Enum. pl. 3: 232. 1841. *Gynestum strictum* Poit., Mém. Mus. Hist. Nat. 9: 391. 1822. Type. French Guiana. Without locality, n.d., *A. Poiteau s.n.* (holotype, P) (Figure 6.68e–f; Plate IIIb).

Stems cespitose or solitary, 0.5–3 m tall, 0.5–2 cm diam, occasionally procumbent and rooting, sometimes forming large clumps. *Leaves* 5–15, spaced all the way up the stem; sheath 1.5–20 cm long; petiole 4–53 cm long, sheath and petiole (and rachis) sometimes densely scurfy-brown-tomentose, glabrescent; rachis 16–49 cm long; blade entire, 14–75 cm long, 4–22 cm wide at top of rachis, deeply or shallowly bifid, the veins forming a 20–30° angle with the rachis, or pinnate and the pinnae 3(–12) per side, ± sigmoid, evenly spaced and well separated, the middle ones 10–41 cm long, 2–11 cm wide, veins forming a 40–60° angle with the rachis. *Inflorescences* interfoliar or infrafoliar, spicate; peduncle 1–13 cm long; prophyll 1–14 cm long; peduncular bract much reduced

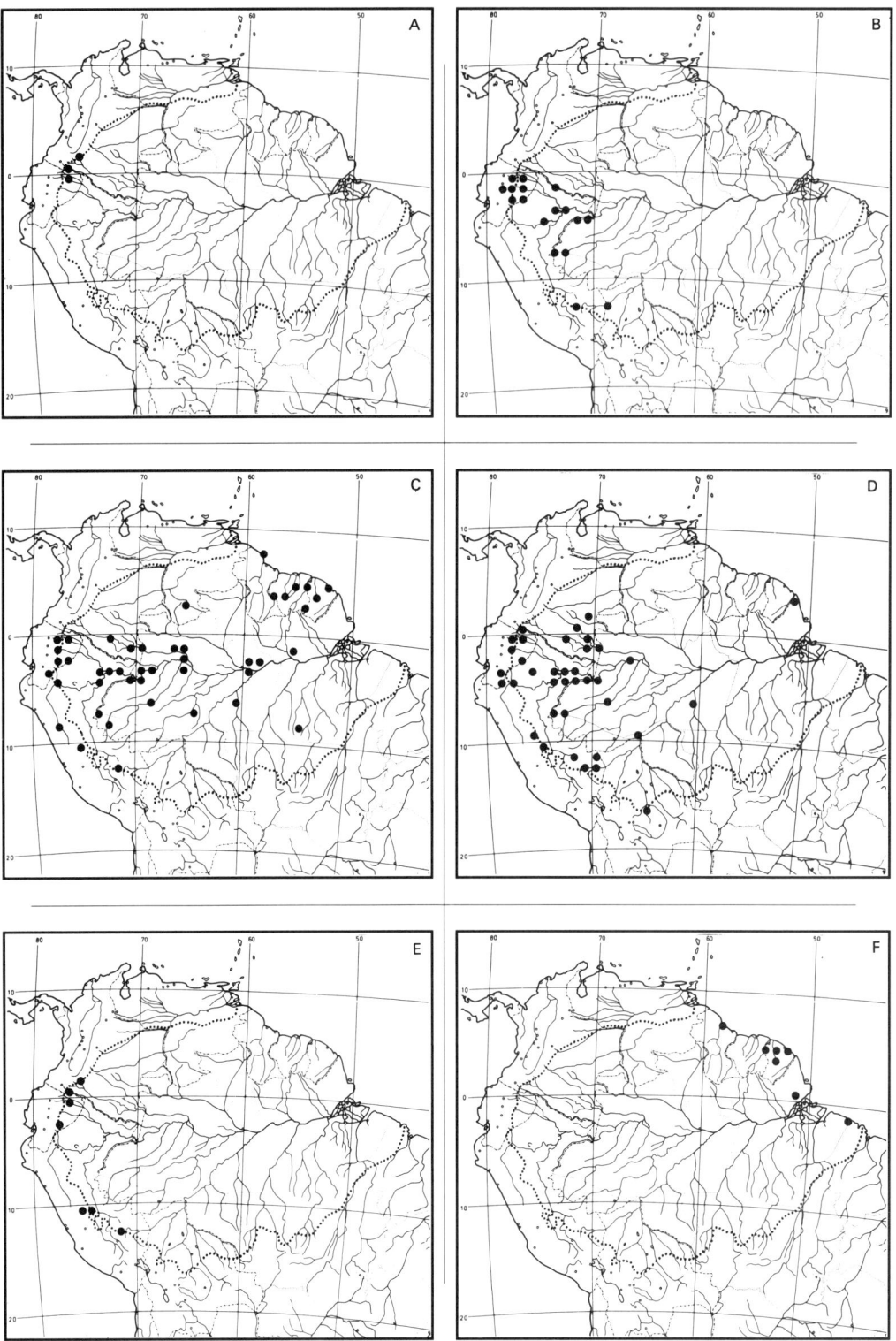

Figure 6.69. (A) *Geonoma polyandra;* **(B)** *Geonoma stricta* var. *piscicauda;* **(C)** *Geonoma stricta* var. *stricta;* **(D)** *Geonoma stricta* var. *trailii;* **(E)** *Geonoma triglochin;* **(F)** *Geonoma umbraculiformis.*

but leaving a scar, occasionally to 6 cm long, or absent; rachilla 1 (very rarely 2–3), 2–30 cm long, 0.3–1.2 cm diam, brown-tomentose, the pits arranged in closely spaced vertical files or almost decussate, lower lip prominent with an entire margin, upper lip scarcely raised; staminate flowers 2.5–6 mm long; sepals ovate or ovate-lanceolate, 2.5–4 mm long; petals ovate, 3.5–4 mm long; filaments very briefly inflexed apically, with 2 long, free thecae; pistillode near top of filament tube; pistillate flowers 4–5 mm long at anthesis (including exserted stigmas); sepals ovate, 3 mm long; petals ovate, 3–4 mm long; staminodial tube crenate apically; *fruits* ovoid or ellipsoid, 7–11 mm long, 5–7 mm diam, black or blue.

Geonoma stricta is a widespread species, and leaf shape and size as well as and inflorescence size are very variable. Leaf shape is here used to divide it into three varieties, but many intermediates occur, and there is no apparent correlation with inflorescence size in this division. Skov (1989) also divided this species into three varieties, but used the younger name *G. pycnostachys*. Leaf shape variation is similar to that in *Geonoma macrostachys* and *G. maxima* (Figure 6.59).

KEY TO THE VARIETIES OF
GEONOMA STRICTA

1. Leaves entire; veins emerging at a 20–35° angle from the rachis; leaf venation linear.
    2. Leaf blade with ± parallel sides, the apex narrower than the middle, shallowly bifid. . . . .
        . . . . 19b. *G. stricta* var. *stricta*.
    2. Leaf blade oblanceolate, the apex much wider than the middle, deeply bifid apically . . . .
        . . . . . 19a. *G. stricta* var. *piscicauda*.
1. Leaves pinnate; veins emerging at a 40–60° angle from the rachis; leaf venation sigmoid
    . . . . 19c. *G. stricta* var. *trailii*.

19a.Geonoma stricta var. **piscicauda** (Dammer) Henderson, stat. nov. *Geonoma piscicauda* Dammer, Verh. Bot. Vereins Prov. Brandenburg 48: 123. 1906 (1907). Type. Brazil. Acre: Rio Juruá, Juruá-mirim, May 1901, *E. Ule 5520* (isotypes, K, MG; F neg. 5466).

Geonoma wittiana Dammer, Verh. Bot. Vereins Prov. Brandenburg 48: 124. 1906 (1907). Type. Brazil. Acre: Rio Juruá, Juruá-mirim, Sep 1901 *E. Ule 5884* (isotype, MG).

Geonoma lanceolata Burret, Notizbl. Bot. Gart. Berlin-Dahlem 11: 7. 1930. Type. Brazil. Amazonas: Rio Içá, May 1930, *C. Lakó 18/G. Huebner 146* (holotype, B).

Geonoma herthae Burret, Notizbl. Bot. Gart. Berlin-Dahlem 14: 325. 1939. Type. Ecuador. Pacayacu, 200 m, 11 Jun 1937, *H. Schultze-Rhonhof 2394* (holotype, B, n.v.).

Stems 1–2 m tall, 0.5–2 cm diam. *Leaves* entire, oblanceolate, long cuneate basally, deeply bifid apically, (23–)55–75 cm long, (9–)16–22 cm wide at apex of rachis, plicate, veins forming a 20–30° angle with the rachis. *Inflorescences* with the rachilla (7–)14–22 cm long, (0.5–)0.8–1.2 cm diam; fruits ellipsoid, 7–8 mm long, 5–6 mm diam, black.

Colombia (Amazonas, Caquetá, Putumayo), Ecuador (Napo, Pastaza, Santiago-Zamora), Peru (Loreto, Madre de Dios), and Brazil (Acre, Amazonas) (Figure 6.69B); lowland rain forest on *terra firme*.

From the east, this variety first appears near the Rio Javari, on the frontier between Brazil and Peru, and from there is common into Colombia, Ecuador, and Peru. It generally does not occur far to the south. Intermediates occur with var. *stricta*. The name *Geonoma piscicauda* was used by Wessels Boer (1968) for pinnate-leaved plants, although the type has an entire leaf.

19b.Geonoma stricta var. **stricta**

Geonoma pycnostachys Mart., Hist. nat. palm. 2: 16. 1823. Type. Brazil. Amazonas: Rio Japurá, n.d., *C. Martius s.n.* (holotype, M; F neg. 18523).

Geonoma maguirei L. H. Bailey, Bull. Torrey Bot. Club 75: 102. 1948. Type. Surinam. Coppename River headwaters, 24 Jul 1944, *B. Maguire 24166* (isotype, NY).

Stems cespitose, 0.5–1.8 m tall, 0.5–1 cm diam. *Leaves* entire, ± parallel sided, slightly narrower at the apex, shallowly bifid apically, 14–60 cm long, 4–23 cm wide at apex of rachis, veins forming at a 20–35° angle with the rachis. *Inflorescences* with rachilla 2–19 cm long, 0.3–

3 cm diam; *fruits* ovoid-ellipsoid, ca. 7 mm long, 5 mm diam, blue or blue-black.

Colombia (Amazonas, Putumayo, and also the Pacific coast in Chocó), Venezuela (Amazonas), the Guianas, Ecuador (Morona-Santiago, Napo, Pastaza), Peru (Huánuco, Loreto, Pasco, Madre de Dios, San Martín), and Brazil (Acre, Amapá, Amazonas, Pará) (Figure 6.69C); lowland rain forest on *terra firme,* occasionally in inundated areas, at elevations below 550 m but to 1500 m, on eastern Andean slopes.

Brazil: *ubim.* Ecuador: *dé-dé* (Secoya), *dijadere* (Siona), *hoja ancha, chontilla, urcu-tauna* (Quichua). Guyana: *hikuriparipia.* Surinam: *makaka ima.* The palm heart is chewed to prevent teeth rotting (Secoya, Ecuador).

Starting from the northeast, there is a diminutive form occurring in the Guianas that has been called *Geonoma stricta.* Leaves of this form are the smallest in the species, and are only 14–27 cm long and 4–8 cm wide (Figure 6.59k). Continuing west, plants tend to get larger, but are still relatively small at least until Manaus (where they have been called *G. pycnostachys).* This diminutive form then disappears except for a few isolated populations in Colombia (Amazonas), Peru (Loreto), and Ecuador (Napo). Somewhere near Tefé a larger form occurs with leaves 30–60 cm long and 10–23 cm wide (Figure 6.59j), and larger inflorescences. This form continues due west to Ecuador, but in general does not occur much to the south.

19c. Geonoma stricta var. **trailii** (Burret) Henderson, stat. nov. *Geonoma trailii* Burret, Bot. Jahrb. Syst. 63: 83. 1930. *Geonoma elegans* Mart. var. *amazonica* Trail, J. Bot. 5: 324. 1876. Type. Brazil. Amazonas: Rio Purus, Barreiras de Mancira, 29 Sep 1874, *J. Trail 1032/CXXXIII* (holotype, K; isotype, P; F neg. 38666)

Geonoma dasystachys Burret, Bot. Jahrb. Syst. 63: 251. 1930. Type. Brazil. Amazonas: Rio Negro, Jauapasse assu, 5 Jul 1874, *J. Trail 981/XC* (holotype, K; F neg. 38646).

Geonoma trauniana Dammer, Verh. Bot. Vereins Prov. Brandenburg 48: 124. 1906 (1907). Type. Brazil. Amazonas (Acre?): Rio Juruá, Fortaleza, Oct 1901, *E. Ule 5946* (isotype, MG; F neg. 5840).

Geonoma raimondii Burret, Bot. Jahrb. Syst. 63. 182. 1930. Type. Peru. Amazonas: without locality, n.d., *A. Raimondi 978* (holotype, B, n.v.).

Geonoma bella Burret, Notizbl. Bot. Gart. Berlin-Dahlem 12: 304. 1935. Type. Brazil. Amazonas: Mun. Tefé, Paranagua, 22 May 1933, *B. Krukoff 4543* (isotypes, MO, NY, US).

Stems cespitose, 1–3(–5) m tall, 0.7–2(–3) cm diam. *Leaves* with sheath and petiole (and rachis) usually densely scurfy-brown-tomentose, glabrescent; pinnae 3(–12) per side, ± sigmoid, evenly spaced and well separated, the middle ones 10–41 cm long, 2–11 cm wide, veins forming a 40–60° angle with the rachis. *Inflorescences* with peduncle 5–13 cm long; rachilla 1(–3), (5–)13–30 cm long, (3–)5–10 mm diam, brown-tomentose; *fruits* ovoid, 8–9 mm long, ca. 7 mm diam, black or rarely blue.

Western Amazon region of Colombia (Amazonas, Vaupés), Ecuador (Morona-Santiago, Napo, Pastaza), Peru (Amazonas, Loreto, Madre de Dios, Pasco, Ucayali), Brazil (Acre, Amazonas), and Bolivia (Cochabamba, Pando), and an outlying population in French Guiana and Brazil (Amapá) (Figure 6.69D); lowland rain forest on *terra firme,* usually at low elevations but occasionally up to 1250 m on eastern Andean slopes.

Brazil: *ubim, ubim juriti.* Colombia: *bo-só-moo-hee* (Makuna), *jiowoohiyui* (Mui), *kaa-ñoo-et* (Maku), *palmicha.* Ecuador: *ya na muyl* (Quechua). Peru: *palmiche blanco, sápap* (Mayna Jívaro), *taco de bajillal.* The leaves are occasionally used for thatching.

This variety, previously known as *G. piscicauda,* becomes common somewhere near Tefé, in Amazonas, Brazil, and is abundant from there west into Colombia, Ecuador, and Peru and also reaching south into Bolivia. Intermediates are found between it and the larger form of var. *stricta.* Although leaves typically have 3 pinnae per side (Figure 6.59i), plants from the Rio Juruá in Amazonas, Brazil, have up to 12. A few collections from Peru (Amazonas, Loreto) *(B. Berlin 1808, R. Kayap 551, S. Knapp*

7692, A. Kujikat 98) are similar to var. *trailii* vegetatively, but have a peduncular bract and 2–3 rachillae. They are here considered a local form of the variety.

20. Geonoma triglochin Burret, Notizbl. Bot. Gart. Berlin-Dahlem 11: 8. 1930. Type. Colombia. Caquetá: Sucre, 10 Jul 1926, G. Woronow & S. Juzepczuk 5858 (holotype, LE, n.v.) (Figure 6.68g–h).

Stems solitary, erect or partly repent, 0.5–4 m tall, 3–5 cm diam. *Leaves* 15–25, spreading umbrellalike; sheath and petiole 30–50 cm long, petiolar part very short; rachis 0.8–1.6 m long, brown-tomentose; pinnae broad, often 3 per side or occasionally intermixed with linear ones, or leaf entire and becoming split with age, bifid apically, cuneate basally, the blade 0.7–1.9 m long, 22–55 cm wide, darker brown abaxially when dry, veins forming a 25–30° angle with the rachis, alternate veins brown-tomentose abaxially. *Inflorescences* interfoliar, branched to 1 or rarely 2 orders, erect; peduncle 26–50 cm long; prophyll 13–33 cm long; peduncular bract 17–40 cm long, exceeding the prophyll apically, both bracts tubular, flattened and persistent; rachis 0–14 cm long; rachillae 2–9, simple or occasionally the basal one(s) bifurcate, 10–30 cm long, (3.5–)8–13 mm diam, the pits closely arranged in vertical files, lower lip prominent and bifid, upper lip slightly raised; staminate flowers 8 mm long at anthesis (including exserted stamens); sepals lanceolate, 3.5 mm long; petals ovate, 4.5 mm long; stamens 6; filaments very briefly inflexed apically, with 2 long, free thecae; pistillode at apex of filament tube; pistillate flowers 3.5 mm long (in bud); sepals lanceolate, 3 mm long; petals ovate, 3 mm long; staminodial tube deeply digitate apically; *fruits* globose to ellipsoid, 1–1.5 cm long, black.

Western Amazon region in Colombia (Caquetá), Ecuador (Morona-Santiago, Napo), and Peru (Junín, Madre de Dios, Pasco) (Figure 6.69E); lowland or premontane rain forest on *terra firme,* in Andean foothills at 300–1150 m elevation. Although relatively widespread it is apparently a rare palm.

Ecuador: *an-hí-da-du* (Siona). The leaves are used for thatching (and populations have probably been greatly reduced by cutting the plants to collect the leaves).

Geonoma triglochin is distinguished by its tall, stout stem, which is often at least partly procumbent; numerous, large, entire or almost entire leaves, which are crowded at the apex of the stem and spread umbrellalike; elongate peduncle with persistent, flattened bracts and few, thick rachillae; staminate flowers with long, free thecae reflexed in relation to the filament; and pistillate flowers with digitately lobed staminodial tube. Wessels Boer (1968) included in synonymy here his *Geonoma umbraculiformis,* and so conceived a species ranging from the Andean foothills of Colombia, Ecuador, and Peru to the eastern Amazon region of Brazil and the Guianas, with nothing in between. There appear, however, to be two populations, a western and an eastern, which are here called *G. triglochin* and *G. umbraculiformis,* respectively. The main difference between them is that the latter is a smaller palm and in particular has a much more slender inflorescence. However, they are very similar and were combined by Skov (1989), who also included *Geonoma oldemanii.*

21. Geonoma umbraculiformis Wess. Boer, Flora of Suriname 5(1): 35. 1965. Type. Surinam. Lawa River near Cottica Mts., n.d., Versteeg 322 (holotype, U, n.v.) (Figure 6.68i–j).

Stems solitary, 0.6–2 m tall, 2–2.5 cm diam. *Leaves* 12–17, spreading; sheath and petiole 15–30 cm long (petiolar part short or absent); rachis 35–90 cm long; blade entire or irregularly split, 0.9–1.2 m long, 20–30 cm wide, bifid apically, cuneate basally, darker brown abaxially when dry, veins forming a 25–30° angle with the rachis, alternate veins tomentose abaxially. *Inflorescences* interfoliar, branched to 1 order; peduncle 25–40 cm long; prophyll 12–20 cm long; peduncular bract 20–25 cm long, exceeding the prophyll; rachis 0–5 cm long; rachillae 2–5, 15–23 cm long, 3–3.5 mm diam at anthesis, the pits arranged in loose spirals ca. 1.5 mm apart, lower lip prominent and bifid, upper lip slightly raised; staminate flowers 3.5 mm long (in bud); sepals linear-lanceolate, 3.5 mm long; petals 3.5 mm long; stamens 6; filaments very briefly inflexed apically, with 2 long, free thecae; pistillode small; pistillate flowers 4.5 mm long at anthesis (including exserted stigmas); sepals ovate, 3 mm long; petals ovate, 4 mm long; staminodial tube deeply digitate apically, the fingers spreading at anthesis; *fruits* globose to ellipsoid-ovoid, 1–1.3 cm long, ca. 0.9 cm diam, black.

Eastern Amazon region in the Guianas and Brazil (Amapá, Pará) (Figure 6.69F); lowland or premontane rain forest, usually on mountain slopes on *terra firme,* at 300–750 m elevation.

Guyana: *dhalebana.*

Geonoma umbraculiformis is distinguished by its stout, solitary stem; numerous entire or almost entire leaves with narrow vein angle and short or absent petiole; long peduncle with peduncular bract exceeding the prophyll; short rachis with few, relatively thin rachillae; staminate flowers with free, inflexed thecae; and pistillate flowers with digitately lobed staminodial tube; see notes under *G. triglochin* for similarities to that species.

PHYTELEPHANTOIDEAE

32. *Phytelephas* **Phytelephas** Ruiz & Pav., Syst. veg. fl. peruv. chil. 299. 1798.

Moderate, dioecious palms. *Stems* solitary or cespitose, short and subterranean, or aerial and then procumbent or erect. *Leaves* pinnate, reduplicate; sheath open and not forming a crownshaft; petiole long; rachis long; pinnae regularly arranged and spreading in 1 plane, linear. *Inflorescences* strongly dimorphic, branched to 1 order; staminate inflorescences interfoliar; peduncle bearing a prophyll and 1 peduncular bract, with a few smaller bracts present; rachis elongate with numerous, short rachillae, or rachillae much reduced; staminate flowers sessile to pedicellate, with much reduced perianth; stamens 150–700; pistillate inflorescences interfoliar with the rachis contracted; pistillate flowers 4–10 per inflorescence; tepals ca. 5; staminodes numerous; gynoecium syncarpous, multiloculate, multiovulate, the styles and stigmas elongate; *fruits* several-seeded, borne in large heads, with obscure stigmatic residue, the epicarp with woody, pointed warts; seeds with homogeneous endosperm and subbasal embryo; germination remote-ligular; eophyll pinnate.

A genus of five species (although Barfod [1991] recognized four) distributed from Panama to Bolivia, usually at low to mid-elevations in wetter areas, on either side of the Andes. Two species occur in the western Amazon region.

The species have an interesting north–south distribution pattern that may be compared with, for example, *Chelyocarpus.* From the south, *Phytelephas macrocarpa* occurs in Amazon Bolivia and Peru; *P. tenuicaulis* in northern Peru, Ecuador, and Colombia; *P. schottii* H. Wendl. in the Magdalena valley of Colombia; and *P. seemanii* O. F. Cook on the Pacific coast of Colombia and in Central America. *Phytelephas aequatorialis* Spruce and *P. tumacana* O. F. Cook occur as outliers on the Pacific coast of Colombia and Ecuador. In fact, the position of these latter two species, as the trans-Andean relatives of the Amazon *P. macrocarpa* and *P. tenuicaulis,* is reminiscent of the situation in *Aiphanes aculeata/eggersi.* Viewed in this light, the genus may have had a southern origin.

Barfod (1991) treated *Phytelephas macrocarpa* and *P. tenuicaulis* as subspecies of a wide-ranging *P. macrocarpa.* In the material examined, however, there are two distinct forms with more or less nonoverlapping ranges, and I recognize them here as separate species.

KEY TO THE SPECIES OF *PHYTELEPHAS*

1. Stems short and subterranean, very thick, rarely to 1.8 m tall, with flowers and fruits borne near ground level; leaf rachis 2.6–7.2 m long; middle pinnae 42–87 cm long, 4–6.5 cm wide . . . . 1. *P. macrocarpa.*
1. Stems longer and aerial, 1.5–7.2 m tall and 8–10 cm diam, with flowers and fruits borne at the apex of the stem; leaf rachis 1.9–4.1 m long; middle pinnae 33–57 cm long, 2.5–3 cm wide. . . . . 2. *P. tenuicaulis.*

1. Phytelephas macrocarpa Ruiz & Pav., Syst. veg. fl. peruv. chil. 301. 1798. *Elephantusia macrocarpa* (Ruiz & Pav.) Willd., Sp. pl. 4: 1156. 1805. Lectotype (Barfod, 1991). Peru. Junín: Vitoc, n.d., *H. Ruíz & J. Pavón s.n.* (isolectotype, K) (Figure 6.70a–b).
Phytelephas microcarpa Ruiz & Pav., Syst. veg. fl. peruv. chil. 302. 1798. *Elephantusia microcarpa* (Ruiz & Pav.) Willd., Sp. pl. 4: 1157. 1805. *Yarina microcarpa* (Ruiz & Pav.) O. F. Cook, J. Wash. Acad. Sci. 17: 223. 1927. Lectotype (Barfod, 1991). Peru. Junin: Vitoc and Pampahermosa, n.d., *H. Ruíz & J. Pavón s.n.* (hololectotype, G, n.v.).

Stems solitary or rarely cespitose, subterranean or aerial and procumbent, then to 1.8 m tall, to 28 cm diam, covered with persistent leaf bases. *Leaves* 12–20; petiole and sheath 1–3 m long, sheath fibrous at apex; rachis 2.6–7.2 m long; pinnae 42–95 per side, regularly arranged and spreading in 1 plane, linear, the middle ones 42–87 cm long, 4–6.5 cm wide. *Inflorescences* strongly dimorphic; staminate inflorescences interfoliar; peduncle ca. 60 cm long; prophyll ca. 40 cm long; peduncular bract ca. 35 cm long; staminate flowers sessile, densely crowded on the rachis; sepals and petals reduced to obscure bracteoles; stamens 150–300; pistillode absent; pistillate inflorescences interfoliar, not seen; *fruits* several, to 9 cm long, densely crowded in a cluster, irregularly obdeltoid, with woody pointed warts especially on apical surface; seeds several per fruit.

Western Amazon region in Peru (Cusco, Huánuco, Junín, Loreto, Madre de Dios, Pasco, Ucayali), Brazil (Acre, Amazonas), and Bolivia (Beni, La Paz, Pando) (Figure 6.71A); lowland rain forest on *terra firme,* often on sloping ground, or seasonally inundated areas near streams, at elevations below 1000 m.

Bolivia: *marfil, palmera marfil.* Brazil: *jarina.* Peru: *polo punta, tagua, yarina.*

The endosperm of immature fruits is eaten; when mature and hard, it is extracted from the fruits and carved into a variety of objects, including rings, buttons, and false teeth. In Peru, the leaves are used for thatching houses. They are used either in a horizontal position *(acomodado)* or in a vertical position *(tejido).* They are said to be more durable than those of other palms, and thatching can last for 30 years.

2. Phytelephas tenuicaulis (Barfod) Henderson, stat. nov. *Phytelephas macrocarpa* subsp. *tenuicaulis* Barfod, Opera bot. 105: 62. 1991. Type. Ecuador. Napo: Añangu, S bank of Río Napo, 95 km downstream from Coca, 0°32'S, 76°23'W, 300 m, 28 Jul 1985, *H. Balslev et al. 60698* (isotype, NY) (Figure 6.70c–d; Plate IIId).

Stems cespitose and forming clumps of 2–8 stems or occasionally solitary, usually leaning, 1.5–7.2 m tall, 8–10 cm diam, rough with old leaf scars, with a mound of roots at base. *Leaves* 8–20; sheath ca. 40 cm long, fibrous at margins; petiole 40–60 cm long; rachis 1.9–4.1 m long; pinnae 35–73 per side, regularly arranged and spreading in 1 plane, linear, the middle ones 33–57 cm long, 2.5–3 cm wide. *Inflorescences* strongly dimorphic; staminate inflorescence interfoliar; peduncle 30–42 cm long; prophyll ca. 40 cm long; peduncular bract 33–40 cm long, with several smaller bracts present distally; rachis 30–45 cm long at anthesis, elongating afterward; staminate flowers densely crowded on rachis; sepals and petals reduced to obscure bracteoles 2–4 mm long; stamens 150–300; pistillode absent; pistillate inflorescences interfoliar; peduncle 20–23 cm long; prophyll ca. 21 cm long; peduncular bract ca. 32 cm long, with other smaller bracts present; rachis ca. 4 cm long; pistillate flowers 4–10, densely crowded on rachis; tepals ca. 5, lanceolate, 5–

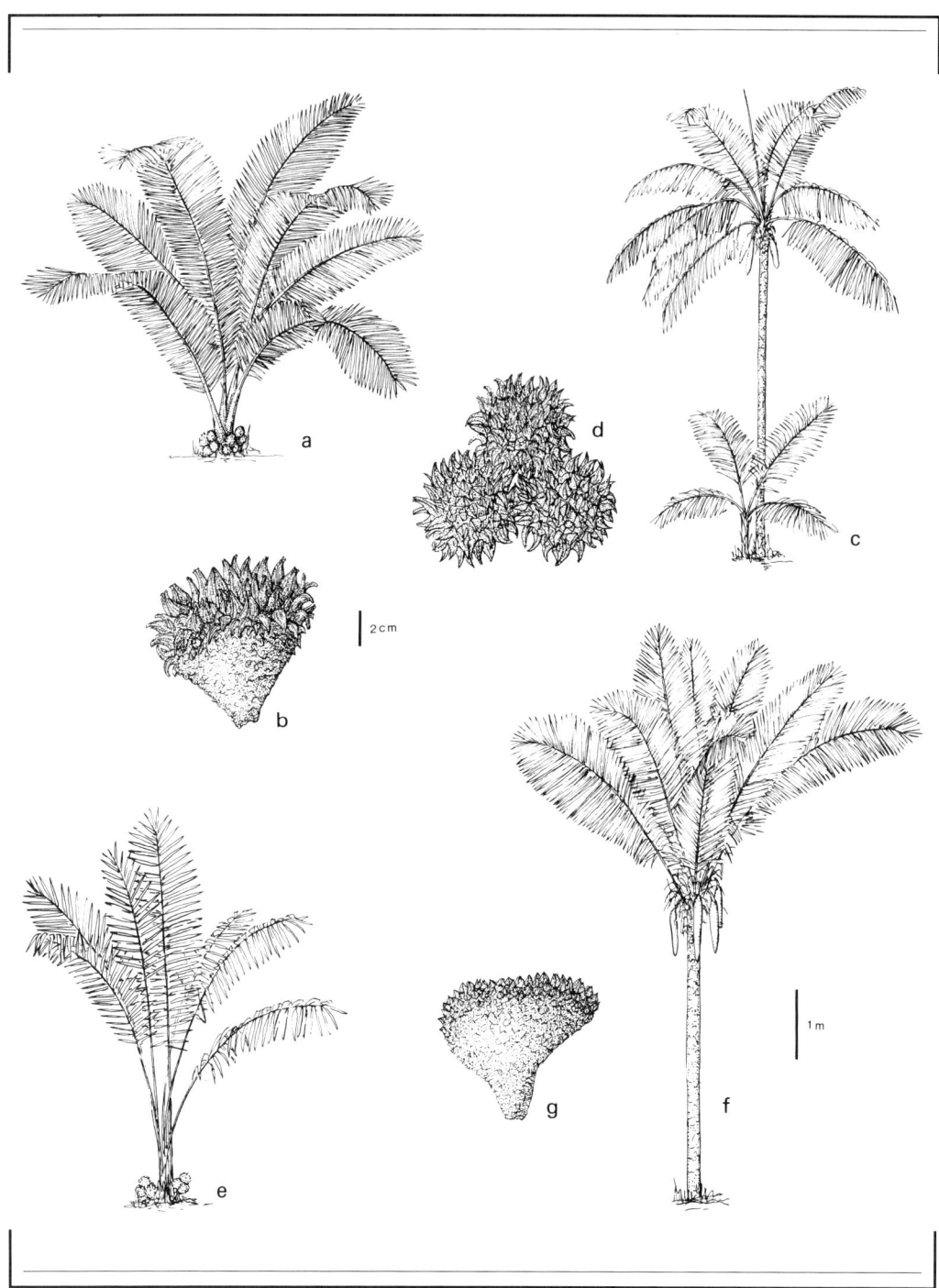

Figure 6.70. **(a)** *Phytelephas macrocarpa,* **(b)** fruit (from *A. Henderson 823*); **(c)** *Phytelephas tenuicaulis,* **(d)** infructescence (from *W. Davis 1016*); **(e)** *Ammandra dasyneura;* **(f)** *Aphandra natalia,* **(g)** fruit (from *W. Davis 997*).

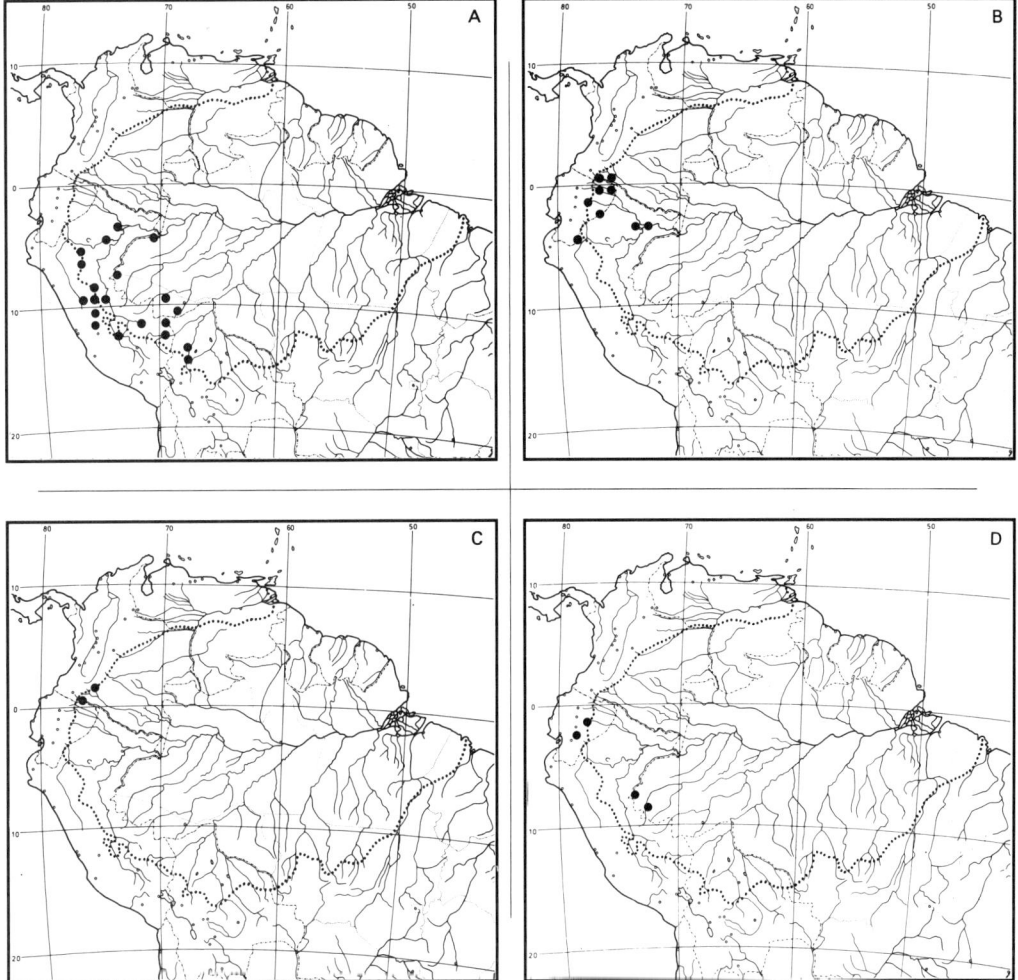

Figure 6.71. (A) *Phytelephas macrocarpa;* **(B)** *Phytelephas tenuicaulis;* **(C)** *Ammandra dasyneura;* **(D)** *Aphandra natalia.*

14 cm long, fleshy; style to 10 cm long; stigmas 5, to 4 cm long; staminodes numerous, stamenlike; *fruits* to 10 in a tight cluster, irregularly obdeltoid, with woody, pointed warts especially on apical surface, 6–9 cm long; seeds ca. 5 per fruit.

Colombia (Putumayo), Ecuador (Napo, Pastaza), and Peru (Amazonas, Loreto) (Figure 6.71B); usually in low-lying, inundated forest near streams and rivers, but occasionally found on *terra firme,* always at low elevations.

Barfod and co-workers (1987) described pollination of this species (as *Phytelephas microcarpa*) in Ecuador, and concluded that the most likely pollinators were curculionid beetles.

Colombia: *shee-she-tsha,* (Kofán), Ecuador: *omacabo* (Huao), *omakabo* (Waorani), *sehua* (Siona), *shishihe* (Kofán), *tagua, yarina*. Peru: *caápi, chápi* (Achual Jívaro), *tagua, yarina.*

The leaves are commonly used to thatch roofs, especially the ridge, and are the preferred palm for thatching; the hearts are occasionally eaten; the immature and still liquid endosperm is drunk; and an oil is extracted from the fruits.

33. *Ammandra*   **Ammandra** O. F. Cook, J. Wash. Acad. Sci. 17: 218. 1927.

Moderate, dioecious palms. *Stems* solitary or cespitose, short and subterranean, or aerial and then procumbent or erect. *Leaves* pinnate, reduplicate; sheath open and not forming a crownshaft; petiole elongate; rachis long; pinnae regularly arranged and spreading in 1 plane, linear. *Inflorescences* strongly dimorphic, branched to 1 order; staminate inflorescence interfoliar; peduncle bearing a prophyll and 1 peduncular bract, with a few smaller bracts present; rachis elongate with numerous short rachillae; staminate flowers borne in clusters of 6–9; perianth much reduced and bractlike; receptacle rounded, with 300–1200 small stamens; pistillode much reduced; pistillate inflorescences interfoliar, with contracted rachis and rachillae; pistillate flowers 6–10 per inflorescence, sessile; tepals 7–10; staminodes numerous; gynoecium syncarpous, multiloculate, multiovulate; styles and stigmas elongate; *fruits* several-seeded, irregularly prismatic, borne in a club-shaped head, covered with woody, pointed warts; seed with homogeneous endosperm and subbasal embryo; germination remote-ligular; eophyll pinnate.

A genus of two species (Barfod, 1991): one in the Pacific coast of Colombia (Chocó), and the other in the northwestern Amazon region of Colombia, Ecuador, Peru, and western Brazil.

1. Ammandra dasyneura (Burret) Barfod, Opera Bot. 105: 43. 1991. *Phytelephas dasyneura* Burret, Notizbl. Bot. Gart. Berlin-Dahlem 11: 5. 1930. Lectotype (Barfod, 1991). Colombia. Caquetá: Getuchá, Río Ortequaza, 30 Jul 1926, *G. Woronow & S. Juzepczuk 6335* (hololectotype, LE, n.v.) (Figure 6.70e).

Stems cespitose, short and subterranean or aerial and ca. 1 m tall, ca. 12 cm diam. *Leaves* 12–20; sheath and petiole 2.7–3.5 m long, not scaly, sheath fibrous at the margins; rachis 2.7–4 m long, not scaly; pinnae 40–60 per side, linear, regularly arranged and spreading in 1 plane, the middle ones 65–85 cm long, 3.5–5 cm wide. *Inflorescences* strongly dimorphic; staminate inflorescences interfoliar; peduncle 40–75 cm long; prophyll 30–55 cm long; peduncular bract 35–50 cm long, with other peduncular bracts present; rachis 30–60 cm long; rachillae 45–70, 0.5–3 cm long, each with 1–9 flowers; staminate flowers on short pedicels, with a much reduced perianth; stamens 800–1200, ca. 2 mm long; pistillate inflorescences interfoliar, with peduncle, prophyll, and peduncular bract similar to those of staminate inflorescence; rachis greatly reduced; pistillate flowers with 7–10 tepals, 7–10 cm long, and ca. 8 stigmas on a 5–7 cm long style; staminodes numerous, stamenlike; *fruits* 4–10 per rachis, closely packed together and irregularly prismatic, 8–10 cm long, 10–12 cm diam, blackish-brown, covered with woody, pointed warts.

Colombia (Caquetá, Putumayo) and Ecuador (Napo) (Figure 6.71C); lowland rain forest on *terra firme* at low elevations.

Ecuador: *patisak'o* (Kofán), *ñume'mba* (Kofán), *tú-te-se-wa* (Siona), *yarina blanca*. The leaves are used for thatch; the immature and still liquid endosperm is drunk.

34. *Aphandra*   **Aphandra** Barfod, Opera Bot. 105: 44. 1991.

Moderate to large, dioecious palms. *Stems* solitary, erect. *Leaves* pinnate, reduplicate; sheath open and not forming a crownshaft, sheath and petiole fibrous on margins; petiole elongate; rachis long; pinnae regularly arranged and spreading in 1 plane, linear. *Inflorescences* strongly dimorphic, branched to 1 order; staminate inflorescence interfoliar; peduncle bearing a prophyll and 1 peduncular bract, with a few smaller bracts present; rachis elongate with numerous short rachillae; staminate flowers borne in clusters of ca. 4; perianth much reduced and bractlike; receptacle rounded, with 400–650 small stamens; pistillate inflorescences interfoliar, with contracted rachis and rachillae; pistillate flowers 24–40 per inflorescence, sessile; tepals 7–9; staminodes numerous; gynoecium syncarpous, 7–8-loculate, 7–8-ovulate; styles and stigmas elongate; *fruits* several-seeded, irregularly prismatic, borne in a club-shaped head, covered with woody, pointed warts; seed with homogeneous endosperm and subbasal embryo; germination remote-ligular; eophyll pinnate.

A genus of one species (Barfod, 1991) distributed in eastern Ecuador, eastern Peru, and western Brazil.

1. **Aphandra natalia** (Balslev & Henderson) Barfod, Opera Bot. 105: 44. 1991. *Ammandra natalia* Balslev & Henderson, Syst. Bot. 12: 501. 1987. Type. Ecuador. Morona-Santiago: road from Mendez to Sucua, km 18, just S of Logroño, 2°35′S, 78°11′W, ca. 800 m, 14 Jul 1985, *H. Balslev & A. Henderson 60651* (isotype, NY) (Figure 6.70f–g).

Stems solitary, 3–11 m tall, 20–22 cm diam, with a mound of roots at the base. *Leaves* 10–20; sheath 0.7–1 m long, fibrous on margins, and with an elongate ligule disintegrating into a mass of fibers, sheath and petiole scaly; petiole 2.3–2.5 m long; rachis 4–5.4 m long, scaly; pinnae 90–120 per side, regularly arranged and spreading in 1 plane, linear, middle pinna 1–1.2 m long, 6–8 cm diam. *Inflorescences* strongly dimorphic; staminate inflorescences interfoliar; peduncle 70–90 cm long; prophyll 40–50 cm long; peduncular bract 80–150 cm long, a few smaller bracts present; rachis 1–1.7 m long; rachillae 200–300, 1.5 cm long; staminate flowers borne on receptacles with much reduced perianth and 200–300 stamens; pistillate inflorescences interfoliar; peduncle 30–45 cm long; prophyll 40–60 cm long; peduncular bract 35–45 cm long, with smaller bracts present; rachis 5–8 cm long; flowers 25–40, crowded on rachis; tepals 7–9, 15–25 cm long, linear, acute; staminodes stamenlike; style 14 cm long; stigmas 6–8, 4.5 cm long; *fruits* densely crowded on rachis, ca. 30 per infructescence, 7–12 cm long, 4–11 cm diam, cone-shaped, irregular, with prominent warty projections.

Western Amazon region in Ecuador (Morona-Santiago, Napo, Pastaza), Peru (Loreto, Ucayali), and Brazil (Acre) (Figure 6.71D); lowland rain forest on *terra firme* at low to mid elevations (300–800 m).

Brazil: *piaçaba, piassaba*. Ecuador: *chilimoyo, tagua, wamowe* (Waorani). The leaf fibers are used to make brooms; the leaves are used for weaving (Waorani, Ecuador). There is considerable interest in the use and management of this species in Ecuador (Borgtoft Pedersen, 1992), where fibers are extracted commercially for broom manufacture. There is also a small industry in Iquitos, Peru, using the fibers for broom manufacture.

APPENDIX

Numerical List of Taxa and Specimens Examined

Taxa

1.1 Chelyocarpus chuco
1.2 Chelyocarpus repens
1.3 Chelyocarpus ulei
2.1 Itaya amicorum
3.1 Raphia taedigera
4.1 Mauritia carana
4.2 Mauritia flexuosa
5.1 Mauritiella aculeata
5.2 Mauritiella armata
6.1a Lepidocaryum tenue
       var. casiquiarense
6.1b Lepidocaryum tenue
       var. gracile
6.1c Lepidocaryum tenue
       var. tenue
7.1 Chamaedorea angustisecta
7.2 Chamaedorea fragrans
7.3 Chamaedorea pauciflora
7.4 Chamaedorea pinnatifrons
8.1a Wendlandiella gracilis
       var. gracilis
8.1b Wendlandiella gracilis
       var. polyclada
8.1c Wendlandiella gracilis
       var. simplicifrons
9.1 Dictyocaryum ptarianum
10.1 Iriartella setigera
10.2 Iriartella stenocarpa
11.1 Iriartea deltoidea
12.1 Socratea exorrhiza
12.2 Socratea salazarii

13.1 Wettinia augusta
13.2 Wettinia drudei
13.3 Wettinia maynensis
14.1 Manicaria saccifera
15.1 Leopoldinia major
15.2 Leopoldinia piassaba
15.3 Leopoldinia pulchra
16.1 Euterpe catinga
       var. catinga
16.2 Euterpe longebracteata
16.3 Euterpe oleracea
16.4a Euterpe precatoria
       var. longevaginata
16.4b Euterpe precatoria
       var. precatoria
17.1 Prestoea schultzeana
18.1 Oenocarpus bacaba
18.2 Oenocarpus balickii
18.3a Oenocarpus bataua
       var. bataua
18.3b Oenocarpus bataua
       var. oligocarpa
18.4 Oenocarpus circumtextus
18.5 Oenocarpus distichus
18.6 Oenocarpus makeru
18.7 Oenocarpus mapora
18.8 Oenocarpus minor
18.9 Oenocarpus simplex
19.1 Hyospathe elegans
20.1 Syagrus cocoides
20.2 Syagrus comosa

20.3	*Syagrus inajai*	26.18	*Bactris glaucescens*
20.4	*Syagrus orinocensis*	26.19a	*Bactris hirta* var. *hirta*
20.5	*Syagrus petraea*	26.19b	*Bactris hirta* var. *mollis*
20.6	*Syagrus sancona*	26.19c	*Bactris hirta* var. *pulchra*
20.7	*Syagrus smithii*	26.20	*Bactris killipii*
20.8	*Syagrus stratincola*	26.21	*Bactris macana*
21.1	*Attalea attaleoides*	26.22	*Bactris macroacantha*
21.2	*Attalea butyracea*	26.23a	*Bactris major* var. *infesta*
21.3	*Attalea dahlgreniana*	26.23b	*Bactris major* var. *major*
21.4	*Attalea eichleri*	26.23c	*Bactris major* var. *megalocarpa*
21.5	*Attalea insignis*	26.23d	*Bactris major* var. *socialis*
21.6	*Attalea luetzelburgii*	26.24a	*Bactris maraja* var. *chaetospatha*
21.7	*Attalea maripa*	26.24b	*Bactris maraja* var. *juruensis*
21.8	*Attalea microcarpa*	26.24c	*Bactris maraja* var. *maraja*
21.9	*Attalea phalerata*	26.25	*Bactris oligocarpa*
21.10	*Attalea racemosa*	26.26	*Bactris oligoclada*
21.11	*Attalea septuagenata*	26.27	*Bactris pliniana*
21.12	*Attalea speciosa*	26.28	*Bactris ptariana*
21.13	*Attalea spectabilis*	26.29	*Bactris rhapidacantha*
21.14	*Attalea tessmannii*	26.30	*Bactris riparia*
22.1	*Barcella odora*	26.31	*Bactris setulosa*
23.1	*Elaeis oleifera*	26.32	*Bactris simplicifrons*
24.1	*Acrocomia aculeata*	26.33	*Bactris syagroides*
25.1	*Aiphanes aculeata*	26.34	*Bactris tefensis*
25.2	*Aiphanes deltoidea*	26.35a	*Bactris tomentosa* var. *sphaerocarpa*
25.3	*Aiphanes ulei*	26.35b	*Bactris tomentosa* var. *tomentosa*
25.4	*Aiphanes weberbaueri*	26.36	*Bactris trailiana*
26.1a	*Bactris acanthocarpa* var. *acanthocarpa*	26.37	*Bactris turbinocarpa*
26.1b	*Bactris acanthocarpa* var. *intermedia*	27.1	*Desmoncus giganteus*
26.2	*Bactris acanthocarpoides*	27.2a	*Desmoncus mitis* var. *leptoclonos*
26.3	*Bactris aubletiana*	27.2b	*Desmoncus mitis* var. *leptospadix*
26.4	*Bactris balanophora*	27.2c	*Desmoncus mitis* var. *mitis*
26.5	*Bactris bidentula*	27.2d	*Desmoncus mitis* var. *rurrenabaquensis*
26.6	*Bactris bifida*	27.2e	*Desmoncus mitis* var. *tenerrimus*
26.7	*Bactris brongniartii*	27.3	*Desmoncus orthacanthos*
26.8	*Bactris campestris*		
26.9	*Bactris coloniata*		
26.10a	*Bactris concinna* var. *concinna*		
26.10b	*Bactris concinna* var. *inundata*		
26.10c	*Bactris concinna* var. *sigmoidea*		
26.11	*Bactris constanciae*		
26.12	*Bactris corossilla*		
26.13	*Bactris cuspidata*		
26.14	*Bactris elegans*		
26.15	*Bactris fissifrons*		
26.16	*Bactris gasipaes*		
26.17	*Bactris gastoniana*		

27.4 *Desmoncus phoenicocarpus*
27.5a *Desmoncus polyacanthos*
 var. *polyacanthos*
27.5b *Desmoncus polyacanthos*
 var. *prunifer*
28.1 *Astrocaryum acaule*
28.2 *Astrocaryum aculeatum*
28.3 *Astrocaryum chambira*
28.4 *Astrocaryum gynacanthum*
28.5 *Astrocaryum huaimi*
28.6 *Astrocaryum jauari*
28.7a *Astrocaryum murumuru*
 var. *ciliatum*
28.7b *Astrocaryum murumuru*
 var. *ferrugineum*
28.7c *Astrocaryum murumuru*
 var. *huicungo*
28.7d *Astrocaryum murumuru*
 var. *javarense*
28.7e *Astrocaryum murumuru*
 var. *macrocalyx*
28.7f *Astrocaryum murumuru*
 var. *murumuru*
28.7g *Astrocaryum murumuru*
 var. *perangustatum*
28.7h *Astrocaryum murumuru*
 var. *urostachys*
28.9 *Astrocaryum paramaca*
28.10 *Astrocaryum sciophilum*
28.11 *Astrocaryum vulgare*
29.1 *Pholidostachys synanthera*
30.1 *Asterogyne guianensis*
31.1 *Geonoma arundinacea*
31.2 *Geonoma aspidiifolia*
31.3 *Geonoma baculifera*
31.4a *Geonoma brevispatha*
 var. *brevispatha*
31.4b *Geonoma brevispatha*
 var. *occidentale*
31.5 *Geonoma brongniartii*

31.6 *Geonoma camana*
31.7 *Geonoma deversa*
31.8a *Geonoma interrupta*
 var. *euspatha*
31.8b *Geonoma interrupta*
 var. *interrupta*
31.9 *Geonoma laxiflora*
31.10 *Geonoma leptospadix*
31.11 *Geonoma longepedunculata*
31.12a *Geonoma macrostachys*
 var. *acaulis*
31.12b *Geonoma macrostachys*
 var. *macrostachys*
31.12c *Geonoma macrostachys*
 var. *poiteauana*
31.13a *Geonoma maxima*
 var. *ambigua*
31.13b *Geonoma maxima*
 var. *chelidonura*
31.13c *Geonoma maxima*
 var. *maxima*
31.13d *Geonoma maxima*
 var. *spixiana*
31.14 *Geonoma myriantha*
31.15 *Geonoma oldemanii*
31.16 *Geonoma oligoclona*
31.17 *Geonoma poeppigiana*
31.18 *Geonoma polyandra*
31.19a *Geonoma stricta*
 var. *piscicauda*
31.19b *Geonoma stricta*
 var. *stricta*
31.19c *Geonoma stricta*
 var. *trailii*
31.20 *Geonoma triglochin*
31.21 *Geonoma umbraculiformis*
32.1 *Phytelephas macrocarpa*
32.2 *Phyleltphas tenuicaulis*
33.1 *Ammandra dasyneura*
34.1 *Aphandra natalia*

Specimens Examined

A. Abraham   107 (26.32), 327 (5.2)
E. Acero   196 (15.3), 202 (4.2)
P. Acevedo   1595 (31.13b), 3369 (26.32), 4880 (31.15)
G. Agostini   222 (26.32)
L. Aguirre-Galviz   1021 (19.1), 1152 (27.2c)
M. Alexiades   35 (26.10b), 66 (26.10b), 269 (21.2), 271 (7.1)

H. Allard   22479 (31.12a)
P. Allen   3252 (28.1), 3277 (20.4), 3357 (26.12)
W. Alverson   81 (18.3a), 82 (16.4a)
I. Amaral   368 (27.5a), 706 (27.5a), 713 (10.1), 902 (7.4), 1015 (31.4a), 1016 (20.5), 1195 (10.1), 1200 (12.1), 1299 (16.2), 1311 (31.12a), 1333 (19.1), 1488 (19.1), 1599 (26.32)

L. Anananch 171 (11.1)

E. Ancuash 51 (26.9), 1039 (19.1)

A. Anderson 93 (15.3), 190 (28.4), 194 (5.2), 196 (28.2), 198 (26.1a), 200 (21.8), 201 (26.7), 216 (15.3), 218 (26.32), 284 (16.4b), 288 (21.12), 292 (21.4), 293 (16.2), 391 (21.12), 1085 (28.8f), 1140 (4.2), 1175 (27.5a), 2041 (27.5a), 2210 (16.3), 10630 (10.1)

W. Anderson 8138 (31.4a), 9625 (27.5a), 9712 (20.2), 10247 (20.5), 10577 (27.5a), 10592 (31.12a), 11127 (31.13b), 35977 (31.4a), 36881 (20.2), 11109 (26.32)

E. Appun 240 (12.1)

I. Araujo 23 (4.2)

W. Archer 2371 (12.1), 8353 (27.3)

A. Argüello 424 (29.1), 521 (26.31), 526 (29.1), 634 (12.1), 635 (13.3), 638 (18.7), 661 (16.4b), 684 (7.3)

L. Aristeguieta s.n. (16.3), 7378 (28.4), 7415 (31.3)

F. Ayala 2362 (19.1), 2366 (31.19a), 2776 (31.13b), 3595 (26.10b), 3877 (26.22), 3885 (10.2), 3886 (6.1c), 3884 (19.1)

G. Aymard 2866 (26.23b), 3091 (31.8b), 3753 (7.4), 3958 (28.4), 3967 (31.10), 4003 (31.13a), 5344 (31.13a), 5813 (10.1), 6117 (31.7), 6123 (10.1), 6953 (12.1), 7214 (26.32), 7230 (16.4b), 7290 (31.3), 7322 (31.10), 7825 (31.3), 7848 (31.7), 7857 (31.7), 7869 (26.7), 7925 (10.1), 8161 (31.12c), 8196 (14.1), 8335 (26.12), 8503 (31.3), 8515 (26.1a), 8599 (16.2), 8610 (26.24c), 8631 (31.7), 8923 (31.13c), 8972 (18.2), 9014 (26.8), 9388 (27.3), 9719 (31.3), 9721 (31.13b), 9733 (10.1), 9770 (16.1), 9850 (26.24c)

L. Bailey s.n. (14.1), 185 (16.3), 216a (23.1), 338 (28.11), 1444 (4.2)

M. Baker 5638 (31.8b), 5911 (31.12b), 5922 (11.1), 5960 (31.12b), 5967 (7.3), 5990 (17.1), 6037 (18.3a), 6047 (26.16), 6243 (13.3), 6342 (7.4), 6368 (31.12b), 6671 (31.8b), 6766 (11.1), 6788 (31.12a), 6805 (25.3), 6812 (19.1), 6827a (26.12), 6905 (31.13c), 6912 (31.19c), 7002 (32.2), 7003 (11.1), 7018 (31.6)

W. Balée 530 (18.5), 572 (18.5), 679 (26.1a), 824 (26.7), 1032 (28.4), 1101 (18.5), 1549 (31.3), 1582 (26.19a), 1591 (31.13c), 1672 (28.4), 1920 (20.1), 1955 (26.7), 2091 (26.35b), 2166 (18.5), 2173 (28.11), 2251 (28.4), 2301 (27.4), 2303 (20.3), 2500 (18.5), 2650 (21.12), 2675 (31.3), 2824 (18.5), 2965 (31.10), 3144 (18.5), 3347 (31.3), 3355 (26.1a), 3376 (21.12), 3377 (21.7), 3385 (26.7), 3481 (28.11), 3484 (28.4), 3525 (26.35b), 3541 (12.1), 3543 (21.12), 3545 (26.23a), 3546 (26.1a)

M. Balick 902 (18.3a), 904 (16.3), 905 (12.1), 906 (28.4), 907 (26.13), 909 (20.3), 910 (18.5), 911 (12.1), 912 (20.2), 913 (5.2), 914 (20.5), 915 (18.5), 916 (4.2), 917 (20.5), 918 (16.2), 919 (28.4), 920 (21.7), 921 (4.2), 922 (28.2), 924 (26.1a), 925 (28.4), 926 (31.7), 928 (31.12a), 929 (18.3a), 930 (18.5), 931 (6.1b), 932 (31.13b), 934 (26.32), 935 (10.1), 936 (31.12c), 937 (26.25), 938 (31.4a), 939 (31.12a), 940 (26.24c), 941 (20.1), 943 (26.24c), 944 (31.12a), 945 (26.32), 947 (31.19b), 950 (26.19a), 951 (27.5a), 953 (18.8), 955 (18.3a), 959 (21.3), 967 (26.14), 974 (5.2), 975 (18.7), 977 (12.1), 987 (18.7), 989 (26.15), 1013 (18.3a), 1043 (31.5), 1045 (31.17), 1046 (31.7), 1066 (1.3), 1131 (31.12a), 1133 (19.1), 1136 (26.19b), 1137 (31.6), 1138 (31.19c), 1139 (31.12b), 1147 (28.8e), 1149 (13.1), 1150 (7.2), 1158 (18.7), 1189 (18.3a), 1192 (20.4), 1193 (16.4b), 1195 (12.1), 1196 (18.7), 1197 (28.1), 1200 (28.4), 1202 (31.7), 1204 (26.24c), 1232 (14.1), 1300 (21.3), 1301 (21.12), 1303 (31.3), 1304 (21.12), 1305 (21.3), 1306 (18.5), 1309 (21.12), 1310 (21.9), 1313 (21.4), 1343 (20.1), 1345 (20.1), 1346 (20.1), 1351 (21.12), 1353 (21.12), 1354 (21.12), 1358 (21.9), 1359 (21.12), 1360 (26.23d), 1361 (12.1), 1362 (18.7), 1364 (31.7), 1365 (26.10b), 1366 (1.1), 1367 (21.12), 1368 (18.3a), 1374 (26.24b), 1375 (26.10c), 1402 (28.5), 1428 (26.23d), 1429 (20.6), 1432 (21.12), 1475 (28.11), 1476 (16.3), 1479 (28.4), 1493 (27.5a), 1527 (18.5), 1528 (21.12), 1555 (21.7), 1556 (24.1), 1578 (21.4), 1579 (21.4), 1580 (21.4), 1587 (31.4a), 1588 (27.3), 1597 (21.4), 1615 (18.5), 1616 (20.1), 1621 (16.3), 1626 (28.11), 1636 (12.1), 1647 (18.3a), 1657 (18.3a), 1659 (18.7), 1660 (16.3), 1675 (21.4), 1678 (18.5), 1715 (27.3), 1946 (21.12), 2689 (16.4a), 7621 (18.3a), 7622 (18.3a)

J. Ballivián 30 (24.1)

H. Balslev 781 (31.20), 1594 (31.12b), 1595 (31.19b), 2302 (19.1), 2330 (17.1), 2335 (31.5), 2385 (17.1), 2391 (31.12b), 2416 (7.4), 3121 (26.31), 3122 (18.7), 4263 (18.7), 4264 (26.10a), 4265 (13.3), 4266 (7.4), 4267 (12.1), 4270 (31.8b), 4275 (7.4), 4276 (31.19a), 4288 (26.31), 4290 (7.4), 4298 (26.31), 4300 (31.7), 4302 (31.19a), 4303 (31.11), 4305 (19.1), 4306 (31.13c), 4309 (18.3a), 4310 (18.7), 4311 (16.4b), 4312 (31.7), 4313 (28.6), 4314 (11.1), 4315 (19.1), 4317 (16.4b), 4318 (12.1), 4319 (31.12b), 4320 (31.19c), 4322 (11.1), 4323 (12.1),4325 (31.12b), 4327 (18.7), 4328 (26.10b), 4338 (32.2), 4339 (21.2), 4342 (12.1), 4343 (7.4), 4344 (28.8h), 4345 (26.30), 4346 (5.2), 4352 (7.4), 4372 (31.12b), 4386 (28.3), 4387 (17.1), 4413 (13.3), 4414 (11.1), 4415 (31.8b), 4420 (31.19b), 4421 (7.4), 4422 (26.12), 4423 (29.1), 4426 (13.3), 4542 (31.12b), 4568 (28.8h), 4569 (31.12b), 4623 (27.1), 4624 (31.6), 4625 (7.3), 4626 (31.19b), 4631 (11.1), 4632 (12.1), 4635 (26.12), 4636 (16.4b), 4639 (18.7), 4775 (27.2c), 4777 (31.18), 4782 (31.19a), 4783 (31.11), 4785 (26.32), 4789 (21.7), 4790 (26.12), 4791 (31.7), 4797 (18.3a), 4811 (31.19c), 4812 (28.3), 4813 (16.4b), 4814 (31.5), 4816 (31.12b), 4817 (28.6), 4825 (31.12b), 4828 (31.5), 4838 (31.19b), 4858 (26.10a), 4860 (17.1), 4863 (31.12b), 4865 (11.1), 4869 (5.2), 4870 (26.12), 4871 (7.4), 4872 (31.12b), 4873 (17.1), 10404 (7.4), 60629 (31.12b), 60500 (19.1), 60506 (31.12b), 60515 (26.24b), 60516 (26.24c), 60517 (31.12b), 60518 (17.1), 60520 (31.13c), 60537 (18.7), 60538 (13.3), 60539 (11.1), 60563 (11.1),60571 (31.5), 60577 (32.2), 60579 (31.12b), 60584 (31.19b), 60589 (7.4), 60592 (21.7), 60603 (17.1), 60606 (27.1), 60607 (21.7), 60609 (31.12b), 60612 (27.1), 60613 (18.3a), 60615 (16.4b), 60618 (26.24c), 60649 (31.8b), 60650 (12.1), 60672 (12.1), 60682 (19.1), 60692 (31.8b), 60693 (26.10a), 60694 (28.3), 60697 (32.2), 60700 (17.1), 60702 (28.8h), 60704 (18.3a), 60730 (18.7), 60731 (31.13c), 60733 (25.3), 60738 (7.3), 60739 (31.18), 60745 (21.7), 60748 (21.2), 60749 (4.2), 60752 (27.3), 62001 (11.1), 62005 (18.3a), 62017 (26.31), 62111 (16.3), 62031 (7.4), 62035 (31.19b), 62039 (26.24b), 62041 (13.3), 62043 (31.8b), 62044 (18.7), 62050 (27.5b), 62051 (26.30), 62052 (26.32), 62053 (31.19a), 62054 (31.7), 62055 (31.11), 62056 (5.2), 62057 (27.3), 62058 (17.1), 62059 (19.1), 62060 (31.19c), 62061 (31.12a), 62062 (26.1a), 62063 (31.12b), 62064 (31.20), 62065 (13.3), 62066 (31.18), 62069 (19.1), 62070 (33.1), 62071 (31.20), 62073 (19.1), 62075 (31.12b), 62081 (31.19a), 62038 (26.12), 62117 (26.31), 62200 (25.3), 62201 (31.19b), 62206 (31.20), 62210 (31.12b), 62213 (17.1), 62211 (31.11), 62216 (31.6), 62217 (28.8h), 62401 (23.1), 62406 (31.11), 62413 (17.1), 62417 (1.3), 62525 (20.6), 62535 (13.3), 69047 (16.4b)

P. Bamps 5344 (6.1c)

M. Bang 1734 (11.1)

P. Barbour 2678 (7.4), 4474 (29.1), 4778 (31.19c), 4798 (31.7)

H. Barclay 9443 (19.1)

A. Barfod 60009 (29.1), 60027 (11.1), 60065 (16.3), 60079 (18.3a), 60103 (16.4a), 60174 (16.4a), 60185 (16.3)

Barnes s.n. (14.1)

H. Beck 60 (19.1), 147 (14.1), 163 (31.15), 165 (20.3), 218 (12.1), 260 (16.3), 273 (28.8f), 278 (18.5), 284 (19.1), 297 (26.7), 372a (18.3a), 396 (21.7), 524 (4.2)

S. Beck 1556 (26.23d), 1623 (31.5), 5247 (26.18), 5945 (26.30), 9992 (28.5), 10090 (5.2), 10178 (21.2), 10181 (4.2), 16480 (27.2d), 18234 (7.4), 18258 (31.5), 19152 (26.10b), 19260 (1.1), 19265 (4.2), 20301 (1.1), 19612 (18.3a)

F. Beekman 28 (27.4)

D. Bell 88–228 (26.19a)

C. Belshaw 3586 (11.1)

B. Bennett 3571 (31.19c), 3683 (29.1), 3878 (4.2), 3991 (11.1), 4116 (7.3), 4118 (31.12a), 4120 (29.1)

C. Berg 699 (31.3), P18133 (19.1), P19464 (10.1)

B. Berlin 262 (13.3), 449 (26.9), 498 (26.12), 527 (28.3), 646 (32.2), 670 (26.9), 688 (17.1), 762 (26.12), 763 (26.27), 831 (28.7c), 832 (17.1), 1804 (16.4a), 1808 (31.19c), 1877 (26.32), 1879 (26.32)

R. Bernal 27 (26.21), 59 (26.24c), 105 (26.16), 115 (18.7), 285 (31.8b), 300 (11.1), 306 (31.8b), 395 (27.3), 397 (21.2), 401 (23.1),

409 (27.3), 457 (18.7), 474 (3.1), 483 (21.2), 494 (18.3a), 507 (27.3), 508 (21.2), 523 (27.3), 629 (24.1), 639 (7.4), 646 (26.7), 649 (26.24c), 686 (26.7), 693 (26.31), 694 (14.1), 800 (18.7), 858 (18.3a), 873 (26.31), 891 (16.3), 906 (29.1), 938 (11.1), 1007 (19.1), 1018 (14.1), 1021 (26.24c), 1025 (26.31), 1037 (18.7), 1088 (26.9), 1091 (19.1), 1096 (26.24c), 1110 (26.24c), 1163 (26.32), 1201 (23.1), 1209 (26.23b), 1359 (20.6), 1379 (26.31), 1382 (26.24c), 1384 (29.1), 1399 (31.10), 1404 (21.2), 1406 (25.3), 1407 (7.3), 1408 (31.12b), 1409 (31.18), 1410 (26.7), 1413 (31.8b), 1415 (31.11), 1421 (31.20), 1425 (32.2), 1547 (21.2), 1549 (24.1), 1960 (24.1)

A. Bernardi   753 (16.4b), 831 (31.8a), 941 (9.1), 1496 (26.32), 2017 (7.4), 2825 (31.7), 6617 (5.2)

P. Berry   654 (28.1), 794 (15.3), 1443 (6.1a), 1463 (26.24c), 1468 (26.4), 1515 (4.1), 1538 (26.19a), 1655 (18.1)

J. Betancur   1273 (31.8b), 1315 (20.4), 2369 (20.4), 2818 (31.8b) 3191 (29.1)

P. Bettella   121 (7.4), 124 (26.21)

D. Bierhorst   S254 (23.1)

F. Billiet   1202 (16.3), 1285 (26.23a), 4413 (26.23a)

G. Black   471745 (15.3), 52–14536 (5.2)

U. Blicher   1 (21.7), 5 (21.2)

Blum   1496 (26.24c)

J. Blydenstein   1073 (28.1), 1688 (20.4)

J. Boeke   1467 (7.4)

F. Bond   63 (27.3), 64 (14.1), 214 (14.1)

B. Boom   1711 (28.10), 1751 (28.10), 1788 (28.10), 1848 (18.1), 1891 (28.10), 2298 (12.1), 2327 (12.1), 2346 (18.1), 2415 (12.1), 2635 (12.1), 4129 (26.1a), 4145 (21.9), 4151 (16.4b), 4152 (18.7), 4154 (28.8f), 4155 (12.1), 4159 (28.2), 4436 (31.13b), 4462 (26.1a), 4509 (26.24b), 4537 (21.7), 4538 (18.3a), 4656 (12.1), 4805 (26.24b), 4836 (18.7), 4984 (26.16), 5423 (10.1), 5474 (10.1), 5540 (31.7), 6213 (24.1), 6616 (20.4), 6720 (16.4b), 7373 (26.26), 8101 (26.26), 9331 (10.1)

R. Borchsenius   91427 (25.4), 91430 (31.19b)

C. Bosque   Y-SN1 (16.4a), 24 (26.31)

B. Boswezen   5620 (26.1b), 6508 (31.19b)

H. Boudet Fernandes   618 (26.19a)

J. Brandbyge   30125 (17.1), 33388 (17.1), 32603 (17.1)

A. Braun   s.n. (7.4)

D. Breedlove   33936 (31.8b)

A. Brenes   13566 (11.1)

N. Britton   278 (16.3), 494 (21.7), 1006 (18.3b), 1559 (27.5a), 1805 (31.8b), 1933 (31.8b), 2278 (31.8b)

W. Broadway   289 (28.11), 309 (27.3), 367 (24.1), 644 (28.11), 645 (26.23a), 754 (25.1), 935 (24.1), 8958 (16.4b), 9750 (16.4b), 9850 (26.8), 9851 (31.8b)

N. Brokaw   155 (31.8b)

O. Buchtein   1247 (11.1), 1249 (11.1)

G. Bunting   4977 (7.4), 6972 (21.2), 7500 (16.4a), 7661 (27.3), 8329 (18.3a), 8512 (31.8b), 8857 (26.23b), 9526 (31.8b), 10179 (26.31), 10189 (16.4a), 10219 (27.3), 10241a (7.4), 10260 (16.4a), 10836 (31.7), 10850 (18.7), 10907 (16.4a), 10945 (18.3a), 11193 (18.7), 11443 (31.7), 12102 (19.1), 12103a (7.4)

W. Burger   4305a (31.8b), 4720 (NY), 5895 (12.1)

Byron   67–17 (16.4b)

I. Cabrera   1820 (28.1), 2002 (31.7), 2005 (21.7), 2406 (31.7), 2703 (14.1), 3054 (27.2b), 3308 (6.1c)

C. Calderon   2709 (27.2c), 2851 (27.2a)

R. Callejas   2386 (31.8b), 3012 (7.4), 3410 (31.10), 4021 (29.1), 4221 (29.1), 4309 (7.4), 4946 (26.24c), 5399 (31.8b), 7963 (16.4a), 8529 (16.4a)

M. Cameron   151 (19.1), 153 (19.1), 156 (12.1), 157 (28.8e)

D. Campbell   14476 (12.1), 14857a (28.9), 14990a (21.8), P20818 (6.1c), P21875 (18.8), P21985 (15.3), P22295 (21.8), P22409 (20.1)

J. Cardiel   112 (31.4b)

F. Cardona   409 (10.1), 922 (31.3)

Carmona   21 (26.31), 173 (26.31)

A. Carvalho   3464 (26.1a), 3469 (26.19a)

F. Casas   8103 (31.7), 8136 (31.12a), 8169 (31.12a), 8212 (16.4b), 8254 (31.13b), 8309 (31.12a), 8436 (26.30)

A. Castellanos   23493 (20.2)

P. Cavalcante   997 (16.3), 1568 (26.17), 1940 (16.3), 2402 (26.17), 3280 (16.4b)

P. Cazalet   7752 (19.1), 7775 (31.19a)

C. Cerón   161 (12.1), 162 (28.3), 322 (16.4b),

330 (13.3), 360 (18.3a), 387 (17.1), 567 (31.19c), 588 (17.1), 662 (32.2), 714 (27.1), 819 (11.1), 948 (18.3a), 2272 (17.1), 2512 (12.1), 2861 (11.1), 3292 (17.1,) 3401 (16.4b), 4264 (18.3a), 5030 (18.3a), 5104 (28.3), 5107 (28.6)

J. Chagas 234 (21.1), 1133 (21.8)

M. Chaparro et al. 29 (19.1)

F. Chávez 103 (26.24c), 577 (7.4), 600 (19.1), 617 (26.24c), 626 (27.5a), 638 (12.2), 641 (11.1), 643 (31.7), 644 (19.1), 647 (11.1), 652 (31.5), 662 (4.2), 673 (18.7), 674 (18.7), 676 (28.7d), 678 (12.1), 679 (21.9), 680 (18.3a), 681 (11.1), 683 (27.5a), 684 (26.24c), 685 (31.8b), 688 (31.12b), 689 (31.12a), 692 (31.5), 693 (31.7), 694 (26.35a), 695 (31.12b), 696 (31.12b), 697 (31.12b), 698 (31.12b), 699 (31.12b), 701 (19.1), 702 (31.8b), 703 (7.1), 705 (31.5), 706 (7.4), 708 (26.24c), 710 (31.12a), 711 (8.1c), 713 (25.1), 714 (26.24c), 715 (26.24c), 716 (26.24c), 718 (31.5), 736 (26.19a), 740 (26.1a), 745 (31.5), 750 (26.7), 752 (26.30), 753 (26.10b), 754 (18.7), 759 (7.4), 761 (31.12a), 763 (28.6)

H. Churchill 5564 (26.9), 6268 (16.4a)

C. Cid 7 (31.10), 83 (19.1), 462 (26.32), 520 (26.19a), 522 (20.1), 530 (31.7), 545 (31.19b), 626 (10.1), 842 (31.19b), 1120 (12.1), 1130 (10.1), 1181 (6.1b), 1479 (31.15), 1555 (20.3), 1643A (31.7), 1803 (31.7), 3299 (27.2c), 4102 (6.1c), 4195 (6.1b), 4578 (7.4), 4636 (27.3), 4794 (12.1), 6065 (20.2), 6240 (20.1), 6413 (20.1), 6530 (20.5), 7089 (31.7), 7714 (6.1b), 8588 (10.1), 8629 (31.19b), 9181 (26.8)

H. Clark 6815 (10.1), 7022 (6.1a), 7493 (15.1), 7982 (27.5a)

C. Clement 501CRC88 (26.21)

L. Coêlho s.n. (16.1), s.n. (27.5a), 20974 (21.1)

M. Collela 1247 (28.4), 1439 (28.4), 1497 (28.4), 1499 (12.1), 1564 (10.1), 1595 (31.7), 1684 (5.1), 1814 (31.3), 2023 (5.2), 2024 6.1a), 2027 (31.13c), 2067 (12.1) 2119 (5.2), 2160 (6.1a)

G. Colonnello 932 (7.4)

E. Contreras 868 (27.3), 976 (27.3), 2166 (31.8b), 2581 (27.3)

O. Cook 14 (16.4a), 37 (16.4a), 59 (16.4a), 151 (16.4a), 176 (16.4a), 899 (7.4)

I. Cordeiro 23 (10.1), 32 (22.1), 199 (27.5a)

M. Córdoba 10 (4.1)

R. Cortés 325 (16.1), 345 (18.3a)

R. Cowan 38340 (26.35b), 38365 (19.1)

J. Cowell 124 (18.7)

G. Cremers 4651 (26.32), 5016 (26.32), 7966 (28.4), 9460 (26.8), 10662 (26.8), 10904 (26.8), 11077 (27.3), 11187 (26.3), 11540 (14.1), 11577 (31.13c), 11578 (28.4), 11622 (31.19b), 11803 (26.3), 11804 (26.17), 11838 (27.4), 11839 (20.3), 11978 (26.36), 12037 (26.32), 12066 (26.17), 12082 (14.1), 12139 (27.5a), 12202 (26.25), 12208 (26.25), 12210 (20.3), 12243 (28.9), 12295 (26.3), 12296 (26.3), 12380 (26.3), 12381 (26.25), 12436 (20.3), 12448 (26.3), 12488 (31.15), 12501 (26.14), 12512 (4.2), 12581 (26.3), 12620 (26.17), 12673 (28.4), 12717 (26.8), 12972 (26.29), 13006 (26.14), 13009 (31.19b), 13020 (26.1b)

T. Croat 5306 (26.9), 5740 (26.23b), 8160 (12.1), 9743 (26.23b), 10735 (26.23b), 14182 (26.24c), 15408 (26.9), 15928 (31.8b), 17529 (26.22), 17605 (12.1), 17621 (31.13b), 17992 (21.2), 18102 (26.22), 18113 (26.24c), 18115 (27.5a), 18434 (26.19b), 18438 (26.12), 18458 (31.17), 18474 (31.17), 18478 (26.1a), 18493 (31.17), 18524 (31.17), 18757 (26.32), 18782 (26.7), 19035 (26.24c), 19329 (26.10b), 20006 (26.32), 20039 (27.2c), 20192 (26.19a), 20295 (6.1c), 20337 (26.24c), 20652 (26.24a), 20901 (26.24c), 59348 (12.1), 67233 (31.8b), 14132 (16.4a), 22845 (16.4a)

L. Croizat 180 (10.1), 369 (31.7), 492 (28.4), 499 (16.2), 499a (26.24c), 584 (12.1), 994 (10.1), 995 (10.1), 995c (26.12)

J. Cruxent 110 (31.7), 112 (19.1), 298 (26.12)

H. Cuadros 2620 (24.1), 2621 (24.1)

J. Cuatrecasas 3860 (20.4), 3861 (12.1), 3977 (27.3), 6806 (10.1), 6874 (19.1), 6941 (21.7), 6992 (16.1), 7071 (4.2), 7138 (26.16), 7181 (26.4), 7181a (14.1), 7187 (20.4), 7221 (18.1), 7243 (26.1a), 7265 (12.1), 7266 (18.3a), 7267 (14.1), 7274 (28.4), 7283 (28.1), 7295 (5.2), 7298 (16.4b), 7301 (11.1), 7309 (5.2), 7368 (18.7), 7434 (26.7), 7475 (18.7), 7487 (20.4), 7539 (18.3a), 7562 (21.7), 7574 (28.3), 7559 (20.4), 7620 (11.1), 7768 (4.2), 8870 (21.7), 8871 (18.3a), 8929 (26.16), 8945 (21.2), 8959 (4.2), 9110 (19.1), 9111 (29.1), 10552 (13.2),

10621 (32.2), 10775 (28.8e), 10786 (13.3), 10794 (26.16), 10817 (32.2), 10849 (18.3a),10858 (21.2), 10860 (4.2), 10868 (28.3), 10883 (17.1), 10895 (17.1), 11023 (13.3), 11159 (13.3), 11332 (31.8b), 13208 (26.12), 13327 (26.16), 13328 (26.16), 13962 (18.3a), 14276 (26.16), 15860 (26.31), 16319 (18.3a), 16428 (26.16), 16729 (14.1), 22053 (16.3), 22971 (20.6), 24200 (16.3), 25377 (7.4), 27195 (13.2)

N. Cuello 255 (16.4a), 439 (20.4), 477 (26.32), 485 (26.5), 566 (15.3), 572 (16.4b), 575 (21.10)

H. Curran 2 (16.3), 144 (23.1), 174 (27.3), 187 (26.23c), 313 (23.1), 354 (21.2)

B. Dahlgren 522 (16.2), 610647 (16.3)

D'Alessandro 5 (14.1)

D. Daly 277 (28.11), 535 (27.5a), 1213 (31.3), 1695 (26.1a), 1844 (31.8a), 4483 (26.32), 5110 (13.1), 5353 (19.1), 5902 (19.1), 6479 (32.1), 6494 (27.2d)

A. da Silva 445 (21.5)

G. Davidse 5295 (20.4), 5683 (4.2), 12426 (15.3), 12443 (28.6), 12611 (5.1), 12721 (27.3), 13115 (5.1), 13837 (5.1), 14549 (20.4), 15225 (28.1), 15357 (21.10), 15368 (28.4), 15646 (15.3), 16286 (16.3), 16528 (31.7), 16846 (15.3), 16849 (26.8), 16915 (26.8), 16973 (28.4), 17025 (21.8), 17316 (31.7), 17386 (5.2), 17708 (26.8), 17711 (28.1), 17769 (5.2), 17831 (27.3), 17988 (27.5a), 18451 (18.7), 18480 (7.4), 18521 (31.8b), 18605 (31.7), 21075 (7.4), 21617 (31.7), 21746 (31.7), 21852 (27.5a), 26599 (31.13c), 26693 (10.1), 26695 (31.13c), 26711 (31.13b), 26715 (31.13b), 26736 (28.1), 26844 (5.1), 27002 (10.1), 27069 (31.7), 27125 (26.32), 27163 (31.7), 27237 (31.13c), 27672 (27.5a), 27712 (27.3), 27714 (15.1), 27845 (15.1)

C. Davidson 3571 (5.2), 10582 (10.1)

W. Davis 843 (19.1), 850 (31.19b), 853 (31.17), 870 (31.12b), 871 (19.1), 872 (31.12a), 929 (11.1), 948 (12.1), 949 (17.1), 960 (31.19a), 961 (31.12b), 963 (21.7), 977 (28.8h), 978 (28.3), 997 (34.1), 1004 (18.3a), 1015 (31.13c) 1016 (32.2), 1173 (26.16), 1174 (4.2), 1185 (4.2), 1188 (21.9), 1318 (7.2)

E. Dawson 15218 (5.2), 15220 (28.5), 15221 (20.2)

A. de Carvalho 627 (27.3)

H. de Foresta 566 (26.3), 571 (26.25)

J.-J. de Granville 46 (31.3), 92 (28.9), 132 (26.13), 164 (26.32), 179 (26.3), 184 (26.3), 186 (27.4), 291 (31.7), 474 (26.29), 514 (26.3), 694 (20.3), 697 (26.32), 736 (26.13), 757 (31.13a), 863 (26.19a), 977 (31.10), 1041 (26.19a), 1113 (26.36), 1149 (26.17), 1215 (26.19a), 1285 (27.4), 1421 (26.13), 1432 (26.17), 1512 (26.24c), 1577 (26.3), 1624 (26.17), 1638 (26.17), 1728 (26.27), 1736 (26.29), 1737 (28.9), 1770 (26.7), 1835 (20.3), 1879 (28.4), 1892 (31.3), 1932 (26.3), 1947 (26.19a), 2087 (21.1), 2210 (31.10), 2259 (27.5a), 2272 (26.19a), 2419 (18.1), 2591 (26.14), 2592 (26.14), 2592 (31.10), 2890 (31.15), 3068 (26.17), 3075 (26.3), 3257 (28.10), 3411 (18.3b), 3444 (18.1), 3543 (26.32), 3585 (27.4), 3750 (4.2), 3774 (31.7), 3775 (26.19a), 3983 (26.11), 4005 (26.19a), 4074 (27.4), 4394 (20.8), 4481 (28.8f), 4496 (26.19a), 4508a (28.10), 4536 (26.25), 4552 (26.17), 4604 (26.14), 4620 (26.3), 4774 (26.3), 4837 (26.2), 4850 (26.11), 4851 (26.29), 4860 (26.3), 4903 (26.17), 4964 (26.29), 4969 (27.4), 5207 (26.14), 5223 (26.29), 5249 (28.4), 5250 (26.29), 5253 (26.17), 5299 (26.32), 5361 (26.8), 5362 (23.1), 5363a (26.23a), 5363 (26.23a), 5391 (26.2), 5392 (26.1b), 5397 (28.8f), 5404 (26.7), 5406 (28.8f), 5407 (28.9), 5408 (20.3), 5413 (26.29), 5426 (26.11), 5566 (21.1), 5583 (27.4), 5824 (26.3), 5889 (26.2), 5961 (26.3), 6001 (26.3), 6174 (26.17), 6175 (26.17), 6179 (26.3), 6187 (26.2), 6188 (26.11), 6196 (27.4), 6238 (26.2), 6261 (26.3), 6382 (31.21), 6461 (26.32), 6895 (26.7), 6955 (26.25), 7061 (26.35b), 7086 (20.3), 7087 (21.1), 7109 (26.29), 7125 (28.8f), 7137 (26.32), 7195 (26.24b), 7206 (26.23a), 7208 (21.8), 7213 (26.11), 7222 (28.8f), 7273 (26.17), 7353 (26.3), 7373 (26.25), 7409 (26.17), 7498 (26.32), 7545 (26.25), 7623 26.13), 7777 (26.17), 7806 (26.14), 7823 (21.1), 7886 (26.24c), 7888 (26.3), 7982 (31.8a), 8223 (28.4), 8250 (20.8), 8327 (27.3), 8340 (28.11), 8352 (26.1b), 8376 (26.25), 8381 (26.1b), 8381 (26.1b), 8422 (31.19b), 8470 (26.32), 8509 (26.1b), 8512 (26.32), 8584 (31.21), 8617 (31.13a), 8618 (26.25), 8673 (26.1b), 8787 (31.19b), 8801 (31.8a), 8854 (26.24c), 8872 (26.24c), 9031 (28.10), 9093 (26.1b),

9095 (26.24b), 9096 (31.15), 9104 (26.3), 9173 (26.17), 9176 (26.3), 9200 (31.19b), 9204 (26.17), 9229 (26.25), 9239 (31.3), 9432 (26.19a), 9432 (26.19a), 9510 (31.13c), 9513 (26.24c), 9692 (31.7), 9693 (26.2), 9694 (26.3), 9913 (20.3), 9968 (26.1b), 9969 (26.19a), 10001 (26.25), 10004 (9.11), 10151 (26.3), 10158 (31.3), 10181 (26.11), 10212 (31.21), 10216 (31.19c), 10273 (26.32), 10292 (26.19a), 10297 (26.32), 10301 (26.19a), 10303 (31.7), 10312 (21.8), 10319 (14.1), 10353 (26.3), 10354 (26.11), 10362 (31.19b), 10381 (26.19a), 10390 (26.14), 10393 (26.25), 10399 (26.17), 10408 (19.1), 10410 (31.7), 10418 (26.29), 10450 (26.24c), 10451 (26.11), 10455 (26.1b), 10469 (26.17), 10474 (26.3), 10477 (28.10), 10484 (31.21), 10514 (31.7), 10554 (26.13), 10621 (31.8a), 10622 (26.32), 10687 (28.4), 10821 (26.13), 10821 (26.13), 10822 (26.32), 10854 (26.24c), 10883 (31.8a), 10884 (26.32), 10893 (19.1), 10945 (28.4), 10984 (26.32), 10995 (26.13), 10996 (26.25), 11032 (21.8), 11039 (26.17), 11073 (28.9), 11078 (26.2), 11079 (20.8), 11118 (26.32), 11137 (26.29), 11148 (26.24b), 11596 (26.25), 11698 (27.4), 11844 (26.17), 13020 (26.1b)

W. de Jong 17 (18.7), 49 (12.1), 134 (21.7), 159 (31.17), 160 (19.1), 161 (31.12b)

J. De La Cruz 3360 (27.4), 4186 (26.26), 4234 (31.13a)

F. Delascio 2236 (21.7), 7964 (27.3), 8198 (27.3), 8626 (28.4)

L. Delgado 571 (18.3a), 665 (16.4b), 774 (31.7), 775 (26.12), 1062 (26.1a)

G. de Nevers 3785 (31.8b), 4457 (31.8b), 4714 (19.1), 4793 (16.4a), 4960 (31.8b), 5043 (19.1), 5234 (19.1), 5343 (26.9), 5741 (16.4a), 5872 (16.4a), 6127 (26.24c), 6224 (16.4a), 6327 (16.4a), 6331 (31.8b), 6350 (16.4a), 6373 (19.1), 6643 (19.1), 6742 (19.1), 6941 (19.1), 7469 (26.16), 7841 (31.7), 8240 (26.23b), 8242 (26.9), 8298 (18.3a), 8331 (26.24c), 8360 (12.1), 8391 (19.1), 8553 (11.1)

E. de Oliveira 3590 (20.1), 3885 (20.1), 4463 (20.1)

C. Díaz 821 (12.1), 3130 (7.4), 3604 (7.4)

W. Diaz 55 (7.4), 319 (27.5a), 321 (26.12)

C. Dodson 9163 (16.4a), 14677 (16.3)

L. Dorr 4817 (7.4), 5201 (7.4), 5659 (7.4), 5874 (11.1), 7153 (26.12), 7179 (26.12), 7181 (31.8b), 7182 (26.31), 7188 (16.4a), 7189 (18.3a), 7194 (31.8b), 7779 (27.5a), 7861 (31.8b)

G. dos Santos 412 (28.4)

C. Doyle s.n. (26.24c), 16 (20.6), 18 (26.16),

E. Dryander s.n. (21.2)

A. Ducke 8695 (26.32), 10675 (26.11), 11762 (6.1b), 11776 (16.4b), 12305 (18.4)

A. Dugand 2506 (27.3), 2897 (18.3a), 2907 (20.4), 5329 (24.1),

J. Duivenvoorden 951 (31.19b), 1429 (26.24c), 2202 (13.1), 2357 (10.2), 2922 (27.5a)

J. Duke 217 (16.4a), 10997 (16.3), 11342 (11.1), 11663 (14.1)

J. Duque 2407 (31.1), 2410 (31.9)

J. Duque-Jaramillo 2142 (7.4), 2310 (7.3), 2332 (26.10b), 2401 (18.3a), 2404 (26.15), 4587 (25.1), 4588 (21.2)

J. Dwyer 4859 (26.24c), 9404 (31.7), 9778 (17.1)

W. Egler 45951 (26.7), 46414 (26.19a), 46691 (31.7)

G. Eiten 5490 (20.2), 9200 (31.4a), 9209 (27.5a),

R. Ek 599 (31.13a)

Bro. Elias 1317 (23.1),

S. Elcoro 66 (19.1)

M. Elvin-Lewis 10072 (19.1)

C. Ely 1 (16.4b), 5 (26.10b), 6 (18.7), 11 (7.1), 13 (21.9), 14 (32.1), 15 (26.24c), 16 (26.10b), 17 (4.2), 19 (28.8f), 25 (12.1), 31 (28.8f), 32 (21.7), 42 (31.13b), 44 (26.24c), 46 (18.7), 48 (16.4b), 59 (18.3a), 66 (21.9), 69 (28.2),

F. Encarnación 1006 (6.1c)

J. Espina 24 (28.4), 62 (31.12c), 224 (15.3), 270 (15.3)

D. Faber-Langendoen 1563 (31.8b)

D. Fairchild 1062 (20.4)

D. Fanshawe 542 (18.1), F573 (31.13a), 2476 (27.3)

M. Fariñas 321 (31.7), 384 (10.1), 491 (31.7)

C. Farney 1875 (15.3), 1904 (5.2),

A. Fendler 460 (3.1), 732 (27.5a), 2465 (7.4), 2468 (27.5a)

A. Fernández 1744 (10.1), 1831 (31.3), 1921 (26.12), 1943 (26.1a), 2410 (27.5a), 2604 (10.1), 2653 (10.1), 2675 (10.1), 3358 (26.19a), 3452 (10.1), 3470 (26.32), 3942 (27.5a), 4110 (16.4b), 4124 (31.3), 4489

(21.7), 4836 (26.32), 6351 (12.1), 6618 (26.7), 7004 (18.3a), 7008 (21.7), 7010 (16.4b), 7106 (31.7), 7502 (31.7), 7825 (26.1a), 7916 (27.5a), 7009 (18.1)

L. Ferreira 56 (15.3), 104 (26.5)

R. Ferreyra 7988 (19.1)

C. Feuillet 647 (26.17), 1144 (26.11), 9926 (19.1), 9933 (31.13a)

T. Filgueiras 1474 (20.5)

M. Fleury 525 (31.13c), 671 (26.7)

E. Foldats 9198 (31.7), 9200 (26.4)

D. Folli 673 (27.3)

E. Forero 1790 (31.8b), 3052 (26.16), 6384 (27.2a), 6387 (7.3), 9080 (27.3),

Forest Department of British Guiana 843 (27.5a), 2349 (16.4b), 3038 (31.19b), 3349 (26.28), 5203 (26.23a), 5212 (27.3), 6335 (31.21), 6781 (26.26), 6888 (26.26), 6899 (31.13a)

M. Foster 1731 (26.21), 1787 (20.6), 1891 (21.2), 2118 (26.16), 2245 (31.1), 2312 (14.1), 8775 (18.3a), 9806 (18.3a)

R. Foster 3255 (19.1), 3581 (17.1), 4338b (26.32), 4353B (31.7), 4457 (31.12b), 5313 (28.7d), 5750 (11.1), 6996 (8.1c), 7168 (27.5a), 7305 (31.19b), 7327 (29.1), 7364 (19.1), 7840 (10.2), 7846 (31.5), 7859 (26.24b), 7904 (31.10), 7908 (26.1a), 7949 (31.12a), 7999 (19.1), 8677 (26.22), 8682 (26.22), 8840 (8.1a), 8889 (11.1), 9318 (31.12b), 9475 (31.20), 9579 (25.1), 9655 (7.1), 9725 (18.7), 9726 (12.1), 9727 (12.2), 9729 (4.2), 9733 (16.4b), 10001 (31.11), 10050 (31.5), 10636 (27.2b), 10661 (7.4), 10759 (31.19a), 10760 (31.20), 10761 (31.20), 10791 (31.13b), 10917 (25.4), 10997 (12.2), 10999 (16.4a), 11288 (26.10a), 11293 (32.1), 11327 (1.3), 11470 (18.7), 11487 (18.7), 11554 (21.14), 11648 (26.10c), 11680 (27.5a), 11764 (26.32), 11764 (26.32), 11904 (31.19c), 11913 (7.3), 12876 (25.1), 13339 (7.1), 13393 (31.5)

L. Fournier 338 (16.4a)

D. Frame 152 (26.14), 154 (7.3), 155 (12.1), 242 (12.1)

P. Franco 2256 (7.4), 2278 (7.4), 2404 (7.4)

R. Fróes 2001 (27.3), 11622 (20.1), 11757 (16.3), 11765 (27.5a), 22368 (15.1), 22665 (22.1), 25887 (26.32), 26042 (28.9), 28315 (22.1), 28741 (21.6), 29155 (14.1), 31323 (15.3), 32027 (5.2), 51651 (26.8)

H. Fuchs 21903 (26.16), 22053 (16.3), 22187 (18.3a)

V. Funk 8106 (31.4b), 8150 (31.4b), 8187 (31.19c), 8233 (7.3), 8272 (31.19c)

G. Galeano 277 (18.7), 693 (7.4), 860 (21.7), 897 (26.36), 899 (26.20), 913 (26.22), 984 (6.1c), 985 (26.1a), 1056 (28.6), 1092 (27.5a), 1142 (4.2), 1143 (5.2), 1148 (29.1), 1157 (5.2), 1175 (16.1), 1206 (6.1c), 1215 (19.1), 1240 (16.4a), 1242 (18.7), 1249 (31.13b), 1257 (24.1), 1273 (21.2), 1280 (26.4), 1281 (9.1), 1284 (31.10), 1286 (26.32), 1289 (16.1), 1290 (4.1), 1292 (18.1), 1293 (31.13c), 1295 (26.19a), 1303 (31.12b), 1304 (31.7), 1310 (31.19c), 1315 (28.7a), 1317 (27.2c), 1320 (26.14), 1321 (19.1), 1323 (31.17), 1329 (16.4b), 1330 (20.7), 1344 (31.12b), 1350 (26.1a), 1444 (10.2) 1445 (10.2), 1448 (6.1c), 1452 (26.19a), 1466 (31.17), 1487 (31.19a), 1492 (31.13d), 1493 (26.24c), 1494 (25.3), 1508 (7.3), 1509 (10.2), 1510 (25.3), 1557 (26.24b), 1562 (7.3), 1563 (7.3), 1564 (31.6), 1568 (18.7), 1611 (26.32), 1619 (31.19b), 1621 (6.1c), 1654 (2.1), 1669 (20.7), 1671 (1.3), 1702 (26.5), 1703 (26.24b), 1735 (31.12a), 1762 (31.12a), 1770 (31.13d), 1797 (26.30), 1811 (31.13d), 1812 (28.7a), 1821 (31.9), 1848 (26.4), 1849 (16.1), 1869 (31.2), 1882 (4.1), 1886 (26.12), 1887 (31.12b), 1890 (31.12a), 1955 (16.4a), 1966 (26.31), 1968 (29.1), 1969 (26.24c), 1973 (31.13c), 1974 (18.4), 1975 (26.32), 1994 (26.19a), 1995 (31.19b), 1996 (26.22), 1997 (18.4), 1999 (19.1), 2023 (26.1a), 2024 (26.15), 2025 (31.13d), 2026 (18.2), 2029 (26.19c), 2032 (26.19a), 2033 (18.2), 2034 (31.1), 2051 (18.8), 2067 (26.12), 2068 (21.2), 2069 (26.24c), 2071 (16.1), 2073 (26.35a), 2074 (31.19c), 2075 (31.13b), 2078 (21.11), 2079 (31.6), 2080 (31.13c), 2081 (18.8), 2094 (18.7), 2095 (26.24c), 2096 (31.12b), 2097 (31.5), 2098 (31.19c), 2099 (6.1c), 2100 (31.19b), 2100 (31.19b), 2104 (19.1), 2106 (31.16), 2107 (26.20), 2109 (31.13b), 2110 (26.14), 2111 (31.7), 2112 (31.7), 2113 (27.5b), 2114 (26.19a), 2116 (28.7a), 2117 (18.2), 2260 (21.10), 2287 (16.1)

H. García Barriga 8443 (25.1), 13652 (31.7), 13772 (6.1c), 13914 (18.1), 13921 (6.1c), 13931 (19.1), 14007 (28.6), 14029 (19.1),

14324 (31.13b), 14395 (31.7), 14471 (6.1c), 14819 (26.10a), 14980 (7.3), 15126 (26.19a), 17061 (20.4), 20947 (21.7), 20993 (25.1)

A. Gely 51 (26.16)

P. Gentle 4883a (31.7), 6363 (31.8b), 6462 (16.4a), 7177 (31.8b), 7558 (31.7), 8201 (31.8b), 8585 (16.4a), 8587 (31.7), 9157 (31.8b), 9169 (31.7), 9289 (31.8b)

A. Gentry 4999 (27.3), 8726 (3.1), 12892 (15.3), 14819 (27.5a), 15810 (6.1c), 15877 (31.12b), 15937 (32.1), 16276 (31.4b), 16664 (26.7), 19020 (6.1c), 19113 (28.6), 20702 (26.22), 21341 (21.8), 21730 (26.12), 21740 (26.22), 21864 (26.30), 22000 (12.1), 22007 (27.5a), 22068 (32.2), 22935 (25.4), 24254 (14.1), 24255 (16.3), 24362 (16.4a), 25038 (26.22), 25513 (26.6), 25521 (13.1), 25785 (27.5a), 25872 (4.2), 26227 (26.30), 26254 (6.1c), 26272 (26.24c), 27100 (12.2), 27511 (7.4), 27724 (31.12a), 27981 (12.1), 28029 (26.24c), 28031 (26.32), 29086 (26.32), 29155 (31.17), 29157 (17.1), 29430 (11.1), 29486 (26.12), 29587 (13.3), 29825 (11.1), 31407 (26.32), 31621 (31.19c), 32104 (26.24c), 32124 (26.30), 35999 (11.1), 36367 (26.12), 36541 (26.22), 37092 (26.32), 37118 (7.4), 37936 (31.8b), 38082 (18.7), 38749 (31.9), 38788 (26.19c), 38798 (31.12b), 39011 (26.22), 39403 (26.35a), 39542 (31.17), 39548 (31.13b), 39571 (31.5), 39595 (31.7), 39626 (31.5), 39653 (26.15), 39715 (26.32), 39721 (31.19c), 41186 (26.23b), 41368 (31.5), 41393 (26.24c), 41556 (31.13b), 41877 (31.11), 42003a (31.11), 42124 (31.19c), 42125 (31.11), 42131 (31.13b), 42155 (26.24c), 42195 (31.7), 42246 (26.19b), 42252 (26.32), 42310 (26.6), 42345 (26.22), 42346 (26.19b), 42361 (26.32), 42591 (26.35a), 42619 (26.15), 43160 (31.9), 43421 (8.1c), 45263 (7.4), 45905 (26.10c), 46074 (26.1a), 46204 (26.1a), 46441 (26.32), 46460 (28.1), 46467 (26.19a), 47507 (26.21), 47512 (26.21), 47554 (16.4a), 47807 (19.1), 47982 (16.4a), 48116 (7.4), 48325 (16.4a), 49086 (21.7), 50232 (28.10), 50245 (26.8), 50276 (26.16), 50296 (28.4), 51143 (26.1a), 52152 (26.24c), 52166 (31.12a), 52179 (31.12a), 52224 (26.10b), 52246 (21.2), 52256 (26.10b), 52270 (26.32), 52306 (21.7), 53024 (19.1), 53769 (11.1), 53796 (12.1), 54253 (28.8e), 54290 (32.2), 54319 (31.13b), 54605 (31.12b), 54605 (31.12b), 54606 (31.5), 55449 (31.8b), 55618 (28.8e), 55702 (31.19b), 55814 (31.12b), 56045 (31.7), 56146 (27.5b), 56188 (26.7), 56265 (31.12a), 56337 (29.1), 56347 (28.7d), 56390 (31.13b), 56540 (26.24a), 57335 (18.3a), 57498 (31.5), 57500 (31.5), 57519 (18.7), 57521 (31.4b), 57528 (7.3), 57590 (26.24c), 57654 (11.1), 57953 (26.1a), 57954 (11.1), 57999 (12.1), 58432 (7.4), 58442A (31.5), 58542 (7.1), 58553 (26.12), 58592 (1.3), 60024 (17.1), 60249 (31.19c), 60885 (27.2b), 61140 (7.4), 61643 (32.2), 61918 (31.7), 61918 (31.7), 61967 (26.32), 61978 (27.2b), 63159 (26.1b), 63160 (26.13), 63189 (26.3), 63342a (31.11), 63541 (13.3), 65642 (26.19b), 65696 (26.24c), 65743 (28.8e), 65782 (26.32), 65792 (8.1b), 65800 (31.12b), 65845 (31.19b), 65881 (21.10), 65889 (31.7), 65911 (18.7), 66024 (31.12a), 68666 (31.12a), 68908 (26.10a), 68908 (26.10b), 69193 (20.3), 69487 (28.5), 69530 (26.19a), 69545 (31.19a), 69546 (31.4b), 69557 (31.7), 70504 (16.4b), 70662 (25.1), 70710 (8.1c), 70719 (7.1), 70783 (13.1), 70851 (7.4), 77678 (18.3a), 77762 (16.4b)

R. Gill 52 (12.1), 57 (18.3a), 54563 (19.1)

L. Gillespie 812 (27.5a), 1050 (27.4), 1226 (27.3), 1262 (26.28), 1357 (26.8), 1439 (26.32), 1588 (31.13a), 2041 (28.4), 2087 (31.3), 2088 (26.32), 2091 (31.3), 2115 (26.26), 2118 (28.4), 2122 (26.24c), 2136 (26.24c), 2161 (26.24c), 2164 (31.3), 2182 (26.24c), 2183 (26.26), 2225 (26.32), 2280 (26.26), 2376 (16.3), 2377 (31.3), 2569 (26.8)

H. Gines 2145 (7.4)

S. Glassman 13049 (20.2), 13092 (21.9)

A. Glaziou 22263 (20.2)

H. Gleason 231 (26.26), 269 (27.4), 797 (4.2)

L. Glenboski C-86 (32.1), C161 (28.3), C200 (6.1c), 214 (11.1), 233 (14.1), C253 (18.3a), 217 (19.1)

L. Gómez 19688 (11.1) 19512 (18.7), 20529 (11.1)

M. Goulding 35 (16.4b)

C. Grández 1692 (31.9), 2233 (14.1)

M. Grant 10555 (19.1)

M. Grayum 8784 (19.1)

P. Grenand 391 (26.14), 1033 (31.13c), 1090

(26.3), 1103 (26.17), 1297 (26.32), 1541 (26.25)
A. Grijalva 501 (31.7)
J. Grimes 3329 (26.17)
P. Grubb 1447 (7.4), 1667 (11.1)
F. Guanchez 3223 (26.4), 3230 (26.1a)
G. Gutierrez 894 (31.7)
W. Hahn 3625 (31.13c), 3630 (26.3), 3737 (26.11), 4203 (26.19a), 4220 (31.13a), 4304 (31.2), 4703 (26.28), 4729 (26.28), 4730 (26.19a), 4800 (27.5a), 5129 (26.11), 5130 (26.28), 5134 (27.4), 5300 (26.26), 5799 (31.13a), 5800 (28.4), 5810 (26.28), 5811 (26.26)
F. Hallé 592 (31.3)
C. Hamilton 2824 (11.1), 2980 (18.7)
B. Hammel 15989 (17.1)
R. Harley 19613 (31.4a)
G. Harling 3260 (31.19a), 3498 (29.1), 3569 (31.12b), 3577 (31.5), 3600 (31.19c), 3762 (31.1)
E. Harris 1094 (21.7), 1129 (27.3)
G. Hatschbach 35907 (31.4a), 44102 (20.5), 46160 (31.4a)
S. Hayes 897 (31.8b)
E. Heinrichs 300 (7.4), 375 (31.12b), 501 (31.7)
A. Henderson 8 (12.1), 9 (31.13c), 10 (26.19a), 11 (18.3a), 12 (16.1), 13 (10.1), 14 (16.4b), 16 (31.7), 17 (11.1), 18 (12.1), 19 (5.2), 20 (10.1), 21 (28.1), 24 (26.32), 25 (26.19a), 33 (11.1), 34 (26.23a), 35 (28.4), 36 (27.5a), 38 (26.19a), 39 (31.13c), 40 (10.1), 41 (5.2), 42 (11.1), 43 (26.16), 44 (12.1), 56 (31.8b), 62 (27.3), 64 (31.8b), 65 (11.1), 66 (12.1), 72 (12.1), 79 (11.1), 80 (12.1), 81 (12.1), 86 (16.4a), 87 (11.1), 89 (26.24c), 95 (31.7), 103 (12.1), 104 (12.1), 105 (12.1), 109 (12.1), 110 (31.8b), 119 (11.1), 131 (12.1), 132 (12.1), 133 (11.1), 136 (12.1), 138 (29.1), 142 (11.1), 149 (12.1), 153 (19.1), 155 (19.1), 161 (29.1), 172 (26.6), 173 (15.3), 174 (12.1), 175 (26.7), 177 (16.4b), 179 (15.3), 180 (26.30), 181 (26.24c), 182 (14.1), 184 (15.3), 186 (28.4), 187 (31.2), 188 (10.1), 189 (31.13d), 191 (31.7), 192 (18.7), 193 (28.4), 194 (12.1), 195 (31.12a), 196 (26.32), 197 (26.19a), 199 (11.1), 200 (31.10), 201 (31.7), 203 (31.13b), 204 (18.7), 219 (26.32), 230 (31.13b), 236 (5.2), 237 (6.1c), 241 (26.19a), 243 (11.1), 246 (26.14), 247 (16.2), 248 (26.19a), 249 (26.6), 250 (31.7), 251 (26.1a), 253 (16.4b), 254 (21.7), 255 (15.3), 256 (12.1), 257 (6.1c), 260 (28.1), 297 (12.1), 298 (26.36), 301 (26.19c), 301A (31.19c), 302 (31.19b), 303 (31.19c), 304 (10.1), 337 (20.1), 481 (12.1), 482 (31.2), 483 (4.1), 484 (22.1), 522 (16.4a), 525 (16.4a), 526 (7.4), 534 (28.8f), 537 (11.1), 539 (12.2), 540 (10.2), 541 (12.2), 542 (10.2), 605 (28.6), 606 (26.7), 618 (24.1), 624 (26.24c), 625 (26.32), 626 (28.4), 632 (10.1), 633 (31.13b), 634 (20.3), 635 (21.8), 636 (16.1), 637 (16.4b), 638 (26.19a), 639 (31.2), 640 (18.8), 641 (26.32), 642 (26.35b), 643 (31.13d), 644 (20.3), 645 (3.2), 647 (10.1), 648 (26.9a), 649 (31.13c), 650 (15.3), 651 (16.1), 654 (26.14), 655 (31.2), 656 (21.1), 657 (26.1a), 658 (26.17), 659 (31.13d), 661 (18.8), 662 (31.19b), 663 (26.32), 664 (31.13c), 665 (26.35b), 666 (26.19a), 668 (21.8), 669 (26.24c), 670 (31.13b), 671 (26.1a), 672 (18.1), 675 (28.4), 676 (19.1), 677 (26.35b), 678 (31.13b), 679 (26.35b), 680 (28.10), 681 (10.1), 682 (15.3), 683 (10.2), 684 (10.2), 685 (8.1b), 687 (28.1), 701 (12.1), 703 (16.4a), 705 (14.1), 707 (19.1), 717 (31.8b), 724 (18.7), 729 (19.1), 730 (26.24c), 750 (2.1), 751 (2.1), 768 (21.9), 811 (21.4), 812 (26.18), 813 (12.1), 814 (26.10b), 815 (26.7), 816 (31.12a), 817 (16.4b), 818 (31.7), 819 (31.19c), 820 (26.24b), 821 (21.5), 822 (28.7d), 823 (32.1), 824 (18.7), 825 (7.3), 827 (19.1), 828 (31.5), 829 (31.12b), 830 (13.2), 832 (10.1), 833 (21.8), 834 (18.2), 835 (31.13b), 836 (31.7), 837 (6.1c), 838 (26.20), 839 (26.32), 840 (7.3), 842 (2.1), 843 (31.6), 844 (26.24c), 845 (31.17), 846 (25.3), 847 (26.24a), 848 (26.32), 849 (26.24b), 850 (31.19b), 851 (31.19b), 852 (28.7d), 853 (31.5), 854 (26.27), 855 (21.8), 856 (21.5), 857 (26.32), 858 (26.32), 859 (26.19a), 860 (31.7), 861 (10.1), 862 (31.12b), 863 (31.6), 864 (21.2), 865 (19.1), 866 (31.1), 867 (26.6), 868 (25.3), 869 (7.3), 870 (7.3), 871 (7.4), 874 (26.1a), 875 (27.2c), 876 (26.32), 877 (31.10), 878 (31.19c), 879 (31.7), 880 (31.7), 881 (26.35a), 882 (26.24c), 883 (26.19a), 884 (2.1), 885 (31.19a), 886 (31.6), 887 (28.3), 888 (26.14), 889 (31.19b), 890 (31.19b), 891 (31.7), 892

(19.1), 893 (31.3), 906 (26.1a), 919 (16.1), 949 (16.1), 973 (10.1), 974 (5.2), 975 (15.2), 976 (31.13c), 977 (27.5a), 978 (26.19a), 995 (26.7), 996 (31.3), 1034 (7.4), 1040 (19.1), 1041 (31.7), 1043 (31.19b), 1044 (26.17), 1045 (26.36), 1046 (20.3), 1047 (28.10), 1048 (31.13b), 1049 (26.24b), 1049 (26.35b), 1050 (31.7), 1051 (26.19a), 1052 (26.1a), 1053 (31.13d), 1054 (10.1), 1055 (31.13c), 1056 (31.13b), 1057 (26.20), 1058 (26.14), 1059 (26.20), 1061 (18.3a), 1062 (26.11), 1063 (26.19c), 1064 (28.4), 1065 (26.32), 1066 (31.19b), 1067 (12.1), 1068 (18.3a), 1069 (26.24c), 1071 (21.1), 1072 (16.4b), 1073 (18.1), 1074 (18.8), 1075 (31.13c), 1077 (27.4), 1078 (26.4), 1079 (31.2), 1100 (31.9), 1101 (27.2a), 1102 (31.9), 1103 (26.6), 1104 (13.1), 1105 (31.19b), 1106 (31.6), 1108 (31.19c), 1109 (26.12), 1110 (31.12a), 1111 (6.1c), 1112 (31.19a), 1113 (7.4), 1114 (19.1), 1115 (9.1), 1120 (2.5.32), 1121 (31.12a), 1122 (31.13b), 1124 (19.1), 1125 (28.7d), 1126 (34.1), 1127 (7.4), 1128 (31.6), 1129 (7.3), 1130 (7.3), 1132 (31.19b), 1133 (25.3), 1134 (8.1a), 1135 (16.4b), 1136 (28.3), 1137 (16.4a), 1138 (16.4a), 1139 (12.2), 1140 (32.1), 1141 (18.7), 1142 (7.4), 1143 (31.14), 1144 (1.3), 1145 (25.1), 1146, (5.2), 1147 (18.2), 1148 (6.1c), 1149 (31.13b), 1150 (10.2), 1151 (26.19a), 1152 (31.2), 1153 (27.4), 1154 (18.8), 1155 (26.14), 1157 (20.3), 1158 (28.10), 1159 (26.1a), 1160 (26.2), 1161 (31.13b), 1162 (26.17), 1163 (31.13c), 1165 (21.8), 1166 (12.1), 1167 (18.3a), 1168 (16.4b), 1169 (16.1), 1170 (21.1), 1171 (26.32), 1172 (26.24c), 1173 (4.2), 1174 (21.7), 1175 (28.2), 1176 (5.2), 1177 (26.32), 1181 (31.8b), 1500 (26.24c), 1501 (26.24b), 1501 (26.35b), 1502 (26.17), 1503 (20.1), 1504 (15.3), 1505 (26.24b), 1506 (28.9), 1507 (26.1a), 1508 (23.1), 1509 (28.8f), 1510 (26.10b), 1511 (26.5), 1512 (31.13b), 1513 (21.12), 1514 (26.20), 1515 (26.19a), 1516 (26.32), 1517 (31.13c), 1517 (31.13b), 1518 (31.13b), 1519 (4.1), 1520 (31.19b), 1521 (31.10), 1522 (26.24b), 1523 (31.12a), 1524 (26.10b), 1525 (26.24c), 1526 (7.4), 1527 (19.1), 1528 (31.19b), 1529 (31.13d), 1530 (16.4b), 1531 (31.19c), 1532 (31.3), 1533 (31.7), 1534 (26.14), 1535 (26.24b), 1536 (31.12a), 1537 (26.32), 1538 (31.10), 1539 (26.19a), 1540 (26.1a), 1541 (16.1), 1542 (18.1), 1543 (10.1), 1544 (26.7), 1545 (26.24c), 1546 (26.6), 1547 (23.1), 1548 (26.6), 1549 (18.1), 1550 (31.7), 1551 (26.32), 1553 (31.10), 1554 (31.13c), 1555 (26.5), 1556 (26.36), 1557 (31.13d), 1558 (31.13d), 1559 (26.1a), 1560 (26.25), 1561 (26.25), 1562 (26.25), 1572 (26.19a), 1573 (26.20), 1575 (26.14), 1576 (26.19c), 1577 (26.24b), 1578 (31.19b), 1579 (31.19b), 1580 (31.19b), 1581 (26.25), 1582 (26.25), 1583 (26.34), 1585 (31.16), 1586 (31.7), 1587 (26.34), 1589 (26.1a), 1593 (28.1), 1594 (18.8), 1595 (31.19c), 1596 (26.32), 1598 (26.25), 1599 (26.1a), 1603 (28.8f), 1604 (26.30), 1605 (26.17), 1606 (26.24c), 1607 (27.5a), 1608 (26.25), 1609 (27.4), 1610 (27.5a), 1611 (26.1b), 1612 (26.32), 1613 (26.36), 1614 (3.1), 1617 (27.3), 1620 (26.32), 1621 (16.4b), 1622 (16.3), 1625 (31.8b), 1629 (26.1a), 1630 (26.10a), 1631 (7.3), 1632 (31.7), 1634 (31.13b), 1636 (31.5), 1637 (31.5), 1638 (31.19c), 1639 (26.32), 1641 (26.19a), 1642 (27.5a), 1643 (21.9), 1644 (21.2), 1645 (21.10), 1646 (26.5), 1647 (26.30), 1648 (21.2), 1649 (31.5), 1650 (19.1), s.n. (21.8), 1651 (31.12a), 1652 (31.5), 1653 (7.4), 1654 (26,10b), 1655 (26.24c), 1656 (21.14), 1657 (34.1), 1658 (26.10c), 1659 (21.9), 1660 (7.3), 1661 (31.6), 1662 (7.3), 1663 (26.21), 1665 (26.6), 1666 (26.35a), 1667 (26.32), 1668 (31.19b), 1669 (1.3), 1671 (26.7), 1672 (26.10c), 1673 (31.7), 1674 (28.8f), 1675 (27.5a), 1676 (26.30), 1677 (26.30), 1678 (26.35a), 1679 (31.14), 1680 (31.12a), 1681 (28.3), 1682 (31.19b), 1683 (25.1), 1684 (31.14), 1685 (28.3), 1686 (26.27), 1687 (31.13b), 1689 (26.27), 1690 (31.6), 1691 (21.14), 1692 (26.12), 1693 (27.2a), 1694 (20.7), 1695 (26.32), 1696 (27.3), 1697 (26.32), 1698 (10.2), 1699 (6.1c), 1700 (31.19a), 1701 (13.1), 1702 (18.3a), 1703 (16.4b), 1704 (26.1a), 1705 (31.7), 1706 (28.3), 1707 (31.12a), 1708 (26.27), 1709 (26.24c), 1710 (11.1), 1711 (21.7), 1712 (31.13b), 1713 (4.2), 1714 (12.1), 1716 (28.1), 1717 (21.14), 1718 (8.1b), 1719 (27.1), 1730 (24.1), 1731 (16.3), 1732 (18.5),

1733 (28.11), 1734 (21.3), 1736 (21.13), 1738 (20.1), 1739 (28.6), 1742 (21.8), 1743 (21.9), 1815 (3.1), s.n. (21.1)

E. Heringer   401 (4.2), 3153 (4.2), 9060 (20.5), 18294 (20.5), 11447 (20.2), 8589/783 (31.4a),

F. Hermann   11208 (21.5)

S. Hill   12958 (6.1b), 13105 (6.1b)

A. Hladik   3027 (27.5a)

W. Hodge   6079 (31.5)

F. Hoehne   5053 (6.1b)

M. Hoff   5798 (31.12c), 5811 (26.32), 5823 (31.12c), 6066 (31.12c), 6424 (26.25), 6425 (26.25), 6427 (26.25), 6446 (26.32), 6454 (31.12c), 6526 (26.32)

B. Hoffman   394 (26.24c), 450 (26.4), 464 (12.1), 799 (27.5a), 824 (31.13a), 1403 (31.13a), 1551 (28.4), 1577 (31.13a), 1907 (26.32)

L. Holm-Nielsen   22002 (17.1), 22704 (17.1), 25229 (16.3)

B. Holst   2368 (31.7), 2433 (18.1), 3082 (26.32), 3359 (9.1), 3364 (16.1)

R. Holm   663 (23.1)

E. Holt   352, 496 (10.1)

M. Hopkins   173 (21.9), 112 (21.12), 158 (21.12)

E. Huamán   s.n. (1.3), 4 (26.19a), 19 (26.19a), 23 (18.7), 24 (31.6)

V. Huashicat   161 (26.12), 358 (7.4), 1042 (26.32), 1178 (26.24c), 1630 (26.32)

C. Hubbuch   1 (26.36), 2 (26.7), 3 (26.10b), 4 (26.24c), 5 (26.23a), 7 (26.1a), 10 (1.1)

J. Huber   2201 (3.1)

O. Huber   2067 (10.1), 2078 (28.4), 3237 (28.1), 3929 (5.2), 4748 (20.4), 6256 (31.8b), 6440 (21.10), 9411 (10.1), 11083 (16.1), 12404 (26.32), 12418 (10.1)

M. Huft   1617 (31.7)

J. Idrobo   506 (27.5a), 679 (20.4), 709 (31.8b), 2498 (31.8a), 4707 (27.2b), 4747 (26.12), 6482 (6.1c), 6851 (6.1c), 8885 (27.2c), 11858 (18.3a)

H. Irwin   5422 (31.4a), 6525 (20.2), 6645 (5.2), 7194 (27.5a), 8851 (31.4a), 12479 (31.4a), 13655 (20.2), 15965 (20.2), 17364 (28.5), 17461 (5.2), 18989 (20.5), 19233 (31.4a), 21810 (31.4a), 24627 (5.2), 24838 (31.4a), 26619 (20.2), 47282 (31.19c), 47290 (19.1), 47351 (31.7), 47455 (31.13c), 47456 (12.1), 47528 (28.9), 47588 (31.7), 47627 (31.12c), 47631 (31.7), 47632 (20.3), 47707 (31.12c), 47714 (31.3), 47715 (28.10), 47716 (19.1), 47795 (31.7), 48082 (31.7), 48083 (26.14), 48171 (26.27), 48172 (16.3), 55422 (4.2), 55587 (31.3)

J. Jangoux   1615 (31.3), 1772 (20.5), 85–001 (31.19c), 85–003 (18.7), 85–019 (13.1), 85–021 (13.1), 85–027 (21.14), 85–029 (31.12a), 85–030 (31.19c), 85–031 (7.3), 85–042 (21.2), 85–050 (31.19c), 85–070 (31.14), 85–075 (31.12a), 85–076 (31.6), 85– 110 (13.1)

M. Jansen-Jacobs   340 (12.1), 368 (31.13a), 384 (31.13a), 442 (16.3), 681 (28.4), 684 (26.7), 823 (31.8a), 824 (31.10), 824 (31.10), 832 (16.2), 895 (18.1), 1018 (12.1), 1019 (28.2), 1058 (16.3), 1060 (21.7), 1090 (28.4), 1230 (19.1), 1293 (18.3b), 1490 (26.2), 1508 (31.3), 1521 (31.10), 1526 (19.1), 1582 (10.1), 1583 (26.2), 1588 (28.10), 1608 (18.3a), 1729 (28.4), 1730 (31.7), 1734 (26.2), 1745 (12.1), 1746 (26.1a), 1862 (4.2), 2383 (26.14), 2385 (26.32), 2386 (31.3), 2387 (18.1), 2388 (26.2), 2389 (21.7), 2390 (28.10), 2391 (16.4b), 2430 (5.2), 2478 (26.7), 2479 (28.2), 2539 (27.5a), 2566 (12.1)

A. Janssen   205 (5.2)

J. Jaramillo   2794 (17.1), 3263 (7.3), 4607 (27.2c), 6874 (31.19c), 31526 (17.1)

R. Jaramillo   438 (31.11), 1258 (20.4)

R. Jaramillo-Mejía   1241 (18.7), 1004 (18.7)

C. Jativa   1182 (16.3)

G. Jenman   527 (21.7), 7516 (14.1), 7517 (4.2), 7575 (16.4b), 7726 (27.3)

A. Juncosa   696 (16.3)

R. Juwa   52 (16.4b), 143 (28.3)

F. Kahn   1681 (31.8b), 1685 (13.3), 1694 (13.1), 1697 (31.19b), 1702 (12.1), 1705 (7.3), 1715 (31.13b), 1718 (10.2), 1721 (6.1c), 1723 (18.2), 1724 (21.8), 1727 (18.7), 1748 (6.1c), 1757 (31.19b), 1771 (26.22), 1776 (28.7d), 1777 (28.7d), 1779 (28.7d), 1780 (28.7d), 1781 (26.32), 1782 (28.8f), 1789 (26.6), 1823 (28.8f), 1838 (1.3), 1842 (11.1), 1852 (16.4b), 1875 (7.3), 1892 (31.6), 1893 (31.6), 1894 (31.17), 1899 (31.10), 1961 (26.6), 1965 (26.22), 1969 (29.1), 1973 (1.2), 1974 (1.2), 1978 (26.32), 1988 (13.1), 2031 (28.7c), 2051 (12.2), 2064 (31.19b), 2065 (7.3), 2073 (7.3), 2074 (7.3), 2094 (28.8f), 2108 (13.1), 2111 (26.1a), 2117 (31.13b), 2119 (31.7), 2128 (28.8f), 2129 (20.6), 2130 (18.7), 2133 (31.4b), 2148 (21.9), 2152

(31.19c), 2154 (31.19c), 2156 (10.2), 2157 (31.13b), 2167 (12.2), 2175 (11.1), 2182 (31.19c), 2230 (16.4b), 2232 (12.1), 2310 (9.1), 2314 (9.1), 2320 (26.24c), 2554 (25.4), 2556 (25.2), 2654 (28.7c)

M. Kawasaki   343 (15.3)

R. Kayap   247 (12.2), 325 (26.27), 370 (26.16), 552 (26.9), 854 (26.32), 1033 (26.12), 1358 (26.12)

S. Keel   32 (26.5)

W. Kellerman   7175 (16.4a)

C. Kelloff   631 (27.5a)

H. Kennedy   660 (31.8b

L. Kenoyer   170 (21.2)

M. Kessler   387 (19.1)

T. Killeen   1409 (4.2), 2212 (21.9), 2733 (12.1), 2761 (5.2), 2829 (26.23d)

E. Killip   8788 (7.4), 11055 (20.6), 11136 (7.4), 14330 (27.3), 14459 (23.1), 14756 (23.1), 14808 (18.7), 14866 (31.7), 14900 (26.24c), 15314 (29.1), 15336 (31.8b), 17040 (7.4), 18996 (7.4), 19452 (7.4), 20413 (7.4), 22760 (7.4), 22864 (7.1), 22899 (31.5), 23006 (31.5), 23999 (18.3a), 24608 (25.4), 24706 (19.1), 24926 (7.1), 24929 (7.4), 25103 (7.1), 25141 (21.9), 25532 (7.4), 25627 (7.4), 26170 (31.10), 26246 (10.2), 26247 (31.11), 26414 (31.10), 26429 (10.2), 26430 (19.1), 26430 (31.7), 26431 (31.20), 26455 (31.5), 26461 (26.1a), 26532 (31.7), 26540 (19.1), 26584 (31.12b), 26717 (1.3), 26722 (31.5), 26770 (19.1), 26807 (31.12b), 26985 (10.2), 26987 (31.13b), 26992 (10.2), 26993 (31.13b), 26994 (10.2), 27007 (26.10b), 27071 (27.2b), 27140 (26.24c), 27171 (4.2), 27295 (31.12b), 27360 (31.12a), 27633 (21.2), 27647 (32.1), 27648 (18.7), 27654 (31.9), 27775 (8.1b), 27977 (31.13b), 28026 (31.17), 28045 (27.5a), 28142 (31.12a), 28145 (31.12a), 28345 (7.3), 28541 (31.17), 28546 (19.1), 28729 (31.17), 28807 (27.2b), 28814 (21.8), 28840 (28.7c), 28845 (18.3a), 28847 (31.12a), 28895 (31.12a), 28946 (32.1), 29299 (7.4), 29333 (32.1), 29333 (32.2), 29338 (26.19b), 29373 (31.12a), 29433 (31.12b), 29439 (26.19b), 29511 (7.4), 29526 (7.4), 29595 (7.4), 29621 (31.6), 29659 (26.32), 29673 (31.12a), 29694 (19.1), 29830 (26.19b), 29831 (6.1c), 29889 (26.32), 29926 (10.2), 29951 (31.13b), 30573 (27.5a), 34267 (16.4b), 34270 (21.5), 34275 (18.7), 34905 (26.16), 35502 (26.16), 37314 (26.32), 37398 (31.10), 37399 (31.12c), 37437 (31.3), 37488a (19.1), 37822 (7.4), 39006 (16.3)

R. King   6053 (17.1), 6065 (19.1), 6149 (31.6), 6204 (31.12b)

S. King   550 (16.3), 656 (18.7), 659 (16.4a), 660 (18.3a), 661 (18.7)

J. Kirkbride   1919 (7.4), 2284 (7.4), 2964 (5.2)

G. Klug   181 (19.1), 395 (8.1b), 448 (31.12a), 537 (31.13b), 1495 (29.1), 1496 (31.13b), 1497 (31.7), 4367 (7.2)

S. Knapp   5693 (12.1), 5919 (16.4a), 6041 (31.8b), 6510 (26.32), 6898 (6.1c), 7113 (31.12a), 7114 (26.32), 7146 (26.32), 7153 (12.2), 7185 (10.2), 7422 (7.2), 7616 (31.1), 7673 (31.19b), 7692 (31.19c),

J. Korning   58733 (17.1)

T. Koyama   7315 (19.1), 7326 (31.8a)

B. Krukoff   1101 (6.1c), 1122 (19.1), 1601 (21.7), 1608 (28.2), 1612 (11.1), 1685 (7.3), 4545 (19.1), 4996 (27.2a), 5361 (31.12a), 5572 (21.5), 5618 (21.5), 5622 (21.5), 5739 (11.1), 5758 (18.3a), 5803 (18.7), 6082 (31.3), 6402 (18.7), 6497 (26.10b), 6517 (26.27), 6765 (28.4), 6772 (31.13c), 7033 (26.36), 7052 (18.7), 7065 (21.7), 7107 (26.32), 7157 (31.13d), 7158 (31.7), 7184 (26.36), 7204 (6.1c), 7284 (15.3), 7552 (14.1), 8079 (19.1), 8127 (26.24c), 8144 (31.12a), 8598 (7.3), 8599 (19.1), 8795 (6.1c), 10481 (7.4), 10806 (7.4), 10807 (7.4), 10808 (31.5), 10836 (31.5), 10982 (31.8a), 10983 (31.7)

J. Kuhlmann   1831 (28.1), 1907 (28.9)

A. Kujikat   98 (31.19c), 337 (19.1)

P. Kukle   125 (18.8)

R. Kummrow   2818 (20.5)

P. Kunkumas   RBAE123 (18.3a)

O. Kuntze   s.n. (7.4), s.n. (21.4)

L. Kvist   167a (26.28), 246 (9.1), 40198 (7.4)

J. Lanjouw   474 (31.3)

J. Lanna   77 (27.3)

C. La Rotta   116 (6.1c), 132 (21.10), 134 (28.3), 137 (27.2c), 143 (18.3a), 409 (6.1c), 539 (6.1c)

D. Larpin   207 (26.3), 313 (26.17), 403 (28.9), 772 (20.8), 839 (20.8)

T. Lasser   1644 (12.1)

P. Lau   19 (18.7)

J. Lawesson   39407 (17.1), 39522 (17.1), 44473 (16.4b)

C. Le Fiell   2 (26.24c)

A. Le Goff   42 (27.3), 43 (26.23a)

K. Lems   5036 (12.1), 64091402 (11.1)

H. Leng   22 (26.19a), 68 (26.32), 135 (26.26), 171 (26.32), 217 (26.24c), 282 (26.19a), 350 (26.27), 356 (26.19a)

J. Leveau   221 (26.12)

G. Lewis   813 (27.3), 1059 (27.5a)

W. Lewis   10073 (29.1), 10095 (31.11), 10114 (31.13b), 10263 (31.12b), 10269 (18.3a), 10331 (26.32), 10332 (31.19c), 10350 (7.3), 10448 (26.16), 11359 (31.11), 11922 (32.2), 11962 (31.12b), 12124 (31.19b), 12153 (31.12b), 12154 (31.19b), 12173 (26.24b), 12405 (31.19b), 12678 (28.3), 13045 (31.7), 13212 (21.14), 13900 (31.10)

R. Liesner   2067 (11.1), 3463 (6.1a), 3619 (6.1a), 3834 (26.4), 5595 (26.7), 8494 (26.24c), 8608 (15.3), 8914 (26.24c), 8940 (27.5a), 9134 (5.2), 9551 (16.4a), 9974 (16.4a), 10439 (18.7), 10570 (31.7), 10628 (26.12), 10896 (7.4), 10943 (31.8b), 10991 (31.7), 11252 (26.1a), 11368 (31.7), 11486 (28.4), 11743 (7.4), 12349 (31.8b), 13986 (26.32), 13995 (31.12c), 13999 (31.3), 16439 (26.32), 17331 (31.7), 17554 (31.7), 17569 (31.13c), 17724 (14.1), 17734 (16.1), 17843 (31.13c), 18194 (10.1), 18355 (5.2), 18846 (31.7), 18928 (19.1), 18942 (26.19a), 19031 (26.36), 19184 (31.10), 19217 (10.1), 19594 (31.3), 20059 (10.1), 20208 (16.2), 20561 (28.4), 20889 (5.2), 20902 (10.1), 22120 (16.1), 22256 (26.4), 22258 (26.32), 22261 (10.1), 22283 (26.32), 23526 (26.32), 24278 (31.7), 24284 (28.4), 24531 (31.13c), 24556 (26.12), 25031 (26.32), 25051 (16.1), 25417 (10.1), 25463 (26.1a), 25587 (31.13c), 25625 (26.12), 25728 (27.5a), 25833 (31.13c)

J. Lima   677 (22.1)

J. Lindeman   58 (26.17), 226 (12.1), 389 (31.7), 443 (31.13a), 676 (26.24c), 4148 (31.3), 4176 (26.17), 6807 (31.7)

P. Lisboa   2755 (26.10b)

C. Listabarth   11–2190 (31.12a), 11–14389 (25.1), 11–30193 (12.1), 12–5689 (7.4)

E. Little   6397b (11.1), 8414 (16.4b), 8464 (21.2), 13764 (24.1), 21236 (11.1)

E. Lleras   2067 (21.9), 2068 (28.5), 2079 (18.5), P16633 (28.4), P16916 (31.12a), P16953 (20.7), P16971 (29.1), P16972 (6.1c), P17168 (31.12a), P17184 (27.5a), P17235 (2.1), P17288 (6.1c), P17313 (31.19b), P17318 (10.1), P17401 (31.19b), P17402 (31.13b), P17443 (10.1)

M. Lobo   109 (21.13)

H. Loomis   48 (25.1)

A. Loureiro   s.n. (26.35a), s.n. (6.1c)

S. Lowrie et al.   690 (31.7)

G. Lozano   455 (28.4)

P. von Luetzelburg   20026 (27.3), 22051 (4.2), 22119 (5.1), 22395 (6.1a), 22350 (15.2), 22663 (15.2), 22894 (15.2), 23046 (4.1), 23047 (15.1), 23071 (10.1)

C. Lundell   16237 (27.3), 16387 (31.8b)

J. Luteyn   4419 (19.1), 8517 (7.4), 8518 (31.5), 8521 (32.2), 8569 (17.1), 8571 (31.5), 8679 (31.12b), 8706 (7.4), 8707 (4.2), 8719 (31.12b), 8519 (19.1), 8672 (17.1), 9060 (19.1), 9392 (26.32)

P. Maas   3559 (27.4), 6668 (31.7), 6846 (31.13c), P12651 (16.4a), P12704 (31.19b), P12706 (19.1), P12879 (27.2a), P13024 (31.12a), P13130 (27.2a), P13268 (31.12a)

F. Macbride   5418 (32.1)

A. Macedo   3346 (31.4a), 3450 (5.2)

U. Maciel   1387 (16.4b)

M. Madison   5348 (32.2), 6642 (6.1a)

B. Maguire   22939 (31.13a), 24094 (31.3), 24116 (19.1), 24117 (31.12c), 24154 (31.3), 24155 (19.1), 24178 (26.17), 24461 (26.8), 27733 (16.1), 28292 (26.12), 28510 (9.1), 28701 (26.19a), 28798 (26.19a), 28799 (26.19a), 28800 (31.16), 29197 (10.1), 29347 (31.7), 29361a (31.13c), 29722 (10.1), 33049a (26.32), 36558 (15.1), 37656 (5.2), 39058 (31.19b), 39161 (26.32), 39205 (26.32), 39294 (26.32), 40599a (26.19c), 40722 (26.32), 40822 (31.12c), 41597 (10.1), 43868 (26.32), 46740 (31.3), 46985 (12.1), 53740 (26.32), 60281 (21.10)

I. Malave   23 (27.3)

N. Marshall   110 (26.3)

G. Martinelli   7155 (6.1b)

E. Martínez   6981 (31.8b)

P. Martínez   22 (26.23b)

N. Mashu   21 (11.1), RBAE0024 (13.3), 25 (16.4b), RBAE3 (18.3a)

W. Maxon   6825 (31.8b)

S. McDaniel   10950 (31.12a), 11032 (31.19c), 16892 (31.13b), 21527 (6.1c), 27438 (25.4), 27840 (26.19c), 29635 (13.2)

T. McDowell   2608 (26.24c), 2668 (26.32), 2696 (31.13a), 2708 (31.8a), 3109 (27.4),

3302 (26.32), 3306 (26.28), 3367 (31.13a), 3441 (31.2), 3446 (26.32), 3487 (31.13a), 3674 (5.2) 3685 (26.24c), 3692 (26.19a), 3730 (27.5a), 3767 (26.19a), 4104 (27.4), 4258 (26.32), 4268 (31.13a), 4398 (31.12c), 4424 (26.24c), 4479 (31.13a), 4482 (26.32), 4755 (28.4), 4914 (31.2)

K. Mejia 89 (13.1), 642 (31.7), 725 (9.1), 727 (10.2), 730 (11.1)

R. Mejía 1258 (20.4)

F. Mello 2957 (21.8)

P. Mendoza 105 (34.1), 106 (28.8h), 118 (32.2)

Y. Mexia 6135 (31.17), 6877 (17.1), 6891 (7.3), 6905 (34.1), 7563 (7.4), 8291 (11.1)

J. Miller 1653 (16.4b), 1742 (31.7), 2180 (17.1)

W. Milliken 460 (31.3), 706 (31.13a), 711 (31.7)

A. Molina 14979 (31.7), 18001 (12.1)

M. Monsalve 972 (12.1), 996 (11.1),

O. Monteiro 1341 (21.1), 1360 (16.4b)

H. Moore 6532 (12.1), 6559 (23.1), 6574 (11.1), 6592 (31.8b), 6593 (16.4a), 6711 (26.24c), 8341 (32.1), 8358 (25.2), 8361 (13.3), 8369 (26.22), 8371 (13.1), 8372 (4.2), 8376 (18.7), 8392 (21.9), 8393 (32.1), 8397 (18.7), 8400 (21.7), 8403 (20.6), 8404 (26.7), 8406 (18.7), 8408 (20.6), 8411 (26.10b), 8416 (28.3), 8420 (28.8e), 8421 (26.35a), 8423 (26.19b), 8427 (26.24c), 8429 (5.2), 8435 (26.19b), 8436 (26.22), 8437 (26.20), 8438 (26.19b), 8442 (28.6), 8443 (26.7), 8449 (27.2c), 8452 (26.19c), 8456 (25.4), 8460 (26.27), 8464 (26.32), 8472 (26.10b), 8473 (26.10a), 8477 (8.1b), 8478 (21.2), 8480 (32.2), 8483 (26.7), 8485 (26.24c), 8486 (26.22), 8488 (6.1c), 8492 (8.1b), 8498 (26.10b), 8500 (28.7c), 8510 (6.1c), 8512 (21.8), 8515 (26.12), 8521 (13.3), 8525 (18.7), 8527 (25.4), 8529 (25.3), 8532 (32.1), 8535 (26.10b), 8537 (20.6), 8540 (26.10a), 8560 (26.10c), 8567 (26.24c), 8571 (26.19a), 8579 (26.10a), 8580 (13.1), 8591 (27.2d), 8595 (26.22), 9449 (14.1), 9453 (14.1), 9496 (32.1), 9506 (26.15), 9523 (28.4), 9524 (21.7), 9525 (15.3), 9526 (26.32), 9527 (28.1), 9528 (20.3), 9529 (26.2), 9531 (21.8), 9532 (18.3a), 9533 (10.1), 9534 (31.13c), 9535 (21.1), 9536 (16.1), 9537 (5.2), 9538 (12.1), 9539 (16.4b), 9540 (18.8), 9541 (28.2), 9542 (26.32), 9543 (31.7), 9544 (26.4), 9546 (4.1), 9547 (16.3), 9551 (28.8f), 9552 (26.1a), 9556 (20.3), 9832 (9.1), 9837 (26.31), 9841 (19.1), 9871 (14.1), 10212 (32.1), 10218 (20.6), 10312 (21.7), 10314 (28.9), 10315 (26.24c), 10316 (28.10), 10317 (4.2), 10319 (18.1), 10320 (16.3), 10321 (26.27), 10322 (28.10), 10323 (28.9), 10324 (28.4), 10326 (26.25), 10327 (26.2), 10332 (28.4), 10334 (20.3), 10335 (26.32), 10336 (26.24c), 10337 (28.10), 10338 (26.29), 10339 (26.29), 10340 (26.17), 10341 (26.24c), 10346 (26.17), 10347 (21.7), 10348 (16.3), 10351 (14.1), 10354 (23.1), 10355 (4.2), 10356 (26.8), 10357 (28.11), 10359 (26.7), 10360 (27.5a), 10361 (26.32), 10362 (23.1), 10510 (27.3)

L. Mora APA358 (18.3a)

M. Moraes 610 (7.1), 792 (20.6), 842 (25.1), 843 (20.6), 845 (27.2d), 846 (7.1), 847 (16.4a), 896 (7.1), 922 (28.8f), 1053 (26.23d), 1091 (26.18), 1094 (26.30)

L. Moreno 3 (26.23d), 6 (11.1), 7 (12.1), 8 (26.30), 14 (26.1a), 15 (26.10b), 16 (27.3), 18 (12.1), 21 (5.2), 22 (28.5), 23 (26.30), 24 (26.18), 25 (28.2), 26 (21.7), 27 (18.5), 31 (20.6), 34 (31.4b), 35 (31.12b), 36 (7.1), 37 (8.1c), 39 (26.24b), 41 (26.24b), 42 (25.1), 43 (31.7), 44 (7.1), 46 (27.3), 47 (27.2d), 48 (32.1), 52 (26.10c), 54 (28.8f), 57 (7.4), 66 (25.1), 71 (25.1), 72 (1.1), 73 (18.7), 77 (26.24b), 81 (7.4), 82 (7.4)

P. Moreno 33 (26.24b), 56 (26.32), 58 (31.13b), 60 (26.10b), 65 (12.2)

S. Mori 5051 (16.4a), 8063 (27.4), 8066 (26.26), 8091 (26.32), 8100 (27.4), 8101 (26.26), 8134 (28.4), 8154 (31.13a), 8155 (31.13a), 8156 (31.13a), 8455 (27.4), 8628 (31.3), 8753 (19.1), 8936 (19.1), 8937 (31.19b), 8990 (26.5), 9012 (31.13c), 9128 (27.5a), 9169 (19.1), 9188 (31.19b), 12538 (31.4a), 14677 (18.3a), 14678 (28.8h), 14680 (28.8h), 14683 (28.3), 14684 (28.3), 14731 (26.3), 14872 (28.9), 14908 (19.1), 14954 (21.8), 15000 (19.1), 15063 (26.29), 15097 (28.4), 15200 (26.25), 15552 (31.13c), 15582 (26.1b), 15604 (27.4), 16771 (20.2), 17197 (26.27), 17204 (31.7), 17287 (26.8), 17421 (21.8), 17466 (20.3), 17548 (28.4), 17608 (26.17), 17611 (31.12c), 17709 (20.3), 17732 (28.6), 17734 (26.24c), 17735 (27.5a), 18083 (26.3), 18106 (19.1), 18164 (26.32), 19053

(26.2), 19122 (28.4), 19603 (26.11), 20206 (10.1), 20516 (12.1), 20586 (26.36), 20608 (26.1a), 21004 (26.3), 21591 (31.13c), 21781 (10.1), 21821 (5.1), 21835 (31.13b), 21960 (26.5), 22195 (26.25), 22393 (26.30), 22756 (19.1), 22846 (20.3)

M. Nee 3615 (26.23b), 7020 (26.9), 7846 (26.24c), 17811 (27.5a), 29977 (27.3), 30323 (11.1), 30885 (31.7), 30947 (16.4b), 31432 (26.23a), 31458 (28.2), 31605 (21.9), 31607 (26.23d), 31617 (31.4b), 31631 (18.2), 31632 (16.4b), 31653 (11.1), 31664 (11.1), 31704 (1.1), 31889 (21.9), 31897 (12.1), 31898 (21.9), 31901 (21.12), 31904 (20.6), 33453 (26.23d), 33730 (26.23d), 34279 (24.1), 34333 (28.4), 34334 (26.14), 34335 (31.7), 34372 (31.7), 34379 (21.12), 34409 (12.1), 34420 (16.4b), 34430 (7.4), 34436 (31.4b), 34457 (26.7), 34477 (18.3a), 34482 (16.4b), 34483 (5.2), 34511 (26.18), 34662 (18.5), 34686 (18.3a), 34691 (16.4b), 34724 (18.5), 34767 (28.4), 34801 (26.1a), 34831 (28.2), 34853 (4.2), 34877 (16.4b), 34883 (5.2), 34884 (28.1), 34886 (18.2), 34903 (31.10), 34905 (27.2a), 34908 (26.24c), 34913 (26.14), 34929 (26.19a), 34953 (26.23a), 34990 (27.2a), 34994 (31.7), 35422 (7.4), 35453 (21.9), 35684 (26.16), 35949 (18.3a), 35966 (11.1), 36029 (7.4), 36030 (7.4), 36034 (28.8f), 36870 (25.1), 37545 (26.18), 38697 (26.18), 38804 (12.1), 38806 (16.4b), 39077 (7.4), 39112 (7.4), 39399 (26.23d), 39611 (27.5a), 40133 (28.5), 40972 (7.4), 41013 (7.4), 41230 (5.2), 41881 (7.4), 42329 (27.4), 42897 (31.19b)

B. Nelson 931–A (5.1), P21099 (5.2)

L. Noblick 4527 (20.5), 4530 (21.4), 4643 (20.5), 4658 (20.2), 4676 (4.2), 4695 (27.3), 4724 (27.5a), 4781 (26.1a), 4785 (26.19a), 4864 (21.9)

P. Núñez 5515 (7.1), 5747 (12.2), 6910 (1.3), 8550 (7.4), 8938 (7.4), 9611 (26.24c), 9616 (18.7), 9798 (28.5), 9832 (26.10b), 9839 (31.7), 9926 (7.1), 9974 (26.10b), 10015 (27.5a), 10114 (25.1), 10147 (7.1), 10488 (7.1), 10590 (31.5), 10645 (12.1), 10673 (18.7), 11059 (18.7), 11394 (26.21), 11408 (18.7), 12242 (31.5), 12934 (11.1), 12935 (32.1),11409 (7.4), 11488 (12.2), 13782 (27.5a), 14748 (1.3), 13595 (19.1), 14976 (21.14)

R. Oldeman 15 (26.11), 95 (12.1), 161 (31.7), 277 (31.7), 851 (26.3), 907 (26.3), 1088 (28.10), 1137 (28.9), 1163 (28.4), 1169 (26.29), 1201 (28.9), 1269 (27.5a), 1914 (14.1), 2487 (26.14), 2581 (26.3), 3164 (31.12c), 3175 (26.14), 3253 (27.5a), 4069 (26.3), 4603 (23.1)

F. Oldenburger 1584 (20.5)

E. Oliveira 3861 (31.13c), 4264 (28.4), 4675 (26.19a), 4676 (31.13c)

B. Ollgaard 9035 (31.19a), 9236 (31.19b), 38944 (17.1), 38990 (17.1), 39076 (17.1), 39213 (17.1)

Omawale 126 (28.2), 220 (21.7)

J. Ongley P21746 (10.1), P21754 (15.3), P21779 (26.32),

A. Ortega 103 (11.1)

R. Ortega 2443 (7.4)

M. Pabón 193 (21.7), 203 (21.10), 204 (18.3a), 256 (5.1), 310 (28.1), 329 (27.3),

P. Palacios 2443 (10.1), 2488 (21.10)

W. Palacios 341 (17.1), 590 (19.1), 702 (31.13c), 1674 (11.1), 2299 (17.1)

R. Pardini 10 (31.13b), 11 (31.6), 12 (26.10b), 13 (31.7), 14 (31.1), 15 (31.12b), 16 (29.1), 17 (31.19c), 18 (26.24c), 19 (31.19b), 20 (21.5), 21 (7.4), 22 (7.4), 23 (31.6), 24 (26.27), 25 (31.12b), 26 (6.1c), 27 (28.4), 28 (31.19b), 29 (31.7), 30 (26.1a), 31 (26.19a), 32 (31.13d), 33 (26.7), 34 (26.10b), 35 (31.9), 36 (21.2), 37 (18.8), 38 (26.24c), 39 (27.5a), 40 (28.3), 42 (26.24c), 43 (26.10b), 44 (31.13b), 45 (31.12a), 46 (26.30), 48 (31.7), 49 (31.12b), 50 (31.7), 51 (31.13c), 52 (31.19c), 53 (31.13d), 54 (5.2), 55 (25.1), 56 (31.19b), 57 (26.19a), 58 (26.10b), 59 (26.24a), 60 (31.9), 61 (26.24c), 62 (26.35a), 63 (31.10), 64 (26.19a), 66 (28.1), 67 (28.11), 69 (20.1), 70 (21.13), 71 (26.22), 72 (27.5a), 73 (26.13)

R. Pardo 7 (3.1)

W. Pariona 919 (11.1)

F. Pennell 8576 (25.1)

T. Pennington 367 (26.28)

A. Persaud 101 (28.4)

A. Perez 20 (31.10)

R. Persaud 111 (9.1), 119 (9.1)

O. Phillips 156 (32.1), 625 (21.2), 632 (21.9), 634 (18.7), 635 (7.3)

W. Philipson 1656 (20.6), 1718 (26.12)

M. Pinard 806 (11.1), 807 (31.12a), 808

(31.7), 820 (28.8f), 834 (12.1), 836 (1.1), 837 (31.13b), 838 (18.7), 839 (32.1), 840 (26.24c), 841 (26.10b), 842 (11.1), 843 (26.1a), 845 (26.21), 846 (26.23a), 847 (25.1), 848 (7.1), 849 (16.4b), 850 (28.2), 851 (21.7), 852 (18.3a), 853 (18.3a), 855 (4.2), 858 (21.9), 859 (31.13b)

C. Pinheiro 89-1 (21.4), 89-2 (21.4), 89-9 (21.3), 89-10 (21.4) 89-11 (21.12), 89-12 (21.3), 89-13 (21.7), 89-14 (18.5), 89-15 (16.3), 89-17 (21.12)

H. Pinkley 95 (26.16), 335a (26.10b), 476 (32.2), 486 (25.3), 524 (33.1), 536 (16.4b)

M. Pinto 1 (16.3)

J. Pipoly 3816 (31.7), 7442 (27.4), 7451 (27.5a), 7729 (5.2), 8145 (26.26), 8848 (26.28), 9236 (27.4), 9956 (27.4), 10674 (26.32), 10946 (31.2), 11231 (9.1), 11411 (26.8), 11651 (16.3), 12207 (6.1c), 12325 (31.9), 12409 (31.19c), 12444 (31.6), 12487 (19.1), 12652 (10.2), 12675 (31.19c), 12714 (26.19c), 12873 (31.5), 12885 (31.19c), 12892 (31.12a), 13001 (31.19b), 13004 (6.1c), 13040 (31.19a), 13059 (31.12b), 13071 (11.1), 13072 (12.1), 13154 (6.1c), 13238 (18.3a), 13292 (31.13b), 13308 (26.35a), 13321 (18.7), 13434 (31.13b), 13454 (31.19a), 13601 (26.15), 13608 (6.1c), 13643 (28.3), 13685 (28.8e), 13739 (31.12a), 13758 (26.24b), 13839 (28.8e), 13888 (21.5), 13897 (26.32), 13908 (31.13b), 13922 (31.19a), 13946 (31.19a), 14019 (31.7), 14144 (31.6), 14309 (31.13b), 14457 (26.15), 14478 (31.7), 14496 (31.13b), 14505 (6.1c), 14658 (26.15), 14659 (31.13b), 14916 (26.32), 14930 (31.12a), 14937 (31.19c), 15001 (27.2c), 15002 (31.19c)

J. Pirani 1248 (20.2), 1258 (31.4a)

J. Pires 886 (4.2), 887 (16.3), 889 (18.3a), 890 (21.7), 2644 (18.3a)

H. Pittier 4251 (14.1), 7076 (25.1), 8988 (12.1), 8029 (7.4), 14833 (4.2), 15142 (16.4b)

T. Plowman 1657 (26.20), 2121 (17.1), 2485 (31.12a), 2516 (26.32), 2589 (31.6), 2590 (26.24c), 3587 (21.2), 3720 (27.3), 4032 (32.2), 4388 (31.11), 7311 (18.3a), 7483 (26.6), 8267 (20.1), 8425 (5.2), 8541 (21.9), 8656 (26.23a), 8977 (21.4), 9535 (28.4), 9547 (27.4), 9586 (31.3), 9797 (26.14), 9880 (26.32), 9881 (27.5a), 11358 (18.2), 11374 (26.32), 11490 (19.1), 11677 (31.5), 12181 (28.4), 12197 (31.19b), 12235 (31.3), 12236 (31.10), 12240 (19.1), 12247 (31.12a), 12275 (26.32), 12418 (26.6), 12536 (28.4), 12610 (26.32), 12637 (26.35b)

M. Polak 4 (26.26), 136 (31.12c)

J. Poole 1648 (26.30), 1652 (31.12c), 2005 (27.3), 2114 (28.1),

S. Prada 107 (26.30)

G. Prance 1303 (20.2), 1347 (26.1a), 1471 (19.1), 1598 (31.3), 1675 (31.3), 1831 (31.21), 1864 (26.24c), 1992 (31.12c), 2155 (21.8), 2187 (18.1), 2236 (10.1), 2239 (31.13c), 2249 (16.1), 2766 (31.9), 2801 (31.12a), 2835 (6.1c), 3012 (6.1c), 3242 (6.1c), 3257 (10.1), 3523 (18.7), 4559 (31.7), 4989 (31.7), 5022 (10.1), 5529 (31.13b), 5636 (27.2a), 5708 (1.1), 5763 (5.2), 5803 (31.7), 5980 (31.13c), 5981 (26.24c), 7319 (31.5), 7684 (7.3), 7936 (7.3), 8239 (6.1c), 8278 (18.8), 8533 (31.19c), 8536 (31.7), 8546 (26.32), 8677 (18.7), 8717 (1.1), 8718 (26.24c), 9362 (31.3), 9586 (10.1), 9968 (26.32), 10036 (31.8a), 10036a (31.10), 10067 (27.5a), 10266 (31.7), 10873 (26.32), 10894 (26.24c), 11802 (10.2), 11810 (19.1), 11821 (6.1c), 11835 (31.12a), 11932 (26.35a), 12038 (31.5), 12092 (31.12a), 12093 (31.19c), 12108 (10.2), 12109 (26.32), 12310 (27.2c), 12322 (10.2), 12490 (27.2c), 13488 (31.9), 13596 (10.1), 13946 (27.2a), 14071 (6.1c), 14094 (6.1c), 14253 (6.1c), 14667 (31.19b), 14830 (21.8), 15005 (26.35b), 15006 (15.3), 15167 (15.1), 15177 (5.1), 15311 (31.19b), 16363 (31.9), 16491 (31.10), 16492 (31.7), 16493 (31.12b), 16725 (26.10a), 16726 (26.10b), 16740 (26.7), 17572 (21.1), 17781 (31.2), 17890 (10.1), 19288 (20.2), 20197 (15.3), 20198 (28.6), 20461 (10.1), 20470 (18.8), 20490 (28.4), 20568 (28.1), 20701 (6.1c), 20704 (10.1), 20727 (6.1c), 21597 (12.1), 22225 (12.1), 22227 (6.1b), 23838 (31.13b), 23847 (31.19a), 23850 (6.1c), 23909 (31.6), 23911 (18.7), 23993 (27.2c), 24055 (31.19c), 24569 (10.1), 24572 (31.13b), 25489 (31.7), 25503 (31.19b), 26109 (21.9), 26233 (26.18), 26519 (31.13c), 28496 (26.32), 28872 (22.1), 29451 (31.7), 29473 (12.1), 29511 (27.5a), 29626 (27.5a), 29705 (26.8), 29776 (22.1), 30213 (31.7), 30629 (19.1), 30655 (26.13), 30660 (26.25), 30662 (28.8f), 58810 (19.1), 59110 (31.4a), 59372 (21.9), 59388 (28.5)

M. Prévost   842 (27.4), 889 (26.24c), 1747 (26.29), 1844 (26.8), 1852 (27.3), 1854 (26.11)

G. Proctor   26968 (31.7)

J. Pruski   3484 (31.14)

A. Pulle   574 (26.32)

F. Putz   174 (28.1), 177 (6.1a), 179 (15.1), 182 (5.2)

B. Rabelo   1861 (31.3), 2035 (26.23c), 2310 (31.10), 2393 (31.10), 2533 (27.5a)

H. Rainer   P11–4888 (18.7), P11–8888 (31.5), P11–9888 (7.4), P21– 1988 (18.3a), P21–4988 (31.8b), P11–19988 (8.1a), P12–19988 (8.1a), P14–18188 (31.13b), P14–19988 (32.1), P14–22188 (13.1), P16–21188 (11.1), P21–25888 (19.1), P21–31888 (31.10), P22–10988 (31.1), P22–13988 (19.1), P22–23888 (31.13b), P23– 19188 (13.1)

J. Ramírez & D. López   1673 (18.3a)

J. Ramos   160 (31.4a), 972 (7.4), 1333 (7.4), 2782 (20.6), P21825 (31.2)

A. Ranghel-Galindo   s.n. (26.15), 28 (26.16), 144a (7.3), 183 (4.2)

J. Rankin   85 (16.4b)

J. Ratter   5735 (27.5a)

R. Read   s.n. (7.1), 8480 (16.4a)

E. Renteria   3975 (3.1)

J. Revilla   341 (27.5a), 957 (27.2c), 2325 (31.12b), 3728 (31.13b), 3756 (31.12a), 1070 (31.9), 1078 (27.5a)

C. Reynel   5010 (31.13b)

A. Ribeiro   1 (14.1)

B. Ribeiro   1602 (19.1), 1660 (31.21)

W. Richardson   733 (26.8)

M. Rimachi   1743 (31.13b), 3150 (31.9), 3935 (31.12a), 5668 (10.2), 1621 (10.2), 7279 (6.1c)

W. Rodrigues   IL5–2 (26.8), 4931 (26.32), 8394 (22.1), 8448 (26.1a), 8464 (16.4b), 8473 (21.8), 9075 (26.35b),

R. Romero-Casteñeda   8313 (18.7)

N. Rosa   665 (27.5a), 1580 (26.8), 1600 (31.13c), 1686 (22.1), 1830 (28.9), 2773 (26.32), 4420 (26.35b)

J. Rosales   53 (26.7)

J. Ruíz   171 (6.1c), 199 (10.2), 220 (31.12b), 1248 (31.12b), 1272 (19.1), 1300 (31.19b)

H. Rusby   410 (26.32), 411 (31.13a), 414 (18.1)

D. Sabatier   83 (27.3)

M. Saldias   223 (24.1), 371 (7.4), 372 (7.4), 424 (19.1), 533 (26.16), 683 (26.16), 689 (26.23d), 693 (28.8f), 694 (26.16), 1004 (26.16), 1032 (26.23d), 1076 (26.21), 1079 (21.12), 1259 (25.1), 1284 (26.30)

J. Salick   7085 (17.1), 7152 (10.2), 7169 (26.32), 7174 (19.1), 7658 (28.8g),

R. Sanders   1740 (21.7), 1745 (4.2), 1749 (14.1), 1751 (16.4b), 1758 (16.4b), 1759 (28.2), 1810 (28.9), 1811 (26.11), 1812 (26.23b), 1813 (26.17), 1814 (26.3), 1815 (26.8), 1816 (26.11), 1817 (28.8f)

E. Sanoja   2964 (16.1), 3060 (26.32), 3096 (16.1)

J. Santos   757 (27.2c)

C. Sastre   3011 (31.1), 3030 (27.2c), 3145 (7.3), 3445 (31.13c), 4168 (26.32), 4708 (27.5a), 8080 (31.19b)

J. Saunders   553 (27.3)

M. Sauvain   554 (26.32)

A. Scariot   5 (28.4), 7 (16.3), 8 (12.1), 10 (26.18), 170 (18.5), 171 (21.7), 173 (28.8f), 178 (21.12), 238 (21.12), 239 (1.1), 240 (26.23a), 241 (5.2), 242 (26.7), 243 (18.3a), 244 (18.7), 245 (16.4b)

R. Schnell   11546 (26.1b)

Schubert   1184 (26.24c)

R. Schultes   3407 (32.2), 3606 (32.2), 3612 (17.1), 3851 (6.1c), 3864 (18.3a), 3869 (28.6), 3872 (26.16), 3885 (28.4), 3938 (26.4), 3960 (6.1a), 5363 (26.15), 5428 (5.1), 5854 (26.32), 5856a (21.6), 6511 (6.1c), 6522a (15.3), 8290 (31.9), 8322 (26.6), 8327 (19.1), 8885 (15.1), 9346 (15.1), 9380b (15.2), 9527 (6.1a), 12484 (6.1c), 12733 (31.7), 12781 (31.7), 12812 (31.12b), 12868 (6.1c), 12930 (6.1c), 12931 (26.32), 13094 (26.19a), 13095 (31.19c), 13105 (31.19c), 13121 (6.1c), 13309 (6.1c), 13665 (31.19c), 13678 (31.12b), 13697 (7.3), 13801 (26.32), 13843 (4.2), 13989 (6.1c), 14093 (6.1c), 14108 (31.19c), 14867 (28.2), 14873 (11.1), 14886 (26.15), 14934a (16.4b), 15198 (6.1c), 15240 (19.1), 15315 (19.1), 15321 (31.19c), 15579 (6.1c), 15756 (6.1c), 15764 (14.1), 15828 (6.1c), 15984 (14.1), 16018 (26.15), 16020 (26.15), 16023 (6.1c), 16031 (26.20), 16032 (31.12b), 16233 (27.5a), 16630 (10.1), 16642 (26.32), 16648 (26.32), 16732 (26.19a), 16754 (18.3a), 17211 (26.32), 17328 (31.19c), 17341 (26.32), 17353 (31.10), 17421 (10.1), 17422 (26.35a), 17439 (26.4), 17518 (21.6), 17614 (26.32),

17655 (10.1), 17656 (26.4), 17657 (10.1), 17878a (26.35a), 18040 (21.10), 18279 (15.3), 18304 (26.4), 18315 (4.1), 18335a (26.32), 18934 (26.24c), 18967 (31.12a), 19199 (5.1), 19207 (21.6), 19216 (4.1), 20021 (31.19c), 20027 (19.1), 20076 (26.24b), 46219 (26.35a), 46314 (26.35a)

J. Schulz   502 (18.7)

J. Schunke   611 (31.12a), 1639 (19.1), 5414 (10.2), 5416 (31.19c), 5505 (31.13b), 5507 (31.10), 5603 (31.19b), 5615 (19.1), 6038 (7.4), 6136 (7.4), 6141 (7.4), 6613 (31.8b), 6674 (26.24c), 7382 (31.7), 7397 (26.32), 6751 (26.12), 7502 (26.32), 7958 (19.1), 8080 (31.12b), 8163 (19.1), 8374 (26.6), 8421 (31.12b), 8537 (7.4), 9167 (7.4), 9385 (19.1), 9477 (7.2), 9640 (7.4), 10219 (31.12b), 10407 (31.12a), 10954 (31.12a)

R. Secco   265 (16.2), 265a (31.8a), 304 (31.13c),

SEF   8517 (11.1), 8529 (28.3), 8741 (28.8h), 9261 (16.4b), 10333 (28.8h)

P. Shanley   5 (16.4b)

D. Shiki   107 (19.1)

A. Silva   1937 (31.4a)

J. Silva   299 (31.7)

M. Silva   64 (20.2), 199 (6.1c), 357 (16.4b), 379 (6.1c), 729 (6.1c), 730 (6.1c), 3136 (26.35b), 4182 (20.5), 5137 (26.35b)

P. Silverstone-Sopkin   468 (31.7), 5045 (26.21), 5182 (31.8b), 5574 (26.21), 5667 (31.8b), 5693 (25.1), 5732 (31.8b), 5907 (20.6), 5919 (31.8b), 6163 (26.21)

P. Sist   144 (28.10)

F. Skov   64719 (31.8b), 64725 (7.3), 64726 (7.3), 64753 (7.4), 64759 (7.4), 64761 (26.32), 64762 (31.19b), 64765 (31.19c), 64770 (25.4), 64776 (25.3), 64778 (7.3), 64786 (21.7), 64789 (21.2), 64790 (31.5), 64808 (31.11), 64810 (31.19c)

A. Skutch   s.n. (3.1), 4557 (7.3)

A. Smith   2127 (27.5a), 2143 (26.32), 2456 (28.1), 2556 (31.13a), 2583 (28.10), 2584 (18.1), 2661 (10.1), 2770 (31.3), 3036 (28.6), 3425 (28.4)

D. Smith   1163 (18.7), 1285 (19.1), 1891 (12.2), 1935 (31.1), 1945 (26.10c), 1960 (26.1a), 1961 (31.12a), 1996 (19.1), 2016 (17.1), 2017 (32.1), 2124 (29.1), 2131 (11.1), 2635 (7.4), 2855a (31.1), 2858a (31.12b), 2940 (13.3), 3219 (7.4), 3681 (10.2), 3688 (12.2), 3725 (18.3a), 3791 (27.5b), 3800 (16.4b), 3948 (31.10), 3964 (31.19b), 3985 (17.1), 4020 (11.1), 4045 (28.8g), 4343 (7.4), 4346 (7.4), 4455 (19.1), 4833 (18.3a), 4834 (29.1), 4835 (16.4a), 4837 (13.3), 4838 (11.1), 5146 (29.1), 5165 (7.4), 5465 (13.2), 8499 (7.4), 12863 (7.4), 12879 (7.1), 12898 (31.8a), 12918 (12.1), 12919 (18.3a), 12920 (11.1), 12921 (16.4b), 12982 (31.5), 13043 (19.1), 13044 (31.7), 13045 (31.19c), 13276 (11.1), 13565 (20.6), 13623 (24.1)

H. Smith   1713 (27.5a), 2339 (27.3), 2477 (7.4), 2478 (16.4a), 8375 (20.7)

S. Smith   113 (31.4b), 124 (7.3), 155 (31.19c), 240 (31.7), 553 (31.19c), 1318 (31.4b), 1384 (7.3), 1392 (27.5a), 1460 (20.4)

C. Sobrevilla   1773 (7.1)

J. Solomon   3241 (7.1), 6120 (26.32), 6122 (12.1), 6216 (31.13c), 6292 (31.10), 7814 (27.2a), 7842 (12.1), 7977 (31.13c), 9538 (16.4a), 10875 (31.9), 14153 (11.1), 14790 (21.9), 17152 (31.9)

C. Sperling   5802 (26.1a), 6412 (31.7), 6428 (31.19c), 6548 (21.12), 6611 (31.12a)

Spichiger   1106 (31.12a)

R. Spruce   44 (31.13b), 56 (4.1)

G. Stahel   s.n. (21.7), s.n. (27.3), s.n. (28.9)

B. Stein   2592 (17.1)

J. Steinbach   5424 (7.4)

B. Stergios   4673 (18.7), 6380 (7.4), 7554 (31.13b), 9152 (6.1a), 9279 (5.1), 9525 (15.2), 9860 (31.7), 9964 (15.3), 10146 (10.1), 10221 (31.7), 10802d (28.4), 12640 (6.1a), 13001 (19.1), 13227b (31.3)

W. Stevens   8385 (16.4a), 12110 (31.7), 23477 (16.4a), 23799 (16.4a), 24495 (11.1), 24559 (12.1), 24560 (16.4a)

D. Stevenson   731 (26.32), 803 (31.13b), 816 (6.1a), 817 (31.16), 832 (26.32), 867 (31.13c), 868 (10.1), 938 (26.32), 1041 (31.7), 1042 (31.7), 1109 (26.24c)

J. Steyermark   269 (9.1), 1320 (5.2), 54639 (7.4), 74633 (10.1), 74634 (31.7), 74703 (26.32), 74750 (31.10), 74787 (28.4), 74790 (12.1), 74791 (16.4b), 75222 (10.1), 75585 (26.19a), 75979 (9.1), 86703 (16.3), 87178 (12.1), 87273 (27.5a), 87274 (31.13a), 87405 (14.1), 87406 (26.7), 87474 (16.3), 87708 (4.2), 87957 (12.1), 87989 (28.4), 88054 (26.32), 88063 (26.32), 88126 (31.13a), 88144 (26.26), 88407 (26.1a), 88408 (16.4b),

88409 (4.2), 88912 (21.7), 88913 (20.6), 89119 (31.13a), 89171 (26.31), 89402 (9.1), 90221 (10.1), 90576 (31.3), 90577 (19.1), 90596 (26.12), 90660 (28.4), 90666 (31.10), 90720 (12.1), 90777 (11.1), 90779 (12.1), 91399 (26.31), 91584 (7.4), 91643 (16.4a), 91661 (16.4a), 91755 (7.4), 91797 (26.31), 91799 (7.4), 91934 (26.31), 92837 (26.32), 92883 (9.1), 94146 (9.1), 94356 (12.1), 95455 (26.12), 95458 (27.5a), 95723 (12.1), 95769 (26.1a), 96016 (26.31), 96519 (20.6), 97783 (5.2), 98917 (7.4), 99128 (26.31), 99584 (31.7), 99604 (18.3a), 99720 (7.4), 99868 (31.8b), 100368 (26.31), 101287 (27.5a), 101301 (12.1), 101326 (31.7), 101380 (18.7), 101415 (26.12), 101707 (7.4), 102000 (27.3), 102440 (26.32), 102454 (15.1), 102647 (26.5), 102653 (15.1), 102752 (10.1), 102784 (21.8), 102831 (5.2), 102940 (10.1), 103029 (31.13c),103030 (10.1), 103202 (14.1), 103337 (19.1), 103363 (16.4a), 103467 (7.4), 104041 (18.3a), 105553 (7.4), 106056 (12.1), 106829 (16.4a), 106832 (12.1), 107000 (26.12), 107025 (26.12), 107188 (16.2), 107596a (18.1), 113041 (19.1), 113323 (26.32), 113825 (26.8), 114169 (19.1), 114650 (26.23b), 114802 (14.1), 117242 (5.2), 117783 (26.1a), 118914 (19.1), 118954 (26.31), 119430 (26.12), 119437 (16.4a), 119446 (12.1), 120417 (31.8b), 120422 (18.7), 120593 (31.8b), 121267 (21.7), 121520 (31.8b), 122179 (27.5a), 122219 (28.1), 122226 (31.7), 122278 (16.4b), 122329 (26.32), 122340 (14.1), 122373 (28.4), 122392 (26.32), 122423 (27.4), 122698 (31.8b), 122709 (26.23b), 123246 (16.4a), 124283 (19.1), 124731 (26.23b), 124808 (7.4), 125072 (7.4), 125642 (31.13c), 125695 (28.4), 125699 (31.7), 125705 (10.1), 125783 (28.1), 125785 (26.12), 125886 (28.4), 127374 (31.13a), 129705 (31.7), 130208 (26.19a), 130552 (14.1), 130877 (26.32), 131457 (28.6), 131532 (26.32), 131589 (14.1), 131602 (21.10), 131605 (15.1), 131716 (28.1), 131988 (18.1)

A. Stoffers   249 (19.1)

J. Strudwick   4572 (14.1), 4667 (26.8), 4668 (5.2), 4669 (28.4), 4671 (31.13c), 4672 (21.7), 4673 (20.3), 4680 (26.23a), 4681 (16.3), 5001 (31.7), 5002 (26.32), 5003 (28.8f), 5007 (31.3), 5008 (26.7), 5009 (31.13c), 5010 (31.7), 5011 (16.3), 5012 (12.1), 5013 (27.5a), 5014 (18.5), 5015 (18.3a)

G. Sullivan   1180 (19.1)

F. Tamayo   2991 (4.2), 3121 (5.2)

G. Tate   49 (10.1), 157 (26.19a), 166 (31.3), 355 (31.13c), 356 (21.7), 435 (19.1)

A. Tavares   74 (26.32)

E. Taylor   1049 (20.1)

J. Taylor   11587 (31.8b)

L. Teixeira   106 (6.1c), 491 (7.4), 961 (26.14), 996 (26.19a), 1018 (26.24c), 1019 (10.1), 1033 (31.7)

E. Tejera   29 (27.3)

G. Tessmann   5082 (18.3a), 5117 (28.8e), 5210 (28.6), 5236 (27.2b), 5324 (12.1), 5381 (4.2), 5483 (26.16)

W. Thomas   428 (28.8f), 2669 (5.2), 4080 (26.1a), 4081 (16.2), 5147 (31.7), 5626 (31.4a), 6026 (27.3), 6210 (16.4b), 6471 (16.4b), 6478 (12.1), 6498 (16.4b), 6523 (12.1), 6598 (28.2), 6634 (28.4), 6777 (27.5b)

S. Tillett   43911 (18.3b), 44841 (9.1), 44921 (31.13a), 44931 (31.2), 45037 (26.28), 45046 (26.32), 45458 (31.13a), 45468 (26.32), 45477 (26.26), 45503 (26.24c), 45508 (26.1a), 45509 (31.13a), 45515 (26.26), 45530 (28.4), 45540 (27.5a), 45614 (31.13a), 45662 (31.8a), 45713 (31.12c), 45725 (26.32), 45769 (12.1), 152–172 (16.1), 672130 (7.3), 752–172a (16.1)

M. Timaná   907 (29.1), 1234 (31.12a), 1272 (27.5a), 1334 (26.10a), 1417 (31.5), 1444 (26.10a), 1588 (31.12a), 1612 (31.5), 1624 (31.12a), 1668 (31.12a),

P. Tomlinson   2 (27.3)

R. Toro   850 (7.4), 1318 (7.4)

J. Torres   107 (28.3), 2994 (31.1), 3132 (31.13c), 3137 (28.4), 3139 (31.7), 3152 (31.6), 3157 (31.19c), 3166 (12.1), 3181 (26.15), 3183 (28.7a), 3185 (27.2c), 3188 (6.1c), 3199 (21.10), 3205 (21.10), 3206 (21.10)

J. Trail   826/CLII (26.24c), 829/XXXVII (26.24c), 830/CLXVIII (26.24c), 837 (26.7), 841/CLXXVIII (26.24c), 848/CXXIII (26.7), 850 (26.13), 851/VI (26.13), 852/CCXXIV (26.13), 853/XXXV (26.13), 856/XII (26.37), 857/LIII (26.19a), 862/CCXIV (26.19a), 863/XI (26.19a), 864/CLXXXIX (26.19a), 865/CCXX (26.19a), 866/LXIX (26.19a), 867/LXIX (26.19a), 868/LXVI

(26.19a), 870/LXIX (26.19a), 874/XXII (26.14), 874/XXII (26.14), 875/CXXXIV (26.14), 877/XLVIII (26.10b), 879/XLIV (26.13), 881/CCXI (26.22), 885/XCIX (26.5), 886/CI (26.5), 887/XIV (26.5), 888/CVI (26.4), 889/XV (26.36), 894/LXXXII (26.30), 895/CLXXIII (26.15), 896/CLXXIX (26.15), 907/LXXXVI (26.6), 908/LV (26.6), 912/XIII (26.11), 913/CXIV (26.11), 916 (26.32), 917/XC (26.32), 919/CLXXXI (26.32), 920/LXIV (26.32), 921 (26.32), 922/CLXIX (26.32), 923/CLXIX (26.32), 925 (26.32), 926 (26.32), 928 (26.32), 929/XXVIII (26.32), 930 (26.32), 931 (26.32), 932 (26.32), 933 (26.32), 934 (26.32), 935 (26.32), 936 (26.32), 937 (26.32), 938 (26.32), 939/LXXVIII (26.32), 941 (26.2), 942/CXXIII (31.13c), 944/XLIX (31.7), 945/LXXIV (31.7), 946/LXX (31.7), 947/LXXI (31.7), 948/LXXII (31.7), 950/LVII (31.7), 951/CXXXI (31.7), 952/CC (31.7), 953/CLIII (31.7), 954/CLI (31.7), 955 (31.7), 959/CLXXXVII (31.7), 960/CLXIII (31.7), 963/CLXXII (31.10), 964/CVIII (31.2), 967/l (31.3), 968/CXXIV (31.3), 969/CLXXX (31.3), 970 (31.3), 971 (31.3), 972/CVII (31.13d), 973/CXII (31.13d), 974/CXXXV (31.13d), 975 (31.13d), 979/CLXXXII (31.6), 981/XC (31.13b), 982/CXLIII (31.13b), 985/XXIX (31.13b), 986/XXV (31.13c), 987/XXVI (31.13c), 988/LVI (31.13c), 988/CXX (31.13b), 990/XLVII (31.13c), 991/CLXX (31.13c), 992/CXVIII (31.13c), 993/LXXXIX (31.13c), 995/CXXXVII (31.13c), 996/CXXXVIII (31.13c), 999/CLXXIV (31.13b),1000/CXXV (31.13b),1001/CLXIV (31.13b), 1003/XLI (31.13b), 1008/CLXXVI (31.13b), 1010/CCVII (31.12b), 1011/CLXXXIV (31.12b), 1012/CCIII (31.12b), 1013/CXXVIII (31.12b), 1014/CXXII (31.12a), 1015/CXXII (31.12a), 1016/CXXIIa (31.12a), 1020/CLXXI (31.16), 1021/CCI (31.16), 1022/LIX (31.9), 1025/XVI (31.19b), 1027/CLXII (31.19b), 1028/CXCVIII (31.19a), 1029/CXCVII (31.19a), 1031/CLXI (31.19c), 1033/CXXXIII (31.19c), 1034/CXXXIII (31.19c), 1053 (12.1)
J. Treacy 483 (4.2)
G. Triana 23 (21.10), 93 (27.5a), 170 (15.3)
S. Tsugaru B-394 (16.3), B1040 (15.3)
S. Tunqui 65 (26.32), 293 (31.13b), 360 (31.13c)
U. G. Bio 88 (4.2)
R. Valencia 84784 (17.1)
J. Valerio 1088 (16.4a)
T. van Andel 108 (7.3), 245 (31.1), 329 (31.19c)
H. van der Werff 5647 (31.8b)
R. Vanegas 1 (21.2)
C. Vargas 7748 (29.1), 7813 (11.1), 14488 (13.1), 14490 (26.24c), 16105 (26.10c), 16304 (4.2), 16656 (11.1), 17751 (13.1), 17375 (25.1), 17582 (29.1), 17896 (19.1), 18616 (26.19a), 18618 (26.19a), 18694 (25.1), 18714 (26.24c), 19163 (7.4), 23145 (7.4), 23810 (7.4)
R. Vásquez 042 (31.9), 238 (31.13b), 336 (18.7), 354 (31.12a), 461 (31.12a), 476 (31.19a), 477 (31.13b), 955 (31.5), 1052 (31.13b), 1106 (12.1), 1396 (26.7), 1426 (31.13b), 1427 (31.7), 1428 (21.7), 1452 (12.1), 1582 (31.12b), 1601 (26.12), 1678 (26.10b), 1779 (31.7), 1797 (31.13b), 1926 (26.12), 2098 (31.12a), 2137 (31.17), 2185 (26.24a), 2228 (31.19a), , 2300 (27.5a), 2360 (28.8e), 2448 (31.13b), 2488 (26.24c), 2646 (27.2b), 2956 (27.2c), 3002 (7.4), 3054 (12.1), 3075 (13.2), 3077 (13.3), 3119 (27.5a), 3255 (10.2), 3381 (27.5b), 3506 (26.24b), 3724 (31.12a), 3751 (18.7), 3935 (26.22), 4499 (18.3a), 4503 (16.1), 4611 (10.2), 4628 (29.1), 4719 (31.6), 4819 (12.1), 5061 (31.7), 5216 (18.7), 5245 (31.13c), 5294 (25.4), 5324 (21.10), 5687 (31.12a), 5696 (16.4b), 6106 (31.12a), 6216 (26.20), 6271 (31.12a), 6626 (31.13c), 6857 (19.1), 7110 (31.7), 7112 (2.1), 7314 (31.19c), 7321 (19.1), 7413 (31.1), 7501 (5.2), 7574 (26.19b), 7575 (26.19c), 7585 (26.20), 7591 (31.13c), 7752 (31.13b), 7756 (21.8), 7757 (31.10), 8329 (27.2b), 8528 (26.24c), 8806 (31.17), 8904 (16.1), 9004 (31.10), 9005 (31.19b), 9056a (31.19c), 9152 (32.2), 9185 (21.7), 9225 (18.7), 9246 (16.4b), 9579 (18.2), 9744 (31.5), 9745 (31.19c), 10002 (26.24a), 10006 (31.10), 10103 (31.12a), 10563 (26.19c), 10594 (26.10b), 10686 (27.5b), 10836 (31.19b), 10885 (31.1), 10966 (5.2), 10968 (16.1), 11055 (26.24c), 11130 (19.1), 11246 (27.5b), 11491 (26.32), 11525 (26.10a), 11535 (31.7),

11674 (19.1), 11713 (26.6), 11773 (31.19a), 11775 (31.12b), 11779 (31.12b), 11814 (19.1), 12015 (26.24a), 12017 (29.1), 12053 (26.19c), 12057 (31.12b), 12059 (31.12a), 12096 (7.4), 12164 (31.12b), 12167 (31.12b), 12237 (31.19a), 12304 (31.7), 12362 (31.12a), 12693 (17.1), 12762 (26.24b), 12797 (26.19a), 12963 (26.32), 12996 (19.1), 13019 (31.12b), 13029 (26.32), 13050 (12.1), 13070 (26.35a), 13073 (19.1), 13110 (31.12b), 13112 (31.19a), 13129 (26.24b), 13147 (8.1b), 13203 (26.22), 13235 (31.12a), 13261 (31.7), 13353 (26.24a), 13353 (26.24b), 13358 (26.22), 13460 (8.1b), 13470 (31.12b), 13529 (27.2b), 13559 (27.5b), 13810 (21.7), 13843 (25.4), 13851 (31.19a), 13904 (6.1c), 14108 (27.2b), 14181 (25.4), 14195 (31.19c), 14323 (21.10), 14327 (31.19c), 14330 (26.22), 14359 (26.35a), 14411 (31.7), 14413 (31.17), 14438 (26.24c), 14439 (31.19c), 14444 (31.7), 14446 (31.13b), 14453 (19.1), 14455 (31.19a), 14456 (31.17), 14516 (28.3), 14532 (25.4), 14542 (25.4), 14568 (6.1c), 14592 (28.8e), 14593 (11.1), 14617 (12.1), 14621 (11.1), 14633 (18.3a), 14801 (28.3), 14828 (16.4b), 15164 (31.19a)

G. Vieira 40 (6.1c), 576 (31.7)

M. Vieira et al. 816 (31.8a)

P. Vincelli 980 (28.4), 1044 (18.1)

C. von Hagen 1368 (27.3)

T. Wachter 29 (26.32), 52 (7.4)

J. Walker 174 (27.3)

B. Wallnöfer 114–8488 (25.4), 12–311287 (31.1)

P. Warner 470 (26.21)

A. Warush RBAE80 (7.3), RBAE97 (31.7)

D. Wasshausen 892 (31.1)

J. Wessels Boer 155 (23.1), 157 (18.2), 163 (28.9), 165 (21.8), 170 (27.3), 172 (26.32), 173 (28.2), 174 (27.5a), 175 (28.2), 176 (31.3), 177 (27.5a), 181 (31.7), 182 (26.17), 183 (28.4), 185 (12.1), 186 (26.32), 189 (26.11), 193 (27.5a), 196 (28.6), 197 (27.3), 198 (26.2), 211 (28.10), 278 (31.3), 279 (26.32), 281 (31.13a), 283 (18.1), 284 (16.3), 288 (21.7), 296 (26.17), 308 (28.9), 309 (28.11), 310 (28.2), 312 (28.2), 313 (28.11), 314 (28.2), 320 (31.7), 324 (26.2), 325 (31.3), 327 (28.10), 329 (16.4b), 332 (28.4), 335 (31.3), 336 (31.7), 338 (28.10), 345 (21.7), 346 (16.3), 347 (21.7), 349 (31.3), 350 (16.3), 353 (16.3), 357 (21.7), 388 (31.3), 392 (21.8), 457 (19.1), 460 (31.10), 467 (26.2), 473 (18.1), 486 (26.23c), 500 (28.2), 509 (18.3b), 521 (28.2), 526 (27.3), 570 (28.6), 571 (16.4b), 575 (27.5a), 619 (27.5a), 644 (26.2), 649 (26.2), 805 (21.3), 810 (21.7), 912 (31.10), 942 (31.3), 970 (27.5a), 1037 (27.5a), 1064 (31.19b), 1065 (26.3), 1066 (26.3), 1091 (26.27), 1168 (26.25), 1203 (26.14), 1204 (21.1), 1224 (26.29), 1226 (28.9), 1329 (21.7), 1335 (27.3), 1336 (28.11), 1365 (21.3), 1371 (21.7), 1374 (26.37), 1379 (18.1), 1392 (26.37), 1428 (26.37), 1493 (21.8), 1495 (26.14), 1496 (19.1), 1512 (19.1), 1513 (31.19b), 1514 (18.3b), 1515 (26.8), 1558 (31.19b), 1584 (26.23a), 1587 (21.3), 1591 (26.14), 1598 (31.19b), 1601 (26.32), 1608 (26.14), 1621 (26.8), 1624 (14.1), 1626 (26.23b), 1632 (27.3), 1633 (26.23b), 1635 (31.8b), 1636 (26.31), 1638 (16.4b), 1639 (27.3), 1640 (26.23b), 1641 (26.23b), 1643 (16.3), 1646 (26.23b), 1650 (16.3), 1651 (26.23b), 1653 (16.3), 1657 (26.23b), 1658 (27.3), 1701 (7.4), 1762 (20.6), 1764 (26.23b), 1798 (26.23c), 1814 (16.4b), 1821 (21.7), 1854 (26.21), 1858 (16.4a), 1865 (16.4a), 1868 (16.4a), 1869 (31.8b), 1874 (28.4), 1876 (28.6), 1891 (18.3a), 1894 (21.10), 1906 (21.10), 1912 (21.2), 1913 (21.8), 1914 (28.1), 1916 (5.1), 1920 (21.7), 1943 (15.3), 1944 (28.1), 1946 (28.6), 1951 (24.1), 1963 (26.31), 2013 (31.8b), 2040 (16.4a), 2044 (16.4a), 2047 (12.1), 2062 (18.3b), 2083 (26.32), 2105 (21.2), 2111 (31.7), 2252 (16.1), 2262 (4.1), 2273 (21.8), 2274 (28.3), 2275 (5.2), 2283 (26.32), 2299 (26.5), 2300 (5.2), 2301 (28.1), 2304 (28.6), 2308 (15.1), 2309 (15.2), 2310 (26.4), 2322 (21.10), 2357 (21.6), 2368 (28.2), 2372 (26.32), 2374 (21.6), 2375 (31.13b), 2376 (27.3), 2377 (27.2c), 2383 (26.32), 2384 (26.12), 2385 (26.19a), 2399 (31.12a), 2402 (28.8f), 2407 (6.1a), 2409 (21.8), 2410 (26.8), 2438 (26.12), 2453 (26.21), 2454 (7.4), 2458 (16.4a), 2459 (31.8b), 2460 (19.1), 6508 (31.19b)

L. Williams 7386 (7.4), 11391 (19.1), 11623 (12.1), 11797 (26.24c), 11888 (21.2), 12605 (28.4), 12626 (24.1), 12665 (4.2), 12724 (27.3), 12994 (18.3a), 13116 (15.3), 13138 (20.4) 14013 (10.1), 14053 (27.3), 14064

(15.2), 14075 (5.1), 14179 (26.24c), 14318 (4.2), 14345 (18.3a), 14379 (5.2), 14393 (10.1), 14521 (18.1), 14574 (26.19a), 14592 (26.16), 14671 (12.1), 14737 (6.1a), 14925 (21.7), 15073 (5.2), 15198 (31.3), 15539 (6.1a), 15586 (18.1), 15750 (26.5), 15841 (10.1)

R. Williams 392 (26.24b), 393 (7.1), 402 (11.1), 657 (27.2d)

P. Wilson 547 (27.3)

J. Wurdack 40852 (20.4), 42734 (15.1), 42940 (15.1), 43188 (31.7), 43282 (5.1), 43486 (27.3)

H. Young 42 (31.4b), 126 (26.24c), 150 (26.10b)

K. Young 49 (4.2), 59 (31.19c), 69 (31.7), 102 (26.19a), 189 (11.1), 985 (31.12b), 1005 (31.5), 1044 (7.3)

T. Yuncker 8809 (27.3)

V. Zak 4850 (11.1)

J. Zarucchi 1065 (26.10b), 1285 (26.32), 1706 (26.15), 1710 (10.1), 1712 (31.7), 1715 (27.1), 1747 (31.7), 1748 (31.19c), 1795 (20.4), 1809 (18.3a), 1834 (18.7), 1838 (18.3a), 1839 (16.4b), 1842 (18.3a), 1849 (18.3a), 1886 (16.4b), 1887 (16.1), 1888 (5.2), 1889 (26.4), 1894 (10.1), 2107 (21.6), 2108 (21.6), 6044 (7.4)

E. Zardini 4090 (20.5)

S. Zent 098502b (16.4b), 098502c (16.4b), 1085–16 (31.7), 128505 (26.12)

S. Zuluaga 20 (27.3)

References

Ab'Sáber, A. 1982. The paleoclimate and paleoecology of Brazilian Amazonia. In G. Prance, ed. Biological Diversification in the Tropics. Columbia University Press, New York, 41–59.

Absy, M., A. Cleef, M. Fournier, L. Martin, M. Servant, A. Sifeddine, M. Ferreira, F. Soubies, K. Suguio, B. Turcq, & T. van der Hammen. 1991. Mise en évidence de quatre phases d'ouverture de la forêt dense dans le sud-est de l'Amazonie au cours des 60,000 dernières années. Première comparaison avec d'autres régions tropicales. C. R. Acad. Sci. Paris 312, ser. 2:673–678.

Allen, P. 1965. *Raphia* in the western world. Principes 9:66–70.

Anderson, A. 1981. White-sand vegetation of Brazilian Amazon. Biotropica 13:199–210.

———. 1983. The biology of *Orbignya martiana* (Palmae), a tropical dry forest dominant in Brazil. Ph.D. diss., University of Florida, Gainesville.

———. 1988. Use and management of native forests dominated by açaí palm (*Euterpe oleracea* Mart.) in the Amazon estuary. Adv. Econ. Bot. 6:144–154.

Anderson, A., & M. Balick. 1988. Taxonomy of the *babassu* complex (*Orbignya* spp.: Palmae). Syst. Bot. 13:32–50.

Anderson, A., P. May, & M. Balick. 1991. The Subsidy from Nature. Columbia University Press, New York.

Anderson, A., W. Overall, & A. Henderson. 1988. Pollination ecology of a forest dominant palm (*Orbignya phalerata* Mart.) in northern Brazil. Biotropica 20:192–205.

Anderson, R., & S. Mori. 1967. A preliminary investigation of *Raphia* palm swamps, Puerto Viejo, Costa Rica. Turrialba 17:221–224.

Ataroff, M., & T. Schwarzkopf. 1992. Leaf production, reproductive patterns, field germination and seedling survival in *Chamaedorea bartlingiana*, a dioecious understory palm. Oecologia 92:250–256.

Aublet, J. 1775. Histoire des plantes de la Guiane française. 4 vols. Didot, Paris.

Ayres, J., & T. Clutton-Brock. 1992. River boundaries and species range size in Amazonian primates. Amer. Nat. 140:531–537.

Bahn, P. 1993. 50,000-year-old Americans of Pedra Furada. Nature 362:114–115.

Bailey, L. 1933. Certain palms of Panama. Gentes Herb. 3:33–116.

———. 1941. *Acrocomia*—Preliminary paper. Gentes Herb. 4:421–476.

———. 1947. Indigenous palms of Trinidad and Tobago. Gentes Herb. 7:353–445.

Balée, W. 1988. Indigenous adaptation to Amazonian palm forests. Principes 32:47–54.

———. 1989. The culture of Amazonian forests. Adv. Econ. Bot. 7:1–21.

———. 1992. People of the fallow: A historical ecology of foraging in lowland South America. In K. Redford & C. Padoch, eds. Conservation of Neotropical Forests: Working from Traditional Resource Use. Columbia University Press, New York, 35–57.

Balick, M. 1980. Wallace, spruce and palm trees of the Amazon: An historical perspective. Bot. Mus. Leafl. 28:263–269.

———. 1986. Systematics and economic botany of the *Oenocarpus–Jessenia* complex. Adv. Econ. Bot. 3:1–140.

———. 1991. A new hybrid palm from Amazonian Brazil, *Oenocarpus* x *andersonii*. Bol. Mus. Paraense Hist. Nat. 7:505–510.

Balick, M., & H. Beck. 1990. Useful Palms of the World. Columbia University Press, New York.

Balick, M., & S. Gershoff. 1990. A nutritional study of *Aiphanes caryotifolia* (Kunth) Wendl. (Palmae) fruit: An exceptional source of vitamin A and high quality protein from tropical America. Adv. Econ. Bot. 8:35–40.

Balick, M., C. Pinheiro, & A. Anderson. 1987. Hybridization in the babassu palm complex: I. *Orbignya phalerata* x *O. eichleri*. Amer. J. Bot. 74:1013–1032.

Balslev, H., J. Luteyn, J. Ollgaard, & L. Holm-Nielsen. 1987. Composition and structure of adjacent unflooded and floodplain forest in Amazonian Ecuador. Opera Bot. 92:37–57.

Barbosa Rodrigues, J. 1875. Enumeratio Palmarum Novarum. Brown & Evaristo, Sebastianopolis.

———. 1879. Protesto-Appendice ao Enumeratio Palmarum Novarum. Typographia Nacional, Rio de Janeiro.

———. 1903. Sertum Palmarum Brasiliensium. 2 vols. Veuve Monnom, Brussels.

Barfod, A. 1991. A monographic study of the subfamily Phytelephantoideae (Arecaceae). Opera Bot. 105:1–73.

Barfod, A., A. Henderson, & H. Balslev. 1987. A note on the pollination of *Phytelephas microcarpa* (Palmae). Biotropica 19:191–192.

Bates, H. 1863. The Naturalist on the River Amazons. Murray, London.

Beach, J. 1984. The reproductive biology of the Peach or "Pejibayé" palm (*Bactris gasipaes*) and a wild congener (*B. porschiana*) in the Atlantic lowlands of Costa Rica. Principes 28:107–119.

Beccari, O. 1918. Asiatic palms—Lepidocaryeae. Part 3. Ann. Roy. Bot. Gard. Calcutta 12:1–231.

Bentham, G., & J. Hooker. 1883. Genera Plantarum. Vol. 3, part 2. Reeve, London.

Bernal, R. 1989. Proposal to conserve *Bactris gasipaes* Kunth against *Bactris ciliata* (R. & P.) Mart. (Palmae). Taxon 38:520–522.

———. 1992. Colombian palm products. In M. Plotkin & L. Famolare, eds. Sustainable Harvest and Marketing of Rain Forest Products. Island Press, Washington, D.C., 158–172.

Bernal, R., G. Galeano, & A. Henderson. 1991. Notes on *Oenocarpus* (Palmae) in the Colombian Amazon. Brittonia 43:154–164.

Bernal, R., & A. Henderson. In press. Nuevas evidencias sobre el origin del chontaduro o pejibaye, *Bactris gasipaes* (Palmae). Caldasia.

Bigarella, J. 1973. Geology of the Amazon and Parnaiba basins. In A. Nairn & F. Stehli, eds. The Ocean Basins and Margins. Vol. 1: The South Atlantic. Plenum, New York, 25–86.

Bigarella, J., & A. Ferreira. 1985. Amazonian geology and the Pleistocene and the Cenozoic environments and paleoclimates. In G. Prance & T. Lovejoy, eds. Amazonia. Pergamon Press, Oxford, 49–71.

Blombery, A., & T. Rodd. 1982. Palms. Angus & Robertson, Sydney.

Bondar, G. 1957. Novo genero e nova especie de palmeiras da tribo Attaleinae. Arq. Jard. Bot. Rio de Janeiro 15:49–55.

Borchsenius, F. 1993. Flowering biology and insect visitation of three Ecuadorean *Aiphanes* species. Principes 37:139–150.

Borchsenius, F., & R. Bernal. In press. *Aiphanes* (Palmae). Flora Neotropica.

Borgtoft Pedersen, H. 1992. Use and management of *Aphandra natalia* (Palmae) in Ecuador. Bull. Inst. Fr. Ét. Andines 21:741–753.

Bradford, D., & C. Smith. 1977. Seed predation and seed number in *Scheelea* palm fruits. Ecology 58:667–673.

Brown, C., & W. Lidstone. 1878. Fifteen Thousand Miles on the Amazon and Its tributaries. Edward Stanford, London.

Brown, K., & G. Brown. 1992. Habitat alteration and species loss in Brazilian forests. In T. Whitmore & J. Sayer, eds. Tropical Deforestation and Species Extinction. Chapman & Hall, London, 119–142.

Brummitt, R., & C. Powell. 1992. Authors of Plant Names. Royal Botanic Gardens, Kew, London.

Buringh, P. 1979. Introduction to the Study of Soils in Tropical and Subtropical Regions. 3rd ed. Center for Agricultural Publishing and Documentation, Wageningen, Netherlands.

Burret, M. 1929a. Die Gattung *Euterpe* Gaertn. Bot. Jahrb. Syst. 63:49–76.

———. 1929b. Die Palmengattungen *Orbignya, Attalea, Scheelea* und *Maximiliana*. Notizbl. Bot. Gart. Berlin-Dahlem 10:498–593, 651–701.

———. 1930. Iriarteae. Notizbl. Bot. Gart. Berlin-Dahlem 10:918–942.

———. 1933–1934. *Bactris* und verwandte Palmengattungen. Repert. Spec. Nov. Regni Veg. 34:167–184, 185–253.

———. 1934a. Die Palmengattung *Desmoncus* Mart. Repert. Spec. Nov. Regni Veg. 36:197–221.

———. 1934b. Die Palmengattung *Astrocaryum* G. F. W. Meyer. Repert. Spec. Nov. Regni Veg. 35:114–158.

———. 1935. Die Palmengattungen *Mauritia* L.f. und *Mauritiella* Burret nov. gen. Notizbl. Bot. Gart. Berlin-Dahlem 12:605–611.

Calzavara, B. 1972. As possibilidades do açaizeiro no estuário amazonico. Bol. Fac. Ci. Agrar. Pará, Belém 5:1–103.

Campbell, D., D. Daly, G. Prance, & U. Maciel. 1986. Quantitative ecological inventory of terra firme and varzea tropical forest on the Rio Xingu, Brazilian Amazon. Brittonia 38:369–393.

Campbell, L. 1986. Comments. Curr. Anthropol. 27:488.

Caputo, M. 1991. Solimoes megashear: Intraplate tectonics in northwest Brazil. Geology 19:246–249.

Castelnau, F. de. 1853. Vues et scènes recueillies pendant l'expédition dans les parties centrales de l'Amérique du Sud, Rio de Janeiro à Lima, et de Lima au Para. Bertrand, Paris.

Clement, C., J. Aguiar, D. Arkcoll, J. Firmino, & R. Leandro. 1989. *Pupunha brava* (*Bactris dahlgreniana* Glassman): Progenitor da pupunha (*B. gasipaes* H.B.K.)? Bol. Mus. Paraense Hist. Nat. 5:39–55.

Clement, C., & J. Mora Urpí. 1987. Pejibaye palm (*Bactris gasipaes,* Arecaceae): Multi-use potential for the lowland humid tropics. Econ. Bot. 41:302–311.

Colinvaux, P. 1987. Amazon diversity in light of the paleoecological record. Quart. Sci. Rev. 6:93–114.

Condit, R., S. Hubbell, & R. Foster. 1992. Short-term dynamics of a neotropical forest. Bioscience 42:822–828.

Cook, O. 1942. A Brazilian origin for the commercial oil palm. Scientific Monthly. June: 577–580.

Cracraft, J., & R. Prum. 1988. Patterns and processes of diversification: Speciation and historical congruence in some neotropical birds. Evolution 42:603–620.

Cronquist, A. 1988. The Evolution and Classification of Flowering Plants. 2nd ed. New York Botanical Garden, New York.

Dahlgren, B. 1940. Travels of Ruiz, Pavón, and Dombey in Peru and Chile (1777–1788). Field Mus. Nat. Hist., Bot. ser. 21:1–372.

———. 1959. Plates. Index of American Palms. Field Mus. Nat. Hist., Bot. ser. 14:1–412.

Daly, D., & G. Prance. 1989. Brazilian Amazon. In D. Campbell & D. Hammond, eds. Floristic Inventory of Tropical Countries. New York Botanical Garden, New York, 401–426.

D'Arcy, W. 1987. Flora of Panama: Checklist and Index. Missouri Botanical Garden, St. Louis.

Darwin, C. 1859. The Origin of Species. Murray, London.

Darwin, S. 1993. Albert Charles Smith—Recipient of the 1992 Asa Gray Award. Syst. Bot. 18:1–5.

de Blank, S. 1952. A reconnaissance of the American oil palm. Trop. Agri. 29:90–101.

de Candolle, A. 1857. Sketch on the life and writings of M. de Martius, secretary to the

Bavarian Academy of Science. Hooker's J. Bot. Kew Gard. Misc. 9:6–10.

de Granville, J.-J. 1977. Notes biologiques sur quelques palmiers guyanais. Cah. ORSTOM, ser. Biol. 12:347–353.

de Granville, J.-J., & A. Henderson. 1988. A new species of *Asterogyne* (Palmae) from French Guiana. Brittonia 40:76–80.

de Lisboa, C. 1968. História dos animais e arvores de Maranhão. Universidade Federal do Paraná, Curitiba, Brazil.

Denevan, W. 1992. The pristine myth: The landscape of the Americas in 1492. Ann. Associ. Amer. Geog. 82:369–385.

Denslow, J. 1987. Tropical rain forest gaps and tree species diversity. Ann. Rev. Ecol. Syst. 18:431–451.

de Steven, D., D. Windsor, F. Putz, & B. de León. 1987. Vegetative and reproductive phenologies of a palm assemblage in Panama. Biotropica 19:342–356.

Devall, M., & R. Kiester. 1987. Notes on *Raphia* at Corcovado. Brenesia 28:89–96.

Dixon, J. 1979. Origin and distribution of reptiles in the lowland tropical rainforests of South America. In W. Duellman, ed. The South American Herpetofauna: Its Origin, Evolution, and Dispersal. Monograph No. 7. Museum of Natural History, University of Kansas, Lawrence, 217–240.

Donnelly, T. 1992. Geological setting and tectonic history of Mesoamerica. In D. Quintero & A. Aiello, eds. Insects of Panama and Mesoamerica. Oxford University Press, New York.

Dransfield, J., D. Johnson, & H. Synge. 1988. The Palms of the New World: A Conservation Census. IUCN, Cambridge.

Drude, O. 1881. Cyclanthaceae et Palmae. In C. Martius, ed. Flora Brasiliensis. Vol. 3, part 2, fasc. 85. Monachii, Leipzig, 225–460.

———. 1882. Palmae part II. In C. Martius ed. Flora Brasiliensis. Vol. 3, part 2, fasc. 86. Monachii, Leipzig, 461–610.

———. 1887. Palmae. In A. Engler & K. Prantl, eds. Die Natürlichen Pflanzenfamilien. Vol. 2, part 3. Engelmann, Leipzig, 1–93.

Ducke, A., & G. Black. 1953. Phytogeographical notes on the Brazilian Amazon. Anais Acad. Brasil. Ci. 25:1–46.

Dugand, A. 1953. Notas sobre el genero *Attalea* (Palmae) en Colombia. Mutisia 18:1–10.

Dumont, J., S. Lamotte, & F. Kahn. 1990. Wetland and upland forest ecosystems in Peruvian Amazonia: Plant species diversity in light of some geological and botanical evidence. Forest Ecol. and Manage. 33–34:125–139.

Eden, M. 1990. Ecology and Land Management in Amazonia. Bellhaven Press, London.

Eiten, G. 1972. The cerrado vegetation of Brazil. Bot. Rev. 38:1–341.

Emmons, L. 1984. Geographic variation in densities and diversities of non-flying mammals in Amazonia. Biotropica 16:210–222.

Endler, J. 1982. Pleistocene forest refuges: Fact or fancy? In G. Prance, ed. Biological Diversification in the Tropics. Columbia University Press, New York, 641–657.

FAO/UNESCO. 1974. Soil Map of the World, 1:50,000. Vol. 4: South America. UNESCO, Paris.

Fearnside, P. 1989. Extractive reserves in Brazilian Amazonia. Bioscience 39:387–393.

Ferreira, A. 1972. Viagem filosófica pela capitanias de Grão-Pará, Rio Negro, Mato Grosso e Cuiabá. Conselho Federal de Cultura, Rio de Janeiro.

Foster, M. 1992. Tent roosts of Macconnell's bat (*Vampyressa macconnelli*). Biotropica 24:447–454.

Frailey, C., E. Lavina, A. Rancy, & J. Pereira. 1988. A proposed Pleistocene/Holocene lake in the Amazon basin and its significance to Amazonian geology and biogeography. Acta Amazônica 18:119–143.

Furch, K. 1984. Water chemistry of the Amazon basin: The distribution of chemical elements among freshwaters. In H. Sioli, ed. The Amazon: Limnology and the Landscape Ecology of a Mighty Tropical River and Its Basin. Junk, Dordecht, 167–199.

Galeano, G. 1991. Las palmas de la región de Araracuara. Tropenbos-Colombia, Bogotá.

Galeano, G., & R. Bernal. 1989. La identidad de *Scheelea insignis* (Palmae). Caldasia 16:10–13.

García, M. 1988. Observaciones de polinización en *Jessenia bataua* (Arecaceae). Departmento de Ciencias Biológicas, Pontificia Universidad Católica del Ecuador, Quito.

Gentry, A. 1982. Patterns of neotropical plant species diversity. Evol. Biol. 15:1–84.

———. 1986. Species richness and floristic composition of Choco region plant communities. Caldasia 15:71–84.

———. 1988. Tree species richness of upper Amazonian forests. Proc. Natl. Acad. Sci. USA 85:156–159.

Gheerbrant, A. 1992. The Amazon: Past, Present and Future. Abrams, New York.

Glanz, W., R. Thorington, J. Giacalone-Madden, & L. Heaney. 1983. Seasonal food use and demographic trends in *Sciurus*. In E. Leigh, A. Rand, & D. Windson, eds. The Ecology of a Tropical Forest. Smithsonian Institution Press, Washington, D.C., 239–252.

Glassman, S. 1965. Geographic distribution of New World palms. Principes 9:132–134.

———. 1972. A Revision of B. E. Dahlgren's Index of American Palms. Cramer, Lehre, Germany.

———. 1977a. Preliminary taxonomic studies in the palm genus *Orbignya* Mart. Phytologia 36:89–115.

———. 1977b. Preliminary taxonomic studies in the palm genus *Scheelea* Karsten. Phytologia 37:219–250.

———. 1977c. Preliminary taxonomic studies in the palm genus *Attalea* H.B.K. Fieldiana, Bot. 38:31–61.

———. 1978. Preliminary taxonomic studies in the palm genus *Maximiliana* Mart. Phytologia 38:161–172.

———. 1987. Revisions of the palm genus *Syagrus* Mart. and other selected genera in the *Cocos* alliance. Ill. Biol. Monogr. 56:1–230.

González, V. 1987. Los morichales de los llanos orientales. Ediciones Corpoven, Caracas.

Görts-van Rijn, A. 1991. Flora of the Guianas: Index of Suriname Plant Collectors. Koeltz Scientific Books, Königstein.

Goulding, M. 1989. Amazon: The Flooded Forest. BBC Books, London.

Goulding, M., M. Carvalho, & E. Ferreira. 1988. Rio Negro: Rich Life in Poor Water. SPB Academic, The Hague.

Grabert, H. 1983. The Amazon shearing system. Tectonophysics 95:329–336.

Greenberg, J., C. Turner, & S. Zegura. 1986. The settlement of the Americas: A comparison of the linguistic, dental, and genetic evidence. Curr. Anthropol. 27:477–488.

Haffer, J. 1969. Speciation in Amazonian forest birds. Science 165:131–137.

———. 1987a. Quaternary history of tropical America. In T. Whitmore & G. Prance, eds. Biogeography and Quaternary History in Tropical Amercia. Clarendon Press, Oxford, 1–18.

———. 1987b. Biogeography of neotropical birds. In T. Whitmore & G. Prance, eds. Biogeography and Quaternary History in Tropical Amercia. Clarendon Press, Oxford, 105–150.

Hahn, W. 1991. Notes on the genus *Acanthococos*. Principes 35:167–171.

Hallé, F. 1977. The longest leaf in palms? Principes 21:18.

Heinen, H., & K. Ruddle. 1974. Ecology, ritual, and economic organization in the distribution of palm starch among the Warao of the Orinoco Delta. J. Anthropol. Res. 30:116–138.

Hemming, J. 1987. Amazon Frontier: The Defeat of the Brazilian Indians. Macmillan, London.

Henderson, A. 1985. Pollination of *Socratea exorrhiza* and *Iriartea ventricosa*. Principes 29:64–71.

———. 1986a. A review of pollination studies in the Palmae. Bot. Rev. 52:221–259.

———. 1986b. *Barcella odora*. Principes 30:74–76.

———. 1990. Arecaceae. Part 1: Introduction and Iriarteinae. Flora Neotrop. Monogr. 53:1–100.

Henderson, A., & M. Balick. 1987. Notes on the palms of Amazônia Legal. Principes 31:116–122.

Henderson, A., & M. Balick. 1991. *Attalea crassispatha*, a rare and endemic Haitian palm. Brittonia 43:189–194.

Henderson, A., & F. Chávez. 1993. *Desmoncus* as a useful palm in the western Amazon basin. Principes 37:184–186.

Henderson, A., S. Churchill, & J. Luteyn. 1991. Neotropical plant diversity. Nature 351:21–22.

Henderson, A., & J. Steyermark. 1986. New

Palms from Venezuela. Brittonia 38:309–313.
Hiepko, P. 1987. The collections of the Botanical Museum Berlin- Dahlem (B) and their history. Englera 7:219–249.
Hodel, D. 1992. Chamaedorea Palms. International Palm Society and Allen Press, Lawrence, Kan.
Holmgren, P., N. Holmgren, & L. Barnett. 1990. Index Herbariorum. 8th ed. New York Botanical Garden, New York.
Hoorn, C. 1990. Evolución de los ambientes sedimentarios durante el Terciario y el Cuaternario en la Amazonia colombiana. Colom. Amazonica 4:97–126.
———. 1991. Nota geológica; la formación Pevas ("Terciario Inferior Amazónico"): Depósitos fluvio-lacustres del Mioceno medio a superior. Colomb. Amazonica 5:119–130.
———. In press. Marine incursions and the influence of Andean tectonics on the Miocene depositional history of northwestern Amazonia: Results of a palynostratigraphic study. Palaeogeog., Palaeoclimatol., Palaeoecol.
Hoyos, J., & A. Braun. 1984. Palmas Tropicales. Sociedad y Fundación La Salle de Ciencias Naturales, Caracas.
Huber, J. 1906. La végétation de la vallée du Rio Purus (Amazone). Bull. Herb. Boissier 6:249–276.
Huber, O. 1989. Shrublands of the Venezuelan Guayana. In L. Holm-Nielsen, I. Nielsen, & H. Balslev, eds. Tropical Forests. Academic Press, London, 271– 285.
Huber, O., & C. Alarcon. 1988. Mapa de vegetación de Venezuela. Ministerio del Ambiente y de los Recursos Naturales Renovables, Caracas.
Huber, O., J. Steyermark, G. Prance, & C. Alès. 1984. The vegetation of the Sierra Parima, Venezuela-Brazil: Some results of recent exploration. Brittonia 36:104–139.
Humboldt, A. von. 1850. Aspects of Nature. Translated by Mrs. Sabine. Lea & Blanchard, Philadelphia.
Humboldt, A. von, A. Bonpland, & C. Kunth. 1815. Nova Genera et Species Plantarum. Librarie Grecque-Latine-Allemande, Paris.
Hunter, M., G. Jacobson, & T. Webb. 1988. Paleoecology and the coarse-filter approach to maintaining biological diversity. Cons. Biol. 2:375–385.
IBGE. 1988. Mapa de vegetacão do Brasil. Fundação Instituto Brasiliero de Geografia e Estatística, Rio de Janeiro.
Imchanitzkaya, N. 1987. Palmae (Arecaceae) in Herbario Instituti Botanici Nomine V. L. Komarovii (Liningrad) Conservatae 2. Specimena authentica taxorum novorum a H. Karstenio Descriptorum. Novitates Syst. Pl. Vasc. 24:26–42.
Instituto Geográfico. 1985. Mapa de bosques de Colombia. Ministerio de Hacienda y Credito Público, Bogotá.
Irion, G. 1989. Quaternary geological history of the Amazon lowlands. In L. Holm-Nielsen, I. Nielsen, & H. Balslev, eds. Tropical Forests. Academic Press, London, 23–34.
Jacobson, G., T. Webb, & E. Grimm. 1987. Patterns and rates of vegetation change during the deglaciation of eastern North America. In W. Ruddiman & H. Wright, eds. North America and Adjacent Oceans During the Last Glaciation. Vol. K-3: The Geology of North America. Geological Society of America, Boulder, Colo. 277–288.
Janzen, D. 1971. The fate of *Scheelea rostrata* fruits beneath the parent tree; predispersal attack by bruchids. Principes 15:89–101.
———. 1974. Tropical blackwater rivers, animals, and mast fruiting by the Dipterocarpaceae. Biotropica 6:69–103.
Jordan, C. 1989. An Amazonian Rainforest. UNESCO, Paris.
Junk, W. 1984. Ecology of the *várzea*, floodplain of Amazonian whitewater rivers. In H. Sioli, ed. The Amazon: Limnology and the Landscape Ecology of a Mighty Tropical River and Its Basin. Junk, Dordecht, 215–243.
Junk, W., & K. Furch. 1985. The physical and chemical properties of Amazonian waters and their relationships with the biota. In G. Prance & T. Lovejoy, eds. Amazonia. Pergamon Press, Oxford, 1–17.
Kahn, F. 1986. Les palmiers des forets tropicales humides du bas Tocantins (Amazonie Bresilienne). Rev. Ecol. (Terre Vie) 41:3–13.
Kahn, F., & J.-J. de Granville. 1992. Palms in Forest Ecosystems of Amazonia. Springer-Verlag, Berlin.

Kahn, F., & K. Mejia. 1987. Notes on the biology, ecology, and use of a small Amazonian palm: *Lepidocaryum tessmannii*. Principes 31:14–19.

Kahn, F., & K. Mejia. 1988. A new species of *Chelyocarpus* (Palmae, Coryphoideae) from Peruvian Amazon. Principes 32:69–72.

Kahn, F., & K. Mejia. 1990. Palm communities in wetland forest ecosystems of Peruvian Amazonia. Forest Ecol. Manage. 33:169–179.

Kahn, F., & B. Millán. 1992. *Astrocaryum* (Palmae, Cocoeae, Bactridinae) in Amazonia. A preliminary treatment. Bull. Inst. Fr. Ét. Andines 21:459–531.

Kiltie, R. 1981. Distribution of palm fruits on a rain forest floor: Why white-lipped peccaries forage near objects. Biotropica 13:141–145.

Klammer, G. 1984. The relief of the extra-Andean Amazon basin. In H. Sioli, ed. The Amazon: Limnology and the Landscape Ecology of a Mighty Tropical River and Its Basin. Junk, Dordecht, 47–83.

Kubitzki, K. 1990. The psammophilous flora of northern South America. Mem. N. Y. Bot. Gard. 64:248–253.

———. 1991. Dispersal and distribution in *Leopoldinia* (Palmae). Nord. J. Bot. 11:429–432.

Lawrence, G., A. Buchheim, G. Daniels, & H. Dolezal. 1968. B-P-H. Hunt Botanical Library, Pittsburgh.

Lentz, D. 1990. *Acrocomia mexicana*: Palm of the ancient Mesoamericans. J. Ethnobiol. 10:183–194.

Lescure, J.-P., L. Emperaire, & C. Franciscon. 1992. *Leopoldinia piassaba* Wallace (Arecaceae): A few biological and economic data from the Rio Negro region (Brazil). Forest Ecol. Manage. 55:83–86.

Lévi-Strauss, C. 1955. Tristes Tropiques. Translated by J. Weightman & D. Weightman. Librarie Plon, Paris.

Linnaeus, C. 1753. Species Plantarum. Impensis Laurentii Salvii, Stockholm.

———. 1781. Supplementum Plantarum. Impensis Orphanotrophei, Braunschweig.

Listabarth, C. 1992. A survey of pollination strategies in the Bactridinae (Palmae). Bull. Inst. Fr. Ét. andines 21:699–714.

———. 1993a. Insect-induced wind pollination of the palm *Chamaedorea pinnatifrons* and pollination in the related *Wendlandiella* sp. Biodivers. Conserv. 2:39–50.

———. 1993b. Pollination in *Geonoma macrostachys* and three congeners, *G. acaulis, G. gracilis,* and *G. interrupta*. Bot. Acta 106:455–514.

Liu, K., & P. Colinvaux. 1985. Forest changes in the Amazon basin during the last glacial maximum. Nature 318:556–557.

Macbride, J. F. 1960. Palmae. Flora of Peru. Field Mus. Nat. Hist., Bot. ser. 13:321–418.

Mahar, D. 1989. Government Policies and Deforestation in Brazil's Amazon Region. World Bank, Washington, D.C.

Malcolm, J. 1990. Estimation of mammalian densities in continuous forest north of Manaus. In A. Gentry, ed. Four Neotropical Forests. Yale University Press, New Haven, 339–357.

Marshall, E. 1990. Clovis counterrevolution. Science 249:738–741.

Martius, C. 1823–1837. Historia Naturalis Palmarum. Vol. 2: Genera et Species. Weigel, Leipzig.

———. 1831–1853. Historia Naturalis Palmarum. Vol. 1: De Palmas Generatim. Weigel, Leipzig.

———. 1842–1847. Palmetum Orbignianum. Vol. 7, part 3 of A. d'Orbigny. Voyage dans l'Amérique méridionale. Bertrand, Paris.

Mayr, E. 1982. The Growth of Biological Thought. Harvard University Press, Cambridge, Mass.

McCourt, W., J. Aspen, & M. Brook. 1984. New geological and geochronological data from the Colombian Andes: Continental growth by multiple accretion. J. Geol. Soc. Lond. 141:831–845.

McCurrach, J. 1960. Palms of the World. Harper & Brothers, New York.

Meggers, B. 1971. Amazonia, Man and Culture in a Counterfeit Paradise. Aldine Atherton, Chicago.

Meggers, B., & C. Evans. 1957. Archaeological investigations at the mouth of the Amazon. Bull. Bur. Amer. Ethnol. 167:1–664.

Merrill, E. 1943. Destruction of the Berlin herbarium. Science 98:490–491.

Molion, L. 1987. On the dynamic climatology of the Amazon basin and associated rain-producing mechanisms. In R. Dickson, ed.

The Geophysiology of Amazonia. Wiley, New York, 391–407.

Moore, H. 1969. The Geonomoid Palms. Taxon 18:230–232.

———. 1972. *Chelyocarpus* and its allies *Cryosophila* and *Itaya* (Palmae). Principes 16:67–88.

———. 1973. The major groups of palms and their distribution. Gentes Herb. 11:27–141.

———. 1977. Endangerment at the specific and generic levels in palms. In G. Prance & T. Elias, eds. Extinction Is Forever. New York Botanical Garden, New York, 267–282.

Moore, H., & J. Dransfield. 1978. A new species of *Wettinia* and notes on the genus. Notes Roy. Bot. Gard. Edinburgh 36:259–267.

Moraes, M., & J. Sarmiento. 1992. Contribución al estudio de biología reproductiva de una especie de *Bactris* (Palmae) en el bosque de galería (Depto. Beni, Bolivia). Bull. Inst. Fr. Ét. andines 21:685–698.

Mora Urpí, J., & E. Solís. 1980. Polinización en *Bactris gasipaes* H.B.K. (Palmae). Rev. Biol. Trop. 28:153–174.

Mori, S. 1991. The Guayana lowland floristic province. C. R. Soc. Biogéog. 67:67–75.

Mori, S., & P. Becker. 1991. Flooding affects survival of Lecythidaceae in *terra firme* forest near Manaus. Biotropica 23:87–90.

Mori, S., & F. Ferreira. 1987. A distinguished Brazilian botanist, João Barbosa Rodrigues (1842–1909). Brittonia 39:73–85.

Moses, T. 1962. Palms of Brazil. Principes 6:26–37.

Munn, C., J. Thomsen, & C. Yamashita. 1989–1990. The hyacinth macaw. Audubon Wildlife Report 1989–1990:405–419.

Nelson, B. In press. Natural forest disturbance and change in the Brazilian Amazon. Remote Sensing Reviews.

Nelson, B., C. Ferreira, M. da Silva, & M. Kawasaki. 1990. Endemism centres, refugia and botanical collection density in Brazilian Amazon. Nature 345:714–716.

Niklas, K. 1992. Plant Biomechanics. University of Chicago Press, Chicago.

Nixon, K., & Q. Wheeler. 1990. An amplification of the phylogenetic species concept. Cladistics 6:211–223.

Noblick, L. 1991. The Indigenous Palms of the State of Bahia, Brazil. Ph.D. diss., University of Illinois, Chicago.

Nobre, C. 1984. The Amazon and climate. In Proceedings of the Climate Conference for Latin America and the Caribbean, Paipa, Colombia. World Meteorological Association, Geneva, 409–416.

Oldeman, R. 1969. Étude biologique des pinotières de la Guyane française. Cah. ORSTOM, ser. Biol. 10:3–18.

Olesen, J., & H. Balslev. 1990. Flower biology and pollinators of the Amazonian monoecious palm, *Geonoma macrostachys:* A case of Bakerian mimicry. Principes 34:181–190.

ONERN (Oficina Nacional de Evaluación de Recursos Naturales). 1976. Inventario, evaluación e integración de los recursos naturales de la zona Iquitos, Nauta, Requena y Colonia Angamos. ONERN, Lima.

Otedoh, M. 1977. The African origin of *Raphia taedigera*—Palmae. Nigerian Field 42:11–16.

———. 1982. A revision of the genus *Raphia* Beauv. (Palmae). J. Nigerian Inst. Oil Palm Res. 6:145–189.

Padoch, C. 1988. Aguaje (*Mauritia flexuosa* L.f.) in the economy of Iquitos, Peru. Adv. Econ. Bot. 6:214–224.

Papavero, N. 1971. Essays on the History of Neotropical Dipterology, with Special Reference to Collectors (1750–1905). Museu de Zoologia, Universidade de São Paulo.

Peres, C. In press. Patterns of abundance and fruiting phenology of arborescent palms in a central-Amazonian *terra firme* forest. Principes.

Peters, C., M. Balick, F. Kahn, & A. Anderson. 1989. Oligarchic forests of economic plants in Amazonia: Utilization and conservation of an important tropical resource. Conserv. Biol. 3:341–349.

Phillips, O. 1993. The potential for harvesting fruits in tropical rainforests: New data from Amazonian Peru. Biodivers. Conserv. 2:18–38.

Piedade, M. 1985. Ecologia e biologia reprodutiva de *Astrocaryum jauari* Mart. (Palmae) como exemplo de populacão adaptada às áreas inundáveis do Rio Negro (igapós). M.S. thesis, Universidade do Amazonas, Manaus, Brazil.

Pinard, M. 1993. Impacts of stem harvesting on populations of *Iriartea deltoidea* (Palmae) in an extractive reserve in Acre, Brazil. Biotropica 25:2–14.

Pinard, M., & F. Putz. 1992. Population matrix models and palm resource management. Bull. Inst. Fr. Ét. Andines 21:637–649.

Pires, J., & G. Prance. 1985. The vegetation types of the Brazilian Amazon. In G. Prance & T. Lovejoy, eds. Amazonia. Pergamon Press, Oxford, 109–145.

Pitman, W., S. Cande, J. LaBrecque, & J. Pindell. 1993. Fragmentation of Gondwana: The separation of Africa from South America. In P. Goldblatt, ed. Biological Relationships Between Africa and South America. Yale University Press. New Haven, 15–34.

Poiteau, P. 1822. Histoire des palmiers de la Guiane française. Mém. Mus. Hist. Nat. 9:385–392.

Potztal, E. 1965. Maximilian Burret, 1883–1964. Willdenowia 4:23–31.

Pouzat, J., H. Bilal, D. Nammour, & M. Pimbert. 1989. A comparative study of the host plant's influence on the sex pheromone dynamics of three bruchid species. Acta Oecologica 10:401–410.

Prance, G. 1971. An index of plant collectors in Brazilian Amazonia. Acta Amazônica 1:25–68.

———. 1977. The phytogeographic subdivisions of Amazonia and their influence on the selection of biological reserves. In G. Prance & T. Elias, eds. Extinction Is Forever. New York Botanical Garden, New York, 195–213.

———. 1982. A review of the phytogeographic evidences for Pleistocene climate changes in the Neotropics. Ann. Missouri Bot. Gard. 69:594–624.

———. 1987. Vegetation. In T. Whitmore, & G. Prance, eds. Biogeography and Quaternary History in Tropical America. Clarendon Press, Oxford, 28–45.

Prance, G., B. Nelson, M. da Silva, & D. Daly. 1984. Projeto Flora Amazônica. Acta Amazônica, suppl. 14:5–29.

Prance, G., & H. Schubart. 1978. Notes on the vegetation of Amazonia. I. A preliminary note of the origin of the open white sand campinas of the lower Rio Negro. Brittonia 30:60–63.

Puhakka, M., R. Kalliola, M. Rajasilta, & J. Salo. 1992. River types, site evolution and successional vegetation patterns in Peruvian Amazon. J. Biogeogr. 19:651–665.

Putz, F. 1979. Biology and human use of *Leopoldinia piassaba*. Principes 23:149–156.

Putzer, H. 1984. The geological evolution of the Amazon basin and its mineral resources. In H. Sioli, ed. The Amazon: Limnology and the Landscape Ecology of a Mighty Tropical River and Its Basin. Junk, Dordecht, 15–46.

Räsänen, M., J. Salo, & R. Kalliola. 1987. Fluvial perturbance in the western Amazon basin: Regulation by long-term sub-Andean tectonics. Science 238:1398–1401.

Raven, P. 1979. Plate tectonics and southern hemisphere biogeography. In K. Larsen & L. Holm-Nielsen, eds. Tropical Botany. Academic Press, London, 3–24.

Read, R. 1979. Palms of the Lesser Antilles. Smithsonian Institution Press, Washington, D.C.

Reichel-Dolmatoff, G. 1989. Biological and social aspects of the Yuruparí complex of the Colombian Vaupés territory. J. Lat. Amer. Lore 15:95–135.

Rioja, G. 1992. The jatata project: The pilot experience of Chimane empowerment. In M. Plotkin & L. Famolare, eds. Sustainable Harvest and Marketing of Rain Forest Products. Island Press, Washington, D.C., 192–196.

Roosevelt, A. 1991. Moundbuilders of the Amazon. Academic Press, San Diego.

Roosevelt, A., R. Housley, M. Imazio, S. Maranca, & R. Johnson. 1991. Eighth millennium pottery from a prehistoric shell midden in the Brazilian Amazon. Science 254:1621–1624.

Roubik, D. 1989. Ecology and Natural History of Tropical Bees. Cambridge University Press, Cambridge.

Salati, E. 1985. The climatology and hydrology of Amazonia. In G. Prance & T. Lovejoy, eds. Amazonia. Pergamon Press, Oxford, 18–48.

Salati, E., & J. Marques. 1984. Climatology of the Amazon region. In H. Sioli, ed. The Amazon: Limnology and the Landscape Ecology of a Mighty Tropical River and Its Basin. Junk, Dordecht, 85–126.

Salo, J. 1987. Pleistocene forest refuges in the Amazon: Evaluation of the biostratigraphical, lithostratigraphical and geomorphological data. Ann. Zool. Fennici 24:203–211.

Salo, J., R. Kalliola, I. Häkkinen, Y. Mäkinen, P. Niemelä, M. Puhakka, & P. Coley. 1986. River dynamics and the diversity of Amazon lowland forest. Nature 322:254–258.

Sanders, R. 1991. Cladistics of *Bactris* (Palmae): Survey of characters and refutation of Burret's classification. Selbyana 12:105–133.

Sanford, R., J. Saldarriaga, K. Clark, C. Uhl, & R. Herrera. 1985. Amazon rain-forest fires. Science 227:53–55.

Scariot, A., E. Lleras, & J. Hay. 1991. Reproductive biology of the palm *Acrocomia aculeata* in central Brazil. Biotropica 23:12–22.

Scariot, A., A. Oliveira, & E. Lleras. 1989. Species richness, density and distribution of palms in an eastern Amazonian seasonally flooded forest. Principes 33:172–179.

Schultes, R. 1990. Taxonomic, nomenclatural and ethnobotanic notes on *Elaeis*. Elaeis 2:172–187.

Silberbauer-Gottsberger, I. 1990. Pollination and evolution in palms. Phyton 30:213–233.

Singer, R., & I. Aguiar. 1986. Litter decomposing and ectomycorrhizal Basidiomycetes in an *igapó* forest. Plant Syst. Evol. 153:107–117.

Sioli, H. 1984. The Amazon and its main affluents: Hydrography, morphology of the river courses, and river types. In H. Sioli, ed. The Amazon: Limnology and the Landscape Ecology of a Mighty Tropical River and Its Basin. Junk, Dordrecht, 127–165.

Sist, P. 1987. Régénération, dynamique des populations et dissémination d'un palmier de Guyane française: *Jessenia bataua* (Mart.) Burret subsp. *oligocarpa* (Griseb. & H. Wendl.) Balick. Bull. Mus. natn. Hist. Nat., Paris 9:317–336.

———. 1989a. Peuplement et phénologie des palmiers en forêt guyanaise (Piste de Saint Elie). Rev. Ecol. (Terre Vie) 4:113–151.

———. 1989b. Demography of *Astrocaryum sciophilum,* an understory palm of French Guiana. Principes 33:142–151.

Skole, D., & C. Tucker. 1993. Tropical deforestation and habitat fragmentation in the Amazon: satellite data from 1978 to 1988. Science 260:1905–1910.

Skov, F. 1989. Hypertaxonomy—a new tool for revisional work and a revision of *Geonoma* (Palmae) in Ecuador. Ph.D. diss., Aarhus University, Denmark.

Skov, F., & H. Balslev. 1989. A revision of *Hyospathe* (Arecaceae). Nord. J. Bot. 9:189–202.

Smith, N. 1980. Anthrosols and human carrying capacity in Amazonia. Ann. Assoc. Amer. Geog. 70:553–566.

Smythe, N. 1989. Seed survival in the palm *Astrocaryum standleyanum:* Evidence for dependence upon its seed dispersers. Biotropica 21:50–56.

Sombroek, W. 1966. Amazon Soils. Center for Agricultural Publishing and Documentation, Wageningen, Netherlands.

———. 1984. Soils of the Amazon region. In H. Sioli, ed. The Amazon: Limnology and the Landscape Ecology of a Mighty Tropical River and Its Basin. Junk, Dordecht, 521–535.

Spix, J., & C. Martius. 1831. Reise in Brasilien in den Jahren 1817 bis 1820. Vol. 3. J. Spix, Munich.

Spruce, R. 1860. On *Leopoldinia piassaba,* Wallace. J. Proc. Linn. Soc., Bot. 4:58–63.

———. 1871. Palmae Amazonicae. J. Proc. Linn. Soc., Bot. 11:65–183.

———. 1908. Notes of a Botanist on the Amazon and Andes. 2 vols. Macmillan, London.

Stafleu, F., & R. Cowan. 1976. Taxonomic Literature. Vol. 1: A–G. Bohn, Scheltema & Holkema, Utrecht.

———. 1979. Taxonomic Literature. Vol. 2: H–Le. Bohn, Scheltema & Holkema, Utrecht.

———. 1981. Taxonomic Literature. Vol. 3: Lh–O. Bohn, Scheltema & Holkema, Utrecht.

———. 1983. Taxonomic Literature. Vol. 4: P–Sak. Bohn, Scheltema & Holkema, Utrecht.

———. 1985. Taxonomic Literature. Vol. 5: Sal–Ste. Bohn, Scheltema & Holkema, Utrecht.

———. 1986. Taxonomic Literature. Vol. 6: Sti–Vuy. Bohn, Scheltema & Holkema, Utrecht.

———. 1988. Taxonomic Literature. Vol. 7: W–Z. Bohn, Scheltema & Holkema, Utrecht.

Steward, J. 1946–1959. Handbook of South American Indians. 7 vols. Smithsonian Institution Press, Washington, D.C.

Steyermark, J. 1986. Speciation and endemism in the flora of the Venezuelan tepuis. In F. Vuilleumier & M. Monasterio, eds. High Altitude Tropical Biogeography. Oxford University Press, New York, 317–373.

St. John, T., & A. Anderson. 1982. A reexamination of plant phenolics as a source of tropical black water rivers. Trop. Ecol. 23:151–154.

Strudwick, J., & G. Sobel. 1988. Uses of *Euterpe oleracea* Mart. in the Amazon estuary, Brazil. Adv. Econ. Bot. 6:225–253.

Taylor, D. 1991. Paleobiogeographic relationships of Andean angiosperms of Cretaceous to Pliocene age. Palaeogeog. Palaeoclimatol., Palaeoecol. 88:69–84.

Terborgh, J. 1983. Five New World Monkeys. Princeton University Press, Princeton.

———. 1986. Keystone plant resources in the tropical forest. In M. Soulé, ed. Conservation Biology. Sinauer, Sunderland, Mass., 330–344.

———. 1992. Diversity and the Tropical Rain Forest. Scientific American Library, New York.

Timm, R. 1987. Tent construction by bats of the genera *Artibeus* and *Uroderma*. Fieldiana, Zool. 39:187–212.

Tomlinson, P. 1961. Anatomy of Monocotyledons. II. Palmae. Oxford University Press, London.

Trail, J. 1876. Descriptions of new species and varieties of palms collected in the valley of the Amazon in north Brazil, in 1874. J. Bot. (Hooker) 14:323–333, 353–359.

———. 1877a. Descriptions of new species and varieties of palms collected in the valley of the Amazon in north Brazil, in 1874. J. Bot. (Hooker) 15:1–10, 40–49, 75–81.

———. 1877b. Some remarks on the synonymy of palms of the Amazon Valley. J. Bot. (Hooker) 15:129–132.

Tuomisto, H., K. Ruokolainen, & J. Salo. 1992. Lago Amazonas: Fact or fancy? Acta Amazonica 22:353–361.

Turner, J. 1982. How do refuges produce biological diversity? Allopatry and parapatry, extinction and gene flow in mimetic butterflies. In G. Prance, ed. Biological Diversification in the Tropics. Columbia University Press, New York, 309–335.

Uhl, N., & J. Dransfield. 1987. Genera Palmarum. L. H. Bailey Hortorium and Allen Press, Lawrence, Kan.

Urdaneta, H. 1981. Planificación silvicultural de los bosques ricos en *palma manaca* (*Euterpe oleracea*), en el delta del Río Orinoco. Universidad de los Andes, Mérida.

Urrego, L. 1987. Estudio preliminar de la fenología de la canangucha (*Mauritia flexuosa* L.f.). Colom. Amazonica 2:57–81.

van der Hammen, T. 1989. History of the montane forests of the northern Andes. Plant System. Evol. 162:109–114.

———. 1992. Palaeoecological background: Neotropics. In N. Myers, ed. Tropical Forests and Climate. Kluwer, Dordrecht, 37–47.

Vásquez, R., & A. Gentry. 1989. Use and misuse of forest-harvested fruits in the Iquitos area. Cons. Biol. 3:350–361.

Veléz, G. 1992. Estudio fenológico de diecinueve frutales silvestres utilizados por las comunidades indígenas de la región de Araracuara—Amazonia Colombiana. Colom. Amazonica 6:135–186.

Wallace, A. 1853. Palm Trees of the Amazon and Their Uses. Van Hoorst, London.

———. 1889. A Narrative of Travels on the Amazon and Rio Negro. Ward, Lock, London.

Webb, T. 1987. The appearance and disappearance of major vegetational assemblages: Long-term vegetational dynamics in eastern North America. Vegetatio 69:177–187.

Wessels Boer, J. 1965. The Indigenous Palms of Suriname. Brill, Leiden.

———. 1968. The Geonomoid Palms. Verh. Kon. Ned. Akad. Wetensch., Afd. Natuurk., Tweede Reeks 58:1–202.

———. 1971a. Clave para las palmas venezolanas. Act. Bot. Venezuelica 6:299–362.

———. 1971b. *Bactris* x *moorei*, a hybrid in palms. Acta Bot. Neerl. 20:167–172.

———. 1976. Fa Joe Kan Tak' Mi No Moi: Suriname Wandelflora. 2 vols. Stinasu, Paramaribo, Surinam.

———. 1988. Palmas indígenas de Venezuela. Pittiera 17:1–332.

Whitmore, T., & G. Prance. 1987. Biogeography and Quaternary History in Tropical America. Clarendon Press, Oxford.

Wilbert, J. 1976. *Manicaria saccifera* and its cultural significance among the Warao Indians of Venezuela. Bot. Mus. Leafl. 24:275–335.

Wilson, D., & D. Janzen. 1972. Predation on *Scheelea* palm seeds by bruchid beetles: Seed density and distance from the parent palm. Ecology 53:954–959.

Worbes, M. 1985. Structural and other adaptations to long-term flooding by trees in Central Amazonia. Amazoniana 9:459–484.

Worbes, M., & W. Junk. 1989. Dating tropical trees by means of ^{14}C from bomb tests. Ecology 70:503–507.

Worbes, M., H. Klinge, J. Revilla, & C. Martius. 1992. On the dynamics, floristic subdivision and geographical distribution of várzea forests in central Amazonia. J. Veg. Sci. 3:553–564.

Wright, S. 1983. The dispersion of eggs by a bruchid beetle among *Scheelea* palm seeds and the effect of distance to the parent palm. Ecology 64:1016–1021.

Yungjohann, J. 1989. White Gold: The Diary of a Rubber Tapper in the Amazon, 1906–1916. Synergetic Press, Oracle, Ariz.

Zona, S., & A. Henderson. 1989. A review of animal-mediated seed dispersal of palms. Selbyana 11:6–21.

Subject Index

A

Acre Basin, 6
acrisols, 8
agoutis, 40
Amazon Basin, 4
Amazon region, definition of, 3–4
Amazon savannas, 14
Amazonas Basin, 6
Amazônia Legal, 3
Amazonia, Man and Culture in a Counterfeit Paradise, (Meggers), 18
Andes, 4, 7, 33
anthropogenic palm forests, 18
Arecaceae, 21
arenosols, 9
Aublet, Jean, 46

B

Bailey, Liberty, 51–52, 53, 54
bana, 15
Barbosa Rodrigues, João, 48, 50, 53, 54; palm regions described by, 17
Barreiras/Alter do Chão Formation, 6
Belterra clays, 6
Berlin herbarium, 51
Bonpland Aimé, 45,46
Brazilian Shield, 4, 33
bruchid beetles, 40
Burret Maximilian, 51, 53, 54

C

caatinga, 15
cambisols, 8
campina, 15
cerrado, 14
Chapada dos Parecís, 6
charcoal, 9
climate, 9, 29, 31; short-term change in, 30
Clovis site, 17
Cuatrecasas, José, 52

D

deforestation, 22, 23
degree square, 22
demography, 44
dispersal, 39
diversity, 22–23
d'Orbigny Alcide, 47
Drude, O., palm regions described by, 17
Dugand, Armando, 52

E

ecological biogeography, 29
ectomycorrhizal fungi, 9
El Niño, 10
elevational range, 30
Enumeratio Palmarum Novarum (Barbosa Rodrigues), 50
environmental determinism, 18

F

ferralsols, 7
Ferreira, Alexandre, 45n
Fitzcarrald Arch, 6
Flora Brasiliensis (Martius), 47
Flora of Suriname (Wessels Boer), 52

flowering plants, diversity of, 22
fluvisols, 8
friagens, 9
friajes, 9

G

gallery forests, 15
Genera Palmarum (Moore), 52
Gentry, Alwyn, 52
geologic history, 4–7
gleyic solonchaks, 9
gleysols, 8
Gondwanaland, breakup of, 4
Guayana Highland, 13
Guayana Shield, 4, 13, 32
Gurupá Arch, 5, 6

H

Handbook of South American Indians (Steward), 18
Historia Naturalis Palmarum (Martius), 47
historical biogeography, 32–34
Holocene, 32
Hopp, Werner, 51
Huber, Jacques 51
Huebner, E.,
humans: cultural development of, in Amazon, 18; occupation of Americas by, 17–19
Humboldt Alexander von, 3, 45, 46, 47
hylaea, 3

I

igapó, 8
Indian black-earth soils, 9
Intertropical Convergence Zone, 10
inundated forests, 15
inundated savanna, 14
Iquitos Arch, 5, 6

K

keystone species, 41
Killip, Ellsworth, 51
Köppen system, 10
Krukoff, Boris, 52
Kunth, Carl 46

L

Lago Amazonas, 6
Lakó, Karl, 51
lingua geral, 24
Linnaeus, Carl (father) 45, 52
Linnaeus, Carl (son), 45
Lisboa, Cristovão de, 46–47
llanos, 14
local names, 24–25
lowland rain forest, 15
Luetzelburg, Philip von, 51

M

macaws, 41
Madre de Dios Basin, 6
Maguire, Bassett, 52
Manu, 40
Marajó Basin, 6
Martius, Carl von, 46, 47, 48
Mexia, Ynes, 51
monkeys, 39–41
Moore, Harold, 51, 52

N

nitosols, 9

P

Pacific accretions, 7
Palm Trees of the Amazon and Their Uses (Wallace), 48
Palmae, 21
Palmae Amazonicae (Spruce), 48

palms: abundance of, 22; botanists, 45–52; collecting density of, 22; collectors of, 45; conservation status of, 24; diversity of, 22; endemism of, 21; fruits of, 25; number of species of, 21; starch in, 25; stems of, 25; uses of, 24, 25–26; variation of, 53
Pastaza–Marañon Basin, 6
Pavón, José, 46
Pebas Formation, 6
peccaries, 41
Pedra Furada site, 17
phenology, 37
physical environment, 37
phytogeographic regions, 14, 16–17
pneumatophores, 16
podzols, 8
Poeppig, Eduard, 47
Poiteau, Pierre, 46
pollen cores, 31
pollination, 38
population of Americas, 18; pre-Columbian, 18
Potztal, Eva, 51
Prance, Ghillean, 52; phyteographic regions of, 17
predation, 39
Pristine Myth, 18
Projeto Flora Amazônica, 52
Protesto Appendice (Barbosa Rodrigues), 50
Purus Arch, 5, 6

Q

Quaternary, 30

R

rainfall: between-year variation in, 9, 10; seasonal distribution of, 9, 10; total annual amount of, 9

Ralegh, Sir Walter 45*n*
Refuge Theory, 30; and palms, 32; botanical evidence for, 3.
reproductive biology, 38
restingas, 15
Rio Negro, 11
Rio Solimões, 12
rivers: annual rise and fall of, 12; as boundaries, 17; blackwater, 12, 13; channel shifting in, 31; clear-water, 12, 13; meandering, 11; systems of, 11–13, 30; whitewater, 12
rodents, 39
Roraima Formation, 5
Ruíz, Hipólito, 46

S

Schultes, Richard, 52
Serra do Cachimbo, 6, 14
Serra do Divisor, 6
Serra do Roncador, 6
Serra dos Carajás, 6, 14
Sertum Palmarum Brasiliensium (Barbosa Rodrigues), 50

Smith, Albert, 51
Soils, 7–9
Solimões Basin, 6
Solimões Formation, 6
species: complexes, 57; concepts, 52–54; number of, 21
Species Plantarum (Linnaeus), 45
Spix, 47
Spruce, Richard, 48, 49; palm regions described by, 16–17
squirrels, 40
Steyermark, Julian, 52
stone lines, 7
surazos, 9

T

tannins, 13
temperature, 9
tepuis, 13
terra firme, 9
terra preta do índio, 9
Tessmann, Günther, 51
topography, 11, 13, 30
Trail, James, 48, 50
transition forests, 15

U

Ucayali Basin, 6
Ule, Ernst, 51

V

várzea, 8; diversity of, 16
Vegetation, types of, 14–16

W

wallaba, 15
Wallace, Alfred Russel, 48, 49, 53
Wessels Boer, Jan Gerard, 52, 53
white sand, 15
World Soil Classification, 7

Y

Yurupari myth, 26

Index of Scientific Names

Accepted generic and specific names, and their principal page reference, are in bold face.

A

Acanthococos, 161
Acanthococos emensis Toledo, 161
Aacanthococos emensis var. *pubifolia* Toledo, 161
Acanthococos hassleri Barb. Rodr., 161
Acanthococos sericea Burret, 161
Acanthorrhiza chuco (Mart.) Drude, 63
Acrocomia Mart., **160**
Acrocomia aculeata (Jacq.) Lodd., 14, 16, 18, **161**
Acrocomia eriocantha Barb. Rodr., 161
Acrocomia lasiospatha Mart., 161
Acrocomia lasiospatha Wallace, 161
Acrocomia mexicana, 162
Acrocomia microcarpa Barb. Rodr., 161
Acrocomia sclerocarpa Mart, 161
Acrocomia sclerocarpa var. *wallaceana* Drude, 161
Acrocomia wallaceana Becc., 161
Aiphanes Willd., **162**
Aiphanes aculeata, 135
Aiphanes aculeata Willd., 30, 39, **163**
Aiphanes caryotifolia (Kunth) H. Wendl., 163
Aiphanes deltoidea Burret, **165**
Aiphanes eggersii, 135, 165
Aiphanes ernesti (Burret) Burret, 163
Aiphanes horrida (Jacq.) Burret, 163
Aiphanes orinocensis Burret, 163
Aiphanes praemorsa (Poepp.) Burret, 163
Aiphanes schultzeana Burret, 165
Aiphanes tessmannii Burret, 167
Aiphanes ulei (Dammer) Burret, **165**
Aiphanes weberbaueri Burret, **167**
Alfonsia oleifera Kunth, 160
Allagoptera, 14
Allagoptera leucocalyx (Drude) Kuntze, 14, 57
Ammandra O. F. Cook, **294**
Ammandra dasyneura (Burret) Barfod, **294**
Ammandra natalia Balslev & Henderson, 295
Amylocarpus acanthocnemis (Mart.) Barb. Rodr., 214
Amylocarpus angustifolius Huber, 216
Amylocarpus arenarius (Barb. Rodr.) Barb. Rodr., 216
Amylocarpus cuspidatus (Mart.) Barb. Rodr., 186
Amylocarpus ericetinus (Barb. Rodr.) Barb. Rodr., 195
Amylocarpus floccosus (Spruce) Barb. Rodr., 186
Amylocarpus formosus (Barb. Rodr.) Barb. Rodr., 196
Amylocarpus geonomoides (Drude) Barb. Rodr., 195
Amylocarpus hirtus (Mart.) Barb. Rodr., 193
Amylocarpus hylophilus (Spruce) Barb. Rodr., 195

Amylocarpus hylophilus var. *glabrescens* (Drude) Barb. Rodr., 196
Amylocarpus hylophilus var. *macrocarpus* (Drude) Barb. Rodr., 196
Amylocarpus hylophilus var. *nanus* (Drude) Barb. Rodr., 196
Amylocarpus inermis (Trail) Barb. Rodr., 216
Amylocarpus linearifolius (Barb. Rodr.) Barb. Rodr., 195
Amylocarpus luetzelburgii (Burret) Burret, 216
Amylocarpus marajay (Barb. Rodr.) Barb. Rodr., 186
Amylocarpus microspathus (Barb. Rodr.) Barb. Rodr., 216
Amylocarpus mitis (Mart.) Barb. Rodr., 186
Amylocarpus pectinatus (Mart.) Barb. Rodr., 195
Amylocarpus platispinus Barb. Rodr., 196
Amylocarpus pulcher (Trail) Barb. Rodr., 196
Amylocarpus settipinnatus (Barb. Rodr.) Barb. Rodr., 195
Amylocarpus simplicifrons (Mart.) Barb. Rodr., 214
Amylocarpus simplicifrons var. *subpinnata* (Trail) Barb. Rodr., 216
Amylocarpus syagroides (Barb. Rodr. & Trail) Barb. Rodr., 218
Amylocarpus tenuissimus (Barb. Rodr.) Barb. Rodr., 216
Amylocarpus xanthocarpus (Barb. Rodr.) Barb. Rodr., 216
Aphandra Barfod, **294**
Aphandra natalia (Balslev & Henderson) Barfod, **295**
Arecastrum, 128
Areceae, 100, 103, 105
Arecoideae, 88, 100, 103, 105, 128, 136, 157, 160, 251
Arikuryroba, 128

Artibeus, 25
Asterogyne H. Wendl. & Hook. f., **253**
Asterogyne guianensis Granv. & Henderson, 32, **253**
Astrocaryum G. Mey., **234**
Astrocaryum acanthopodium Barb. Rodr., 247
Astrocaryum acaule Mart., **236**
Astrocaryum aculeatissimum (Schott) Burret, 234
Astrocaryum aculeatum G. Mey., **238**
Astrocaryum aculeatum of Barb. Rodr., 247
Astrocaryum aculeatum of Wallace, 175
Astrocaryum alatum Loomis, 235
Astrocaryum awarra de Vriese, 250
Astrocaryum campestre Mart., 57, 234
Astrocaryum carnosum Kahn & Millán, 245
Astrocaryum caudescens Barb. Rodr., 238
Astrocaryum chambira Burret, 25, **238**
Astrocaryum chonta Mart., 246
Astrocaryum ciliatum Kahn & Millán, 245
Astrocaryum confertum H. Wendl., 234
Astrocaryum cuatrecasanum Dugand, 246
Astrocaryum farinosum Barb. Rodr., 249
Astrocaryum ferrugineum Kahn & Millán, 245
Astrocaryum giganteum Barb. Rodr., 236
Astrocaryum gratum Kahn & Millán, 246
Astrocaryum guara Burret, 242
Astrocaryum guianense Splitg., 250
Astrocaryum gymnopus Burret, 239
Astrocaryum gynacanthum Mart., 22, **239**

Astrocaryum gynacanthum var. *dasychaetum* Burret, 239
Astrocaryum gynacanthum var. *munbaca* Trail, 239
Astrocaryum horridum Barb. Rodr., 246
Astrocaryum huami Mart., **241**
Astrocaryum huebneri Burret, 236
Astrocaryum huicungo Dammer, 245
Astrocaryum humile Wallace, 173
Astrocaryum jauari Mart., 13, 16, **242**
Astrocaryum javarense (Trail) Drude, 246
Astrocaryum leiospatha Barb. Rodr., 242
Astrocaryum luetzelburgii Burret, 236
Astrocaryum macrocalyx Burret, 246
Astrocaryum macrocarpum Huber, 238
Astrocaryum malybo H. Karst., 234
Astrocaryum manaoense Barb. Rodr., 238
Astrocaryum mexicanum Liebm., 234
Astrocaryum minus Trail, 239
Astrocaryum minus var. *terrafirme* Drude, 239
Astrocaryum munbaca Mart., 25, 239
Astrocaryum murumuru Mart., 8, 33, 39, 40, 41, **242**
Astrocaryum murumuru var. **ciliatum** (Kahn & Millán) Henderson, **245**
Astrocaryum murumuru var. **ferrugineum** (Kahn & Millán) Henderson, **245**
Astrocaryum murumuru var. **huicungo** (Dammer) Henderson, **245**
Astrocaryum murumuru var. **javarense** (Trail) Henderson, **246**
Astrocaryum murumuru var.

macrocalyx (Burret) Henderson, **246**
Astrocaryum murumuru var. **murumuru**, **246**
Astrocaryum murumuru var. **perangustatum** (Kahn & Millán) Henderson, **246**
Astrocaryum murumuru var. **urostachys** (Burret) Henderson, **247**
Astrocaryum paramaca Mart., 25, **247**
Astrocaryum paramaca var. *javarense* Trail, 246
Astrocaryum paramaca var. *platyacantha* Drude, 247
Astrocaryum perangustatum Kahn & Millán, 247
Astrocaryum plicatum Drude, 249
Astrocaryum princeps Barb. Rodr., 238
Astrocaryum princeps var. *aurantiacum* Barb. Rodr., 238
Astrocaryum princeps var. *flavum* Barb. Rodr., 238
Astrocaryum princeps var. *sulphureum* Barb. Rodr., 238
Astrocaryum princeps var. *vitellinum* Barb. Rodr., 238
Astrocaryum rodriguesii Trail, 247
Astrocaryum rodriguesii var. *minus* (Trail) Barb. Rodr., 239
Astrocaryum sciophilum (Miq.) Pulle, 33, 41, 42, **249**
Astrocaryum scopatum Kahn & Millán, 245
Astrocaryum segregatum Drude, 250
Astrocaryum sociale Barb. Rodr., 249
Astrocaryum standleyanum L. H. Bailey, 40, 234
Astrocaryum triandrum Galeano, Bernal & Kahn, 235
Astrocaryum tucuma Mart., 25, 238
Astrocaryum tucuma of Wallace, 250

Astrocaryum tucumoides Drude, 250
Astrocaryum ulei Burret, 246
Astrocaryum urostachys Burret, 247
Astrocaryum vulgare Mart., 18, 41, **250**
Astrocaryum vulgare of Wallace, 238
Astrocaryum vulgare of Warburg, 239
Astrocaryum yauaperyense Barb. Rodr., 246
Atitara aerea (Drude) Barb. Rodr., 232
Atitara ataxacantha (Barb. Rodr.) Kuntze, 229
Atitara horrida (Splitg.) Kuntze, 229
Atitara leptospadix (Mart.) Kuntze, 228
Atitara macroacantha (Mart.) Kuntze, 232
Atitara macrocarpa (Barb. Rodr.) Barb. Rodr., 229
Atitara macrodon (Barb. Rodr.) Barb. Rodr., 230
Atitara mitis (Mart.) Kuntze, 227
Atitara nemorosa (Barb. Rodr.) Barb. Rodr., 230
Atitara oligacantha (Barb. Rodr.) Kuntze, 232
Atitara orthacantha (Mart.) Barb. Rodr., 228
Atitara palustris (Trail) Kuntze, 229
Atitara paraensis Barb. Rodr., 233
Atitara phenogophylla (Drude) Kuntze, 233
Atitara philippiana (Barb. Rodr.) Barb. Rodr., 233
Atitara phoenicocarpa (Barb. Rodr.) Kuntze, 230
Atitara polyacantha (Mart.) Kuntze, 232
Atitara prunifera (Poepp.) Kuntze, 233
Atitara pumila (Trail) Kuntze, 228

Atitara riparia (Spruce) Kuntze, 232
Atitara setosa (Mart.) Kuntze, 232
Attalea Kunth, 14, 21, 46, **136**
Attalea agrestis Barb. Rodr., 147
Attalea attaleoides (Barb. Rodr.) Wess. Boer, 33, 41, 137, **139**
Attalea blepharopus Mart., 149
Attalea butyracea (Mutis) Wess. Boer, 8, 12, 16, 39, **139**, 143
Attalea cephalotes Poepp., 139
Attalea cryptanthera Wess. Boer, 146
Attalea dahlgreniana (Bondar) Wess. Boer, **143**
Attalea eichleri (Drude) Henderson, 14, **143**, 144
Attalea excelsa Mart., 149, 151
Attalea ferruginea Burret, 151
Attalea goeldiana Huber, 144
Attalea hoehnei Burret, 149
Attalea humboldtiana Spruce, 142
Attalea humilis, 144
Attalea insignis (Mart.) Drude, **144**
Attalea luetzelburgii (Burret) Wess. Boer, **144**
Attalea lydiae (Drude) Barb. Rodr., 153
Attalea macrolepis (Burret) Wess. Boer, 142
Attalea macropetala (Burret) Wess. Boer, 146
Attalea maripa (Aubl.) Mart., 4, 18, 24, **146**
Attalea microcarpa Mart., 137, **147**, 157
Attalea monosperma Barb. Rodr., 155
Attalea phalerata Mart., 40, 41, **149**, 151
Attalea pixuna Barb. Rodr., 153
Attalea princeps Mart., 149
Attalea pycnocarpa Wess. Boer, 142
Attalea racemosa Spruce, 137, **151**

Attalea regia (Mart.) Wess. Boer, 146
Attalea sagotii (Trail) Wess. Boer, 147
Attalea septuagenata Dugand, 14, 137, **153**
Attalea speciosa Mart., 15, 19, 41, 42, 137, **153**, 144, 155, 157
Attalea spectabilis Mart., 137, **155**, 149, 157
Attalea spectabilis var. *monosperma* (Barb. Rodr.) Drude, 155
Attalea spectabilis var. *polyandra* Drude, 153
Attalea spectabilis var. *typica* Drude, 155
Attalea tessmannii Burret, 51, **157**
Attalea transitiva Barb. Rodr., 139
Attalea wallisii Huber, 142
Attaleinae, 136
Augustinea major (Jacq.) Oerst., 200

B

Bactridinae, 160
Bactris Jacq., 58, **167**
Bactris acanthocarpa Mart., **171**
Bactris acanthocarpa subsp. *trailiana* (Barb. Rodr.) Trail, 222
Bactris acanthocarpa var. **acanthocarpa, 173**
Bactris acanthocarpa var. *crispata* Drude, 174
Bactris acanthocarpa var. *excapa* Barb. Rodr., 173
Bactris acanthocarpa var. **intermedia** Henderson, **174**
Bactris acanthocarpoides Barb. Rodr., **174**
Bactris acanthocnemis Mart., 214
Bactris acanthospatha (Trail) Trail, 200

Bactris actinoneura Drude & Trail, 207
Bactris aculeifera Drude, 173
Bactris amoena Burret, 216
Bactris angustifolia Dammer, 221
Bactris arenaria Barb. Rodr., 216
Bactris aristata Mart., 190
Bactris armata Barb. Rodr., 206
Bactris arundinacea (Trail) Drude, 221
Bactris atrox Burret, 196
Bactris aubletiana Trail, **175**
Bactris balanophora Spruce, **175**
Bactris bella Burret, 205
Bactris bicuspidata Spruce, 173
Bactris bidentula Spruce, 13, 16, 33, **177**
Bactris bifida Mart., 33, **179**
Bactris bifida var. *humaitensis* Trail, 179
Bactris bifida var. *puruensis* Trail, 179
Bactris bijugata Burret, 206
Bactris brevifolia Spruce, 214
Bactris brongniartii Mart., 34, **179**
Bactris burretii Glassman, 180
Bactris campestris Poepp., 14, **180**
Bactris capillacea (Trail) Drude, 221
Bactris capinensis Huber, 222
Bactris carolensis Spruce, 214
Bactris caryotaefolia, 193
Bactris chaetochlamys Burret, 207
Bactris chaetorachis Mart., 203
Bactris chaetospatha Mart., 205
Bactris chaetospatha var. *macrophylla* Drude, 206
Bactris chapadensis Barb. Rodr., 203
Bactris chloracantha Poepp., 206
Bactris chlorocarpa Burret, 206
Bactris ciliata (Ruiz & Pav.) Mart., 191
Bactris coccinea Barb. Rodr., 213

Bactris coloniata L. H. Bailey, **181**
Bactris concinna Mart., **181**
Bactris concinna subsp. *depauperata* Trail, 185
Bactris concinna var. **concinna, 184**
Bactris concinna var. *depauperata* (Trail) Drude, 185
Bactris concinna var. **inundata** Spruce, **185**
Bactris concinna var. **sigmoidea** Henderson, **185**
Bactris confluens var. *acanthospatha* Trail, 200
Bactris constanciae Barb. Rodr., 50, **185**
Bactris corossilla H. Karst., **185**
Bactris curuena (Trail) Trail, 201
Bactris cuspidata Mart., **186**
Bactris cuspidata var. *angustipinnata* Trail, 186
Bactris cuspidata var. *coriacea* Trail, 186
Bactris cuspidata var. *marajay* (Barb. Rodr.) Drude, 186
Bactris cuspidata var. *mitis* (Mart.) Drude, 186
Bactris cuspidata var. *tenuis* (Wallace) Drude, 214
Bactris dahlgreniana Glassman, 198
Bactris dakamana (L. H. Bailey) Glassman, 217
Bactris demerarana L. H. Bailey, 203
Bactris devia H. E. Moore, 174
Bactris duidae Steyerm., 185
Bactris duplex H. E. Moore, 185
Bactris elatior Wallace, 206
Bactris elegans Barb. Rodr., 50, **189**
Bactris elegantissima Burret, 189
Bactris ericetina Barb. Rodr., 195
Bactris erostrata Burret, 207
Bactris essequiboensis (L. H. Bailey) Glassman, 217
Bactris eumorpha Trail, 221

Bactris eumorpha subsp. *arundinacea* Trail, 221
Bactris eumorpha subsp. *eumorpha*, 221
Bactris exaltata Barb. Rodr., 201
Bactris exscapa (Barb. Rodr.) Barb. Rodr., 173
Bactris fissifrons Mart., **190**
Bactris fissifrons var. *robusta* Trail, 190
Bactris floccosa Spruce, 186
Bactris formosa Barb. Rodr., 196
Bactris gasipaes Kunth, **190**
Bactris gastoniana Barb. Rodr., **191**
Bactris gaviona (Trail) Trail, 201
Bactris geonomoides Drude, 195
Bactris geonomoides var. *setosa* Drude, 195
Bactris glaucescens Drude, **193**
Bactris gleasonii (L. H. Bailey) Glassman, 217
Bactris gracilis Barb. Rodr., 216
Bactris granariuscarpa Barb. Rodr., 206
Bactris guineensis, 179
Bactris gymnospatha Burret, 207
Bactris hirta Mart., **193**
Bactris hirta subsp. *pulchra* Trail, 196
Bactris hirta var. **hirta, 195**
Bactris hirta var. **mollis** (Dammer) Henderson, **196**
Bactris hirta var. **pulchra** (Trail) Henderson, **196**
Bactris hoppii Burret, 51, 196
Bactris huberiana Burret, 51, 216
Bactris huebneri Burret, 51, 196
Bactris humilis (Wallace) Burret, 173
Bactris hylophila Spruce, 195
Bactris hylophila var. *glabrescens* Drude, 196
Bactris hylophila var. *macrocarpa* Drude, 196
Bactris hylophila var. *nana* Drude, 196
Bactris incommoda Trail, 205
Bactris inermis Trail, 216

Bactris inermis var. *tenuissimis* Barb. Rodr., 216
Bactris infesta Mart., 201
Bactris insignis (Mart.) Baillon, 190
Bactris integrifolia Wallace, 195
Bactris interruptepinnata Barb. Rodr., 173
Bactris inundata Mart., 213
Bactris juruensis Trail, 205
Bactris juruensis var. *lissospatha* Trail, 205
Bactris kamarupa Steyerm., 207
Bactris killipii Burret, 51, **198**
Bactris krichana Barb. Rodr., 205
Bactris kuhlmannii Burret, 217
Bactris kuhlmannii var. *aculeata* Burret, 217
Bactris lakoi Burret, 51, 196
Bactris lanceolata Burret, 180
Bactris leptocarpa Trail, 180
Bactris leptochaete Burret, 174
Bactris leptospadix Burret, 207
Bactris leptotricha Burret, 207
Bactris linearifolia Barb. Rodr., 195
Bactris littoralis Barb. Rodr., 213
Bactris longicuspis Burret, 207
Bactris longifrons Mart., 213
Bactris longipes Poepp., 195
Bactris longipes var. *exilis* Trail, 195
Bactris longisecta Burret, 207
Bactris luetzelburgii Burret, 51, 216
Bactris luetzelburgii var. *anacantha* Burret, 216
Bactris macana (Mart.) Pittier, 30, 33, **198**
Bactris macroacantha Mart., **200**
Bactris macrocalyx Burret, 174
Bactris macrocarpa Wallace, 206
Bactris maguirei (L. H. Bailey) Steyerm., 217
Bactris major Jacq., **200**
Bactris major var. **infesta** (Mart.) Drude, **201**

Bactris major var. **major, 203**
Bactris major var. **megalocarpa** (Trail) Henderson, **203**
Bactris major var. **socialis** Drude, **204**
Bactris maraja Mart., 15, 47, **204**
Bactris maraja subsp. *limnaia* Trail, 206
Bactris maraja subsp. *maraja* Trail, 206
Bactris maraja subsp. *sobralensis* Trail, 206
Bactris maraja var. **chaetospatha** (Mart.) Henderson, **205**
Bactris maraja var. **juruensis** (Trail) Henderson, **205**
Bactris maraja var. *limnaia* (Trail) Drude, 206
Bactris maraja var. **maraja, 206**
Bactris maraja var. *sobralensis* (Trail) Drude, 206
Bactris maraja var. *trailii* A. D. Hawkes, 206
Bactris marajaacu Barb. Rodr., 180
Bactris marajay Barb. Rodr., 186
Bactris mattogrossensis Barb. Rodr., 203
Bactris megalocarpa Trail, 203
Bactris megistocarpa Burret, 224
Bactris microcalyx Burret, 174
Bactris microcarpa Spruce, 195
Bactris microspadix Burret, 206
Bactris microspatha Barb. Rodr., 216
Bactris minax Miq., 203
Bactris mitis Mart., 186
Bactris mitis subsp. *inermis* Trail, 216
Bactris mitis subsp. *mitis* Trail, 186
Bactris mitis subsp. *tenuis* (Wallace) Trail, 214
Bactris mitis subsp. *uaupensis* (Spruce) Trail, 216
Bactris mollis Dammer, 196
Bactris monticola Barb. Rodr., 206

Bactris multiramosa Burret, 218
Bactris naevia Poepp., 216
Bactris negrensis Spruce, 216
Bactris negrensis var. *carolensis* (Spruce) Burret, 214
Bactris negrensis var. *minor* Spruce, 216
Bactris nemorosa Barb. Rodr., 201
Bactris nigrispina Barb. Rodr., 177
Bactris obovata Burret, 216
Bactris oligocarpa Barb. Rodr. & Trail, **209**
Bactris oligocarpa var. *brachycaulis* Trail, 209
Bactris oligoclada Burret, **209**
Bactris pallidispina Mart., 179
Bactris palustris Barb. Rodr., 177
Bactris paraensis Splitg., 247
Bactris paucijuga Barb. Rodr., 206
Bactris paucisecta Burret, 216
Bactris pectinata Mart., 195
Bactris pectinata subsp. *hylophila* (Spruce) Trail, 195
Bactris pectinata subsp. *hylophila* var. *setipinnata* (Barb. Rodr.) Trail, 195
Bactris pectinata subsp. *hylophila* var. *subintegrifolia* Trail, 195
Bactris pectinata subsp. *microcarpa* (Spruce) Trail, 195
Bactris pectinata subsp. *microcarpa* var. *nana* Trail, 195
Bactris pectinata subsp. *turbinata* (Spruce) Trail, 195
Bactris pectinata subsp. *turbinata* var. *spruceana* Trail, 195
Bactris penicillata Barb. Rodr., 205
Bactris pinnatisecta Burret, 174
Bactris piranga Trail, 205
Bactris piscatorum Weddell, 180
Bactris platyacantha Burret, 200

Bactris platyspinus (Barb. Rodr.) Burret, 196
Bactris pliniana Granv. & Henderson, **211**
Bactris praemorsa Poepp., 163
Bactris ptariana Steyerm., **211**
Bactris pulchella Burret, 205
Bactris pulchra (Trail) Trail, 196
Bactris pulchra var. *inermis* Dammer, 196
Bactris pyrenoglyphoides A. D. Hawkes, 203
Bactris rhapidacantha Wess. Boer, **213**
Bactris riparia Mart., 13, **213**
Bactris rivularis Barb. Rodr., 179
Bactris schultesii (L. H. Bailey) Glassman, 217
Bactris sciophila Miq., 249
Bactris setiflora Burret, 200
Bactris setipinnata Barb. Rodr., 195
Bactris setulosa H. Karst., 33, **214**
Bactris simplex Burret, 216
Bactris simplicifrons Mart., 34, **214**
Bactris simplicifrons of Spruce, 195
Bactris simplicifrons var. *acanthocnemis* (Mart.) Drude, 214
Bactris simplicifrons var. *brevifolia* (Spruce) Trail, 214
Bactris simplicifrons var. *carolensis* (Spruce) Trail, 214
Bactris simplicifrons var. *negrensis* (Spruce) Trail, 216
Bactris simplicifrons var. *subpinnata* Trail, 216
Bactris sobralensis (Trail) Barb. Rodr. var. *limnaia* (Trail) Barb. Rodr., 206
Bactris socialis Mart., 204
Bactris socialis subsp. *curuena* Trail, 201
Bactris socialis subsp. *gaviona* Trail, 201
Bactris soropanae Steyerm., 217
Bactris sp. Wallace, 177, 221
Bactris sphaerocarpa Trail, 221

Bactris sphaerocarpa subsp. *pinnatisecta* Trail, 221
Bactris sphaerocarpa var. *ensifolia* Trail, 221
Bactris sphaerocarpa var. *minor* Trail, 221
Bactris sphaerocarpa var. *pinnatisecta* (Trail) A. D. Hawkes, 221
Bactris sphaerocarpa var. *platyphylla* Trail, 221
Bactris sphaerocarpa var. *schizophylla* Drude, 221
Bactris stahelii (L. H. Bailey) Glassman, 217
Bactris syagroides Barb. Rodr. & Trail, **218**
Bactris sylvatica Barb. Rodr., 206
Bactris tarumanensis Barb. Rodr., 173
Bactris tefensis Henderson, 32, **220**
Bactris tenera (H. Karst.) H. Wendl., 180
Bactris tenerrima Mart., 228
Bactris tenuis Wallace, 214
Bactris tenuis var. *inermis* (Trail) Burret, 216
Bactris tenuissimis (Barb. Rodr.) Barb. Rodr., 216
Bactris tomentosa Mart., **220**
Bactris tomentosa subsp. *capillacea* Trail, 221
Bactris tomentosa subsp. *tomentosa* Trail, 220
Bactris tomentosa var. *negrensis* A. D. Hawkes, 220
Bactris tomentosa var. **sphaerocarpa** (Trail) Henderson, **221**
Bactris tomentosa var. **tomentosa, 221**
Bactris trailiana Barb. Rodr., **222**
Bactris trichospatha Trail, 207
Bactris trichospatha subsp. *jurutensis* Trail, 207
Bactris trichospatha subsp. *trichospatha* Trail, 207
Bactris trichospatha subsp.

trichospatha var. *elata* Trail, 207
Bactris trichospatha subsp. *trichospatha* var. *robusta* Trail, 207
Bactris trichospatha var. *cararaucensis* A. D. Hawkes, 207
Bactris trichospatha var. *jurutensis* (Trail) Drude, 207
Bactris trichospatha var. *patens* Drude, 207
Bactris trichospatha var. *robusta* (Trail) Drude, 207
Bactris turbinata Spruce, 195
Bactris turbinocarpa Barb. Rodr., **222**
Bactris uaupensis Spruce, 216
Bactris ulei Burret, 51, 198
Bactris umbraticola Barb. Rodr., 206
Bactris umbrosa Barb. Rodr., 206
Bactris unaensis Barb. Rodr., 196
Bactris venezuelensis Steyerm., 185
Bactris vexans Burret, 224
Bactris x *moorei* Wess. Boer, 173
Bactris xanthocarpa Barb. Rodr., 216
Balanophora, 177
Barbosa, 128
Barcella Drude, **157**
Barcella odora (Trail) Drude, 9, 15, **159**
Borassus pinnatifrons Jacq., 84
Butia, 14
Butiinae, 128

C

Calameae, 68
Calamoideae, 68, 70
Calyptrogyne robusta (Trail) Burret, 251
Calyptrogyne synanthera (Mart.) Burret, 251
Calyptronoma robusta Trail, 251

Calyptronoma synanthera (Mart.) L. H. Bailey, 251
Caryota horrida Jacq., 163
Catis martiana O. F. Cook, 109
Catoblastus H. Wendl., 21, 97
Catoblastus drudei O. F. Cook & Doyle, 98
Catoblastus maynensis (Spruce) Drude, 100
Catoblastus pubescens var. *krinocarpa* (Trail) Drude, 98
Cebus albifrons, 40
Cebus apella, 40
Ceroxyloideae, 79
Chamaedorea Willd., 17, **79**
Chamaedorea amazonica Dammer, 84
Chamaedorea angustisecta Burret, **80**
Chamaedorea bartlingiana, 85
Chamaedorea boliviensis Dammer, 84
Chamaedorea cataractarum Hort. (non Mart.), 82
Chamaedorea depauperata Dammer, 84
Chamaedorea fragrans (Ruiz & Pav.) Mart., **82**
Chamaedorea gratissima Hort., 82
Chamaedorea integrifolia (Trail) Dammer, 82
Chamaedorea lanceolata (Ruiz & Pav.) Kunth, 84
Chamaedorea lechleriana H. Wendl., 82
Chamaedorea leonis H. E. Moore, 80
Chamaedorea linearis (Ruiz & Pav.) Mart., 82
Chamaedorea pauciflora Mart., 33, **82**
Chamaedorea pavoniana H. Wendl., 82
Chamaedorea pinnatifrons (Jacq.) Oerst., 39, **84**
Chamaedorea poeppigiana (Mart.) Gentry, 82
Chamaedorea ruizii H. Wendl., 82
Chamaedorea verschaffeltii Hort., 82

Chelyocarpus, 67
Chelyocarpus Dammer, **62**
Chelyocarpus chuco (Mart.) H. E. Moore, **63**
Chelyocarpus repens Kahn & Mejia, **65**
Chelyocarpus ulei Dammer, **65**
Chrysalidosperma smithii H. E. Moore, 135
Chrysallidosperma, 128
Coccothrinax, 67
Cocoeae, 128, 136, 157, 160
Cocos, 128
Cocos aculeatus Jacq., 161
Cocos aequatorialis Barb. Rodr., 131
Cocos butyracea Mutis, 139
Cocos chavesiana Barb. Rodr., 131
Cocos comosa Mart., 131
Cocos inajai (Spruce) Trail, 131
Cocos orinocensis Spruce, 133
Cocos petraea Mart., 133
Cocos purusana Huber, 135
Cocos sancona (H. Karst.) Hook. f., 133
Cocos speciosa Barb. Rodr., 131
Cocos syagrus Drude, 129
Cocos syagrus var. *linearifolia* Barb. Rodr., 129
Cocos venatorum Poepp., 111
Colossoma bidens, 242
Colossoma macropomum, 242
Copernicia alba Morong., 57
Copernicia tectorum (Kunth) Mart., 57
Corozo oleifera (Kunth) L. H. Bailey, 160
Corypheae, 62
Coryphoideae, 62
Cryosophila, 63
Cuatrecasana spruceana (Barb. Rodr.) Dugand, 91
Cuatrecasea vaupesana Dugand, 91
Curculionidae, 119

D

Dahlgrenia ptariana Steyerm., 89

Dasyprocta punctata, 40
Deckeria ventricosa (Mart.) H. Karst., 93
Derelomini, 119
Desmoncus Mart., 39, **224**
Desmoncus aereus Drude, 232
Desmoncus angustisectus Burret, 229
Desmoncus ataxacanthus Barb. Rodr., 229
Desmoncus brevisectus Burret, 233
Desmoncus cirrhiferus Gentry & Zardini, 224
Desmoncus demeraranus L. H. Bailey & H. E. Moore, 229
Desmoncus duidensis Steyerm., 233
Desmoncus giganteus Henderson, **225**
Desmoncus horridus Splitg., 229
Desmoncus huebneri Burret, 229
Desmoncus kaieteurensis L. H. Bailey, 230
Desmoncus kuhlmanii Burret, 229
Desmoncus leptoclonos Dammer, 227
Desmoncus leptoclonos Drude, 227
Desmoncus leptospadix Mart., 228
Desmoncus luetzelburgii Burret, 229
Desmoncus macroacanthos Mart., 232
Desmoncus macrocarpus Barb. Rodr., 229
Desmoncus macrodon Barb. Rodr., 230
Desmoncus maguirei L. H. Bailey, 233
Desmoncus melanacanthos Drude, 229
Desmoncus mitis Mart., 33, **227**
Desmoncus mitis var. **leptoclonos** Henderson, **227**
Desmoncus mitis var. **leptospadix** (Mart.) Henderson, **228**
Desmoncus mitis var. **mitis, 228**
Desmoncus mitis var. **rurrenabaquensis** Henderson, **228**
Desmoncus mitis var. **tenerrimus** (Mart.) Henderson, **228**
Desmoncus multijugus Steyerm., 229
Desmoncus nemorosus Barb. Rodr., 230
Desmoncus oligacanthus Barb. Rodr., 232
Desmoncus orthacanthos Mart., **228**
Desmoncus orthacanthos var. *trailianus* Drude, 229
Desmoncus palustris Trail, 229
Desmoncus paraensis (Barb. Rodr.) Barb. Rodr., 233
Desmoncus parvulus L. H. Bailey, 230
Desmoncus phengophyllus Drude, 233
Desmoncus philippianus Barb. Rodr., 233
Desmoncus phoenicocarpus Barb. Rodr., **230**
Desmoncus polyacanthos Mart., **232**
Desmoncus polyacanthos var. *angustifolia* Drude, 232
Desmoncus polyacanthos var. **polyacanthos, 232**
Desmoncus polyacanthos var. **prunifer** (Mart.) Henderson, **233**
Desmoncus prunifer Poepp., 233
Desmoncus pumilus Trail, 228
Desmoncus riparius Spruce, 232
Desmoncus rudentum Mart., 229
Desmoncus setosus Mart., 232
Desmoncus setosus var. *mitescens* Drude, 228
Desmoncus stans Grayum & de Nevers, 224
Desmoncus tenerrimus (Mart.) Mart., 228
Desmoncus ulei Dammer, 233

Desmoncus vacivus L. H. Bailey, 228
Dictyocaryum H. Wendl., **88**
Dictyocaryum ptarianum (Steyerm.) H. E. Moore & Steyerm., 30, 34, **89**

E

Elaeidinae, 157
Elaeis Jacq., 21, **159**
Elaeis guineensis Jacq., 159, 160
Elaeis melanococca Gaertn., 160
Elaeis melanococca Mart. non Gaertn., 160
Elaeis odora Trail, 159
Elaeis oleifera (Kunth) Cortés, 18, **160**
Elephantusia macrocarpa (Ruiz & Pav.) Willd., 291
Elephantusia microcarpa (Ruiz & Pav.) Willd., 291
Englerophoenix attaleoides (Barb. Rodr.) Barb. Rodr., 139
Englerophoenix insignis (Mart.) Kuntze, 144
Englerophoenix longirostrata (Barb. Rodr.) Barb. Rodr., 146
Englerophoenix maripa (Corrêa) Kuntze, 146
Englerophoenix regia (Mart.) Kuntze, 146
Englerophoenix tetrasticha (Drude) Barb. Rodr., 146
Euterpe Mart., **105**
Euterpe aculeata (Willd.) Spreng., 163
Euterpe aurantiaca H. E. Moore, 107
Euterpe badiocarpa Barb. Rodr., 109
Euterpe caatinga Wallace, 107
Euterpe catinga Wallace, 15, 30, **107**
Euterpe catinga Barb. Rodr., 107
Euterpe catinga var. *aurantiaca* Drude, 107

Euterpe catinga var. **catinga**, 107
Euterpe catinga var. *roraimae*, 107
Euterpe concinna Burret, 107
Euterpe controversa Barb. Rodr., 107
Euterpe edulis Mart., 109, 110, 112
Euterpe jatapuensis Barb. Rodr., 111
Euterpe longebracteata Barb. Rodr., **109**
Euterpe longevaginata Mart., 111
Euterpe longispathacea Barb. Rodr., 109
Euterpe mollissima Spruce, 107
Euterpe oleracea Engel, 111
Euterpe oleracea Mart., 9, 16, 34, 70, **109**
Euterpe petiolata Burret, 111
Euterpe precatoria Mart., 8, 12, 16, 25, **110**
Euterpe precatoria var. **longevaginata** (Mart.) Henderson, **111**
Euterpe precatoria var. **precatoria, 111**
Euterpe schultzeana Burret, 113
Euterpe stenophylla Trail, 111
Euterpe subruminata Burret, 111
Euterpeinae, 105

G

Gastrococos, 161
Gaussia, 87
Geonoma Willd., 46, 58, **253**
Geonoma acaulis (Poit.) Burret, 277
Geonoma acaulis Mart., 274
Geonoma acaulis subsp. *tapajotensis* Trail, 274
Geonoma acutiflora Mart., 260
Geonoma ambigua Spruce, 278
Geonoma appuniana Spruce, 30, 57, 254

Geonoma arundinacea Mart., **258**
Geonoma aspidiifolia Spruce, **260**
Geonoma baculifera (Poit.) Kunth, **260**
Geonoma baculifera var. *macrospatha* (Spruce) Drude, 260
Geonoma bartlettii Dammer, 267
Geonoma beccariana Barb. Rodr., 271
Geonoma bella Burret, 288
Geonoma bijugata Barb. Rodr., 279
Geonoma brachyfoliata Barb. Rodr., 279
Geonoma brevispatha Barb. Rodr., **262**
Geonoma brevispatha var. **brevispatha,** 14, **264**
Geonoma brevispatha var. **occidentale** Henderson, **264**
Geonoma brongniartii Mart., 39, **264**
Geonoma camana Trail, **265**
Geonoma camptoneura Burret, 281
Geonoma capanemae Barb. Rodr., 280
Geonoma chaunostachys Burret, 277
Geonoma chelidonura Spruce, 279
Geonoma chococola Wess. Boer, 285
Geonoma cuneata, 265
Geonoma cuneifolia Burret, 264
Geonoma dammeri Huber, 277
Geonoma dasystachys Burret, 288
Geonoma densiflora Spruce, 279
Geonoma densiflora var. *monticola* Spruce, 279
Geonoma desmarestii Mart., 267
Geonoma deversa (Poit.) Kunth, **267**
Geonoma discolor Spruce, 280
Geonoma elegans Mart. var. *amazonica* Trail, 288

Geonoma estevaniana Burret, 260
Geonoma euspatha Burret, 270
Geonoma falcata Barb. Rodr., 279
Geonoma furcifolia Barb. Rodr., 279
Geonoma fusca Wess. Boer, 260
Geonoma gracilis, 265
Geonoma grandisecta Burret, 281
Geonoma herthae Burret, 287
Geonoma hexasticha Spruce, 280
Geonoma huebneri Burret, 281
Geonoma interrupta (Ruiz & Pav.) Mart., 39, **268**
Geonoma interrupta var. **euspatha** (Burret) Henderson, **270**
Geonoma interrupta var. **interrupta,** 271
Geonoma juruana Dammer, 280
Geonoma jussieuana Mart., 57, 265
Geonoma karuaiana Steyerm., 270
Geonoma killipii Burret, 267
Geonoma lagesiana Dammer, 265
Geonoma lakoi Burret, 280
Geonoma lanceolata Burret, 287
Geonoma latisecta Burret, 281
Geonoma laxiflora Mart., 33, **271**
Geonoma laxiflora var. *depauperata* Trail, 271
Geonoma lehmanii, 265
Geonoma leptospadix Trail, **271**
Geonoma leptostachys Burret, 267
Geonoma longepedunculata Burret, **272**
Geonoma longisecta Burret, 280
Geonoma macropoda Burret, 267
Geonoma macrospatha Spruce, 260
Geonoma macrostachys Mart., 39, **274**

Geonoma macrostachys var. **acaulis** Skov, **274**
Geonoma macrostachys var. **macrostachys, 276**
Geonoma macrostachys var. **poiteauana** (Kunth) Henderson, **277**
Geonoma maguirei L. H. Bailey, 287
Geonoma major Burret, 267
Geonoma maxima (Poit.) Kunth, 34, **277**
Geonoma maxima var. **ambigua** (Spruce) Henderson, **278**
Geonoma maxima var. **chelidonura** (Spruce) Henderson, **279**
Geonoma maxima var. **maxima, 280**
Geonoma maxima var. **spixiana** (Mart.) Henderson, **281**
Geonoma microspatha Spruce, 267
Geonoma microspatha var. *pacimoensis* Spruce, 267
Geonoma multiflora Mart., 280
Geonoma myriantha Dammer, **281**
Geonoma negrensis Spruce, 280
Geonoma oldemanii, 289
Geonoma oldemanii Granv., **283**
Geonoma oligoclada Burret, 285
Geonoma oligoclona Trail, **283**
Geonoma oxycarpa, 270
Geonoma palustris Barb. Rodr., 279
Geonoma paniculigera Mart., 267
Geonoma paniculigera subvar. *gramineifolia* Trail, 267
Geonoma paniculigera var. *cosmiophylla* Trail, 267
Geonoma paniculigera var. *microspatha* (Spruce) Trail, 267
Geonoma paniculigera var. *papyracea* Trail, 267
Geonoma paraensis Spruce, 280
Geonoma parvisecta Burret, 280
Geonoma pauciflora Mart., 254

Geonoma personata Spruce, 279
Geonoma pinnatifrons, 270
Geonoma piscicauda Dammer, 287
Geonoma poeppigiana Mart., 47, **283**
Geonoma poiteana Mart., 277
Geonoma poiteauana Kunth, 277
Geonoma polyandra Skov, **285**
Geonoma procumbens, 265
Geonoma pycnostachys Mart., 287
Geonoma raimondii Burret, 288
Geonoma rectifolia Wallace, 267
Geonoma robusta Burret, 279
Geonoma saramaccana L. H. Bailey, 272
Geonoma schomburgkiana Spruce, 279
Geonoma simplicifrons Willd., 57, 254
Geonoma sodiroi, 265
Geonoma speciosa Barb. Rodr., 279
Geonoma spixiana Mart., 281
Geonoma spruceana Trail, 279
Geonoma spruceana subsp. *intermedia* var. *compta* Trail, 281
Geonoma spruceana subsp. *intermedia* var. *intermedia* Trail, 281
Geonoma spruceana subsp. *intermedia* var. *major* Trail, 279
Geonoma spruceana subsp. *intermedia* var. *tuberculata* Trail, 279
Geonoma spruceana subsp. *spruceana* var. *heptasticha* Trail, 279
Geonoma spruceana subsp. *spruceana* var. *micra* Trail, 279
Geonoma spruceana subsp. *spruceana* var. *spruceana* Trail, 279
Geonoma stricta (Poit.) Kunth, 34, **285**
Geonoma stricta var.

piscicauda (Dammer) Henderson, **287**
Geonoma stricta var. **stricta, 287**
Geonoma stricta var. **trailii** (Burret) Henderson, **288**
Geonoma synanthera Mart., 251
Geonoma tamandua Trail, 276
Geonoma tapajotensis (Trail) Drude, 274
Geonoma tessmannii Burret, 267
Geonoma trailii Burret, 288
Geonoma trauniana Dammer, 288
Geonoma triglochin Burret, **289**
Geonoma trijugata Barb. Rodr., 267
Geonoma tuberculata Spruce, 279
Geonoma uleana Dammer, 258
Geonoma uliginosa Barb. Rodr., 280
Geonoma umbraculiformis Wess. Boer, **289**
Geonoma werdermannii Burret, 264
Geonoma wittiana Dammer, 287
Geonoma woronowii Burret, 276
Geonoma yauaperyensis Barb. Rodr., 267
Geonomeae, 251
Guilielma gasipaes (Kunth) L. H. Bailey, 190
Guilielma insignis Mart., 190
Guilielma macana Mart., 198
Guilielma mattogrossensis Barb. Rodr., 213
Guilielma microcarpa Huber, 198
Guilielma speciosa Mart., 190
Guilielma speciosa var. *coccinea* Barb. Rodr., 190
Guilielma speciosa var. *flava* Barb. Rodr., 190
Guilielma speciosa var. *mitis* Drude, 190

Guilielma speciosa var. *ochracea* Barb. Rodr., 190
Gynestum acaule Poit., 277
Gynestum baculiferum Poit., 260
Gynestum deversum Poit., 267
Gynestum maximum Poit., 277
Gynestum strictum Poit., 285

H

Hyophorbe, 87
Hyophorbeae, 79
Hyophorbeae, 87
Hyospathe Mart., **125**
Hyospathe brevipedunculata Dammer, 127
Hyospathe elegans Mart., **127**
Hyospathe filiformis H. Wendl., 127
Hyospathe gracilis H. Wendl., 127
Hyospathe micropetala Burret, 127
Hyospathe pallida H. E. Moore, 127
Hyospathe tessmannii Burret, 127

I

Iriartea Ruiz & Pav., 17, 24, 46, **93**
Iriartea deltoidea Ruiz & Pav., 26, **93**
Iriartea exorrhiza Mart., 95
Iriartea exorrhiza var. *orbigniana* Drude, 95
Iriartea orbigniana Mart., 95
Iriartea philonotia Barb. Rodr., 95
Iriartea pubescens var. *krinocarpa* Trail, 98
Iriartea setigera Mart., 91
Iriartea spruceana Barb. Rodr., 91
Iriartea stenocarpa (Burret) MacBride, 91

Iriartea ventricosa Mart., 93
Iriarteeae, 88
Iriartella H. Wendl., 24, **89**
Iriartella ferreyrae H. E. Moore, 91
Iriartella pruriens (Barb. Rodr.) Barb. Rodr., 91
Iriartella setigera (Mart.) H. Wendl., 34, **91**
Iriartella setigera var. *pruriens* Barb. Rodr., 91
Iriartella spruceana (Barb. Rodr.) Barb. Rodr., 91
Iriartella stenocarpa Burret, **91**
Itaya H. E. Moore, **67**
Itaya amicorum H. E. Moore, **67**

J

Jessenia H. Karst., 113
Jessenia bataua (Mart.) Burret, 118
Jessenia bataua subsp. *bataua*, 118
Jessenia bataua subsp. *oligocarpa* (Griseb. & H. Wendl.) Balick, 120
Jessenia oligocarpa Griseb. & H. Wendl., 120
Jessenia polycarpa H. Karst., 119
Jessenia weberbaueri Burret, 119
Jubaea, 128
Jubaeopsis, 128

L

Leopoldinia Mart., 13, 47, **103**
Leopoldinia insignis Mart., 104
Leopoldinia major Wallace, 16, **103**
Leopoldinia piassaba Wallace, 16, 29, 32, 48, **104**
Leopoldinia pulchra Mart., 16, 30, **104**
Leopoldiniinae, 103

Lepidocaryeae, 70
Lepidocaryum Mart., **77**
Lepidocaryum allenii Dugand, 79
Lepidocaryum casiquiarense (Spruce) Drude, 78
Lepidocaryum enneaphyllum Barb. Rodr., 79
Lepidocaryum gracile Mart., 79
Lepidocaryum guainiense (Spruce) Drude, 78
Lepidocaryum gujanense Becc., 79
Lepidocaryum macrocarpum (Drude) Becc., 79
Lepidocaryum quadripartitum (Spruce) Drude, 79
Lepidocaryum sexpartitum Trail & Barb. Rodr., 79
Lepidocaryum sexpartitum var. *macrocarpum* Drude, 79
Lepidocaryum sexpartitum var. *microcarpum* Drude, 79
Lepidocaryum tenue Mart., 9, **77**
Lepidocaryum tenue var. **casiquiarense** (Spruce) Henderson, 15, 33, **78**
Lepidocaryum tenue var. **gracile** (Mart.) Henderson, **79**
Lepidocaryum tenue var. *sexpartitum* (Trail & Barb. Rodr.) Trail, 79
Lepidocaryum tenue var. **tenue,** **79**
Lepidocaryum tessmannii Burret, 79
Lepidococcus aculeatus (Kunth) H. Wendl. & Drude, 75
Lepidococcus armatus (Mart.) H. Wendl. & Drude, 76
Lepidococcus duckei (Burret) H. Wendl. & Drude, 76
Lepidococcus huebneri (Burret) H. Wendl. & Drude, 76
Lepidococcus intermedius (Burret) H. Wendl. & Drude, 76
Lepidococcus martianus (Spruce) H. Wendl. & Drude, 76

Lepidococcus peruvianus (Becc.) H. Wendl. & Drude, 76
Lepidococcus pumilus (Wallace) H. Wendl. & Drude, 76
Lepidococcus subinermis (Spruce) A. D. Hawkes, 76

M

Manicaria Gaertn., **100**
Manicaria atricha Burret, 101
Manicaria martiana Burret, 101
Manicaria saccifera, 70
Manicaria saccifera Gaertn., 9, 26, **101**
Manicaria saccifera var. *mediterranea* Trail, 101
Manicariinae, 100
Marara aculeata (Willd.) H. Karst., 163
Marara caryotifolia (Kunth) H. Karst., 163
Markleya dahlgreniana Bondar, 143
Martinezia aculeata (Willd.) Klotzsch, 163
Martinezia aiphanes Mart., 163
Martinezia caryotifolia Kunth, 163
Martinezia ernesti Burret, 163
Martinezia interrupta Ruiz & Pav., 268
Martinezia lanceolata Ruiz & Pav., 84
Martinezia ulei Dammer, 163, 165
Mauritia L. f., **70**
Mauritia aculeata Kunth, 75
Mauritia aculeata of Mart., 76
Mauritia amazonica Barb. Rodr., 75
Mauritia armata Mart., 76
Mauritia carana Wallace, 15, 16, 48, **70**
Mauritia casiquiarensis Spruce, 78
Mauritia flexuosa L. f., 4, 8, 9, 12, 14, 16, 25, 26, 29, 37, 38, 45, **72**
Mauritia flexuosa var. *venezuelana* Steyerm., 72
Mauritia gracilis (Mart.) Spruce, 79
Mauritia gracilis Wallace, 75
Mauritia guainiensis Spruce, 78
Mauritia huebneri Burret, 76
Mauritia intermedia Burret, 76
Mauritia linophila Barb. Rodr., 75
Mauritia martiana Spruce, 76
Mauritia minor Burret, 72
Mauritia peruviana Becc., 76
Mauritia pumila Wallace, 76
Mauritia quadripartita Spruce, 79
Mauritia sphaerocarpa Burret, 72
Mauritia subinermis Spruce, 76
Mauritia tenuis (Mart.) Spruce, 77
Mauritia vinifera Mart., 72
Mauritiella Burret, **74**
Mauritiella aculeata (Kunth) Burret, 13, 15, 16, 33, **75**
Mauritiella armata (Mart.) Burret, 4, 14, **76**
Mauritiella campylostachys Burret, 76
Mauritiella cataractarum Dugand, 75
Mauritiella duckei Burret, 76
Mauritiella huebneri (Burret) Burret, 76
Mauritiella intermedia (Burret) Burret, 76
Mauritiella martiana (Spruce) Burret, 76
Mauritiella nannostachys Burret, 76
Mauritiella peruviana (Becc.) Burret, 76
Mauritiella pumila (Wallace) Burret, 76
Maximiliana Mart., 21, 47, 136, 137
Maximiliana attaleoides Barb. Rodr., 139
Maximiliana elegans H. Karst., 146
Maximiliana inajai Spruce, 131
Maximiliana insignis Mart., 144
Maximiliana longirostrata Barb. Rodr., 146
Maximiliana macrogyne Burret, 146
Maximiliana macropetala Burret, 146
Maximiliana maripa (Aubl.) Drude, 146
Maximiliana martiana H. Karst., 146
Maximiliana regia Mart., 146
Maximiliana stenocarpa Burret, 146
Maximiliana tetrasticha Drude, 146
Maximiliana venatorum (Poepp.) H. Wendl., 111
Metroxylon taedigerum (Mart.) Spreng., 68
Monogynanthus, 234
Morenia integrifolia Trail, 82
Morenia integrifolia var. *nigricans* Trail, 82
Morenia pauciflora (Mart.) Drude, 82
Moreniopsis, 82
Mystrops, 38, 119
Mystrops mexicana, 155

N

Nitidulidae, 119
Nunnezharia fragrans Ruiz & Pav., 82
Nunnezharia integrifolia (Trail) Kuntze, 82
Nunnezharia lanceolata (Ruiz & Pav.) Kuntze, 84
Nunnezharia pauciflora (Mart.) Kuntze, 82
Nunnezharia pinnatifrons (Jacq.) Kuntze, 84

O

Oenocarpus Mart., **113**
Oenocarpus bacaba Mart., 17, **115**

Oenocarpus bacaba var. *bacaba* Wess. Boer, 115
Oenocarpus bacaba var. *grandis* Wess. Boer, 115
Oenocarpus bacaba var. *parvus* Wess. Boer, 118
Oenocarpus bacaba var. *xanthocarpa* Trail, 115
Oenocarpus baccata in Cuervo Marquez, 115
Oenocarpus balickii Kahn, **118**
Oenocarpus bataua Mart., 9, 12, 26, 30, 37, 38, 42, **118**
Oenocarpus bataua var. **bataua, 119**
Oenocarpus bataua var. **oligocarpa** (Griseb. & H. Wendl.) Henderson, **120**
Oenocarpus circumtextus Mart., **120**
Oenocarpus discolor Barb. Rodr., 121
Oenocarpus distichus Mart., 17, 33, **121**
Oenocarpus grandis Burret, 115
Oenocarpus hoppii Burret, 115
Oenocarpus huebneri Burret, 123
Oenocarpus intermedius Burret, 123
Oenocarpus macrocalyx Burret, 123
Oenocarpus makeru Bernal, Galeano & Henderson, 14, **121**
Oenocarpus mapora H. Karst., 30, 33, **121**
Oenocarpus mapora subsp. *mapora* Balick, 121
Oenocarpus microspadix Burret, 123
Oenocarpus minor Mart., **123**
Oenocarpus minor subsp. *intermedius* (Burret) Balick, 123
Oenocarpus minor subsp. *minor* Balick, 123
Oenocarpus multicaulis Spruce, 123
Oenocarpus oligocarpa (Griseb. & H. Wendl.) Wess. Boer, 120
Oenocarpus simplex Bernal, Galeano & Henderson, 14, **125**
Oenocarpus tarampabo Mart., 121
Oenocarpus x *andersonii* Balick, 118
Orbignya Mart., 21, 136, 137
Orbignya agrestis (Barb. Rodr.) Burret, 147
Orbignya barbosiana Burret, 153
Orbignya campestris Barb. Rodr., 144
Orbignya eichleri Drude, 143
Orbignya huebneri Burret, 153
Orbignya humilis Mart., 143, 144
Orbignya longibracteata Barb. Rodr., 144
Orbignya luetzelburgii Burret, 144
Orbignya lydiae Drude, 153
Orbignya macrocarpa Barb. Rodr., 144
Orbignya macropetala Burret, 153
Orbignya martiana Barb. Rodr., 153
Orbignya microcarpa (Mart.) Burret, 147
Orbignya phalerata Mart., 153
Orbignya pixuna (Barb. Rodr.) Barb. Rodr., 153
Orbignya polysticha Burret, 147
Orbignya racemosa (Spruce) Drude, 151
Orbignya sabulosa Barb. Rodr., 147
Orbignya sagotii Trail, 147
Orbignya speciosa (Mart.) Barb. Rodr., 153
Orbignya spectabilis (Mart.) Burret, 155
Orbignya teixeriana Bondar, 144
Orophoma, 72
Orophoma carana (Wallace) Spruce, 70
Orophoma subinermis (Spruce) Drude, 76

P

Palma maripa Aubl., 146
Palma maripa Corrêa, 146
Parajubaea, 128
Parascheelea anchistropetala Dugand, 146
Parascheelea luetzelburgii (Burret) Dugand, 144
Pholidostachys H. Wendl., **251**
Pholidostachys synanthera (Mart.) H. E. Moore, **251**
Phyllotrox, 38, 193
Phytelephantoideae, 290
Phytelephas Ruiz & Pav., 46, **290**
Phytelephas aequatorialis, 290
Phytelephas dasyneura Burret, 294
Phytelephas macrocarpa Ruiz & Pav., 290, **291**
Phytelephas macrocarpa subsp. *tenuicaulis* Barfod, 291
Phytelephas microcarpa Ruiz & Pav., 291
Phytelephas schottii, 290
Phytelephas seemanii, 290
Phytelephas tenuicaulis (Barfod) Henderson, 8, 32, 290, **291**
Phytelephas tumacana, 290
Phytotribus, 193
Pleiogynanthus, 234
Prestoea Hook. f., **112**
Prestoea asplundii H. E. Moore, 113
Prestoea ensiformis (Ruiz & Pav.) H. E. Moore, 113
Prestoea schultzeana (Burret) H. E. Moore, 32, **113**
Prestoea tenuiramosa (Dammer) H. E. Moore, 30, 57
Pyrenoglyphis, 179
Pyrenoglyphis aristata (Mart.) Burret, 190

Pyrenoglyphis bicuspidata (Spruce) Burret, 173
Pyrenoglyphis bifida (Mart.) Burret, 179
Pyrenoglyphis bifida var. *humaitensis* (Trail) Burret, 179
Pyrenoglyphis bifida var. *puruensis* (Trail) Burret, 179
Pyrenoglyphis brongniartii (Mart.) Burret, 179
Pyrenoglyphis chaetorachis (Mart.) Burret, 203
Pyrenoglyphis concinna (Mart.) Burret, 181
Pyrenoglyphis concinna var. *depauperata* (Trail) Burret, 185
Pyrenoglyphis concinna var. *inundata* (Spruce) Burret, 185
Pyrenoglyphis curuena (Trail) Burret, 201
Pyrenoglyphis exaltata (Barb. Rodr.) Burret, 201
Pyrenoglyphis gastoniana (Barb. Rodr.) Burret, 191
Pyrenoglyphis gaviona (Trail) Burret, 201
Pyrenoglyphis hoppii Burret, 203
Pyrenoglyphis infesta (Mart.) Burret, 201
Pyrenoglyphis major (Jacq.) H. Karst., 200
Pyrenoglyphis maraja (Mart.) Burret, 204
Pyrenoglyphis microcarpa Burret, 180
Pyrenoglyphis nemorosa (Barb. Rodr.) Burret, 201
Pyrenoglyphis oligocarpa (Barb. Rodr. & Trail) Burret, 209
Pyrenoglyphis pallidispina (Mart.) Burret, 179
Pyrenoglyphis rivularis (Barb. Rodr.) Burret, 179
Pyrenoglyphis socialis (Mart.) Burret, 204
Pyrenoglyphis turbinocarpa (Barb. Rodr.) Burret, 222

R

Raphia P. Beauv., 21, **68**
Raphia farinifera, 70
Raphia taedigera (Mart.) Mart., 16, 34, 47, **68**
Raphia vinifera var. *taedigera* Drude, 68
Raphiinae, 68
Rhynchophorus palmarum, 101
Rhyticocos, 128
Roystonea oleracea (Jacq.) O. F. Cook, 57

S

Sabal mauritiiformis (H. Karst.) Griseb. & H. Wendl., 57
Sagus taedigera Mart., 68
Scheelea H. Karst., 21, 136, 137
Scheelea anisitsiana Barb. Rodr., 151
Scheelea attaleoides H. Karst., 144
Scheelea bassleriana Burret, 142
Scheelea blepharopus (Mart.) Burret, 149
Scheelea brachyclada Burret, 142
Scheelea butyracea (Mutis) H. Karst., 139
Scheelea cephalotes (Poepp.) H. Karst., 139
Scheelea goeldiana (Huber) Burret, 144
Scheelea huebneri Burret, 142
Scheelea humboldtiana (Spruce) Burret, 142
Scheelea insignis (Mart.) H. Karst., 144
Scheelea macrolepis Burret, 142
Scheelea martiana Burret, 149
Scheelea microspadix Burret, 149
Scheelea parviflora (Barb. Rodr.) Barb. Rodr., 151
Scheelea passargei Burret, 142
Scheelea phalerata (Mart.) Burret, 149
Scheelea princeps (Mart.) H. Karst., 149
Scheelea princeps var. *corumbaensis* Barb. Rodr., 151
Scheelea quadrisperma Barb. Rodr., 151
Scheelea quadrisulcata Barb. Rodr., 151
Scheelea stenorhyncha Burret, 142
Scheelea tessmannii Burret, 142
Scheelea tetrasticha (Drude) Burret, 146
Scheelea wallisii (Huber) Burret, 142
Scheelea weberbaueri Burret, 149
Schippia, 63
Sciurus aestuans, 250
Sciurus granatensis, 40
Socratea H. Karst., 24, **94**
Socratea albolineata Steyerm., 95
Socratea exorrhiza (Mart.) H. Wendl., 12, 16, 25, 30, **95**
Socratea gracilis Burret, 95
Socratea microchlamys Burret, 95
Socratea orbigniana (Mart.) H. Karst., 95
Socratea philonotia (Barb. Rodr.) Hook. f., 95
Socratea salazarii H. E. Moore, **95**
Syagrus Mart., 14, **128**
Syagrus aequatorialis (Barb. Rodr.) Barb. Rodr., 131
Syagrus allenii Glassman, 133
Syagrus brachyrhyncha Burret, 129
Syagrus chavesiana (Barb. Rodr.) Barb. Rodr., 131
Syagrus cocoides Mart., 33, **129**
Syagrus cocoides var. *linearifolia* Barb. Rodr., 129
Syagrus comosa (Mart.) Mart., 33, **131**
Syagrus flexuosa, 128

Syagrus inajai (Spruce) Becc., **131**
Syagrus orinocensis (Spruce) Burret, **133**
Syagrus petraea (Mart.) Becc., 14, **133**
Syagrus purusana (Huber) Frambach, 135
Syagrus sancona H. Karst., 30, 33, **133**
Syagrus smithii (H. E. Moore) Glassman, **135**
Syagrus speciosa (Barb. Rodr.) Barb. Rodr., 131
Syagrus stratincola Wess. Boer, 32, **135**
Syagrus tessmannii Burret, 135
Synechanthus, 87

Tessmanniodoxa, 51
Tessmanniodoxa chuco (Mart.) Burret, 63
Tessmanniophoenix, 51
Tessmanniophoenix chuco (Mart.) Burret, 63
Tessmanniophoenix longibracteata Burret, 65
Thrinacinae, 63
Thrinacineae, 62
Thrinax, 67
Thrinax? chuco Mart., 63
Tilmia caryotifolia (Kunth) O. F. Cook, 163
Trithrinax, 63
Trithrinax chuco (Mart.) Walp., 63

Wendlandiella polyclada Burret, 88
Wendlandiella simplicifrons Burret, 88
Wettinella maynensis (Spruce) O. F. Cook & Doyle, 100
Wettinia Poepp., 17, 21, **97**
Wettinia augusta Poepp. & Endl., **98**
Wettinia drudei (O. F. Cook & Doyle) Henderson, 32, **98**
Wettinia illaqueans Spruce, 100
Wettinia maynensis Spruce, **100**
Wettinia poeppigii Kunth, 98
Wettinia praemorsa, 97

T

Taenianthera acaulis (Mart.) Burret, 274
Taenianthera camana (Trail) Burret, 265
Taenianthera gracilis Burret, 274
Taenianthera lagesiana (Dammer) Burret, 265
Taenianthera lakoi Burret, 277
Taenianthera macrostachys (Mart.) Burret, 274
Taenianthera minor Burret, 276
Taenianthera oligosticha Burret, 274
Taenianthera tamandua (Trail) Burret, 276
Taenianthera tapajotensis (Trail) Burret, 274

V

Vampyressa, 25
Vampyressa macconnelli, 246
Voanioala, 128

W

Wendlandiella Dammer, **85**
Wendlandiella gracilis Dammer, 33, 39, **87**
Wendlandiella gracilis var. **gracilis**, 88
Wendlandiella gracilis var. **polyclada** (Burret) Henderson, **88**
Wendlandiella gracilis var. **simplicifrons** (Burret) Henderson, **88**

Y

Yarina microcarpa (Ruiz & Pav.) O. F. Cook, 291
Yuyba dakamana L. H. Bailey, 217
Yuyba essequiboensis L. H. Bailey, 217
Yuyba gleasonii L. H. Bailey, 217
Yuyba maguirei L. H. Bailey, 217
Yuyba schultesii L. H. Bailey, 217
Yuyba simplicifrons (Mart.) L. H. Bailey, 214
Yuyba simplicifrons var. *acanthocnemis* (Mart.) A. D. Hawkes, 214
Yuyba stahelii L. H. Bailey, 217

Index of Common Names

A

a-ibcom-ba, 67
aacaiba, 84
abadek, 147
açaí, 110, 112
açaí branco, 110
açaí chumbo, 109
açaí da mata, 109, 112
açairana, 135
açaizeiro, 110
açaizinho, 107
agee, 78
aguajales, 74
aguaje, 70, 74
aguajillo, 77
akamin-ékoenari, 189
albarico, 242
amaint-nak, 241
amankayo, 157
ambakai, 94
ampakay, 94
an-hí-da-du, 289
anaccu tssatssa'vo', 95
anaja, 147
anajá, 147
ankú, 112
ansepara, 184
anuya wili, 193
anuyawili, 209

arandacê, 131
aricuri, 142
arimkwe, 112
ariry, 129
aroira palm, 250
asaí de sabana, 107
asaí paso, 107, 112
assaí, 24
assaí chumbinho, 107
assaí cubinha, 107
assaí da mata, 112
assaí de caatinga, 107
assaí de catinga, 107
assai-rana, 267
assaizeiro, 110
assay da terra firme, 109
awara, 238, 250
awarra, 250
awëke, 241
awilâ, 272
ayry, 25

B

baba maripa, 147
babaçu, 155
babassu, 155
babassú, 153

baboenpina, 110
baby-ité, 77
bacabão, 118
bacaba, 115, 121, 123
bacába, 24
bacabeira, 115
bacabi, 123
bacabinha, 123, 125
bacabiña, 123
bacurí, 151
bamba maka, 233
bambakka, 233
bambamaka, 230
bango palm, 180
baraboro, 262
barahuasca, 228
baro, 268
barobaro, 262
baru baru, 278
barubaru, 276
batái, 239
batú, 119
bejuco alcalde, 233
bo-só-moo-hee, 288
boba, 95
boegroe maka, 250
bombona paso, 89
bonbon, 95
bongie-bongie, 175
boo b(o) me g(e), 177

boo-han-ñee-kaw-ne, 91
boyon, 153
buba, 95
bubúmemëku, 174
buçu, 101
bulishi, 250
bunyashiri, 209
buritales, 74
buriti, 74
buriticillo, 77
buritirana, 75, 77
burityzinho, 78
bussú, 101

C

ca'hue, 113
ca-mu-vé, 230, 233
ca-újico, 190
caápi, 293
cabaya, 78
cachepai montañero, 180
cachuda barriguda, 94
cadanarite, 75
cadena, 78
cadotodek, 94
caiaué, 157
calson-panga, 276
calzón panga, 277
camanará, 78
camona blanca, 94
camonilla, 93, 97, 100
camuare, 230
camuari, 230
canahuaníma, 174
canambo, 142, 143
canangucha, 74
cananguchales, 74
cananguchillo, 77
canangucho de sabana, 70
canangucho paso, 70
caña negra, 180
caraná, 74, 77, 78
caraná do matto, 74
caranai, 78
caranaí, 65, 75, 77
caranatinga, 77
caraña, 77, 78
carnaubinha, 65

carana-y, 78
caruto, 112
casha pona, 95
casha ponita, 93
cashapona, 24
cashapona de altura, 97
cashipana, 85
casicusi, 147
catarina, 149
catherina, 149
catolé, 135
cau, 78
ccu'ye, 100
ceguera, 177
cemau, 238
ceyacepan, 217
céyacépan, 113
chambira, 25, 239
chambira ñeja, 209
chambirilla, 242
chápi, 293
chapil, 119
chellochellpan, 127
chicyorah, 123
chilimoyo, 295
chima, 191
chimbo, 123
chincha, 113
chiqui-chiqui, 104
chiquichiqui, 104
chiquichiquito, 104
chonta, 191, 238, 244
chonta de castilla, 191
chonta de nutria, 93
chonta durillo, 214
chonta duro de tintin, 217
chonta loro, 244
chonta pala, 191
chontadura, 191
chontadurillo, 214
chontaduro de los peces, 174
chontaduro de rana, 194
chontaduro de rana de rastrojo, 217
chontaduro paso, 177
chontilla, 85, 87, 177, 184, 190, 193, 200, 209, 214, 217, 242, 288
chontilla blanca, 85
chontilla conguillo, 209
chontilla de llana-muncu, 127

chontillo, 268
chontillo de loma, 268
chontillo de monte, 268
choó-no-hee, 127
chuchana, 244
churrguay, 133
churruái, 133
churubay, 133
cinamillo, 123
citaganomecu, 84
cocão, 157
cocito, 133
cocito yagua, 153
coco, 135, 144, 153
coco curuá, 148
coco-curúa, 144
coco de mono, 239
coco palha preta, 139
coco pancha, 144
codime, 153
cola de pava, 91
cola de pescado, 95
comou, 115
conta, 157
contillo, 144
coquinha, 193
coquino, 144
coquita, 193
coquito, 133, 186
coroba, 143
corocito, 236
corombolo, 239
corozo, 157, 238
cosá, 119
counana, 249, 250
counana agouti, 241, 249
crepísh, 94
cubarillo, 218
cubarra, 180, 241
cubarro, 174, 177, 180, 181, 190, 194, 241
cucurita, 24, 147
cucurito, 147, 204
cucurrito, 105
cudídi, 115
cui'cui, 174
cullulí, 265
cumare, 239
cuparú, 186
curuá, 155
curúa, 146

curuaí, 149
curuaraua, 146
cusí, 147
cusí, 155
cusimacho, 147
cutata fedié, 127

D

dalibana, 277
dé-dé, 288
de-de-hueoco, 127
dédé-bui, 285
derehue'co, 127
devéke, 241
dhalebana, 290
dhalli, 278
dhallibanna, 278
dí-dí, 285
dija-dere, 288
dois por dois, 143
du, 186

E

eérü, 189
enredadera, 230, 233
espera i, 233
espina, 209, 236
espina de sardina, 177
espinita, 194

F

falso bombonaje, 67
faux wi blanc, 175
feuille chasseur, 131
firichí, 189
foodá, 228
fuidorda, 93

G

gabiroba catolé, 131
geke, 276

gerua, 101
giyikabemo, 113
go-güi-n(e), 272
go-güi-re de centro de monte, 268
grua, 146
guacamaya-panga, 276
guacamayopanga, 276
guajo, 24, 112, 147
guasai, 112
guichire, 24, 147
guie, 249

H

haai maba, 175
haijowa-'i, 262
halago, 84
hanaimaka, 175, 193, 211
he-bu-ca-nu, 119
highland kiskismaka, 241
hikuriparipia, 288
hikuripipia, 278
ho-ta-mo-hee', 78
ho-tá-ñe, 177
hoja ancha, 288
hoja de llana-muncu, 127
hoja grande, 65
hoja redonda, 65
hua-so-e-ne, 217
huachapona, 95
huacrapona, 24, 94, 95
huan-só-dé-dé, 276
huanima, 191
huasai, 112
huasai de varillal, 107
huasipanga, 285
huasoëne, 268
huicosa, 123
huicungo, 244
huiririma, 242
hungurahui, 119
hunguravi, 119

I

icaí, 264
iika, 244

imáp, 97
imayá, 24, 147
ina yuga, 147
inajá, 147
inajai, 147
inaja-'y, 147
inayá, 147
inayo, 147
inayuga, 147
inaza, 147
ini-bue, 112
iñéjhe, 74
irapai, 78
irapaillo, 276
iraparpukwaha, 230
irapay, 78
iriri, 129
ite palm, 74
itsama, 119

J

jací, 142
jaciarana, 135
jacitara, 227, 228, 230, 233
jará, 104, 105
jarevá, 131
jarina, 291
jarive, 147
jassitara, 233
jatata, 264, 268
jatata grande, 278
jatatilla, 85
jauarí, 242
jawi, 175
jawpe, 194
jetahu-'y, 155
ji, 203
jiowoohiyui, 288
jiyui, 285
jo-da-jime-ru, 217
jodaj mena, 84
ju, 241
jû-kû-fê-nâ, 89
jubatí, 47
juçara, 110, 112
juduaro, 186
jupatí, 70
jupihu, 241

jussara, 77, 110
juzyba, 110

K

ka-be-re, 110
ká-roo, 78
kaa-ñoo-et, 288
kaarú, 78
kamancá, 181
kamancha, 186, 200
kamanchá, 218
kamáncha, 221, 260
kamawa, 230
kamawarri, 233
kampana, 268
kamuari, 230
kan-tine-é, 77
karugiri, 78
karwari, 230
katirina, 157
kauwaya, 77
kidallibanaro, 211
kikyura, 74
kiskis pina, 110
kiskismaka, 194, 211, 218
ko-rü'-ne, 242
kokerite, 147
komora, 180
kopayan, 133
koua'hm', 213
kubina, 91
kudina, 153
kuík, 135
kukarit, 147
kumri, 239
kumu, 115
kunkúk, 119
kunkupij, 127
kunuana, 149
kupat, 95
kúpat, 95
kwere'i, 174

L

liba awara, 242
lu, 115

M

ma-na-cáy, 112
ma-tav-icú-li, 236
maá-kan, 91
mabaco, 149, 153
mabe, 91
mábi, 91
macába, 115
macana, 214
macanilla, 91, 95
macanillo, 186
macaúba, 157
macoupi, 149
macoupi blanc, 139
majalie, 143
majo, 119
maka kow maka, 175
makaka ima, 288
makeru, 121
mana'i, 151
manac, 112
manaca, 107, 112
manáca-man-ash-quili, 95
manácam, 95
manaco, 112
manai maka, 175
manaka, 110
manamazu, 104
manása, 268
manasha, 268
manicole, 112
manicoli, 105
manni koemboe, 115
mantas, 278
mapanaré, 142
mapora, 112
mapure, 115
maradai, 180
maraja, 217
marajá, 24, 174, 180, 184, 186, 189, 193, 194, 198, 200, 203, 205, 209, 211, 214, 221, 241
maraja'i, 180
maraja'y, 180
marajá de cacho, 180
marajá do igapó, 177
marajá do jacaré, 177
marajá pupunha, 180, 209, 214
marajaú, 184, 214
marajazinha, 221

maraju, 241
maramapé, 104
marark'y, 131
marayaú, 203, 204
marayua, 241
marazaiwa-ran, 194
marfil, 291
maria-ci, 180
maria-wa, 180
marím ipa, 67
maripa, 143, 147
maritaryna, 155
marrashemapar, 93, 127
masanke, 244
maswa, 211
mataa, 239
mavaco, 101, 147, 149
mayo, 119
mecuá-bak, 101
meena, 262
mil peso, 119
milpesillo, 115, 123
milpesillo de sabana, 120
milpesos, 119
miriti, 74
miriti-rn, 70
mo, 276
moeroekoe, 189
molinillo, 85
momogaik-t-t-co, 93
momoigiak-u, 93
moo-ee-a, 78
moo-heé, 78
moretillo, 77
morichales, 74
moriche, 74, 75
morichita, 104
morichito, 75, 77, 78
morichito de tierra firme, 77
moriquito, 78
mormuru, 244
motacú, 151
motacusillo, 147
mountain maripa, 149
mourou-mourou, 244
mucuja, 157
mucujazeiro, 157
muin, 78
muiriká, 78
mulumulusi, 250
mumbaca, 25

mumbaca branca, 181
mumbacucu, 25
mumu, 250
munbaca, 174, 185, 241
munbaca da preta, 241
muriti, 74
muru muru, 250
murumuru, 25, 244, 250
murumuru da terra firme, 249
mutacú, 151

N

na-í, 113
nai, 113
naicá, 113
naimacca, 193
nain, 74
naja'i, 147
nao maka, 175
naya-huë-dadu, 276
ne, 74
ne-e-da, 112
nenea, 112
ní-ní, 265
niejilla, 184
nomkie muruku pina, 112
non, 74
novaco, 153
nu-que, 184
ña-k-r, 127
ñai-cü-r(ü), 127
ñai-gú-ru, 276
ñeeinó, 200
ñeja, 209, 214
ñeja negra, 179
ñejilla, 180, 184, 186, 198, 221, 127
ñejilla de canto de cocha, 214
ñicó, 95
ñobea, 95
ñu-quan-ne, 268
ñucua-ëné, 85
ñume'mba, 294

O

o-có-pui, 277
oco-be-to, 242

ohauimon, 268
omacabo, 293
omakabo, 293
onipa, 95
oo-chee-an, 177
oo-ree-ñaw, 217
oomawe, 278
oompa, 147
or-ó-boto, 133
ora, 94
orocori, 244
ouricuri, 151
owi, 262
owi-ran, 272

P

pa'i ine, 214
pa-pa, 143
pachiubarana, 89
pachiubina, 91
pachiubinha de espinha, 167
pachuba, 94
pachuda zancona, 95
pachuvilla, 95
paipigu, 184
palha branca, 139, 149
palha de flecha, 144
palha preta, 155
palha rasgada, 144
palha redonda, 65
palha vermelha, 149
palheira, 155
palhera, 139
palhera branca, 139
palhera vermelha, 149
palla, 142
palma coco, 239
palma cola de pescado, 91, 262
palma de coco, 153
palma de cumare, 239
palma de pantano, 113
palma de San Pablo, 278
palma de tintas, 127
palma gorere, 84
palma letchera, 119
palma mabaca, 153
palma real, 24, 74, 142, 147
palma real de loma, 271

palma San Pablo, 262, 268
palma seje, 119, 120
palma temiche, 101
palma vermelha, 149
palma yagua, 143
palmeira, 78, 82, 149
palmeirinha, 82
palmera, 95, 184
palmera marfil, 291
palmicha, 198, 253, 288
palmiche, 105, 127, 253, 264, 268, 271, 272, 277, 285
palmiche blanco, 288
palmiche grande, 253
palmiche negro, 278
palmier bache, 74
palmier toulouri, 101
palmilla, 77
palmita, 127
palmita tijereta, 84
palmito, 105, 112
palmito manaca, 112
pambil, 94, 95
pamiwa, 112
pan si noha, 167
panabí, 112
pani, 244
panima, 238
paramaka, 249
paramaka x boefro maka, 249
parapi-balli, 217
parena, 157
pata de gallo, 78
pata de grillo, 78
patabá, 115, 119
patauá, 119
patauá branca, 119
patauá roxa, 119
patawa, 120
pati, 129, 131
patigua, 94
patisak'o, 294
patoá, 119
paxi'y, 95
paxiuba, 24, 95
paxiúba barrigouda, 94
paxiuba barriguda, 24
paxiúba barriguda, 94
paxiubão, 94
paxiubarana, 91
paxiubinha, 24, 91, 95, 100

Index of Common Names 359

paxiúbinha de macaco, 98
paxiubinha do macaco, 93
pe-co-r, 253
pëpë, 131
perenao, 143
perinão, 143
petit ouara, 213
piaçaba, 295
piaçava, 144
piassaba, 104, 157, 295
piassaba preta, 157
piassava, 144
pifaia, 89
pijacicara, 230
pijiguao, 191
pijiwau de monte, 177
pijuaito, 218
pijuayo, 191
pijuayo del monte, 200
pina, 110
pindiwa'y, 121
pinduwa'ywa, 121
pinglo, 167
pinot, 110
pinuwa-pihun, 110
pinuwa-'y, 121
pinuwa'yw, 121
piri'a-hu, 203
pirima, 131
piririma, 129, 131
piritiumë, 209
piritu, 209, 213
pishuallo rojo, 233
po-ta'-me, 78
polo punta, 291
pona, 93, 94, 100
pona colorada, 89
pona de altura, 97
pona lisa, 94
ponilla, 24, 87, 93, 98, 127, 214, 271, 276, 278
posuí, 123
prasara, 110, 112
pui, 78
púi, 271
pui ocho hojas, 78
pupunha, 165, 191
pupunha brava, 129, 165, 200
pupunha-brava, 131
pupunha de mata, 174

pupunha mança wa', 174
pupunha silvestre, 133
pupunharana, 131
pusuy, 123
puy, 78

Q

quëboitsama, 123
quidíja, 238
quirigua, 100

R

rayador, 95
rayhoo, 112
rohn-di, 250
rotetenpar, 218
ruan-rie, 144
rui-re'-gö, 241

S

saápap, 127
sacha aguajillo, 67
sacurí, 149
sadke, 112
sake, 112
san pablo, 127, 262, 265, 271
San Paulo, 268
sangapilla, 82, 84
sangapilla masha, 272
sangapilla sacha, 85
sápap, 127, 272, 277, 288
sauarai, 242
saupak, 127
sehua, 293
seje, 115, 118, 119, 120
seje grande, 119
seje hembra, 119
seje pequeño, 115
sejito, 115
shapaha, 143
shapaja, 143, 151

shapajilla, 143, 149
shebon, 143, 153
shebon enano, 153
sheboncita, 143
shee-she-tsha, 293
shimbu, 123
shinishëoxo, 209
shiquita, 95
shishihe, 293
shiwamuyo, 119
sialla, 82
síi, 184
siname, 119
sinami, 123
sinamillo, 123
sirá, 244
sitanó, 190
siyaiye, 82
siyeyi, 82
spinho preto, 241
suma-yuca, 84
sumuke, 135
sumuqué, 135
suté-dé-dé, 268

T

taco de altura, 278
taco de bajillal, 288
tado, 78
tagua, 291, 293, 295
takamã, 250
taki-taki, 249
takone, 239
tamutubë, 193
tananë, 278
tas, 262
tassie, 262
te-ere, 264
tegpayaje, 78
temiche, 101
tepa, 94
teren, 100
tevy, 78
ti aacaiba, 84
ti-wara, 241
tiemaka, 249
titara, 233

too'-ee, 70
toókee, 135
totai, 157
toulouri, 101
troolie, 101
truli, 101
tsao-he-tsi, 285
tú-te-se-wa, 294
tuamo, 94
tucum, 193, 238, 250
tucúm, 239
tucum bravo, 180
tucum de indio, 133
tucumã, 238, 250
tucumã branco, 249
tucuma, 157, 238, 239, 250
tucuma-í, 236
tucumazeiro, 250
tuee-tee', 78
tuinfa, 239
tukum 'y, 250
tunci sake, 112
tupí, 242
tupina'i, 209
turiri, 101
turu palm, 120
turuji, 268
turúji, 272

U

ü-da-ti-ño-rü, 285
u-ma-chú-ku-su, 217
ubi, 262
ubí, 101
ubim, 24, 85, 253, 260, 262, 264, 267, 268, 272, 276, 278, 283, 288
ubím, 262
ubim açaí, 278
ubim brava, 272
ubim cavalo, 278
ubim comum, 268
ubim da várzea, 271
ubim de espinho, 179, 209
ubim do igapó, 91
ubim galope, 276
ubím grande, 262

ubim juriti, 288
ubim-mirim, 217
ubim no céu, 278
ubim rana, 194
ubim rasteiro, 276
ubim uassú, 253
ubimaçu, 283
ubimrana, 278
ubimzinho, 276
ubi-ran, 278
ubussu, 101
ubussuhy, 222
uhahik, 211
uichira, 24, 147
ujagkit, 211
uksha, 272
ulukpana, 175
ümeh, 143
unabai, 236
unamo, 119
ungurabe morado, 119
unguragua, 119
ungurahua, 119
ungurahui, 119
ungurawi, 119
upa, 95
urcu-tauna, 288
urishi, 241
urpi-chunda, 227
urucuri, 151
urucurizeiro, 151
usahua, 244
utata-jididi, 127
úun shigki, 181
uva de montaña, 209
uwí, 191
uwinim, 218
uyainim, 218

V

vá-va-ra, 268
vacavilla, 123
vaí, 268
vara casha, 228, 233
vara-casha, 227
varo-varo, 268
vo-ti, 129

volador, 230
voladora, 233

W

wa-heé, 101
wá-hee, 101
wa-hó, 147
wa-tos-uné, 217
wa-'y, 155
wabo-yaka, 112
waï, 262, 277
walte, 100
wamowe, 295
wapoe, 110
wapu, 110
wasei, 110
washí, 101
wayúr, 253
wi-ni-co, 100
wilã si, 175
wili, 189
woy, 191
wulo, 241

X

xëbichoqui, 147
xëbini, 151
xila, 67

Y

ya na muyl, 288
ya-yo-(e)r(u), 177
yachoro, 190
yadua, 91
yagua, 144
yajuji, 101
yampuna ujuk, 277
yará, 104, 105
yarina, 291, 293
yarina blanca, 294
yasita, 230

yasitara, 233
yauary, 25
yaun, 84
yaún, 84
yavarí, 242

yija-déré, 277
yu-'y, 241
yunkúp, 278
yurua, 211
yuyba, 175, 218

Z

zagrinette, 203, 213
zagrinette forêt, 213
zancona, 95

DATE DUE

DUE DATE SUBJECT TO CHANGE
IF A RECALL IS REQUESTED

JAN 22 1997

MAY 23 1997

SEP 6 1997

RETURNED MAY 23 1997

DEMCO 12801830